1635: The Boston Public Latin School, now the oldest public school in the U.S., is founded.

1637: Descartes introduces the idea of a coordinate system.

1638: The first American Indian reservation is established in Connecticut.

1647: Fermat states that $x^n + y^n = z^n$ has no integer solutions for $n < 2$. This becomes known as Fermat's Last Theorem.

1654: Fermat and Pascal lay the groundwork for probability theory.

1665: Newton and Leibnitz begin work on calculus.

1678: Elena Piscopia becomes the first woman to earn a PhD. She teaches mathematics at the University of Padua until 1684.

1679: Leibnitz develops binary arithmetic but does not publish the results until 1701.

1698: The first steam engine is patented.

1706: William Jones introduces the symbol π to represent the ratio of the circumference to the diameter of a circle.

1727: Euler introduces the symbol *e* for the base of the natural logarithms.

1735: Euler introduces the notation $f(x)$.

1736: Euler solves the Konisberg bridge problem.

1741: *American Magazine* is the first magazine published in the United States.

1746: Lucy Terry publishes *Bar Fight*, the earliest existing poem by an African-American.

1748: Maria Gaëtana Agnesi writes the most respected calculus text in Italy.

1755: Samuel Johnson publishes the first dictionary of the English language.

1776: The Declaration of Independence is signed.

1781: Caroline Herschel and her brother William discover Uranus.

1785: Marie Jean Condorcet publishes a treatise on voting theory.

1786: Caroline Herschel becomes the first woman to discover a comet.

1789: George Washington becomes the first President of the United States.

1796: Edward Jenner discovers a smallpox vaccination.

1800 – 1899

1808: Marie Sophie Germain defines Germain prime numbers, which are important to cryptography.

1813: Jane Austen writes *Pride and Prejudice.*

1816: The first bicycle is invented by Karl von Drais.

1824: Beethoven composes the Ninth Symphony.

1829: Louis Braille publishes the Braille method of writing.

1837: Concepts in non-Euclidean geometry are investigated by Nikolai Lobachevsky.

1847: George Boole shows that rules of logic can be treated mathematically, which lays the foundation for computer logic.

1849: Elizabeth Blackwell becomes the first woman physician in the U.S.

1851: *The New York Times* newspaper debuts.

1858: Augustus Möbius invents the Möbius strip.

1861: The American Civil War begins.

1865: The 13th Amendment to the U.S. Constitution is ratified, abolishing slavery.

1869: Susan B. Anthony founds the National Woman Suffrage Association.

1874: Cantor develops his set theory.

1876: Alexander Graham Bell invents the telephone.

1879: Thomas Edison invents the light bulb.

1881: Venn diagrams are introduced.

1893: Karl Pearson writes his first paper on statistics and is considered one of the founders of statistics.

1896: Nobel prizes are established.

1900 – Present

1901: Theodore Roosevelt renames the President's residence The White House.

1903: The Wright brothers make their first flight.

1914: The Panama Canal is opened.

1920: The 19th Amendment to the U.S. Constitution is ratified, giving women the right to vote.

1928: The first scheduled television show is broadcast and Amelia Earhart is the first woman to fly across the Atlantic.

1936: The first Fields Medal is awarded. It is the mathematician's equivalent to the Nobel prize.

1945: The first fully electronic computer is built.

1952: Grace Murray Hopper develops A-O, the precursor to modern computer programming languages.

1953: Watson and Crick establish the double helix of a DNA molecule.

1955: Rosa Parks is arrested for not giving up her seat on a bus.

1963: Valentina Tereshkova is the first woman to orbit Earth.

1963: Martin Luther King, Jr., gives his "I have a dream" speech.

1967: Christian Barnard performs the first heart transplant.

1968: String theory, a mathematical theory used in physics, begins.

1969: Neil Armstrong walks on the moon.

1971: Ray Tomlinson creates the ARPANET e-mail system and introduces the "@" symbol to connect the user and the host computer.

1973: Martin Cooper invents the cell phone.

1975: Benoit Mandelbrot begins his investigations of fractals.

1976: The four-color conjecture is proved.

1981: IBM introduces the personal computer.

1984: The World Wide Web is invented.

1991: The first web browser is created.

1994: Andrew Wiles proves Fermat's Last Theorem.

1996: The DVD is invented.

1996: The Great Internet Mersenne Prime Search begins. An award of $100,000 is offered to the person or group of people who find the first prime number with 10 million or more digits.

2000: The Clay Mathematics Institute states the Millennium math problems. Any person or group of people who solves one of these problems will receive $1,000,000.

2003: The Human Genome Project is completed.

2003: The Recording Industry Association of America (RIAA) sues 261 individuals for allegedly distributing copyrighted music files using the Internet.

2006: In 2002, Grigori Perelman presented a proof of the Poincare conjecture, one of the seven Millenimum problems. It is not until 2006 that mathematicians finally verify Perelman's proof.

Mathematical Thinking and Quantitative Reasoning

Richard N. Aufmann
Palomar College, California

Joanne S. Lockwood
New Hampshire Community Technical College, New Hampshire

Richard D. Nation
Palomar College, California

Daniel K. Clegg
Palomar College, California

HOUGHTON MIFFLIN COMPANY
Boston   New York

Publisher: Richard Stratton
Senior Sponsoring Editor: Lynn Cox
Senior Marketing Manager: Katherine Greig
Marketing Associate: Naveen Hariprasad
Development Editor: Lisa Collette
Associate Editor: Noel Kamm
Editorial Assistant: Laura Ricci
Associate Project Editor: Susan Miscio
Editorial Assistant: Joanna Carter
Art and Design Manager: Gary Crespo
Cover Design Manager: Anne Katzeff
Photo Editor: Jennifer Meyer Dare
Composition Buyer: Chuck Dutton
New Title Project Manager: James Lonergan

Cover Photograph: © Royalty-Free/CORBIS

Photo credits for front endpapers: Pyramid of Cheops: CORBIS; abacus: Getty Images; Great Wall of China: CORBIS; *Mona Lisa*: Getty Images; telescope: CORBIS; Leonhard Euler: Kean Collection/Getty Images; Declaration of Independence: CORBIS; bicycle: Getty Images; Panama Canal; Getty Images; Grace Hopper: Bettman/CORBIS; Neil Armstrong: Getty Images.

Additional photo credits are found immediately after the answer section in the back of the book.

Printed in the U.S.A.

Library of Congress Catalog Card Number: 2006923761

ISBNs:
Instructor's Annotated Edition:
ISBN 13: 978-0-618-77738-9
ISBN 10: 0-618-77738-5

For orders, use student text ISBNs:
ISBN 13: 978-0-618-77737-2
ISBN 10: 0-618-77737-7

123456789-CRK-11 10 09 08 07

CONTENTS

Algebraic Models 121

Measurement and Geometric Models 202

CHAPTER

The Mathematics of Finance *415*

CHAPTER

Probability and Statistics *494*

CHAPTER

9

Apportionment and Voting *602*

CHAPTER

10

The Mathematics of Graphs *666*

Web Appendix: Algebra Review (Available only online at this textbook's
Online Study Center at: **hmco.college.com/pic/aufmannMTQR.**)

PREFACE

Mathematical Thinking and Quantitative Reasoning presents an analytical investigation of topics and concepts that are relevant to modern society. Our goal is to demonstrate the power of mathematics and quantitative reasoning in solving contemporary problems.

Mathematical Thinking and Quantitative Reasoning provides glimpses into how mathematics is used to solve real-life problems. Students will learn how prime numbers are used to encrypt information sent across the Internet, the role of modular arithmetic in verifying credit card numbers, how to determine whether to lease or buy a car, how statistics is used to predict the outcome of elections, and how mathematics can be used to evaluate voting systems.

Two features that we have incorporated in the text are Math Matters and Investigations. Math Matters are vignettes of interesting applications related to the topic being discussed. Each section of the text ends with an Investigation, which is an extension of one of the topics presented in that section. For instance, one Investigation extends the ideas of formal logic to logic gates in computers; another Investigation examines how to determine whether a die is fair.

The exercise sets in *Mathematical Thinking and Quantitative Reasoning* have been carefully selected to reinforce and extend the concepts developed in each section. The exercises range from drill-and-practice to interesting challenges. Some of the exercise sets include outlines for further explorations, suggestions for essays, critical thinking problems, and cooperative learning activities. In all cases, the exercises were chosen to illustrate the many facets of the topic under discussion.

The purpose of this text is to strengthen students' quantitative reasoning skills by having them solve a variety of real-world problems. Although we assume that the reader has an intermediate algebra background, each topic is carefully developed, and appropriate material is reviewed whenever necessary. When deciding on the depth of coverage, our singular criterion was to make mathematics accessible.

Student Success

Mathematical Thinking and Quantitative Reasoning is designed to foster student success through an integrated text and media program.

AIM for Success Student Preface

This "how to use this text" preface explains what is required of a student to be successful and how this text has been designed to foster student success. *AIM for Success* can be used as a lesson on the first day of class or as a project for students to complete to strengthen their study skills.

page xvii

An Interactive Method

Mathematical Thinking and Quantitative Reasoning is written in a style that encourages the student to interact with the textbook. Each section contains a variety of worked examples. Each example is given a title so that the student can see at a glance the type of problem that is being solved. Most examples include annotations that assist the student in moving from step to step, and the final answer is in color in order to be readily identifiable.

Check Your Progress Exercises

Following each worked example is a *Check Your Progress* exercise for the student to work. By solving this exercise, the student actively practices concepts as they are presented in the text. For each *Check Your Progress* exercise, there is a detailed solution in the Solutions appendix.

page 128

Student Success, *continued*

Question/Answer Feature

At various places throughout the text, a *Question* is posed about the topic that is being developed. This question encourages students to pause, think about the current discussion, and answer the question. Students can immediately check their understanding by referring to the *Answer* to the question provided in a footnote on the same page. This feature creates another opportunity for the student to interact with the textbook.

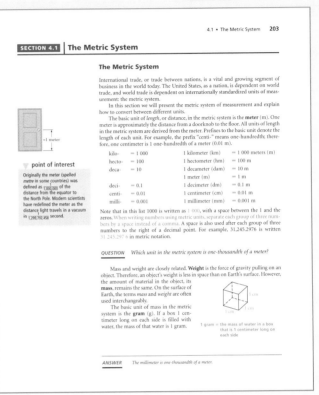

page 203

page 151

Investigations

Each section ends with an *Investigation* along with corresponding *Investigation* Exercises. These activities engage students in the mathematics of the section. Some *Investigations* are designed as in-class cooperative learning activities that lend themselves to a hands-on approach. They can also be assigned as projects or extra credit assignments. The *Investigations* are a unique and important feature of this text. They provide opportunities for students to take an active role in the learning process.

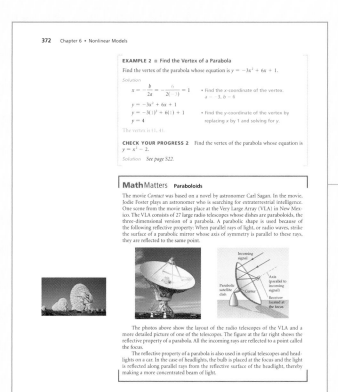

page 372

Conceptual Understanding

Mathematical Thinking and Quantitative Reasoning helps students understand the course concepts through features in the exposition.

Math Matters

This feature of the text typically contains an interesting sidelight about mathematics, its history, or its applications.

Historical Note

These margin notes provide historical background information related to the concept under discussion or vignettes of individuals who were responsible for major advancements in their fields of expertise.

Point of Interest

These notes provide interesting information related to the topics under discussion. Many of these are of a contemporary nature and, as such, they provide students with the needed motivation for studying concepts that may at first seem abstract and obscure.

Take Note

These notes alert students to a point requiring special attention or are used to amplify the concepts that are currently being developed.

Calculator Note

These notes provide information about how to use the various features of a calculator.

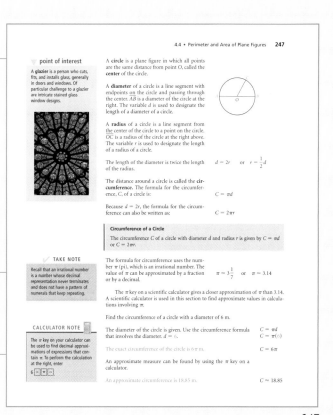

page 247

Conceptual Understanding, *continued*

Mathematical Thinking and Quantitative Reasoning helps students understand course concepts through features in the exercise sets.

Exercise Sets

The exercise sets were carefully written to provide a wide variety of exercises that range from drill-and-practice to interesting challenges. Exercise sets emphasize skill building, skill maintenance, concepts, and applications. Icons are used to identify various types of exercise.

The *Think About It* exercises are conceptual in nature. They ask students to think about the concepts studied in a section, make generalizations, and apply them to more abstract problems. The focus is on mental mathematics, not calculation or computation. These exercises are meant to help students synthesize concepts.

Writing exercises

Data analysis exercises

Graphing calculator exercises

Internet exercises

Think About It exercises

---

152   Chapter 3 • Algebraic Models

Exercise Set 3.2

1. Provide two examples of situations in which unit rates are used.

2. Explain why unit rates are used to describe situations involving units such as miles per gallon.

3. What is the purpose of exchange rates in international trade?

4. Provide two examples of situations in which ratios are used.

5. Explain why ratios are used to describe situations involving information such as student–teacher ratios.

6. What does the phrase "the cross products are equal" mean?

7. Explain why the product of the means in a proportion equals the product of the extremes.

In Exercises 8–13, write the expression as a unit rate.

8. 582 miles in 12 hours

9. 138 miles on 6 gallons of gasoline

10. 544 words typed in 8 minutes

11. 100 meters in 8 seconds

12. $9100 for 350 shares of stock

13. 1000 square feet of wall covered with 2.5 gallons of paint

14. A rate of 288 miles in 6 hours is closest to which unite rate?
   a. 100 mph   b. 50 mph   c. 200 mph   d. 25 mph

15. A rate of $123.75 in 11 hours is closest to which unite rate?
   a. $20 per hour   b. $1 per hour   c. $10 per hour   d. $120 per hour

Solve Exercises 16–21.

16. **Wages** A machinist earns $490 for working a 35-hour week. What is the machinist's hourly rate of pay?

17. **Space Vehicles** Each of the Space Shuttle's solid rocket motors burns 680,400 kilograms of propellant in 2.5 minutes. How much propellant does each motor burn in 1 minute?

18. **Photography** During filming, an IMAX camera uses 65-mm film at a rate of 5.6 feet per second.
   a. At what rate per minute does the camera go through film?
   b. How quickly does the camera use a 500-foot roll of 65-mm film? Round to the nearest second.

19. **Consumerism** Which is the more economical purchase, a 32-ounce jar of mayonnaise for $2.79 or a 24-ounce jar of mayonnaise for $2.09?

20. **Consumerism** Which is the more economical purchase, an 18-ounce box of corn flakes for $2.89 or a 24-ounce box of corn flakes for $3.89?

21. **Wages** You have a choice of receiving a wage of $34,000 per year, $2840 per month, $650 per week, or $16.50 per hour. Which pay choice would you take? Assume a 40-hour work week and 52 weeks of work per year.

22. **Baseball** Baseball statisticians calculate a hitter's at-bats per home run by dividing the number of times the player has been at bat by the number of home runs the player has hit.
   a. Calculate the at-bats per home run ratio for each player in the table on the following page. Round to the nearest tenth.
   b. Which player has the lowest rate of at-bats per home run? Which player has the second lowest rate?

Babe Ruth

page 152

---

684   Chapter 10 • The Mathematics of Graphs

41. **Transportation** For the train routes given in Exercise 5, find a route that visits each city and returns to the starting city without visiting any city twice.

42. **Transportation** For the direct air flights given in Exercise 6, find a route that visits each city and returns to the starting city without visiting any city twice.

43. **Architecture** In Exercise 7, you were asked to draw a graph that represents a museum floor plan. Describe what a Hamiltonian circuit in the graph would correspond to in the museum.

44. **Transportation** Consider a subway map, like the one given in Exercise 32. If we draw a graph in which each vertex represents a train junction, and an edge between vertices means that a train travels between those two junctions, what does a Hamiltonian circuit correspond to in regard to the subway?

Extensions
CRITICAL THINKING

45. **Route Planning** A security officer patrolling a city neighborhood needs to drive every street each night. The officer has drawn a graph representing the neighborhood, in which the edges represent the streets and the vertices correspond to street intersections. Would the most efficient way to drive the streets correspond to an Euler circuit, a Hamiltonian circuit, or neither? (The officer must return to the starting location when finished.) Explain your answer.

46. **Route Planning** A city engineer needs to inspect the traffic signs at each street intersection of a neighborhood. The engineer has drawn a graph representing the neighborhood, where the edges represent the streets and the vertices correspond to street intersections. Would the most efficient route correspond to an Euler circuit, a Hamiltonian circuit, or neither? (The engineer must return to the starting location when finished.) Explain your answer.

47. Is there an Euler circuit in the graph below? Is there an Euler walk? Is there a Hamiltonian circuit? Justify your answer. (You do not need to find any of the circuits or paths.)

48. Is there an Euler circuit in the graph below? Is there an Euler walk? Is there a Hamiltonian circuit? Justify your answer. (You do not need to find any of the circuits or paths.)

COOPERATIVE LEARNING

49. a. Draw a connected graph with six vertices that has no Euler circuit and no Hamiltonian circuit.
   b. Draw a graph with six vertices that has a Hamiltonian circuit but no Euler circuit.
   c. Draw a graph with five vertices that has an Euler circuit but no Hamiltonian circuit.

50. **Travel** A map of South America is shown below.

Venezuela Guyana
Suriname
French
Guiana
Colombia
Ecuador
Brazil
Peru
Bolivia
Paraguay
Chile
Uruguay
Argentina

   a. Draw a graph in which the vertices represent the 13 countries of South America, and two vertices are joined by an edge if the corresponding countries share a common border.
   b. Two friends are planning a driving tour of South America. They would like to drive across every border on the continent. Is it possible to plan such a route that never crosses the same border twice? What would the route correspond to on the graph?
   c. Find a route the friends can follow that will start and end in Venezuela and that crosses every border while recrossing the fewest borders possible. *Hint:* On the graph, add multiple edges corresponding to border crossings that allow an Euler circuit.

page 684

Extensions

Extension exercises are placed at the end of each exercise set. These exercises are designed to extend concepts. In most cases these exercises are more challenging and require more time and effort than the preceding exercises. The *Extension* exercises always include at least two of the following types of exercises:

Critical Thinking
Cooperative Learning
Explorations

Some *Critical Thinking* exercises require the application of two or more procedures or concepts.

The *Cooperative Learning* exercises are designed for small groups of two to four students.

Many of the *Exploration* exercises require students to search for information on the Internet or through reference materials in a library.

Assessment and Review

Mathematical Thinking and Quantitative Reasoning is designed to foster student success through practice and review.

Chapter Summary

At the end of each chapter there is a *Chapter Summary* that includes *Key Terms* and *Essential Concepts* that were covered in the chapter. These chapter summaries provide a single point of reference as the student prepares for an examination. Each key word references the page number where the word was first introduced.

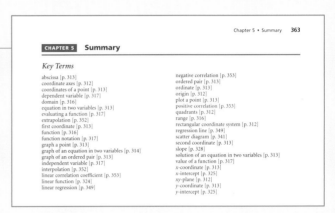

page 363

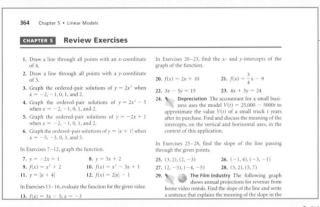

page 364

Chapter Review Exercises

Review Exercises are found near the end of each chapter. These exercises were selected to help the student integrate the major topics presented in the chapter. The answers to all *Review Exercises* appear in the answer section, along with a section reference for each exercise. These section references indicate the section or sections in which a student can locate the concepts needed to solve each exercise.

Chapter Test

The *Chapter Test* exercises are designed to emulate a possible test of the material in the chapter. The answers to all *Chapter Test* exercises appear in the answer section, along with a section reference for each exercise. The section references indicate the section or sections in which a student can locate the concepts needed to solve each exercise.

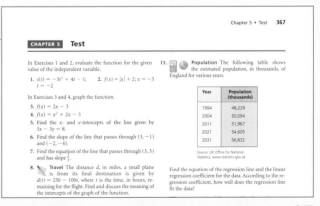

page 367

Supplements for the Instructor

Mathematical Thinking and Quantitative Reasoning has an extensive support package for the instructor that includes:

Instructor's Annotated Edition (IAE): The *Instructor's Annotated Edition* is an exact replica of the student textbook with the following additional text-specific items for the instructor: answers to *all* of the end-of-section and end-of-chapter exercises, answers to *all* Investigation and Exploration exercises, Instructor Notes, Suggested Assignments, and Ⓟ icons denoting tables and art that appear in PowerPoint® slides.

Online Teaching Center

Online Teaching Center: This free companion website contains an abundance of instructor resources such as solutions to all exercises in the text, digital art and tables, suggested course syllabi, Chapter Tests, a Graphing Calculator Guide, and Microsoft® Excel spreadsheets. Visit **hmco.college.com/pic/aufmannMTQR** and click on the Online Teaching Center icon.

Online Instructor's Solutions Manual: The *Online Instructor's Solutions Manual* offers worked-out solutions to *all* of the exercises in each exercise set as well as solutions to the Investigation and Exploration exercises.

HM Testing CD-ROM (powered by Diploma™): *HM Testing* (powered by *Diploma*) offers instructors a flexible and powerful tool for test generation and test management. Now supported by the Brownstone Research Group's market-leading *Diploma* software, this new version of *HM Testing* significantly improves on functionality and ease of use by offering all the tools needed to create, author, deliver, and customize multiple types of tests—including authoring and editing algorithmic questions.

Eduspace®: Eduspace, powered by Blackboard®, is Houghton Mifflin's customizable and interactive online learning tool.

Eduspace provides instructors with online courses and content. By pairing the widely recognized tools of Blackboard with quality, text-specific content from Houghton Mifflin Company, Eduspace makes it easy for instructors to create all or part of a course online. This online learning tool also contains ready-to-use homework exercises, quizzes, tests, tutorials, and supplemental study materials.

Visit **eduspace.com** for more information.

Supplements for the Student

Mathematical Thinking and Quantitative Reasoning has an extensive support package for the student that includes:

Student Solutions Manual: The *Student Solutions Manual* contains complete, worked-out solutions to *all* odd-numbered exercises and *all* of the solutions to the Chapter Reviews and Chapter Tests in the text.

Online Study Center **Online Study Center:** This free companion website contains an abundance of student resources such as an Algebra Review appendix, a Graphing Calculator Guide, and Microsoft® Excel spreadsheets. Visit **hmco.college.com/pic/aufmannMTQR** and click on the Online Study Center icon.

Online CLAST Preparation Guide: The CLAST Preparation Guide is a competency-based study guide that reviews and offers preparatory material for the CLAST (College Level Academic Skills Test) objectives required by the State of Florida for mathematics. The guide includes a correlation of the CLAST objectives to the *Mathematical Thinking and Quantitative Reasoning* text, worked-out examples, practice examples, cumulative reviews, and sample diagnostic tests with grading sheets.

Houghton Mifflin Instructional DVDs: These text-specific DVDs, professionally produced by Dana Mosely, provide explanations of key concepts, examples, and exercises in a lecture-based format. They offer students a valuable resource for further instruction and review. They also provide support for students in online courses.

Eduspace®: Eduspace, powered by Blackboard®, is Houghton Mifflin's customizable and interactive online learning tool for instructors and students. Eduspace is a text-specific, web-based learning environment that instructors can use to offer students a combination of practice exercises, multimedia tutorials, video explanations, online algorithmic homework, and more. Specific content is available 24 hours a day to help students succeed in their course.

SMARTHINKING® Live, Online Tutoring: Houghton Mifflin has partnered with SMARTHINKING to provide an easy-to-use, effective, online tutorial service. Through state-of-the-art tools and a two-way whiteboard, students communicate in real-time with qualified e-structors who can help them understand difficult concepts and guide them through the problem-solving process while studying or completing homework.

Three levels of service are offered to students.

- **Live Tutorial Help** provides real-time, one-on-one instruction.

- **Question submission** allows students to submit questions to the tutor outside the scheduled hours and receive a response within 24 hours.

- **Independent Study Resources** connects students around-the-clock to additional educational resources, ranging from interactive websites to Frequently Asked Questions.

Visit **smarthinking.com** for more information. *Limits apply; terms and hours of SMARTHINKING service are subject to change.*

Acknowledgments

The authors would like to thank the people who have reviewed this manuscript and provided many valuable suggestions.

Nkechi Madonna Agwu, *Borough of Manhattan Community College, City University of New York*

Robert L. Bernhardt, *East Carolina University*

Dr. Cherlyn L. Converse, *Vanguard University*

Mindy Diesslin, *Wright State University*

Dr. Vincent Dimiceli, *Oral Roberts University*

Fred Feldon, *Coastline Community College*

Allen G. Fuller, Ph.D., *Gordon College*

Troy L. Goodsell, *Brigham Young University—Idaho*

Louise Hainline, *Brooklyn College of CUNY*

Philip D. Harris, *Wright State University—Lake Campus*

Kathryn Lavelle, *Westchester Community College*

Kyong-Hee M. Lee, *Colby-Sawyer College*

Mary K. Patton, *University of Illinois at Springfield*

Judith A. Silver, *Marshall University*

Angela M. Tisi, *University of Northern Colorado*

Christinia Vertullo, *Marist College*

Denise A. Widup, *University of Wisconsin—Parkside*

Fred Worth, *Henderson State University*

AIM FOR SUCCESS

Welcome to *Mathematical Thinking and Quantitative Reasoning*. As you begin this course, we know two important facts: (1) You want to succeed. (2) We want you to succeed. In order to accomplish these goals, an effort is required from each of us. For the next few pages, we are going to show you what is required of you to achieve your goal and how we have designed this text to help you succeed.

✓ **TAKE NOTE**

Motivation alone will not lead to success. For instance, suppose a person who cannot swim is placed in a boat, taken out to the middle of a lake, and then thrown overboard. That person has a lot of motivation to swim, but there is a high likelihood the person will drown without some help. Motivation gives us the desire to learn but is not the same as learning.

Motivation

One of the most important keys to success is motivation. We can try to motivate you by offering interesting or important ways that you can benefit from mathematics. But, in the end, the motivation must come from you. On the first day of class it is easy to be motivated. Eight weeks into the term, it is harder to keep that motivation.

To stay motivated, there must be outcomes from this course that are worth your time, money, and energy. List some reasons you are taking this course. Do not make a mental list—actually write them out. Do this now.

Although we hope that one of the reasons you listed was an interest in mathematics, we know that many of you are taking this course because it is required to graduate. Although you may not agree that this course should be necessary, it is! If you are motivated to graduate or complete the requirements for your major, then use that motivation to succeed in this course. Do not become distracted from your goal of completing your education!

Commitment

To be successful, you must make a commitment to succeed. This means devoting time to math so that you achieve a better understanding of the subject.

List some activities (sports, hobbies, talents such as dance, art, or music) that you enjoy and at which you would like to become better. Do this now.

Next to these activities, put the number of hours each week that you spend practicing these activities.

Whether you listed surfing or sailing, aerobics or restoring cars, or any other activity you enjoy, note how many hours a week you spend on each activity. To succeed in math, you must be willing to commit the same amount of time. Success requires some sacrifice.

The "I Can't Do Math" Syndrome

There may be things you cannot do, for instance, lift a two-ton boulder. You can, however, do math. It is much easier than lifting the two-ton boulder. When you first learned the activities you listed above, you probably could not do them well. With practice, you got better. With practice, you will be better at math. Stay focused, motivated, and committed to success.

It is difficult for us to emphasize how important it is to overcome the "I Can't Do Math" syndrome. If you listen to interviews of very successful athletes after a particularly bad performance, you will note that they focus on the positive aspect of what they did, not the negative. Sports psychologists encourage athletes to always be positive—to have a "Can Do" attitude. You need to develop this attitude toward math.

Strategies for Success

Know the Course Requirements   To do your best in this course, you must know exactly what your instructor requires. Course requirements may be stated in a syllabus, which is a printed outline of the main topics of the course, or they may be presented orally. When they are listed in a syllabus or on other printed pages, keep them in a safe place. When they are presented orally, make sure to take complete notes. In either case, it is important that you understand them completely and follow them exactly. Be sure you know the answer to each of the following questions.

1. What is your instructor's name?

2. Where is your instructor's office?

3. At what times does your instructor hold office hours?

4. Besides the textbook, what other materials does your instructor require?

5. What is your instructor's attendance policy?

6. If you must be absent from a class meeting, what should you do before returning to class? What should you do when you return to class?

7. What is the instructor's policy regarding collection or grading of homework assignments?

8. What options are available if you are having difficulty with an assignment? Is there a math tutoring center?

9. If there is a math lab at your school, where is it located? What hours is it open?

10. What is the instructor's policy if you miss a quiz?

11. What is the instructor's policy if you miss an exam?

12. Where can you get help when studying for an exam?

Remember: Your instructor wants to see you succeed. If you need help, ask! Do not fall behind. If you were running a race and fell behind by 100 yards, you might be able to catch up, but it would require more effort than if you had not fallen behind.

Time Management   We know that there are demands on your time. Family, work, friends, and entertainment all compete for your time. We do not want to see you receive poor job evaluations because you are studying math. However, it is also true that we do not want to see you receive poor math test scores because you devoted too much time to work. When several competing and important tasks require your time and energy, the only way to manage the stress of being successful at each is to manage your time efficiently.

Instructors often advise students to spend twice the amount of time outside of class studying as they spend in the classroom. Time management is important if you are to accomplish this goal and succeed in school. The following activity is intended to help you structure your time more efficiently.

✔ **TAKE NOTE**

For time management to work, you must have realistic ideas of how much time is available. There are very few people who can *successfully* work full-time and go to school full-time. If you work 40 hours a week, take 15 units, spend the recommended time studying given at the right, and sleep 8 hours a day, you will use over 80% of the available hours in a week. That leaves less than 20% of the hours in a week for family, friends, eating, recreation, and other activities.

Take out a sheet of paper and list the names of each course you are taking this term, the number of class hours each course meets, and the number of hours you should spend outside of class studying course materials. Now create a weekly calendar with the days of the week across the top and each hour of the day in a vertical column. Fill in the calendar with the hours you are in class, the hours you spend at work, and other commitments such as sports practice, music lessons, or committee meetings. Then fill in the hours that are more flexible—for example, study time, recreation, and meal times.

	Monday	Tuesday	Wednesday	Thursday	Friday	Saturday	Sunday
10–11 a.m.	History	Rev Spanish	History	Rev Span Vocab	History	Jazz Band	
11–12 p.m.	Rev History	Spanish	Study Group	Spanish	Math Tutor	Jazz Band	
12–1 p.m.	Lunch		Lunch		Lunch		Soccer

We know that many of you must work. If that is the case, realize that working 10 hours a week at a part-time job is equivalent to taking a three-unit class. If you must work, consider letting your education progress at a slower rate to allow you to be successful at both work and school. There is no rule that says you must finish school in a certain time frame.

Schedule Study Time   As we encouraged you to do by filling out the time management form, schedule a certain time to study. You should think of this time as being at work or class. Reasons for "missing study time" should be as compelling as reasons for missing work or class. "I just didn't feel like it" is not a good reason to miss your scheduled study time. Although this may seem like an obvious exercise, list a few reasons you might want to study. Do this now.

Of course we have no way of knowing the reasons you listed, but from our experience one reason given quite frequently is "To pass the course." There is nothing wrong with that reason. If that is the most important reason for you to study, then use it to stay focused.

One method of keeping to a study schedule is to form a **study group**. Look for people who are committed to learning, who pay attention in class, and who are punctual. Ask them to join your group. Choose people with similar educational goals but different methods of learning. You can gain from seeing the material from a new perspective. Limit groups to four or five people; larger groups are unwieldy.

There are many ways to conduct a study group. Begin with the following suggestions and see what works best for your group.

1. Test each other by asking questions. Each group member might bring two or three sample test questions to each meeting.

2. Practice teaching each other. Many of us who are teachers learned a lot about our subject when we had to explain it to someone else.

3. Compare class notes. You might ask other students about material in your notes that is difficult for you to understand.

4. Brainstorm test questions.

5. Set an agenda for each meeting. Set approximate time limits for each agenda item and determine a quitting time.

And now, probably the most important aspect of studying is that it should be done in relatively small chunks. If you can only study three hours a week for this course (probably not enough for most people), do it in blocks of one hour on three separate days, preferably after class. Three hours of studying on a Sunday is not as productive as three hours of paced study.

Features of This Text That Promote Success

Preparing for Class   Before the class meeting in which your instructor begins a new section, you should read the title of the section. Next, browse through the chapter material, being sure to note each word in bold type. These words indicate important concepts that you must know to learn the material. Do not worry about trying to understand all the material. Your instructor is there to assist you with that endeavor. The purpose of browsing through the material is so that your brain will be prepared to accept and organize the new information when it is presented to you. Turn to page 416. Write down the title of Section 7.1.

Write down the words in the section that are in bold print. It is not necessary for you to understand the meanings of these words. You are in this class to learn their meanings.

Math Is Not a Spectator Sport   To learn mathematics, you must be an active participant. Listening and watching your instructor do mathematics is not enough. Mathematics requires that you interact with the lesson you are studying. If you have been writing down the things we have asked you to, you have been interactive. This textbook has been designed so that you can be an active learner.

Check Your Progress   One of the key instructional features of this text is a completely worked-out example followed by a *Check Your Progress*.

EXAMPLE 4 ■ Calculate Simple Interest

Calculate the simple interest due on a 45-day loan of $3500 if the annual interest rate is 8%.

Solution

Use the simple interest formula. Substitute the following values into the formula: $P = 3500$, $r = 8\% = 0.08$, and $t = \dfrac{\text{number of days}}{360} = \dfrac{45}{360}$.

$$I = Prt$$

$$I = 3500(0.08)\left(\frac{45}{360}\right)$$

$$I = 35$$

The simple interest due is $35.

CHECK YOUR PROGRESS 4   Calculate the simple interest due on a 120-day loan of $7000 if the annual interest rate is 5.25%.

Solution   *See page S25.*

    The simple interest formula can be used to find the interest rate on a loan when the interest, principal, and time period of the loan are known. An example is given below.

page 418

TAKE NOTE

If you have difficulty with a particular algebra topic, visit the textbook's Online Student Center at **hmco.college.com/pic/ aufmannMTQR** and study the Algebra Review appendix to refresh your skills.

Note that each Example is completely worked out and that the *Check Your Progress* following the example is not. Study the worked-out example carefully by working through each step. You should do this with paper and pencil.

Now work the *Check Your Progress*. If you get stuck, refer to the page number following the word *Solution*, which directs you to the page on which the *Check Your Progress* is solved—a complete worked-out solution is provided. Try to use the given solution to get a hint about the step you are stuck on. Then try to complete your solution.

When you have completed the solution, check your work against the solution we provide.

CHECK YOUR PROGRESS 4, *page 418*

$P = 7000, r = 5.25\% = 0.0525$

$$t = \frac{\text{number of days}}{360} = \frac{120}{360}$$

$I = Prt$

$$I = 7000(0.0525)\left(\frac{120}{360}\right)$$

$I = 122.5$

The simple interest due is $122.50.

---

page S25

Be aware that there is frequently more than one way to solve a problem. Your answer, however, should be the same as the given answer. If you have any question as to whether your method will "always work," check with your instructor or with someone in the math center.

Browse through the textbook and write down the page numbers on which two other paired example features occur.

Remember: Be an active participant in your learning process. When you are sitting in class watching and listening to an explanation, you may think that you understand. However, until you actually try to do it, you will have no confirmation of the new knowledge or skill. Most of us have had the experience of sitting in class thinking we knew how to do something only to get home and realize we didn't.

Rule Boxes   Pay special attention to rules placed in boxes. These rules tell you why certain types of problems are solved the way they are. When you see a rule, try to rewrite the rule in your own words.

Simple Interest Formula

The simple interest formula is

$I = Prt$

where *I* is the interest, *P* is the principal, *r* is the interest rate, and *t* is the time period.

Chapter Exercises   When you have completed studying a section, do the section exercises. Math is a subject that needs to be learned in small sections and practiced continually in order to be mastered. Doing the exercises in each exercise set will help you master the problem-solving techniques necessary for success. As you work through the exercises, check your answers to the odd-numbered exercises against those given in the back of the book.

Preparing for a Test   There are important features of this text that can be used to prepare for a test.

- Chapter Summary

- Chapter Review Exercises

- Chapter Test

After completing a chapter, read the Chapter Summary. (See page 489 for the Chapter 7 Summary.) This summary highlights the important topics covered in the chapter. The page number following each topic refers you to the page in the text on which you can find more information about the concept.

Following the Chapter Summary are Chapter Review Exercises (see page 490). Doing the review exercises is an important way of testing your understanding of the chapter. The answer to each review exercise is given at the back of the book, along with, in brackets, the section reference from which the question was taken (see page A36). After checking your answers, restudy any section from which a question you missed was taken. It may be helpful to retry some of the exercises for that section to reinforce your problem-solving techniques.

Each chapter ends with a Chapter Test (see page 492). This test should be used to prepare for an exam. We suggest that you try the Chapter Test a few days before your actual exam. Take the test in a quiet place and try to complete the test in the same amount of time you will be allowed for your exam. When taking the Chapter Test, practice the strategies of successful test takers: 1) scan the entire test to get a feel for the questions; 2) read the directions carefully; 3) work the problems that are easiest for you first; and perhaps most importantly, 4) try to stay calm.

When you have completed the Chapter Test, check your answers (see page A36). Next to each answer is, in brackets, the reference to the section from which the question was taken. If you missed a question, review the material in that section and re-work some of the exercises from that section. This will strengthen your ability to perform the skills learned in that section.

Your career goal goes here. →

Is it difficult to be successful? YES! Successful music groups, artists, professional athletes, teachers, sociologists, chefs, and _____ have to work very hard to achieve their goals. They focus on their goals and ignore distractions. The things we ask you to do to achieve success take time and commitment. We are confident that if you follow our suggestions, you will succeed.

CHAPTER ①

Problem Solving

1.3 Problem Solving Using Sets

Most occupations require good problem-solving skills. For instance, architects and engineers must solve many complicated problems as they design and construct modern buildings that are aesthetically pleasing, functional, and that meet stringent safety requirements. Two goals of this chapter are to help you become a better problem solver and to demonstrate that problem solving can be an enjoyable experience.

One problem that many have enjoyed is the Monty Hall (host of the game show *Let's Make a Deal*) problem, which is stated as follows. The grand prize in *Let's Make a Deal* is behind one of three curtains. Less desirable prizes (for instance, a goat and a box of candy) are behind the other two curtains. You select one of the curtains, say curtain A. Monty Hall reveals one of the less desirable prizes behind one of the other curtains. You are then given the opportunity either to stay with your original choice or to choose the remaining closed curtain.

Curtain A	Curtain B	Curtain C

Marilyn vos Savant, author of the "Ask Marilyn" column featured in *Parade Magazine,* analyzed this problem,[1] claiming that you *double* your chances of winning the grand prize by switching to the other closed curtain. Many readers, including some mathematicians, responded with arguments that contradicted Marilyn's analysis.

What do you think? Do you have a better chance of winning the grand prize by switching to the other closed curtain or staying with your original choice?

Of course there is also the possibility that it does not matter, if the chances of winning are the same with either strategy.

Discuss the Monty Hall problem with some of your friends and classmates. Is everyone in agreement? Additional information on this problem is given in Exploration Exercise 56 on page 15.

---

1. "Ask Marilyn," *Parade Magazine*, September 9, 1990, p. 15.

Online Study Center

For online student resources, visit this textbook's Online Study Center at **college.hmco.com/pic/aufmannMTQR.**

| # Inductive and Deductive Reasoning

Inductive Reasoning

The type of reasoning that forms a conclusion based on the examination of specific examples is called *inductive reasoning*. The conclusion formed by using inductive reasoning is often called a **conjecture,** since it may or may not be correct.

> **Inductive Reasoning**
>
> **Inductive reasoning** is the process of reaching a general conclusion by examining specific examples.

When you examine a list of numbers and predict the next number in the list according to some pattern you have observed, you are using inductive reasoning.

EXAMPLE 1 ▪ Use Inductive Reasoning to Predict a Number

Use inductive reasoning to predict the most probable next number in each of the following lists.

a. 3, 6, 9, 12, 15, ?   **b.** 1, 3, 6, 10, 15, ?

Solution

a. Each successive number is 3 larger than the preceding number. Thus we predict that the most probable next number in the list is 3 larger than 15, which is 18.

b. The first two numbers differ by 2. The second and the third numbers differ by 3. It appears that the difference between any two numbers is always 1 more than the preceding difference. Since 10 and 15 differ by 5, we predict that the next number in the list will be 6 larger than 15, which is 21.

CHECK YOUR PROGRESS 1   Use inductive reasoning to predict the most probable next number in each of the following lists.

a. 5, 10, 15, 20, 25, ?   **b.** 2, 5, 10, 17, 26, ?

Solution   *See page S1.*

Inductive reasoning is not used just to predict the next number in a list. In Example 2 we use inductive reasoning to make a conjecture about an arithmetic procedure.

EXAMPLE 2 ▪ Use Inductive Reasoning to Make a Conjecture

Consider the following procedure: Pick a number. Multiply the number by 8, add 6 to the product, divide the sum by 2, and subtract 3.

    Complete the above procedure for several different numbers. Use inductive reasoning to make a conjecture about the relationship between the size of the resulting number and the size of the original number.

Solution

Suppose we pick 5 as our original number. Then the procedure would produce the following results:

Original number: 5
Multiply by 8: $8 \times 5 = 40$
Add 6: $40 + 6 = 46$
Divide by 2: $46 \div 2 = 23$
Subtract 3: $23 - 3 = 20$

> ✔ **TAKE NOTE**
>
> In Example 5, we will use a deductive method to verify that the procedure in Example 2 always yields a result that is four times the original number.

We started with 5 and followed the procedure to produce 20. Starting with 6 as our original number produces a final result of 24. Starting with 10 produces a final result of 40. Starting with 100 produces a final result of 400. In each of these cases the resulting number is four times the original number. We *conjecture* that following the given procedure will produce a resulting number that is four times the original number.

CHECK YOUR PROGRESS 2   Consider the following procedure: Pick a number. Multiply the number by 9, add 15 to the product, divide the sum by 3, and subtract 5.

Complete the above procedure for several different numbers. Use inductive reasoning to make a conjecture about the relationship between the size of the resulting number and the size of the original number.

Solution   *See page S1.*

---

historical note

Galileo Galilei
(găl-ə-lā′-ē) entered the University of Pisa to study medicine at the age of 17, but he soon realized that he was more interested in the study of astronomy and the physical sciences. Galileo's study of pendulums assisted in the development of pendulum clocks. ∎

Scientists often use inductive reasoning. For instance, Galileo Galilei (1564–1642) used inductive reasoning to discover that the time required for a pendulum to complete one swing, called the *period* of the pendulum, depends on the length of the pendulum. Galileo did not have a clock, so he measured the periods of pendulums in "heartbeats." The following table shows some results obtained for pendulums of various lengths. For the sake of convenience, a length of 10 inches has been designated as 1 unit.

Length of Pendulum, in Units	Period of Pendulum, in Heartbeats
1	1
4	2
9	3
16	4

The period of a pendulum is the time it takes for the pendulum to swing from left to right and back to its original position.

EXAMPLE 3 ■ Use Inductive Reasoning to Solve an Application

Use the data in the table on the preceding page and inductive reasoning to answer each of the following.

a. If a pendulum has a length of 25 units, what is its period?

b. If the length of a pendulum is quadrupled, what happens to its period?

Solution

a. In the table, each pendulum has a period that is the square root of its length. Thus we conjecture that a pendulum with a length of 25 units will have a period of 5 heartbeats.

b. In the table, a pendulum with a length of 4 units has a period that is twice that of a pendulum with a length of 1 unit. A pendulum with a length of 16 units has a period that is twice that of a pendulum with a length of 4 units. It appears that quadrupling the length of a pendulum doubles its period.

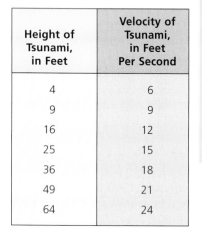

Height of Tsunami, in Feet	Velocity of Tsunami, in Feet Per Second
4	6
9	9
16	12
25	15
36	18
49	21
64	24

CHECK YOUR PROGRESS 3   A tsunami is a sea wave produced by an underwater earthquake. The velocity of a tsunami as it approaches land depends on the height of the tsunami. Use the table at the left and inductive reasoning to answer each of the following questions.

a. What happens to the height of a tsunami when its velocity is doubled?

b. What should be the height of a tsunami if its velocity is 30 feet per second?

Solution   *See page S1.*

Conclusions based on inductive reasoning may be incorrect. As an illustration, consider the circles shown below. For each circle, all possible line segments have been drawn to connect each dot on the circle with all the other dots on the circle.

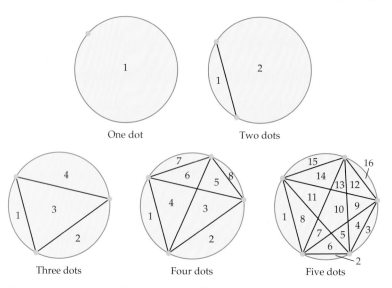

The maximum numbers of regions formed by connecting dots on a circle

For each circle, count the number of regions formed by the line segments that connect the dots on the circle. Your results should agree with the results in the following table.

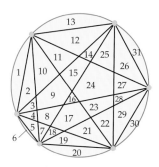

The line segments connecting six dots on a circle yield a maximum of 31 regions.

Number of dots	1	2	3	4	5	6
Maximum number of regions	1	2	4	8	16	?

There appears to be a pattern. Each additional dot seems to double the number of regions. Guess the maximum number of regions you expect for a circle with six dots. Check your guess by counting the maximum number of regions formed by the line segments that connect six dots on a *large* circle. Your drawing will show that for six dots, the maximum number of regions is 31 (see the figure at the left), not 32 as you may have guessed. With seven dots the maximum number of regions is 57. This is a good example to keep in mind. Just because a pattern holds true for a few cases, it does not mean the pattern will continue. When you use inductive reasoning, you have no guarantee that your conclusion is correct.

Counterexamples

A statement is a true statement if and only if it is true in all cases. If you can find *one case* for which a statement is not true, called a **counterexample,** then the statement is a false statement. In Example 4 we verify that each statement is a false statement by finding a counterexample for each.

EXAMPLE 4 ■ Find a Counterexample

Verify that each of the following statements is a false statement by finding a counterexample.

For all numbers x:

a. $|x| > 0$ **b.** $x^2 > x$ **c.** $\sqrt{x^2} = x$

Solution

A statement may have many counterexamples, but we need only find one counterexample to verify that the statement is false.

a. Let $x = 0$. Then $|0| = 0$. Because 0 is not greater than 0, we have found a counterexample. Thus "for all x, $|x| > 0$" is a false statement.

b. For $x = 1$ we have $1^2 = 1$. Since 1 is not greater than 1, we have found a counterexample. Thus "for all x, $x^2 > x$" is a false statement.

c. Consider $x = -3$. Then $\sqrt{(-3)^2} = \sqrt{9} = 3$. Since 3 is not equal to -3, we have found a counterexample. Thus "for all x, $\sqrt{x^2} = x$" is a false statement.

CHECK YOUR PROGRESS 4   Verify that each of the following statements is a false statement by finding a counterexample for each.

For all numbers x:

a. $\dfrac{x}{x} = 1$ **b.** $\dfrac{x + 3}{3} = x + 1$ **c.** $\sqrt{x^2 + 16} = x + 4$

Solution   See page S1.

QUESTION   *How many counterexamples are needed to prove that a statement is false?*

Deductive Reasoning

Another type of reasoning is called *deductive reasoning*. Deductive reasoning is distinguished from inductive reasoning in that it is the process of reaching a conclusion by applying general principles and procedures.

> **Deductive Reasoning**
>
> **Deductive reasoning** is the process of reaching a conclusion by applying general assumptions, procedures, or principles.

EXAMPLE 5 ■ Use Deductive Reasoning to Establish a Conjecture

Use deductive reasoning to show that the following procedure produces a number that is four times the original number.

   Procedure: Pick a number. Multiply the number by 8, add 6 to the product, divide the sum by 2, and subtract 3.

Solution

Let *n* represent the original number.

Multiply the number by 8:	$8n$
Add 6 to the product:	$8n + 6$
Divide the sum by 2:	$\dfrac{8n + 6}{2} = 4n + 3$
Subtract 3:	$4n + 3 - 3 = 4n$

We started with *n* and ended with 4*n*. The procedure given in this example produces a number that is four times the original number.

CHECK YOUR PROGRESS 5   Use deductive reasoning to show that the following procedure produces a number that is three times the original number.

   Procedure: Pick a number. Multiply the number by 6, add 10 to the product, divide the sum by 2, and subtract 5. *Hint:* Let *n* represent the original number.

Solution   See page S1.

ANSWER   *One*

MathMatters   **The MYST® Adventure Games
and Inductive Reasoning**

Most games have several rules, and the players are required to use a combination of deductive and inductive reasoning to play the game. However, the MYST® computer/video adventure games have no specific rules. Thus your only option is to explore and make use of inductive reasoning to discover the clues needed to solve the game.

MYST® III: *EXILE*

Inductive Reasoning vs. Deductive Reasoning

In Example 6 we analyze arguments to determine whether they use inductive or deductive reasoning.

EXAMPLE 6 ■ Determine Types of Reasoning

Determine whether each of the following arguments is an example of inductive reasoning or deductive reasoning.

a. During the past 10 years, a tree has produced plums every other year. Last year the tree did not produce plums, so this year the tree will produce plums.

b. All home improvements cost more than the estimate. The contractor estimated my home improvement will cost $35,000. Thus my home improvement will cost more than $35,000.

Solution

a. This argument reaches a conclusion based on specific examples, so it is an example of inductive reasoning.

b. Because the conclusion is a specific case of a general assumption, this argument is an example of deductive reasoning.

CHECK YOUR PROGRESS 6   Determine whether each of the following arguments is an example of inductive reasoning or deductive reasoning.

a. All Janet Evanovich novels are worth reading. The novel *To the Nines* is a Janet Evanovich novel. Thus *To the Nines* is worth reading.

b. I know I will win a jackpot on this slot machine in the next 10 tries, because it has not paid out any money during the last 45 tries.

Solution   *See page S1.*

Logic Puzzles

Some logic puzzles, similar to the one in Example 7, can be solved by using deductive reasoning and a chart that enables us to display the given information in a visual manner.

EXAMPLE 7 ■ Solve a Logic Puzzle

Each of four neighbors, Sean, Maria, Sarah, and Brian, has a different occupation (editor, banker, chef, or dentist). From the following clues, determine the occupation of each neighbor.

1. Maria gets home from work after the banker but before the dentist.
2. Sarah, who is the last to get home from work, is not the editor.
3. The dentist and Sarah leave for work at the same time.
4. The banker lives next door to Brian.

Solution

From clue 1, Maria is not the banker or the dentist. In the following chart, write X1 (which stands for "ruled out by clue 1") in the Banker and the Dentist columns of Maria's row.

	Editor	Banker	Chef	Dentist
Sean				
Maria		X1		X1
Sarah				
Brian				

From clue 2, Sarah is not the editor. Write X2 (ruled out by clue 2) in the Editor column of Sarah's row. We know from clue 1 that the banker is not the last to get home, and we know from clue 2 that Sarah is the last to get home; therefore, Sarah is not the banker. Write X2 in the Banker column of Sarah's row.

	Editor	Banker	Chef	Dentist
Sean				
Maria		X1		X1
Sarah	X2	X2		
Brian				

From clue 3, Sarah is not the dentist. Write X3 for this condition. There are now X's for three of the four occupations in Sarah's row; therefore, Sarah must be the chef. Insert a $\checkmark$ in that box. Since Sarah is the chef, none of the other three people can be the chef. Write X3 for these conditions. There are now Xs for three of the four occupations in Maria's row; therefore, Maria must be the editor. Insert a $\checkmark$ to indicate that Maria is the editor, and write X3 twice to indicate that neither Sean nor Brian is the editor.

	Editor	Banker	Chef	Dentist
Sean	X3		X3	
Maria	$\checkmark$	X1	X3	X1
Sarah	X2	X2	$\checkmark$	X3
Brian	X3		X3	

From clue 4, Brian is not the banker. Write X4 for this condition. Since there are three Xs in the Banker column, Sean must be the banker. Insert a $\checkmark$ in that box. Thus Sean cannot be the dentist. Write X4 in that box. Since there are 3 Xs in the Dentist column, Brian must be the dentist. Insert a $\checkmark$ in that box.

	Editor	Banker	Chef	Dentist
Sean	X3	$\checkmark$	X3	X4
Maria	$\checkmark$	X1	X3	X1
Sarah	X2	X2	$\checkmark$	X3
Brian	X3	X4	X3	$\checkmark$

Sean is the banker, Maria is the editor, Sarah is the chef, and Brian is the dentist.

CHECK YOUR PROGRESS 7   Brianna, Ryan, Tyler, and Ashley were recently elected as the new class officers (president, vice president, secretary, treasurer) of the sophomore class at Summit College. From the following clues, determine which position each holds.

1. Ashley is younger than the president but older than the treasurer.
2. Brianna and the secretary are both the same age, and they are the youngest members of the group.
3. Tyler and the secretary are next door neighbors.

Solution   *See page S1.*

Investigation

The Game of *Sprouts* by John H. Conway

The mathematician John H. Conway has created several games that are easy to play but complex enough to provide plenty of mental stimulation. For instance, in 1967 Conway, along with Michael Paterson, created the two-person, paper-and-pencil game of *Sprouts*. After more than three decades, the game of *Sprouts* has not been completely analyzed. Here are the rules for *Sprouts*.

John H. Conway
Princeton University

Rules for Sprouts

1. A few spots (dots) are drawn on a piece of paper.

2. The players alternate turns. A turn (move) consists of drawing a curve between two spots or drawing a curve that starts at a spot and ends at the same spot. The active player then places a new spot on the new curve. **No curve may pass through a previously drawn spot. No curve may cross itself or a previously drawn curve.**

3. A spot with no rays emanating from it has three lives. A spot with one ray emanating from it has two lives. A spot with two rays emanating from it has one life. A spot is dead and cannot be used when it has three rays that emanate from it. See the figure at the left.

4. The winner is the last player who is able to draw a curve.

Here is an example of a game of *2-Spot Sprouts*.

A spot with three lives.

A spot with two lives.

A spot with one life.

A dead spot.

The status of a spot in the game of *Sprouts*

A Game of 2-Spot Sprouts

Start

First Move

The first player connects *A* with *B* and draws spot *C*.

Second Move

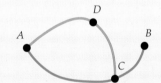

The second player connects *A* with *C* and draws spot *D*.

Third Move

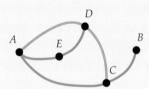

The first player connects *D* with *A* and draws spot *E*.

Fourth Move

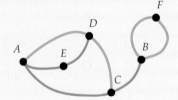

The second player connects *B* to itself and draws spot *F*. The second player is the winner because the first player cannot draw another curve.

(continued)

Note in the previous game that the spots **E** and **F** both have one life, but they cannot be connected because no curve can cross a previously drawn curve.

Investigation Exercises

1. Play a game of *1-Spot Sprouts.* Explain how you know that it is not possible for the first player to win a game of *1-Spot Sprouts.*

2. Every *n-Spot Sprouts* game has a maximum of $3n - 1$ moves. To verify this, we use the following deductive argument.

    At the start of a game, each spot has three lives. Thus an *n-Spot Sprouts* game starts with $3n$ lives. Each time a move is completed, two lives are killed and one life is created. Thus each move decreases the number of lives by one. The game cannot continue when only one life remains. Thus the maximum number of moves is $3n - 1$.

    It can also be shown that an *n-Spot Sprouts* game must have at least $2n$ turns. Play several *2-Spot Sprouts* games. Did each *2-Spot Sprouts* game you played have at least $2(2) = 4$ moves and at most $3(2) - 1 = 5$ moves?

3. Play several *3-Spot Sprouts* games. Did each *3-Spot Sprouts* game you played have at least $2(3) = 6$ moves and at most $3(3) - 1 = 8$ moves?

4. In a *2-Spot Sprouts* game, the second player can always play in a manner that will guarantee a win. In the following exercises, you are asked to illustrate a winning strategy for the second player in three situations.

    a. In a *2-Spot Sprouts* game, the first player makes the first move as shown at the upper left. What move can the second player make to guarantee a win? Explain how you know this move will guarantee that the second player will win.

    b. In a *2-Spot Sprouts* game, the first player makes the first move as shown at the lower left. What move can the second player make to guarantee a win? Explain how you know this move will guarantee that the second player will win.

    c. In a *2-Spot Sprouts* game, the first player draws a curve from *A* to *B* and the second player draws a curve from *B* back to *A* as shown at the right.

    At this point, the game can progress in several different ways. However, it is possible to show that regardless of how the first player responds on the third move, the second player can win the game on the fourth move. Choose, *at random,* a next move for the first player and demonstrate how the second player can win the game on the fourth move.

    The first two moves in a game of *2-Spot Sprouts*

    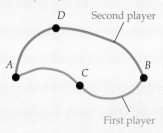

5. In 1990, David Applegate, Daniel Sleator, and Guy Jacobson used a computer program to try to analyze the game of *n-Spot Sprouts.* From their work, they *conjectured* that the first player has a winning strategy when the number of spots divided by 6 leaves a remainder of 3, 4, or 5. Assuming that this conjecture is correct, determine in which *n-Spot Sprouts* games, with $1 \leq n \leq 11$, the first player has a winning strategy.

The first move in a game of *2-Spot Sprouts*

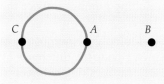

The first move in a game of *2-Spot Sprouts*

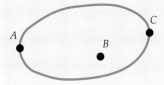

Exercise Set 1.1

In Exercises 1–10, use inductive reasoning to predict the most probable next number in each list.

1. 4, 8, 12, 16, 20, 24, ?

2. 5, 11, 17, 23, 29, 35, ?

3. 3, 5, 9, 15, 23, 33, ?

4. 1, 8, 27, 64, 125, ?

5. 1, 4, 9, 16, 25, 36, 49, ?

6. 80, 70, 61, 53, 46, 40, ?

7. $\dfrac{3}{5}, \dfrac{5}{7}, \dfrac{7}{9}, \dfrac{9}{11}, \dfrac{11}{13}, \dfrac{13}{15}, ?$

8. $\dfrac{1}{2}, \dfrac{2}{3}, \dfrac{3}{4}, \dfrac{4}{5}, \dfrac{5}{6}, \dfrac{6}{7}, ?$

9. 2, 7, −3, 2, −8, −3, −13, −8, −18, ?

10. 1, 5, 12, 22, 35, ?

In Exercises 11–16, use inductive reasoning to decide whether the conclusion for each argument is correct. *Note:* The numbers 1, 2, 3, 4, . . . are called **natural numbers** or **counting numbers**. The numbers . . . , −3, −2, −1, 0, 1, 2, 3, . . . are called **integers**.

11. The sum of any two even numbers is an even number.

12. If a number with three or more digits is divisible by 4, then the number formed by the last two digits of the number is divisible by 4.

13. The product of an odd integer and an even integer is always an even number.

14. The cube of an odd integer is always an odd number.

15. Pick any counting number. Multiply the number by 6. Add 8 to the product. Divide the sum by 2. Subtract 4 from the quotient. The resulting number is twice the original number.

16. Pick any counting number. Multiply the number by 8. Subtract 4 from the product. Divide the difference by 4. Add 1 to the quotient. The resulting number is twice the original number.

Experimental Data Galileo used inclines similar to the one shown below to measure the distance balls of various weights would travel in equal time intervals. The conclusions that Galileo reached from these experiments were contrary to the prevailing Aristotelian theories on the subject, and he lost his post at the University of Pisa because of them.

An experiment with an incline and three balls produced the following results. The three balls are each the same size; however, ball A has a mass of 20 grams, ball B has a mass of 40 grams, and ball C has a mass of 80 grams.

Time, in Seconds	Distance Traveled, in Inches		
	Ball A (20 grams)	Ball B (40 grams)	Ball C (80 grams)
1	6	6	6
2	24	24	24
3	54	54	54
4	96	96	96

In Exercises 17–24, use the above table and inductive reasoning to answer each question.

17. If the weight of a ball is doubled, what effect does this have on the distance it rolls in a given time interval?

18. If the weight of a ball is quadrupled, what effect does this have on the distance it rolls in a given time interval?

19. How far will ball A travel in 5 seconds?

20. How far will ball A travel in 6 seconds?

21. If a particular time is doubled, what effect does this have on the distance a ball travels?

22. If a particular time is tripled, what effect does this have on the distance a ball travels?

23. How much time is required for one of the balls to travel 1.5 inches?

24. How far will one of the balls travel in 1.5 seconds?

In Exercises 25–32, determine whether the argument is an example of inductive reasoning or deductive reasoning.

25. Andrea enjoyed reading the *Dark Tower* series by Stephen King, so I know she will like his next novel.

26. All pentagons have exactly five sides. Figure A is a pentagon. Therefore, Figure A has exactly five sides.

27. Every English setter likes to hunt. Duke is an English setter, so Duke likes to hunt.

28. Cats don't eat tomatoes. Scat is a cat. Therefore, Scat does not eat tomatoes.

29. A number is a "neat" number if the sum of the cubes of its digits equals the number. Therefore, 153 is a "neat" number.

30. The Atlanta Braves have won five games in a row. Therefore, the Atlanta Braves will win their next game.

31. Since

$$11 \times (1)(101) = 1111$$
$$11 \times (2)(101) = 2222$$
$$11 \times (3)(101) = 3333$$
$$11 \times (4)(101) = 4444$$
$$11 \times (5)(101) = 5555$$

we know that the product of 11 and a multiple of 101 is a number in which every digit is the same.

32. The following equations show that $n^2 - n + 11$ is a prime number for all counting numbers $n = 1, 2, 3, 4, \dots$.

$$n = 1 \quad (1)^2 - 1 + 11 = 11$$
$$n = 2 \quad (2)^2 - 2 + 11 = 13$$
$$n = 3 \quad (3)^2 - 3 + 11 = 17$$
$$n = 4 \quad (4)^2 - 4 + 11 = 23$$

Note: A **prime number** is a counting number greater than 1 that has no counting number factors other than itself and 1. The first 15 prime numbers are 2, 3, 5, 7, 11, 13, 17, 19, 23, 29, 31, 37, 41, 43, and 47.

In Exercises 33–42, find a counterexample to show that the statement is false.

33. $x > \dfrac{1}{x}$

34. $x + x > x$

35. $x^3 \geq x$

36. $|x + y| = |x| + |y|$

37. $-x < x$

38. $\dfrac{(x + 1)(x - 1)}{(x - 1)} = x + 1$ *Hint:* Division by zero is undefined.

39. If the sum of two natural numbers is even, then the product of the two natural numbers is even.

40. If the product of two natural numbers is even, then both of the numbers are even numbers.

41. Pick any three-digit counting number. Reverse the digits of the original number. The difference of these two numbers has a tens digit of 9.

42. If a counting number with two or more digits remains the same with its digits reversed, then the counting number is a multiple of 11.

43. Use deductive reasoning to show that the following procedure always produces a number that is equal to the original number.
Procedure: Pick a number. Multiply the number by 6 and add 8. Divide the sum by 2, subtract twice the original number, and subtract 4.

44. Use deductive reasoning to show that the following procedure always produces the number 5.
Procedure: Pick a number. Add 4 to the number and multiply the sum by 3. Subtract 7 and then decrease this difference by the triple of the original number.

45. **Stocks** Each of four siblings (Anita, Tony, Maria, and Jose) is given $5000 to invest in the stock market. Each chooses a different stock. One chooses a utility stock, another an automotive stock, another a technology stock, and the other an oil stock. From

the following clues, determine which sibling bought which stock.

a. Anita and the owner of the utility stock purchased their shares through an online brokerage, whereas Tony and the owner of the automotive stock did not.

b. The gain in value of Maria's stock is twice the gain in value of the automotive stock.

c. The technology stock is traded on NASDAQ, whereas the stock that Tony bought is traded on the New York Stock Exchange.

46. Gourmet Chefs The Changs, Steinbergs, Ontkeans, and Gonzaleses were winners in the All-State Cooking Contest. There was a winner in each of four categories: soup, entrée, salad, and dessert. From the following clues, determine in which category each family was the winner.

a. The soups were judged before the Ontkeans' winning entry.

b. This year's contest was the first for the Steinbergs and for the winner in the dessert category. The Changs and the winner of the soup category entered last year's contest.

c. The winning entrée took 2 hours to cook, whereas the Steinbergs' entrée required no cooking at all.

47. Collectibles The cities of Atlanta, Chicago, Philadelphia, and Seattle held conventions this summer for collectors of coins, stamps, comic books, and baseball cards. From the following clues, determine which collectors met in which city.

a. The comic book collectors convention was in August, as was the convention held in Chicago.

b. The baseball card collectors did not meet in Philadelphia, and the coin collectors did not meet in Seattle or Chicago.

c. The convention in Atlanta was held during the week of July 4, whereas the coin collectors convention was held the week after that.

d. The convention in Chicago had more collectors attending it than did the stamp collectors convention.

48. Map Coloring The following map shows eight states in the central time zone of the United States. Four colors have been used to color the states such that no two bordering states are the same color.

a. Can this map be colored, using only three colors, such that no two bordering states are the same color? Explain.

b. Can this map be colored, using only two colors, such that no two bordering states are the same color? Explain.

49. Driving Time You need to buy groceries at the supermarket, deposit a check at the credit union, and purchase a book at the bookstore. You can complete the errands in any order; however, you must start and end at your home. The driving time, in minutes, between each of these locations is given in the following figure.

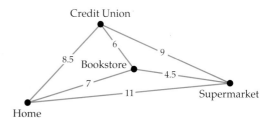

Find a route whose total driving time is less than 30 minutes.

50. Driving Time Suppose, in Exercise 49, that you need to go to the supermarket after you have completed the other two errands. What route should you take to minimize your travel time?

Extensions

CRITICAL THINKING

51. Use inductive reasoning to predict the next letter in the following list.

O, T, T, F, F, S, S, E, . . .

Hint: Look for a pattern that involves letters from words used for counting.

52. Use inductive reasoning to predict the next symbol in the following list.

Ⅿ, ♡, 8, Ⅿ, ⌀, . . .

Hint: Look for a pattern that involves counting numbers and symmetry about a line.

53. The Foucault Pendulum For the World's Fair in 1850, the French physicist Jean Bernard Foucault (*foo-ko*) installed a pendulum in the Pantheon in Paris. Foucault's pendulum had a period of about 16.4 seconds. If a pendulum with a length of 0.25 meter has a period of 1 second, determine which of the following lengths best approximates the length of Foucault's pendulum. *Hint:* Use the results of Example 3b.

    a. 7 meters   **b.** 27 meters

    c. 47 meters   **d.** 67 meters

54. Counterexamples Find a counterexample to prove that the inductive argument in

    a. Exercise 31 is incorrect.

    b. Exercise 32 is incorrect.

EXPLORATIONS

55. The Game of Life The Game of Life was invented by the mathematician John. H. Conway in 1970. It is not really a game in that there are no players, and no one wins or loses. The game starts by placing "pieces" on a grid. Once the pieces are in place, the rules determine everything that happens. It may sound like a boring game, but many people have found it to be an intriguing game that is full of surprises! Write a short essay that explains how the Game of Life is played and some of the reasons why it continues to be such a popular game.

Many websites have applets that allow you to play the Game of Life. Use one of these applets to play the Game of Life before you write your essay. *Note:* Conway's Game of Life is not the same game as the Game of Life board game with the pink and blue babies in the back seat of a plastic car.

56. The Monty Hall Problem Redux You can use the Internet to perform an experiment to determine the best strategy for playing the Monty Hall problem, which was stated in the Chapter 1 opener on page 1. Here is the procedure.

    a. Use a search engine to find a website that provides a simulation of the Monty Hall problem. This problem is also known as the three-door problem, so search under both of these titles. Once you locate a site that provides a simulation, play the simulation 30 times using the strategy of not switching. Record the number of times you win the grand prize. Now play the simulation 30 times using the strategy of switching. How many times did you win the grand prize by not switching? How many times did you win the grand prize by switching?

    b. On the basis of this experiment, which strategy seems to be the best strategy for winning the grand prize? What type of reasoning have you used?

SECTION 1.2 | # Problem-Solving Strategies

Polya's Problem-Solving Strategy

Ancient mathematicians such as Euclid and Pappus were interested in solving mathematical problems, but they were also interested in *heuristics*, the study of the methods and rules of discovery and invention. In the seventeenth century, the mathematician and philosopher René Descartes (1596–1650) contributed to the field of heuristics. He tried to develop a universal problem-solving method. Although he did not achieve this goal, he did publish some of his ideas in *Rules for the Direction of the Mind* and his better-known work *Discourse de la Methode.*

Another mathematician and philosopher, Gottfried Wilhelm Leibniz (1646–1716), planned to write a book on heuristics titled *Art of Invention.* Of the problem-solving process, Leibniz wrote, "Nothing is more important than to see the sources of invention which are, in my opinion, more interesting than the inventions themselves."

historical note

George Polya
After a brief stay at Brown University, George Polya (pōl'yə) moved to Stanford University in 1942 and taught there until his retirement. While at Stanford, he published 10 books and a number of articles for mathematics journals. Of the books Polya published, *How to Solve It* (1945) is one of his best known. In this book, Polya outlines a strategy for solving problems from virtually any discipline.

"A great discovery solves a great problem but there is a grain of discovery in the solution of any problem. Your problem may be modest; but if it challenges your curiosity and brings into play your inventive faculties, and if you solve it by your own means, you may experience the tension and enjoy the triumph of discovery." ■

One of the foremost recent mathematicians to make a study of problem solving was George Polya (1887–1985). He was born in Hungary and moved to the United States in 1940. The basic problem-solving strategy that Polya advocated consisted of the following four steps.

Polya's Four-Step Problem-Solving Strategy

1. Understand the problem.
2. Devise a plan.
3. Carry out the plan.
4. Review the solution.

Polya's four steps are deceptively simple. To become a good problem solver, it helps to examine each of these steps and determine what is involved.

Understand the Problem This part of Polya's four-step strategy is often overlooked. You must have a clear understanding of the problem. To help you focus on understanding the problem, consider the following questions.

- Can you restate the problem in your own words?
- Can you determine what is known about these types of problems?
- Is there missing information that, if known, would allow you to solve the problem?
- Is there extraneous information that is not needed to solve the problem?
- What is the goal?

Devise a Plan Successful problem solvers use a variety of techniques when they attempt to solve a problem. Here are some frequently-used procedures.

- Make a list of the known information.
- Make a list of information that is needed.
- Draw a diagram.
- Make an organized list that shows all the possibilities.
- Make a table or a chart.
- Work backwards.
- Try to solve a similar but simpler problem.
- Look for a pattern.
- Write an equation. If necessary, define what each variable represents.
- Perform an experiment.
- Guess at a solution and then check your result.
- Use indirect reasoning.

Carry Out the Plan Once you have devised a plan, you must carry it out.

- Work carefully.
- Keep an accurate and neat record of all your attempts.
- Realize that some of your initial plans will not work and that you may have to devise another plan or modify your existing plan.

Review the Solution   Once you have found a solution, check the solution.

- Ensure that the solution is consistent with the facts of the problem.
- Interpret the solution in the context of the problem.
- Ask yourself whether there are generalizations of the solution that could apply to other problems.

In Example 1 we apply Polya's four-step problem-solving strategy to solve a problem involving the number of routes between two points.

EXAMPLE 1 ■ Apply Polya's Strategy   *(Solve a Similar but Simpler Problem)*

Consider the map shown in Figure 1.1. Allison wishes to walk along the streets from point A to point B. How many direct routes can Allison take?

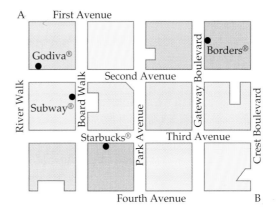

Figure 1.1 *City Map*

Solution

Understand the Problem   We would not be able to answer the question if Allison retraced her path or traveled away from point B. Thus we assume that on a direct route, she always travels along a street in a direction that gets her closer to point B.

Devise a Plan   The map in Figure 1.1 has many extraneous details. Thus we make a diagram that allows us to concentrate on the essential information. See the figure at the left.

Because there are many routes, we consider the similar but simpler diagrams shown below. The number at each street intersection represents the number of routes from point A to that particular intersection.

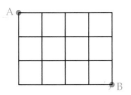

A simple diagram of the street map in Figure 1.1

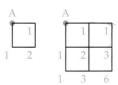

Simple Street Diagrams

✔ **TAKE NOTE**

The strategy of working a similar but simpler problem is an important problem-solving strategy that can be used to solve many problems.

A street diagram with the number of routes to each intersection labeled

Look for a pattern. It appears that the number of routes to an intersection is the *sum* of the number of routes to the adjacent intersection to its left and the number of routes to the intersection directly above. For instance, the number of routes to the intersection labeled 6 is the sum of the number of routes to the intersection to its left, which is three, and the number of routes to the intersection directly above, which is also three.

Carry Out the Plan Using the pattern discovered above, we see from the figure at the left that the number of routes from point A to point B is $20 + 15 = 35$.

Review the Solution Ask yourself whether a result of 35 seems reasonable. If you were required to draw each route, could you devise a scheme that would enable you to draw each route without missing a route or duplicating a route?

CHECK YOUR PROGRESS 1 Consider the street map in Figure 1.1. Allison wishes to walk directly from point A to point B. How many different routes can she take if she wants to go past Starbucks on Third Avenue?

Solution *See page S2.*

Example 2 illustrates the technique of using an organized list.

EXAMPLE 2 ■ **Apply Polya's Strategy** *(Make an Organized List)*

A baseball team won two out of their last four games. In how many different orders could they have two wins and two losses in four games?

Solution

Understand the Problem There are many different orders. The team may have won two straight games and lost the last two (WWLL). Or maybe they lost the first two games and won the last two (LLWW). Of course there are other possibilities, such as WLWL.

Devise a Plan We will make an *organized list* of all the possible orders. An organized list is a list that is produced using a system that ensures that each of the different orders will be listed once and only once.

Carry Out the Plan Each entry in our list must contain two Ws and two Ls. We will use a strategy that makes sure each order is considered, with no duplications. One such strategy is to always write a W unless doing so will produce too many Ws or a duplicate of one of the previous orders. If it is not possible to write a W, then and only then do we write an L. This strategy produces the six different orders shown below.

1. WWLL (Start with two wins)
2. WLWL (Start with one win)
3. WLLW
4. LWWL (Start with one loss)
5. LWLW
6. LLWW (Start with two losses)

Review the Solution   We have made an organized list. The list has no duplicates and the list considers all possibilities, so we are confident that there are six different orders in which a baseball team can win exactly two out of four games.

CHECK YOUR PROGRESS 2   A true-false quiz contains five questions. In how many ways can a student answer the questions if the student answers two of the questions with "false" and the other three with "true"?

Solution   *See page S2.*

In Example 3 we make use of several problem-solving strategies to solve a problem involving the total number of games to be played.

EXAMPLE 3 ■ Apply Polya's Strategy   *(Solve a Similar but Simpler Problem)*

In a basketball league consisting of 10 teams, each team plays each of the other teams exactly three times. How many league games will be played?

Solution

Understand the Problem   There are 10 teams in the league and each team plays exactly three games against each of the other teams. The problem is to determine the total number of league games that will be played.

Devise a Plan   Try the strategy of working a similar but simpler problem. Consider a league with only four teams (denoted by A, B, C, and D) in which each team plays each of the other teams only once. The diagram at the left illustrates that the games can be represented by line segments that connect the points A, B, C, and D.

Since each of the four teams will play a game against each of the other three, we might conclude that this would result in $4 \cdot 3 = 12$ games. However, the diagram shows only six line segments. It appears that our procedure has counted each game twice. For instance, when team A plays team B, team B also plays team A. To produce

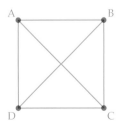

The possible pairings of a league with only four teams

the correct result, we must divide our previous result, 12, by 2. Hence, four teams can play each other once in $\frac{4 \cdot 3}{2} = 6$ games.

Carry Out the Plan Using the process developed above, we see that 10 teams can play each other *once* in a total of $\frac{10 \cdot 9}{2} = 45$ games. Since each team plays each opponent exactly three times, the total number of games is $45 \cdot 3 = 135$.

Review the Solution We could check our work by making a diagram that includes all 10 teams represented by dots labeled A, B, C, D, E, F, G, H, I, and J. Because this diagram would be somewhat complicated, let's try the method of making an organized list. The figure at the left shows an organized list in which the notation BC represents a game between team B and team C. The notation CB is not shown because it also represents a game between team B and team C. This list shows that 45 games are required for each team to play each of the other teams once. Also notice that the first row has nine items, the second row has eight items, the third row has seven items, and so on. Thus 10 teams require

$$9 + 8 + 7 + 6 + 5 + 4 + 3 + 2 + 1 = 45$$

games if each team plays every other team once and $45 \cdot 3 = 135$ games if each team plays exactly three games against each opponent.

AB AC AD AE AF AG AH AI AJ
BC BD BE BF BG BH BI BJ
CD CE CF CG CH CI CJ
DE DF DG DH DI DJ
EF EG EH EI EJ
FG FH FI FJ
GH GI GJ
HI HJ
IJ

An organized list of all possible games

CHECK YOUR PROGRESS 3 If six people greet each other at a meeting by shaking hands with one another, how many handshakes will take place?

Solution *See page S3.*

In Example 4 we make use of a table to solve a problem.

EXAMPLE 4 ▩ **Apply Polya's Strategy** *(Make a Table and Look for a Pattern)*

Determine the digit 100 places to the right of the decimal point in the decimal representation of $\frac{7}{27}$.

Solution

Understand the Problem Express the fraction $\frac{7}{27}$ as a decimal and look for a pattern that will enable you to determine the digit 100 places to the right of the decimal point.

Devise a Plan Dividing 27 into 7 by long division or by using a calculator produces the decimal 0.259259259.... Since the decimal representation repeats the digits 259 over and over forever, we know that the digit located 100 places to the right of the decimal point is either a 2, a 5, or a 9. A table may help us to see a pat-

CALCULATOR NOTE

Some calculators display $\frac{7}{27}$ as 0.25925925926. However, the last digit 6 is not correct. It is a result of the rounding process. The actual decimal representation of $\frac{7}{27}$ is the decimal 0.259259... or $0.\overline{259}$, in which the digits continue to repeat the 259 pattern forever.

tern and enable us to determine which one of these digits is in the 100th place. Since the decimal digits repeat every three digits, we use a table with three columns.

The first 15 decimal digits of $\frac{7}{27}$

Column 1		Column 2		Column 3	
Location	**Digit**	**Location**	**Digit**	**Location**	**Digit**
1st	2	2nd	5	3rd	9
4th	2	5th	5	6th	9
7th	2	8th	5	9th	9
10th	2	11th	5	12th	9
13th	2	14th	5	15th	9
⋮		⋮		⋮	

Carry Out the Plan Only in column 3 is each of the decimal digit *locations* evenly divisible by 3. From this pattern we can tell that the 99th decimal digit (because 99 is evenly divisible by 3) must be a 9. Since a 2 always follows a 9 in the pattern, the 100th decimal digit must be a 2.

Review the Solution The above table illustrates additional patterns. For instance, if each of the location numbers in column 1 is divided by 3, a remainder of 1 is produced. If each of the location numbers in column 2 is divided by 3, a remainder of 2 is produced. Thus we can find the decimal digit in any location by dividing the location number by 3 and examining the remainder. For instance, to find the digit in the 3200th decimal place of $\frac{7}{27}$, merely divide 3200 by 3 and examine the remainder, which is 2. Thus, the digit 3200 places to the right of the decimal point is a 5.

CHECK YOUR PROGRESS 4 Determine the ones digit of 4^{200}.

Solution *See page S3.*

Example 5 illustrates the method of working backwards. In problems in which you know a final result, this method may require the least effort.

EXAMPLE 5 ▪ Apply Polya's Strategy *(Work Backwards)*

In consecutive turns of a Monopoly game, Stacy first paid $800 for a hotel. She then lost half her money when she landed on Boardwalk. Next, she collected $200 for passing GO. She then lost half her remaining money when she landed on Illinois Avenue. Stacy now has $2500. How much did she have just before she purchased the hotel?

✔ **TAKE NOTE**

Example 5 can also be worked by using algebra. Let *A* be the amount of money Stacy had just before she purchased the hotel. Then

$$\frac{1}{2}\left[\frac{1}{2}(A - 800) + 200\right] = 2500$$

$$\frac{1}{2}(A - 800) + 200 = 5000$$

$$\frac{1}{2}(A - 800) = 4800$$

$$A - 800 = 9600$$

$$A = 10{,}400$$

Which do you prefer, the algebraic method or the method of working backwards?

Solution

Understand the Problem   We need to determine the number of dollars that Stacy had just prior to her $800 hotel purchase.

Devise a Plan   We could guess and check, but we might need to make several guesses before we found the correct solution. An algebraic method might work, but setting up the necessary equation could be a challenge. Since we know the end result, let's try the method of working backwards.

Carry Out the Plan   Stacy must have had $5000 just before she landed on Illinois Avenue; $4800 just before she passed GO; and $9600 prior to landing on Boardwalk. This means she had $10,400 just before she purchased the hotel.

Review the Solution   To check our solution we start with $10,400 and proceed through each of the transactions. $10,400 less $800 is $9600. Half of $9600 is $4800. $4800 increased by $200 is $5000. Half of $5000 is $2500.

CHECK YOUR PROGRESS 5   Melody picks a number. She doubles the number, squares the result, divides the square by 3, subtracts 30 from the quotient, and gets 18. What are the possible numbers that Melody could have picked? What operation does Melody perform that prevents us from knowing with 100% certainty which number she picked?

Solution   See page S3.

Some problems can be solved by making guesses and checking. Your first few guesses may not produce a solution, but quite often they will provide additional information that will lead to a solution.

EXAMPLE 6 ■ **Apply Polya's Strategy**   *(Guess and Check)*

The product of the ages, in years, of three teenagers is 4590. None of the teens are the same age. What are the ages of the teenagers?

Solution

Understand the Problem   We need to determine three distinct whole numbers, from the list 13, 14, 15, 16, 17, 18, and 19, that have a product of 4590.

Devise a Plan   If we represent the ages by *x*, *y*, and *z*, then $xyz = 4590$. We are unable to solve this equation, but we notice that 4590 ends in a zero. Hence, 4590 has a factor of 2 and a factor of 5, which means that at least one of the numbers we seek must be an even number and at least one number must have 5 as a factor. The only number in our list that has 5 as a factor is 15. Thus 15 is one of the numbers and at least one of the other numbers must be an even number. At this point we try to solve by *guessing and checking.*

Carry Out the Plan

$15 \cdot 16 \cdot 18 = 4320$	• No. This product is too small.
$15 \cdot 16 \cdot 19 = 4560$	• No. This product is too small.
$15 \cdot 17 \cdot 18 = 4590$	• Yes. This is the correct product.

The ages of the teenagers are 15, 17, and 18.

Review the Solution   Because $15 \cdot 17 \cdot 18 = 4590$ and each of the ages represents the age of a teenager, we know our solution is correct. None of the numbers 13, 14, 16, and 19 is a factor (divisor) of 4590, so there are no other solutions.

CHECK YOUR PROGRESS 6   Nothing is known about the personal life of the ancient Greek mathematician Diophantus except for the information in the following epigram. "Diophantus passed $\frac{1}{6}$ of his life in childhood, $\frac{1}{12}$ in youth, and $\frac{1}{7}$ more as a bachelor. Five years after his marriage was born a son who died four years before his father, at $\frac{1}{2}$ his father's (final) age."

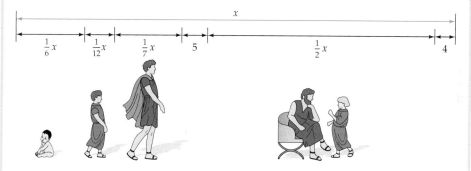

A diagram of the data, where *x* represents the age of Diophantus when he died

How old was Diophantus when he died? *Hint:* Although an equation can be used to solve this problem, the method of guessing and checking will probably require less effort. Also assume that his age when he died is a natural number.

Solution   See page S3.

QUESTION   *Is the process of guessing at a solution and checking your result one of Polya's problem-solving strategies?*

ANSWER   *Yes*

Karl Friedrich Gauss
(1777–1855)

Math Matters   A Mathematical Prodigy

Karl Friedrich Gauss (gous) was a German scientist and mathematician. His work encompassed several disciplines, including number theory, differential geometry, analysis, astronomy, geodesy, and optics. He is often called the "Prince of Mathematicians" and is known for having shown remarkable mathematical prowess as early as age three. It is reported that soon after Gauss entered elementary school, his teacher assigned an arithmetic problem designed to keep the students occupied for a lengthy period of time. The problem consisted of finding the sum of the first 100 natural numbers. Gauss was able to determine the correct sum in a matter of a few seconds. The following solution shows the thought process that Gauss applied as he solved the problem.

Understand the Problem   The sum of the first 100 natural numbers is represented by

$$1 + 2 + 3 + \cdots + 98 + 99 + 100$$

Devise a Plan   Adding the first 100 natural numbers from left to right would produce the desired sum, but would be time consuming and laborious. Gauss considered another method. He added 1 and 100 to produce 101. He noticed that 2 and 99 have a sum of 101, and that 3 and 98 have a sum of 101. Thus the 100 numbers could be thought of as 50 pairs, each with a sum of 101.

$$1 + 2 + 3 + \cdots + 98 + 99 + 100$$

101
101
101

Carry Out the Plan   To find the sum of the 50 pairs, each with a sum of 101, Gauss computed $50 \cdot 101$ and arrived at 5050 as the solution.

Review the Solution   Because the addends in an addition problem can be placed in any order without changing the sum, Gauss was confident that he had the correct solution.

An Extension   The solution of one problem often leads to solutions of additional problems. For instance, the sum $1 + 2 + 3 + \cdots + (n - 2) + (n - 1) + n$ can be found by using the following formula.

A Summation Formula for the First *n* Natural Numbers:

$$1 + 2 + 3 + \cdots + (n - 2) + (n - 1) + n = \frac{n(n + 1)}{2}$$

Some problems are deceptive. After reading one of these problems, you may think that the solution is obvious or impossible. These deceptive problems generally require that you carefully read the problem several times and that you check your solution to make sure it satisfies all the conditions of the problem.

EXAMPLE 7 ■ Solve a Deceptive Problem

A hat and a jacket together cost $100. The jacket costs $90 more than the hat. What are the cost of the hat and the cost of the jacket?

Solution

Understand the Problem   After reading the problem for the first time, you may think that the jacket costs $90 and the hat costs $10. The sum of these costs is $100, but the cost of the jacket is only $80 more than the cost of the hat. We need to find two dollar amounts that differ by $90 and whose sum is $100.

Devise a Plan   Write an equation using h for the cost of the hat and $h + 90$ for the cost of the jacket.

$$h + h + 90 = 100$$

Solve the above equation for h.

Carry Out the Plan

$$2h + 90 = 100 \qquad \text{• Collect like terms.}$$
$$2h = 10 \qquad \text{• Solve for } h.$$
$$h = 5$$

The cost of the hat is $5 and the cost of the jacket is $90 + $5 = $95.

Review the Solution   The sum of the costs is $5 + $95 = $100 and the cost of the jacket is $90 more than the cost of the hat. This check confirms that the hat costs $5 and the jacket costs $95.

CHECK YOUR PROGRESS 7   Two U.S. coins have a total value of 35¢. One of the coins is not a quarter. What are the two coins?

Solution   *See page S3.*

Reading and Interpreting Graphs

Graphs are often used to display numerical information in a visual format that allows the reader to see pertinent relationships and trends quickly. Three of the most common types of graphs are the *bar graph*, the *circle graph*, and the *broken-line graph*.

Figure 1.2 is a bar graph that displays the average U.S. movie theatre ticket prices for the years from 1996 to 2004. The years are displayed on the horizontal axis. Each vertical bar is used to display the average ticket price for a given year. The higher the bar, the greater the average ticket price for that year.

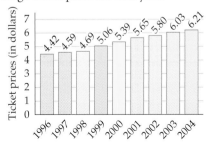

Figure 1.2 *Average U.S. Movie Theatre Ticket Prices*

Source: National Association of Theatre Owners

Figure 1.3 *Types of Automobile Accidents in Twin Falls in 2005*

Figure 1.3 is a circle graph that uses circular sectors to display the percent of automobile accidents of a particular type that occurred in the city of Twin Falls in 2005. The size of a particular sector is proportional to the percent of accidents that occurred in that category.

Figure 1.4 shows two broken-line graphs. The maroon broken-line graph displays the average age at first marriage for men for selected years from 1960 to 2000. The green broken-line graph displays the average age at first marriage for women for selected years during this same time period. The ⌇ symbol on the vertical axis indicates that the ages between 0 and 20 are not displayed. This break in the vertical axis allows the graph to be displayed in a compact form. The line segments that connect points on the graph indicate trends. Increasing trends are indicated by line segments that rise as they move to the right, and decreasing trends are indicated by line segments that fall as they move to the right. The blue arrows in Figure 1.4 show that the average age at which men married for the first time in the year 1990 was about 26 years, rounded to the nearest quarter of a year.

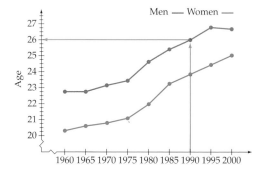

Figure 1.4 *U.S. Average Age at First Marriage*

Source: Bureau of the Census

EXAMPLE 8 ■ Use Graphs to Solve Problems

a. Use Figure 1.2 to determine the minimum average U.S. movie theatre ticket price for the years from 1996 to 2004.

b. Use Figure 1.3 to estimate the number of rear-end collisions that occurred in Twin Falls in the year 2005. *Note:* The total number of accidents in Twin Falls for the year 2005 was 4300.

c. Use Figure 1.4 to estimate the average age at which women married for the first time in the year 2000. Round to the nearest quarter of a year.

Solution

a. The minimum of the average ticket prices is displayed by the shortest vertical bar in Figure 1.2. Thus the minimum average U.S. movie theatre ticket price for the years from 1996 to 2004 was $4.42, in the year 1996.

b. Figure 1.3 indicates that in 2005, 29% of the 4300 automobile accidents in Twin Falls were rear-end collisions. Thus $0.29 \cdot 4300 = 1247$ of the accidents were accidents involving rear-end collisions.

c. To estimate the average age at which women married for the first time in the year 2000, locate 2000 on the horizontal axis of Figure 1.4 and then move directly upward to a point on the green broken-line graph. The height of this point represents the average age at first marriage for women in the year 2000, and it can be estimated by moving horizontally to the vertical axis on the left. Thus the average

age at first marriage for women in the year 2000 was about 25 years, rounded to the nearest quarter of a year.

CHECK YOUR PROGRESS 8

a. Use Figure 1.2 to determine the maximum average U.S. movie theatre ticket price for the years from 1996 to 2004.

b. Use Figure 1.3 to determine the number of lane change accidents that occurred in Twin Falls in the year 2005.

c. Use Figure 1.4 to estimate the average ages at first marriage for men and for women in the year 1975. Round to the nearest quarter of a year.

Solution   *See page S4.*

Investigation

Routes on a Probability Demonstrator

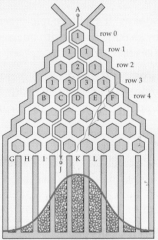

The object shown at the left is called a *Galton board* or *probability demonstrator*. It was invented by the English statistician Francis Galton (1822–1911). This particular board has 256 small red balls that are released so that they fall through an array of hexagons. The board is designed such that when a ball falls on a vertex of one of the hexagons, it is equally likely to fall to the left or to the right. As the ball continues its downward path, it strikes a vertex of a hexagon in the next row, where the process of falling to the left or to the right is repeated. After the ball passes through all the rows of hexagons, it falls into one of the bins at the bottom. In most cases the balls will form a *bell shape*, as shown by the green curve. Examine the numbers displayed in the hexagons in rows 0 through 3. Each number indicates the number of different routes a ball can take from point A to the top of that particular hexagon.

Investigation Exercises

1. How many routes can a ball take as it travels from point A to point B, from A to C, from A to D, from A to E, and from A to F? *Hint:* This problem is similar to Example 1 on page 17.

2. How many routes can a ball take as it travels from point A to points G, H, I, J, and K?

3. Explain how you know that the number of routes from point A to point J is the same as the number of routes from point A to point L.

4. Explain why the greatest number of balls tend to fall into the center bin.

5. The probability demonstrator shown at the lower left has nine rows of hexagons. Determine how many routes a ball can take as it travels from A to B, from A to C, from A to D, from A to E, from A to F, and from A to G.

Exercise Set 1.2

Use Polya's four-step problem-solving strategy and the problem-solving procedures presented in this section to solve each of the following exercises.

1. **Number of Girls** There are 364 first-grade students in Park Elementary School. If there are 26 more girls than boys, how many girls are there?

2. **Heights of Ladders** If two ladders are placed end to end, their combined height is 31.5 feet. One ladder is 6.5 feet shorter than the other ladder. What are the heights of the two ladders?

3. How many squares are in the following figure?

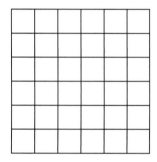

4. What is the 44th decimal digit in the decimal representation of $\frac{1}{11}$?

$$\frac{1}{11} = 0.09090909\ldots$$

5. **Cost of a Shirt** A shirt and a tie together cost $50. The shirt costs $30 more than the tie. What is the cost of the shirt?

6. **Number of Games** In a basketball league consisting of 12 teams, each team plays each of the other teams exactly twice. How many league games will be played?

7. **Number of Routes** Consider the following map. Tyler wishes to walk along the streets from point A to point B. How many direct routes (no backtracking) can Tyler take?

8. **Number of Routes** Use the map in Exercise 7 to answer each of the following.
   a. How many direct routes are there from A to B if you want to pass by Starbucks?
   b. How many direct routes are there from A to B if you want to stop at Subway for a sandwich?
   c. How many direct routes are there from A to B if you want to stop at Subway and at Starbucks?

9. **True-False Test** In how many ways can you answer a 12-question true-false test if you answer each question with either a "true" or a "false"?

10. **A Puzzle** A frog is at the bottom of a 17-foot well. Each time the frog leaps it moves up the side of the wall a distance of 3 feet. If the frog has not reached the top of the well, then the frog slides back 1 foot before it is ready to make another leap. How many leaps will the frog need to reach the top of the well?

11. **Number of Handshakes** If eight people greet each other at a meeting by shaking hands with one another, how many handshakes take place?

12. **Number of Line Segments** Twenty-four points are placed around a circle. A line segment is drawn between each pair of points. How many line segments are drawn?

13. **Number of Ducks and Pigs** The number of ducks and pigs in a field totals 35. The total number of legs among them is 98. Assuming each duck has exactly two legs and each pig has exactly four legs, determine how many ducks and how many pigs are in the field.

14. ✎ **Racing Strategies** Carla and Allison are sisters. They are on their way from school to home. Carla runs half the time and walks half the time. Allison runs half the distance and walks half the distance. If Carla and Allison walk at the same speed and run at the same speed, which one arrives home first? Explain.

15. **Change for a Quarter** How many ways can you make change for 25¢ using dimes, nickels, and/or pennies?

16. **Carpet for a Room** A room measures 12 feet by 15 feet. How many 3-foot-by-3-foot squares of carpet are needed to cover the floor of this room?

17. Determine the units digit of 4^{7022}.

18. Determine the units digit of 2^{6543}.

19. Determine the units digit of $3^{11,707}$.

20. Determine the units digit of 8^{8985}.

21. Find the following sums without using a calculator or a formula. *Hint:* Apply the procedure used by Gauss. (See the Math Matters on page 24.)

   a. $1 + 2 + 3 + 4 + \cdots + 397 + 398 + 399 + 400$

   b. $1 + 2 + 3 + 4 + \cdots + 547 + 548 + 549 + 550$

   c. $2 + 4 + 6 + 8 + \cdots + 80 + 82 + 84 + 86$

22. Explain how you could modify the procedure used by Gauss (see the Math Matters on page 24) to find the following sum.

$$1 + 2 + 3 + 4 + \cdots + 62 + 63 + 64 + 65$$

In Exercises 23–26, use only mental computations and/or estimations to answer each question.

23. The distance from Boston, Massachusetts to Los Angeles, California is about 2595 miles. Is it possible to walk this distance in 6 months?

24. Are there more than or less than 50,000 seconds in a day?

25. The area of a rectangle is found by multiplying its width by its length. For instance, a photograph that measures 4 inches by 5 inches has an area of $4 \times 5 = 20$ square inches. Doubling the width and the length of this photograph produces a new photograph which measures 8 inches by 10 inches. How many times as large is the area of this new photograph compared with the area of the original photograph?

26. The regular price of a watch is $50.00. During a sale, the price of this watch is reduced by 10% ($10\% = 10/100 = 1/10$) of its regular price. After the sale the price of this watch is increased by 10% of the sale price. Is the current price of the watch equal to its regular price, less than its regular price, or more than its regular price?

27. Palindromic Numbers A **palindromic number** is a whole number that remains unchanged when its digits are written in reverse order. Find all palindromic numbers that have exactly

   a. three digits and are the square of a natural number.

   b. four digits and are the cube of a natural number.

28. Speed of a Car A car has an odometer reading of 15,951 miles, which is a palindromic number. (See Exercise 27.) After 2 hours of continuous driving at a constant speed, the odometer reading is the next palindromic number. How fast, in miles per hour, was the car being driven during these 2 hours?

29. A Puzzle Three volumes of the series *Mathematics: Its Content, Methods, and Meaning* are on a shelf with no space between the volumes. Each volume is 1 inch thick without its covers. Each cover is $\frac{1}{8}$ inch thick. See the following figure. A bookworm bores horizontally from the first page of Volume 1 to the last page of Volume III. How far does the bookworm travel?

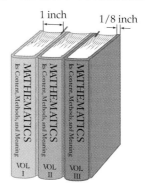

30. Connect the Dots Nine dots are arranged as shown. Is it possible to connect the nine dots with exactly four lines if you are not allowed to retrace any part of a line and you are not allowed to remove your pencil from the paper? If it can be done, demonstrate with a drawing.

31. Admissions The following bar graph shows the numbers of U.S. movie theatre admissions for the years from 1994 to 2004.

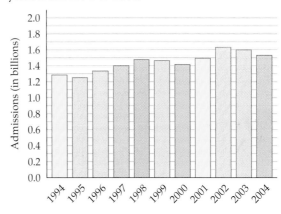

U.S. Movie Theatre Admissions

Source: National Association of Theatre Owners

   a. Estimate the numbers of admissions for the years 1996, 2001, and 2003. Round each estimate to the nearest tenth of a billion.

   b. Which year had the fewest admissions?

c. Which year had the greatest number of admissions?

d. During which two consecutive years did the greatest increase in admissions occur?

32. **Revenues** The following circle graph shows the percents of refreshment revenues that a movie theatre complex received from various types of refreshments on a given day.

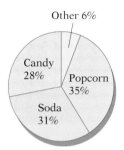

Other 6%

Candy 28%

Popcorn 35%

Soda 31%

Total Revenues from Refreshments: $3910.25

a. Determine the revenue the theatre earned from candy sales for the given day.

b. By how much did the popcorn revenue exceed the soda revenue for the given day?

33. **Box Office Grosses** The following broken-line graph shows the U.S. box office grosses, in billions of dollars, for the years from 1994 to 2004.

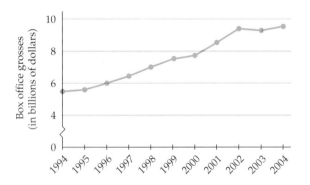

U.S. Movie Theatre Box Office Grosses

Source: National Association of Theatre Owners

a. Which year had the least box office gross?

b. Which year had the greatest box office gross?

c. Movie theatre box office grosses declined from 2002 to 2003, even though the average ticket price increased from $5.80 in 2002 to $6.03 in 2003. Explain how this is possible. *Hint:* See Exercise 31.

34. **Votes in an Election** In a school election, one candidate for class president received more than 94%, but less than 100%, of the votes cast. What is the least possible number of votes cast?

35. **Floor Design** A square floor is tiled with congruent square tiles. The tiles on the two diagonals of the floor are blue. The rest of the tiles are green. If 101 blue tiles are used, find the total number of tiles on the floor. *Note:* The following figure is not drawn to scale.

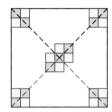

36. **Number of Children** How many children are there in a family wherein each girl has as many brothers as sisters, but each boy has twice as many sisters as brothers?

37. **Brothers and Sisters** I have two more sisters than brothers. Each of my sisters has two more sisters than brothers. How many more sisters than brothers does my youngest brother have?

38. **A Coin Problem** If you take 22 pennies from a pile of 57 pennies, how many pennies do you have?

39. **Bacterial Growth** The bacteria in a petri dish grow in a manner such that each day, the number of bacteria doubles. On what day will the number of bacteria be half of the number present on the 12th day?

40. **Number of River Crossings** Four people on one side of a river need to cross the river in a boat that can carry a maximum load of 180 pounds. The weights of the people are 80, 100, 150, and 170 pounds.

a. Explain how the people can use the boat to get everyone to the opposite side of the river.

b. What is the minimum number of crossings that must be made by the boat?

41. **Examination Scores** On three examinations, Dana received scores of 82, 91, and 76. What score does Dana need on the fourth examination to raise his average to 85?

42. Puzzle from a Movie In the movie *Die Hard: With a Vengeance,* Bruce Willis and Samuel L. Jackson are given a 5-gallon jug and a 3-gallon jug and they must put *exactly* 4 gallons of water on a scale to keep a bomb from exploding. Explain how they could accomplish this feat. Take your time, you have 2 minutes.

43. Find the Fake Coin You have eight coins. They all look identical, but one is a fake and is slightly lighter than the others. Explain how you can use a balance scale to determine which coin is the fake in exactly

a. three weighings.

b. two weighings.

Problems from the Mensa Workout Mensa is a society that welcomes people from every walk of life whose IQ's are in the top 2% of the population. The multiple-choice Exercises 44–47 are from the *Mensa Workout,* which is posted on the Internet at

http://www.mensa.org

44. If it were two hours later, it would be half as long until midnight as it would be if it were an hour later. What time is it now?

a. 18:30   **b.** 20:00   **c.** 21:00   **d.** 22:00   **e.** 23:30

45. Sally likes 225 but not 224; she likes 900 but not 800; she likes 144 but not 145. Which of the following does she like?

a. 1600   **b.** 1700

46. There are 1200 elephants in a herd. Some have pink and green stripes, some are all pink, and some are all blue. One third are pure pink. Is it true that 400 elephants are definitely blue?

a. Yes   **b.** No

47. Following the pattern shown in the number sequence below, what is the missing number?

$$1 \quad 8 \quad 27 \quad ? \quad 125 \quad 216$$

a. 36   **b.** 45   **c.** 46   **d.** 64   **e.** 99

Extensions

CRITICAL THINKING

48. A Famous Puzzle The mathematician Augustus De Morgan wrote that he had the distinction of being x years old in the year x^2. He was 43 in the year 1849.

a. Explain why people born in the year 1980 might share the distinction of being x years old in the year x^2. *Note:* Assume x is a natural number.

b. What is the next year after 1980 for which people born in that year might be x years old in the year x^2?

49. Verify a Procedure Select a two-digit number between 50 and 100. Add 83 to your number. From this number form a new number by adding the digit in the hundreds place to the number formed by the other two digits (the digits in the tens place and the ones place). Now subtract this newly formed number from your original number. Your final result is 16. Use a deductive approach to show that the final result is always 16 regardless of which number you start with.

50. Numbering Pages How many digits does it take in total to number a book from page 1 to page 240?

COOPERATIVE LEARNING

51. The Four 4s Problem The object of this exercise is to create a mathematical expression for every natural number from 1 to 20, using only common mathematical symbols and exactly four 4s (no other digits are allowed). You are allowed to use the following mathematical symbols: $+, -, \times, \div, \sqrt{\ }, ($ and $)$. For example,

$$\frac{4}{4} + \frac{4}{4} = 2, \quad 4^{(4-4)} + 4 = 5, \quad \text{and}$$

$$4 - \sqrt{4} + 4 \times 4 = 18$$

52. A Cryptarithm The following puzzle is a famous *cryptarithm.*

$$
\begin{array}{r}
\text{SEND} \\
+ \ \text{MORE} \\
\hline
\text{MONEY}
\end{array}
$$

Each letter in the cryptarithm represents one of the digits 0 through 9. The leading digits, represented by S and M, are not zero. Determine which digit is represented by each of the letters so that the addition is correct. *Note:* A letter that is used more than once, such as M, represents the same digit in each position in which it appears.

EXPLORATIONS

53. **The Collatz Problem** There are many unsolved problems in mathematics. One famous unsolved problem is known as the *Collatz problem,* or the $3n + 1$ problem. This problem was created by L. Collatz in 1937. Although the procedures in the Collatz problem are easy to understand, the problem remains unsolved. Search the Internet or a library to find information on the Collatz problem.

a. Write a short report that explains the Collatz problem. In your report, explain the meaning of a "hailstone" sequence.

b. Show that for each of the natural numbers 2, 3, 4, ..., 10, the Collatz procedure does generate a sequence that "returns" to 1.

54. **Paul Erdos** Paul Erdos (1913–1996) was a mathematician known for his elegant solutions of problems in number theory, combinatorics, discrete mathematics, and graph theory. He loved to solve mathematical problems, and for those problems he could not solve, he offered financial rewards, up to $10,000, to the person who could provide a solution.

Write a report on the life of Paul Erdos. In your report, include information about the type of problem that Erdos considered the most interesting. What did Erdos have to say about the Collatz problem mentioned in Exploration Exercise 53?

55. Graphing Expenses Most spreadsheet programs, such as *Excel,* can be used to create bar graphs, circle graphs, and line graphs. For instance, in the figure below, Aileen has entered the months of the year in column A and the dollar amounts of her water bills in column B. She then uses the computer's mouse to highlight the cells containing the data she wishes to graph. To create a vertical bar graph (called a column chart in *Excel*) of her monthly expenditures for water, she selects "Chart" from the "Insert" menu. She clicks the "Next" button and then the "Finish" button. *Note:* During this process, you can select other options that allow you to enhance the graph with special artistic effects, but these can also be added later.

Use a spreadsheet program to enter the data shown in columns A and B above.

a. Use the program to create a vertical bar graph, a circle graph (called a pie chart in *Excel*), and a line graph of the data.

b. Which of the three graphs do you think provides the best pictorial representation of the expenditures? Explain.

| **Problem Solving Using Sets**

Sets

The constellation Scorpius is a set of stars.

In an attempt to better understand the universe, ancient astronomers classified certain groups of stars as constellations. Today we still find it extremely helpful to classify items into groups that enable us to find order and meaning in our complicated world.

Any group or collection of objects is called a **set.** The objects that belong in a set are the **elements,** or **members,** of the set. For example, the set consisting of the four seasons has spring, summer, fall, and winter as its elements.

The following two methods are often used to designate a set.

- Describe the set using words.
- List the elements of the set inside a pair of braces, { }. This method is called the **roster method.** Commas are used to separate the elements.

For instance, let's use S to represent the set consisting of the four seasons. Using the roster method, we would write

$$S = \{\text{spring, summer, fall, winter}\}$$

The order in which the elements of a set are listed is not important. Thus the set consisting of the four seasons can also be written as

$$S = \{\text{winter, spring, fall, summer}\}$$

Table 1.1 gives two examples of sets, where each set is designated by a word description and also by using the roster method.

Table 1.1 *Define Sets by Using a Word Description and the Roster Method*

Description	Roster Method
The set of denominations of U.S. paper currency in production at this time	{$1, $2, $5, $10, $20, $50, $100}
The set of states in the United States that border the Pacific Ocean	{California, Oregon, Washington, Alaska, Hawaii}

▼ **point of interest**

Paper currency in denominations of $500, $1000, $5000, and $10,000 is still in circulation, but production of these bills ended in 1945. If you just happen to have some of these bills, you can still cash them for their face value.

The following sets of numbers are used extensively in many areas of mathematics.

Basic Number Sets

Natural Numbers or Counting Numbers $N = \{1, 2, 3, 4, 5, \ldots\}$

Whole Numbers $W = \{0, 1, 2, 3, 4, 5, \ldots\}$

Integers $I = \{\ldots, -4, -3, -2, -1, 0, 1, 2, 3, 4, \ldots\}$

Rational Numbers Q = the set of all terminating or repeating decimals

Irrational Numbers $\mathscr{I}$ = the set of all nonterminating, nonrepeating decimals

Real Numbers R = the set formed by combining the rational numbers and the irrational numbers

The set of natural numbers is also called the set of counting numbers. The three dots ... are called an **ellipsis** and indicate that the elements of the set continue in a manner suggested by the elements that are listed.

The integers ..., $-4, -3, -2, -1$ are **negative integers.** The integers $1, 2, 3, 4, ...$ are **positive integers.** Note that the natural numbers and the positive integers are the same set of numbers. The integer zero is neither a positive nor a negative integer.

If a number in decimal form terminates or repeats a block of digits, then the number is a rational number. Rational numbers can also be written in the form $\dfrac{p}{q}$, where p and q are integers and $q \neq 0$. For example,

$$\frac{1}{4} = 0.25 \quad \text{and} \quad \frac{3}{11} = 0.\overline{27}$$

are rational numbers. The bar over the 27 means that the block of digits 27 repeats without end; that is, $0.\overline{27} = 0.27272727...$.

A decimal that neither terminates nor repeats is an irrational number. For instance, $0.35335333533335...$ is a nonterminating, nonrepeating decimal and thus is an irrational number.

Every real number is either a rational number or an irrational number.

EXAMPLE 1 ■ **Use the Roster Method to Represent a Set of Numbers**

Use the roster method to write each of the given sets.

a. The set of natural numbers less than 5

b. The solution set of $x + 5 = -1$

c. The set of negative integers greater than -4

Solution

a. The set of natural numbers is given by $\{1, 2, 3, 4, 5, 6, 7, ...\}$. The natural numbers less than 5 are 1, 2, 3, and 4. Using the roster method, we write this set as $\{1, 2, 3, 4\}$.

b. Adding -5 to each side of the equation produces $x = -6$. The solution set of $x + 5 = -1$ is $\{-6\}$.

c. The set of negative integers greater than -4 is $\{-3, -2, -1\}$.

CHECK YOUR PROGRESS 1   Use the roster method to write each of the given sets.

a. The set of whole numbers less than 4

b. The set of counting numbers larger than 11 and less than or equal to 19

c. The set of negative integers between -5 and 7

Solution   See page S4.

Definitions Regarding Sets

A set is a **well-defined set** if it is possible to determine whether any given item is an element of the set. For instance, the set of letters of the English alphabet is a well-defined set. The set of *great songs* is not a well-defined set, because there is no universally accepted method for deciding whether any given song is an element of the set.

The statement "4 is an element of the set of natural numbers" can be written using mathematical notation as $4 \in N$. The symbol $\in$ is read "is an element of." To state that "-3 is not an element of the set of natural numbers," we use the "is not an element of" symbol, $\notin$, and write $-3 \notin N$.

A set is **finite** if the number of elements in the set is a whole number. The **cardinal number** of a finite set is the number of elements in the set. The cardinal number of a finite set A is denoted by the notation $n(A)$. For instance, if $A = \{1, 4, 6, 9\}$, then $n(A) = 4$. In this case, A has a cardinal number of 4, which is sometimes stated as "A has a *cardinality* of 4."

EXAMPLE 2 ■ The Cardinality of a Set

Find the cardinality of each of the following sets.

a. $J = \{2, 5\}$ **b.** $S = \{3, 4, 5, 6, 7, \ldots, 31\}$

Solution

a. Set J contains exactly two elements, so J has a cardinality of 2. Using mathematical notation, we state this as $n(J) = 2$.

b. Only a few elements are actually listed. The number of natural numbers from 1 to 31 is 31. If we omit the numbers 1 and 2, then the number of natural numbers from 3 to 31 must be $31 - 2 = 29$. Thus $n(S) = 29$.

CHECK YOUR PROGRESS 2   Find the cardinality of each of the following sets.

a. $C = \{-1, 5, 4, 11, 13\}$ **b.** $D = \{0\}$

Solution   See page S4.

The following definition is important in our work with sets.

Equal Sets

Set A is **equal** to set B, denoted by $A = B$, if and only if A and B have exactly the same elements.

For instance, the set of natural numbers is equal to the set of counting numbers and {nickel, dime, quarter } = {dime, quarter, nickel}.

The Universal Set and the Complement of a Set

In complex problem-solving situations and even in routine daily activities, we need to understand the set of all elements that are under consideration. For instance, when an instructor assigns letter grades, the possible choices may include A, B, C, D, F, and I. In this case the letter H is not a consideration. When you place a telephone call, you know that the area code is given by a natural number with three digits. In this instance, a rational number such as $\frac{2}{3}$ is not a consideration. The set of all elements that are being considered is called the **universal set.** We will use the letter U to denote the universal set.

The Complement of a Set

The **complement** of a set A, denoted by A', is the set of all elements of the universal set U that are not elements of A.

EXAMPLE 3 ■ Find the Complement of a Set

Riley is shopping for a digital camera at a store that sells the brand names in the universal set

$U = \{\text{Nikon, Canon, Sony, Olympus, JVC, Panasonic, Pentax, Kodak}\}$

Riley's friend Jordan recommends the brand names in $J = \{\text{Nikon, Canon, Pentax}\}$. Riley's sister Elle recommends the brand names in $E = \{\text{Canon, Sony, JVC, Kodak}\}$. Find each of the following sets.

a. J' **b.** E'

Solution

a. Set J' contains all the elements of U that are not elements of J. Because J contains the elements Nikon, Canon, and Pentax, these three elements cannot belong to J'. Thus, $J' = \{\text{Sony, Olympus, JVC, Panasonic, Kodak}\}$.

b. Set E' contains all the elements of U that are not elements of E, so $E' = \{\text{Nikon, Olympus, Panasonic, Pentax}\}$.

CHECK YOUR PROGRESS 3 A seafood restaurant offers the entrées listed in the universal set

$U = \{\text{sea bass, halibut, salmon, tuna, cod, rockfish}\}$

Let $S = \{\text{sea bass, salmon}\}$ and $M = \{\text{halibut, salmon, cod, rockfish}\}$. Find each of the following sets.

a. S' **b.** M'

Solution   *See page S4.*

Empty Set

The **empty set,** or **null set,** is the set that contains no elements.

The symbol $\varnothing$ or $\{\}$ is used to represent the empty set. As an example of the empty set, consider the set of natural numbers that are negative integers.

There are two fundamental results concerning the universal set and the empty set. Because the universal set contains all elements under consideration, the complement of the universal set is the empty set. Conversely, the complement of the empty set is the universal set, because the empty set has no elements and the universal set contains all the elements under consideration. Using mathematical notation, we state these fundamental results as follows:

> **The Complement of the Universal Set and the Complement of the Empty Set**
>
> $U' = \varnothing$ and $\varnothing' = U$

Subsets

Consider the set of letters in the alphabet and the set of vowels {a, e, i, o, u}. Every element of the set of vowels is an element of the set of letters in the alphabet. The set of vowels is said to be a *subset* of the set of letters in the alphabet. We will often find it useful to examine subsets of a given set.

> **A Subset of a Set**
>
> Set A is a **subset** of set B, denoted by $A \subseteq B$, if and only if every element of A is also an element of B.

Here are two fundamental subset relationships.

> **Subset Relationships**
>
> $A \subseteq A$, for any set A
> $\varnothing \subseteq A$, for any set A

A set is always a subset of itself, and the empty set is a subset of any set.

The notation $A \not\subseteq B$ is used to denote that A is *not* a subset of B. To show that A is not a subset of B, it is necessary to find at least one element of A that is not an element of B.

TAKE NOTE

Recall that W represents the set of whole numbers and N represents the set of natural numbers.

EXAMPLE 4 ■ True or False

Determine whether each statement is true or false.

a. $\{5, 10, 15, 20\} \subseteq \{0, 5, 10, 15, 20, 25, 30\}$ **b.** $W \subseteq N$

c. $\{2, 4, 6\} \subseteq \{2, 4, 6\}$ **d.** $\varnothing \subseteq \{1, 2, 3\}$

Solution

a. True; every element of the first set is an element of the second set.

b. False; 0 is a whole number, but 0 is not a natural number.

c. True; every set is a subset of itself.

d. True; the empty set is a subset of every set.

CHECK YOUR PROGRESS 4   Determine whether each statement is true or false.

a. $\{1, 3, 5\} \subseteq \{1, 5, 9\}$

b. The set of counting numbers is a subset of the set of natural numbers.

c. $\varnothing \subseteq U$

d. $\{-6, 0, 11\} \subseteq I$

Solution   See page S4.

U

A

A Venn diagram

The English logician John Venn (1834–1923) developed diagrams, which we now refer to as *Venn diagrams,* that can be used to illustrate sets and relationships between sets. In a **Venn diagram,** the universal set is represented by a rectangular region and subsets of the universal set are generally represented by oval or circular regions drawn inside the rectangle. The Venn diagram at the left shows a universal set and one of its subsets, labeled as set A. The size of the circle is not a concern. The region outside of the circle but inside of the rectangle represents the set A'.

QUESTION   *What set is represented by $(A')'$?*

> **Proper Subset**
>
> Set A is a **proper subset** of set B, denoted by $A \subset B$, if every element of A is an element of B, and $A \neq B$.

To illustrate the difference between subsets and proper subsets, consider the following two examples.

1. Let $R = \{$Mars, Venus$\}$ and $S = \{$Mars, Venus, Mercury$\}$. The first set R is a subset of the second set S because every element of R is an element of S. In addition, R is also a proper subset of S, because $R \neq S$.

2. Let $T = \{$Europe, Africa$\}$ and $V = \{$Africa, Europe$\}$. The first set T is a subset of the second set V; however, T is *not* a proper subset of V because $T = V$.

Some counting problems in the study of probability require that we find all the subsets of a given set. One way to find all the subsets of a given set is to use the method of making an organized list. First list the empty set, which has no elements. Next list all the sets that have exactly one element, followed by all the sets that contain exactly two elements, followed by all the sets that contain exactly three elements, and so on. This process is illustrated in the following example.

EXAMPLE 5 ■ List All the Subsets of a Set

List all the subsets of $\{1, 2, 3, 4\}$.

Solution

An organized list produces the following subsets.

$\{\,\}$ • Subsets with 0 elements

$\{1\}, \{2\}, \{3\}, \{4\}$ • Subsets with 1 element

ANSWER   *The set A' contains the elements of U that are not in A. By definition, the set $(A')'$ contains only the elements of U that are elements of A. Thus $(A')' = A$.*

$\{1, 2\}, \{1, 3\}, \{1, 4\}, \{2, 3\}, \{2, 4\}, \{3, 4\}$ • Subsets with 2 elements

$\{1, 2, 3\}, \{1, 2, 4\}, \{1, 3, 4\}, \{2, 3, 4\}$ • Subsets with 3 elements

$\{1, 2, 3, 4\}$ • Subsets with 4 elements

CHECK YOUR PROGRESS 5   List all of the subsets of {a, b, c, d, e}.

Solution   *See page S4.*

The counting techniques developed in Section 1.2 can be used to verify the following statement.

> **The Number of Subsets of a Set**
>
> A set with n elements has 2^n subsets and $2^n - 1$ proper subsets.

According to the statement above, the set $\{1, 2, 3, 4\}$, with four elements, should have $2^4 = 16$ subsets and $2^4 - 1 = 15$ proper subsets. Examine the subsets listed in Example 5 to confirm that $\{1, 2, 3, 4\}$ does indeed have 16 subsets and 15 proper subsets.

The empty set has zero elements. Its only subset $(2^0 = 1)$ is itself.

Intersection and Union of Sets

In everyday usage, the word *intersection* refers to the *common region* where two streets cross. See the figure at the left. The intersection of two sets is defined in a similar manner.

> **Intersection of Sets**
>
> The **intersection** of sets A and B, denoted by $A \cap B$, is the set of elements common to both A and B.

In the figure at the left, the region shown in blue represents the intersection of sets A and B.

U

$A \cap B$

✔ **TAKE NOTE**

It is a mistake to write

$\{1, 5, 9\} \cap \{3, 5, 9\} = 5, 9$

The intersection of two sets is a set. Thus

$\{1, 5, 9\} \cap \{3, 5, 9\} = \{5, 9\}$

EXAMPLE 6 ■ Find Intersections

Let $A = \{1, 4, 5, 7\}$, $B = \{2, 3, 4, 5, 6\}$, and $C = \{3, 6, 9\}$. Find

a. $A \cap B$ **b.** $A \cap C$

Solution

a. The elements common to both sets are 4 and 5.

$A \cap B = \{1, 4, 5, 7\} \cap \{2, 3, 4, 5, 6\}$

$= \{4, 5\}$

b. Sets A and C have no common elements. Thus $A \cap C = \varnothing$.

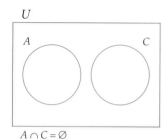

U

$A \cap C = \varnothing$

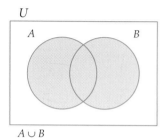

U

$A \cup B$

CHECK YOUR PROGRESS 6   Let $D = \{0, 3, 8, 9\}$, $E = \{3, 4, 8, 9, 11\}$, and $F = \{0, 2, 6, 8\}$. Find

a. $D \cap E$ **b.** $D \cap F$

Solution   *See page S4.*

Two sets are **disjoint** if their intersection is the empty set. The sets A and C in Example 6b are disjoint. The Venn diagram at the upper left illustrates two disjoint sets.

In everyday usage, the word *union* refers to the act of uniting or joining together. The union of two sets has a similar meaning.

Union of Sets

The **union** of sets A and B, denoted by $A \cup B$, is the set that contains all the elements that belong to A or to B or to both.

In the Venn diagram at the lower left, the region shown in blue represents the union of sets A and B.

EXAMPLE 7 ▪ Find Unions

Let $A = \{1, 4, 5, 7\}$, $B = \{2, 3, 4, 5, 6\}$, and $C = \{3, 6, 9\}$. Find

a. $A \cup B$ **b.** $A \cup C$

Solution

a. List all the elements of set A, which are 1, 4, 5, and 7. Then add to your list the elements of set B that have not already been listed—in this case, 2, 3, and 6. Enclose all elements with a pair of braces. Thus

$$A \cup B = \{1, 4, 5, 7\} \cup \{2, 3, 4, 5, 6\}$$
$$= \{1, 2, 3, 4, 5, 6, 7\}$$

b. $A \cup C = \{1, 4, 5, 7\} \cup \{3, 6, 9\}$
$$= \{1, 3, 4, 5, 6, 7, 9\}$$

CHECK YOUR PROGRESS 7   Let $D = \{0, 4, 8, 9\}$, $E = \{1, 4, 5, 7\}$, and $F = \{2, 6, 8\}$. Find

a. $D \cup E$ **b.** $D \cup F$

Solution   *See page S4.*

In mathematical problems that involve sets, the word *and* is interpreted to mean *intersection*. For instance, the phrase "the elements of A and B" means the elements of $A \cap B$. Similarly, the word *or* is interpreted to mean *union*. The phrase "the elements of A or B" means the elements of $A \cup B$.

✔ **TAKE NOTE**

Would you like soup or salad?

In a sentence, the word *or* can mean one or the other, but not both. For instance, if a menu states that you can have soup or salad with your meal, this generally means that you can have either soup or salad for the price of the meal, but not both. In this case the word *or* is said to be an *exclusive or*. In the mathematical statement "A or B," the *or* is an *inclusive or*. It means A or B, or both.

Venn Diagrams and Equality of Sets

The equality $A \cup B = A \cap B$ is true for some sets A and B, but not for all sets A and B. For instance, if $A = \{1, 2\}$ and $B = \{1, 2\}$, then $A \cup B = A \cap B$. However, we can prove that, in general, $A \cup B \neq A \cap B$ by finding an example for which the expressions are not equal. One such example is $A = \{1, 2, 3\}$ and $B = \{2, 3\}$. In this case $A \cup B = \{1, 2, 3\}$, whereas $A \cap B = \{2, 3\}$. This example is called a counter-example. The point to remember is that if you wish to show that two set expressions are *not equal*, then you need to find just one counterexample.

In the next example, we present a technique that uses Venn diagrams to determine whether two set expressions are equal.

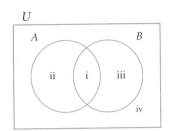

EXAMPLE 8 ■ **Equality of Sets**

Use Venn diagrams to determine whether $(A \cup B)' = A' \cap B'$ for all sets A and B.

Solution

Draw a Venn diagram that shows the two sets A and B, as in the figure at the left. Label the four regions as shown. To determine what region(s) represents $(A \cup B)'$, first note that $A \cup B$ consists of regions i, ii, and iii. Thus $(A \cup B)'$ is represented by region iv. See Figure 1.5.

Draw a second Venn diagram. To determine what region(s) represents $A' \cap B'$, we shade A' (regions iii and iv) with a diagonal up pattern and we shade B' (regions ii and iv) with a diagonal down pattern. The intersection of these shaded regions, which is region iv, represents $A' \cap B'$. See Figure 1.6.

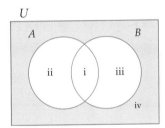

Figure 1.5

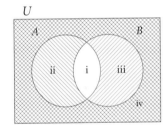

Figure 1.6

Because both $(A \cup B)'$ and $A' \cap B'$ are represented by the same region, we know that $(A \cup B)' = A' \cap B'$ for all sets A and B.

CHECK YOUR PROGRESS 8 Use Venn diagrams to determine whether $(A \cap B)' = A' \cup B'$ for all sets A and B.

Solution *See page S4.*

The equalities that were verified in Example 8 and Check Your Progress 8 are known as **De Morgan's laws.**

De Morgan's Laws

For all sets A and B,

$(A \cup B)' = A' \cap B'$ and $(A \cap B)' = A' \cup B'$

De Morgan's law $(A \cup B)' = A' \cap B'$ can be stated as "the complement of the union of two sets is the intersection of the complements of the sets." De Morgan's law $(A \cap B)' = A' \cup B'$ can be stated as "the complement of the intersection of two sets is the union of the complements of the sets."

In the next example, we extend the Venn diagram procedure illustrated in the previous examples to expressions that involve three sets.

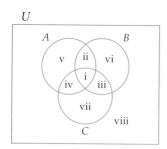

Figure 1.7

EXAMPLE 9 ■ Equality of Set Expressions

Use Venn diagrams to determine whether $A \cup (B \cap C) = (A \cup B) \cap C$ for all sets A, B, and C.

Solution
Draw a Venn diagram that shows the sets A, B, and C and the eight regions they form. See Figure 1.7. To determine what region(s) represents $A \cup (B \cap C)$, we first consider $(B \cap C)$, represented by regions i and iii, because it is in parentheses. Set A is represented by the regions i, ii, iv, and v. Thus $A \cup (B \cap C)$ is represented by all of the listed regions (namely, i, ii, iii, iv, and v), as shown in Figure 1.8.

Draw a second Venn diagram showing the three sets A, B, and C, as in Figure 1.7. To determine what region(s) represents $(A \cup B) \cap C$, we first consider $(A \cup B)$ (regions i, ii, iii, iv, v, and vi) because it is in parentheses. Set C is represented by regions i, iii, iv, and vii. Therefore, the intersection of $(A \cup B)$ and C is represented by the overlap, or regions i, iii, and iv. See Figure 1.9.

Because the sets $A \cup (B \cap C)$ and $(A \cup B) \cap C$ are represented by different regions, we conclude that $A \cup (B \cap C) \neq (A \cup B) \cap C$.

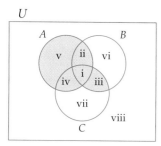

Figure 1.8

CHECK YOUR PROGRESS 9 Use Venn diagrams to determine whether $A \cup (B \cap C) = (A \cup B) \cap (A \cup C)$ for all sets A, B, and C.

Solution   *See page S4.*

Venn diagrams can be used to verify each of the following properties.

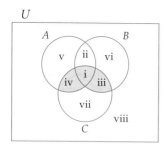

Figure 1.9

Properties of Sets

For all sets A and B,
Commutative Properties
$A \cap B = B \cap A$ Commutative property of intersection
$A \cup B = B \cup A$ Commutative property of union

For all sets A, B, and C,
Associative Properties
$(A \cap B) \cap C = A \cap (B \cap C)$ Associative property of intersection
$(A \cup B) \cup C = A \cup (B \cup C)$ Associative property of union
Distributive Properties
$A \cap (B \cup C) = (A \cap B) \cup (A \cap C)$ Distributive property of intersection over union

$A \cup (B \cap C) = (A \cup B) \cap (A \cup C)$ Distributive property of union over intersection

---

QUESTION   *Does $(B \cup C) \cap A = (A \cap B) \cup (A \cap C)$?*

ANSWER   *Yes. The commutative property of intersection allows us to write $(B \cup C) \cap A$ as $A \cap (B \cup C)$, and $A \cap (B \cup C) = (A \cap B) \cup (A \cap C)$ by the distributive property of intersection over union.*

historical note

The Nobel Prize is an award granted to people who have made significant contributions to society. Nobel Prizes are awarded annually for achievements in physics, chemistry, physiology or medicine, literature, peace, and economics. The prizes were first established in 1901 by the Swedish industrialist Alfred Nobel, who invented dynamite. ∎

Math Matters   Venn Diagrams, Blood Groups, and Blood Types

Karl Landsteiner won a Nobel Prize in 1930 for his discovery of the four different human blood groups. He discovered that the blood of each individual contains exactly one of the following combinations of antigens.

- Only A antigens (blood group A)

- Only B antigens (blood group B)

- Both A and B antigens (blood group AB)

- No A antigens and no B antigens (blood group O)

These four blood groups are shown by the Venn diagram in Figure 1.10.

In 1941, Landsteiner and Alexander Wiener discovered that human blood may or may not contain an Rh, or rhesus, factor. Blood with this factor is called Rh-positive and denoted by Rh+. Blood without this factor is called Rh-negative and is denoted by Rh−.

The Venn diagram in Figure 1.11 illustrates the eight blood types (A+, B+, AB+, O+, A−, B−, AB−, O−) that are possible if we consider antigens and the Rh factor.

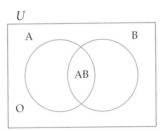

Figure 1.10 *The four blood groups*

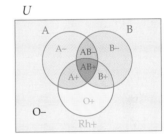

Figure 1.11 *The eight blood types*

Mixing blood from two individuals can lead to blood clumping, or agglutination, which can have fatal consequences. Karl Landsteiner discovered that blood clumping is an immunological reaction that occurs when the receiver of a blood transfusion does not have an antigen that is present in the donor's blood. This discovery has made it possible for blood transfusions to be carried out safely.

Surveys: An Application of Sets

Some applications that involve surveys can be solved by using Venn diagrams. The basic procedure used to solve these applications consists of drawing a Venn diagram that illustrates the sets involved and using the given numerical information to determine the cardinality of each region in the Venn diagram. This procedure is illustrated in Examples 10 and 11.

EXAMPLE 10 ■ A Survey of Preferences

A movie company is making plans for future movies it wishes to produce. The company has done a random survey of 1000 people. The results of the survey are shown below.

695 people like action adventures.

340 people like comedies.

180 people like both action adventures and comedies.

Of the people surveyed, how many people:

a. like action adventures but not comedies?

b. like comedies but not action adventures?

c. do not like either of these types of movies?

Solution

A Venn diagram can be used to illustrate the results of the survey. We use two overlapping circles (see Figure 1.12). One circle represents the set of people who like action adventures, and the other represents the set of people who like comedies. The region i where the circles intersect represents the set of people who like both types of movies.

We start with the information that 180 people like both types of movies and write 180 in region i. See Figure 1.13.

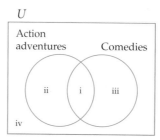

Figure 1.12

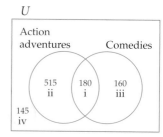

Figure 1.13

a. Regions i and ii have a total of 695 people. So far we have accounted for 180 of these people in region i. Thus the number of people in region ii, which is the set of people who like action adventures but do not like comedies, is 695 − 180 = 515.

b. Regions i and iii have a total of 340 people. Thus the number of people in region iii, which is the set of people who like comedies but do not like action adventures, is 340 − 180 = 160.

c. The number of people who do not like action adventure movies or comedies is represented by region iv. The number of people in region iv must be the total number of people, which is 1000, less the number of people accounted for in regions i, ii, and iii, which is 855. Thus the number of people who do not like either type of movie is 1000 − 855 = 145.

CHECK YOUR PROGRESS 10   The athletic director of a school has surveyed 200 students. The survey results are shown below.

> 140 students like volleyball.
>
> 120 students like basketball.
>
> 85 students like both volleyball and basketball.

Of the students surveyed, how many students:

a. like volleyball but not basketball?

b. like basketball but not volleyball?

c. do not like either of these sports?

Solution   See page S5.

In the next example we consider a more complicated survey that involves three types of music.

EXAMPLE 11 ▪ A Music Survey

A music teacher has surveyed 495 students. The results of the survey are listed below.

> 320 students like rap music.
>
> 395 students like rock music.
>
> 295 students like heavy metal music.
>
> 280 students like both rap music and rock music.
>
> 190 students like both rap music and heavy metal music.
>
> 245 students like both rock music and heavy metal music.
>
> 160 students like all three.

How many students:

a. like exactly two of the three types of music?

b. like only rock music?

c. like only one of the three types of music?

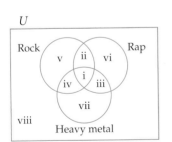

Solution

The Venn diagram at the upper left shows three overlapping circles. Region i represents the set of students who like all three types of music. Each of the regions v, vi, and vii represent the students who like only one type of music.

a. The survey shows that 245 students like rock and heavy metal music, so the numbers we place in regions i and iv must have a sum of 245. Since region i has 160 students, we see that region iv must have 245 − 160 = 85 students. In a similar manner, we can determine that region ii has 120 students and region iii has 30 students. Thus 85 + 120 + 30 = 235 students like exactly two of the three types of music.

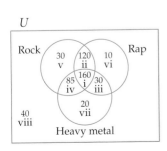

b. The sum of the students represented by regions i, ii, iv, and v must be 395. The number of students in region v must be the difference between this total and the sum of the numbers of students in regions i, ii, and iv. Thus the number of students who like only rock music is 395 − (160 + 120 + 85) = 30. See the Venn diagram at the lower left.

c. Using the same reasoning as in part b, we find that region vi has 10 students and region vii has 20 students. To find the number of students who like only one type of music, find the sum of the numbers of students in regions v, vi, and vii, which is $30 + 10 + 20 = 60$. See the Venn diagram on the preceding page.

CHECK YOUR PROGRESS 11   An activities director for a cruise ship has surveyed 240 passengers. Of the 240 passengers,

135 like swimming.	80 like swimming and dancing.
150 like dancing.	40 like swimming and games.
65 like games.	25 like dancing and games.
	15 like all three activities.

How many passengers:

a. like exactly two of the three types of activities?

b. like only swimming?

c. like none of these activities?

Solution   See page S5.

Investigation

Voting Systems

There are many types of voting systems. When people are asked to vote for or against a resolution, a one-person, one-vote *majority system* is often used to decide the outcome. In this type of voting system, each voter receives one vote, and the resolution passes only if it receives *most* of the votes.

In any voting system, the number of votes required to pass a resolution is called the **quota**. A **coalition** is a set of voters each of whom votes the same way, either for or against a resolution. A **winning coalition** is a set of voters the sum of whose votes is greater than or equal to the quota. A **losing coalition** is a set of voters the sum of whose votes is less than the quota.

Sometimes you can find all the winning coalitions in a voting process by making an organized list. For instance, consider the committee consisting of Alice, Barry, Cheryl, and Dylan. To decide on any issue, they use a one-person, one-vote majority voting system. Since each of the four voters has a single vote, the quota for this majority voting system is 3. The winning coalitions consist of all subsets of the voters that have three or more people. We list these winning coalitions in the left table on page 47, where A represents Alice, B represents Barry, C represents Cheryl, and D represents Dylan.

A **weighted voting system** is one in which some voters' votes carry more weight regarding the outcome of an election. As an example, consider a selection committee that consists of four people, designated by A, B, C, and D. Voter A's vote has a weight of 2, and the vote of each other member of the committee has a weight of 1. The quota for this weighted voting system is 3. A winning coalition must have a weighted voting sum of at least 3. The winning coalitions are listed in the right table on page 47.

(continued)

Winning Coalition	Sum of the Votes
{A, B, C}	3
{A, B, D}	3
{A, C, D}	3
{B, C, D}	3
{A, B, C, D}	4

Winning Coalition	Sum of the Weighted Votes
{A, B}	3
{A, C}	3
{A, D}	3
{B, C, D}	3
{A, B, C}	4
{A, B, D}	4
{A, C, D}	4
{A, B, C, D}	5

A **minimal winning coalition** is a winning coalition that has no proper subset that is a winning coalition. In a minimal winning coalition, each voter is said to be a **critical voter** because if any of the voters leaves the coalition, the coalition will then become a losing coalition. In the table at the right above, the minimal winning coalitions are {A, B}, {A, C}, {A, D}, and {B, C, D}. If any single voter leaves one of these coalitions, the coalition will become a losing coalition. The coalition {A, B, C, D} is not a minimal winning coalition because it contains at least one proper subset, for instance {A, B, C}, that is a winning coalition.

Investigation Exercises

1. A selection committee consists of Ryan, Susan, and Trevor. To decide on issues, they use a one-person, one-vote majority voting system.
   a. Find all winning coalitions.
   b. Find all losing coalitions.

2. A selection committee consists of three people, designated by M, N, and P. M's vote has a weight of 3, N's vote has a weight of 2, and P's vote has a weight of 1. The quota for this weighted voting system is 4. Find all winning coalitions.

3. Determine the minimal winning coalitions for the voting system in Investigation Exercise 2.

Additional information on the applications of mathematics to voting systems is given in Chapter 9.

Exercise Set 1.3

In Exercises 1–8, use the roster method to represent each of the sets.

1. The set of U.S. coins with values of less than 50¢

2. The set of months of the year with names that end with the letter y

3. The set of planets in our solar system with names that start with the letter M

4. The set of months with exactly 30 days

5. The set of negative integers greater than −6

6. The set of whole numbers less than 8

7. The set of integers x that satisfy $2x − 1 = −11$

8. The set of whole numbers x that satisfy $x − 1 < 4$

In Exercises 9–14, write a description of each set.

9. {Tuesday, Thursday}

10. {Libra, Leo}

11. {Mercury, Venus}

12. {penny, nickel, dime}

13. {1, 2, 3, 4, 5, 6, 7, 8, 9}

14. {2, 4, 6, 8}

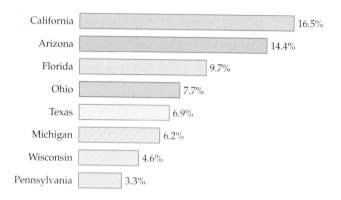

Charter Schools During recent years the number of U.S. charter schools has increased dramatically. The following horizontal bar graph shows the eight states with the greatest percents of U.S. charter schools, as of December 2005.

States with the Greatest Percent of U.S. Charter Schools

Source: U.S. Charter Schools web page, **http://www.uscharterschools.org**

Use the data in the above graph and the roster method to represent each of the sets in Exercises 15 and 16.

15. The set of states with more than 14% of all U.S. charter schools

16. The set of states that have between 5% and 10% of all U.S. charter schools

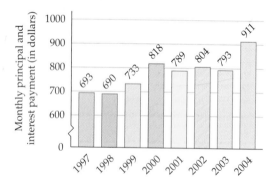

Mortgage Payments The following bar graph shows the monthly principal and interest payments needed to purchase average-priced existing homes in the United States for the years from 1997 to 2004.

Monthly Principal and Interest Payments for Average-Priced Existing Homes

Source: National Association of REALTORS® as reported in the *World Almanac*, 2005, p. 483

Use the data in the above graph and the roster method to represent each of the sets in Exercises 17 and 18.

17. The set of years in which the monthly principal and interest payments for average-priced existing homes exceeded $800

18. The set of years in which the monthly principal and interest payments for average-priced existing homes were between $700 and $800

Ticket Prices The following table shows the average U.S. movie theatre ticket prices for the years from 1985 to 2004.

Year	Price	Year	Price
1985	$3.55	1995	$4.35
1986	3.71	1996	4.42
1987	3.91	1997	4.59
1988	4.11	1998	4.69
1989	3.99	1999	5.06
1990	4.22	2000	5.39
1991	4.21	2001	5.65
1992	4.15	2002	5.80
1993	4.14	2003	6.03
1994	4.08	2004	6.21

Average U.S. Movie Theatre Ticket Prices

Source: National Association of Theatre Owners, http://www.natoonline.org/statisticstickets.htm

Use the information in the above table and the roster method to represent the sets in Exercises 19 and 20.

19. The set of years in which the average ticket prices were less than $4.00

20. The set of years in which the average ticket prices were greater than $4.25 but less than $6.00

In Exercises 21–26, find the cardinality of each set.

21. $A = \{2, 4, 6, 8, 10, 12, 14, 16, 18, 20, 22\}$

22. $B = \{7, 14, 21, 28, 35, 42, 49, 56\}$

23. D = the set of all dogs that can spell "elephant"

24. S = the set of all states in the United States

25. $\{3, 6, 9, 12, 15, \ldots, 363\}$

26. $\{7, 11, 15, 19, 23, 27, \ldots, 407\}$

A Cardinality Formula

For any finite sets A and B,
$$n(A \cup B) = n(A) + n(B) - n(A \cap B)$$

In Exercises 27–30, use this cardinality formula to solve each exercise. Do not use a Venn diagram or a calculator.

27. If $n(A) = 30$, $n(B) = 20$, and $n(A \cap B) = 5$, find $n(A \cup B)$.

28. If $n(A) = 65$, $n(B) = 45$, and $n(A \cap B) = 15$, find $n(A \cup B)$.

29. If $n(A \cup B) = 60$, $n(A) = 40$, and $n(A \cap B) = 8$, find $n(B)$.

30. If $n(A \cup B) = n(A) + n(B)$, what must be the relationship between sets A and B?

In Exercises 31–34, find the complement of the set given that $U = \{0, 1, 2, 3, 4, 5, 6, 7, 8\}$.

31. $\{2, 4, 6, 7\}$ 32. $\{3, 6\}$

33. The set of odd natural numbers less than 8

34. The set of even counting numbers less than 10

In Exercises 35–40, determine whether the first set is a subset of the second set.

35. $\{a, b, c, d\}$; $\{a, b, c, d, e, f, g\}$

36. $\{3, 5, 7\}$; $\{3, 4, 5, 6\}$

37. The set of integers; the set of rational numbers

38. The set of real numbers; the set of integers

39. The set of all sandwiches; the set of all hamburgers

40. $\{2, 4, 6, \ldots, 5000\}$; the set of even whole numbers

41. A class of 16 students has 2^{16} subsets. Determine how long (to the nearest hour) it would take you to write all the subsets, assuming you can write each subset in 1 second.

42. A class of 32 students has 2^{32} subsets. Determine how long (to the nearest year) it would take you to write all the subsets, assuming you can write each subset in 1 second.

43. **Sandwich Choices** A delicatessen makes a roast-beef-on-sour-dough sandwich for which you can choose from eight condiments.

   a. How many different types of roast-beef-on-sour-dough sandwiches can the delicatessen prepare?

   b. What is the minimum number of condiments the delicatessen must have available if it wishes to offer at least 2000 different types of roast-beef-on-sour-dough sandwiches?

44. **Omelet Choices** A restaurant provides a brunch where the omelets are individually prepared. Each guest is allowed to choose from 10 different ingredients.

   a. How many different types of omelets can the restaurant prepare?

   b. What is the minimum number of ingredients that must be available if the restaurant wants to advertise that it offers over 4000 different omelets?

45. **Truck Options** A truck company makes a pickup truck with 12 upgrade options. Some of the options are air conditioning, chrome wheels, and a CD player.

   a. How many different versions of this truck can the company produce?

   b. What is the minimum number of upgrade options the company must be able to provide if it wishes to offer at least 14,000 different versions of this truck?

46. **Voting Coalitions** Five people, designated A, B, C, D, and E, serve on a committee. To pass a motion, at least three of the committee members must vote for the motion. In such a situation, any set of three or more voters is called a **winning coalition** because if this set of people votes for a motion, the motion will pass. Any nonempty set of two or fewer voters is called a **losing coalition.**

   a. List all the winning coalitions.

   b. List all the losing coalitions.

In Exercises 47–60, let $U = \{1, 2, 3, 4, 5, 6, 7, 8\}$, $A = \{2, 4, 6\}$, $B = \{1, 2, 5, 8\}$, and $C = \{1, 3, 7\}$. Find each of the following.

47. $A \cup B$ 48. $A \cap B$

49. $A \cap B'$ 50. $B \cap C'$

51. $(A \cup B)'$ 52. $(A' \cap B)'$

53. $A \cup (B \cup C)$ 54. $A \cap (B \cup C)$

55. $A \cap (B \cap C)$ 56. $A' \cup (B \cap C)$

57. $(A \cup C') \cap (B \cup A')$ 58. $(A \cup C') \cup (B \cup A')$

59. $(A \cup B) \cap (B \cap C')$ 60. $(B \cap A') \cup (B' \cup C)$

In Exercises 61–66, use a Venn diagram to show the following sets.

61. $A \cap B'$ 62. $(A \cap B)'$

63. $(A \cup B)'$ 64. $(A' \cap B) \cup B'$

65. $A \cap (B \cup C')$ 66. $A \cap (B' \cap C)$

In Exercises 67–70, draw two Venn diagrams to determine whether the given expressions are equal for all sets A and B.

67. $A \cap B'$; $A' \cup B$ 68. $A' \cap B$; $A \cup B'$

69. $A \cup (A' \cap B)$; $A \cup B$

70. $A' \cap (B \cup B')$; $A' \cup (B \cap B')$

In Exercises 71–74, draw two Venn diagrams to determine whether the given expressions are equal for all sets A, B, and C.

71. $(A \cup C) \cap B'$; $A' \cup (B \cup C)$

72. $A' \cap (B \cap C)$; $(A \cup B') \cap C$

73. $(A' \cap B) \cup C$; $(A' \cap C) \cap (A' \cap B)$

74. $A' \cup (B' \cap C)$; $(A' \cup B') \cap (A' \cup C)$

Computers and televisions make use of *additive color mixing*. The following figure shows that when the *primary colors* red R, green G, and blue B are all three mixed together using additive color mixing, they produce white, W. Using set notation, we state this as $R \cap B \cap G = W$. The colors yellow Y, cyan C, and magenta M are called *secondary colors*. A secondary color is produced by mixing exactly two of the primary colors.

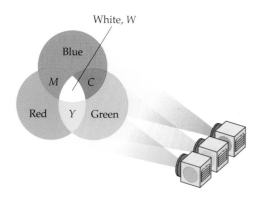

Additive Color Mixing In Exercises 75–77, determine which color is represented by each of the following sets. Assume the colors are being mixed using additive color mixing. (Use R for red, G for green, and B for blue.)

75. $(R \cap G) \cap B'$

76. $(R \cap G') \cap B$

77. $(R' \cap G) \cap B$

Artists who paint with pigments use *subtractive color mixing* to produce different colors. In a subtractive color mixing system, the primary colors are cyan C, magenta M, and yellow Y. The following figure shows that when the three primary colors are mixed in equal amounts using subtractive color mixing, they form black, K. Using set notation, we state this as $C \cap M \cap Y = K$. In subtractive color mixing, the colors red R, blue B, and green G are called *secondary colors*. A secondary color is produced by mixing equal amounts of exactly two of the primary colors.

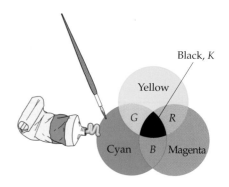

Subtractive Color Mixing In Exercises 78–80, determine what color is represented by each of the following sets. Assume the colors are being mixed using subtractive color mixing. (Use C for cyan, M for magenta, and Y for yellow.)

78. $(C \cap M) \cap Y'$

79. $(C' \cap M) \cap Y$

80. $(C \cap M') \cap Y$

In Exercises 81 and 82, draw a Venn diagram with each of the given elements placed in the correct region. *Hint:* See the Math Matters about *Venn Diagrams, Blood Groups, and Blood Types* on page 43.

81. $U = \{$Sue, Bob, Al, Jo, Ann, Herb, Eric, Mike, Sal$\}$

$A = \{$Sue, Herb$\}$

$B = \{$Sue, Eric, Jo, Ann$\}$

$Rh+ = \{$Eric, Sal, Al, Herb$\}$

82. $U = \{$Hal, Marie, Rob, Armando, Joel, Juan, Melody$\}$

$A = \{$Marie, Armando, Melody$\}$

$B = \{$Rob, Juan, Hal$\}$

$Rh+ = \{$Hal, Marie, Rob, Joel, Juan, Melody$\}$

In Exercises 83–90, let $U = \{$English, French, History, Math, Physics, Chemistry, Psychology, Drama$\}$. Also,

$A = \{$English, History, Psychology, Drama$\}$,

$B = \{$Math, Physics, Chemistry, Psychology, Drama$\}$, and

$C = \{$French, History, Chemistry$\}$.

Find each of the following.

83. $n(B \cup C)$ **84.** $n(A \cup B)$

85. $n(B) + n(C)$ **86.** $n(A) + n(B)$

87. $n((A \cup B) \cup C)$ **88.** $n(A \cap B)$

89. $n(A) + n(B) + n(C)$ **90.** $n((A \cap B) \cap C)$

91. **Investors** In a survey of 600 investors, it was reported that 380 had invested in stocks, 325 had invested in bonds, and 75 had not invested in either stocks or bonds.

a. How many investors had invested in both stocks and bonds?

b. How many investors had only invested in stocks?

92. **Medical Treatments** A team physician has determined that of all the athletes who were treated for minor back pain, 72% responded to an analgesic, 59% responded to a muscle relaxant, and 44% responded to both forms of treatment.

a. What percent of the athletes who were treated responded to the muscle relaxant but not to the analgesic?

b. What percent of the athletes who were treated did not respond to either form of treatment?

93. **Hand Guns** A special-interest group has conducted a survey concerning a ban on hand guns. *Note:* A rifle is a gun, but it is not a hand gun. The survey yielded the following results for the 1000 households that responded.

271 own a hand gun.
437 own a rifle.
497 support the ban on hand guns.
140 own both a hand gun and a rifle.
202 own a rifle but no hand gun and do not support the ban on hand guns.
74 own a hand gun and support the ban on hand guns.
52 own both a hand gun and a rifle and also support the ban on hand guns.

How many of the surveyed households:

a. own only a hand gun and do not support the ban on hand guns?

b. do not own a gun and support the ban on hand guns?

c. do not own a gun and do not support the ban on hand guns?

94. **Music** A survey of college students was taken to determine how the students acquired music. The survey showed the following results.

365 students acquired music from CDs.
298 students acquired music from the Internet.
268 students acquired music from cassettes.
212 students acquired music from both CDs and cassettes.
155 students acquired music from both CDs and the Internet.
36 students acquired music from cassettes, but not from CDs or the Internet.
98 students acquired music from CDs, cassettes, and the Internet.

Of those surveyed,

a. how many acquired music from CDs but not from the Internet or cassettes?

b. how many acquired music from the Internet but not from CDs or cassettes?

c. how many acquired music from CDs or the Internet?

d. how many acquired music from the Internet and cassettes?

95. **Advertising** A computer company advertises its computers in *PC World*, in *PC Magazine*, and on television. A survey of 770 customers finds that the numbers of customers that are familiar with the company's computers because of the different forms of advertising are as follows:

305, *PC World*
290, *PC Magazine*
390, television
110, *PC World* and *PC Magazine*
135, *PC Magazine* and television
150, *PC World* and television
85, all three sources

How many of the surveyed customers know about the computers because of:

a. exactly one of these forms of advertising?

b. exactly two of these forms of advertising?

c. *PC World* and neither of the other two forms of advertising?

96. Gratuities The management of a hotel conducted a survey. It found that of the 2560 guests who were surveyed:

1785 tip the wait staff.
1219 tip the luggage handlers.
831 tip the maids.
275 tip the maids and the luggage handlers.
700 tip the wait staff and the maids.
755 tip the wait staff and the luggage handlers.
245 tip all three services.
210 do not tip these services.

How many of the surveyed guests tip:

a. exactly two of the three services?

b. only the wait staff?

c. only one of the three services?

Extensions

CRITICAL THINKING

97.  **Blood Types** A report from the *American Association of Blood Banks* shows that in the United States:

44% of the population has the A antigen.
15% of the population has the B antigen.
84% of the population has the Rh+ factor.
34% of the population is A+.
9% of the population is B+.
4% of the population has the A antigen and the B antigen.
3% of the population is AB+.

See the Math Matters on pg 43.

Source: http://www.aabb.org/
All_About_Blood/FAQs/aabb_
faqs.htm

Find the percent of the U.S. population that is:

a. A−   **b.** O+   **c.** O−

98. Given that set *A* has 47 elements and set *B* has 25 elements, determine each of the following.

a. The maximum possible number of elements in $A \cup B$

b. The minimum possible number of elements in $A \cup B$

c. The maximum possible number of elements in $A \cap B$

d. The minimum possible number of elements in $A \cap B$

EXPLORATIONS

99. Search Engines The following Venn diagram displays *U* parceled into 16 distinct regions by four sets.

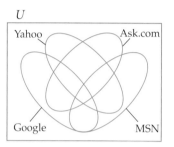

A survey of 1250 Internet users shows the following results concerning the use of the search engines Google, MSN, Yahoo!, and Ask.com.

585 use Google.
620 use Yahoo!.
560 use Ask.com.
450 use MSN.
100 use only Google, Yahoo!, and Ask.com.
41 use only Google, Yahoo!, and MSN.
50 use only Google, Ask.com, and MSN.
80 use only Yahoo!, Ask.com, and MSN.
55 use only Google and Yahoo!.
34 use only Google and Ask.com.
45 use only Google and MSN.
50 use only Yahoo! and Ask.com.
30 use only Yahoo! and MSN.
45 use only Ask.com and MSN.
60 use all four.

Use the above Venn diagram and the given information to answer the following questions.

How many of the Internet users:

a. use only Google?

b. use exactly three of the four search engines?

c. do not use any of the four search engines?

Summary

Key Terms

additive/subtractive color mixing [p. 50]
bar graph [p. 25]
broken-line graph [p. 25]
cardinal number of a finite set [p. 35]
circle graph [p. 25]
Collatz problem [p. 32]
complement of a set [p. 36]
conjecture [p. 2]
counterexample [p. 5]
counting number [p. 12]
deductive reasoning [p. 6]
De Morgan's laws [p. 41]
disjoint sets [p. 40]
element (member) of a set [p. 33]
ellipsis [p. 34]
empty set or null set [p. 36]
equal sets [p. 35]
finite set [p. 35]
inductive reasoning [p. 2]
integer [p. 12]
intersection [p. 39]
irrational number [p. 33]
natural number [p. 12]
palindromic number [p. 29]
prime number [p. 13]
proper subset [p. 38]
rational number [p. 33]
real number [p. 33]
roster method [p. 33]
set [p. 33]
subset [p. 37]
union [p. 40]
universal set [p. 36]
Venn diagram [p. 38]
well-defined set [p. 34]
whole number [p. 33]
winning/losing coalition [p. 46]

Essential Concepts

■ **Inductive Reasoning**
Inductive reasoning is the process of reaching a general conclusion by examining specific examples. A conclusion based on inductive reasoning is called a *conjecture*. A conjecture may or may not be correct.

■ **Deductive Reasoning**
Deductive reasoning is the process of reaching a conclusion by applying general assumptions, procedures, or principles.

■ **True and False Statements**
A statement is a *true statement* if and only if it is true in all cases. If you can find one case in which a statement is not true, then the statement is a *false statement*.

■ **Polya's Problem-Solving Strategy**
Many problems can be solved by applying *Polya's problem-solving strategy*:
1. Understand the problem.
2. Devise a plan.
3. Carry out the plan.
4. Review the solution.

■ **Subset of a Set**
Set A is a *subset* of set B, denoted by $A \subseteq B$, if and only if every element of A is also an element of B.

■ **Proper Subset of a Set**
Set A is a *proper subset* of set B, denoted by $A \subset B$, if and only if every element of A is also an element of B, and $A \neq B$.

■ **Number of Subsets of a Set**
A set with n elements has 2^n subsets and $2^n - 1$ proper subsets.

■ **Intersection of Sets**
The *intersection* of sets A and B, denoted by $A \cap B$, is the set of elements common to both A and B.

■ **Union of Sets**
The *union* of sets A and B, denoted by $A \cup B$, is the set that contains all the elements that belong to A or to B or to both.

■ **De Morgan's Laws**
$$(A \cup B)' = A' \cap B' \qquad (A \cap B)' = A' \cup B'$$

■ **Commutative Properties of Sets**
$$A \cap B = B \cap A \qquad A \cup B = B \cup A$$

■ **Associative Properties of Sets**
$$(A \cap B) \cap C = A \cap (B \cap C)$$
$$(A \cup B) \cup C = A \cup (B \cup C)$$

■ **Distributive Properties of Sets**

$$A \cap (B \cup C) = (A \cap B) \cup (A \cap C)$$
$$A \cup (B \cap C) = (A \cup B) \cap (A \cup C)$$

■ **A Cardinality Formula**
For any finite sets A and B,

$$n(A \cup B) = n(A) + n(B) - n(A \cap B)$$

CHAPTER 1 Review Exercises

In Exercises 1–4, determine whether the argument is an example of inductive reasoning or deductive reasoning.

1. All books written by John Grisham make the best-seller list. The book *The Last Juror* is a John Grisham book. Therefore, *The Last Juror* made the best-seller list.

2. Samantha got an A on each of her first four math tests, so she will get an A on the next math test.

3. We had rain yesterday, so there is less chance of rain today.

4. All amoeba multiply by dividing. I have named the amoeba shown in my microscope Amelia. Therefore, Amelia multiplies by dividing.

5. Find a counterexample to show that the following conjecture is false.

   Conjecture: For all numbers x, $x^4 > x$.

6. Find a counterexample to show that the following conjecture is false.

   Conjecture: For all counting numbers n, $\dfrac{n^3 + 5n + 6}{6}$

   is an even number.

7. Find a counterexample to show that the following conjecture is false.

   Conjecture: For all numbers x, $(x + 4)^2 = x^2 + 16$.

8. Find a counterexample to show that the following conjecture is false.

   Conjecture: For numbers a and b, $(a + b)^3 = a^3 + b^3$.

Polya's Problem-Solving Strategy In Exercises 9–14, solve each problem using Polya's four-step problem-solving strategy. Label your work so that each of Polya's four steps is identified.

9. A rancher decides to enclose a rectangular region by using an existing fence along one side of the region and 2240 feet of new fence on the other three sides. The rancher wants the length of the rectangular region to be five times its width. What will be the dimensions of the rectangular region?

10. In how many ways can you answer a 15-question test if you answer each question with either a "true," a "false," or an "always false"?

11. The skyboxes at a large sports arena are equally spaced around a circle. The 11th skybox is directly opposite the 35th skybox. How many skyboxes are there in the sports arena?

12. A rancher needs to get a dog, a rabbit, and a basket of carrots across a river. The rancher has a small boat that will only stay afloat carrying the rancher and one of the critters or the rancher and the carrots. The rancher cannot leave the dog alone with the rabbit because the dog will eat the rabbit. The rancher cannot leave the rabbit alone with the carrots because the rabbit will eat the carrots. How can the rancher get across the river with the critters and the carrots?

13. An investor bought 20 shares of stock for a total cost of $1200 and then sold all the shares for $1400. A few months later the investor bought 25 shares of the same stock for a total cost of $1800 and then sold all the shares for $1900. How much money did the investor earn on these investments?

14. If 15 people greet each other at a meeting by shaking hands with one another, how many handshakes will take place?

15. List five strategies that are included in Polya's second step (devise a plan).

16. List three strategies that are included in Polya's fourth step (review the solution).

17. **Match Students with Their Majors** Michael, Clarissa, Reggie, and Ellen are attending Florida State University (FSU). One student is a computer science major, one is a chemistry major, one is a business major, and one is a biology major. From the following clues, determine which major each student is pursuing.

   a. Michael and the computer science major are next door neighbors.

   b. Clarissa and the chemistry major have attended FSU for 2 years. Reggie has attended FSU for

3 years, and the biology major has attended FSU for 4 years.

c. Ellen has attended FSU for fewer years than Michael.

d. The business major has attended FSU for 2 years.

18. **Little League Baseball** Each of the Little League teams in a small rural community is sponsored by a different local business. The names of the teams are the Dodgers, the Pirates, the Tigers, and the Giants. The businesses that sponsor the teams are the bank, the supermarket, the service station, and the drug store. From the following clues, determine which business sponsors each team.

a. The Tigers and the team sponsored by the service station have winning records this season.

b. The Pirates and the team sponsored by the bank are coached by parents of the players, whereas the Giants and the team sponsored by the drug store are coached by the director of the Community Center.

c. Jake is the pitcher for the team sponsored by the supermarket and coached by his father.

d. The game between the Tigers and the team sponsored by the drug store was rained out yesterday.

19. **Map Coloring** The following map shows seven countries in the Indian subcontinent. Four colors have been used to color the countries such that no two bordering countries are the same color.

a. Can this map be colored, using only three colors, such that no two bordering countries are the same color? Explain.

b. Can this map be colored, using only two colors, such that no two bordering countries are the same color? Explain.

20. **Find a Route** The following map shows the 10 bridges and 3 islands between the suburbs of North Bay and South Bay.

a. During your morning workout, you decide to jog over each bridge exactly once. Draw a route that you can take. Assume that you want to start from North Bay and that your workout concludes after you jog over the 10th bridge.

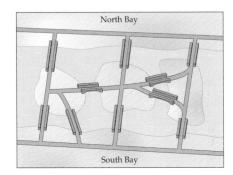

b. Assume you want to start your jog from South Bay. Can you find a route that crosses each bridge exactly once?

21. **Areas of Rectangles** Two perpendicular line segments partition the interior of a rectangle into four smaller rectangles. The areas of these smaller rectangles are x, 2, 5, and 10 square inches. Find all possible values of x.

22. **A Cryptarithm** In the following addition problem, each letter represents one of the digits 0, 1, 2, 3, 4, 5, 6, 7, 8, or 9. The leading digits represented by A and B are nonzero digits. What digit is represented by each letter?

$$\begin{array}{r} A \\ + \ B\ B \\ \hline A\ D\ D \end{array}$$

23. **Make Change** In how many different ways can change be made for a dollar using only quarters and/or nickels?

24. **Counting Problem** In how many different orders can a basketball team win exactly three out of its last five games?

25. What is the units digit of 7^{56}?

26. What is the units digit of 23^{85}?

27. **Verify a Conjecture** Use deductive reasoning to show that the following procedure always produces a number that is twice the original number.
Procedure: Pick a number. Multiply the number by 4, add 12 to the product, divide the sum by 2, and subtract 6.

28. Explain why 2004 nickels are worth more than 100 dollars.

29. College Graduates The following bar graph shows the percents of the U.S. population, age 25 and over, who have attained bachelor's degrees or higher for selected years from 1940 to 2000.

Percent of U.S. Population, Age 25 and Over, with a Bachelor's Degree or Higher

Source: U.S. Census Bureau

a. During which 10-year period did the percent of bachelor's degree or higher recipients increase the most?

b. What was the amount of that increase?

30. The Film Industry The following circle graph categorizes, by their ratings, the 655 films released during a recent year.

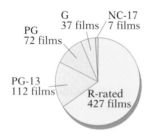

Ratings of Films Released

Source: MPA Worldwide Market Research

a. How many times as many PG-13 films as NC-17 films were released?

b. How many times as many R-rated films as NC-17 films were released?

31. SAT Scores The following broken-line graphs show the average SAT math scores and the average SAT verbal scores for the years from 1999 to 2004.

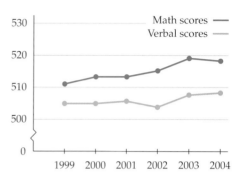

Average SAT Scores

Source: The College Board

a. In which year did the math scores increase and the verbal scores decrease from the previous year?

b. In which year did the verbal scores increase and the math scores decrease from the previous year?

32. Palindromic Numbers Recall that palindromic numbers read the same from left to right as they read from right to left. For instance, 37,573 is a palindromic number. Find the smallest palindromic number larger than 1000 that is a multiple of 5.

33. Narcissistic Numbers A **narcissistic number** is a two-digit natural number that is equal to the sum of the squares of its digits. Find all narcissistic numbers.

34. Number of Intersections Two different lines can intersect in at most one point. Three different lines can intersect in at most three points, and four different lines can intersect in at most six points.

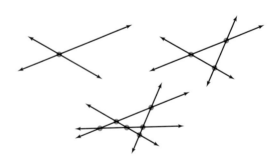

a. Determine the maximum number of intersections for five different lines.

b. Does it appear, by inductive reasoning, that the maximum number of intersection points I_n for n different lines is given by $I_n = \dfrac{n(n-1)}{2}$?

In Exercises 35–38, use the roster method to write each set.

35. The set of whole numbers less than 8

36. The set of integers that satisfy $x^2 = 64$

37. The set of natural numbers that satisfy $x + 3 \leq 7$

38. The set of counting numbers larger than -3 and less than or equal to 6

In Exercises 39–46, let $U = \{2, 6, 8, 10, 12, 14, 16, 18\}$, $A = \{2, 6, 10\}$, $B = \{6, 10, 16, 18\}$, and $C = \{14, 16\}$. Find each of the following.

39. $A \cap B$

40. $A \cup B$

41. $A' \cap C$

42. $B \cup C'$

43. $A \cup (B \cap C)$

44. $(A \cup C)' \cap B'$

45. $(A \cap B')'$

46. $(A \cup B \cup C)'$

In Exercises 47–50, determine whether the first set is a proper subset of the second set.

47. The set of natural numbers; the set of whole numbers

48. The set of integers; the set of real numbers

49. The set of counting numbers; the set of natural numbers

50. The set of real numbers; the set of rational numbers

In Exercises 51–54, list all the subsets of the given set.

51. $\{I, II\}$

52. $\{s, u, n\}$

53. $\{$penny, nickel, dime, quarter$\}$

54. $\{A, B, C, D, E\}$

In Exercises 55–58, find the number of subsets of the given set.

55. The set of the four musketeers

56. The set of the letters of the English alphabet

57. The set of the letters of "uncopyrightable," which is the longest English word with no repeated letters

58. The set of the seven dwarfs

In Exercises 59–62, draw a Venn diagram to represent the given set.

59. $A \cap B'$

60. $A' \cup B'$

61. $(A \cup B) \cup C'$

62. $A \cap (B' \cup C)$

In Exercises 63–66, draw Venn diagrams to determine whether the expressions are equal for all sets A, B, and C.

63. $A' \cup (B \cup C);\quad (A' \cup B) \cup (A' \cup C)$

64. $(A \cap B) \cap C';\quad (A' \cup B') \cup C$

65. $A \cap (B' \cap C);\quad (A \cup B') \cap (A \cup C)$

66. $A \cap (B \cup C);\quad A' \cap (B \cup C)$

In Exercises 67 and 68, draw a Venn diagram with each of the given elements placed in the correct region.

67. $U = \{e, h, r, d, w, s, t\}$
$A = \{t, r, e\}$
$B = \{w, s, r, e\}$
$C' = \{s, r, d, h\}$

68. $U = \{\alpha, \beta, \Gamma, \gamma, \Delta, \delta, \varepsilon, \theta\}$
$A' = \{\beta, \Delta, \theta, \gamma\}$
$B = \{\delta, \varepsilon\}$
$C = \{\beta, \varepsilon, \Gamma\}$

In Exercises 69 and 70, state the cardinality of each set.

69. $\{5, 6, 7, 8, 6\}$

70. $\{4, 6, 8, 10, 12, \ldots, 22\}$

71. A Computer Survey In a survey of 300 college students, 170 students indicated that they owned a laptop (portable) computer, 94 indicated that they owned a desktop computer, and 35 indicated that they owned both a laptop and a desktop computer.

How many of the students in the survey:

a. owned only a laptop computer?

b. owned only a desktop computer?

c. did not own either a laptop or a desktop computer?

72. An Exercise Survey In a survey at a health club, 208 members indicated that they enjoyed aerobic exercises, 145 indicated that they enjoyed weight training, 97 indicated that they enjoyed both aerobics and weight training, and 135 indicated that they did not enjoy either of these types of exercise. How many members were surveyed?

73. Coffee Preferences A gourmet coffee bar conducted a survey to determine the preferences of its customers. Of the customers surveyed,

221 liked espresso.

127 liked cappuccino and chocolate-flavored coffee.

182 liked cappuccino.

136 liked espresso and chocolate-flavored coffee.

209 liked chocolate-flavored coffee.

96 liked all three types of coffee.

116 liked espresso and cappuccino.

82 liked none of these types of coffee.

How many of the customers in the survey:

a. liked only chocolate-flavored coffee?

b. liked cappuccino and chocolate-flavored coffee but not espresso?

c. liked espresso and cappuccino but not chocolate-flavored coffee?

d. liked exactly one of the three types of coffee?

74. A Television Survey A survey of 250 families in a housing development was taken to determine how they acquired their television service. The survey found that the families acquired their television service through a cable service, through a satellite service, or by using an antenna. Of the families surveyed:

155 used a cable service.

142 used a satellite service.

80 used an antenna.

64 used both a cable service and a satellite service.

26 used both a cable service and an antenna.

55 used both a satellite service and an antenna.

14 used all three methods to acquire their television service.

How many of the families in the survey:

a. used only a cable service to acquire their television service?

b. used only a satellite service to acquire their television service?

c. did not use any of the three methods?

CHAPTER 1   Test

1. Determine whether each of the following is an example of inductive reasoning or deductive reasoning.

a. All novels by Sidney Sheldon are gruesome. The novel *Are You Afraid of the Dark?* was written by Sidney Sheldon. Therefore, *Are You Afraid of the Dark?* is a gruesome novel.

b. Ashlee Simpson's last album made the top-ten list, so her next album will also make the top-ten list.

c. Two computer programs, a *bubble sort* and a *shell sort*, are used to sort data. In each of 10 experiments, the shell sort program took less time to sort the data than did the bubble sort program. Thus the shell sort program is the faster of the two sorting programs.

d. If a geometric figure is a rectangle, then it is a parallelogram. Figure A is a rectangle. Therefore, Figure A is a parallelogram.

2. Counting Problem In how many different orders can a basketball team win exactly four out of its last six games?

3. What is the units digit of 3^{4513}?

4. Vacation Money Shelly has saved some money for a vacation. Shelly spends half of her vacation money on an airline ticket; she then spends $50 for sunglasses, $22 for a taxi, and one-third of her remaining money for a room with a view. After her sister repays her a loan of $150, Shelly finds that she has $326. How much vacation money did Shelly have at the start of her vacation?

5. Number of Routes How many different routes are there from point A to point B in the following figure?

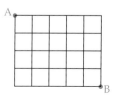

6. **Number of Games** In a league of nine football teams, each team plays every other team in the league exactly once. How many league games will take place?

7. **Ages of Children** The four children in the Rivera family are Reynaldo, Ramiro, Shakira, and Sasha. The ages of the two teenagers are 13 and 15. The ages of the younger children are 5 and 7. From the following clues, determine the age of each of the children.

   a. Reynaldo is older than Ramiro.

   b. Sasha is younger than Shakira.

   c. Sasha is 2 years older than Ramiro.

   d. Shakira is older than Reynaldo.

8. **Palindromic Numbers** Find the smallest palindromic number larger than 600 that is a multiple of 3.

9. Find a counterexample to show that the following conjecture is false.

   Conjecture: For all numbers x, $\dfrac{(x-4)(x+3)}{(x-4)} = x + 3$

10. **Navigation Systems** The following bar graph shows the numbers of new vehicles sold with navigation systems for the years 1998 to 2003.

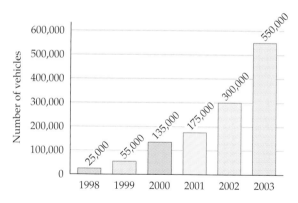

New Vehicles Sold with Navigation Systems
Source: J. D. Power and Associates

   a. Between which two years did the number of new vehicles sold with navigation systems increase the most?

   b. What was the amount of that increase?

11. Use the roster method to represent the set of whole number less than 6.

12. Let $U = \{1, 2, 3, 4, 5, 6, 7, 8, 9, 10\}$, $A = \{3, 5, 7, 8\}$, $B = \{2, 3, 8, 9, 10\}$, and

$C = \{1, 4, 7, 8\}$. Use the roster method to write each of the following sets.

   a. $A \cup B$

   b. $A' \cap B$

   c. $A \cap (B' \cup C)$

13. Let $U =$ {Ford, Hummer, Lexus, BMW, Pontiac, Chrysler, Mercedes, Mazda, Chevrolet}, $A =$ {Ford, Pontiac, Chrysler, Chevrolet}, $B =$ {Lexus, Mazda}, and $C =$ {Hummer, Pontiac}.

Find each of the following:

   a. $n(B \cup C')$

   b. $n((A \cup B) \cap C)$

14. List all the subsets of {a, d, p}.

15. Determine the number of subsets of a set with 21 elements.

16. Draw a Venn diagram and shade in the region represented by $(A \cup B') \cap C$.

17. According to one of De Morgan's laws, the set $(A \cup B)'$ can also be written as _____.

18. **Motorcycle Options** A motorcycle manufacturer makes a motorcycle with five upgrade options. Each option is independent of the other options.

   a. How many different versions of this motorcycle can the company produce?

   b. What is the minimum number of upgrade options the company must be able to provide if it wishes to offer at least 250 different versions of this motorcycle?

19. **A Survey** In the town of LeMars, 385 families have a CD player, 142 families have a DVD player, 41 families have both a CD player and a DVD player, and 55 families do not have a CD player or a DVD player. How many families live in LeMars?

20. News Survey A survey of 1000 households was taken to determine how they obtained news about current events. The survey considered only television, newspapers, and the Internet as sources for news. Of the households surveyed,

724 obtained news from television.

545 obtained news from newspapers.

280 obtained news from the Internet.

412 obtained news from both television and newspapers.

185 obtained news from both television and the Internet.

105 obtained news from television, newspapers, and the Internet.

64 obtained news from the Internet but not from television or newspapers.

Of those households that were surveyed,

a. how many obtained news from television but not from newspapers or the Internet?

b. how many obtained news from newspapers but not from television or the Internet?

c. how many obtained news from television or newspapers?

d. how many did not obtain news from television, newspapers, or the Internet?

CHAPTER 2

Logic and Its Applications

I know what you're thinking about," said Tweedledum (to Alice), "but it isn't so, nohow."

"Contrariwise," continued Tweedledee, "if it was so, it might be; and if it were so, it would be; but as it isn't, it ain't. That's logic."

This excerpt, from Lewis Carroll's *Through the Looking-Glass,* sums up Tweedledee's understanding of logic. In today's complex world, it is not so easy to summarize the topic of logic. For lawyers and business people, logic is the science of correct reasoning. They often use logic to construct valid arguments, analyze legal contracts, and solve complicated problems. The principles of logic can also be used as a production tool. For example, programmers use logic to design computer software, engineers use logic to design the electronic circuits in computers, and mathematicians use logic to solve problems and construct mathematical proofs.

In this chapter, you will encounter several facets of logic. Specifically, you will use logic to

- analyze information and relationships between statements,

- determine the validity of arguments,

- determine valid conclusions based on given assumptions, and

- analyze electronic circuits.

Online Study Center

For online student resources, visit this textbook's Online Study Center at **college.hmco.com/pic/aufmannMTQR.**

SECTION 2.1 | # Logic Statements and Quantifiers

Logic Statements

George Boole
(bool) was born in
1815 in Lincoln,
England. He was
raised in poverty,
but he was very
industrious and
had learned Latin and Greek by
the age of 12. Later he mastered
German, French, and Italian. His
first profession, at the young age
of 16, was that of an assistant
school teacher. At the age of 20
he started his own school.

In 1849 Boole was ap-
pointed the chairperson of mathematics at Queens College in
Cork, Ireland. He was known as a
dedicated professor who gave
detailed lectures. He continued to
teach at Queens College until he
died of pneumonia in 1864.

Many of Boole's mathematical ideas, such as Boolean algebra, have applications in the
areas of computer programming
and the design of telephone
switching devices. ■

One of the first mathematicians to make a serious study of symbolic logic was
Gottfried Wilhelm Leibniz (1646–1716). Leibniz tried to advance the study of logic
from a merely philosophical subject to a formal mathematical subject. Leibniz never
completely achieved this goal; however, several mathematicians, such as Augustus
De Morgan (1806–1871) and George Boole (1815–1864), contributed to the ad-
vancement of symbolic logic as a mathematical discipline.

Boole published *The Mathematical Analysis of Logic* in 1848. In 1854 he published the more extensive work *An Investigation of the Laws of Thought.* Concerning
this document, the mathematician Bertrand Russell stated, "Pure mathematics was
discovered by Boole in a work which is called *The Laws of Thought.*" Although some
mathematicians feel this is an exaggeration, the following paragraph, extracted from
An Investigation of the Laws of Thought, gives some insight into the nature and scope
of this document.

> The design of the following treatise is to investigate the fundamental laws of those
> operations of the mind by which reasoning is performed; to give expression to
> them in the language of a Calculus, and upon this foundation to establish the sci-
> ence of Logic and construct its method; to make the method itself the basis of a
> general method for the application of the mathematical doctrine of probabilities;
> and finally, to collect from the various elements of truth brought to view in the
> course of these inquiries some probable intimations concerning the nature and
> constitution of the human mind.[1]

Every language contains different types of sentences, such as statements, questions, and commands. For instance,

"Is the test today?" is a question.

"Go get the newspaper" is a command.

"This is a nice car" is an opinion.

"Denver is the capital of Colorado" is a statement of fact.

The symbolic logic that Boole was instrumental in creating applies only to sentences that are *statements* as defined below.

Statement

A **statement** is a declarative sentence that is either true or false, but not both
simultaneously.

---

1. Bell, E. T. *Men of Mathematics.* New York: Simon and Schuster, Inc., Touchstone Books, Reissue edition,
1986.

It may not be necessary to determine whether a sentence is true or false to determine whether it is a statement. For instance, the following sentence is either true or false:

Every even number greater than 2 can be written as the sum of two prime numbers.

At this time mathematicians have not determined whether the sentence is true or false, but they do know that it is either true or false and that it is not both true and false. Thus the sentence is a statement.

✔ **TAKE NOTE**

The following sentence is a famous paradox:

This is a false sentence.

It is not a statement, because if we assume it to be a true sentence, then it is false, and if we assume it to be a false sentence, then it is true. Statements cannot be true and false at the same time.

EXAMPLE 1 ■ Identify Statements

Determine whether each sentence is a statement.

a. Florida is a state in the United States.

b. The word *dog* has four letters.

c. How are you?

d. $9^{(9^9)} + 2$ is a prime number.

e. $x + 1 = 5$

Solution

a. Florida is one of the 50 states in the United States, so this sentence is true and it is a statement.

b. The word *dog* consists of exactly three letters, so this sentence is false and it is a statement.

c. The sentence "How are you?" is a question; it is not a declarative sentence. Thus it is not a statement.

d. You may not know whether $9^{(9^9)} + 2$ is a prime number; however, you do know that it is a whole number larger than 1, so it is either a prime number or it is not a prime number. The sentence is either true or false, and it is not both true and false simultaneously, so it is a statement.

e. $x + 1 = 5$ is a statement. It is known as an *open statement*. It is true for $x = 4$, and it is false for any other value of x. For any given value of x, it is true or false but not both.

CHECK YOUR PROGRESS 1   Determine whether each sentence is a statement.

a. Open the door.

b. 7055 is a large number.

c. $4 + 5 = 8$

d. In the year 2019, the president of the United States will be a woman.

e. $x > 3$

Solution   See page S5.

Charles Dodgson
(Lewis Carroll)

MathMatters   **Charles Dodgson**

One of the best-known logicians is Charles Dodgson (1832–1898). Dodgson was educated at Rugby and Oxford, and in 1861 he became a lecturer in mathematics at Oxford. His mathematical works include *A Syllabus of Plane Algebraical Geometry, The Fifth Book of Euclid Treated Algebraically,* and *Symbolic Logic.* Although Dodgson was a distinguished mathematician in his time, he is best known by his pen name Lewis Carroll, which he used when he published *Alice's Adventures in Wonderland* and *Through the Looking-Glass.*

Queen Victoria of the United Kingdom enjoyed *Alice's Adventures in Wonderland* to the extent that she told Dodgson she was looking forward to reading another of his books. He promptly sent her *A Syllabus of Plane Algebraical Geometry,* and it was reported that she was less than enthusiastic about the latter book.

Compound Statements

Connecting statements with words and phrases such as *and, or, not, if … then,* and *if and only if* creates a **compound statement.** For instance, "I will attend the meeting or I will go to school" is a compound statement. It is composed of the two **component statements** "I will attend the meeting" and "I will go to school." The word *or* is a **connective** for the two component statements.

George Boole used symbols such as *p, q, r,* and *s* to represent statements and the symbols $\wedge, \vee, \sim, \rightarrow,$ and $\leftrightarrow$ to represent connectives. See Table 2.1.

Table 2.1 *Logic Symbols*

Original Statement	Connective	Statement in Symbolic Form	Type of Compound Statement
not *p*	not	$\sim p$	negation
p and *q*	and	$p \wedge q$	conjunction
p or *q*	or	$p \vee q$	disjunction
If *p*, then *q*	If … then	$p \rightarrow q$	conditional
p if and only if *q*	if and only if	$p \leftrightarrow q$	biconditional

QUESTION   *What connective is used in a conjunction?*

_____

ANSWER   *The connective and.*

The Truth Table for ~p

p	*~p*
T	F
F	T

> **Truth Value and Truth Tables**
>
> The **truth value** of a statement is true (T) if the statement is true and false (F) if the statement is false. A **truth table** is a table that shows the truth values of a statement for all possible truth values of its components.

The **negation** of the statement "Today is Friday" is the statement "Today is not Friday." In symbolic logic, the tilde symbol ~ is used to denote the negation of a statement. If a statement p is true, its negation $\sim p$ is false, and if a statement p is false, its negation $\sim p$ is true. See the table at the left. The negation of the negation of a statement is the original statement. Thus, $\sim(\sim p)$ can be replaced by p in any statement.

EXAMPLE 2 ■ Write the Negation of a Statement

Write the negation of each statement.

a. Bill Gates has a yacht.

b. The number 10 is a prime number.

c. The Dolphins lost the game.

Solution

a. Bill Gates does not have a yacht.

b. The number 10 is not a prime number.

c. The Dolphins did not lose the game.

CHECK YOUR PROGRESS 2   Write the negation of each statement.

a. 1001 is divisible by 7.

b. 5 is an even number.

c. The fire engine is not red.

Solution   See page S5.

We will often find it useful to write compound statements in symbolic form.

EXAMPLE 3 ■ Write Compound Statements in Symbolic Form

Consider the following statements.

p: Today is Friday.

q: It is raining.

r: I am going to a movie.

s: I am not going to the basketball game.

Write the following compound statements in symbolic form.

a. Today is Friday and it is raining.

b. It is not raining and I am going to a movie.

c. I am going to the basketball game or I am going to a movie.

d. If it is raining, then I am not going to the basketball game.

Solution

a. $p \wedge q$ **b.** $\sim q \wedge r$ **c.** $\sim s \vee r$ **d.** $q \rightarrow s$

CHECK YOUR PROGRESS 3   Use p, q, r, and s as defined in Example 3 to write the following compound statements in symbolic form.

a. Today is not Friday and I am going to a movie.
b. I am going to the basketball game and I am not going to a movie.
c. I am going to a movie if and only if it is raining.
d. If today is Friday, then I am not going to a movie.

Solution   *See page S5.*

In the next example, we translate symbolic logic statements into English sentences.

EXAMPLE 4 ■ Translate Symbolic Statements

Consider the following statements.

   p: The game will be played in Atlanta.
   q: The game will be shown on CBS.
   r: The game will not be shown on ESPN.
   s: The Dodgers are favored to win.

Write each of the following symbolic statements in words.

a. $q \wedge p$ **b.** $\sim r \wedge s$ **c.** $s \leftrightarrow \sim p$

Solution
a. The game will be shown on CBS and the game will be played in Atlanta.
b. The game will be shown on ESPN and the Dodgers are favored to win.
c. The Dodgers are favored to win if and only if the game will not be played in Atlanta.

CHECK YOUR PROGRESS 4   Consider the following statements.

   e: All men are created equal.
   t: I am trading places.
   a: I get Abe's place.
   g: I get George's place.

Use the information above to translate the dialogue in the following speech bubbles.

Solution   *See page S5.*

The Truth Table for $p \wedge q$

p	q	$p \wedge q$
T	T	T
T	F	F
F	T	F
F	F	F

If you order cake *and* ice cream in a restaurant, the waiter will bring *both* cake and ice cream. In general, the **conjunction** $p \wedge q$ is true if both p and q are true, and the conjunction is false if either p or q is false. The truth table at the left shows the four possible cases that arise when we form a conjunction of two statements.

> **Truth Value of a Conjunction**
>
> The conjunction $p \wedge q$ is true if and only if both p and q are true.

Sometimes the word *but* is used in place of the connective *and* to form a conjunction. For instance, "My local phone company is *SBC*, but my long-distance carrier is Sprint" is equivalent to the conjunction "My local phone company is *SBC* and my long-distance carrier is Sprint."

Any **disjunction** $p \vee q$ is true if p is true or q is true or both p and q are true. The truth table at the left shows that the disjunction p or q is false if both p and q are false; however, it is true in all other cases.

The Truth Table for $p \vee q$

p	q	$p \vee q$
T	T	T
T	F	T
F	T	T
F	F	F

> **Truth Value of a Disjunction**
>
> The disjunction $p \vee q$ is true if p is true, if q is true, or if both p and q are true.

EXAMPLE 5 ■ Determine the Truth Value of a Statement

Determine whether each statement is true or false.

a. $7 \geq 5$

b. 5 is a whole number and 5 is an even number.

c. 2 is a prime number and 2 is an even number.

Solution

a. $7 \geq 5$ means $7 > 5$ or $7 = 5$. Because $7 > 5$ is true, the statement $7 \geq 5$ is a true statement.

b. This is a false statement because 5 is not an even number.

c. This is a true statement because each component statement is true.

CHECK YOUR PROGRESS 5   Determine whether each statement is true or false.

a. 21 is a rational number and 21 is a natural number.

b. $4 \leq 9$

c. $-7 \geq -3$

Solution   See page S6.

Truth tables for the conditional and biconditional are given in Section 2.3.

Quantifiers and Negation

In a statement, the word *some* and the phrases *there exists* and *at least one* are called **existential quantifiers.** Existential quantifiers are used as prefixes to assert the existence of something.

In a statement, the words *none, no, all,* and *every* are called **universal quantifiers.** The universal quantifiers *none* and *no* deny the existence of something, whereas the universal quantifiers *all* and *every* are used to assert that every element of a given set satisfies some condition.

Recall that the negation of a false statement is a true statement and the negation of a true statement is a false statement. It is important to remember this fact when forming the negation of a quantified statement. For instance, what is the negation of the false statement, "All dogs are mean"? You may think that the negation is "No dogs are mean," but this is also a false statement. Thus the statement "No dogs are mean" is not the negation of "All dogs are mean." The negation of "All dogs are mean," which is a false statement, is in fact "Some dogs are not mean," which is a true statement. The statement "Some dogs are not mean" can also be stated as "At least one dog is not mean" or "There exists a dog that is not mean."

What is the negation of the false statement "No doctors write in a legible manner"? Whatever the negation is, we know it must be a true statement. The negation cannot be "All doctors write in a legible manner," because this is also a false statement. The negation is "Some doctors write in a legible manner." This can also be stated as "There exists at least one doctor who writes in a legible manner."

Table 2.2 summarizes the concepts needed to write the negations of statements that contain one of the quantifiers *all*, *none*, or *some*.

Table 2.2 *The Negation of a Statement That Contains a Quantifier*

Original Statement	Negation
All _____ are _____ .	Some _____ are not _____ .
No(ne) _____ .	Some _____ .
Some _____ are not _____ .	All _____ are _____ .
Some _____ .	No(ne) _____ .

EXAMPLE 6 ■ Write the Negation of a Quantified Statement

Write the negation of each of the following statements.

a. Some baseball players are worth a million dollars.

b. All movies are worth the price of admission.

c. No odd numbers are divisible by 2.

Solution

a. No baseball player is worth a million dollars.

b. Some movies are not worth the price of admission.

c. Some odd numbers are divisible by 2.

CHECK YOUR PROGRESS 6 Write the negations of the following statements.

a. All bears are brown.

b. No math class is fun.

c. Some vegetables are not green.

Solution See page S6.

Investigation

Claude E. Shannon

Switching Networks

In 1939, Claude E. Shannon (1916–2001) wrote a thesis on an application of symbolic logic to *switching networks.* A switching network consists of wires and switches that can open or close. Switching networks are used in many electrical appliances, telephone equipment, and computers. Figure 2.1 shows a switching network that consists of a single switch P that connects two terminals. An electric current can flow from one terminal to the other terminal provided the switch P is in the closed position. If P is in the open position, then the current cannot flow from one terminal to the other. If a current can flow between the terminals we say that a network is closed, and if a current cannot flow between the terminals we say that the network is open. We designate this network by the letter P. There exists an analogy between a network P and a statement p in that a network is either open or closed, and a statement is either true or false.

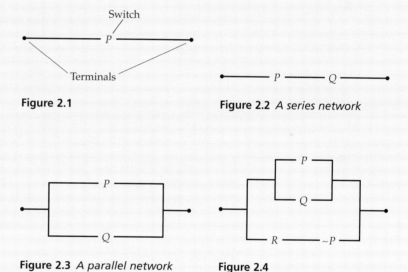

Figure 2.1

Figure 2.2 *A series network*

Figure 2.3 *A parallel network*

Figure 2.4

Figure 2.2 shows two switches P and Q connected in **series.** This series network is closed if and only if both switches are closed. We will use $P \wedge Q$ to denote this series network because it is analogous to the logic statement $p \wedge q$, which is true if and only if both p and q are true.

Figure 2.3 shows two switches P and Q connected in **parallel.** This parallel network is closed if either P or Q is closed. We will designate this parallel network by $P \vee Q$ because it is analogous to the logic statement $p \vee q$, which is true if p is true or if q is true.

Series and parallel networks can be combined to produce more complicated networks, as shown in Figure 2.4.

The network shown in Figure 2.4 is closed provided P or Q is closed or provided both R and $\sim P$ are closed. Note that the switch $\sim P$ is closed if P is open, and $\sim P$ is open if P is closed. We use the symbolic statement $(P \vee Q) \vee (R \wedge \sim P)$ to represent this network.

If two switches are always open at the same time and always closed at the same time, then we will use the same letter to designate both switches.

(continued)

Investigation Exercises

Write a symbolic statement to represent each of the networks in Investigation Exercises 1–6.

1.

2.
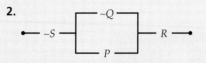

3.
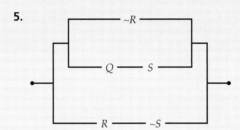

4.
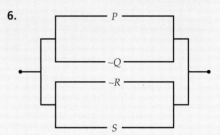

5.

6.

7. Which of the networks in Investigation Exercises 1–6 are closed networks, given that P is closed, Q is open, R is closed, and S is open?

8. Which of the networks in Investigation Exercises 1–6 are closed networks, given that P is open, Q is closed, R is closed, and S is closed?

In Investigation Exercises 9–14, draw a network to represent each statement.

9. $(\sim P \vee Q) \wedge (R \wedge P)$

10. $P \wedge [(Q \wedge \sim R) \vee R]$

11. $(\sim P \wedge Q \wedge R) \vee (P \wedge R)$

12. $(Q \vee R) \vee (S \vee \sim P)$

13. $[(\sim P \wedge R) \vee Q] \vee (\sim R)$

14. $(P \vee Q \vee R) \wedge S \wedge (\sim Q \vee R)$

Warning Circuits The circuits shown in Investigation Exercises 15 and 16 include a switching network, a warning light, and a battery. In each circuit the warning light will turn on only when the switching network is closed.

15. Consider the following circuit.

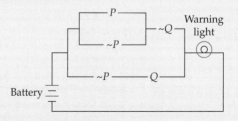

For each of the following conditions, determine whether the warning light will be on or off.

a. P is closed and Q is open.

b. P is closed and Q is closed.

c. P is open and Q is closed.

d. P is open and Q is open.

(continued)

16. An engineer thinks that the following circuit can be used in place of the circuit shown in Investigation Exercise 15. Do you agree? Explain.

Exercise Set 2.1

In Exercises 1–10, determine whether each sentence is a statement.

1. West Virginia is west of the Mississippi River.
2. 1031 is a prime number.
3. The area code for Storm Lake, Iowa, is 512.
4. Some negative numbers are rational numbers.
5. Have a fun trip.
6. Do you like to read?
7. All hexagons have exactly five sides.
8. If x is a negative number, then x^2 is a positive number.
9. Mathematics courses are better than history courses.
10. Every real number is a rational number.

In Exercises 11–18, determine the components of each compound statement.

11. The principal will attend the class on Tuesday or Wednesday.
12. 5 is an odd number and 6 is an even number.
13. A triangle is an acute triangle if and only if it has three acute angles.
14. Some birds can swim and some fish can fly.
15. I ordered a salad and a cola.
16. If this is Saturday, then tomorrow is Sunday.
17. $5 + 2 \geq 6$
18. $9 - 1 \leq 8$

In Exercises 19–22, write the negation of each statement.

19. The Giants lost the game.
20. The lunch was served at noon.
21. The game did not go into overtime.
22. The game was not shown on ABC.

In Exercises 23–32, write each sentence in symbolic form. Represent each component of the sentence with the letter indicated in parentheses. Also state whether the sentence is a conjunction, a disjunction, a negation, a conditional, or a biconditional.

23. If today is Wednesday (w), then tomorrow is Thursday (t).
24. It is not true that Sue took the tickets (t).
25. All squares (s) are rectangles (r).
26. I went to the post office (p) and the bookstore (s).
27. A triangle is an equilateral triangle (l) if and only if it is an equiangular triangle (a).
28. A number is an even number (e) if and only if it has a factor of 2 (t).
29. If it is a dog (d), it has fleas (f).
30. Polynomials that have exactly three terms (p) are called trinomials (t).
31. I will major in mathematics (m) or computer science (c).
32. All pentagons (p) have exactly five sides (s).

In Exercises 33–38, write each symbolic statement in words. Use p, q, r, s, t, and u as defined below.

p: The tour goes to Italy.

q: The tour goes to Spain.

r: We go to Venice.

s: We go to Florence.

t: The hotel fees are included.

u: The meals are not included.

33. $p \wedge \sim q$ **34.** $r \vee s$

35. $r \rightarrow \sim s$ **36.** $p \rightarrow r$

37. $s \leftrightarrow \sim r$ **38.** $\sim t \wedge u$

39. If p is a false statement and the compound statement $p \vee q$ is a true statement, what can be said about the q statement?

40. If p is a false statement and the compound statement $p \wedge q$ is a false statement, must q also be a false statement?

41. If p is a true statement, what can be said about the compound statement $p \vee q$?

42. If the statement "All ps are qs" is a true statement, what can be said about the statement "Some ps are not qs"?

In Exercises 43–54, use the definitions presented in Table 2.2, page 68, to write the negation of each quantified statement.

43. Some cats do not have claws.

44. Some dogs are not friendly.

45. All classic movies were first produced in black and white.

46. Everybody enjoyed the dinner.

47. None of the numbers were even numbers.

48. At least one student received an A.

49. No irrational number can be written as a terminating decimal.

50. All cameras use film.

51. All cars run on gasoline.

52. None of the students took my advice.

53. Every item is on sale.

54. All of the telephone lines are not busy.

In Exercises 55–68, determine whether each statement is true or false.

55. 7 < 5 or 3 > 1.

56. $3 \leq 9$

57. $(-1)^{50} = 1$ and $(-1)^{99} = -1$.

58. $7 \neq 3$ or 9 is a prime number.

59. $-5 \geq -11$

60. $4.5 \leq 5.4$

61. 2 is an odd number or 2 is an even number.

62. 5 is a natural number and 5 is a rational number.

63. There exists an even prime number.

64. The square of any real number is a positive number.

65. Some real numbers are irrational.

66. All irrational numbers are real numbers.

67. Every integer is a rational number.

68. Every rational number is an integer.

Extensions

CRITICAL THINKING

Write Quotations in Symbolic Form In Exercises 69–74, translate each quotation into symbolic form. For each component, indicate what letter you used to represent the component.

69. If you can count your money, you don't have a billion dollars.   *J. Paul Getty*

70. If you aren't fired with enthusiasm, then you will be fired with enthusiasm.   *Vince Lombardi*

71. Those who do not learn from history are condemned to repeat it.   *George Santayana*

72. We don't like their sound, and guitar music is on the way out.   *Decca Recording Company,* rejecting the Beatles in 1962

73. If people concentrated on the really important things in life, there'd be a shortage of fishing poles.   *Doug Larson*

74. If you're killed, you've lost a very important part of your life.   *Brooke Shields*

Write Statements in Symbolic Form In Exercises 75–80, translate each mathematical statement into symbolic form. For each component, indicate what letter you used to represent the component.

75. An angle is a right angle if and only if its measure is 90°.

76. Any angle inscribed in a semicircle is a right angle.

77. If two sides of a triangle are equal in length, the angles opposite those sides are congruent.

78. The sum of the measures of the three angles of any triangle is 180°.

79. All squares are rectangles.

80. If the corresponding sides of two triangles are proportional, then the triangles are similar.

EXPLORATIONS

81.  **Raymond Smullyan** is a logician, a philosopher, a professor, and an author of many books on logic and puzzles. Some of his fans rate his puzzle books as the best ever written. Search the Internet to find information on the life of Smullyan and his work in the area of logic. Write a few paragraphs that summarize your findings.

SECTION 2.2 | **Truth Tables and Applications**

Truth Tables

In Section 2.1, we defined truth tables for the negation of a statement, the conjunction of two statements, and the disjunction of two statements. Each of these truth tables is shown below for review purposes.

Negation

p	$\sim p$
T	F
F	T

Conjunction

p	q	$p \wedge q$
T	T	T
T	F	F
F	T	F
F	F	F

Disjunction

p	q	$p \vee q$
T	T	T
T	F	T
F	T	T
F	F	F

p	q	**Given Statement**
T	T	
T	F	
F	T	
F	F	

Standard truth table form for a given statement that involves only the two simple statements p and q

In this section, we consider methods of constructing truth tables for a statement that involves a combination of conjunctions, disjunctions, and/or negations. If the given statement involves only two simple statements, then start with a table with four rows (see the table at the left), called the **standard truth table form,** and proceed as shown in Example 1.

EXAMPLE 1 ■ **Truth Tables**

a. Construct a table for $\sim(\sim p \vee q) \vee q$.

b. Use the truth table from part a to determine the truth value of $\sim(\sim p \vee q) \vee q$, given that p is true and q is false.

Solution

a. Start with the standard truth table form and then include a $\sim p$ column.

p	q	$\sim p$
T	T	F
T	F	F
F	T	T
F	F	T

Now use the truth values from the $\sim p$ and q columns to produce the truth values for $\sim p \vee q$, as shown in the following table.

p	q	$\sim p$	$\sim p \vee q$
T	T	F	T
T	F	F	F
F	T	T	T
F	F	T	T

Negate the truth values in the $\sim p \vee q$ column to produce the following.

p	q	$\sim p$	$\sim p \vee q$	$\sim(\sim p \vee q)$
T	T	F	T	F
T	F	F	F	T
F	T	T	T	F
F	F	T	T	F

As our last step, we form the disjunction of $\sim(\sim p \vee q)$ with q and place the results in the rightmost column of the table. See the following table. The shaded column is the truth table for $\sim(\sim p \vee q) \vee q$.

p	q	$\sim p$	$\sim p \vee q$	$\sim(\sim p \vee q)$	$\sim(\sim p \vee q) \vee q$	
T	T	F	T	F	T	Row 1
T	F	F	F	T	T	Row 2
F	T	T	T	F	T	Row 3
F	F	T	T	F	F	Row 4

b. In row 2 of the above truth table, we see that when p is true and q is false, the statement $\sim(\sim p \vee q) \vee q$ in the rightmost column is true.

CHECK YOUR PROGRESS 1

a. Construct a truth table for $(p \wedge {\sim}q) \vee ({\sim}p \vee q)$.

b. Use the truth table that you constructed in part a to determine the truth value of $(p \wedge {\sim}q) \vee ({\sim}p \vee q)$, given that p is true and q is false.

Solution   *See page S6.*

p	q	r	Given Statement
T	T	T	
T	T	F	
T	F	T	
T	F	F	
F	T	T	
F	T	F	
F	F	T	
F	F	F	

Standard truth table form for a statement that involves the three simple statements p, q, and r

Compound statements that involve exactly three simple statements require a standard truth table form with $2^3 = 8$ rows, as shown at the left.

EXAMPLE 2 ■ Truth Tables

a. Construct a truth table for $(p \wedge q) \wedge ({\sim}r \vee q)$.

b. Use the truth table from part a to determine the truth value of $(p \wedge q) \wedge ({\sim}r \vee q)$, given that p is true, q is true, and r is false.

Solution

a. Using the procedures developed in Example 1, we can produce the following table. The shaded column is the truth table for $(p \wedge q) \wedge ({\sim}r \vee q)$. The numbers in the squares below the columns denote the order in which the columns were constructed. Each truth value in the column numbered 4 is the conjunction of the truth values to its left in the columns numbered 1 and 3.

p	q	r	$p \wedge q$	${\sim}r$	${\sim}r \vee q$	$(p \wedge q) \wedge ({\sim}r \vee q)$	
T	T	T	T	F	T	T	Row 1
T	T	F	T	T	T	T	Row 2
T	F	T	F	F	F	F	Row 3
T	F	F	F	T	T	F	Row 4
F	T	T	F	F	T	F	Row 5
F	T	F	F	T	T	F	Row 6
F	F	T	F	F	F	F	Row 7
F	F	F	F	T	T	F	Row 8
			1	2	3	4	

b. In row 2 of the above truth table, we see that $(p \wedge q) \wedge ({\sim}r \vee q)$ is true when p is true, q is true, and r is false.

CHECK YOUR PROGRESS 2

a. Construct a truth table for $({\sim}p \wedge r) \vee (q \wedge {\sim}r)$.

b. Use the truth table that you constructed in part a to determine the truth value of $({\sim}p \wedge r) \vee (q \wedge {\sim}r)$, given that p is false, q is true, and r is false.

Solution   *See page S6.*

Plugger caller I.D.

Plugger Logic

Alternative Procedure for the Construction of a Truth Table

In Example 3 we use an *alternative procedure* to construct a truth table. **This alternative procedure generally requires less writing, less time, and less effort than the procedure explained in Examples 1 and 2.**

Alternative Procedure for the Construction of a Truth Table

If the given statement has n simple statements, then use a standard form that has 2^n rows.

1. In each row, enter the truth values for each *simple* statement and for any negation of a simple statement.

2. Use the truth values from Step 1 to enter the truth value under each connective within a pair of grouping symbols (parentheses ( ), brackets [ ], braces { }). If some grouping symbols are nested inside other grouping symbols, then work from the inside out.

3. Use the truth values from Step 2 to determine the truth values under the remaining connectives.

✔ **TAKE NOTE**

In a symbolic statement, grouping symbols are generally used to indicate the order in which logical connectives are applied. If grouping symbols are not used to specify the order in which logical connectives are applied, then we use the following **Order of Precedence Agreement:** First apply the negations from left to right, then apply the conjunctions from left to right, and finally apply the disjunctions from left to right.

EXAMPLE 3 ■ **Use the Alternative Procedure to Construct a Truth Table**

Construct a truth table for $p \vee [\sim(p \wedge \sim q)]$.

Solution

The given statement $p \vee [\sim(p \wedge \sim q)]$ has the two simple statements p and q. Thus we start with a standard form that has $2^2 = 4$ rows.

Step 1. In each column, enter the truth values for the statements p and $\sim q$, as shown in the columns numbered 1, 2, and 3 of the following table.

p	q	p	$\vee$	$[\sim$	$(p$	$\wedge$	$\sim q)]$
T	T	T			T		F
T	F	T			T		T
F	T	F			F		F
F	F	F			F		T
		1			2		3

Step 2. Use the truth values in columns 2 and 3 to determine the truth values to enter under the "and" connective. See the column numbered 4. Now negate the truth values in the column numbered 4 to produce the truth values in the column numbered 5.

p	q	p	$\vee$	$[\sim$	$(p$	$\wedge$	$\sim q)]$
T	T	T		T	T	F	F
T	F	T		F	T	T	T
F	T	F		T	F	F	F
F	F	F		T	F	F	T
		1		5	2	4	3

Step 3. Use the truth values in the columns numbered 1 and 5 to determine the truth values to enter under the "or" connective. See the column numbered 6, which is the truth table for $p \vee [\sim(p \wedge \sim q)]$.

p	q	p	$\vee$	$[\sim$	$(p$	$\wedge$	$\sim q)]$
T	T	T	T	T	T	F	F
T	F	T	T	F	T	T	T
F	T	F	T	T	F	F	F
F	F	F	T	T	F	F	T
		1	6	5	2	4	3

CHECK YOUR PROGRESS 3 Construct a truth table for $\sim p \vee (p \wedge q)$.

Solution   *See page S6.*

MathMatters   **A Three-Valued Logic**

In traditional logic, either a statement is true or it is false. Many mathematicians have tried to extend traditional logic so that sentences that are *partially* true are assigned a truth value other than T or F. Jan Lukasiewicz was one of the first mathematicians to consider a three-valued logic in which a statement is true, false, or "somewhere between true and false." In his three-valued logic, Lukasiewicz classified the truth value of a statement as true (T), false (F), or maybe (M). The following table shows truth values for negation, conjunction, and disjunction in this three-valued logic.

p	q	Negation $\sim p$	Conjunction $p \wedge q$	Disjunction $p \vee q$
T	T	F	T	T
T	M	F	M	T
T	F	F	F	T
M	T	M	M	T
M	M	M	M	M
M	F	M	F	M
F	T	T	F	T
F	M	T	F	M
F	F	T	F	F

historical note

Jan Lukasiewicz (lōō-kä-shä-věch) (1878–1956) was the Polish Minister of Education in 1919 and served as a professor of mathematics at Warsaw University from 1920–1939. Most of Lukasiewicz's work was in the area of logic. He is well known for developing *polish notation,* which was first used in logic to eliminate the need for parentheses in symbolic statements. Today *reverse polish notation* is used by many computers and calculators to perform computations without the need to enter parentheses. ∎

✔ **TAKE NOTE**

In the remaining sections of this chapter, the ≡ symbol will often be used to denote that two statements are equivalent.

Equivalent Statements

Two statements are **equivalent** if they both have the same truth values for all possible truth values of their component statements. Equivalent statements have identical truth values in the final columns of their truth tables. The notation $p \equiv q$ is used to indicate that the statements p and q are equivalent.

EXAMPLE 4 ■ **Verify That Two Statements Are Equivalent**

Show that $\sim(p \vee \sim q)$ and $\sim p \wedge q$ are equivalent statements.

Solution

Construct two truth tables and compare the results. The truth tables below show that $\sim(p \vee \sim q)$ and $\sim p \wedge q$ have the same truth values for all possible truth values of their component statements. Thus the statements are equivalent.

p	q	$\sim(p \vee \sim q)$
T	T	F
T	F	F
F	T	T
F	F	F

p	q	$\sim p \wedge q$
T	T	F
T	F	F
F	T	T
F	F	F

Identical truth values

Thus $\sim(p \vee \sim q) \equiv \sim p \wedge q$.

CHECK YOUR PROGRESS 4 Show that $p \vee (p \wedge \sim q)$ and p are equivalent.

Solution *See page S7.*

The truth tables in Table 2.3 show that $\sim(p \vee q)$ and $\sim p \wedge \sim q$ are equivalent statements. The truth tables in Table 2.4 show that $\sim(p \wedge q)$ and $\sim p \vee \sim q$ are equivalent statements.

Table 2.3

p	q	$\sim(p \vee q)$	$\sim p \wedge \sim q$
T	T	F	F
T	F	F	F
F	T	F	F
F	F	T	T

Table 2.4

p	q	$\sim(p \wedge q)$	$\sim p \vee \sim q$
T	T	F	F
T	F	T	T
F	T	T	T
F	F	T	T

These equivalences are known as **De Morgan's laws for statements.**

De Morgan's Laws for Statements

For any statements p and q,

$$\sim(p \vee q) \equiv \sim p \wedge \sim q$$
$$\sim(p \wedge q) \equiv \sim p \vee \sim q$$

De Morgan's laws can be used to restate certain English sentences in equivalent forms.

EXAMPLE 5 ■ **State an Equivalent Form**

Use one of De Morgan's laws to restate the following sentence in an equivalent form.

It is not the case that I graduated or I got a job.

Solution

Let p represent the statement "I graduated." Let q represent the statement "I got a job." In symbolic form, the original sentence is $\sim(p \vee q)$. One of De Morgan's laws states that this is equivalent to $\sim p \wedge \sim q$. Thus a sentence that is equivalent to the original sentence is "I did not graduate and I did not get a job."

CHECK YOUR PROGRESS 5   Use one of De Morgan's laws to restate the following sentence in an equivalent form.

It is not true that I am going to the dance and I am going to the game.

Solution   See page S7.

Tautologies and Self-Contradictions

A **tautology** is a statement that is always true. A **self-contradiction** is a statement that is always false.

EXAMPLE 6 ■ **Verify Tautologies and Self-Contradictions**

Show that $p \vee (\sim p \vee q)$ is a tautology.

Solution

Construct a truth table as shown below.

p	q	p	$\vee$	$(\sim p$	$\vee$	$q)$
T	T	T	T	F	T	T
T	F	T	T	F	F	F
F	T	F	T	T	T	T
F	F	F	T	T	T	F
		1	5	2	4	3

The table shows that $p \vee (\sim p \vee q)$ is always true. Thus $p \vee (\sim p \vee q)$ is a tautology.

CHECK YOUR PROGRESS 6   Show that $p \wedge (\sim p \wedge q)$ is a self-contradiction.

Solution   See page S7.

QUESTION   *Is the statement $x + 2 = 5$ a tautology or a self-contradiction?*

ANSWER   *Neither. The statement is not true for all values of x, and it is not false for all values of x.*

Investigation

Switching Networks—Part II

The Investigation in Section 2.1 introduced the application of symbolic logic to switching networks. This Investigation makes use of *closure tables* to determine under what conditions a switching network is open or closed. **In a closure table, we use a 1 to designate that a switch or switching network is closed and a 0 to indicate that it is open.**

Figure 2.5 shows a switching network that consists of the single switch P and a second network that consists of the single switch $\sim P$. The table below shows that the switching network $\sim P$ is open when P is closed and is closed when P is open.

Negation Closure Table

P	$\sim P$
1	0
0	1

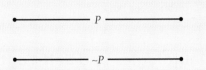

Figure 2.5

Figure 2.6 shows switches P and Q connected to form a series network. The table below shows that this series network is closed if and only if both P and Q are closed.

Series Network Closure Table

P	Q	$P \wedge Q$
1	1	1
1	0	0
0	1	0
0	0	0

Figure 2.6 *A series network*

Figure 2.7 shows switches P and Q connected to form a parallel network. The table below shows that this parallel network is closed if P is closed or if Q is closed.

Parallel Network Closure Table

P	Q	$P \vee Q$
1	1	1
1	0	1
0	1	1
0	0	0

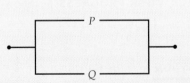

Figure 2.7 *A parallel network*

(continued)

Now consider the network shown in Figure 2.8. To determine the required conditions under which this network is closed, we first write a symbolic statement that represents the network, and then we construct a closure table.

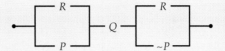

Figure 2.8

A symbolic statement that represents the network in Figure 2.8 is

$$[(R \lor P) \land Q] \land (R \lor \sim P)$$

The closure table for this network is shown below.

P	Q	R	[(R	∨	P)	∧	Q]	∧	(R	∨	∼P)	
1	1	1	1	1	1	1	1	1	1	1	0	Row 1
1	1	0	0	1	1	1	1	0	0	0	0	Row 2
1	0	1	1	1	1	0	0	0	1	1	0	Row 3
1	0	0	0	1	1	0	0	0	0	0	0	Row 4
0	1	1	1	1	0	1	1	1	1	1	1	Row 5
0	1	0	0	0	0	0	1	0	0	1	1	Row 6
0	0	1	1	1	0	0	0	0	1	1	1	Row 7
0	0	0	0	0	0	0	0	0	0	1	1	Row 8

1	6	2	7	3	9	4	8	5

The rows numbered 1 and 5 of the above table show that the network is closed whenever

- *P* is closed, *Q* is closed, and *R* is closed, or
- *P* is open, *Q* is closed, and *R* is closed.

Thus the switching network in Figure 2.8 is closed provided *Q* is closed and *R* is closed. The switching network is open under all other conditions.

Investigation Exercises

Construct a closure table for each of the following switching networks. Use the closure table to determine the required conditions for the network to be closed.

1.

2.

3.

4.

(continued)

5.

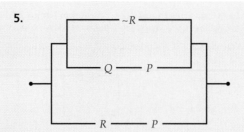

6.

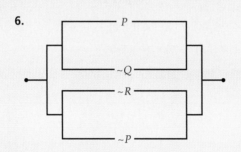

7. *Warning Circuits*

    a. The following circuit shows a switching network used in an automobile. The warning buzzer will buzz only when the switching network is closed. Construct a closure table for this switching network.

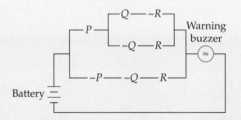

$$\{P \wedge [(Q \wedge \sim R) \vee (\sim Q \wedge R)]\} \vee [(\sim P \wedge \sim Q) \wedge R]$$

    b. An engineer thinks that the following circuit can be used in place of the circuit in part a. Do you agree? *Hint:* Construct a closure table for this switching network and compare your closure table with the closure table in part a.

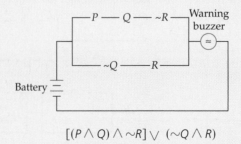

$$[(P \wedge Q) \wedge \sim R] \vee (\sim Q \wedge R)$$

Exercise Set 2.2

1. If $\sim(p \vee q)$ is a true statement, what can be said about the statement $\sim p \wedge \sim q$?

2. Use one of De Morgan's laws to rewrite $\sim(\sim p \wedge q)$ as an equivalent statement.

3. Assume x is a real number. Is the statement $x = x + 1$ a tautology or a self-contradiction?

4. How many rows are needed to construct a truth table for the statement $(p \wedge \sim q) \vee (\sim r \vee q)$?

In Exercises 5–14, determine the truth value of the compound statement given that p is a false statement, q is a true statement, and r is a true statement.

5. $p \vee (\sim q \vee r)$

6. $r \wedge \sim(p \vee r)$

7. $(p \wedge q) \vee (\sim p \wedge \sim q)$

8. $(p \wedge q) \vee [(\sim p \wedge \sim q) \vee q]$

9. $[\sim(p \wedge \sim q) \vee r] \wedge (p \wedge \sim r)$

10. $(p \wedge \sim q) \vee [(p \wedge \sim q) \vee r]$

11. $[(p \wedge \sim q) \vee \sim r] \wedge (q \wedge r)$

12. $(\sim p \wedge q) \wedge [(p \wedge \sim q) \vee r]$

13. $[(p \wedge q) \wedge r] \vee [p \vee (q \wedge \sim r)]$

14. $\{[(\sim p \wedge q) \wedge r] \vee [(p \wedge q) \wedge \sim r]\} \vee [p \wedge (q \wedge r)]$

15. **a.** Given that p is a false statement, what can be said about $p \wedge (q \vee r)$?

    b. Explain why it is not necessary to know the truth values of q and r to determine the truth value of $p \wedge (q \vee r)$ in part a above.

16. **a.** Given that q is a true statement, what can be said about $q \vee \sim r$?

    b. Explain why it is not necessary to know the truth value of r to determine the truth value of $q \vee \sim r$ in part a above.

In Exercises 17–32, construct a truth table for each compound statement.

17. $\sim p \vee q$

18. $(q \wedge \sim p) \vee \sim q$

19. $p \wedge \sim q$

20. $p \vee [\sim(p \wedge \sim q)]$

21. $(p \wedge \sim q) \vee [\sim(p \wedge q)]$

22. $(p \vee q) \wedge [\sim(p \vee \sim q)]$

23. $\sim(p \vee q) \wedge (\sim r \vee q)$

24. $[\sim(r \wedge \sim q)] \vee (\sim p \vee q)$

25. $(p \wedge \sim r) \vee [\sim q \vee (p \wedge r)]$

26. $[r \wedge (\sim p \vee q)] \wedge (r \vee \sim q)$

27. $[(p \wedge q) \vee (r \wedge \sim p)] \wedge (r \vee \sim q)$

28. $(p \wedge q) \wedge \{[\sim(\sim p \vee r)] \wedge q\}$

29. $q \vee [\sim r \vee (p \wedge r)]$

30. $\{[\sim(p \vee \sim r)] \wedge \sim q\} \vee r$

31. $(\sim q \wedge r) \vee [p \wedge (q \wedge \sim r)]$

32. $\sim[\sim p \wedge (q \wedge r)]$

In Exercises 33–40, use two truth tables to show that the given compound statements are equivalent.

33. $p \vee (p \wedge r)$; p

34. $q \wedge (q \vee r)$; q

35. $p \wedge (q \vee r)$; $(p \wedge q) \vee (p \wedge r)$

36. $p \vee (q \wedge r)$; $(p \vee q) \wedge (p \vee r)$

37. $p \vee (q \wedge \sim p)$; $p \vee q$

38. $\sim[p \vee (q \wedge r)]$; $\sim p \wedge (\sim q \vee \sim r)$

39. $[(p \wedge q) \wedge r] \vee [p \wedge (q \wedge \sim r)]$; $p \wedge q$

40. $[(\sim p \wedge \sim q) \wedge r] \vee [(p \wedge q) \wedge \sim r] \vee [p \wedge (q \wedge r)]$;
    $(p \wedge q) \vee [(\sim p \wedge \sim q) \wedge r]$

In Exercises 41–46, make use of one of De Morgan's laws to write the given statement in an equivalent form.

41. It is not the case that it rained or it snowed.

42. I did not pass the test and I did not complete the course.

43. She did not visit France and she did not visit Italy.

44. It is not true that I bought a new car and I moved to Florida.

45. It is not true that she received a promotion or that she received a raise.

46. It is not the case that the students cut classes or took part in the demonstration.

In Exercises 47–52, use a truth table to determine whether the given statement is a tautology.

47. $p \lor \sim p$
48. $q \lor [\sim (q \land r) \land \sim q]$
49. $(p \lor q) \lor (\sim p \lor q)$
50. $(p \land q) \lor (\sim p \lor \sim q)$
51. $(\sim p \lor q) \lor (\sim q \lor r)$
52. $\sim [p \land (\sim p \lor q)] \lor q$

In Exercises 53–58, use a truth table to determine whether the given statement is a self-contradiction.

53. $\sim r \land r$
54. $\sim (p \lor \sim p)$
55. $p \land (\sim p \land q)$
56. $\sim [(p \lor q) \lor (\sim p \lor q)]$
57. $[p \land (\sim p \lor q)] \lor q$
58. $\sim [p \lor (\sim p \lor q)]$
59. Explain why the statement $7 \le 8$ is a disjunction.
60. **a.** Why is the statement $5 \le 7$ true?
    b. Why is the statement $7 \le 7$ true?

Extensions

CRITICAL THINKING

61. How many rows are needed to construct a truth table for the statement $[p \land (q \lor \sim r)] \lor (s \land \sim t)$?
62. Explain why no truth table can have exactly 100 rows.

COOPERATIVE LEARNING

In Exercises 63 and 64, construct a truth table for the given compound statement. *Hint:* Use a table with 16 rows.

63. $[(p \land \sim q) \lor (q \land \sim r)] \land (r \lor \sim s)$
64. $s \land [\sim (\sim r \lor q) \lor \sim p]$

EXPLORATIONS

65. **Disjunctive Normal Form** Read about the *disjunctive normal form* of a statement in a logic text.

    a. What is the disjunctive normal form of a statement that has the following truth table?

p	q	r	Given Statement
T	T	T	T
T	T	F	F
T	F	T	T
T	F	F	F
F	T	T	F
F	T	F	F
F	F	T	T
F	F	F	F

    b. Explain why the disjunctive normal form is a valuable concept.

66. **Conjunctive Normal Form** Read about the *conjunctive normal form* of a statement in a logic text. What is the conjunctive normal form of the statement defined by the truth table in Exercise 65a?

| **The Conditional and Related Statements**

Conditional Statements

If you don't get in that plane, you'll regret it. Maybe not today, maybe not tomorrow, but soon, and for the rest of your life.

This quotation is from the movie *Casablanca*. Rick, played by Humphrey Bogart, is trying to convince Ilsa, played by Ingrid Bergman, to get on the plane with Laszlo. The sentence "If you don't get in that plane, you'll regret it" is a *conditional*

Humphrey Bogart and Ingrid Bergman star in *Casablanca* (1942).

statement. **Conditional statements** can be written in *if p, then q* form or in *if p, q* form. For instance, all of the following are conditional statements.

If we order pizza, then we can have it delivered.

If you go to the movie, you will not be able to meet us for dinner.

If *n* is a prime number greater than 2, then *n* is an odd number.

In any conditional statement represented by "If *p*, then *q*" or by "If *p*, *q*", the *p* statement is called the **antecedent** and the *q* statement is called the **consequent.** For instance, in the conditional statement,

If our school was this nice, I would go there more than once a week,[2]

the *antecedent* is "our school was this nice" and the *consequent* is "I would go there more than once a week."

Arrow Notation

The conditional statement "If *p*, then *q*" can be written using the **arrow notation** $p \rightarrow q$. The arrow notation $p \rightarrow q$ is read as "if *p*, then *q*" or as "*p* implies *q*."

To determine the truth table for $p \rightarrow q$, consider the advertising slogan for a web authoring software product that states "If you can use a word processor, you can create a web page." This slogan is a conditional statement. The antecedent is *p*, "you can use a word processor," and the consequent is *q*, "you can create a web page." Now consider the truth value of $p \rightarrow q$ for each of the four possibilities in Table 2.5.

Table 2.5

p: you can use a word processor	*q*: you can create a web page	$p \rightarrow q$	
T	T	?	Row 1
T	F	?	Row 2
F	T	?	Row 3
F	F	?	Row 4

Row 1: Antecedent T, consequent T You can use a word processor, and you can create a web page. In this case the truth value of the advertisement is true. To complete Table 2.5, we place a T in place of the question mark in row 1.

Row 2: Antecedent T, consequent F You can use a word processor, but you cannot create a web page. In this case the advertisement is false. We put an F in place of the question mark in row 2 of Table 2.5.

Row 3: Antecedent F, consequent T You cannot use a word processor, but you can create a web page. Because the advertisement does not make any statement about what you might or might not be able to do if you cannot use a word processor, we cannot state that the advertisement is false, and we are compelled to place a T in place of the question mark in row 3 of Table 2.5.

---

2. A quote from the movie *The Basketball Diaries*, 1995.

Row 4: Antecedent F, consequent F You cannot use a word processor, and you cannot create a web page. Once again we must consider the truth value in this case to be true because the advertisement does not make any statement about what you might or might not be able to do if you cannot use a word processor. We place a T in place of the question mark in row 4 of Table 2.5.

The truth table for the conditional $p \rightarrow q$ is given in Table 2.6.

Table 2.6 *The Truth Table for $p \rightarrow q$*

p	q	$p \rightarrow q$
T	T	T
T	F	F
F	T	T
F	F	T

Truth Value of the Conditional $p \rightarrow q$

The conditional $p \rightarrow q$ is false if p is true and q is false. It is true in all other cases.

EXAMPLE 1 ■ **Find the Truth Value of a Conditional**

Determine the truth value of each of the following.

a. If 2 is an integer, then 2 is a rational number.

b. If 3 is a negative number, then $5 > 7$.

c. If $5 > 3$, then $2 + 7 = 4$.

Solution
a. Because the consequent is true, this is a true statement.

b. Because the antecedent is false, this is a true statement.

c. Because the antecedent is true and the consequent is false, this is a false statement.

CHECK YOUR PROGRESS 1 Determine the truth value of each of the following.

a. If $4 \geq 3$, then $2 + 5 = 6$.

b. If $5 > 9$, then $4 > 9$.

c. If Tuesday follows Monday, then April follows March.

Solution See page S7.

EXAMPLE 2 ■ **Construct a Truth Table for a Statement Involving a Conditional**

Construct a truth table for $[p \wedge (q \vee \sim p)] \rightarrow \sim p$.

Solution
Using the alternative procedure for truth table construction, we produce the following table.

p	q	$[p$	$\wedge$	$(q$	$\vee$	$\sim p)]$	$\rightarrow$	$\sim p$
T	T	T	T	T	T	F	F	F
T	F	T	F	F	F	F	T	F
F	T	F	F	T	T	T	T	T
F	F	F	F	F	T	T	T	T
		1	6	2	5	3	7	4

CHECK YOUR PROGRESS 2   Construct a truth table for $[\,p \wedge (\,p \rightarrow q)\,] \rightarrow q$.

Solution   *See page S7.*

CALCULATOR NOTE

TI-84 Plus

Program FACTOR

```
0→dim (L1)
Prompt N
1→S: 2→F:0→E
√(N)→M
While F≤M
While fPart (N/F)=0
E+1→E:N/F→N
End
If E>0
Then
F→L1(S)
E→L1(S+1)
S+2→S:0→E
√(N)→M
End
If F=2
Then
3→F
Else
F+2→F
End: End
If N≠1
Then
N→L1(S)
1→L1(S+1)
End
If S=1
Then
Disp N, " IS PRIME"
Else
Disp L1
```

Math Matters — Use Conditional Statements to Control a Calculator Program

Computer and calculator programs use conditional statements to control the flow of a program. For instance, the "If…Then" instruction in a TI-83 or TI-84 Plus calculator program directs the calculator to execute a group of commands if a condition is true and to skip to the End statement if the condition is false. See the program steps below.

> If *condition*
> Then (skip to End if *condition* is false)
> *command* if *condition* is true
> *command* if *condition* is true
> End
> *command*

The TI-83 or TI-84 Plus program FACTOR shown at the left factors a number N into its prime factors. Note the use of the "If…Then" instructions highlighted in red.

Equivalent Forms of the Conditional

Every conditional statement can be stated in many equivalent forms. It is not even necessary to state the antecedent before the consequent. For instance, the conditional "If I live in Boston, then I must live in Massachusetts" can also be stated as

> I must live in Massachusetts, if I live in Boston.

Table 2.7 lists some of the various forms that may be used to write a conditional statement.

Table 2.7 *Common Forms of* $p \rightarrow q$

Every conditional statement $p \rightarrow q$ can be written in the following equivalent forms.	
If p, then q.	Every p is a q.
If p, q.	q, if p.
p only if q.	q provided p.
p implies q.	q is a necessary condition for p.
Not p or q.	p is a sufficient condition for q.

EXAMPLE 3 ■ Write a Statement in an Equivalent Form

Write each of the following in "If p, then q" form.

a. The number is an even number provided it is divisible by 2.

b. Today is Friday, only if yesterday was Thursday.

Solution

a. The statement "The number is an even number provided it is divisible by 2" is in "q provided p" form. The antecedent is "it is divisible by 2," and the consequent is "the number is an even number." Thus its "If p, then q" form is

  If it is divisible by 2, then the number is an even number.

b. The statement "Today is Friday, only if yesterday was Thursday" is in "p only if q" form. The antecedent is "today is Friday." The consequent is "yesterday was Thursday." Its "If p, then q" form is

  If today is Friday, then yesterday was Thursday.

CHECK YOUR PROGRESS 3   Write each of the following in "If p, then q" form.

a. Every square is a rectangle.

b. Being older than 30 is sufficient to show I am at least 21.

Solution   *See page S7.*

The Converse, the Inverse, and the Contrapositive

Every conditional statement has three related statements. They are called the *converse*, the *inverse*, and the *contrapositive*.

Statements Related to the Conditional Statement

The **converse** of $p \rightarrow q$ is $q \rightarrow p$.

The **inverse** of $p \rightarrow q$ is $\sim p \rightarrow \sim q$.

The **contrapositive** of $p \rightarrow q$ is $\sim q \rightarrow \sim p$.

The above definitions show the following:

■ The converse of $p \rightarrow q$ is formed by interchanging the antecedent p with the consequent q.

■ The inverse of $p \rightarrow q$ is formed by negating the antecedent p and negating the consequent q.

■ The contrapositive of $p \rightarrow q$ is formed by negating both the antecedent p and the consequent q and interchanging these negated statements.

EXAMPLE 4 ■ **Write the Converse, Inverse, and Contrapositive of a Conditional**

Write the converse, inverse, and contrapositive of

If I get the job, then I will rent the apartment.

Solution
Converse: If I rent the apartment, then I get the job.
Inverse: If I do not get the job, then I will not rent the apartment.
Contrapositive: If I do not rent the apartment, then I did not get the job.

CHECK YOUR PROGRESS 4   Write the converse, inverse, and contrapositive of

If we have a quiz today, then we will not have a quiz tomorrow.

Solution   See page S7.

Table 2.8 shows that any conditional statement is equivalent to its contrapositive, and that the converse of a conditional statement is equivalent to the inverse of the conditional statement.

Table 2.8 *Truth Tables for the Conditional and Related Statements*

p	q	Conditional $p \rightarrow q$	Converse $q \rightarrow p$	Inverse $\sim p \rightarrow \sim q$	Contrapositive $\sim q \rightarrow \sim p$
T	T	T	T	T	T
T	F	F	T	T	F
F	T	T	F	F	T
F	F	T	T	T	T

$$q \rightarrow p \equiv \sim p \rightarrow \sim q$$
$$p \rightarrow q \equiv \sim q \rightarrow \sim p$$

EXAMPLE 5 ■ **Determine Whether Related Statements Are Equivalent**

Determine whether the given statements are equivalent.

a. If a number ends with a 5, then the number is divisible by 5.
   If a number is divisible by 5, then the number ends with a 5.

b. If two lines in a plane do not intersect, then the lines are parallel.
   If two lines in a plane are not parallel, then the lines intersect.

Solution

a. The second statement is the converse of the first. The statements are not equivalent.

b. The second statement is the contrapositive of the first. The statements are equivalent.

CHECK YOUR PROGRESS 5 Determine whether the given statements are equivalent.

a. If $a = b$, then $a \cdot c = b \cdot c$.
If $a \neq b$, then $a \cdot c \neq b \cdot c$.

b. If I live in Nashville, then I live in Tennessee.
If I do not live in Tennessee, then I do not live in Nashville.

Solution See page S7.

In mathematics, it is often necessary to prove statements that are in "If p, then q" form. If a proof cannot readily be produced, mathematicians often try to prove the contrapositive, "If $\sim q$, then $\sim p$." Because a conditional and its contrapositive are equivalent statements, a proof of either statement also establishes the proof of the other statement.

QUESTION *A mathematician wishes to prove the following statement about the integer x.*

> *If x^2 is an odd integer, then x is an odd integer.* (I)

If the mathematician is able to prove the statement, "If x is an even integer, then x^2 is an even integer," does this also prove statement (I)?

EXAMPLE 6 ■ Use the Contrapositive to Determine a Truth Value

Write the contrapositive of each statement and use the contrapositive to determine whether the original statement is true or false.

a. If $a + b$ is not divisible by 5, then a and b are not both divisible by 5.

b. If x^3 is an odd integer, then x is an odd integer. (Assume x is an integer.)

c. If a geometric figure is not a rectangle, then it is not a square.

Solution

a. If a and b are both divisible by 5, then $a + b$ is divisible by 5. This is a true statement, so the original statement is also true.

b. If x is an even integer, then x^3 is an even integer. This is a true statement, so the original statement is also true.

c. If a geometric figure is a square, then it is a rectangle. This is a true statement, so the original statement is also true.

ANSWER *Yes, because the second statement is the contrapositive of (I).*

CHECK YOUR PROGRESS 6 Write the contrapositive of each statement and use the contrapositive to determine whether the original statement is true or false.

a. If $3 + x$ is an odd integer, then x is an even integer. (Assume x is an integer.)

b. If two triangles are not similar triangles, then they are not congruent triangles. *Note:* Similar triangles have the same shape. Congruent triangles have the same size and shape.

c. If today is not Wednesday, then tomorrow is not Thursday.

Solution   *See page S7.*

The Biconditional

The statement $(p \rightarrow q) \wedge (q \rightarrow p)$ is called a **biconditional** and is denoted by $p \leftrightarrow q$, which is read as "p if and only if q."

Table 2.9 *The Truth Table for $p \leftrightarrow q$*

p	q	$p \leftrightarrow q$
T	T	T
T	F	F
F	T	F
F	F	T

The Biconditional $p \leftrightarrow q$

$$p \leftrightarrow q \equiv [(p \rightarrow q) \wedge (q \rightarrow p)]$$

Table 2.9 shows that $p \leftrightarrow q$ is true only when the components p and q have the same truth value.

EXAMPLE 7 ■ Determine the Truth Value of a Biconditional

State whether each biconditional is true or false.

a. $x + 4 = 7$ if and only if $x = 3$.

b. $x^2 = 36$ if and only if $x = 6$.

Solution

a. Both components are true when $x = 3$ and both are false when $x \neq 3$. Both components have the same truth value for any value of x, so this is a true statement.

b. If $x = -6$, the first component is true and the second component is false. Thus, this is a false statement.

CHECK YOUR PROGRESS 7   State whether each biconditional is true or false.

a. $x > 7$ if and only if $x > 6$.

b. $x + 5 > 7$ if and only if $x > 2$.

Solution   *See page S8.*

Rear Admiral Grace Hopper

Math Matters   Grace Hopper

Grace Hopper (1906–1992) was a visionary in the field of computer programming. She was a mathematics professor at Vassar from 1931 to 1943, but retired from teaching to start a career in the U.S. Navy at the age of 37.

The Navy assigned Hopper to the Bureau of Ordnance Computation at Harvard University. It was here that she was given the opportunity to program computers. It has often been reported that she was the third person to program the world's first large-scale digital computer. Grace Hopper had a passion for computers and computer programming. She wanted to develop a computer language that would be user-friendly and enable people to use computers in a more productive manner.

Grace Hopper had a long list of accomplishments. She designed some of the first computer compilers, she was one of the first to introduce English commands into computer languages, and she wrote the precursor to the computer language COBOL.

Grace Hopper retired from the Navy (for the first time) in 1966. In 1967 she was recalled to active duty and continued to serve in the Navy until 1986, at which time she was the nation's oldest active duty officer.

In 1951, the UNIVAC I computer that Grace Hopper was programming started to malfunction. The malfunction was caused by a moth that had become lodged in one of the computer's relays. Grace Hopper pasted the moth into the UNIVAC I logbook with a label that read, "computer bug." Since then computer programmers have used the word *bug* to indicate any problem associated with a computer program. Modern computers use logic gates (see below) instead of relays to process information, so actual bugs are not a problem; however, bugs such as the "Year 2000 bug" can cause serious problems.

Investigation

Logic Gates

Modern digital computers use *gates* to process information. These gates are designed to receive two types of electronic impulses, which are generally represented as a 1 or a 0. Figure 2.9 shows a *NOT gate.* It is constructed so that a stream of impulses that enter the gate will exit the gate as a stream of impulses in which each 1 is converted to a 0 and each 0 is converted to a 1.

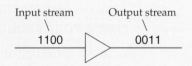

Figure 2.9 *NOT gate*

(continued)

Note the similarity between the logical connective *not* and the logic gate NOT. The *not* connective converts the sequence of truth values T F to F T. The NOT gate converts the input stream 1 0 to 0 1. If the 1's are replaced with T's and the 0's with F's, then the NOT logic gate yields the same results as the *not* connective.

Many gates are designed so that two input streams are converted to one output stream. For instance, Figure 2.10 shows an *AND gate.* The AND gate is constructed so that a 1 is the output if and only if both input streams have a 1. In any other situation a 0 is produced as the output.

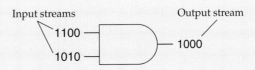

Input streams 1100 1010   Output stream 1000

Figure 2.10 *AND gate*

Note the similarity between the logical connective *and* and the logic gate AND. The *and* connective combines the sequence of truth values T T F F with the truth values T F T F to produce T F F F. The AND gate combines the input stream 1 1 0 0 with the input stream 1 0 1 0 to produce 1 0 0 0. If the 1's are replaced with T's and the 0's with F's, then the AND logic gate yields the same result as the *and* connective.

The *OR gate* is constructed so that its output is a 0 if and only if both input streams have a 0. All other situations yield a 1 as the output. See Figure 2.11.

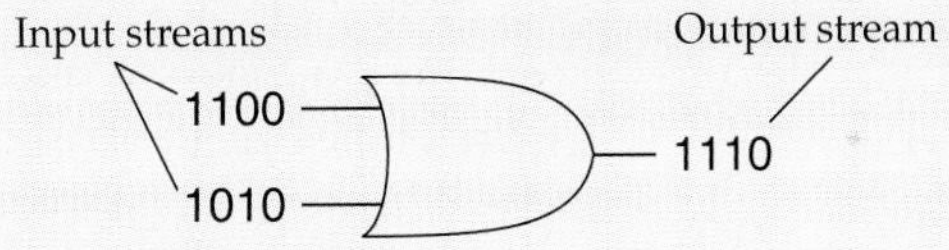

Input streams 1100 1010   Output stream 1110

Figure 2.11 *OR gate*

Figure 2.12 shows a network that consists of a NOT gate and an AND gate.

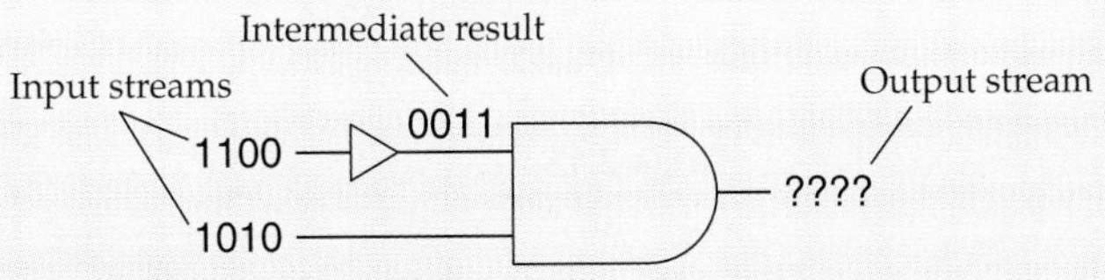

Intermediate result

Input streams 1100 1010   0011   Output stream ????

Figure 2.12

QUESTION   *What is the output stream for the network in Figure 2.12?*

ANSWER   *0 0 1 0*

(continued)

Investigation Exercises

1. For each of the following, determine the output stream for the given input streams.

a. Input streams

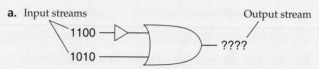

Output stream

b. Input streams

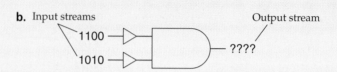

Output stream

c. Input streams

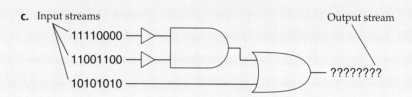

Output stream

2. Construct a network using NOT, AND, and OR gates, as needed, that accepts the two input streams 1 1 0 0 and 1 0 1 0 and produces the output stream 0 1 1 1.

Table 2.10 *Sheffer's stroke*

p	q	p|q
T	T	F
T	F	T
F	T	T
F	F	T

In 1913, the logician Henry M. Sheffer created a connective that we now refer to as *Sheffer's stroke* (or *NAND*). This connective is often denoted by the symbol |. Table 2.10 shows that $p|q$ is equivalent to $\sim(p \wedge q)$. Sheffer's stroke $p|q$ is false when both p and q are true and true in all other cases.

Any logic statement can be written using only Sheffer's stroke connectives. For instance, Table 2.11 shows that $p|p \equiv \sim p$ and $(p|p)|(q|q) \equiv p \vee q$.

Figure 2.13 shows a logic gate called a NAND gate. This gate models the Sheffer's stroke connective in that its output is 0 when both input streams are 1 and its output is 1 in all other cases.

Table 2.11

p	q	p|p	(p|p)|(q|q)
T	T	F	T
T	F	F	T
F	T	T	T
F	F	T	F

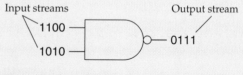

Figure 2.13 *NAND gate*

3. a. Complete a truth table for $p|(q|q)$.

   b. Use the results of Investigation Exercise 3a to determine an equivalent statement for $p|(q|q)$.

(continued)

4. a. Complete a truth table for $(p|q)|(p|q)$.

    b. Use the results of Investigation Exercise 4a to determine an equivalent statement for $(p|q)|(p|q)$.

5. a. Determine the output stream for the following network of NAND gates. *Note:* In a network of logic gates, a solid circle • is used to indicate a connection. A symbol such as $\nleftarrow$ is used to indicate "no connection."

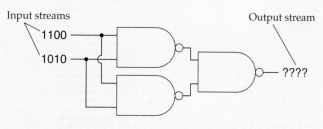

Figure 2.14

    b. What logic gate is modeled by the network in Figure 2.14?

6. NAND gates are functionally complete in that any logic gate can be constructed using only NAND gates. Construct a network of NAND gates that would produce the same output stream as an OR gate.

Exercise Set 2.3

In Exercises 1–4, identify the antecedent and the consequent of each conditional statement.

1. If I had the money, I would buy the painting.

2. If Shelly goes on the trip, she will not be able to take part in the graduation ceremony.

3. If they had a guard dog, then no one would trespass on their property.

4. If I don't get to school before 7:30, I won't be able to find a parking place.

In Exercises 5–10, determine the truth value of the given statement.

5. If x is an even integer, then x^2 is an even integer.

6. If all cats are black, then I am a millionaire.

7. If $4 < 3$, then $7 = 8$.

8. If $x < 2$, then $x + 5 < 7$.

9. If $|x| = 6$, then $x = 6$.

10. If $\pi = 3$, then $2\pi = 6$.

In Exercises 11–20, construct a truth table for the given statement.

11. $(p \wedge \sim q) \rightarrow [\sim(p \wedge q)]$

12. $[(p \rightarrow q) \wedge p] \rightarrow p$

13. $[(p \rightarrow q) \wedge p] \rightarrow q$

14. $(\sim p \vee \sim q) \rightarrow \sim(p \wedge q)$

15. $[r \wedge (\sim p \vee q)] \rightarrow (r \vee \sim q)$

16. $[(p \rightarrow \sim r) \wedge q] \rightarrow \sim r$

17. $[(p \rightarrow q) \vee (r \wedge \sim p)] \rightarrow (r \vee \sim q)$

18. $\{p \wedge [(p \rightarrow q) \wedge (q \rightarrow r)]\} \rightarrow r$

19. $[\sim(p \rightarrow \sim r) \wedge \sim q] \rightarrow r$

20. $[p \wedge (r \rightarrow \sim q)] \rightarrow (r \vee q)$

In Exercises 21–24, let v represent "I will take a vacation," let p represent "I get the promotion," and let t represent "I am transferred." Write each of the following statements in symbolic form.

21. If I get the promotion, I will take a vacation.

22. If I am not transferred, I will take a vacation.

23. If I am transferred, then I will not take a vacation.

24. If I will not take a vacation, then I will not be transferred and I get the promotion.

In Exercises 25–30, construct a truth table for each statement to determine if the statements are equivalent.

25. $p \rightarrow \sim r$; $r \vee \sim p$

26. $p \rightarrow q$; $q \rightarrow p$

27. $\sim p \rightarrow (p \vee r)$; r

28. $p \rightarrow q$; $\sim q \rightarrow \sim p$

29. $p \rightarrow (q \vee r)$; $(p \rightarrow q) \vee (p \rightarrow r)$

30. $\sim q \rightarrow p$; $p \vee q$

In Exercises 31–36, write each statement in "If p, then q" form.

31. We will be in good shape for the ski trip provided we take the aerobics class.

32. We can get a dog only if we install a fence around the back yard.

33. He can join the band if he has the talent to play a keyboard.

34. I will be able to prepare for the test only if I have the textbook.

35. I will be able to receive my credential provided Education 147 is offered in the spring semester.

36. Being in excellent shape is a necessary condition for running the Boston marathon.

37. Write the converse of $r \rightarrow s$.

38. Write the contrapositive of $p \rightarrow (\sim q)$.

39. If a conditional statement is a true statement, must its contrapositive also be a true statement?

40. If a conditional statement is a true statement, must its converse also be a true statement?

In Exercises 41–54, write the **a.** converse, **b.** inverse, and **c.** contrapositive of the given statement.

41. If I were rich, I would quit this job.

42. If we had a car, then we would be able to take the class.

43. If she does not return soon, we will not be able to attend the party.

44. I will be in the talent show only if I can do the same comedy routine I did for the banquet.

45. Every parallelogram is a quadrilateral.

46. If you get the promotion, you will need to move to Denver.

47. I would be able to get current information about astronomy provided I had access to the Internet.

48. You need four-wheel drive to make the trip to Death Valley.

49. We will not have enough money for dinner if we take a taxi.

50. If you are the president of the United States, then your age is at least 35.

51. She will visit Kauai only if she can extend her vacation for at least two days.

52. In a right triangle, the acute angles are complementary.

53. Two lines perpendicular to a given line are parallel.

54. If $x + 5 = 12$, then $x = 7$.

In Exercises 55–58, determine whether the given statements are equivalent.

55. If Kevin wins, we will celebrate.
If we celebrate, then Kevin will win.

56. If I save $1000, I will go on the field trip.
If I go on the field trip, then I saved $1000.

57. If she attends the meeting, she will make the sale.
If she does not make the sale, then she did not attend the meeting.

58. If you understand algebra, you can remember algebra.
If you do not understand algebra, you cannot remember algebra.

In Exercises 59–64, write the contrapositive of the statement and use the contrapositive to determine whether the original statement is true or false.

59. If $3x - 7 = 11$, then $x \neq 7$.

60. If $x \neq 3$, then $5x + 7 \neq 22$.

61. If $a \neq 3$, then $|a| \neq 3$.

62. If $a + b$ is divisible by 3, then a is divisible by 3 and b is divisible by 3.

63. If $\sqrt{a + b} \neq 5$, then $a + b \neq 25$.

64. Assume x is an integer. If x^2 is an even integer, then x is an even integer.

In Exercises 65–70, state whether the given biconditional is true or false. Assume that x and y are real numbers.

65. $x^2 = 9$ if and only if $x = 3$.

66. x is a positive number if and only if $x > 0$.

67. $|x|$ is a positive number if and only if $x \neq 0$.

68. $|x + y| = x + y$ if and only if $x + y > 0$.

69. $4 = 7$ if and only if $2 = 3$.

70. Today is March 1 if and only if yesterday was February 28.

Extensions
CRITICAL THINKING

71. Give an example of a true conditional statement whose
   a. converse is true.   **b.** converse is false.

72. Give an example of a true conditional statement whose
   a. inverse is true.   **b.** inverse is false.

In Exercises 73–76, determine the original statement if the given statement is related to the original in the manner indicated.

73. *Converse:* If you can do it, you can dream it.

74. *Inverse:* If I did not have a dime, I would not spend it.

75. *Contrapositive:* If I were a singer, I would not be a dancer.

76. *Negation:* Pigs have wings and pigs cannot fly.

77. Explain why it is not possible to find an example of a true conditional statement whose contrapositive is false.

78. If a conditional statement is false, must its converse be true? Explain.

79. A Puzzle Lewis Carroll (Charles Dodgson) wrote many puzzles, many of which he recorded in his diaries. Solve the following puzzle, which appears in one of his diaries.

The Dodo says that the Hatter tells lies.
The Hatter says that the March Hare tells lies.
The March Hare says that both the Dodo and the Hatter tell lies.
Who is telling the truth?[3]

Hint: Consider the three different cases in which only one of the characters is telling the truth. In only one of these cases can all three of the statements be true.

EXPLORATIONS

80. Puzzles Use a library or the Internet to find puzzles created by Lewis Carroll. For the puzzle that you think is most interesting, write an explanation of the puzzle and give its solution.

81. A Factor Program If you have access to a TI-83 or a TI-84 Plus calculator, enter the program FACTOR on page 87 into the calculator and demonstrate the program to a classmate.

SECTION 2.4 | **Arguments**

Arguments

In this section we consider methods of analyzing arguments to determine whether they are *valid* or *invalid*. For instance, consider the following argument.

> If Aristotle was human, then Aristotle was mortal. Aristotle was human. Therefore, Aristotle was mortal.

To determine whether the above argument is a valid argument, we must first define the terms *argument* and *valid argument*.

---

3. From *Lewis Carroll's Games and Puzzles*, edited by Edward Wakeling.

historical note

Aristotle
(ăr′ĭ-stŏt′l)
(384–322 B.C.)
was an ancient
Greek philosopher
who studied
under Plato. He
wrote about many subjects, including logic, biology, politics, astronomy, metaphysics, and ethics. His ideas about logic and the reasoning process have had a major impact on mathematics and philosophy. ■

Argument and Valid Argument

An **argument** consists of a set of statements called **premises** and another statement called the **conclusion.** An argument is **valid** if the conclusion is true whenever all the premises are assumed to be true. An argument is **invalid** if it is not a valid argument.

In the argument about Aristotle, the two premises and the conclusion are shown below. It is customary to place a horizontal line between the premises and the conclusion.

First Premise: If Aristotle was human, then Aristotle was mortal.

Second Premise: Aristotle was human.

Conclusion: Therefore, Aristotle was mortal.

Arguments can be written in **symbolic form.** For instance, if we let h represent the statement "Aristotle was human" and m represent the statement "Aristotle was mortal," then the argument can be expressed as

$$h \rightarrow m$$
$$h$$
$$\therefore m$$

The three dots $\therefore$ are a symbol for "therefore."

EXAMPLE 1 ■ Write an Argument in Symbolic Form

Write the following argument in symbolic form.

The fish is fresh or I will not order it. The fish is fresh. Therefore I will order it.

Solution

Let f represent the statement "The fish is fresh." Let o represent the statement "I will order it." The symbolic form of the argument is

$$f \vee \sim o$$
$$f$$
$$\therefore o$$

CHECK YOUR PROGRESS 1   Write the following argument in symbolic form.

If she doesn't get on the plane, she will regret it. She does not regret it. Therefore, she got on the plane.

Solution   See page S8.

Arguments and Truth Tables

The following truth table procedure can be used to determine whether an argument is valid or invalid.

> **Truth Table Procedure to Determine the Validity of an Argument**
>
> **1.** Write the argument in symbolic form.
>
> **2.** Construct a truth table that shows the truth value of each premise and the truth value of the conclusion for all combinations of truth values of the component statements.
>
> **3.** If the conclusion is true in every row of the truth table in which all the premises are true, the argument is valid. If the conclusion is false in any row in which all the premises are true, the argument is invalid.

We will now use the above truth table procedure to determine the validity of the argument about Aristotle.

1. Once again we let h represent the statement "Aristotle was human" and m represent the statement "Aristotle was mortal." In symbolic form, the argument is

$h \rightarrow m$	First premise
h	Second premise
$\therefore m$	Conclusion

2. Construct a truth table as shown below.

h	m	First Premise $h \rightarrow m$	Second Premise h	Conclusion m	
T	T	T	T	T	Row 1
T	F	F	T	F	Row 2
F	T	T	F	T	Row 3
F	F	T	F	F	Row 4

3. Row 1 is the only row in which all the premises are true, so it is the only row that we examine. Because the conclusion is true in row 1, the argument is valid.

In Example 2, we use the truth table method to determine the validity of a more complicated argument.

EXAMPLE 2 ■ Determine the Validity of an Argument

Determine whether the following argument is valid or invalid.

If it rains, then the game will not be played. It is not raining. Therefore, the game will be played.

Solution

If we let r represent "it rains" and g represent "the game will be played," then the symbolic form is

$r \rightarrow \sim g$
$\sim r$
$\therefore g$

The truth table for this argument is as follows:

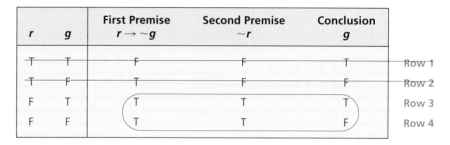

r	g	First Premise $r \rightarrow \sim g$	Second Premise $\sim r$	Conclusion g	
T	T	F	F	T	Row 1
T	F	T	F	F	Row 2
F	T	T	T	T	Row 3
F	F	T	T	F	Row 4

QUESTION   *Why do we need to examine only rows 3 and 4?*

Because the conclusion in row 4 is false and the premises are both true, we know the argument is invalid.

CHECK YOUR PROGRESS 2   Determine the validity of the following argument.

If the stock market rises, then the bond market will fall.

The bond market did not fall.

∴The stock market did not rise.

Solution   *See page S8.*

The argument in Example 3 involves three statements. Thus we use a truth table with $2^3 = 8$ rows to determine the validity of the argument.

EXAMPLE 3 ▪ Determine the Validity of an Argument

Determine whether the following argument is valid or invalid.

If I am going to run the marathon, then I will buy new shoes.

If I buy new shoes, then I will not buy a television.

∴If I buy a television, I will not run the marathon.

Solution
Label the statements

m: I am going to run the marathon.

s: I will buy new shoes.

t: I will buy a television.

The symbolic form of the argument is

$m \rightarrow s$

$s \rightarrow \sim t$

∴$t \rightarrow \sim m$

ANSWER   *Rows 3 and 4 are the only rows in which all of the premises are true.*

The truth table for this argument is as follows:

m	s	t	First Premise $m \rightarrow s$	Second Premise $s \rightarrow \sim t$	Conclusion $t \rightarrow \sim m$	
T	T	T	T	F	F	Row 1
T	T	F	T	T	T	Row 2
T	F	T	F	T	F	Row 3
T	F	F	F	T	T	Row 4
F	T	T	T	F	T	Row 5
F	T	F	T	T	T	Row 6
F	F	T	T	T	T	Row 7
F	F	F	T	T	T	Row 8

The only rows in which both premises are true are rows 2, 6, 7, and 8. Because the conclusion is true in each of these rows, the argument is valid.

CHECK YOUR PROGRESS 3   Determine whether the following argument is valid or invalid.

> If I arrive before 8 A.M., then I will make the flight.
> If I make the flight, then I will give the presentation.
> ∴If I arrive before 8 A.M., then I will give the presentation.

Solution   *See page S8.*

Standard Forms

Some arguments can be shown to be valid if they have the same symbolic form as an argument that is known to be valid. For instance, we have shown that the argument

$$h \rightarrow m$$
$$\underline{h}$$
$$\therefore m$$

is valid. This symbolic form is known as **modus ponens** or the **law of detachment.** All arguments that have this symbolic form are valid. Table 2.12 shows four symbolic forms and the name used to identify each form. Any argument that has a symbolic form identical to one of these symbolic forms is a valid argument.

Table 2.12 *Standard Forms of Four Valid Arguments*

Modus Ponens	Modus Tollens	Law of Syllogism	Disjunctive Syllogism
$p \rightarrow q$	$p \rightarrow q$	$p \rightarrow q$	$p \lor q$
$\underline{p}$	$\underline{\sim q}$	$\underline{q \rightarrow r}$	$\underline{\sim p}$
$\therefore q$	$\therefore \sim p$	$\therefore p \rightarrow r$	$\therefore q$

✔ **TAKE NOTE**

In logic, the ability to identify standard forms of arguments is an important skill. If an argument has one of the standard forms in Table 2.12, then it is a valid argument. If an argument has one of the standard forms in Table 2.13, then it is an invalid argument. The standard forms can be thought of as laws of logic. Concerning the laws of logic, the logician Gottlob Frege (frā′gə) (1848–1925) stated, "The laws of logic are not like the laws of nature. They…are laws of the laws of nature."

The law of syllogism can be extended to include more than two conditional premises. For example, if the premises of an argument are $a \to b, b \to c, c \to d, \ldots, y \to z$, then a valid conclusion for the argument is $a \to z$. We will refer to any argument of this form with more than two conditional premises as the **extended law of syllogism.**

Table 2.13 shows two symbolic forms associated with invalid arguments. Any argument that has one of these symbolic forms is invalid.

Table 2.13 *Standard Forms of Two Invalid Arguments*

Fallacy of the Converse	Fallacy of the Inverse
$p \to q$	$p \to q$
q	$\sim p$
$\therefore p$	$\therefore \sim q$

EXAMPLE 4 ■ Use a Standard Form to Determine the Validity of an Argument

Use a standard form to determine whether the following argument is valid or invalid.

The program is interesting or I will watch the basketball game.

The program is not interesting.

∴I will watch the basketball game.

Solution
Label the statements.

    i: The program is interesting.

    w: I will watch the basketball game.

In symbolic form the argument is:

$i \lor w$

$\sim i$

$\therefore w$

This symbolic form matches the standard form known as disjunctive syllogism. Thus the argument is valid.

CHECK YOUR PROGRESS 4   Use a standard form to determine whether the following argument is valid or invalid.

If I go to Florida for spring break, then I will not study.

I did not go to Florida for spring break.

∴I studied.

Solution   *See page S9.*

Consider an argument with the following symbolic form.

$q \to r$	Premise 1
$r \to s$	Premise 2
$\sim t \to \sim s$	Premise 3
q	Premise 4
$\therefore t$	

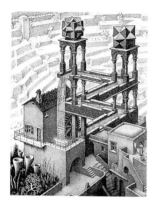

Waterfall by M. C. Escher

M. C. Escher (1898–1972) created many works of art that defy logic. In this lithograph, the water completes a full cycle even though the water is always traveling downward.

To determine whether the argument is valid or invalid using a truth table, we would require a table with $2^4 = 16$ rows. It would be time-consuming to construct such a table and, with the large number of truth values to be determined, we might make an error. Thus we consider a different approach that makes use of a sequence of valid arguments to arrive at a conclusion.

$q \rightarrow r$	Premise 1
$\underline{r \rightarrow s}$	Premise 2
$\therefore q \rightarrow s$	Law of syllogism
$q \rightarrow s$	The previous conclusion
$\underline{s \rightarrow t}$	Premise 3 expressed in an equivalent form
$\therefore q \rightarrow t$	Law of syllogism
$q \rightarrow t$	The previous conclusion
$\underline{q}$	Premise 4
$\therefore t$	Modus ponens

This sequence of valid arguments shows that t is a valid conclusion for the original argument.

EXAMPLE 5 ■ Determine the Validity of an Argument

Determine whether the following argument is valid.

> If the movie was directed by George Lucas (l), then I want to see it (w). If I want to see a movie, then the movie's production costs must have exceeded 20 million dollars (c). The movie's production costs were less than 20 million dollars. Therefore, the movie was not directed by George Lucas.

Solution
In symbolic form the argument is:

$l \rightarrow w$	Premise 1
$w \rightarrow c$	Premise 2
$\underline{\sim c}$	Premise 3
$\therefore \sim l$	Conclusion

Applying the law of syllogism to Premises 1 and 2 produces

$l \rightarrow w$	Premise 1
$\underline{w \rightarrow c}$	Premise 2
$\therefore l \rightarrow c$	Law of syllogism

Combining the above conclusion $l \rightarrow c$ with Premise 3 gives us

$l \rightarrow c$	Conclusion from above
$\underline{\sim c}$	Premise 3
$\therefore \sim l$	Modus tollens

This sequence of valid arguments has produced the conclusion given in the original argument. Thus the original argument is a valid argument.

CHECK YOUR PROGRESS 5   Determine whether the following argument is valid.

I start to fall asleep if I read a math book. I drink a soda whenever I start to fall asleep. If I drink a soda, then I must eat a candy bar. Therefore, I eat a candy bar whenever I read a math book.

Hint: p whenever *q* is equivalent to $q \rightarrow p$.

Solution   See page S9.

Math Matters   **The Paradox of the Unexpected Hanging**

The following paradox, known as "the paradox of the unexpected hanging," has proved difficult to analyze.

The man was sentenced on Saturday. "The hanging will take place at noon," said the judge to the prisoner, "on one of the seven days of next week. But you will not know which day it is until you are so informed on the morning of the day of the hanging."

The judge was known to be a man who always kept his word. The prisoner, accompanied by his lawyer, went back to his cell. As soon as the two men were alone the lawyer broke into a grin. "Don't you see?" he exclaimed. "The judge's sentence cannot possibly be carried out."

"I don't see," said the prisoner.

"Let me explain. They obviously can't hang you next Saturday. Saturday is the last day of the week. On Friday afternoon you would still be alive and you would know with absolute certainty that the hanging would be on Saturday. You would know this *before* you were told so on Saturday morning. That would violate the judge's decree."

"True," said the prisoner.

"Saturday, then, is positively ruled out," continued the lawyer. "This leaves Friday as the last day they can hang you. But they can't hang you on Friday because by Thursday afternoon only two days would remain: Friday and Saturday. Since Saturday is not a possible day, the hanging would have to be on Friday. Your knowledge of that fact would violate the judge's decree again. So Friday is out. This leaves Thursday as the last possible day. But Thursday is out because if you're alive Wednesday afternoon, you'll know that Thursday is to be the day."

"I get it," said the prisoner, who was beginning to feel much better. "In exactly the same way I can rule out Wednesday, Tuesday, and Monday. That leaves only tomorrow. But they can't hang me tomorrow because I know it today!"

In brief, the judge's decree seems to be self-refuting. There is nothing logically contradictory in the two statements that make up his decree; nevertheless, it cannot be carried out in practice.

He [the prisoner] is convinced, by what appears to be unimpeachable logic, that he cannot be hanged without contradicting the conditions specified in his sentence. Then, on Thursday morning, to his great surprise, the hangman arrives. Clearly he did not expect him. What is more surprising, the judge's decree is now seen to be perfectly correct. The sentence can be carried out exactly as stated.[4]

---

4. Reprinted with the permission of Simon & Schuster from *The Unexpected Hanging and Other Mathematical Diversions* by Martin Gardner. Copyright © 1969 by Martin Gardner.

Investigation

Fallacies

Any argument that is not valid is called a **fallacy.** Ancient logicians enjoyed the study of fallacies and took pride in their ability to analyze and categorize different types of fallacies. In this Investigation we consider the four fallacies known as *circulus in probando,* the fallacy of experts, the fallacy of equivocation, and the fallacy of accident.

Circulus in Probando

A fallacy of *circulus in probando* is an argument that uses a premise as the conclusion. For instance, consider the following argument.

> The Chicago Bulls are the best basketball team because there is no basketball team that is better than the Chicago Bulls.

> The fallacy of *circulus in probando* is also known as *circular reasoning* or *begging the question.*

Fallacy of Experts

A fallacy of experts is an argument that uses an expert (or a celebrity) to lend support to a product or an idea. Often the product or idea is outside the expert's area of expertise. The following endorsements may qualify as fallacy of experts arguments.

> Tiger Woods for Rolex watches

> Lindsey Wagner for Ford Motor Company

Fallacy of Equivocation

A fallacy of equivocation is an argument that uses a word with two interpretations in two different ways. The following argument is an example of a fallacy of equivocation.

> The highway sign read $268 fine for littering,

> so I decided fine, for $268, I will litter.

Fallacy of Accident

The following argument is an example of a fallacy of accident.

> Everyone should visit Europe.

> Therefore, prisoners on death row should be allowed to visit Europe.

Using more formal language, we can state the argument as follows.

> If you are a prisoner on death row (d), then you are a person (p).

> If you are a person (p), then you should be allowed to visit Europe (e).

> ∴If you are a prisoner on death row, then you should be allowed to visit Europe.

The symbolic form of the argument is

$$d \rightarrow p$$
$$p \rightarrow e$$
$$\therefore d \rightarrow e$$

This argument appears to be a valid argument because it has the standard form of the law of syllogism. Common sense tells us the argument is not valid, so where have we gone wrong in our analysis of the argument?

(continued)

The problem occurs with the interpretation of the word "everyone." Often, when we say "everyone," we really mean "most everyone." A fallacy of accident may occur whenever we use a statement that is often true in place of a statement that is always true.

Investigation Exercises

1. Write an argument that is an example of *circulus in probando*.

2. Give an example of an argument that is a fallacy of experts.

3. Write an argument that is an example of a fallacy of equivocation.

4. Write an argument that is an example of a fallacy of accident.

5. Algebraic arguments often consist of a list of statements. In a valid algebraic argument, each statement (after the premises) can be deduced from the previous statements. The following argument that $1 = 2$ contains exactly one step that is not valid. Identify the step and explain why it is not valid.

$$
\begin{array}{ll}
\text{Let} \quad a = b & \bullet \text{ Premise} \\
a^2 = ab & \bullet \text{ Multiply each side by } a. \\
a^2 - b^2 = ab - b^2 & \bullet \text{ Subtract } b^2 \text{ from each side.} \\
(a + b)(a - b) = b(a - b) & \bullet \text{ Factor each side.} \\
a + b = b & \bullet \text{ Divide each side by } (a - b). \\
b + b = b & \bullet \text{ Substitute } b \text{ for } a. \\
2b = b & \bullet \text{ Collect like terms.} \\
2 = 1 & \bullet \text{ Divide each side by } b.
\end{array}
$$

Exercise Set 2.4

In Exercises 1–8, use the indicated letters to write each argument in symbolic form.

1. If you can read this bumper sticker (r), you're too close (c). You can read the bumper sticker. Therefore, you're too close.

2. If Lois Lane marries Clark Kent (m), then Superman will get a new uniform (u). Superman does not get a new uniform. Therefore, Lois Lane did not marry Clark Kent.

3. If the price of gold rises (g), the stock market will fall (s). The price of gold did not rise. Therefore, the stock market did not fall.

4. I am going shopping (s) or I am going to the museum (m). I went to the museum. Therefore, I did not go shopping.

5. If we search the Internet (s), we will find information on logic (i). We searched the Internet. Therefore, we found information on logic.

6. If we check the sports results on the Excite channel (c), we will know who won the match (w). We know who won the match. Therefore, we checked the sports results on the Excite channel.

7. If the power goes off ($\sim p$), then the air conditioner will not work ($\sim a$). The air conditioner is working. Therefore, the power is not off.

8. If it snowed (s), then I did not go to my chemistry class ($\sim c$). I went to my chemistry class. Therefore, it did not snow.

In Exercises 9–24, use a truth table to determine whether the argument is valid or invalid.

9. $p \vee \sim q$
$\underline{\sim q}$
$\therefore p$

10. $\sim p \wedge q$
$\underline{\sim p}$
$\therefore q$

11. $p \rightarrow \sim q$
$\underline{\sim q}$
$\therefore p$

12. $p \rightarrow \sim q$
$\underline{p}$
$\therefore \sim q$

13. $\sim p \rightarrow \sim q$
$\underline{\sim p}$
$\therefore \sim q$

14. $\sim p \rightarrow q$
$\underline{p}$
$\therefore \sim q$

15. $(p \rightarrow q) \wedge (\sim p \rightarrow q)$
$\underline{q}$
$\therefore p$

16. $(p \vee q) \wedge (p \wedge q)$
$\underline{p}$
$\therefore q$

17. $\dfrac{(p \wedge \sim q) \vee (p \to q)}{q \vee p}$
$\therefore \sim p \wedge q$

18. $\dfrac{(p \wedge \sim q) \to (p \vee q)}{q \to \sim p}$
$\therefore p \to q$

19. $\dfrac{(p \wedge \sim q) \vee (p \vee r)}{r}$
$\therefore p \vee q$

20. $\dfrac{(p \to q) \to (r \to \sim q)}{p}$
$\therefore \sim r$

21. $\dfrac{\begin{array}{c}p \leftrightarrow q \\ p \to r\end{array}}{\therefore \sim r \to \sim p}$

22. $\dfrac{\begin{array}{c}p \wedge r \\ p \to \sim q\end{array}}{\therefore r \to q}$

23. $\dfrac{\begin{array}{c}p \wedge \sim q \\ p \leftrightarrow r\end{array}}{\therefore q \vee r}$

24. $\dfrac{\begin{array}{c}p \to r \\ r \to q\end{array}}{\therefore \sim p \to \sim q}$

In Exercises 25–30, use the indicated letters to write the argument in symbolic form. Then use a truth table to determine whether the argument is valid or invalid.

25. If you finish your homework (h), you may attend the reception (r). You did not finish your homework. Therefore, you cannot go to the reception.

26. The X Games will be held in Oceanside (o) if and only if the city of Oceanside agrees to pay \$100,000 in prize money (a). If San Diego agrees to pay \$200,000 in prize money (s), then the city of Oceanside will not agree to pay \$100,000 in prize money. Therefore, if the X Games were held in Oceanside, then San Diego did not agree to pay \$200,000 in prize money.

27. If I can't buy the house ($\sim b$), then at least I can dream about it (d). I can buy the house or at least I can dream about it. Therefore, I can buy the house.

28. If the winds are from the east (e), then we will not have a big surf ($\sim s$). We do not have a big surf. Therefore, the winds are from the east.

29. If I master college algebra (c), then I will be prepared for trigonometry (t). I am prepared for trigonometry. Therefore, I mastered college algebra.

30. If it is a blot (b), then it is not a clot ($\sim c$). If it is a zlot (z), then it is a clot. It is a blot. Therefore, it is not a zlot.

In Exercises 31–34, use a standard form to determine whether each argument is valid or invalid.

31. 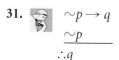 $\dfrac{\begin{array}{c}\sim p \to q \\ \sim p\end{array}}{\therefore q}$

32. 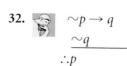 $\dfrac{\begin{array}{c}\sim p \to q \\ \sim q\end{array}}{\therefore p}$

33. 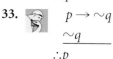 $\dfrac{\begin{array}{c}p \to \sim q \\ \sim q\end{array}}{\therefore p}$

34. 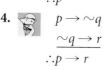 $\dfrac{\begin{array}{c}p \to \sim q \\ \sim q \to r\end{array}}{\therefore p \to r}$

In Exercises 35–44, determine whether the argument is valid or invalid by comparing its symbolic form with the standard symbolic forms given in Tables 2.12 and 2.13. For each valid argument, state the name of its standard form.

35. If you take Art 151 in the fall, you will be eligible to take Art 152 in the spring. You were not eligible to take Art 152 in the spring. Therefore, you did not take Art 151 in the fall.

36. He attended Stanford or Yale. He did not attend Yale. Therefore, he attended Stanford.

37. If I had a nickel for every logic problem I have solved, then I would be rich. I have not received a nickel for every logic problem I have solved. Therefore, I am not rich.

38. If it is a dog, then it has fleas. It has fleas. Therefore, it is a dog.

39. If we serve salmon, then Vicky will join us for lunch. If Vicky joins us for lunch, then Marilyn will not join us for lunch. Therefore, if we serve salmon, Marilyn will not join us for lunch.

40. If I go to college, then I will not be able to work for my Dad. I did not go to college. Therefore, I went to work for my Dad.

41. If my cat is left alone in the apartment, then she claws the sofa. Yesterday I left my cat alone in the apartment. Therefore, my cat clawed the sofa.

42. If I wish to use the new software, then I cannot continue to use this computer. I don't wish to use the new software. Therefore, I can continue to use this computer.

43. If Rita buys a new car, then she will not go on the cruise. Rita went on the cruise. Therefore, Rita did not buy a new car.

44. If Hideo Nomo pitches, then I will go to the game. I did not go to the game. Therefore, Hideo Nomo did not pitch.

In Exercises 45–50, use a sequence of valid arguments to show that each argument is valid.

45. $\dfrac{\begin{array}{c}\sim p \to r \\ r \to t \\ \sim t\end{array}}{\therefore p}$

46. $\dfrac{\begin{array}{c}r \to \sim s \\ \sim s \to \sim t \\ r\end{array}}{\therefore \sim t}$

47. If we sell the boat (s), then we will not go to the river ($\sim r$). If we don't go to the river, then we will go camping (c). If we do not buy a tent ($\sim t$), then we will not go camping. Therefore, if we sell the boat, then we will buy a tent.

48. If it is an ammonite (a), then it is from the Cretaceous period (c). If it is not from the Mesozoic era ($\sim m$), then it is not from the Cretaceous period. If it is from the Mesozoic era, then it is at least 65 million years old (s). Therefore, if it is an ammonite, then it is at least 65 million years old.

49. If the computer is not operating ($\sim o$), then I will not be able to finish my report ($\sim f$). If the office is closed (c), then the computer is not operating. Therefore, if I am able to finish my report, then the office is open.

50. If he reads the manuscript (r), he will like it (l). If he likes it, he will publish it (p). If he publishes it, then you will get royalties (m). You did not get royalties. Therefore, he did not read the manuscript.

Extensions

CRITICAL THINKING

51. An Argument by Lewis Carroll The following argument is from *Symbolic Logic* by Lewis Carroll, written

in 1896. Determine whether the argument is valid or invalid.

> Babies are illogical.
> Nobody is despised who can manage a crocodile.
> Illogical persons are despised.
> Hence, babies cannot manage crocodiles.

EXPLORATIONS

52. **Fallacies** Consult a logic text or search the Internet for information on fallacies. Write a report that includes examples of at least three of the following fallacies.

Ad hominem

Ad populum

Ad baculum

Ad vercundiam

Non sequitur

Fallacy of false cause

Pluriam interrogationem

SECTION 2.5 | **Euler Diagrams**

Euler Diagrams

Many arguments involve sets whose elements are described using the quantifiers *all*, *some*, and *none*. The mathematician Leonhard Euler (laônhärt oi′lər) used diagrams to determine whether arguments that involved quantifiers were valid or invalid. The following figures show Euler diagrams that illustrate the four possible relationships that can exist between two sets.

All Ps are Qs.　No Ps are Qs.　Some Ps are Qs.　Some Ps are not Qs.

Euler diagrams

Euler used diagrams to illustrate logic concepts. Some 100 years later, John Venn extended the use of Euler's diagrams to illustrate many types of mathematics. In this section, we will construct diagrams to determine the validity of arguments. We will refer to these diagrams as Euler diagrams.

EXAMPLE 1 ■ Use an Euler Diagram to Determine the Validity of an Argument

Use an Euler diagram to determine whether the following argument is valid or invalid.

All college courses are fun.

This course is a college course.

∴This course is fun.

Solution

The first premise indicates that the set of college courses is a subset of the set of fun courses. We illustrate this subset relationship with an Euler diagram, as shown in Figure 2.15. The second premise tells us that "this course" is an element of the set of college courses. If we use c to represent "this course," then c must be placed inside the set of college courses, as shown in Figure 2.16.

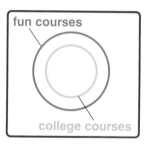

Figure 2.15

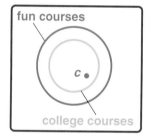

Figure 2.16

Figure 2.16 illustrates that c must also be an element of the set of fun courses. Thus the argument is valid.

CHECK YOUR PROGRESS 1 Use an Euler diagram to determine whether the following argument is valid or invalid.

All lawyers drive BMWs.

Susan is a lawyer.

∴Susan drives a BMW.

Solution See page S9.

If an Euler diagram can be drawn so that the conclusion does not necessarily follow from the premises, then the argument is invalid. This concept is illustrated in the next example.

This impressionist painting, *Dance at Bougival* by Renoir, is on display at the Museum of Fine Arts, Boston.

EXAMPLE 2 ■ **Use an Euler Diagram to Determine the Validity of an Argument**

Use an Euler diagram to determine whether the following argument is valid or invalid.

> Some impressionist paintings are Renoirs.
> *Dance at Bougival* is an impressionist painting.
> ∴*Dance at Bougival* is a Renoir.

Solution

The Euler diagram in Figure 2.17 illustrates the premise that some impressionist paintings are Renoirs. Let *d* represent the painting *Dance at Bougival*. Figures 2.18 and 2.19 show that *d* can be placed in one of two regions.

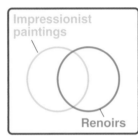
Figure 2.17

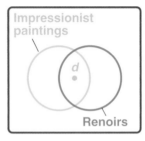
Figure 2.18

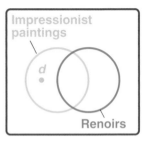
Figure 2.19

Although Figure 2.18 supports the argument, Figure 2.19 shows that the conclusion does not necessarily follow from the premises, and thus the argument is invalid.

CHECK YOUR PROGRESS 2   Use an Euler diagram to determine whether the following argument is valid or invalid.

> No prime numbers are negative.
> The number 7 is not negative.
> ∴The number 7 is a prime number.

Solution   *See page S9.*

✔ **TAKE NOTE**

Even though the conclusion in Example 2 is true, the argument is invalid.

QUESTION   *If one particular example can be found for which the conclusion of an argument is true when its premises are true, must the argument be valid?*

Some arguments can be represented by an Euler diagram that involves three sets, as shown in Example 3.

ANSWER   *No. To be a valid argument, the conclusion must be true whenever the premises are true. Just because the conclusion is true for one specific example, it does not mean the argument is a valid argument.*

EXAMPLE 3 ▦ **Use an Euler Diagram to Determine the Validity of an Argument**

Use an Euler diagram to determine whether the following argument is valid or invalid.

> No psychologist can juggle.
>
> All clowns can juggle.
> _____
>
> ∴No psychologist is a clown.

Solution

The Euler diagram in Figure 2.20 shows that the set of psychologists and the set of jugglers are disjoint sets. Figure 2.21 shows that because the set of clowns is a subset of the set of jugglers, no psychologists p are elements of the set of clowns. Thus the argument is valid.

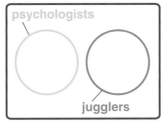

Figure 2.20

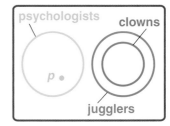

Figure 2.21

CHECK YOUR PROGRESS 3   Use an Euler diagram to determine whether the following argument is valid or invalid.

> No mathematics professors are good-looking.
>
> All good-looking people are models.
> _____
>
> ∴No mathematics professor is a model.

Solution   *See page S10.*

Math Matters   A Famous Puzzle

Three men decide to rent a room for one night. The regular room rate is $25; however, the desk clerk charges the men $30 because it will be easier for each man to pay one-third of $30 than it would be for each man to pay one-third of $25. Each man pays $10 and the porter shows them to their room.

After a short period of time, the desk clerk starts to feel guilty and gives the porter $5, along with instructions to return the $5 to the three men.

On the way to the room the porter decides to give each man $1 and pocket $2. After all, the men would find it difficult to split $5 evenly.

Thus each man has paid $10 and received a refund of $1. After the refund, the men have paid a total of $27. The porter has $2. The $27 added to the $2 equals $29.

QUESTION   *Where is the missing dollar? (See Answer on the following page.)*

Euler Diagrams and the Extended Law of Syllogism

Example 4 uses Euler diagrams to visually illustrate the extended law of syllogism from Section 2.4.

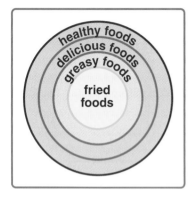

✔ TAKE NOTE

Although the conclusion in Example 4 is false, the argument in Example 4 is valid.

EXAMPLE 4 ■ **Use an Euler Diagram to Determine the Validity of an Argument**

Use an Euler diagram to determine whether the following argument is valid or invalid.

> All fried foods are greasy.
>
> All greasy foods are delicious.
>
> All delicious foods are healthy.
>
> ∴All fried foods are healthy.

Solution
The figure at the left illustrates that every fried food is an element of the set of healthy foods, so the argument is valid.

CHECK YOUR PROGRESS 4   Use an Euler diagram to determine whether the following argument is valid or invalid.

> All squares are rhombi.
>
> All rhombi are parallelograms.
>
> All parallelograms are quadrilaterals.
>
> ∴All squares are quadrilaterals.

Solution   *See page S10.*

Using Euler Diagrams to Form Conclusions

In Example 5, we make use of an Euler diagram to determine a valid conclusion for an argument.

ANSWER   *The $2 the porter kept was added to the $27 the men spent to produce a total of $29. The fact that this amount just happens to be close to $30 is a coincidence. All the money can be located if we total the $2 the porter has, the $3 that was returned to the men, and the $25 the desk clerk has, to produce $30.*

EXAMPLE 5 ■ **Use an Euler Diagram to Determine the Conclusion for an Argument**

Use an Euler diagram and all of the premises in the following argument to determine a valid conclusion for the argument.

> All *M*s are *N*s.
>
> No *N*s are *P*s.
>
> ∴?

Solution

The first premise indicates that the set of *M*s is a subset of the set of *N*s. The second premise indicates that the set of *N*s and the set of *P*s are disjoint sets. The following Euler diagram illustrates these set relationships. An examination of the Euler diagram allows us to conclude that no *M*s are *P*s.

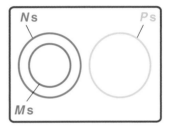

CHECK YOUR PROGRESS 5   Use an Euler diagram and all of the premises in the following argument to determine a valid conclusion for the argument.

> Some rabbits are white.
>
> All white animals like tomatoes.
>
> ∴?

Solution   *See page S10.*

Investigation

Using Logic to Solve Puzzles

Many puzzles can be solved by making an assumption and then checking to see if the assumption is supported by the conditions (premises) associated with the puzzle. For instance, consider the following addition problem in which each letter represents a digit from 0 through 9, and different letters represent different digits.

✔ TAKE NOTE

When working with cryptarithms, we assume that the leading digit of each number is a nonzero digit.

(continued)

Note that the T in T E E is a carry from the middle column. Because the sum of any two single digits plus a previous carry of at most 1 is 19 or less, the T in T E E must be a 1. Replacing all the T's with 1's produces:

$$
\begin{array}{r}
1\,A \\
+\,B\,1 \\
\hline
1\,E\,E
\end{array}
$$

Now B must be an 8 or a 9, because these are the only digits that would produce a carry into the leftmost column.

Case 1: Assume B is a 9. Then A must be an 8 or smaller, and A + 1 does not produce a carry into the middle column. The sum of the digits in the middle column is 10, and thus E is a 0. This presents a dilemma because the units digit of A + 1 must also be a 0, which requires A to be a 9. The assumption that B is a 9 is not supported by the conditions of the problem; thus we reject the assumption that B is a 9.

Case 2: Assume B is an 8. To produce the required carry into the leftmost column, there must be a carry from the column on the right. Thus A must be a 9, and we have the result shown below.

$$
\begin{array}{r}
1\,9 \\
+\,8\,1 \\
\hline
1\,0\,0
\end{array}
$$

A check shows that this solution satisfies all the conditions of the problem.

Investigation Exercises

Solve the following cryptarithms. Assume that no leading digit is a 0.
(*Source:* http://www.geocities.com/Athens/Agora/2160/puzzles.html)[5]

1.
$$
\begin{array}{r}
S\,O \\
+\,S\,O \\
\hline
T\,O\,O
\end{array}
$$

2.
$$
\begin{array}{r}
U\,S \\
+\,A\,S \\
\hline
A\,L\,L
\end{array}
$$

3.
$$
\begin{array}{r}
C\,O\,C\,A \\
+\,C\,O\,L\,A \\
\hline
O\,A\,S\,I\,S
\end{array}
$$

4.
$$
\begin{array}{r}
A\,T \\
E\,A\,S\,T \\
+\,W\,E\,S\,T \\
\hline
S\,O\,U\,T\,H
\end{array}
$$

5. Copyright © 2002 by Jorge A. C. B. Soares.

Exercise Set 2.5

In Exercises 1–4, draw an Euler diagram that illustrates the relationship between the given sets. Also use a dot to show an element of the first set that satisfies the given relationship.

1. All cats (C) are nimble (N).

2. Some mathematicians (M) are extroverts (E).

3. Some actors (A) are not famous (F).

4. No alligators (A) are trustworthy (T).

In Exercises 5–24, use an Euler diagram to determine whether the argument is valid or invalid.

5. All frogs are poetical.
   Kermit is a frog.
   ∴Kermit is poetical.

6. All Oreo cookies have a filling.
   All Fig Newtons have a filling.
   ∴All Fig Newtons are Oreo cookies.

7. Some plants have flowers.
   All things that have flowers are beautiful.
   ∴Some plants are beautiful.

8. No squares are triangles.
   Some triangles are equilateral.
   ∴No squares are equilateral.

9. No rocker would do the Mariachi.
   All baseball fans do the Mariachi.
   ∴No rocker is a baseball fan.

10. Nuclear energy is not safe.
    Some electric energy is safe.
    ∴No electric energy is nuclear energy.

11. Some birds bite.
    All things that bite are dangerous.
    ∴Some birds are dangerous.

12. All fish can swim.
    That barracuda can swim.
    ∴That barracuda is a fish.

13. All men behave badly.
    Some hockey players behave badly.
    ∴Some hockey players are men.

14. All grass is green.
    That ground cover is not green.
    ∴That ground cover is not grass.

15. Most teenagers drink soda.
    No CEOs drink soda.
    ∴No CEO is a teenager.

16. Some students like history.
    Vern is a student.
    ∴Vern likes history.

17. No mathematics test is fun.
    All fun things are worth your time.
    ∴No mathematics test is worth your time.

18. All prudent people shun sharks.
    No accountant is imprudent.
    ∴No accountant fails to shun sharks.

19. All candidates without a master's degree will not be considered for the position of director.
    All candidates who are not considered for the position of director should apply for the position of assistant.
    ∴All candidates without a master's degree should apply for the position of assistant.

20. Some whales make good pets.
    Some good pets are cute.
    Some cute pets bite.
    ∴Some whales bite.

21. All prime numbers are odd.
    2 is a prime number.
    ∴2 is an odd number.

22. All Lewis Carroll arguments are valid.
    Some valid arguments are syllogisms.
    ∴Some Lewis Carroll arguments are syllogisms.

23. All aerobics classes are fun.
    Jan's class is fun.
    ∴Jan's class is an aerobics class.

24. No sane person takes a math class.
    Some students that take a math class can juggle.
    ∴No sane person can juggle.

In Exercises 25–30, use all of the premises in each argument to determine a valid conclusion for the argument.

25. All Reuben sandwiches are good.
All good sandwiches have pastrami.
All sandwiches with pastrami need mustard.
∴?

26. All cats are strange.
Boomer is not strange.
∴?

27. All multiples of 11 end with a 5.
1001 is a multiple of 11.
∴?

28. If it isn't broken, then I do not fix it.
If I do not fix it, then I do not get paid.
∴?

29. Some horses are frisky.
All frisky horses are grey.
∴?

30. If we like to ski, then we will move to Vail.
If we move to Vail, then we will not buy a house.
If we do not buy a condo, then we will buy a house.
∴?

31. Examine the following three premises:

1. All people who have Xboxes play video games.

2. All people who play video games enjoy life.

3. Some mathematics professors enjoy life.

Now consider each of the following six conclusions. For each conclusion, determine whether the argument formed by the three premises and the conclusion is valid or invalid.

a. ∴Some mathematics professors have Xboxes.

b. ∴Some mathematics professors play video games.

c. ∴Some people who play video games are mathematics professors.

d. ∴Mathematics professors never play video games.

e. ∴All people who have Xboxes enjoy life.

f. ∴Some people who enjoy life are mathematics professors.

32. Examine the following three premises:

1. All people who drive pickup trucks like Willie Nelson.

2. All people who like Willie Nelson like country western music.

3. Some people who like heavy metal music like Willie Nelson.

Now consider each of the following five conclusions. For each conclusion, determine whether the argument formed by the three premises and the conclusion is valid or invalid.

a. ∴ Some people who like heavy metal music drive pickup trucks.

b. ∴Some people who like heavy metal music like country western music.

c. ∴Some people who like Willie Nelson like heavy metal music.

d. ∴All people who drive pickup trucks like country western music.

e. ∴People who like heavy metal music never drive pickup trucks.

Extensions

CRITICAL THINKING

33. A Crossnumber Puzzle In the following *crossnumber puzzle*, each square holds a single digit from 0 through 9. Use the clues under the *Across* and *Down* headings to solve the puzzle.

Across

1. One-fourth of 3 across

3. Two more than 1 down with its digits reversed

Down

1. Larger than 20 and less than 30

2. Half of 1 down

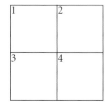

EXPLORATIONS

34. **Bilateral Diagrams** Lewis Carroll (Charles Dodgson) devised a *bilateral diagram* (two-part board) to analyze syllogisms. His method has some advantages over Euler diagrams and Venn diagrams. Use a library or the Internet to find information on Carroll's method of analyzing syllogisms. Write a few paragraphs that explain his method and its advantages.

Summary

Key Terms

antecedent [p. 85]
argument [p. 98]
arrow notation [p. 85]
biconditional [p. 91]
component statement [p. 64]
compound statement [p. 64]
conclusion [p. 98]
conditional statement [p. 85]
conjunction [p. 66]
connective [p. 64]
consequent [p. 85]
contrapositive [p. 88]
converse [p. 88]
disjunction [p. 67]
disjunctive syllogism [p. 101]
equivalent statements [p. 77]
Euler diagram [p. 108]
existential quantifier [p. 67]
extended law of syllogism [p. 102]
fallacy of the converse [p. 102]
fallacy of the inverse [p. 102]
invalid argument [p. 98]
inverse [p. 88]
law of detachment [p. 101]
law of syllogism [p. 101]
modus ponens [p. 101]
modus tollens [p. 101]
negation [p. 65]
premise [p. 98]
quantifier [p. 67]
self-contradiction [p. 79]
standard forms of arguments [p. 101]
standard truth table form [p. 73]
statement [p. 62]
symbolic form [p. 98]
tautology [p. 79]
truth table [p. 65]
truth value [p. 65]
universal quantifier [p. 67]
valid argument [p. 98]

Essential Concepts

- **Truth Values**
  $\sim p$ is true if and only if p is false.
  $p \wedge q$ is true if and only if both p and q are true.
  $p \vee q$ is true if and only if p is true, q is true, or both p and q are true.
  The *conditional* $p \rightarrow q$ is false if p is true and q is false. It is true in all other cases.

- **Order of Precedence Agreement**
  If grouping symbols are not used to specify the order in which logical connectives are applied, then we use the following *Order of Precedence Agreement*.
  First apply the negations from left to right, then apply the conjunctions from left to right, and finally apply the disjunctions from left to right.

- **De Morgan's Laws for Statements**
  $\sim(p \wedge q) \equiv \sim p \vee \sim q$ and $\sim(p \vee q) \equiv \sim p \wedge \sim q$

- **Statements Related to the Conditional Statement**
  The *converse* of $p \rightarrow q$ is $q \rightarrow p$.
  The *inverse* of $p \rightarrow q$ is $\sim p \rightarrow \sim q$.
  The *contrapositive* of $p \rightarrow q$ is $\sim q \rightarrow \sim p$.

- **Equivalent Form of the Biconditional**
  $p \leftrightarrow q \equiv [(p \rightarrow q) \wedge (q \rightarrow p)]$

- **Valid Arguments**

Modus ponens	Modus tollens	Law of syllogism	Disjunctive syllogism
$p \rightarrow q$	$p \rightarrow q$	$p \rightarrow q$	$p \vee q$
$\underline{p \qquad}$	$\underline{\sim q}$	$\underline{q \rightarrow r}$	$\underline{\sim p}$
$\therefore q$	$\therefore \sim p$	$\therefore p \rightarrow r$	$\therefore q$

- **Invalid Arguments**

Fallacy of the converse	Fallacy of the inverse
$p \rightarrow q$	$p \rightarrow q$
$\underline{q \qquad}$	$\underline{\sim p}$
$\therefore p$	$\therefore \sim q$

Review Exercises

In Exercises 1–6, determine whether each sentence is a statement. Assume that a and b are real numbers.

1. How much is a ticket to London?
2. 91 is a prime number.
3. $a > b$
4. $a^2 \geq 0$
5. Lock the car.
6. Clark Kent is Superman.

In Exercises 7–10, write each sentence in symbolic form. Represent each component of the sentence with the letter indicated in parentheses. Also state whether the sentence is a conjunction, a disjunction, a negation, a conditional, or a biconditional.

7. Today is Monday (m) and it is my birthday (b).
8. If x is divisible by 2 (d), then x is an even number (e).
9. I am going to the dance (g) if and only if I have a date (d).
10. All triangles (t) have exactly three sides (s).

In Exercises 11–16, write the negation of each quantified statement.

11. Some dogs bite.
12. Every dessert at the Cove restaurant is good.
13. All winners receive prizes.
14. Some cameras do not use film.
15. None of the students received an A.
16. At least one person enjoyed the story.

In Exercises 17–22, determine whether each statement is true or false.

17. $5 > 2$ or $5 = 2$.
18. $3 \neq 5$ and 7 is a prime number.
19. $4 \leq 7$
20. $-3 < -1$
21. Every repeating decimal is a rational number.
22. There exists a real number that is not positive and not negative.

In Exercises 23–28, determine the truth value of the statement given that p is true, q is false, and r is false.

23. $(p \wedge q) \vee (\sim p \vee q)$
24. $(p \rightarrow \sim q) \leftrightarrow \sim(p \vee q)$
25. $(p \wedge \sim q) \wedge (\sim r \vee q)$
26. $(r \wedge \sim p) \vee [(p \vee \sim q) \leftrightarrow (q \rightarrow r)]$
27. $[p \wedge (r \rightarrow q)] \rightarrow (q \vee \sim r)$
28. $(\sim q \vee \sim r) \rightarrow [(p \leftrightarrow \sim r) \wedge q]$

In Exercises 29–36, construct a truth table for the given statement.

29. $(\sim p \rightarrow q) \vee (\sim q \wedge p)$
30. $\sim p \leftrightarrow (q \vee p)$
31. $\sim(p \vee \sim q) \wedge (q \rightarrow p)$
32. $(p \leftrightarrow q) \vee (\sim q \wedge p)$
33. $(r \leftrightarrow \sim q) \vee (p \rightarrow q)$
34. $(\sim r \vee \sim q) \wedge (q \rightarrow p)$
35. $[p \leftrightarrow (q \rightarrow \sim r)] \wedge \sim q$
36. $\sim(p \wedge q) \rightarrow (\sim q \vee \sim r)$

In Exercises 37–40, make use of De Morgan's laws to write the given statement in an equivalent form.

37. It is not true that Bob failed the English proficiency test and he registered for a speech course.
38. Ellen did not go to work this morning and she did not take her medication.
39. Wendy will go to the store this afternoon or she will not be able to prepare her fettuccine al pesto recipe.
40. Gina enjoyed the movie, but she did not enjoy the party.

In Exercises 41–44, use a truth table to show that the given pairs of statements are equivalent.

41. $\sim p \rightarrow \sim q; \ p \vee \sim q$
42. $\sim p \vee q; \ \sim(p \wedge \sim q)$
43. $p \vee (q \wedge \sim p); \ p \vee q$
44. $p \leftrightarrow q; \ (p \wedge q) \vee (\sim p \wedge \sim q)$

In Exercises 45–48, use a truth table to determine whether the given statement is a tautology or a self-contradiction.

45. $p \wedge (q \wedge \sim p)$

46. $(p \wedge q) \vee (p \rightarrow \sim q)$

47. $[\sim(p \rightarrow q)] \leftrightarrow (p \wedge \sim q)$

48. $p \vee (p \rightarrow q)$

In Exercises 49–52, identify the antecedent and the consequent of each conditional statement.

49. If he has talent, he will succeed.

50. If I had a teaching credential, I could get the job.

51. I will follow the exercise program provided I join the fitness club.

52. I will attend only if it is free.

In Exercises 53–58, determine whether the given statement is true or false. Assume that x and y are real numbers.

53. $x = y$ if and only if $|x| = |y|$.

54. $x > y$ if and only if $x - y > 0$.

55. If $x + y = 2x$, then $y = x$.

56. If $x > y$, then $\frac{1}{x} > \frac{1}{y}$.

57. If $x^2 > 0$, then $x > 0$.

58. If $x^2 = y^2$, then $x = y$.

In Exercises 59–62, write each statement in "If p, then q" form.

59. Every nonrepeating, nonterminating decimal is an irrational number.

60. Being well known is a necessary condition for a politician.

61. I could buy the house provided I could sell my condominium.

62. Being divisible by 9 is a sufficient condition for being divisible by 3.

In Exercises 63–68, write the **a.** converse, **b.** inverse, and **c.** contrapositive of the given statement.

63. If $x + 4 > 7$, then $x > 3$.

64. All recipes in this book can be prepared in less than 20 minutes.

65. If a and b are both divisible by 3, then $(a + b)$ is divisible by 3.

66. If you build it, they will come.

67. Every trapezoid has exactly two parallel sides.

68. If they like it, they will return.

69. What is the inverse of the contrapositive of $p \rightarrow q$?

70. What is the converse of the contrapositive of the inverse of $p \rightarrow q$?

In Exercises 71–74, determine the original statement if the given statement is related to the original statement in the manner indicated.

71. *Converse:* If $x > 2$, then x is an odd prime number.

72. *Negation:* The senator will attend the meeting and she will not vote on the motion.

73. *Inverse:* If their manager will not contact me, then I will not purchase any of their products.

74. *Contrapositive:* If Ginny can't rollerblade, then I can't rollerblade.

In Exercises 75–78, use a truth table to determine whether the argument is valid or invalid.

75. $(p \wedge \sim q) \wedge (\sim p \rightarrow q)$
$\underline{\quad p \qquad\qquad\qquad\qquad}$
$\therefore \sim q$

76. $p \rightarrow \sim q$
$\underline{\quad q \qquad\qquad}$
$\therefore \sim p$

77. r
$p \rightarrow \sim r$
$\underline{\sim p \rightarrow q}$
$\therefore p \wedge q$

78. $(p \vee \sim r) \rightarrow (q \wedge r)$
$\underline{\quad r \wedge p \qquad\qquad\qquad}$
$\therefore p \vee q$

In Exercises 79–84, determine whether the argument is valid or invalid by comparing its symbolic form with the symbolic forms in Tables 2.12 and 2.13 on pages 101 and 102.

79. We will serve either fish or chicken for lunch. We did not serve fish for lunch. Therefore, we served chicken for lunch.

80. If Mike is a CEO, then he will be able to afford to make a donation. If Mike can afford to make a donation, then he loves to ski. Therefore, if Mike does not love to ski, he is not a CEO.

81. If we wish to win the lottery, we must buy a lottery ticket. We did not win the lottery. Therefore, we did not buy a lottery ticket.

82. Robert can charge it on his MasterCard or his Visa. Robert does not use his MasterCard. Therefore, Robert charged it on his Visa.

83. If we are going to have a caesar salad, then we need to buy some eggs. We did not buy eggs. Therefore, we are not going to have a caesar salad.

84. If we serve lasagna, then Eva will not come to our dinner party. We did not serve lasagna. Therefore, Eva came to our dinner party.

In Exercises 85–88, use an Euler diagram to determine whether the argument is valid or invalid.

85. No wizard can yodel.
All lizards can yodel.

∴No wizard is a lizard.

86. Some dogs have tails.
Some dogs are big.

∴Some big dogs have tails.

87. All Italian villas are wonderful. It is not wise to invest in expensive villas. Some wonderful villas are expensive. Therefore, it is not wise to invest in Italian villas.

88. All logicians like to sing "It's a small world after all." Some logicians have been presidential candidates. Therefore, some presidential candidates like to sing "It's a small world after all."

CHAPTER 2 **Test**

---

1. Determine whether each sentence is a statement.

a. Look for the cat.

b. Clark Kent is afraid of the dark.

2. Write the negation of each statement.

a. Some trees are not green.

b. None of the kids had seen the movie.

3. Determine whether each statement is true or false.

a. $5 \leq 4$

b. $-2 \geq -2$

4. Determine the truth value of each statement given that p is true, q is false, and r is true.

a. $(p \vee \sim q) \wedge (\sim r \wedge q)$

b. $(r \vee \sim p) \vee [(p \vee \sim q) \leftrightarrow (q \rightarrow r)]$

In Exercises 5 and 6, construct a truth table for the given statement.

5. $\sim(p \wedge \sim q) \vee (q \rightarrow p)$ **6.** $(r \leftrightarrow \sim q) \wedge (p \rightarrow q)$

7. Use one of De Morgan's laws to write the following in an equivalent form.

Elle did not eat breakfast and she did not take a lunch break.

8. ✎ What is a tautology?

9. Construct a truth table for each of the following statements to determine if the statements are equivalent.

$\sim(p \rightarrow q); p \wedge \sim q$

10. Determine whether the given statement is true or false. Assume that x, y, and z are real numbers.

a. $x = y$ if $|x| = |y|$. **b.** If $x > y$, then $xz > yz$.

11. Write the **a.** converse, **b.** inverse, and **c.** contrapositive of the following statement.

If $x + 7 > 11$, then $x > 4$.

12. Write the standard form known as modus ponens.

13. Write the standard form known as the law of syllogism.

In Exercises 14 and 15, use a truth table to determine whether the argument is valid or invalid.

14. $(p \wedge \sim q) \wedge (\sim p \rightarrow q)$
p

∴$\sim q$

15. r
$p \rightarrow \sim r$
$\sim p \rightarrow q$

∴$p \wedge q$

In Exercises 16–20, determine whether the argument is valid or invalid. Explain how you made your decision.

16. If we wish to win the talent contest, we must practice. We did not win the contest. Therefore, we did not practice.

17. Gina will take a job in Atlanta or she will take a job in Kansas City. Gina did not take a job in Atlanta. Therefore, Gina took a job in Kansas City.

18. No wizard can glow in the dark.
Some lizards can glow in the dark.

∴No wizard is a lizard.

19. Some novels are worth reading.
War and Peace is a novel.

∴*War and Peace* is worth reading.

20. If I cut my night class, then I will go to the party. I went to the party. Therefore, I cut my night class.

CHAPTER 3

Algebraic Models

In your study of mathematics, you probably noticed that as you advanced the problems became less concrete and more abstract. Problems that are concrete provide information pertaining to a specific instance. Abstract problems are theoretical; they are stated without reference to a specific instance. Here's an example of a concrete problem:

If one candy bar costs 25 cents, how many candy bars can be purchased with 2 dollars?

To solve this problem, you need to calculate the number of cents in 2 dollars (multiply 2 by 100), and divide the result by the cost per candy bar (25 cents).

$$\frac{100 \cdot 2}{25} = \frac{200}{25} = 8$$

If one candy bar costs 25 cents, 8 candy bars can be purchased with 2 dollars.

Here is a related abstract problem:

If one candy bar costs c cents, how many candy bars can be purchased with d dollars?

Use the same procedure to solve the related abstract problem. Calculate the number of cents in d dollars (multiply d by 100), and divide the result by the cost per candy bar (c cents).

$$\frac{100 \cdot d}{c} = \frac{100d}{c}$$

If one candy bar costs c cents, $\frac{100d}{c}$ candy bars can be purchased with d dollars.

It is the variables in the problem above that make it abstract. At the heart of the study of algebra is the use of variables. Variables enable us to generalize situations and state relationships among quantities. These relationships are often stated in the form of equations. In this chapter, we will be using equations to solve applications.

Online Study Center

For online student resources, visit this textbook's Online Study Center at **college.hmco.com/pic/aufmannMTQR**.

| # First-Degree Equations and Formulas

Solving First-Degree Equations

The fuel economy, in miles per gallon, of a particular car traveling at a speed of v miles per hour can be calculated using the variable expression $-0.02v^2 + 1.6v + 3$, where $10 \leq v \leq 75$. For example, suppose the speed of a car is 30 miles per hour. We can calculate the fuel economy by substituting 30 for v in the variable expression and then using the Order of Operations Agreement to evaluate the resulting numerical expression.

$$-0.02v^2 + 1.6v + 3$$
$$-0.02(30)^2 + 1.6(30) + 3 = -0.02(900) + 1.6(30) + 3$$
$$= -18 + 48 + 3$$
$$= 33$$

The fuel economy is 33 miles per gallon.

The **terms** of a variable expression are the addends of the expression. The expression $-0.02v^2 + 1.6v + 3$ has three terms. The terms $-0.02v^2$ and $1.6v$ are **variable terms** because each contains a variable. The term 3 is a **constant term**; it does not contain a variable.

Each variable term is composed of a **numerical coefficient** and a **variable part** (the variable or variables and their exponents). For the variable term $-0.02v^2$, -0.02 is the coefficient and v^2 is the variable part.

Like terms of a variable expression are terms with the same variable part. Constant terms are also like terms. Examples of like terms are

$4x$ and $7x$

$9y$ and y

$5x^2y$ and $6x^2y$

8 and -3

An **equation** expresses the equality of two mathematical expressions. Each of the following is an equation.

$8 + 5 = 13$

$4y - 6 = 10$

$x^2 - 2x + 1 = 0$

$b = 7$

Each of the equations below is a **first-degree equation in one variable**, where *first-degree* means that the variable has an exponent of 1.

$x + 11 = 14$

$3z + 5 = 8z$

$2(6y - 1) = 34$

QUESTION *Which of the following are first-degree equations in one variable?*

a. $5y + 4 = 9 - 3(2y + 1)$ *b.* $\sqrt{x} + 9 = 16$

c. $p = -14$ *d.* $2x - 5 = x^2 - 9$

e. $3y + 7 = 4z - 10$

A **solution of an equation** is a number that, when substituted for the variable, results in a true equation.

3 is a solution of the equation $x + 4 = 7$ because $3 + 4 = 7$.

9 is not a solution of the equation $x + 4 = 7$ because $9 + 4 \neq 7$.

To **solve an equation** means to find all solutions of the equation. The following properties of equations are often used to solve equations.

Properties of Equations

Addition Property
The same number can be added to each side of an equation without changing the solution of the equation.

If $a = b$, then $a + c = b + c$.

Subtraction Property
The same number can be subtracted from each side of an equation without changing the solution of the equation.

If $a = b$, then $a - c = b - c$.

Multiplication Property
Each side of an equation can be multiplied by the same *nonzero* number without changing the solution of the equation.

If $a = b$ and $c \neq 0$, then $ac = bc$.

Division Property
Each side of an equation can be divided by the same *nonzero* number without changing the solution of the equation.

If $a = b$ and $c \neq 0$, then $\dfrac{a}{c} = \dfrac{b}{c}$.

✔ **TAKE NOTE**

In the Multiplication Property, it is necessary to state $c \neq 0$ so that the solutions of the equation are not changed. For example, if $\frac{1}{2}x = 4$, then $x = 8$. But if we multiply each side of the equation by 0, we have

$$0 \cdot \frac{1}{2}x = 0 \cdot 4$$

$$0 = 0$$

The solution $x = 8$ is lost.

In solving a first-degree equation in one variable, the goal is to rewrite the equation with the variable alone on one side of the equation and a constant term on the other side of the equation. The constant term is the solution of the equation.

ANSWER *The equations in **a** and **c** are first-degree equations in one variable. The equation in **b** is not a first-degree equation in one variable because it contains the square root of a variable. The equation in **d** contains a variable with an exponent other than 1. The equation in **e** contains two variables.*

✔ **TAKE NOTE**

You should always check the solution of an equation. The check for the example at the right is shown below.

$$t + 9 = -4$$
$$\overline{-13 + 9\ |\ -4}$$
$$-4 = -4$$

This is a true equation. The solution -13 checks.

For example, to solve the equation $t + 9 = -4$, use the Subtraction Property to subtract the constant term 9 from each side of the equation.

$$t + 9 = -4$$
$$t + 9 - 9 = -4 - 9$$
$$t = -13$$

Now the variable (t) is alone on one side of the equation and a constant term (-13) is on the other side. The solution is -13.

To solve the equation $-5q = 120$, use the Division Property. Divide each side of the equation by the coefficient -5.

$$-5q = 120$$
$$\frac{-5q}{-5} = \frac{120}{-5}$$
$$q = -24$$

Now the variable (q) is alone on one side of the equation and a constant (-24) is on the other side. The solution is -24.

✔ **TAKE NOTE**

An equation has some properties that are similar to those of a balance scale. For instance, if a balance scale is in balance and equal weights are added to each side of the scale, then the balance scale remains in balance. If an equation is true, then adding the same number to each side of the equation produces another true equation.

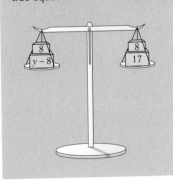

EXAMPLE 1 ■ Solve a First-Degree Equation Using One of the Properties of Equations

Solve.

a. $y - 8 = 17$ **b.** $4x = -2$ **c.** $-5 = 9 + b$ **d.** $-a = -36$

Solution

a. Because 8 is subtracted from y, use the Addition Property to add 8 to each side of the equation.

$$y - 8 = 17$$
$$y - 8 + 8 = 17 + 8$$
$$y = 25 \qquad \text{• A check will show that 25 is a solution.}$$

The solution is 25.

b. Because x is multiplied by 4, use the Division Property to divide each side of the equation by 4.

$$4x = -2$$
$$\frac{4x}{4} = \frac{-2}{4}$$
$$x = -\frac{1}{2} \qquad \text{• A check will show that } -\frac{1}{2} \text{ is a solution.}$$

The solution is $-\dfrac{1}{2}$.

c. Because 9 is added to b, use the Subtraction Property to subtract 9 from each side of the equation.

$$-5 = 9 + b$$

$$-5 - 9 = 9 - 9 + b$$

$$-14 = b$$

The solution is -14.

d. The coefficient of the variable is -1. Use the Multiplication Property to multiply each side of the equation by -1.

$$-a = -36$$

$$-1(-1a) = -1(-36)$$

$$a = 36$$

The solution is 36.

> ✔ **TAKE NOTE**
>
> When the coefficient of a variable is 1 or negative 1, the 1 is usually not written; $1a$ is written as a, and $-1a$ is written as $-a$.

CHECK YOUR PROGRESS 1   Solve.

a. $c - 6 = -13$ **b.** $4 = -8z$ **c.** $22 + m = -9$ **d.** $5x = 0$

Solution   *See page S10.*

When solving more complicated first-degree equations in one variable, use the following sequence of steps.

Steps for Solving a First-Degree Equation in One Variable

1. If the equation contains fractions, multiply each side of the equation by the least common multiple (LCM) of the denominators to clear the equation of fractions.

2. Use the Distributive Property to remove parentheses.

3. Combine any like terms on the right side of the equation and any like terms on the left side of the equation.

4. Use the Addition Property or the Subtraction Property to rewrite the equation with only one variable term and only one constant term.

5. Use the Multiplication Property or the Division Property to rewrite the equation with the variable alone on one side of the equation and a constant term on the other side of the equation.

If one of the above steps is not needed to solve a given equation, proceed to the next step. Remember that the goal is to rewrite the equation with the variable alone on one side of the equation and a constant term on the other side of the equation.

EXAMPLE 2 ■ Solve a First-Degree Equation Using the Properties of Equations

Solve.

a. $5x + 9 = 23 - 2x$ **b.** $8x - 3(4x - 5) = -2x + 6$

c. $\dfrac{3x}{4} - 6 = \dfrac{x}{3} - 1$

Solution

a. There are no fractions (Step 1) or parentheses (Step 2). There are no like terms on either side of the equation (Step 3). Use the Addition Property to rewrite the equation with only one variable term (Step 4). Add $2x$ to each side of the equation.

$$5x + 9 = 23 - 2x$$
$$5x + 2x + 9 = 23 - 2x + 2x$$
$$7x + 9 = 23$$

Use the Subtraction Property to rewrite the equation with only one constant term (Step 4). Subtract 9 from each side of the equation.

$$7x + 9 - 9 = 23 - 9$$
$$7x = 14$$

Use the Division Property to rewrite the equation with the x alone on one side of the equation (Step 5). Divide each side of the equation by 7.

$$\frac{7x}{7} = \frac{14}{7}$$
$$x = 2$$

The solution is 2.

b. There are no fractions (Step 1). Use the Distributive Property to remove parentheses (Step 2).

$$8x - 3(4x - 5) = -2x + 6$$
$$8x - 12x + 15 = -2x + 6$$

Combine like terms on the left side of the equation (Step 3). Then rewrite the equation with the variable alone on one side and a constant on the other.

$$-4x + 15 = -2x + 6 \qquad \bullet \text{ Combine like terms.}$$
$$-4x + 2x + 15 = -2x + 2x + 6 \qquad \bullet \text{ The Addition Property}$$
$$-2x + 15 = 6$$
$$-2x + 15 - 15 = 6 - 15 \qquad \bullet \text{ The Subtraction Property}$$
$$-2x = -9$$
$$\frac{-2x}{-2} = \frac{-9}{-2} \qquad \bullet \text{ The Division Property}$$
$$x = \frac{9}{2}$$

The solution is $\dfrac{9}{2}$.

historical note

The letter x is used universally as the standard letter for a single unknown, which is why x-rays were so named. The scientists who discovered them did not know what they were, and so labeled them the "unknown rays," or x-rays. ■

c. The equation contains fractions (Step 1); multiply each side of the equation by the LCM of the denominators. Then rewrite the equation with the variable alone on one side and a constant on the other.

$$\frac{3x}{4} - 6 = \frac{x}{3} - 1$$

$$12\left(\frac{3x}{4} - 6\right) = 12\left(\frac{x}{3} - 1\right)$$ • The Multiplication Property

$$12 \cdot \frac{3x}{4} - 12 \cdot 6 = 12 \cdot \frac{x}{3} - 12 \cdot 1$$ • The Distributive Property

$$9x - 72 = 4x - 12$$

$$9x - 4x - 72 = 4x - 4x - 12$$ • The Subtraction Property

$$5x - 72 = -12$$

$$5x - 72 + 72 = -12 + 72$$ • The Addition Property

$$5x = 60$$

$$\frac{5x}{5} = \frac{60}{5}$$ • The Division Property

$$x = 12$$

The solution is 12.

CHECK YOUR PROGRESS 2   Solve.

a. $4x + 3 = 7x + 9$ **b.** $7 - (5x - 8) = 4x + 3$ **c.** $\dfrac{3x - 1}{4} + \dfrac{1}{3} = \dfrac{7}{3}$

Solution   *See page S10.*

Math Matters   **The Hubble Space Telescope**

The Hubble Space Telescope missed the stars it was targeted to photograph during the second week of May, 1990, because it was pointing in the wrong direction. The telescope was off by about one-half of one degree as a result of an arithmetic error—an addition instead of a subtraction.

Applications

In some applications of equations, we are given an equation that can be used to solve the application. This is illustrated in Example 3.

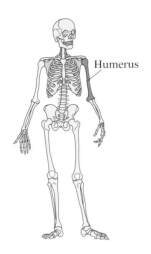

Humerus

EXAMPLE 3 ■ Solve an Application

Forensic scientists have determined that the equation $H = 2.9L + 78.1$ can be used to approximate the height H, in centimeters, of an adult on the basis of the length L, in centimeters, of the adult's humerus (the bone extending from the shoulder to the elbow).

a. Use this equation to approximate the height of an adult whose humerus measures 36 centimeters.

b. According to this equation, what is the length of the humerus of an adult whose height is 168 centimeters?

Solution

a. Substitute 36 for L in the given equation. Solve the resulting equation for H.

$$H = 2.9L + 78.1$$
$$H = 2.9(36) + 78.1$$
$$H = 104.4 + 78.1$$
$$H = 182.5$$

The adult's height is approximately 182.5 centimeters.

b. Substitute 168 for H in the given equation. Solve the resulting equation for L.

$$H = 2.9L + 78.1$$
$$168 = 2.9L + 78.1$$
$$168 - 78.1 = 2.9L + 78.1 - 78.1$$
$$89.9 = 2.9L$$
$$\frac{89.9}{2.9} = \frac{2.9L}{2.9}$$
$$31 = L$$

The length of the adult's humerus is approximately 31 centimeters.

CHECK YOUR PROGRESS 3   The amount of garbage generated by each person living in the United States has been increasing and is approximated by the equation $P = 0.05Y - 95$, where P is the number of pounds of garbage generated per person per day and Y is the year.

a. Find the amount of garbage generated per person per day in 1990.

b. According to the equation, in what year will 5.6 pounds of garbage be generated per person per day?

Solution   *See page S11.*

In many applied problems, we are not given an equation that can be used to solve the problem. Instead, we must use the given information to write an equation whose solution answers the question stated in the problem. This is illustrated in Examples 4 and 5.

EXAMPLE 4 ■ Solve an Application of First-Degree Equations

The cost of electricity in a certain city is \$.08 for each of the first 300 kWh (kilowatt-hours) and \$.13 for each kilowatt-hour over 300 kWh. Find the number of kilowatt-hours used by a family that receives a \$51.95 electric bill.

Solution

Let k = the number of kilowatt-hours used by the family. Write an equation and then solve the equation for k.

\$.08 for each of the first 300 kWh + \$.13 for each kilowatt-hour over 300	=	\$51.95

$$0.08(300) + 0.13(k - 300) = 51.95$$
$$24 + 0.13k - 39 = 51.95$$
$$0.13k - 15 = 51.95 \qquad \text{• Simplify } 24 - 39.$$
$$0.13k - 15 + 15 = 51.95 + 15 \quad \text{• Add 15 to each side.}$$
$$0.13k = 66.95$$
$$\frac{0.13k}{0.13} = \frac{66.95}{0.13} \qquad \text{• Divide each side by 0.13.}$$
$$k = 515$$

The family used 515 kWh of electricity.

CHECK YOUR PROGRESS 4   For a classified ad, a newspaper charges \$11.50 for the first three lines and \$1.50 for each additional line for an ad that runs for three days, or \$17.50 for the first three lines and \$2.50 for each additional line for an ad that runs for seven days. You want your ad to run for seven days. Determine the number of lines you can place in the ad for \$30.

Solution   *See page S11.*

✔ **TAKE NOTE**

If the family uses 500 kilowatt-hours of electricity, they are billed \$.13 per kilowatt-hour for 200 kilowatt-hours (500 − 300). If they use 650 kilowatt-hours, they are billed \$.13 per kilowatt-hour for 350 kilowatt-hours (650 − 300). If they use k kilowatt-hours, $k > 300$, they are billed \$.13 per kilowatt-hour for (k − 300) kilowatt-hours.

EXAMPLE 5 ■ Solve an Application of First-Degree Equations

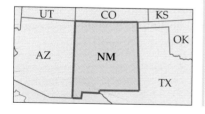

In January, 1990, the population of New Mexico was 1,515,100 and the population of West Virginia was 1,793,500. During the 1990s, New Mexico's population increased at an average rate of 21,235 people per year while West Virginia's population decreased at an average rate of 15,600 people per year. If these rate changes remained stable, in what year would the populations of New Mexico and West Virginia have been the same? Round to the nearest year.

▼ **point of interest**

Is the population of your state increasing or decreasing? You can find out by checking a reference such as the *Information Please Almanac,* which was the source for the data in Example 5 and Check Your Progress 5.

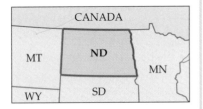

Solution

Let $n =$ the number of years until the populations are the same. Write an equation and then solve the equation for n.

The 1990 population of New Mexico plus the annual increase times n	=	the 1990 population of West Virginia minus the annual decrease times n

$$1{,}515{,}100 + 21{,}235n = 1{,}793{,}500 - 15{,}600n$$
$$1{,}515{,}100 + 21{,}235n + 15{,}600n = 1{,}793{,}500 - 15{,}600n + 15{,}600n$$
$$1{,}515{,}100 + 36{,}835n = 1{,}793{,}500$$
$$1{,}515{,}100 - 1{,}515{,}100 + 36{,}835n = 1{,}793{,}500 - 1{,}515{,}100$$
$$36{,}835n = 278{,}400$$
$$\frac{36{,}835n}{36{,}835} = \frac{278{,}400}{36{,}835}$$
$$n \approx 8$$

The variable n represents the number of years after 1990. Add 8 to the year 1990.

$$1990 + 8 = 1998$$

To the nearest year, the populations would have been the same in 1998.

CHECK YOUR PROGRESS 5   In January, 1990, the population of North Dakota was 638,800 and the population of Vermont was 562,576. During the 1990s, North Dakota's population decreased at an average rate of 1370 people per year while Vermont's population increased at an average rate of 5116 people per year. If these rate changes remained stable, in what year would the populations of North Dakota and Vermont have been the same? Round to the nearest year.

Solution   See page S11.

Literal Equations

A **literal equation** is an equation that contains more than one variable. Examples of literal equations are:

$$2x + 3y = 6$$
$$4a - 2b + c = 0$$

A **formula** is a literal equation that states a relationship between two or more quantities in an application problem. Examples of formulas are shown below. These formulas are taken from physics, mathematics, and business.

$$\frac{1}{R_1} + \frac{1}{R_2} = \frac{1}{R}$$
$$s = a + (n - 1)d$$
$$A = P + Prt$$

QUESTION *Which of the following are literal equations?*

a. $5a - 3b = 7$

b. $a^2 + b^2 = c^2$

c. $a_1 + (n - 1)d$

d. $3x - 7 = 5 + 4x$

The addition, subtraction, multiplication, and division properties of equations can be used to solve some literal equations for one of the variables. In solving a literal equation for one of the variables, the goal is to rewrite the equation so that the letter being solved for is alone on one side of the equation and all numbers and other variables are on the other side. This is illustrated in Example 6.

EXAMPLE 6 ■ **Solve a Literal Equation**

a. Solve $A = P(1 + i)$ for i.

b. Solve $I = \dfrac{E}{R + r}$ for R.

Solution

a. The goal is to rewrite the equation so that i is alone on one side of the equation and all other numbers and letters are on the other side. We will begin by using the Distributive Property on the right side of the equation.

$$A = P(1 + i)$$
$$A = P + Pi$$

Subtract P from each side of the equation.

$$A - P = P - P + Pi$$
$$A - P = Pi$$

Divide each side of the equation by P.

$$\frac{A - P}{P} = \frac{Pi}{P}$$
$$\frac{A - P}{P} = i$$

b. The goal is to rewrite the equation so that R is alone on one side of the equation and all other variables are on the other side of the equation. Because the equation contains a fraction, we will first multiply both sides of the equation by the denominator $R + r$ to clear the equation of fractions.

$$I = \frac{E}{R + r}$$
$$(R + r)I = (R + r)\frac{E}{R + r}$$
$$RI + rI = E$$

ANSWER *The literal equations are* **a** *and* **b**. **c** *is not an equation.* **d** *does not have more than one variable.*

Subtract from the left side of the equation the term that does not contain a capital R.

$$RI + rI - rI = E - rI$$
$$RI = E - rI$$

Divide each side of the equation by I.

$$\frac{RI}{I} = \frac{E - rI}{I}$$
$$R = \frac{E - rI}{I}$$

CHECK YOUR PROGRESS 6

a. Solve $s = \dfrac{A + L}{2}$ for L.

b. Solve $L = a(1 + ct)$ for c.

Solution   *See page S11.*

Investigation

Body Mass Index

Body mass index, or **BMI,** expresses the relationship between a person's height and weight. It is a measurement for gauging a person's weight-related level of risk for high blood pressure, heart disease, and diabetes. A BMI value of 25 or less indicates a low risk; a BMI value of 25 to 30 indicates a moderate risk; a BMI of 30 or more indicates a high risk.

The formula for body mass index is

$$B = \frac{705W}{H^2}$$

where B is the BMI, W is weight in pounds, and H is height in inches.

To determine how much a woman who is 5'4" should weigh in order to have a BMI of 24, first convert 5'4" to inches.

$$5'4'' = 5(12)'' + 4'' = 60'' + 4'' = 64''$$

(continued)

Substitute 24 for *B* and 64 for *H* in the body mass index formula. Then solve the resulting equation for *W*.

$$B = \frac{705W}{H^2}$$

$$24 = \frac{705W}{64^2}$$ • **B = 24, H = 64**

$$24 = \frac{705W}{4096}$$

$$4096(24) = 4096\left(\frac{705W}{4096}\right)$$ • **Multiply each side of the equation by 4096.**

$$98{,}304 = 705W$$

$$\frac{98{,}304}{705} = \frac{705W}{705}$$ • **Divide each side of the equation by 705.**

$$139 \approx W$$

A woman who is 5′4″ should weigh about 139 pounds in order to have a BMI of 24.

Investigation Exercises

1. Amy is 140 pounds and 5′8″ tall. Calculate Amy's BMI. Round to the nearest tenth. Rank Amy as a low, moderate, or high risk for weight-related disease.

2. Carlos is 6′1″ and weighs 225 pounds. Calculate Carlos's BMI. Round to the nearest tenth. Would you rank Carlos as a low, moderate, or high risk for weight-related disease?

3. Roger is 5′11″. How much should he weigh in order to have a BMI of 25? Round to the nearest pound.

4. Brenda is 5′3″. How much should she weigh in order to have a BMI of 24? Round to the nearest pound.

5. Bohdan weighs 185 pounds and is 5′9″. How many pounds must Bohdan lose in order to reach a BMI of 23? Round to the nearest pound.

6. Pat is 6′3″ and weighs 245 pounds. Calculate the number of pounds Pat must lose in order to reach a BMI of 22. Round to the nearest pound.

7. Zack weighs 205 pounds and is 6′0″. He would like to lower his BMI to 20.

   a. By how many points must Zack lower his BMI? Round to the nearest tenth.

   b. How many pounds must Zack lose in order to reach a BMI of 20? Round to the nearest pound.

8. Felicia weighs 160 pounds and is 5′7″. She would like to lower her BMI to 20.

   a. By how many points must Felicia lower her BMI? Round to the nearest tenth.

   b. How many pounds must Felicia lose in order to reach a BMI of 20? Round to the nearest pound.

Exercise Set 3.1

1. What is the difference between an expression and an equation? Provide an example of each.

2. What is the solution of the equation $x = 8$? Use your answer to explain why the goal in solving an equation is to get the variable alone on one side of the equation.

3. Explain how to check the solution of an equation.

In Exercises 4–14, solve the equation.

4. $x + 7 = -5$

5. $9 + b = 21$

6. $-9 = z - 8$

7. $b - 11 = 11$

8. $-3x = 150$

9. $-48 = 6z$

10. $-9a = -108$

11. $-\dfrac{3}{4}x = 15$

12. $\dfrac{5}{2}x = -10$

13. $-\dfrac{x}{4} = -2$

14. $\dfrac{2x}{5} = -8$

15. If $x - 679 = 841$, is x a positive or a negative number? Why?

16. If $-400b = 1200$, is b a positive or a negative number? Why?

17. Given $\dfrac{5}{6}y = 30$, state whether $y > 30$ or $y < 30$.

18. Given $\dfrac{4}{3}d = 24$, state whether $d > 24$ or $d < 24$.

In Exercises 19–45, solve the equation.

19. $4 - 2b = 2 - 4b$

20. $4y - 10 = 6 + 2y$

21. $5x - 3 = 9x - 7$

22. $10z + 6 = 4 + 5z$

23. $3m + 5 = 2 - 6m$

24. $6a - 1 = 2 + 2a$

25. $5x + 7 = 8x + 5$

26. $2 - 6y = 5 - 7y$

27. $4b + 15 = 3 - 2b$

28. $2(x + 1) + 5x = 23$

29. $9n - 15 = 3(2n - 1)$

30. $7a - (3a - 4) = 12$

31. $5(3 - 2y) = 3 - 4y$

32. $9 - 7x = 4(1 - 3x)$

33. $2(3b + 5) - 1 = 10b + 1$

34. $2z - 2 = 5 - (9 - 6z)$

35. $4a + 3 = 7 - (5 - 8a)$

36. $5(6 - 2x) = 2(5 - 3x)$

37. $4(3y + 1) = 2(y - 8)$

38. $2(3b - 5) = 4(6b - 2)$

39. $3(x - 4) = 1 - (2x - 7)$

40. $\dfrac{2y}{3} - 4 = \dfrac{y}{6} - 1$

41. $\dfrac{x}{8} + 2 = \dfrac{3x}{4} - 3$

42. $\dfrac{2x - 3}{3} + \dfrac{1}{2} = \dfrac{5}{6}$

43. $\dfrac{2}{3} + \dfrac{3x + 1}{4} = \dfrac{5}{3}$

44. $\dfrac{1}{2}(x + 4) = \dfrac{1}{3}(3x - 6)$

45. $\dfrac{3}{4}(x - 8) = \dfrac{1}{2}(2x + 4)$

Car Payments The monthly car payment on a 60-month car loan at a 9 percent rate is calculated by using the formula $P = 0.02076L$, where P is the monthly car payment and L is the loan amount. Use this formula for Exercises 46 and 47.

46. If you can afford a maximum monthly car payment of $300, what is the maximum loan amount you can afford? Round to the nearest cent.

47. If the maximum monthly car payment you can afford is $350, what is the maximum loan amount you can afford? Round to the nearest cent.

Cassette Tape The music you hear when listening to a magnetic cassette tape is the result of the magnetic tape passing over magnetic heads, which read the magnetic information on the tape. The length of time a tape will play depends on the length of the tape and the operating speed of the tape player. The formula is $T = \frac{L}{S}$, where T is the time in seconds, L is the length of the tape in inches, and S is the operating speed in inches per second. Use this formula for Exercises 48–50.

48. How long a tape does a marine biologist need to record 16 seconds of whale songs at an operating speed of $1\frac{7}{8}$ inches per second?

49. How long a tape does a police officer need to record a 3-minute confession at an operating speed of $7\frac{1}{2}$ inches per second?

50. A U2 cassette tape takes 50 minutes to play on a system with an operating speed of $3\frac{3}{4}$ inches per second. Find the length of the tape.

Deep-Sea Diving The pressure on a diver can be calculated using the formula $P = 15 + \frac{1}{2}D$, where P is the pressure in pounds per square inch and D is the depth in feet. Use this formula for Exercises 51 and 52.

51. Find the depth of a diver when the pressure on the diver is 45 pounds per square inch.

52. Find the depth of a diver when the pressure on the diver is 55 pounds per square inch.

Foot Races The world-record time for a 1-mile race can be approximated by $t = 17.08 - 0.0067y$, where y is the year of the race, $1950 \le y \le 2006$, and t is the time, in minutes, of the race. Use this formula for Exercises 53 and 54.

53. Approximate the year in which the first "4-minute mile" was run. The actual year was 1954.

54. In 1985, the world-record time for a 1-mile race was 3.77 minutes. For what year does the equation predict this record time?

Black Ice Black ice is an ice covering on roads that is especially difficult to see and therefore extremely dangerous for motorists. The distance a car traveling at 30 miles per hour will slide, on black ice, after its brakes are applied is related to the outside temperature by the formula $C = \frac{1}{4}D - 45$, where C is the Celsius temperature and D is the distance, in feet, that the car will slide. Use this formula for Exercises 55 and 56.

55. Determine the distance a car will slide on black ice when the outside air temperature is $-3°C$.

56. How far will a car slide on black ice when the outside air temperature is $-11°C$?

Crickets The formula $N = 7C - 30$ approximates N, the number of times per minute a cricket chirps when the air temperature is C degrees Celsius. Use this formula for Exercises 57 and 58.

57. What is the approximate air temperature when a cricket chirps 100 times per minute? Round to the nearest tenth.

58. Determine the approximate air temperature when a cricket chirps 140 times per minute. Round to the nearest tenth.

Bowling In order to equalize all the bowlers' chances of winning, some players in a bowling league are given a handicap, or a bonus of extra points. Some leagues use the formula $H = 0.8(200 - A)$, where H is the handicap and A is the bowler's average score in past games. Use this formula for Exercises 59 and 60.

59. A bowler has a handicap of 20. What is the bowler's average score?

60. Find the average score of a bowler who has a handicap of 25.

In Exercises 61–73, write an equation as part of solving the problem.

61. **Depreciation** As a result of depreciation, the value of a car is now $13,200. This is three-fifths of its original value. Find the original value of the car.

62. **Computers** The operating speed of a personal computer is 100 gigahertz. This is one-third the speed of a newer model. Find the speed of the newer model personal computer.

63. **College Tuition** The graph below shows average tuition and fees at private four-year colleges for selected years.

    a. For the 1988–89 school year, the average combined tuition and fees at private four-year colleges was $171 more than five times the average combined tuition and fees at public four-year colleges. Determine the average tuition and fees at public four-year colleges for the school year 1988–89.

    b. For the 2003–04 school year, the average combined tuition and fees at private four-year colleges was $934 more than four times the average combined tuition and fees at public four-year colleges. Find the average tuition and fees at public four-year colleges for the school year 2003–04.

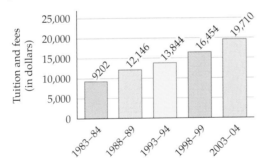

Tuition and Fees at Private Four-Year Colleges

Source: The College Board

64. **Adoption** In a recent year, Americans adopted 21,616 children from foreign countries. The graph on the next page shows the top three countries where the children were born.

a. The number of children adopted from China was 1489 more than three times the number adopted from South Korea. Determine the number of children adopted from South Korea that year.

b. The number of children adopted from Russia was 259 more than six times the number adopted from Kazakhstan. Determine the number of children from Kazakhstan that Americans adopted that year.

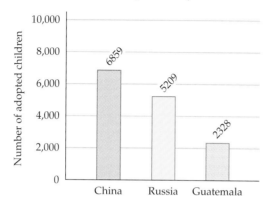

Birth Countries of Adopted American Children
Source: U.S. State Department

65. **Installment Purchases** The purchase price of a big-screen TV, including finance charges, was $3276. A down payment of $450 was made, and the remainder was paid in 24 equal monthly installments. Find the monthly payment.

66. **Auto Repair** The cost to replace a water pump in a sports car was $600. This included $375 for the water pump and $45 per hour for labor. How many hours of labor were required to replace the water pump?

67. **Space Vehicles** The table below provides statistics on the space shuttle *Discovery*, on which John Glenn was a crew member in 1998. His previous space flight was on the *Friendship 7* in 1962. (*Source: Time* magazine, August 17, 1998)

a. The lift-off thrust of the *Discovery* was 85,000 kilograms less than 20 times the lift-off thrust of the *Friendship 7*. Find the lift-off thrust of the *Friendship 7*.

b. The weight of the *Discovery* was 290 kilograms greater than 36 times the weight of the *Friendship 7*. Find the weight of the *Friendship 7*.

The Space Shuttle *Discovery*	
Crew size	7
Crew work area	66 cubic meters
Windows	5
Computers	5
Toggle switches	856
Lift-off thrust	3,175,000 kilograms
Weight	69,770 kilograms

68. College Staffing A university employs a total of 600 teaching assistants and research assistants. There are three times as many teaching assistants as research assistants. Find the number of research assistants employed by the university.

69. Wages A service station attendant is paid time-and-a-half for working over 40 hours per week. Last week the attendant worked 47 hours and earned $631.25. Find the attendant's regular hourly wage.

70. Investments An investor deposited $5000 in two accounts. Two times the smaller deposit is $1000 more than the larger deposit. Find the amount deposited in each account.

71. Computers A computer screen consists of tiny dots of light called *pixels*. In a certain graphics mode, there are 1040 vertical pixels. This is 400 more than one-half the number of horizontal pixels. Find the number of horizontal pixels.

72. Shipping An overnight mail service charges $5.60 for the first 6 ounces and $.85 for each additional ounce or fraction of an ounce. Find the weight, in ounces, of a package that cost $10.70 to deliver.

73. Telecommunications The charges for a long-distance telephone call are $1.42 for the first 3 minutes and $.65 for each additional minute or fraction of a minute. If charges for a call were $10.52, for how many minutes did the phone call last?

In Exercises 74–91, solve the formula for the indicated variable.

74. $A = \dfrac{1}{2}bh$; h (Geometry)

75. $P = a + b + c$; b (Geometry)

76. $d = rt$; t (Physics)

77. $E = IR$; R (Physics)

78. $PV = nRT$; R (Chemistry)

79. $I = Prt$; r (Business)

80. $P = 2L + 2W$; W (Geometry)

81. $F = \dfrac{9}{5}C + 32$; C (Temperature conversion)

82. $P = R - C$; C (Business)

83. $A = P + Prt$; t (Business)

84. $S = V_0 t - 16t^2$; V_0 (Physics)

85. $T = fm - gm$; f (Engineering)

86. $P = \dfrac{R - C}{n}$; R (Business)

87. $R = \dfrac{C - S}{t}$; S (Business)

88. $V = \dfrac{1}{3}\pi r^2 h$; h (Geometry)

89. $A = \dfrac{1}{2}h(b_1 + b_2)$; b_2 (Geometry)

90. $a_n = a_1 + (n - 1)d$; d (Mathematics)

91. $S = 2\pi r^2 + 2\pi rh$; h (Geometry)

In Exercises 92 and 93, solve the equation for y.

92. $2x - y = 4$

93. $4x + 3y = 6$

In Exercises 94 and 95, solve the equation for x.

94. $ax + by + c = 0$

95. $y - y_1 = m(x - x_1)$

Extensions

CRITICAL THINKING

In Exercises 96–99, solve the equation.

96. $3(4x + 2) = 7 - 4(1 - 3x)$

97. $9 - (6 - 4x) = 3(1 - 2x)$

98. $4(x + 5) = 30 - (10 - 4x)$

99. $7(x - 20) - 5(x - 22) = 2x - 30$

100. Use the numbers 5, 10, and 15 to make equations by filling in the boxes: $x + \square = \square - \square$. Each equation must use all three numbers.

a. What is the largest possible solution of these equations?

b. What is the smallest possible solution of these equations?

101. Solve the equation $ax + b = cx$ for x. Is your solution valid for all numbers a, b, and c? Explain.

COOPERATIVE LEARNING

102. Some equations have no solution. For instance, $x = x + 1$ has no solution. If we subtract x from each side of the equation, the result is $0 = 1$, which is not a true statement. One possible interpretation of the equation $x = x + 1$ is "A number is equal to one more than itself." Because there is no number that is equal to one more than itself, the equation has no solution.

Now consider the equation $ax + b = cx + d$. Determine what conditions on a, b, c, and d will result in an equation with no solution.

EXPLORATIONS

103. Recall that formulas state relationships among quantities. When we know there is an explicit relationship between two quantities, often we can write a formula to express the relationship.

For example, the Hooksett tollbooth on Interstate 93 in New Hampshire collects a toll of $.75 from each car that passes through the tollbooth. We can write a formula that describes the amount of money collected from passenger cars on any given day.

Let A be the total amount of money collected, and let c be the number of cars that pass through the tollbooth on a given day. Then

$$A = \$.75c$$

is a formula that expresses the total amount of money collected from passenger cars on any given day.

a. How much money is collected from passenger cars on a day on which 5500 passenger cars pass through the tollbooth?

b. How many passenger cars passed through the tollbooth on a day on which $3243.75 was collected in tolls from passenger cars?

Write formulas for each of the following situations. Include as part of your answer a list of variables that are used, and state what each variable represents.

c. Write a formula to represent the total cost of renting a copier from a company that charges $325 per month plus $.04 per copy made.

d. Suppose you buy a used car with 30,000 miles on it. You expect to drive the car about 750 miles per month. Write a formula to represent the total number of miles the car has been driven after you have owned it for m months.

e. A parking garage charges $2.50 for the first hour and $1.75 for each additional hour. Write a formula to represent the parking charge for parking in this garage for h hours, where h is a whole number.

f. Write a formula to represent the total cost of renting a car from a company that rents cars for $19.95 per day plus 25¢ for every mile driven over 100 miles.

g. Think of a mathematical relationship that can be modeled by a formula. Identify the variables in the relationship and write a formula that models the relationship.

| **Rate, Ratio, and Proportion**

Rates

The word *rate* is used frequently in our everyday lives. It is used in such contexts as unemployment rate, tax rate, interest rate, hourly rate, infant mortality rate, school dropout rate, inflation rate, and postage rate.

A **rate** is a comparison of two quantities. A rate can be written as a fraction. A car travels 135 miles on 6 gallons of gas. The miles-to-gallons rate is written

$$\frac{135 \text{ miles}}{6 \text{ gallons}}$$

Note that the units (miles and gallons) are written as part of the rate.

▼ **point of interest**

Unit rates are used in a wide variety of situations. One unit rate you may not be familiar with is used in the airline industry: cubic feet of air per minute per person. Typical rates are: economy class, 7 cubic feet/minute/person; first class, 50 cubic feet/minute/person; cockpit, 150 cubic feet/minute/person.

A **unit rate** is a rate in which the number in the denominator is 1. To find a unit rate, divide the number in the numerator of the rate by the number in the denominator of the rate. For the preceding example,

$$135 \div 6 = 22.5$$

The unit rate is $\dfrac{22.5 \text{ miles}}{1 \text{ gallon}}$.

This rate can be written 22.5 miles/gallon or 22.5 miles per gallon, where the word *per* has the meaning "for every."

Unit rates make comparisons easier. For example, if you travel 37 miles per hour and I travel 43 miles per hour, we know that I am traveling faster than you are. It is more difficult to compare speeds if we are told that you are traveling $\dfrac{111 \text{ miles}}{3 \text{ hours}}$ and I am traveling $\dfrac{172 \text{ miles}}{4 \text{ hours}}$.

EXAMPLE 1 ■ Calculate a Unit Rate

A dental hygienist earns $780 for working a 40-hour week. What is the hygienist's hourly rate of pay?

Solution
The hygienist's rate of pay is $\dfrac{\$780}{40 \text{ hours}}$.
To find the hourly rate of pay, divide 780 by 40.

$$780 \div 40 = 19.5$$

$$\frac{\$780}{40 \text{ hours}} = \frac{\$19.50}{1 \text{ hour}} = \$19.50/\text{hour}$$

The hygienist's hourly rate of pay is $19.50 per hour.

CHECK YOUR PROGRESS 1   You pay $4.92 for 1.5 pounds of hamburger. What is the cost per pound?

Solution   *See page S11.*

EXAMPLE 2 ■ Solve an Application of Unit Rates

A teacher earns a salary of $34,200 per year. Currently the school year is 180 days. If the school year were extended to 220 days, as is proposed in some states, what annual salary should the teacher be paid if the salary is based on the number of days worked per year?

Solution
Find the current salary per day.

$$\frac{\$34,200}{180 \text{ days}} = \frac{\$190}{1 \text{ day}} = \$190/\text{day}$$

Multiply the salary per day by the number of days in the proposed school year.

$$\frac{\$190}{1 \text{ day}} \cdot 220 \text{ days} = \$190(220) = \$41,800$$

The teacher's annual salary should be $41,800.

CHECK YOUR PROGRESS 2   In July 2004, the federal minimum wage was $5.15 per hour, and the minimum wage in California was $6.75. How much greater is an employee's pay for working 35 hours and earning the California minimum wage rather than the federal minimum wage?

Solution   *See page S11.*

Grocery stores are required to provide customers with unit price information. The **unit price** of a product is its cost per unit of measure. Unit pricing is an application of unit rate.

The price of a 2-pound box of spaghetti is $1.89. The unit price of the spaghetti is the cost per pound. To find the unit price, write the rate as a unit rate.

The numerator is the price and the denominator is the quantity. Divide the number in the numerator by the number in the denominator.

$$\frac{\$1.89}{2 \text{ pounds}} = \frac{\$.945}{1 \text{ pound}}$$

The unit price of the spaghetti is $.945 per pound.

Unit pricing is used by consumers to answer the question "Which is the better buy?" The answer is that the product with the lower unit price is the more economical purchase.

EXAMPLE 3 ■ **Determine the More Economical Purchase**

Which is the more economical purchase, an 18-ounce jar of peanut butter priced at $3.49 or a 12-ounce jar of peanut butter priced at $2.59?

Solution
Find the unit price for each item.

$$\frac{\$3.49}{18 \text{ ounces}} \approx \frac{\$.194}{1 \text{ ounce}} \qquad \frac{\$2.59}{12 \text{ ounces}} \approx \frac{\$.216}{1 \text{ ounce}}$$

Compare the two prices per ounce.

$$\$.194 < \$.216$$

The item with the lower unit price is the more economical purchase.
The more economical purchase is the 18-ounce jar priced at $3.49.

CHECK YOUR PROGRESS 3   Which is the more economical purchase, 32 ounces of detergent for $2.99 or 48 ounces of detergent for $3.99?

Solution   *See page S12.*

Rates such as crime statistics or data on fatalities are often written as rates per hundred, per thousand, per hundred thousand, or per million. For example, the table below shows bicycle deaths per million people in a recent year in the states with the highest and lowest rates. (*Source:* Environmental Working Group)

Rates of Bicycle Fatalities (Deaths per Million People)			
Highest		**Lowest**	
Florida	8.8	North Dakota	1.7
Arizona	7.0	Oklahoma	1.6
Louisiana	5.9	New Hampshire	1.4
South Carolina	5.4	West Virginia	1.2
North Carolina	4.5	Rhode Island	1.1

The rates in this table are easier to read than they would be if they were unit rates. As a comparison, consider that the bicycle fatalities in North Carolina would be written as 0.0000045 as a unit rate. Also, it is easier to understand that 7 out of every million people living in Arizona die in bicycle accidents than to consider that 0.000007 out of every person in Arizona dies in a bicycle accident.

QUESTION   *What does the rate given for Oklahoma mean?*

Another application of rates is in the area of international trade. Suppose a company in France purchases a shipment of sneakers from an American company. The French company must exchange euros, which is France's currency, for U.S. dollars in order to pay for the order. The number of euros that is equivalent to one U.S. dollar is called the *exchange rate*. The table below shows the exchange rates per U.S. dollar for three foreign countries and the European Union on December 2, 2005. Use this table for Example 4 and Check Your Progress 4.

Exchange Rates per U.S. Dollar	
British Pound	0.5770
Canadian Dollar	1.1616
Japanese Yen	120.43
Euro	0.8540

EXAMPLE 4 ■ Solve an Application Using Exchange Rates

a. How many Japanese yen are needed to pay for an order costing $10,000?

b. Find the number of British pounds that would be exchanged for $5000.

Solution

a. Multiply the number of yen per $1 by 10,000.

$$10,000(120.43) = 1,204,300$$

An order costing $10,000 would require 1,204,300 yen.

ANSWER   *Oklahoma's rate of 1.6 means that 1.6 out of every million people living in Oklahoma die in bicycle accidents.*

b. Multiply the number of pounds per $1 by 5000.

$$5000(0.5770) = 2885$$

The number of British pounds that would be exchanged for $5000 is 2885.

CHECK YOUR PROGRESS 4

a. How many Canadian dollars would be needed to pay for an order costing $20,000?

b. Find the number of euros that would be exchanged for $25,000.

Solution   *See page S12.*

Solution   *See page S12.*

point of interest

It is believed that billiards was invented in France during the reign of Louis XI (1423–1483). In the United States, the standard billiard table is 4 feet 6 inches by 9 feet. This is a ratio of 1 : 2. The same ratio holds for carom and snooker tables, which are 5 feet by 10 feet.

Ratios

A **ratio** is the comparison of two quantities that have the same units. A ratio can be written in three different ways:

1. As a fraction $\dfrac{2}{3}$

2. As two numbers separated by a colon (:)   2 : 3

3. As two numbers separated by the word *to*   2 to 3

Although units, such as hours, miles, or dollars, are written as part of a rate, units are not written as part of a ratio.

According to the most recent census, there are 50 million married women in the United States, and 30 million of these women work in the labor force. The ratio of the number of married women who are employed in the labor force to the total number of married women in the country is calculated below. Note that the ratio is written in simplest form.

$$\frac{30{,}000{,}000}{50{,}000{,}000} = \frac{3}{5} \quad \text{or} \quad 3:5 \quad \text{or} \quad 3 \text{ to } 5$$

The ratio 3 to 5 tells us that 3 out of every 5 married women in the United States are part of the labor force.

Given that 30 million of the 50 million married women in the country work in the labor force, we can calculate the number of married women who do not work in the labor force.

$$50 \text{ million} - 30 \text{ million} = 20 \text{ million}$$

The ratio of the number of married women who are not in the labor force to the number of married women who are is:

$$\frac{20{,}000{,}000}{30{,}000{,}000} = \frac{2}{3} \quad \text{or} \quad 2:3 \quad \text{or} \quad 2 \text{ to } 3$$

The ratio 2 to 3 tells us that for every 2 married women who are not in the labor force, there are 3 married women who are in the labor force.

EXAMPLE 5 ■ **Determine a Ratio in Simplest Form**

A survey revealed that, on average, eighth-graders watch approximately 21 hours of television each week. Find the ratio, as a fraction in simplest form, of the number of hours spent watching television to the total number of hours in a week.

Solution

A ratio is the comparison of two quantities with the same units. In this example we are given both hours and weeks. We must first convert 1 week to hours.

$$\frac{24 \text{ hours}}{1 \text{ day}} \cdot 7 \text{ days} = (24 \text{ hours})(7) = 168 \text{ hours}$$

Write in simplest form the ratio of the number of hours spent watching television to the number of hours in 1 week.

$$\frac{21 \text{ hours}}{1 \text{ week}} = \frac{21 \text{ hours}}{168 \text{ hours}} = \frac{21}{168} = \frac{1}{8}$$

The ratio is $\frac{1}{8}$.

CHECK YOUR PROGRESS 5

a. According to the National Low Income Housing Coalition, a minimum-wage worker ($5.15 per hour) living in New Jersey would have to work 120 hours per week to afford the rent on an average two-bedroom apartment and be within the federal standard of 30% of income for housing. Find the ratio, as a fraction in simplest form, of the number of hours a minimum-wage worker would spend working per week to the total number of hours in a week.

b. Although a minimum-wage worker in New Jersey would have to work 120 hours per week to afford a two-bedroom rental, the national average is 60 hours of work per week. For this "average" worker, find the ratio, written using the word *to*, of the number of hours per week spent working to the number of hours spent not working.

Solution   See page S12.

A **unit ratio** is a ratio in which the number in the denominator is 1. One situation in which a unit ratio is used is student–faculty ratios. The table below shows the numbers of full-time men and women undergraduates, as well as the numbers of full-time faculty, at two universities in the Pacific 10. Use this table for Example 6 and Check Your Progress 6. (*Source: Barron's Profile of American Colleges,* 26th edition, c. 2005)

University	Men	Women	Faculty
Oregon State University	7509	6478	1352
University of Oregon	6742	7710	798

EXAMPLE 6 ■ Determine a Unit Ratio

 Calculate the student–faculty ratio at Oregon State University. Round to the nearest whole number. Write the ratio using the word *to*.

Solution

Add the number of male undergraduates and the number of female undergraduates to determine the total number of students.

$$7509 + 6478 = 13{,}987$$

Write the ratio of the total number of students to the number of faculty. Divide the numerator and denominator by the denominator. Then round the numerator to the nearest whole number.

$$\frac{13{,}987}{1352} \approx \frac{10.3454}{1} \approx \frac{10}{1}$$

The ratio is approximately 10 to 1.

CHECK YOUR PROGRESS 6

 Calculate the student–faculty ratio at the University of Oregon. Round to the nearest whole number. Write the ratio using the word *to*.

Solution   *See page S12.*

MathMatters   **Scale Model Buildings**

George E. Slye of Tuftonboro, New Hampshire has built scale models of more than 100 of the best-known buildings in the United States and Canada. From photographs, floor plans, roof plans, and architectural drawings, Slye has constructed a replica of each building using a scale of 1 inch per 200 feet.

The buildings and landmarks are grouped together on an 8-foot-by-8-foot base as if they were all erected in a single city. Slye's city includes such well-known landmarks as New York's Empire State Building, Chicago's Sears Tower, Boston's John Hancock Tower, Seattle's Space Needle, and San Francisco's Golden Gate Bridge.

Proportions

historical note

Proportions were studied by the earliest mathematicians. Clay tablets uncovered by archeologists show evidence of the use of proportions in Egyptian and Babylonian cultures dating from 1800 B.C. ■

Thus far in this section we have discussed rates and ratios. Now that you have an understanding of rates and ratios, you are ready to work with proportions. A **proportion** is an equation that states the equality of two rates or ratios. The following are examples of proportions.

$$\frac{250 \text{ miles}}{5 \text{ hours}} = \frac{50 \text{ miles}}{1 \text{ hour}} \qquad \frac{3}{6} = \frac{1}{2}$$

The first example above is the equality of two rates. Note that the units in the numerators (miles) are the same and the units in the denominators (hours) are the same. The second example is the equality of two ratios. Remember that units are not written as part of a ratio.

The definition of a proportion can be stated as follows: If $\frac{a}{b}$ and $\frac{c}{d}$ are equal ratios or rates, then $\frac{a}{b} = \frac{c}{d}$ is a proportion.

Each of the four members in a proportion is called a **term.** Each term is numbered as shown below.

First term ⟶ $\frac{a}{b} = \frac{c}{d}$ ⟵ Third term

Second term ⟶   ⟵ Fourth term

The second and third terms of the proportion are called the **means** and the first and fourth terms are called the **extremes.**

If we multiply both sides of the proportion by the product of the denominators, we obtain the following result.

$$\frac{a}{b} = \frac{c}{d}$$

$$bd\left(\frac{a}{b}\right) = bd\left(\frac{c}{d}\right)$$

$$ad = bc$$

Note that ad is the product of the extremes and bc is the product of the means. In any proportion, the product of the means equals the product of the extremes. This is sometimes phrased, "the cross products are equal."

In the proportion $\frac{3}{4} = \frac{9}{12}$, the cross products are equal.

$\frac{3}{4} \Large\supset\!\!\subset \normalsize \frac{9}{12}$ ⟶ $4 \cdot 9 = 36$ ⟵ Product of the means

         ⟶ $3 \cdot 12 = 36$ ⟵ Product of the extremes

QUESTION   For the proportion $\frac{5}{8} = \frac{10}{16}$, **a.** name the first and third terms, **b.** write the product of the means, and **c.** write the product of the extremes.

ANSWER   **a.** The first term is 5. The third term is 10. **b.** The product of the means is $8(10) = 80$. **c.** The product of the extremes is $5(16) = 80$.

Sometimes one of the terms in a proportion is unknown. In this case, it is necessary to solve the proportion for the unknown number. The **cross-products method,** which is based on the fact that the product of the means equals the product of the extremes, can be used to solve the proportion.

Cross-Products Method of Solving a Proportion

If $\dfrac{a}{b} = \dfrac{c}{d}$, then $ad = bc$.

EXAMPLE 7 ■ Solve a Proportion

Solve: $\dfrac{8}{5} = \dfrac{n}{6}$

Solution

Use the cross-products method of solving a proportion: the product of the means equals the product of the extremes. Then solve the resulting equation for n.

$$\frac{8}{5} = \frac{n}{6}$$

$8 \cdot 6 = 5 \cdot n$ • The cross-products are equal.

$48 = 5n$

$\dfrac{48}{5} = \dfrac{5n}{5}$ • Divide each side by 5.

$9.6 = n$

The solution is 9.6.

TAKE NOTE

Be sure to check the solution.

$$\frac{8}{5} = \frac{9.6}{6}$$
$$8 \cdot 6 = 5 \cdot 9.6$$
$$48 = 48$$

The solution checks.

CHECK YOUR PROGRESS 7 Solve: $\dfrac{42}{x} = \dfrac{5}{8}$

Solution *See page S12.*

Proportions are useful for solving a wide variety of application problems. Remember that when we use the given information to write a proportion involving two rates, the units in the numerators of the rates need to be the same and the units in the denominators of the rates need to be the same. It is helpful to keep in mind that when we write a proportion, we are stating that two rates or ratios are equal.

✔ TAKE NOTE

We have written a proportion with the unit "miles" in the numerators and the unit "gallons" in the denominators. It would also be correct to have "gallons" in the numerators and "miles" in the denominators.

EXAMPLE 8 ■ Solve an Application Using a Proportion

If you travel 290 miles in your car on 15 gallons of gasoline, how far can you travel in your car on 12 gallons of gasoline under similar driving conditions?

Solution

Let $x =$ the unknown number of miles.
Write a proportion and then solve the proportion for x.

$$\frac{290 \text{ miles}}{15 \text{ gallons}} = \frac{x \text{ miles}}{12 \text{ gallons}}$$

$$\frac{290}{15} = \frac{x}{12}$$

$$290 \cdot 12 = 15 \cdot x$$

$$3480 = 15x$$

$$232 = x$$

You can travel 232 miles on 12 gallons of gasoline.

CHECK YOUR PROGRESS 8   On a map, a distance of 2 centimeters represents 15 kilometers. What is the distance between two cities that are 7 centimeters apart on the map?

Solution   *See page S12.*

EXAMPLE 9 ■ Solve an Application Using a Proportion

The table below shows three of the universities in the Big Ten Conference and their student–faculty ratios. (*Source: Barron's Profile of American Colleges*, 26th edition, c. 2005) There are approximately 31,100 full-time undergraduate students at Michigan State University. Approximate the number of faculty at Michigan State University.

University	Student–Faculty Ratio
Michigan State University	13 to 1
University of Illinois	15 to 1
University of Iowa	11 to 1

Solution

Let F = the number of faculty members.

Write a proportion and then solve the proportion for F.

$$\frac{13 \text{ students}}{1 \text{ faculty}} = \frac{31{,}100 \text{ students}}{F \text{ faculty}}$$

$$13 \cdot F = 1(31{,}100)$$

$$13F = 31{,}100$$

$$\frac{13F}{13} = \frac{31{,}100}{13}$$

$$F \approx 2392$$

There are approximately 2392 faculty members at Michigan State University.

CHECK YOUR PROGRESS 9   The profits of a firm are shared by its two partners in the ratio $7:5$. If the partner receiving the larger amount of this year's profits receives \$28,000, what amount does the other partner receive?

Solution   *See page S12.*

EXAMPLE 10 ■ Solve an Application Using a Proportion

In the United States, the average annual number of deaths per million people aged 5 to 34 from asthma is 3.5. Approximately how many people aged 5 to 34 die from asthma each year in this country? Use a figure of 150,000,000 for the number of U.S. residents who are 5 to 34 years old. (*Source:* National Center for Health Statistics)

Solution

Let d = the number of people aged 5 to 34 who die each year from asthma in the United States. Write and solve a proportion. One rate is 3.5 deaths per million people.

$$\frac{3.5 \text{ deaths}}{1{,}000{,}000 \text{ people}} = \frac{d \text{ deaths}}{150{,}000{,}000 \text{ people}}$$

$$3.5(150{,}000{,}000) = 1{,}000{,}000 \cdot d$$

$$525{,}000{,}000 = 1{,}000{,}000d$$

$$\frac{525{,}000{,}000}{1{,}000{,}000} = \frac{1{,}000{,}000d}{1{,}000{,}000}$$

$$525 = d$$

In the United States, approximately 525 people aged 5 to 34 die each year from asthma.

CHECK YOUR PROGRESS 10

New York City has the highest death rate from asthma in the United States. The average annual number of deaths per million people aged 5 to 34 from asthma in New York City is 10.1. Approximately how many people aged 5 to 34 die from asthma each year in New York City? Use a figure of 4,000,000 for the number of residents 5 to 34 years old in New York City. Round to the nearest whole number. (*Source:* National Center for Health Statistics)

Solution   *See page S12.*

MathMatters   Scale Models for Special Effects

Special-effects artists use scale models to create dinosaurs, exploding spaceships, and aliens. A scale model is produced by using ratios and proportions to determine the size of the scale model.

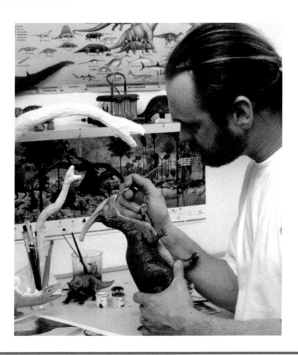

Investigation

Earned Run Average

One measure of a pitcher's success is earned run average. **Earned run average (ERA)** is the number of earned runs a pitcher gives up for every nine innings pitched. The definition of an earned run is somewhat complicated, but basically an earned run is a run that is scored as a result of hits and base running that involves no errors on the part of the pitcher's team. If the opposing team scores a run on an error (for example, a fly ball that should be caught in the outfield is fumbled), then that run is not an earned run.

A proportion is used to calculate a pitcher's ERA. Remember that the statistic involves the number of earned runs per *nine innings*. The answer is always rounded to the nearest hundredth. Here is an example.

During the 2005 baseball season, Pedro Martinez gave up 68 earned runs and pitched 217 innings for the New York Mets. To calculate Pedro Martinez's ERA, let x equal the number of earned runs for every nine innings pitched. Write a proportion and then solve it for x.

$$\frac{68 \text{ earned runs}}{217 \text{ innings}} = \frac{x}{9 \text{ innings}}$$

$$68 \cdot 9 = 217 \cdot x$$

$$612 = 217x$$

$$\frac{612}{217} = \frac{217x}{217}$$

$$2.82 \approx x$$

Pedro Martinez's ERA for the 2005 season was 2.82.

Investigation Exercises

1. In 1979, his rookie year, Jeff Reardon pitched 21 innings for the New York Mets and gave up four earned runs. Calculate Reardon's ERA for 1979.

2. Roger Clemens's first year with the Boston Red Sox was 1984. During that season, he pitched 133.1 innings and gave up 64 earned runs. Calculate Clemens's ERA for 1984.

3. During the 2003 baseball season, Ben Sheets of the Milwaukee Brewers pitched 220.2 innings and gave up 109 earned runs. During the 2004 season, he gave up 71 earned runs and pitched 237.0 innings. During which season was his ERA lower? How much lower?

4. In 1987, Nolan Ryan had the lowest ERA of any pitcher in the major leagues. He gave up 65 earned runs and pitched 211.2 innings for the Houston Astros. Calculate Ryan's ERA for 1987.

5. Find the necessary statistics for a pitcher on your "home team," and calculate that pitcher's ERA.

Earned Run Average Leaders		
Year	Player, club	ERA
	National League	
1990	Danny Darwin, Houston	2.21
1991	Dennis Martinez, Montreal	2.39
1992	Bill Swift, San Francisco	2.08
1993	Greg Maddux, Atlanta	2.36
1994	Greg Maddux, Atlanta	1.56
1995	Greg Maddux, Atlanta	1.63
1996	Kevin Brown, Florida	1.89
1997	Pedro Martinez, Montreal	1.90
1998	Greg Maddux, Atlanta	2.22
1999	Randy Johnson, Arizona	2.48
2000	Kevin Brown, Los Angeles	2.58
2001	Randy Johnson, Arizona	2.49
2002	Randy Johnson, Arizona	2.32
2003	Jason Schmidt, San Francisco	2.34
2004	Jake Peavy, San Diego	2.27
2005	Roger Clemens, Houston	1.87
	American League	
1990	Roger Clemens, Boston	1.93
1991	Roger Clemens, Boston	2.62
1992	Roger Clemens, Boston	2.41
1993	Kevin Appier, Kansas City	2.56
1994	Steve Ontiveros, Oakland	2.65
1995	Randy Johnson, Seattle	2.48
1996	Juan Guzman, Toronto	2.93
1997	Roger Clemens, Toronto	2.05
1998	Roger Clemens, Toronto	2.65
1999	Pedro Martinez, Boston	2.07
2000	Pedro Martinez, Boston	1.74
2001	Freddy Garcia, Seattle	3.05
2002	Pedro Martinez, Boston	2.26
2003	Pedro Martinez, Boston	2.22
2004	Johan Santana, Minnesota	2.61
2005	Kevin Millwood, Cleveland	2.86

Exercise Set 3.2

1. Provide two examples of situations in which unit rates are used.

2. Explain why unit rates are used to describe situations involving units such as miles per gallon.

3. What is the purpose of exchange rates in international trade?

4. Provide two examples of situations in which ratios are used.

5. Explain why ratios are used to describe situations involving information such as student–teacher ratios.

6. What does the phrase "the cross products are equal" mean?

7. Explain why the product of the means in a proportion equals the product of the extremes.

In Exercises 8–13, write the expression as a unit rate.

8. 582 miles in 12 hours

9. 138 miles on 6 gallons of gasoline

10. 544 words typed in 8 minutes

11. 100 meters in 8 seconds

12. $9100 for 350 shares of stock

13. 1000 square feet of wall covered with 2.5 gallons of paint

14. A rate of 288 miles in 6 hours is closest to which unite rate?
   a. 100 mph   b. 50 mph   c. 200 mph   d. 25 mph

15. A rate of $123.75 in 11 hours is closest to which unite rate?
   a. $20 per hour   b. $1 per hour   c. $10 per hour
   d. $120 per hour

Solve Exercises 16–21.

16. **Wages** A machinist earns $490 for working a 35-hour week. What is the machinist's hourly rate of pay?

17. **Space Vehicles** Each of the Space Shuttle's solid rocket motors burns 680,400 kilograms of propellant in 2.5 minutes. How much propellant does each motor burn in 1 minute?

18. **Photography** During filming, an IMAX camera uses 65-mm film at a rate of 5.6 feet per second.
   a. At what rate per minute does the camera go through film?
   b. How quickly does the camera use a 500-foot roll of 65-mm film? Round to the nearest second.

19. **Consumerism** Which is the more economical purchase, a 32-ounce jar of mayonnaise for $2.79 or a 24-ounce jar of mayonnaise for $2.09?

20. **Consumerism** Which is the more economical purchase, an 18-ounce box of corn flakes for $2.89 or a 24-ounce box of corn flakes for $3.89?

21. **Wages** You have a choice of receiving a wage of $34,000 per year, $2840 per month, $650 per week, or $16.50 per hour. Which pay choice would you take? Assume a 40-hour work week and 52 weeks of work per year.

22. **Baseball** Baseball statisticians calculate a hitter's at-bats per home run by dividing the number of times the player has been at bat by the number of home runs the player has hit.
   a. Calculate the at-bats per home run ratio for each player in the table on the following page. Round to the nearest tenth.
   b. Which player has the lowest rate of at-bats per home run? Which player has the second lowest rate?

Babe Ruth

c. Why is this rate used for comparison rather than the number of home runs a player has hit?

Players with 50 or More Home Runs per Season

Year	Baseball Player	Number of Times at Bat	Number of Home Runs Hit	Number of At-Bats per Home Run
1921	Babe Ruth	540	59	_____
1927	Babe Ruth	540	60	_____
1930	Hack Wilson	585	56	_____
1932	Jimmie Foxx	585	58	_____
1938	Hank Greenberg	556	58	_____
1961	Roger Maris	590	61	_____
1961	Mickey Mantle	514	54	_____
1964	Willie Mays	558	52	_____
1977	George Foster	615	52	_____
1998	Mark McGwire	509	70	_____
1998	Sammy Sosa	643	66	_____
2001	Barry Bonds	476	73	_____
2002	Alex Rodriguez	624	57	_____

23. **Population Density** The table below shows the populations and areas of three countries. The population density of a country is the number of people per square mile.

a. Which country has the lowest population density?

b. How many more people per square mile are there in India than in the United States? Round to the nearest whole number.

Country	Population	Area (in square miles)
Australia	20,090,000	2,938,000
India	1,080,264,000	1,146,000
United States	295,734,000	3,535,000

24. **E-mail** Forrester Research, Inc., compiled the following estimates on consumer use of e-mail in the United States.

a. Complete the last column of the table on the following page by calculating the estimated number of messages per day that each user receives. Round to the nearest tenth.

b. The predicted number of messages per person per day in 2005 is how many times the estimated number in 1993?

Year	Number of Users (in millions)	Messages per Day (in millions)	Messages per Person per Day
1993	8	17	_____
1997	55	150	_____
2001	135	500	_____
2005	170	5000	_____

Exchange Rates The table below shows the exchange rates per U.S. dollar for four foreign countries on December 2, 2005. Use this table for Exercises 25 to 28.

Exchange Rates per U.S. Dollar	
Australian Dollar	1.3359
Danish Krone	6.3652
Indian Rupee	46.151
Mexican Peso	10.472

25. How many Danish kroner are equivalent to $10,000?

26. Find the number of Indian rupees that would be exchanged for $45,000.

27. Find the cost, in Mexican pesos, of an order of American computer hardware costing $38,000.

28. Calculate the cost, in Australian dollars, of an American car costing $29,000.

Health Researchers at the Centers for Disease Control and Prevention estimate that 5 million young people living today will die of tobacco-related diseases. Almost one-third of children who become regular smokers will die of a smoking-related illness such as heart disease or lung cancer. The table below shows, for eight states in our nation, the numbers of children under 18 who are expected to become smokers and the numbers who are projected to die of smoking-related illnesses. Use the table for Exercises 29 and 30.

State	Projected Number of Smokers	Projected Number of Deaths
Alabama	260,639	83,404
Alaska	56,246	17,999
Arizona	307,864	98,516
Arkansas	155,690	49,821
California	1,446,550	462,896
Colorado	271,694	86,942
Connecticut	175,501	56,160
Delaware	51,806	16,578

29. Find the ratio of the projected number of smokers in each state listed to the projected number of deaths. Round to the nearest thousandth. Write the ratio using the word *to*.

30. a. Did the researchers calculate different probabilities of death from smoking-related illnesses for each state?

    b. If the projected number of smokers in Florida is 928,464, what would you expect the researchers to project as the number of deaths from smoking-related illnesses in Florida?

Student–Faculty Ratios The table below shows the numbers of full-time men and women undergraduates and the numbers of full-time faculty at universities in the Big East. Use this table for Exercises 31 to 34. Round ratios to the nearest whole number. (*Source: Barron's Profile of American Colleges*, 26th edition, c. 2005)

University	Men	Women	Faculty
Boston College	6292	9756	1283
Georgetown University	2940	3386	655
Syracuse University	4722	6024	815
University of Connecticut	6762	7489	842
West Virginia University	8878	7665	1289

31. Calculate the student–faculty ratio at Syracuse University. Write the ratio using a colon and using the word *to*. What does this ratio mean?

32. Which school listed has the lowest student–faculty ratio?

33. Which school listed has the highest student–faculty ratio?

34. Which schools listed have the same student–faculty ratio?

35. Debt–Equity Ratio A bank uses the ratio of a borrower's total monthly debt to total monthly income to determine eligibility for a loan. This ratio is called the debt–equity ratio. First National Bank requires that a borrower have a debt–equity ratio that is less than $\frac{2}{5}$. Would the homeowner whose monthly income and debt are given below qualify for a loan using these standards?

Monthly Income (in dollars)		Monthly Debt (in dollars)	
Salary	3400	Mortgage	1800
Interest	83	Property tax	104
Rent	640	Insurance	27
Dividends	34	Credit cards	354
		Car loan	199

Solve Exercises 36–51. Round to the nearest hundredth.

36. $\dfrac{3}{8} = \dfrac{x}{12}$

37. $\dfrac{3}{y} = \dfrac{7}{40}$

38. $\dfrac{7}{12} = \dfrac{25}{d}$

39. $\dfrac{16}{d} = \dfrac{25}{40}$

40. $\dfrac{15}{45} = \dfrac{72}{c}$

41. $\dfrac{120}{c} = \dfrac{144}{25}$

42. $\dfrac{65}{20} = \dfrac{14}{a}$

43. $\dfrac{4}{a} = \dfrac{9}{5}$

44. $\dfrac{0.5}{2.3} = \dfrac{b}{20}$

45. $\dfrac{1.2}{2.8} = \dfrac{b}{32}$

46. $\dfrac{0.7}{1.2} = \dfrac{6.4}{x}$

47. $\dfrac{2.5}{0.6} = \dfrac{165}{x}$

48. $\dfrac{y}{6.25} = \dfrac{16}{87}$

49. $\dfrac{y}{2.54} = \dfrac{132}{640}$

50. $\dfrac{1.2}{0.44} = \dfrac{m}{14.2}$

51. $\dfrac{12.5}{m} = \dfrac{102}{55}$

52. Given the proportion $\dfrac{x}{34} = \dfrac{8}{11}$, state whether $x > 8$ or $x < 8$.

53. Given the proportion $\dfrac{7}{12} = \dfrac{3}{t}$, state whether $t > 3$ or $t < 3$.

Solve Exercises 54–65.

54. **Gravity** The ratio of weight on the moon to weight on Earth is 1 : 6. How much would a 174-pound person weigh on the moon?

55. **Management** A management consulting firm recommends that the ratio of middle-management salaries to management trainee salaries be 5 : 4. Using this recommendation, what is the annual middle-management salary if the annual management trainee salary is $36,000?

56. **Gardening** A gardening crew uses 2 pounds of fertilizer for every 100 square feet of lawn. At this rate, how many pounds of fertilizer does the crew use on a lawn that measures 2500 square feet?

57. **Medication** The dosage of a cold medication is 2 milligrams for every 80 pounds of body weight. How many milligrams of this medication are required for a person who weighs 220 pounds?

58. **Fuel Consumption** If your car can travel 70.5 miles on 3 gallons of gasoline, how far can the car travel on 14 gallons of gasoline?

59. **Scale Drawings** The scale on the architectural plans for a new house is 1 inch equals 3 feet. Find the length and width of a room that measures 5 inches by 8 inches on the drawing.

60. **Scale Drawings** The scale on a map is 1.25 inches equals 10 miles. Find the distance between Carlsbad and Del Mar, which are 2 inches apart on the map.

61. **Elections** A pre-election survey showed that two out of every three eligible voters would cast ballots in the county election. There are 240,000 eligible voters in the county. How many people are expected to vote in the election?

62. **Interior Decorating** A paint manufacturer suggests using 1 gallon of paint for every 400 square feet of wall. At this rate, how many gallons of paint would be required to paint a room that has 1400 square feet of wall?

63. **Mileage** Amanda Chicopee bought a new car and drove 7000 miles in the first four months. At the same rate, how many miles will Amanda drive in 3 years?

64. **Lotteries** Three people put their money together to buy lottery tickets. The first person put in $25, the second person put in $30, and the third person put in $35. One of their tickets was a winning ticket. If they won $4.5 million, what was the first person's share of the winnings? Assume the winnings are distributed in proportion to the amount each person contributed.

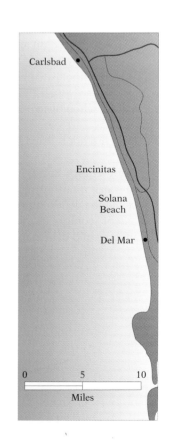

65. **Nutrition** A pancake 4 inches in diameter contains 5 grams of fat. How many grams of fat are in a pancake 6 inches in diameter? Explain how you arrived at your answer.

Extensions

CRITICAL THINKING

66. **Homicide Rates** For U.S. cities with populations over 100,000, those with the highest homicide rates in a recent year are listed below, along with the number of murders in each city and the homicide rate per 100,000 people. (*Source:* FBI Uniform Crime Reports)

  a. Explain what the statistics for Atlanta mean.

  b. What information not provided here was needed to calculate the homicide rates given for each city?

City	Total Number of Murders	Homicide Rate per 100,000 People
Washington	397	73
New Orleans	351	72
Richmond, VA	112	55
Atlanta	196	47
Baltimore	328	46

In Exercises 67 and 68, assume each denominator is a nonzero real number.

67. Determine whether the statement is true or false.
  a. The quotient $a \div b$ is a ratio.   **b.** If $\dfrac{a}{b} = \dfrac{c}{d}$, then $\dfrac{b}{a} = \dfrac{d}{c}$.
  c. If $\dfrac{a}{b} = \dfrac{c}{d}$, then $\dfrac{a}{c} = \dfrac{b}{d}$.   **d.** If $\dfrac{a}{b} = \dfrac{c}{d}$, then $\dfrac{a}{d} = \dfrac{c}{b}$.

68. If $\dfrac{a}{b} = \dfrac{c}{d}$, show that $\dfrac{a + b}{b} = \dfrac{c + d}{d}$.

COOPERATIVE LEARNING

69. **Advertising** Advertising rates for most daytime television programs are lower than for shows aired during prime time. Advertising rates for widely popular sporting events are higher than for contests in which fewer people are interested. In 2006, the cost of a 30-second advertisement during the Super Bowl was approximately $2.5 million! The rates vary because they are based on cost per thousand viewers.

  a. Explain why different time slots and different shows demand different advertising rates.

  b. Why are advertisers concerned about the gender and age of the audience watching a particular program?

  c. How do Nielsen ratings help determine what advertising rates can be charged for an ad airing during a particular program?

70. 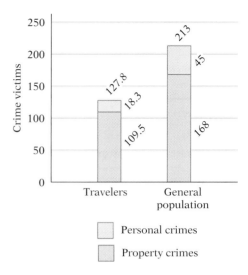 **Crime Rates** According to a recent study, the crime rate against travelers in the United States is lower than that against the general population. The article reporting this study included a bar graph similar to the one at the right.

a. Why are the figures reported based on crime victims per 1000 adults per year?

b. Use the given figures to write the proportion

$$\frac{\text{personal crimes against travelers}}{\text{property crimes against travelers}} = $$

$$\frac{\text{personal crimes against general population}}{\text{property crimes against general population}}$$

Is the proportion true?

c. Why might the crime rate against travelers be lower than that against the general population?

Crime Victims per 1000 Adults per Year

Source: Travel Industry Association of America

EXPLORATIONS

71. The House of Representatives The 50 states in the United States are listed below by their USPS two-letter abbreviations. The number following each state is the figure provided by the U.S. Bureau of the Census for the population of that state in 2000. Figures are in millions and are rounded to the nearest hundred thousand.

Populations of the 50 States in the United States, 2000 Census (in millions)				
AL 4.4	HI 1.2	MA 6.3	NM 1.8	SD 0.8
AK 0.6	ID 1.3	MI 9.9	NY 19.0	TN 5.7
AZ 5.1	IL 12.4	MN 4.9	NC 8.0	TX 20.9
AR 2.7	IN 6.1	MS 2.8	ND 0.6	UT 2.2
CA 33.9	IA 2.9	MO 5.6	OH 11.4	VT 0.6
CO 4.3	KS 2.7	MT 0.9	OK 3.5	VA 7.1
CT 3.4	KY 4.0	NE 1.7	OR 3.4	WA 5.9
DE 0.8	LA 4.5	NV 2.0	PA 12.3	WV 1.8
FL 16.0	ME 1.3	NH 1.2	RI 1.0	WI 5.4
GA 8.2	MD 5.3	NJ 8.4	SC 4.0	WY 0.5

The House of Representatives has a total of 435 members. These members represent the 50 states in proportion to each state's population. As stated in Amendment XIV, Section 2, of the Constitution of the United States, "Representatives shall be apportioned among the several states according to their respective numbers, counting the whole number of persons in each state." Based on the populations for the states given above, determine how many representatives each state should elect to Congress.

Compare your list against the actual number of representatives each state has. These numbers can be found in most almanacs.

Percents

An understanding of percent is vital to comprehending the events that take place in our world today. We are constantly confronted with phrases such as "unemployment of 5%," "annual inflation of 7%," "6% increase in fuel prices," "25% of the daily minimum requirement," and "increase in tuition and fees of 10%."

Percent means "for every 100." Therefore, unemployment of 5% means that 5 out of every 100 people are unemployed. An increase in tuition of 10% means that tuition has gone up $10 for every $100 it cost previously.

▼ **point of interest**

Of all the errors made on federal income tax returns, the four most common errors account for 76% of the mistakes. These errors include an omitted entry (30.7%), an incorrect entry (19.1%), an error in mathematics (17.4%), and an entry on the wrong line (8.8%).

QUESTION *When adults were asked to name their favorite cookie, 52% said chocolate chip. What does this statistic mean? (Source: WEAREVER)*

A percent is a ratio of a number to 100. Thus $\frac{1}{100} = 1\%$, $\frac{50}{100} = 50\%$, and $\frac{99}{100} = 99\%$. Because $1\% = \frac{1}{100}$ and $\frac{1}{100} = 0.01$, we can also write 1% as 0.01.

$$1\% = \frac{1}{100} = \mathbf{0.01}$$

The equivalence $1\% = 0.01$ is used to write a percent as a decimal or to write a decimal as a percent.

To write 17% as a decimal:

$$17\% = 17(1\%) = 17(0.01) = 0.17$$

Note that this is the same as removing the percent sign and moving the decimal point two places to the left.

To write 0.17 as a percent:

$$0.17 = 17(0.01) = 17(1\%) = 17\%$$

Note that this is the same as moving the decimal point two places to the right and writing a percent sign to the right of the number.

ANSWER *This statistic means that 52 out of every 100 people surveyed responded that their favorite cookie was chocolate chip. (In the same survey, the following responses were also given: oatmeal raisin, 10%; peanut butter, 9%; oatmeal, 7%; sugar, 4%; molasses, 4%; chocolate chip oatmeal, 3%.)*

EXAMPLE 1 ▣ Write a Percent as a Decimal

Write each percent as a decimal.

a. 24%   **b.** 183%   **c.** 6.5%   **d.** 0.9%

Solution
To write a percent as a decimal, remove the percent sign and move the decimal point two places to the left.

a. 24% $= 0.24$

b. 183% $= 1.83$

c. 6.5% $= 0.065$

d. 0.9% $= 0.009$

CHECK YOUR PROGRESS 1   Write each percent as a decimal.

a. 74%   **b.** 152%   **c.** 8.3%   **d.** 0.6%

Solution   *See page S12.*

EXAMPLE 2 ▣ Write a Decimal as a Percent

Write each decimal as a percent.

a. 0.62   **b.** 1.5   **c.** 0.059   **d.** 0.008

Solution
To write a decimal as a percent, move the decimal point two places to the right and write a percent sign to the right of the number.

a. 0.62 $= 62\%$

b. 1.5 $= 150\%$

c. 0.059 $= 5.9\%$

d. 0.008 $= 0.8\%$

CHECK YOUR PROGRESS 2   Write each decimal as a percent.

a. 0.3   **b.** 1.65   **c.** 0.072   **d.** 0.004

Solution   *See page S12.*

The equivalence $1\% = \frac{1}{100}$ is used to write a percent as a fraction. To write 16% as a fraction:

$$16\% = 16(1\%) = 16\left(\frac{1}{100}\right) = \frac{16}{100} = \frac{4}{25}$$

Note that this is the same as removing the percent sign and multiplying by $\frac{1}{100}$. The fraction is written in simplest form.

EXAMPLE 3 ■ Write a Percent as a Fraction

Write each percent as a fraction.

a. 25%   **b.** 120%   **c.** 7.5%   **d.** $33\frac{1}{3}\%$

Solution

To write a percent as a fraction, remove the percent sign and multiply by $\frac{1}{100}$. Then write the fraction in simplest form.

a. $25\% = 25\left(\dfrac{1}{100}\right) = \dfrac{25}{100} = \dfrac{1}{4}$

b. $120\% = 120\left(\dfrac{1}{100}\right) = \dfrac{120}{100} = 1\dfrac{20}{100} = 1\dfrac{1}{5}$

c. $7.5\% = 7.5\left(\dfrac{1}{100}\right) = \dfrac{7.5}{100} = \dfrac{75}{1000} = \dfrac{3}{40}$

d. $33\dfrac{1}{3}\% = \dfrac{100}{3}\% = \dfrac{100}{3}\left(\dfrac{1}{100}\right) = \dfrac{1}{3}$

CHECK YOUR PROGRESS 3   Write each percent as a fraction.

a. 8%   **b.** 180%   **c.** 2.5%   **d.** $66\frac{2}{3}\%$

Solution   *See page S13.*

> **✔ TAKE NOTE**
>
> To write a fraction as a decimal, divide the number in the numerator by the number in the denominator. For example,
>
> $$\frac{4}{5} = 4 \div 5 = 0.8$$

To write a fraction as a percent, first write the fraction as a decimal. Then write the decimal as a percent.

EXAMPLE 4 ■ Write a Fraction as a Percent

Write each fraction as a percent.

a. $\dfrac{3}{4}$ **b.** $\dfrac{5}{8}$ **c.** $\dfrac{1}{6}$ **d.** $1\dfrac{1}{2}$

Solution

To write a fraction as a percent, write the fraction as a decimal. Then write the decimal as a percent.

a. $\dfrac{3}{4} = 0.75 = 75\%$

b. $\dfrac{5}{8} = 0.625 = 62.5\%$

c. $\dfrac{1}{6} = 0.16\overline{6} = 16.\overline{6}\%$

d. $1\dfrac{1}{2} = 1.5 = 150\%$

CHECK YOUR PROGRESS 4   Write each fraction as a percent.

a. $\dfrac{1}{4}$ **b.** $\dfrac{3}{8}$ **c.** $\dfrac{5}{6}$ **d.** $1\dfrac{2}{3}$

Solution   *See page S13.*

Math Matters   College Graduates' Job Expectations

The table below shows the expectations of a recent class of college graduates. (*Source:* Yankelovich Partners for Phoenix Home Life Mutual Insurance)

	Men	Women
Expect a job by graduation	31%	23%
Expect a job within 6 months	32%	23%
Expect a starting pay of $30,000 or more	55%	43%
Expect to be richer than their parents	64%	59%

Percent Problems: The Proportion Method

Finding the solution of an application problem involving percent generally requires writing and solving an equation. Two methods of writing the equation will be developed in this section—the *proportion method* and the *basic percent equation*. We will present the proportion method first.

The proportion method of solving a percent problem is based on writing two ratios. One ratio is the percent ratio, written $\dfrac{\text{Percent}}{100}$. The second ratio is the amount-to-base ratio, written $\dfrac{\text{amount}}{\text{base}}$. These two ratios form the proportion used to solve percent problems.

> **The Proportion Used to Solve Percent Problems**
>
> $$\frac{\text{Percent}}{100} = \frac{\text{amount}}{\text{base}}$$

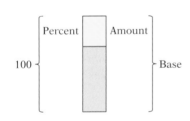

Diagram of the Proportion Method of Solving Percent Problems

The proportion method can be illustrated by a diagram. The rectangle at the left is divided into two parts. On the left, the whole rectangle is represented by 100 and the yellow part by *percent*. On the right, the whole rectangle is represented by the *base* and the yellow part by the *amount*. The ratio of percent to 100 is equal to the ratio of the amount to the base.

When solving a percent problem, first identify the percent, the base, and the amount. It is helpful to know that the base usually follows the phrase "percent of."

QUESTION *In the statement "15% of 40 is 6," which number is the percent? Which number is the base? Which number is the amount?*

EXAMPLE 5 ■ **Solve a Percent Problem for the Base Using the Proportion Method**

The average size of a house in 2003 was 2137 square feet. This is approximately 130% of the average size of a house in 1979. What was the average size of a house in 1979? Round to the nearest whole number.

Solution

We want to answer the question "130% of what number is 2137?" Write and solve a proportion. The percent is 130%. The amount is 2137. The base is the average size of a house in 1979.

$$\frac{\text{Percent}}{100} = \frac{\text{amount}}{\text{base}}$$

$$\frac{130}{100} = \frac{2137}{B}$$ • The percent is **130**. The amount is **2137**.

$$130 \cdot B = 100(2137)$$ • The cross products are equal.

$$130B = 213{,}700$$

$$\frac{130B}{130} = \frac{213{,}700}{130}$$

$$B \approx 1644$$

The average size of a house in 1979 was 1644 square feet.

CHECK YOUR PROGRESS 5   A used Chevrolet Blazer was purchased for $22,400. This was 70% of the cost of the Blazer when new. What was the cost of the Blazer when it was new?

Solution   *See page S13.*

▼ **point of interest**

According to the U.S. Department of Agriculture, of the 356 billion pounds of food produced annually in the United States, about 96 billion pounds are wasted. This is approximately 27% of all the food produced in the United States.

EXAMPLE 6 ■ **Solve a Percent Problem for the Percent Using the Proportion Method**

During 1996, Texas suffered through one of its longest droughts in history. Of the $5 billion in losses caused by the drought, $1.1 billion was direct losses to ranchers. What percent of the total losses was direct losses to ranchers?

ANSWER   *The percent is 15. The base is 40. (It follows the phrase "percent of.") The amount is 6.*

Solution

We want to answer the question "What percent of $5 billion is $1.1 billion?" Write and solve a proportion. The base is $5 billion. The amount is $1.1 billion. The percent is unknown.

$$\frac{\text{Percent}}{100} = \frac{\text{amount}}{\text{base}}$$

$$\frac{p}{100} = \frac{1.1}{5}$$ • The amount is **1.1**. The base is **5**.

$$p \cdot 5 = 100(1.1)$$ • The cross products are equal.

$$5p = 110$$

$$\frac{5p}{5} = \frac{110}{5}$$

$$p = 22$$

Direct losses to ranchers represent 22% of the total losses.

CHECK YOUR PROGRESS 6

Of the approximately 1,300,000 enlisted women and men in the U.S. military, 416,000 are over the age of 30. What percent of the enlisted people are over the age of 30?

Solution *See page S13.*

EXAMPLE 7 ■ Solve a Percent Problem for the Amount Using the Proportion Method

In a recent year, Blockbuster Video customers rented 24% of the approximately 3.7 billion videos rented that year. How many million videos did Blockbuster Video rent that year?

Solution

We want to answer the question "24% of 3.7 billion is what number?" Write and solve a proportion. The percent is 24%. The base is 3.7 billion. The amount is the number of videos Blockbuster Video rented during the year.

$$\frac{\text{Percent}}{100} = \frac{\text{amount}}{\text{base}}$$

$$\frac{24}{100} = \frac{A}{3.7}$$ • The percent is **24**. The base is **3.7**.

$$24(3.7) = 100(A)$$ • The cross products are equal.

$$88.8 = 100A$$

$$\frac{88.8}{100} = \frac{100A}{100}$$

$$0.888 = A$$

The number 0.888 is in billions. We need to convert it to millions.

0.888 billion = 888 million

Blockbuster Video rented approximately 888 million videos that year.

CHECK YOUR PROGRESS 7   A Ford buyers' incentive program offered a 3.5% rebate on the selling price of a new car. What rebate would a customer receive who purchased a $32,500 car under this program?

Solution   *See page S13.*

Percent Problems: The Basic Percent Equation

A second method of solving a percent problem is to use the basic percent equation.

> **The Basic Percent Equation**
>
> $PB = A$, where P is the percent, B is the base, and A is the amount.

When solving a percent problem using the proportion method, we first have to identify the percent, the base, and the amount. The same is true when solving percent problems using the basic percent equation. Remember that the base usually follows the phrase "percent of."

When using the basic percent equation, the percent must be written as a decimal or a fraction. This is illustrated in Example 8.

EXAMPLE 8 ▪ **Solve a Percent Problem for the Amount Using the Basic Percent Equation**

A real estate broker receives a commission of 3% of the selling price of a house. Find the amount the broker receives on the sale of a $275,000 home.

Solution
We want to answer the question "3% of $275,000 is what number?" Use the basic percent equation. The percent is 3% = 0.03. The base is 275,000. The amount is the amount the broker receives on the sale of the home.

$$PB = A$$
$$0.03(275,000) = A$$
$$8250 = A$$

The real estate broker receives a commission of $8250 on the sale.

CHECK YOUR PROGRESS 8   New Hampshire public school teachers contribute 5% of their wages to the New Hampshire Retirement System. What amount is contributed during one year by a teacher whose annual salary is $32,685?

Solution   *See page S13.*

EXAMPLE 9 ■ Solve a Percent Problem for the Base Using the Basic Percent Equation

An investor received a payment of $480, which was 12% of the value of the investment. Find the value of the investment.

Solution
We want to answer the question "12% of what number is 480?" Use the basic percent equation. The percent is 12% = 0.12. The amount is 480. The base is the value of the investment.

$$PB = A$$
$$0.12B = 480$$
$$\frac{0.12B}{0.12} = \frac{480}{0.12}$$
$$B = 4000$$

The value of the investment is $4000.

CHECK YOUR PROGRESS 9 A real estate broker receives a commission of 3% of the selling price of a house. Find the selling price of a home on whose sale the broker received a commission of $14,370.

Solution See page S13.

EXAMPLE 10 ■ Solve a Percent Problem for the Percent Using the Basic Percent Equation

If you answer 96 questions correctly on a 120-question exam, what percent of the questions did you answer correctly?

Solution
We want to answer the question "What percent of 120 questions is 96 questions?" Use the basic percent equation. The base is 120. The amount is 96. The percent is unknown.

$$PB = A$$
$$P \cdot 120 = 96$$
$$\frac{P \cdot 120}{120} = \frac{96}{120}$$
$$P = 0.8$$
$$P = 80\%$$

You answered 80% of the questions correctly.

CHECK YOUR PROGRESS 10 If you answer 63 questions correctly on a 90-question exam, what percent of the questions did you answer correctly?

Solution See page S13.

The table below shows the average cost in the United States for five of the most popular home remodeling projects and the average percent of that cost recouped when the home is sold. Use this table for Example 11 and Check Your Progress 11. (*Source:* National Association of Home Builders)

Home Remodeling Project	Average Cost	Percent Recouped
Addition to the master suite	$36,472	84%
Attic bedroom	$22,840	84%
Major kitchen remodeling	$21,262	90%
Bathroom addition	$11,645	91%
Minor kitchen remodeling	$8,507	94%

EXAMPLE 11 ■ Solve an Application Using the Basic Percent Equation

 Find the difference between the cost of adding an attic bedroom to your home and the amount by which the addition increases the sale price of your home.

Solution

The cost of building the attic bedroom is $22,840, and the sale price increases by 84% of that amount. We need to find the difference between $22,840 and 84% of $22,840.

Use the basic percent equation to find 84% of $22,840. The percent is 84% = 0.84. The base is 22,840. The amount is unknown.

$$PB = A$$
$$0.84(22,840) = A$$
$$19,185.60 = A$$

Subtract 19,185.60 (the amount of the cost that is recouped when the home is sold) from 22,840 (the cost of building the attic bedroom).

$$22,840 - 19,185.60 = 3654.40$$

The difference between the cost of the addition and the increase in value of your home is $3654.40.

CHECK YOUR PROGRESS 11

Find the difference between the cost of a major kitchen remodeling in your home and the amount by which the remodeling increases the sale price of your home.

Solution See page S13.

Percent Increase

When a family moves from one part of the country to another, they are concerned about the difference in the cost of living. Will food, housing, and gasoline cost more in that part of the country? Will they need a larger salary in order to make ends meet?

We can use one number to represent the increased cost of living from one city to another so that no matter what salary you make, you can determine how much you will need to earn in order to maintain the same standard of living. That one number is a percent.

For example, look at the information in the table below. (*Source:* www. homefair.com/homefair/calc/salcalc.html)

If you live in	and are moving to	you will need to make this percent of your current salary
Cincinnati, Ohio	San Francisco, California	236%
St. Louis, Missouri	Boston, Massachusetts	213%
Denver, Colorado	New York, New York	239%

A family in Cincinnati living on $60,000 per year would need 236% of their current income to maintain the same standard of living in San Francisco. Likewise, a family living on $150,000 per year would need 236% of their current income.

$$60{,}000(2.36) = 141{,}600 \qquad 150{,}000(2.36) = 354{,}000$$

The family from Cincinnati living on $60,000 would need an annual income of $141,600 in San Francisco to maintain their standard of living. The family living on $150,000 would need an annual income of $354,000 in San Francisco to maintain their standard of living.

We have used one number, 236%, to represent the increase in the cost of living from Cincinnati to San Francisco. No matter what a family's present income, they can use 236% to determine their necessary comparable income.

QUESTION   *How much would a family in Denver, Colorado living on $55,000 per year need in New York City to maintain a comparable lifestyle? Use the table above.*

Amount of increase

Original value

New value

The percent used to determine the increase in the cost of living is a *percent increase*. **Percent increase** is used to show how much a quantity has increased over its original value. Statements that illustrate the use of percent increase include "sales volume increased by 11% over last year's sales volume" and "employees received an 8% pay increase."

The **federal debt** is the amount the government owes after borrowing the money it needs to pay for its expenses. It is considered a good measure of how much of the government's spending is financed by debt as opposed to taxation. The bar graph below shows the federal debt, according to the U.S. Department of the Treasury, at the end of the fiscal years 1980, 1985, 1990, 1995, and 2000. A fiscal year is the 12-month period that the annual budget spans, from October 1 to September 30. Use the bar graph for Example 12 and Check Your Progress 12.

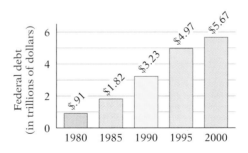

ANSWER   *In New York City, the family would need $55,000(2.39) = $131,450 per year to maintain a comparable lifestyle.*

historical note

The largest percent increase, for a single day, in the Dow Jones Industrial Average occurred on October 6, 1931. The Dow gained approximately 15% of its value. ∎

EXAMPLE 12 ▪ Solve an Application Involving Percent Increase

 Find the percent increase in the federal debt from 1980 to 1995. Round to the nearest tenth of a percent.

Solution

Calculate the amount of increase in the federal debt from 1980 to 1995.

$$4.97 - 0.91 = 4.06$$

We will use the basic percent equation. (The proportion method could also be used.) The base is the debt in 1980. The amount is the amount of increase in the debt. The percent is unknown.

$$PB = A$$
$$P \cdot 0.91 = 4.06 \qquad \text{• The base is 0.91. The amount is 4.06.}$$
$$\frac{P \cdot 0.91}{0.91} = \frac{4.06}{0.91} \qquad \text{• Divide each side by 0.91.}$$
$$P \approx 4.462$$

The percent increase in the federal debt from 1980 to 1995 was 446.2%.

CHECK YOUR PROGRESS 12

 Find the percent increase in the federal debt from 1985 to 2000. Round to the nearest tenth of a percent.

Solution *See page S14.*

Notice in Example 12 that the percent increase is a measure of the *amount of increase* over an *original value*. Therefore, in the basic percent equation, the amount *A* is the *amount of increase* and the base *B* is the *original value*, in this case the debt in 1980.

Percent Decrease

The federal debt is not the same as the federal deficit. The **federal deficit** is the amount by which government spending exceeds the federal budget. The table below shows projected federal deficits. (*Source:* U.S. Government Office of Management and Budget) Note that the deficit listed for 2006 is less than the deficit listed for 2005.

Year	Federal Deficit
2005	$363.570 billion
2006	$267.632 billion
2007	$241.272 billion
2008	$238.969 billion
2009	$237.076 billion

The decrease in the federal deficit can be expressed as a percent. First find the amount of decrease in the deficit from 2005 to 2006.

$$363.570 - 267.632 = 95.938$$

We will use the basic percent equation to find the percent. The base is the deficit in 2005. The amount is the amount of decrease.

$$PB = A$$
$$P \cdot 363.57 = 95.938$$
$$\frac{P \cdot 363.57}{363.57} = \frac{95.938}{363.57}$$
$$P \approx 0.264$$

The federal deficit is projected to decrease by 26.4% from 2005 to 2006.

The percent used to measure the decrease in the federal deficit is a *percent decrease*. **Percent decrease** is used to show how much a quantity has decreased from its original value. Statements that illustrate the use of percent decrease include "the president's approval rating has decreased 9% over the last month" and "there has been a 15% decrease in the number of industrial accidents."

Note in the deficit example above that the percent decrease is a measure of the *amount of decrease* in comparison with an *original value*. Therefore, in the basic percent equation, the amount A is the *amount of decrease* and the base B is the *original value*, in this case the deficit in 2005.

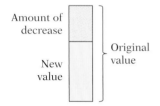

Amount of decrease

New value

Original value

EXAMPLE 13 ■ Solve an Application Involving Percent Decrease

In 1998, GM plants took an average of 32 hours to produce one vehicle. From 1998 to 2005, that time decreased 28.125%. (*Source: Time,* December 5, 2005) Find the average time for GM to produce one vehicle in 2005.

Solution

We will write and solve a proportion. (The basic percent equation could also be used.) The percent is 28.125%. The base is 32. The amount is unknown.

$$\frac{\text{Percent}}{100} = \frac{\text{amount}}{\text{base}}$$

$$\frac{28.125}{100} = \frac{A}{32}$$ • The percent is **28.125**. The base is **32**.

$$28.125(32) = 100 \cdot A$$

$$900 = 100A$$

$$\frac{900}{100} = \frac{100A}{100}$$ • Divide each side by **100**.

$$9 = A$$ • This is the amount of decrease.

Subtract the decrease in time from the 1998 time.

$$32 - 9 = 23$$

In 2005, GM plants took an average of 23 hours to produce one vehicle.

CHECK YOUR PROGRESS 13

 The number of passenger car fatalities in the United States in 2002 was 20,416. This number decreased 3.81% in 2003. (*Source: Time*, May 10, 2004) Find the number of passenger car fatalities in the United States in 2003.

Solution   *See page S14.*

Investigation

Federal Income Tax

Income taxes are the chief source of revenue for the federal government. If you are employed, your employer probably withholds some money from each of your paychecks for federal income tax. At the end of each year, your employer sends you a **Wage and Tax Statement Form** (**W-2 form**), which states the amount of money you earned that year and how much was withheld for taxes.

Every employee is required by law to prepare an income tax return by April 15 of each year and send it to the Internal Revenue Service (IRS). On the income tax return, you must report your total income, or **gross income.** Then you subtract from the gross income any adjustments (such as deductions for charitable contributions or exemptions for people who are dependent on your income) to determine your **adjusted gross income.** You use your adjusted gross income and either a tax table or a tax rate schedule to determine your **tax liability,** or the amount of income tax you owe to the federal government.

After calculating your tax liability, compare it with the amount withheld for federal income tax, as shown on your W-2 form. If the tax liability is less than the amount withheld, you are entitled to a tax refund. If the tax liability is greater than the amount withheld, you owe the IRS money; you have a **balance due.**

The 2006 Tax Rate Schedules table is shown on page 172. To use this table for the exercises that follow, first classify the taxpayer as single, married filing jointly, or married filing separately. Then determine into which range the adjusted gross income falls. Then perform the calculations shown to the right of that range to determine the tax liability.

For example, consider a taxpayer who is single and has an adjusted gross income of $48,720. To find this taxpayer's tax liability, use the portion of the table headed "Schedule X" for taxpayers whose filing status is single.

An income of $48,720 falls in the range $30,650 to $74,200. The tax is $4220.00 + 25% of the amount over $30,650. Find the amount over $30,650.

$48,720 − $30,650 = $18,070

(continued)

Calculate the tax liability:

$$\$4220.00 + 25\%(\$18{,}070) = \$4220.00 + 0.25(\$18{,}070)$$
$$= \$4220.00 + \$4517.50$$
$$= \$8737.50$$

The taxpayer's liability is $8737.50.

Tax Rate Schedules

Single—Schedule X

If line 5 is:		The tax is:	of the amount
Over—	But not over—		over—
$0	$7,550	... 10%	$0
7,550	30,650	**$755.00 + 15%**	7,550
30,650	74,200	4,220.00 + 25%	30,650
74,200	154,800	15,107.50 + 28%	74,200
154,800	336,550	37,675.50 + 33%	154,800
336,550	...	97,653.00 + 35%	336,550

Married Filing Jointly or Qualifying Widow(er)—Schedule Y-1

If line 5 is:		The tax is:	of the amount
Over—	But not over—		over—
$0	$15,100	... 10%	$0
15,100	61,300	**$1,510.00 + 15%**	15,100
61,300	123,700	8,440.00 + 25%	61,300
123,700	188,450	24,040.00 + 28%	123,700
188,450	336,550	42,170.00 + 33%	188,450
336,550	...	91,043.00 + 35%	336,550

Married Filing Separately—Schedule Y-2

If line 5 is:		The tax is:	of the amount
Over—	But not over—		over—
$0	$7,550	... 10%	$0
7,550	30,650	**$755.00 + 15%**	7,550
30,650	61,850	4,220.00 + 25%	30,650
61,850	94,225	12,020.00 + 28%	61,850
94,225	168,275	21,085.00 + 33%	94,225
168,275	...	45,521.50 + 35%	168,275

(continued)

Exercise Set 3.3

Investigation Exercises

Use the 2006 Tax Rate Schedules to solve Exercises 1–8.

1. Joseph Abruzzio is married and filing separately. He has an adjusted gross income of $63,850. Find Joseph's tax liability.

2. Angela Lopez is single and has an adjusted gross income of $31,680. Find Angela's tax liability.

3. Dee Pinckney is married and filing jointly. She has an adjusted gross income of $58,120. The W-2 form shows the amount withheld as $7124. Find Dee's tax liability and determine her tax refund or balance due.

4. Jeremy Littlefield is single and has an adjusted gross income of $78,800. His W-2 form lists the amount withheld as $18,420. Find Jeremy's tax liability and determine his tax refund or balance due.

5. Does a taxpayer in the 33% tax bracket pay 33% of his or her earnings in income tax? Explain your answer.

6. A single taxpayer has an adjusted gross income of $159,000. On what amount of the $159,000 does the taxpayer pay 33% to the Internal Revenue Service?

7. In the table for single taxpayers, how were the figures $755.00 and $4220.00 determined?

8. In the table for married persons filing jointly, how were the figures $1510.00 and $8440.00 determined?

Exercise Set 3.3

1. Name three situations in which percent is used.

2. Explain why multiplying a number by 100% does not change the value of the number.

3. Multiplying a number by 300% is the same as multiplying it by what whole number?

4. Describe each ratio in the proportion used to solve a percent problem.

5. Employee A's annual salary was less than Employee B's annual salary before each employee was given a 5% raise. Which of the two employees now has the higher annual salary?

6. Each of three employees earned the same annual salary before Employee A was given a 3% raise, Employee B was given a 6% raise, and Employee C was given a 4.5% raise. Which of the three employees now has the highest annual salary?

Complete the table of equivalent fractions, decimals, and percents.

	Fraction	Decimal	Percent
7.	$\dfrac{1}{2}$		
8.		0.75	
9.			40%
10.	$\dfrac{3}{8}$		
11.		0.7	
12.			56.25%
13.	$\dfrac{11}{20}$		
14.		0.52	
15.			15.625%
16.	$\dfrac{9}{50}$		

17. Given that 25% of x equals y, $x > 0$, and $y > 0$, state whether $x < y$ or $x > y$.

18. Given that 200% of x equals y, $x > 0$, and $y > 0$, state whether $x < y$ or $x > y$.

19. **Baseball** In 1997, for the first time in Major League Baseball history, interleague baseball games were played during the regular season. According to a Harris Poll, 73% of fans approved and 20% of fans disapproved of the change.

   a. How many fans, out of every 100 surveyed, approved of interleague games?

   b. Did more fans approve or disapprove of the change?

   c. Fans surveyed gave one of three responses: approve, disapprove, or don't know. What percent of the fans surveyed responded that they didn't know? Explain how you calculated this percent.

Solve Exercises 20–36.

20. **Income Tax** In 2004, 34.2 million accountants e-filed income tax returns. This was 114% of the number who e-filed in 2003. (*Source:* Internal Revenue Service) Find the number of accountants who e-filed income tax returns in 2003.

21. **Charitable Contributions** During a recent year, charitable contributions in the United States totaled $190 billion. The graph at the right shows the types of organizations to which this money was donated. Determine how much money was donated to educational organizations. (*Source:* Giving USA 2000/AAFRC Trust for Philanthropy)

22. **Health Insurance** Of the 44 million people in the United States who do not have health insurance, 13.2 million are between the ages of 18 and 24. What percent of the people in the United States who do not have health insurance are between the ages of 18 and 24? (*Source:* U.S. Census Bureau)

23. **Motorists** A survey of 1236 adults nationwide asked, "What irks you most about the actions of other motorists?" The response "tailgaters" was given by 293 people. What percent of those surveyed were most irked by tailgaters? Round to the nearest tenth of a percent. (*Source:* Reuters/Zogby)

24. **Television** A survey by the *Boston Globe* questioned elementary and middle-school students about television. Sixty-eight students, or 42.5% of those surveyed, said that they had televisions in their bedrooms at home. How many students were included in the survey?

25. **Vacations** According to the annual Summer Vacation Survey conducted by Myvesta, a nonprofit consumer education organization, the average summer vacation costs $2252. If 82% of this amount is charged on credit cards, what amount of the vacation cost is charged?

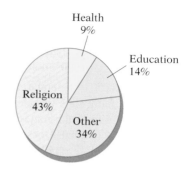

Recipients of Charitable Contributions in the United States

26. **Pets** The average costs associated with owning a dog over its average 11-year life span are shown in the graph at the right. These costs do not include the price of the puppy when purchased. The category labeled "Other" includes such expenses as fencing and repairing furniture damaged by the pet.

   a. Calculate the total of all the expenses.

   b. What percent of the total does each category represent? Round to the nearest tenth of a percent.

   c. If the price of the puppy were included in these data, how would it affect the percents you calculated in part b?

   d. What does it mean to say that these are average costs?

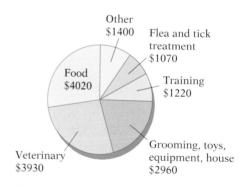

Costs of Owning a Dog

Source: American Kennel Club, *USA Today* research

27. **Time Management** The two circle graphs show how surveyed employees actually spend their time and how they would prefer to spend their time. Assume that employees have 112 hours a week that are not spent sleeping. Round answers to the nearest tenth of an hour. (*Source: Wall Street Journal* Supplement from *Families and Work Institute*)

   a. What is the actual number of hours per week that employees spend with family and friends?

   b. What is the number of hours that employees would prefer to spend on their jobs or careers?

   c. What is the difference between the number of hours employees would prefer to spend on themselves and the actual amount of time employees spend on themselves?

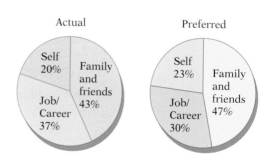

28.   **Prison Inmates**   The graph below shows the numbers of prison inmates in the United States for the years 1990, 1995, and 2000.

   a. The number of federal inmates was what percent of the number of state inmates in 1990? in 2000? Round to the nearest tenth of a percent.

   b. If the ratio of federal inmates to state inmates in 1990 had remained constant, how many state inmates would there have been in 2000, when there were 145,416 federal inmates? Is this more or less than the actual number of state inmates in 2000? Does this mean that the number of federal inmates is growing at a more rapid rate or that the number of state inmates is growing at a more rapid rate?

   c.   Explain how parts a and b are two methods of measuring the same change.

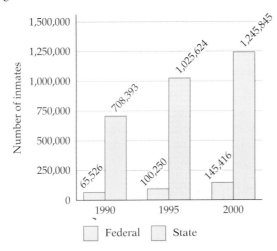

Number of Inmates in the United States

Source: Bureau of Justice Statistics

29.   **Telecommuting**   The graph below shows the growth in the number of telecommuters.

   a. During which 2-year period was the percent increase in the number of telecommuters the greatest?

   b. During which 2-year period was the percent increase in the number of telecommuters the lowest?

   c. During which period was the growth in telecommuting more rapid, from 1998 to 2002 or from 2002 to 2006?

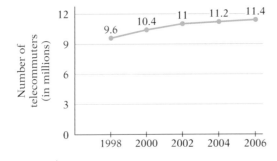

Growth in Telecommuting

30. **Highway Fatalities** In a recent year, the states listed below had the highest rates of truck-crash fatalities. (*Source:* Citizens for Reliable and Safe Highways)

State	Number of Truck-Crash Fatalities	Number of Truck-Crash Fatalities per 100,000 People
Alabama	160	3.76
Arkansas	102	4.11
Idaho	38	3.27
Iowa	86	3.10
Mississippi	123	4.56
Montana	30	3.45
West Virginia	53	2.90
Wyoming	17	3.54

a. What state has the highest rate of truck-crash deaths?

For each of the following states, find the population to the nearest thousand people. Calculate the percent of the population of each state that died in truck accidents. Round to the nearest hundred thousandth of a percent.

b. Mississippi

c. West Virginia

d. Montana

e. Compare the percents in parts b, c, and d with the numbers of truck-crash fatalities per 100,000 people listed in the table. Based on your observations, what percent of the population of Idaho do you think was killed in truck accidents?

f. Explain the relationship between the rates in the table and the percents you calculated.

31. **High-Tech Employees** The table below lists the states with the highest rates of high-tech employees per 1000 workers. High-tech employees are those employed in the computer and electronics industries. Also provided is the size of the civilian labor force for each of these states. (*Source:* American Electronics Association; U.S. Bureau of Labor Statistics)

State	High-Tech Employees per 1000 Workers	Labor Force
Arizona	53	2,229,300
California	62	10,081,300
Colorado	75	2,200,500
Maryland	51	2,798,400
Massachusetts	75	3,291,100
Minnesota	55	4,173,500
New Hampshire	78	663,000
New Jersey	55	4,173,500
Vermont	52	332,200
Virginia	52	3,575,000

  a. Find the number of high-tech employees in each state listed. Round to the nearest whole number.

  b. Which state has the highest rate of high-tech employment? Which state has the largest number of high-tech employees?

  c. The computer and electronics industries employ approximately 4 million workers. Do more or less than half of the high-tech employees work in the 10 states listed in the table?

  d. What percent of the state's labor force is employed in the computer and electronics industries in Massachusetts? in New Jersey? in Virginia? Round to the nearest tenth of a percent.

  e. Compare your answers to part d with the numbers of high-tech employees per 1000 workers listed in the table. Based on your observations, what percent of the labor force in Arizona is employed in the high-tech industries?

  f. Explain the relationship between the rates in the table and the percents you calculated.

32. Consumption of Eggs During the last 40 years, the consumption of eggs in the United States has decreased by 35%. Forty years ago, the average consumption was 400 eggs per person per year. What is the average consumption of eggs today?

33. **Millionaire Households** The table at the right shows the estimated numbers of millionaire households in the United States for selected years.

Year	Number of Households Containing Millionaires
1975	350,000
1997	3,500,000
2005	5,600,000

  a. What was the percent increase in the estimated number of millionaire households from 1975 to 1997?

  b. Find the percent increase in the estimated number of millionaire households from 1997 to 2005.

  c. Find the percent increase in the estimated number of millionaire households from 1975 to 2005.

  d. Provide an explanation for the dramatic increase in the estimated number of millionaire households from 1975 to 2005.

34. **Demographics** The graph at the right shows the projected growth of the number of Americans aged 85 and older.

  a. What is the percent increase in the population of this age group from 1995 to 2030?

  b. What is the percent increase in the population of this age group from 2030 to 2050?

  c. What is the percent increase in the population of this age group from 1995 to 2050?

  d. How many times larger is the population in 2050 than in 1995? How could you determine this number from the answer to part c?

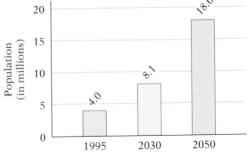

Projected Growth (in millions) of the Population of Americans Aged 85 and Older

Source: U.S. Census Bureau

35. **Occupations** The Bureau of Labor Statistics provides information on the fastest-growing occupations in the United States. These occupations are listed in the table below, along with the predicted percent increases in employment from 2000 to 2010.

Occupation	Employment in 2000	Percent Increase 2000–2010
Software engineers (applications)	380,000	100%
Computer support specialists	506,000	97%
Software engineers (systems)	317,000	90%
Network and computer systems administrators	229,000	82%
Network and data communications analysts	119,000	77%
Desktop publishers	38,000	67%
Database administrators	106,000	66%
Personal and home care aides	414,000	62%
Computer systems analysts	431,000	60%
Medical assistants	329,000	57%
Social and human service assistants	271,000	54%
Physician assistants	58,000	53%
Medical records/health information technicians	136,000	49%
Computer and information systems managers	313,000	48%
Home health aides	615,000	47%
Physical therapist aides	36,000	46%
Occupational therapist aides	9,000	45%
Physical therapist assistants	44,000	45%
Audiologists	13,000	45%
Fitness trainers and aerobics instructors	158,000	40%
Computer and information scientists (research)	28,000	40%
Occupational therapist assistants	17,000	40%
Veterinary and laboratory animal assistants	55,000	40%
Speech–language pathologists	88,000	39%
Mental health/substance abuse social workers	83,000	39%
Dental assistants	247,000	37%
Dental hygienists	147,000	37%
Teachers	234,000	37%
Pharmacy technicians	190,000	36%

a. Which occupation employed the largest number of people in 2000?

b. Which occupation is expected to have the largest percent increase in employment between 2000 and 2010?

c. What increase is expected in the number of people employed as fitness trainers and aerobics instructors from 2000 to 2010?

d. How many people are expected to be employed as teachers in 2010?

e. Is it expected that more people will be employed as home health aides or as computer support specialists in 2010?

f. ✎ Why can't the answer to part e be determined by simply comparing the percent increases for the two occupations?

g. From the list, choose an occupation that interests you. Calculate the projected number of people who will be employed in that occupation in 2010.

36. ● **Elections** According to the Committee for the Study of the American Electorate, the voter turnout in the 2004 presidential election was higher in all but four states than it was in the 2000 election. The four states with lower voter turnout are listed below.

a. How many Maine voters turned out to vote in the 2004 presidential election?

b. How many people in Arizona voted in the 2004 election?

c. How many more people voted in the 2000 election than in the 2004 election in the state of Alaska?

State	2000 Voter Turnout	Percent Decrease in Voter Turnout in 2004 Election
Alaska	280,000	12.4%
Arizona	1,632,000	2.3%
Maine	664,000	0.9%
Montana	433,000	0.1%

Extensions

CRITICAL THINKING

37. **Salaries** Your employer agrees to give you a 5% raise after one year on the job, a 6% raise the next year, and a 7% raise the following year. Is your salary after the third year greater than, less than, or the same as it would be if you had received a 6% raise each year?

38. ● **Work Habits** Approximately 73% of Americans who work in large offices work on weekends, either at home or in the office. The table below shows the average number of hours these workers report they work on weekends. Approximately what percent of Americans who work in large offices work 11 or more hours on weekends?

Numbers of Hours Worked on Weekends	Percent
0 – 1	3%
2 – 5	32%
6 – 10	42%
11 or more	23%

39. Car Purchases In a survey of consumers, approximately 43% said they would be willing to pay $1500 more for a new car if the car had an EPA rating of 80 miles per gallon. If your car currently gets 28 miles per gallon and you drive approximately 10,000 miles per year, in how many months would your savings on gasoline pay for the increased cost of such a car? Assume that the average cost of gasoline is $3.00 per gallon. Round to the nearest whole number.

COOPERATIVE LEARNING

40. **Demography** The statistical study of human populations is referred to as **demography.** Many groups are interested in the sizes of certain segments of the population and projections of population growth. For example, public school administrators want estimates of the number of school-age children who will be living in their districts 10 years from now.

The U.S. government provides estimates of the future U.S. population. You can find these projections at the Census Bureau website at **www.census.gov.** Three different projections are provided: a lowest series, a middle series, and a highest series. These series reflect different theories on how fast the population of this country will grow.

The table below contains data from the Census Bureau website. The figures are from the middle series of projections.

	Age	Under 5	5 – 17	18 – 24	25 – 34	35 – 44	45 – 54	55 – 64	65 – 74	75 & older
2010	Male	9,712	26,544	13,338	18,535	22,181	18,092	11,433	8,180	6,165
	Female	9,274	25,251	12,920	18,699	22,478	18,938	12,529	9,956	10,408
2050	Male	13,877	35,381	18,462	24,533	23,352	21,150	20,403	16,699	19,378
	Female	13,299	33,630	17,871	24,832	24,041	22,344	21,965	18,032	24,751

For the following exercises, round all percents to the nearest tenth of a percent.

a. Which of the age groups listed are of interest to public school officials?

b. Which age groups are of interest to nursing home administrators?

c. Which age groups are of concern to accountants determining benefits to be paid out by the Social Security Administration during the next decade?

d. Which age group is of interest to manufacturers of disposable diapers?

e. Which age group is of primary concern to college and university admissions officers?

f. In which age groups do males outnumber females? In which do females outnumber males?

g. What percent of the projected population aged 75 and older in the year 2010 is female? Does this percent decrease in 2050? If so, by how much?

h. Find the difference between the percent of the population that will be 65 or over in 2010 and the percent that will be 65 or older in 2050.

i. Assume that the work force consists of people aged 25 to 64. What percent increase is expected in this population from 2010 to 2050?

j. What percent of the population is expected to be in the work force in 2010? What percent of the population is expected to be in the work force in 2050?

k. Why are the answers to parts h, i, and j of concern to the Social Security Administration?

l. Describe any patterns you see in the table.

m. Calculate a statistic based on the data in the table and explain why it would be of interest to an institution (such as a school system) or a manufacturer of consumer goods (such as disposable diapers).

41. Nielsen Ratings Nielsen Media Research surveys television viewers to determine the numbers of people watching particular shows. There are an estimated 105.5 million U.S. households with televisions. Each **rating point** represents 1% of that number, or 1,055,000 households. Therefore, for instance, if *60 Minutes* received a rating of 5.8, then 5.8% of all U.S. households with televisions, or $(0.058)(105,500,000) = 6,119,000$ households, were tuned to that program.

A rating point does not mean that 1,055,000 people are watching a program. A rating point refers to the number of households with television sets tuned to that program; there may be more than one person watching a television set in the household.

Nielsen Media Research also describes a program's share of the market. **Share** is the percent of households with television sets in use that are tuned to a program. Suppose that the same week that *60 Minutes* received 5.8 rating points, the show received a share of 11%. This would mean that 11% of all households with televisions *turned on* were turned to *60 Minutes*, whereas 5.8% of all households with televisions were turned to the program.

a. If *CSI* received a Nielsen rating of 10.1 and a share of 17, how many TV households watched the program that week? How many TV households were watching television during that hour? Round to the nearest hundred thousand.

b. Suppose *60 Minutes* received a rating of 5.6 and a share of 11. How many TV households watched the program that week? How many TV households were watching television during that hour? Round to the nearest hundred thousand.

c. Suppose *Two and a Half Men* received a rating of 7.5 during a week in which 19,781,000 people were watching the show. Find the average number of people per TV household who watched the program. Round to the nearest tenth.

Nielsen Media Research has a website on the Internet. You can locate the site by using a search engine. The site does not list rating points or market share, but these statistics can be found on other websites by using a search engine.

d. Find the top two prime-time broadcast television shows for last week. Calculate the numbers of TV households that watched these programs. Compare the figures with the top two cable TV programs for last week.

Direct and Inverse Variation

Direct Variation

An equation of the form $y = kx$ describes many important relationships in business, science, and engineering. The equation $y = kx$, where k is a constant, is an example of a **direct variation**. The constant k is called the **constant of variation** or the **constant of proportionality.** The equation $y = kx$ is the general form of a direct variation equation. We read $y = kx$ as "y varies directly as x."

The circumference C of a circle varies directly as the diameter d. The direct variation equation is written $C = \pi d$. The constant of variation is π.

A nurse earns $30 per hour. The total wage w of the nurse is directly proportional to the number of hours h worked. The direct variation is $w = 30h$. The constant of proportionality is 30.

> **QUESTION:** *The equation $d = 55t$ gives the distance d, in miles, traveled in t hours when traveling at 55 mph. What is the constant of variation in this direct variation equation?*

EXAMPLE 1 ■ Find a Direct Variation Equation

Find the constant of variation if y varies directly as x and $y = 35$ when $x = 5$. Then write the specific direct variation equation that relates y and x.

Solution

$y = kx$	• Write the general form of a direct variation equation.
$35 = k \cdot 5$	• Substitute the given values for y and x.
$7 = k$	• Solve for k by dividing each side of the equation by 5.

The constant of variation is 7.

$y = 7x$	• Write the specific direct variation equation that relates y and x by substituting 7 for k.

The direct variation equation is $y = 7x$.

CHECK YOUR PROGRESS 1   Find the constant of variation if y varies directly as x and $y = 120$ when $x = 8$. Then write the specific direct variation equation that relates y and x.

Solution   See page S14.

EXAMPLE 2 ■ **Solve an Application Using a Direct Variation Equation of the Form $y = kx$**

The distance d sound travels varies directly as the time t it travels. If sound travels 8920 feet in 8 seconds, find the distance that sound travels in 3 seconds.

Solution

$d = kt$	• Write the general form of a direct variation equation.
$8920 = k \cdot 8$	• Substitute 8920 for d and 8 for t.
$1115 = k$	• Solve for k by dividing each side of the equation by 8. The constant of variation is **1115**.
$d = 1115t$	• Write the specific direct variation equation by substituting **1115** for k.
$d = 1115 \cdot 3$	• Find d when $t = 3$.
$d = 3345$	

Sound travels 3345 feet in 3 seconds.

CHECK YOUR PROGRESS 2 The tension T in a spring varies directly as the distance x it is stretched. If $T = 8$ pounds when $x = 2$ inches, find T when $x = 4$ inches.

Solution   See page S14.

A direct variation equation can be written in the form $y = kx^n$, where n is a positive number. For example, the equation $y = kx^2$ is read "y varies directly as the square of x."

The area A of a circle varies directly as the square of the radius r of the circle. The direct variation equation is $A = \pi r^2$. The constant of variation is π.

EXAMPLE 3 ■ **Solve an Application Using a Direct Variation Equation of the Form $y = kx^2$**

The load L that a horizontal beam can safely support is directly proportional to the square of the depth d of the beam. A beam with a depth of 8 inches can safely support 800 pounds. Find the load that a similar beam with a depth of 6 inches can safely support. Assume both beams span the same distance.

Solution

✔ **TAKE NOTE**

The phrase "the square of the depth d" translates to d^2.

$L = kd^2$	• Write the general form of a direct variation equation.
$800 = k \cdot 8^2$	• Substitute 800 for L and 8 for d.
$800 = k \cdot 64$	
$12.5 = k$	• Solve for k by dividing each side of the equation by 64. The constant of variation is **12.5**.
$L = 12.5d^2$	• Write the specific direct variation equation by substituting **12.5** for k.
$L = 12.5 \cdot 6^2$	• Find L when $d = 6$.
$L = 12.5 \cdot 36$	
$L = 450$	

A beam with a depth of 6 inches can safely support a load of 450 pounds.

CHECK YOUR PROGRESS 3 The distance d required for a car to stop varies directly as the square of the velocity v of the car. If a car traveling at 40 mph requires 130 feet to stop, find the stopping distance for a car traveling at 60 mph.

Solution *See page S14.*

Inverse Variation

The equation $y = \dfrac{k}{x}$, where k is a constant, is the general form of an **inverse varia-**

tion equation. The equation $y = \dfrac{k}{x}$ is read "y varies inversely as x" or "y is inversely proportional to x."

The number of items N that can be purchased for a given amount of money is inversely proportional to the cost C of the item. The inverse variation equation that relates N and C is $N = \dfrac{k}{C}$.

It is important to note that in the direct variation equation $y = kx$, as the quantity x increases, y increases. For example, if you are buying power bars, twice as many power bars will cost twice as much money. In the inverse variation equation $y = \dfrac{k}{x}$, as the quantity x increases, y decreases; or as x decreases, y increases. For example, as the distance between you and a light source decreases, the intensity of the light increases; as the distance between you and the light source increases, the intensity of the light decreases.

QUESTION: *The loudness of the music heard on a stereo speaker is 50 decibels at a distance of 20 feet from the speaker. Are these two quantities inversely proportional?*

EXAMPLE 4 ■ Find an Inverse Variation Equation

Find the constant of variation if y varies inversely as x and $y = 15$ when $x = 3$. Then write the specific inverse variation equation that relates y and x.

Solution

$y = \dfrac{k}{x}$ • Write the general form of an inverse variation equation.

$15 = \dfrac{k}{3}$ • Substitute the given values for y and x.

$45 = k$ • Solve for k by multiplying each side of the equation by 3.

The constant of variation is 45.

$y = \dfrac{45}{x}$ • Write the specific inverse variation equation that relates y and x by substituting 45 for k.

The inverse variation equation is $y = \dfrac{45}{x}$.

ANSWER *Yes. As the distance from the speaker increases, the number of decibels decreases; or as the distance from the speaker decreases, the number of decibels increases.*

CHECK YOUR PROGRESS 4   Find the constant of variation if P varies inversely as R and $P = 20$ when $R = 5$. Then write the specific inverse variation equation.

Solution   *See page S14.*

EXAMPLE 5 ■ **Solve an Application Using an Inverse Variation Equation of the Form** $y = \dfrac{k}{x}$

A company that produces personal computers has determined that the number of computers it can sell per month, s, is inversely proportional to the price P of each computer. Two thousand computers can be sold per month when the price is $2500. How many computers can be sold per month when the price is reduced to $2000?

Solution

TAKE NOTE

The inverse variation equation $s = \dfrac{k}{P}$ means that the higher the price, the lower the number of computers sold; the lower the price, the higher the number of computers sold.

$$s = \frac{k}{P}$$
• Write the general form of an inverse variation equation.

$$2000 = \frac{k}{2500}$$
• Substitute 2000 for s and 2500 for P.

$$5{,}000{,}000 = k$$
• Solve for k by multiplying each side of the equation by 2500. The constant of variation is 5,000,000.

$$s = \frac{5{,}000{,}000}{P}$$
• Write the specific inverse variation equation by substituting 5,000,000 for k.

$$s = \frac{5{,}000{,}000}{2000}$$
• Find s when $P = 2000$.

$$s = 2500$$

When the price is $2000, 2500 computers can be sold each month.

CHECK YOUR PROGRESS 5   At an assembly plant, the number of hours h it takes to complete the daily quota of plastic molds is inversely proportional to the number of assembly machines operating, m. If five assembly machines can complete the daily quota in 9 hours, how many hours does it take for four assembly machines to complete the daily quota of plastic molds?

Solution   *See page S14.*

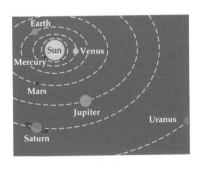

In general, an inverse variation equation can be written as $y = \dfrac{k}{x^n}$, where n is a positive integer. For example, the equation $y = \dfrac{k}{x^2}$ is read "y varies inversely as the square of x."

The gravitational force F between two planets is inversely proportional to the square of the distance d between the planets. This inverse variation equation is written $F = \dfrac{k}{d^2}$ and means that as the distance between two planets increases, the force between them decreases; conversely, as the distance between two planets decreases, the force between them increases.

Math Matters Gravity

Gravity is one of the fundamental forces in the universe. Originally described by Newton, gravity is essentially the natural force of attraction between two objects.

Two factors determine the magnitude of the gravitational force between two objects: their masses and the distance between them. The magnitude of the force is proportional to the product of the masses of the two objects. For example, the force doubles in magnitude if either of the masses is increased to twice its mass. On the other hand, the force grows weaker if the two objects are moved farther apart. As stated in the inverse variation equation $F = \dfrac{k}{d^2}$, the force is inversely proportional to the square of the distance between the objects; so if the distance is tripled, the force is only one-ninth as strong.

The gravitational force of the sun acting on Earth keeps Earth in its orbit. The gravitational force of Earth acting on us holds us to Earth's surface. The gravitational attraction between a person and Earth is proportional to the person's mass and inversely proportional to the square of Earth's radius. The number that describes this gravitational attraction is your weight.

EXAMPLE 6 ■ **Solve an Application Using an Inverse Variation Equation of the Form $y = \dfrac{k}{x^2}$**

The resistance R to the flow of electric current in a wire of fixed length varies inversely as the square of the diameter d of the wire. One wire of diameter 0.01 centimeter has a resistance of 0.5 ohm. Find the resistance in a second wire that is 0.02 centimeter in diameter. Assume the second wire is the same length and made of the same material as the first wire.

Solution

$$R = \frac{k}{d^2}$$
• Write the general form of an inverse variation equation.

$$0.5 = \frac{k}{0.01^2}$$
• Substitute 0.5 for R and 0.01 for d.

$$0.5 = \frac{k}{0.0001}$$

$$0.00005 = k$$
• Solve for k by multiplying each side of the equation by 0.0001. The constant of variation is **0.00005**.

$$R = \frac{0.00005}{d^2}$$
• Write the specific inverse variation equation by substituting 0.00005 for k.

$$R = \frac{0.00005}{0.02^2}$$
• Find R when $d = 0.02$.

$$R = \frac{0.00005}{0.0004}$$

$$R = 0.125$$

The resistance in the second wire is 0.125 ohm.

CHECK YOUR PROGRESS 6 The intensity *I* of a light source is inversely proportional to the square of the distance *d* from the source. If the intensity is 20 foot-candles at a distance of 8 feet, what is the intensity when the distance is 5 feet?

Solution *See page S15.*

Investigation

Gears and Pulleys

The relationship between the size of a gear or pulley and the speed with which it rotates is an inverse variation.

In the diagram at the right, gear A has twice as many teeth as gear B. Therefore, when A makes one turn, B will make two turns.

Suppose gear A has three times as many teeth as gear B. Then when A makes one turn, B will make three turns. If gear A has four times as many teeth as gear B, then A will make one turn while B will make four turns. The fewer teeth gear B has compared with gear A, the more turns it will make for each one turn gear A makes. The speed of a gear that is being rotated by another gear is inversely proportional to the number of teeth on the gear. We can express this relationship by the inverse variation equation

$$R = \frac{k}{T}$$

where *R* is the speed of the gear in revolutions per minute (rpm) and *T* is the number of teeth.

Pulleys operate in much the same way. For the pulley system at the left, the diameter of pulley A is twice that of pulley B. Therefore, when pulley A makes one turn, pulley B will make two turns.

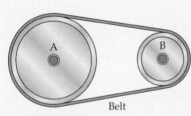

Suppose pulley A has a diameter three times that of pulley B. Then when A makes one turn, B will make three turns. The smaller the diameter of pulley B as compared to the diameter of pulley A, the more turns it will make for each one turn of pulley A. The speed of a pulley that is being rotated by another pulley is inversely proportional to its diameter. We can express this relationship by the inverse variation equation

$$R = \frac{k}{d}$$

where *R* is the speed of the pulley in revolutions per minute (rpm) and *d* is the diameter of the pulley.

(continued)

EXAMPLE

Pulley A is being rotated by pulley B. Pulley B is 16 inches in diameter and is rotating at 240 rpm. Find the speed of pulley A if its diameter is 20 inches.

Solution

$R = \dfrac{k}{d}$ • Write the general form of the inverse variation equation for pulleys.

$240 = \dfrac{k}{16}$ • The speed of pulley B is 240 rpm; $R = 240$. Its diameter is 16 inches; $d = 16$.

$3840 = k$ • Solve for k by multiplying each side of the equation by 16. The constant of variation is **3840**.

$R = \dfrac{3840}{d}$ • Write the specific inverse variation equation by substituting 3840 for k.

$R = \dfrac{3840}{20}$ • The diameter of pulley A is 20 inches; $d = 20$.

$R = 192$

The speed of pulley A is 192 rpm.

Investigation Exercises

1. If gear A and gear B are interlocked, and gear A has 48 teeth and gear B has 12 teeth, how many complete revolutions will gear B make for every one complete revolution gear A makes?

2. Gear A is being rotated by gear B. If gear A has 40 teeth and makes 15 revolutions per minute, how many revolutions per minute will gear B, with 32 teeth, make?

3. Gear A has 20 teeth, turns at a rate of 240 rpm, and is being rotated by gear B. Gear B has 48 teeth. How many revolutions per minute does gear B make?

4. A 12-tooth gear mounted on a motor shaft drives a larger gear. The motor shaft rotates at 1450 rpm. The larger gear has 40 teeth. Find the speed, in revolutions per minute, of the larger gear.

5. Pulley B is being driven by pulley A. Pulley A has a diameter of 12 inches and turns at a rate of 300 rpm. Pulley B has a diameter of 15 inches. How many revolutions per minute does pulley B make?

6. A pulley on a drill press has a diameter of 9 inches and rotates at 1260 rpm. It is belted to a smaller pulley on an electric motor. The smaller pulley has a diameter of 5 inches. Find the speed of the smaller pulley.

7. In a car, the speed of the drive shaft, in revolutions per minute, is converted to rear axle rotation by a ring-and-pinion gear system. See the photograph at the left. The drive shaft is driven by the teeth on the pinion gear. The teeth on the ring gear are on the rear axle. For a pinion gear with 9 teeth and a ring gear with 40 teeth, what is the rear axle speed when the drive shaft is turning at 1200 rpm?

8. The description of pulleys refers to the diameters of circles.
   a. Does the circumference of a circle vary directly with its diameter? Explain.
   b. How is the circumference of a circle affected when the diameter is doubled?

Exercise Set 3.4

1. **a.** When are two quantities directly proportional?

   b. When are two quantities inversely proportional?

2. State whether the two quantities are directly proportional or inversely proportional. Explain your answer.

   a. One acre planted with wheat will produce 45 bushels of wheat.

   b. The intensity of a light is 15 foot-candles at a distance of 10 meters.

   c. A truck travels 17 miles on 1 gallon of fuel.

3. Find the constant of variation when y varies directly as x, and $y = 15$ when $x = 2$.

4. Find the constant of variation when n varies directly as the square of m, and $n = 64$ when $m = 2$.

5. Find the constant of proportionality when T varies inversely as S, and $T = 0.2$ when $S = 8$.

6. Find the constant of variation when W varies inversely as the square of V, and $W = 5$ when $V = 0.5$.

7. If y varies directly as x, and $x = 10$ when $y = 4$, find y when $x = 15$.

8. Given that L varies directly as P, and $L = 24$ when $P = 21$, find P when $L = 80$.

9. Given that P varies directly as R, and $P = 20$ when $R = 5$, find P when $R = 6$.

10. Given that M is directly proportional to P, and $M = 15$ when $P = 30$, find M when $P = 20$.

11. Given that W is directly proportional to the square of V, and $W = 50$ when $V = 5$, find W when $V = 12$.

12. If A varies directly as the square of r, and $A = \frac{22}{7}$ when $r = 1$, find A when $r = 7$.

13. If y varies inversely as x, and $y = 500$ when $x = 4$, find y when $x = 10$.

14. If L varies inversely as the square of d, and $L = 25$ when $d = 2$, find L when $d = 5$.

In Exercises 15–18, do not do any calculations. Determine the answer by thinking about the relationship between the variables x and y.

15. If $y = 5x$, as x grows larger, does y grow larger or smaller?

16. If $y = \frac{100}{x}$, as x grows larger, does y grow larger or smaller?

17. Given the variation equation $y = \frac{6.7}{x}$, which number is closest to the value of y when $x = 3.2$?

    a. 2   **b.** $\frac{1}{2}$ **c.** 20

18. Given the variation equation $y = \frac{101}{x}$, which number is closest to the value of x when $y = 26$?

    a. 4   **b.** $\frac{1}{4}$ **c.** 40

19. **Compensation** A worker's wage w is directly proportional to the number of hours h worked. If \$82 is earned for working 8 hours, how much is earned for working 30 hours?

20. **Physics** The distance d a spring will stretch varies directly as the force F applied to the spring. If a force of 12 pounds is required to stretch a spring 3 inches, what force is required to stretch the spring 5 inches?

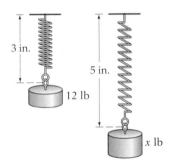

21. **Scuba Diving** The pressure P on a diver in the water varies directly as the depth d. If the pressure is 2.25 pounds per square inch when the depth is 5 feet, what is the pressure when the depth is 12 feet?

22. **Clerical Work** The number of words typed w is directly proportional to the time t spent typing. A typist can type 260 words in 4 minutes. Find the number of words typed in 15 minutes.

23. **Electricity** The current I varies directly as the voltage V in an electric circuit. If the current is 4 amperes when the voltage is 100 volts, find the current when the voltage is 75 volts.

24. **Travel** The distance traveled d varies directly as the time t of travel, assuming that the speed is constant. If it takes 45 minutes to travel 50 miles, how many hours will it take to travel 180 miles?

25. Automotive Technology The distance d required for a car to stop varies directly as the square of the velocity v of the car. If a car traveling at 50 mph requires 170 feet to stop, find the stopping distance for a car traveling at 65 mph.

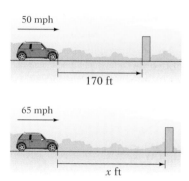

26. Physics The distance d an object falls is directly proportional to the square of the time t of the fall. If an object falls a distance of 8 feet in 0.5 second, how far will the object fall in 5 seconds?

27. Physics The distance s a ball will roll down an inclined plane is directly proportional to the square of the time t. If the ball rolls 5 feet in 1 second, how far will it roll in 4 seconds?

28. Consumerism The number of items N that can be purchased for a given amount of money is inversely proportional to the cost C of the item. If 390 items can be purchased when the cost per item is $.50, how many items can be purchased when the cost per item is $.20?

29. Geometry The length L of a rectangle of fixed area varies inversely as the width W. If the length of the rectangle is 8 feet when the width is 5 feet, find the length of the rectangle when the width is 4 feet.

30. Travel The time t of travel of an automobile trip varies inversely as the speed v. At an average speed of 65 mph, a trip took 4 hours. The return trip took 5 hours. Find the average speed of the return trip.

31. Electricity The current I in an electric circuit is inversely proportional to the resistance R. If the current is 0.25 ampere when the resistance is 8 ohms, find the resistance when the current is 1.2 amperes.

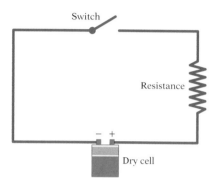

32. Physics At a constant temperature, the pressure P of a gas varies inversely as the volume V. If the pressure is 25 pounds per square inch when the volume is 400 cubic feet, find the pressure when the volume is 150 cubic feet.

33. Physics At a constant temperature, the volume V of a gas varies inversely as the pressure P on the gas. If the volume of the gas is 12 cubic feet when the pressure is 15 pounds per square foot, find the volume of the gas when the pressure is 4 pounds per square foot.

34. Business A computer company that produces personal computers has determined that the number of computers it can sell S per month is inversely proportional to the price P of the computer. Eighteen hundred computers can be sold per month if the price is $1800. How many computers can be sold per month if the price is $1500?

35. Light The intensity I of a light source is inversely proportional to the square of the distance d from the source. If the intensity is 80 foot-candles at a distance of 4 feet, what is the intensity when the distance is 10 feet?

36. Magnetism The repulsive force f between the north poles of two magnets is inversely proportional to the square of the distance d between them. If the repulsive force is 18 pounds when the distance is 3 inches, find the repulsive force when the distance is 1.2 inches.

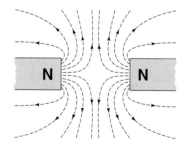

37. **Sound** The loudness L, measured in decibels, of a stereo speaker is inversely proportional to the square of the distance d from the speaker. The loudness is 20 decibels at a distance of 10 feet. What is the loudness at a distance of 6 feet from the speaker?

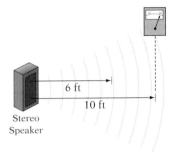

6 ft

10 ft

Stereo
Speaker

Extensions

CRITICAL THINKING

38. **Optics** The distance d a person can see to the horizon from a point above the surface of Earth varies directly as the square root of the height H. If, at a height of 500 feet, the horizon is 19 miles away, how far is the horizon from a point that is 800 feet high? Round to the nearest hundredth of a mile.

39. **Physics** The period p of a pendulum, or the time it takes for the pendulum to make one complete swing, varies directly as the square root of the length L of the pendulum. If the period of a pendulum is 1.5 seconds when the length is 2 feet, find the period when the length is 4.5 feet. Round to the nearest hundredth of a second.

40. Determine whether the statement is true or false.
   a. In the direct variation equation $y = kx$, if x increases, then y increases.
   b. In the inverse variation equation $y = \dfrac{k}{x}$, if x increases, then y increases.
   c. In the direct variation equation $y = kx^2$, if x doubles, then y doubles.
   d. If x varies inversely as y, then when x is doubled, y is doubled.
   e. If a varies inversely as b, then ab is a constant.
   f. If the area of a rectangle is held constant, then the area of the rectangle varies directly as the width.
   g. If the area of a rectangle is held constant, then the length varies directly as the width.

41. a. The variable y varies directly as the cube of x. If x is doubled, by what factor is y increased?
   b. The variable y varies inversely as the cube of x. If x is doubled, by what factor is y decreased?

COOPERATIVE LEARNING

42. You order an extra large pizza, cut into 12 slices, to be delivered to your home.
   a. If there are six people to share the pizza equally, how many slices does each person get?
   b. If two people leave before the pizza arrives, how many slices does each person get?
   c. Is the relationship between the number of people and the number of slices a direct variation or an inverse variation?
   d. What is the constant of variation?
   e. Write the equation that represents the variation.

EXPLORATIONS

43. Joint Variation A variation may involve more than two variables. If a quantity varies directly as the product of two or more variables, it is known as a **joint variation.**

The weight of a rectangular metal block is directly proportional to the volume of the block, given by length times width times height. The variation equation is written

$$\text{Weight} = kLWH$$

The weight of a block with $L = 24$ inches, $W = 12$ inches, and $H = 12$ inches is 72 pounds. Find the weight of another block with $L = 18$ inches, $W = 9$ inches, and $H = 18$ inches.

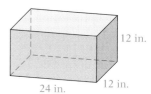

$$\text{Weight} = kLWH$$

$$72 = k(24)(12)(12)$$ • Use the values of the first block.

$$\frac{72}{(24)(12)(12)} = \frac{k(24)(12)(12)}{(24)(12)(12)}$$ • Solve for k.

$$\frac{1}{48} = k$$ • The constant of variation is $\frac{1}{48}$.

$$\text{Weight} = \frac{1}{48}LWH$$ • Write the variation equation, substituting $\frac{1}{48}$ for k.

$$\text{Weight} = \frac{1}{48}(18)(9)(18)$$ • Use the values of the second block.

$$\text{Weight} = 60.75$$

The weight of the other block is 60.75 pounds.

The force on a flat surface that is perpendicular to the wind is directly proportional to the product of the area of the surface and the square of the speed of the wind.

a. Write the joint variation.

b. What effect does doubling the area of the surface have on the force of the wind?

c. What effect does doubling the speed of the wind have on the force of the wind?

d. When the wind is blowing at 30 mph, the force on a 10-square-foot area is 45 pounds. Find the force on this area when the wind is blowing at 60 mph.

The power P in an electric circuit is directly proportional to the product of the current I and the square of the resistance R.

e. If the power is 100 watts when the current is 4 amperes and the resistance is 5 ohms, find the power when the current is 2 amperes and the resistance is 10 ohms.

The pressure p on a liquid varies directly as the product of the depth d and the density D of the liquid.

f. If the pressure is 37.5 pounds per square inch when the depth is 100 inches and the density is 1.2, find the pressure when the density remains the same and the depth is 60 inches.

CHAPTER 3   Summary

Key Terms

constant of proportionality [p. 183]
constant term [p. 122]
constant of variation [p. 183]
direct variation [p. 183]
equation [p. 122]
extremes of a proportion [p. 146]
first-degree equation in one variable [p. 122]
formula [p. 130]
inverse variation [p. 185]
like terms [p. 122]
literal equation [p. 130]
means of a proportion [p. 146]
numerical coefficient [p. 122]
percent [p. 159]
percent decrease [p. 170]
percent increase [p. 168]
proportion [p. 146]
rate [p. 139]
ratio [p. 143]
solution of an equation [p. 123]
solve an equation [p. 123]
term of a proportion [p. 146]
terms of a variable expression [p. 122]
unit price [p. 141]
unit rate [p. 140]
unit ratio [p. 144]
variable part [p. 122]
variable term [p. 122]

Essential Concepts

■ **Properties of Equations**

Addition Property
The same number can be added to each side of an equation without changing the solution of the equation.
If $a = b$, then $a + c = b + c$.

Subtraction Property
The same number can be subtracted from each side of an equation without changing the solution of the equation.
If $a = b$, then $a - c = b - c$.

Multiplication Property
Each side of an equation can be multiplied by the same nonzero number without changing the solution of the equation.
If $a = b$ and $c \neq 0$, then $ac = bc$.

Division Property
Each side of an equation can be divided by the same nonzero number without changing the solution of the equation.
If $a = b$ and $c \neq 0$, then $\frac{a}{c} = \frac{b}{c}$.

■ **Steps for Solving a First-Degree Equation in One Variable**

1. If the equation contains fractions, multiply each side of the equation by the least common multiple (LCM) of the denominators to clear the equation of fractions.

2. Use the Distributive Property to remove parentheses.

3. Combine any like terms on the right side of the equation and any like terms on the left side of the equation.

4. Use the Addition Property or the Subtraction Property to rewrite the equation with only one variable term and only one constant term.

5. Use the Multiplication Property or the Division Property to rewrite the equation with the variable alone on one side of the equation and a constant term on the other side of the equation.

■ **Solve a Literal Equation for One of the Variables**
The goal is to rewrite the equation so that the letter being solved for is alone on one side of the equation and all numbers and other variables are on the other side.

■ **Calculate a Unit Rate**
Divide the number in the numerator of the rate by the number in the denominator of the rate.

■ **Write a Ratio**
A ratio can be written in three different ways: as a fraction, as two numbers separated by a colon (:), or as two numbers separated by the word *to*. Although units, such as hours, miles, or dollars, are written as part of a rate, units are not written as part of a ratio.

■ **Cross-Products Method of Solving a Proportion**
If $\frac{a}{b} = \frac{c}{d}$, then $ad = bc$.

■ **Write a Percent as a Decimal**
Remove the percent sign and move the decimal point two places to the left.

■ **Write a Decimal as a Percent**
Move the decimal point two places to the right and write a percent sign to the right of the number.

■ **Write a Percent as a Fraction**
Remove the percent sign and multiply by $\frac{1}{100}$.

■ **Write a Fraction as a Percent**
First write the fraction as a decimal. Then write the decimal as a percent.

■ **Proportion Used to Solve Percent Problems**
$\frac{\text{Percent}}{100} = \frac{\text{amount}}{\text{base}}$

■ **Basic Percent Equation**
$PB = A$, where P is the percent, B is the base, and A is the amount.

■ **Direct Variation Equation**
$y = kx$, where k is the constant of variation.

■ **Inverse Variation Equation**
$y = \frac{k}{x}$, where k is the constant of variation.

CHAPTER 3 **Review Exercises**

In Exercises 1–5, solve the equation.

1. $5x + 3 = 10x - 17$

2. $3x + \frac{1}{8} = \frac{1}{2}$

3. $6x + 3(2x - 1) = -27$

4. $\frac{5}{12} = \frac{n}{8}$

5. $4y + 9 = 0$

In Exercises 6 and 7, solve the formula for the given variable.

6. $4x + 3y = 12$; y

7. $f = v + at$; t

8. If y varies directly as x, and $x = 20$ when $y = 5$, find y when $x = 12$.

9. If y varies inversely as x, and $y = 400$ when $x = 5$, find y when $x = 20$.

10. Meteorology In June, the temperatures at various elevations of the Grand Canyon can be approximated by the equation $T = -0.005x + 113.25$, where T is the temperature in degrees Fahrenheit and x is the elevation (distance above sea level) in feet. Use this equation to find the elevation at Inner Gorge, the bottom of the canyon, where the temperature is 101°F.

11. Falling Objects Find the time it takes for the velocity of a falling object to increase from 4 feet per second to 100 feet per second. Use the equation $v = v_0 + 32t$, where v is the final velocity of the falling object, v_0 is the initial velocity, and t is the time it takes for the object to fall.

12. Chemistry A chemist mixes 100 grams of water at 80°C with 50 grams of water at 20°C. Use the formula $m_1(T_1 - T) = m_2(T - T_2)$ to find the final temperature of the water after mixing. In this equation, m_1 is the quantity of water at the higher temperature, T_1 is the temperature of the hotter water, m_2 is the quantity of water at the lower temperature, T_2 is the temperature of the cooler water, and T is the final temperature of the water after mixing.

13. **Computer Bulletin Board Service** A computer bulletin board service charges $4.25 per month plus $.08 for each minute over 30 minutes that the service is used. For how many minutes did a subscriber use this service during a month in which the monthly charge was $4.97?

14. **Fuel Consumption** An automobile was driven 326.6 miles on 11.5 gallons of gasoline. Find the number of miles driven per gallon of gas.

15. **Real Estate** A house with an original value of $280,000 increased in value to $350,000 in 5 years. Write, as a fraction in simplest form, the ratio of the increase in value to the original value of the house.

16. **Farm Cropland** The graph below shows the average values per acre of U.S. farm cropland in the years 2000 to 2003. The average value per acre of farm cropland in 2003 was $800 less than twice the average value in 1997. Find the average value per acre of U.S. farm cropland in 1997.

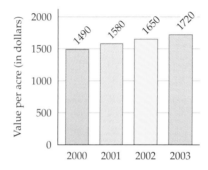

Average Value Per Acre of U.S. Farm Cropland

Source: National Agricultural Statistics Service

17. **City Populations** The table below shows the populations and areas of the five most populous cities in the United States.

   a. The cities are listed in the table according to population, from largest to smallest. Rank the cities according to population density, from largest to smallest.

   b. How many more people per square mile are there in New York City than in Houston? Round to the nearest whole number.

City	Population	Area (in square miles)
New York	8,008,000	321.8
Los Angeles	3,695,000	467.4
Chicago	2,896,000	228.469
Houston	1,954,000	594.03
Philadelphia	1,519,000	136

18. **Student–Faculty Ratios** The following table shows the numbers of full-time men and women undergraduates, as well as the numbers of full-time faculty, at the six colleges in Arizona. In parts a, b, and c, round ratios to the nearest whole number. (*Source: Barron's Profile of American Colleges*, 26th edition, © 2005)

University	Men	Women	Faculty
Arizona State University	14,875	16,054	1,722
Embry-Riddle Aeronautical University	1,181	244	87
Grand Canyon University	450	880	96
Northern Arizona University	4,468	6,566	711
Prescott College	305	435	49
University of Arizona	11,283	12,822	1,495

a. Calculate the student–faculty ratio at Prescott College. Write the ratio using a colon and using the word *to*. What does this ratio mean?

b. Which school listed has the lowest student–faculty ratio? the highest?

c. Which schools listed have the same student–faculty ratio?

19. Advertising The Randolph Company spent $350,000 for advertising last year. Department A and Department B share the cost of advertising in the ratio 3 : 7. Find the amount allocated to each department.

20. Gardening Three tablespoons of a liquid plant fertilizer are to be added to every 4 gallons of water. How many tablespoons of fertilizer are required for 10 gallons of water?

21. Federal Expenditures The table below shows how each dollar of projected spending by the federal government for a recent year was distributed. (*Source:* Office of Management and Budget) Of the items listed, defense is the only discretionary spending by the federal government; all other items are fixed expenditures. The government predicted total expenses of $2,230 billion for the year.

a. Is at least one-fifth of federal spending discretionary spending?

b. Find the ratio of the fixed expenditures to the discretionary spending.

c. Find the amount of the budget to be spent on fixed expenditures.

d. Find the amount of the budget to be spent on Social Security.

How Your Federal Tax Dollar Is Spent	
Health care	23 cents
Social Security	22 cents
Defense	18 cents
Other social aid	15 cents
Remaining government agencies and programs	14 cents
Interest on national debt	8 cents

22. Demographics According to the U.S. Bureau of the Census, the populations of males and females in the United States in 2025 and 2050 are projected to be as shown in the following table.

Year	Males	Females
2025	164,119,000	170,931,000
2050	193,234,000	200,696,000

a. What percent of the projected population in 2025 is female? Round to the nearest tenth of a percent.

b. Does the projected female population in 2050 differ by more or less than 1 percent from the projected female population in 2025?

23. **Populations** According to the Scarborough Report, San Francisco is the city that has the highest percent of people with current U.S. passports: 38.6% of the population, or approximately 283,700 people, have U.S. passports. Estimate the population of San Francisco. Round to the nearest hundred.

24. **Organ Transplants** The graph at the right shows the numbers of people, in thousands, who are listed on the national patient waiting list for various types of organ transplants. What percent of those listed are waiting for kidney transplants? Round to the nearest tenth of a percent.

25. **Nutrition** The table at the right shows the fat, saturated fat, cholesterol, and calorie contents of a 90-gram ground-beef burger and a 90-gram soy burger.

a. Compared with the beef burger, by what percent is the fat content decreased in the soy burger?

b. What is the percent decrease in cholesterol in the soy burger compared with the beef burger?

c. Calculate the percent decrease in calories in the soy burger compared with the beef burger.

26. **Music Sales** According to Nielsen SoundScan, album sales fell from 785.1 million in 2000 to 656.3 million in 2003. What percent decrease does this represent? Round to the nearest tenth of a percent.

27. **Soccer** The chart below shows, by age group, the numbers of girls playing youth soccer. Also shown are the percents of total youth soccer players who are girls. (*Source:* American Youth Soccer Organization)

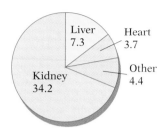

Numbers of People on the National Waiting List for Organ Transplants (in thousands)

Source: United Network for Organ Sharing

	Beef Burger	Soy Burger
Fat	24 g	4 g
Saturated fat	10 g	1.5 g
Cholesterol	75 mg	0 mg
Calories	280	140

Age group	5 – 6	7 – 8	9 – 10	11 – 12	13 – 14	15 – 16	17 – 18
Number of girls playing	23,805	45,181	46,758	39,939	26,157	11,518	4,430
Percent of all players	33%	33%	36%	40%	41%	42%	38%

a. Which age group has the greatest number of girls playing youth soccer?

b. Which age group has the largest percent of girls participating in youth soccer?

c. What percent of the girls playing youth soccer are ages 7 to 10? Round to the nearest tenth of a percent. Is this more or less than half of all the girls playing?

d. How many boys ages 17 to 18 play youth soccer? Round to the nearest ten.

e. The girls playing youth soccer represent 36% of all youth soccer players. How many young people play youth soccer? Round to the nearest hundred.

28. **Partnerships** Last year the profits for a partnership were $140,000. The two partners share the profits in the ratio 4:3. Find the amount received by each partner.

29. **Compensation** A worker's wage w is directly proportional to the number of hours h worked. If $82 is earned for working 4 hours, how much is earned for working 20 hours?

30. **Travel** The time t of travel of an automobile trip varies inversely as the speed v. Traveling at an average speed of 55 mph, a trip took 4 hours. The return trip took 5 hours. Find the average speed of the return trip.

CHAPTER 3   **Test**

In Exercises 1–4, solve the equation.

1. $\dfrac{x}{4} - 3 = \dfrac{1}{2}$

2. $x + 5(3x - 20) = 10(x - 4)$

3. $\dfrac{7}{16} = \dfrac{x}{12}$

4. $x = 12x - 22$

In Exercises 5 and 6, solve the formula for the given variable.

5. $x - 2y = 15;\ y$

6. $C = \dfrac{5}{9}(F - 32);\ F$

7. **Geysers** Old Faithful is a geyser in Yellowstone National Park. It is so named because of its regular eruptions for the past 100 years. An equation that can predict the approximate time until the next eruption is $T = 12.4L + 32$, where T is the time, in minutes, until the next eruption and L is the duration, in minutes, of the last eruption. Use this equation to determine the duration of the last eruption when the time between two eruptions is 63 minutes.

8. **Fines** A library charges a fine for each overdue book. The fine is 15¢ for the first day plus 7¢ a day for each additional day the book is overdue. If the fine for a book is 78¢, how many days overdue is the book?

9. **Rate of Speed** You drive 246.6 miles in 4.5 hours. Find your average rate in miles per hour.

10. **Parks** The table below lists the largest city parks in the United States. The land acreage of Griffith Park in Los Angeles is 3 acres more than five times the acreage of New York's Central Park. What is the acreage of Central Park?

City Park	Land Acreage
Cullen Park (Houston)	10,534
Fairmont Park (Philadelphia)	8,700
Griffith Park (Los Angeles)	4,218
Eagle Creek Park (Indianapolis)	3,800
Pelham Bay Park (Bronx, NY)	2,764
Mission Bay Park (San Diego)	2,300

11. **Baseball** The table below shows six Major League Baseball lifetime record holders for batting. (*Source: Information Please Almanac*)

    a. Calculate the number of at-bats per home run for each player in the table. Round to the nearest thousandth.

    b. The players are listed in the table alphabetically. Rank the players according to the number of at-bats per home run, starting with the best rate.

Baseball Player	Number of Times at Bat	Number of Home Runs Hit	Number of At-Bats per Home Run
Ty Cobb	11,429	4,191	_____
Billy Hamilton	6,284	2,163	_____
Rogers Hornsby	8,137	2,930	_____
Joe Jackson	4,981	1,774	_____
Tris Speaker	10,195	3,514	_____
Ted Williams	7,706	2,654	_____

12. **Golf** In a recent year, the gross revenues from television rights, merchandise, corporate hospitality, and tickets for the four major men's golf tournaments were as shown in the table at the right. What is the ratio, as a fraction in simplest form, of the gross revenue from the British Open to the gross revenue from the U.S. Open?

Golf Championship	Gross Revenue (in millions)
U. S. Open	$35
PGA Championship	$30.5
The Masters	$22
British Open	$20

13. **Partnerships** The two partners in a partnership share the profits of their business in the ratio 5 : 3. Last year the profits were $180,000. Find the amount received by each partner.

14. **Gardening** The directions on a bag of plant food recommend one-half pound for every 50 square feet of lawn. How many pounds of plant food should be used on a lawn that measures 275 square feet?

15. **Crime Rates** The table below lists the U.S. cities with populations over 100,000 that had the highest numbers of violent crimes per 1000 residents per year. The violent crimes include murder, rape, aggravated assault, and robbery. (*Source:* FBI Uniform Crime Reports)

City	Violent Crimes per 1000 People
Baltimore	13.4
Baton Rouge	13.9
Gainesville, Florida	14.2
Lawton, Oklahoma	13.3
Los Angeles - Long Beach	14.2
Miami - Dade	18.9
New Orleans	13.3
New York	13.9

a. Which city had the highest rate of violent crimes?

b. The population of Baltimore is approximately 703,000. Estimate the number of violent crimes committed in that city. Round to the nearest whole number.

16.   **Pets** During a recent year, nearly 1.2 million dogs or litters were registered with the American Kennel Club. The most popular breed was the Labrador retriever, with 172,841 registered. What percent of the registrations were Labrador retrievers? Round to the nearest tenth of a percent. (*Source:* American Kennel Club)

17.   **Digital Camera Sales** The graph at the right shows the projected worldwide sales of digital cameras for 2004 to 2007.

a. What percent of the total number of digital cameras expected to be sold during the four years is the number of digital cameras sold in 2004? Round to the nearest tenth of a percent.

b. Between which two consecutive years shown in the graph is the percent increase the greatest?

c. What is the percent increase in the projected number of digital cameras sold from 2004 to 2007? Round to the nearest tenth of a percent.

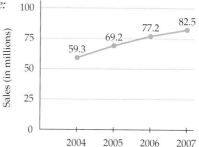

Projected Worldwide Sales of Digital Cameras

Source: IDC

18.   **Working Farms** The number of working farms in the United States in 1977 was 2.5 million. In 1987, there were 2.2 million working farms, and in 1997, there were 2.0 million working farms. (*Source:* CNN)

a. Find the percent decrease in the number of working farms from 1977 to 1997.

b. If the percent decrease in the number of working farms from 1997 to 2017 is the same as it was from 1977 to 1997, how many working farms will there be in the United States in 2017?

c.   Provide an explanation for the decrease in the number of working farms in the United States.

19. **Physics** The distance d a ball rolls down a ramp is directly proportional to the square of the time t the ball rolls. If a ball rolls a distance of 6 feet in 1 second, how far will the ball roll in 3 seconds?

20. **Sound** The loudness L, measured in decibels, of a stereo speaker is inversely proportional to the square of the distance d from the speaker. The loudness is 15 decibels at a distance of 10 feet. What is the loudness at a distance of 3 feet from the speaker? Round to the nearest decibel.

CHAPTER 4

Measurement and Geometric Models

Zero Defects, Zero Accidents, **Total Commitment**

4.3   9-65 E00, 71 ✓
4.4   5-77 E00 ✓
4.5   5-43 E00
4.6   1-61 E00

KAPL, Inc. **On the Path to Mission Success**

4.1	The Metric System	4.5	Properties of Triangles
4.2	The U.S. Customary System	4.6	Volume and Surface Area
4.3	Basic Concepts of Euclidean Geometry	4.7	Introduction to Trigonometry
4.4	Perimeter and Area of Plane Figures		

Look closely at each of the four rectangles shown below. Which rectangle seems to have the most pleasant appearance?

Rectangle A

Rectangle B

Rectangle C

Rectangle D

The length of rectangle A is three times its width; the ratio of the length to the width is 3:1. Rectangle B is a square, so its length is equal to its width; the ratio of the length to the width is 1:1. The ratio of the length of rectangle C to its width is 1.2:1. The shape of rectangle D is referred to as a *golden rectangle*. In a **golden rectangle,** the ratio of the length to the width is approximately 1.6 to 1. The ratio is exactly

$$\frac{\text{length}}{\text{width}} = \frac{1 + \sqrt{5}}{2}$$

The early Greeks considered golden rectangles to be the most pleasing to the eye. They used them extensively in their art and architecture. Many people believe golden rectangles are exhibited in the Parthenon in Athens, Greece. It was built around 435 B.C. as a temple to the goddess Athena.

The World Wide Web is a wonderful source of information about golden rectangles. Simply enter "golden rectangle" in a search engine. It will respond with thousands of websites where you can learn more about the use of golden rectangles in art, architecture, music, nature, and mathematics. Be sure to research Leonardo da Vinci's use of the golden rectangle in his painting *Mona Lisa*.

Online Study Center

For online student resources, visit this textbook's Online Study Center at **college.hmco.com/pic/aufmannMTQR.**

The Metric System

The Metric System

International trade, or trade between nations, is a vital and growing segment of business in the world today. The United States, as a nation, is dependent on world trade, and world trade is dependent on internationally standardized units of measurement: the metric system.

In this section we will present the metric system of measurement and explain how to convert between different units.

The basic unit of *length*, or distance, in the metric system is the **meter** (m). One meter is approximately the distance from a doorknob to the floor. All units of length in the metric system are derived from the meter. Prefixes to the basic unit denote the length of each unit. For example, the prefix "centi-" means one-hundredth; therefore, one centimeter is 1 one-hundredth of a meter (0.01 m).

≈1 meter

▼ **point of interest**

Originally the meter (spelled *metre* in some countries) was defined as $\frac{1}{1,000,000}$ of the distance from the equator to the North Pole. Modern scientists have redefined the meter as the distance light travels in a vacuum in $\frac{1}{1,299,792,458}$ second.

kilo-	= 1 000	1 kilometer (km)	= 1 000 meters (m)
hecto-	= 100	1 hectometer (hm)	= 100 m
deca-	= 10	1 decameter (dam)	= 10 m
		1 meter (m)	= 1 m
deci-	= 0.1	1 decimeter (dm)	= 0.1 m
centi-	= 0.01	1 centimeter (cm)	= 0.01 m
milli-	= 0.001	1 millimeter (mm)	= 0.001 m

Note that in this list 1000 is written as 1 000, with a space between the 1 and the zeros. When writing numbers using metric units, separate each group of three numbers by a space instead of a comma. A space is also used after each group of three numbers to the right of a decimal point. For example, 31,245.2976 is written 31 245.297 6 in metric notation.

QUESTION   *Which unit in the metric system is one-thousandth of a meter?*

Mass and weight are closely related. **Weight** is the force of gravity pulling on an object. Therefore, an object's weight is less in space than on Earth's surface. However, the amount of material in the object, its **mass**, remains the same. On the surface of Earth, the terms *mass* and *weight* are often used interchangeably.

The basic unit of mass in the metric system is the **gram** (g). If a box 1 centimeter long on each side is filled with water, the mass of that water is 1 gram.

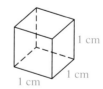

1 gram = the mass of water in a box that is 1 centimeter long on each side

---

ANSWER   *The millimeter is one-thousandth of a meter.*

The units of mass in the metric system have the same prefixes as the units of length.

1 kilogram (kg)	= 1 000 grams (g)
1 hectogram (hg)	= 100 g
1 decagram (dag)	= 10 g
1 gram (g)	= 1 g
1 decigram (dg)	= 0.1 g
1 centigram (cg)	= 0.01 g
1 milligram (mg)	= 0.001 g

The gram is a small unit of mass. A paperclip weighs about 1 gram. In many applications, the kilogram (1 000 grams) is a more useful unit of mass. This textbook weighs about 1 kilogram.

Weight ≈ 1 gram

QUESTION *Which unit in the metric system is equal to one thousand grams?*

Liquid substances are measured in units of *capacity*. The basic unit of capacity in the metric system is the **liter** (L). One liter is defined as the capacity of a box that is 10 centimeters long on each side.

The units of capacity in the metric system have the same prefixes as the units of length.

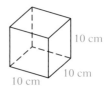

1 liter = the capacity of a box that is
10 centimeters long on each side

1 kiloliter (kl)	= 1 000 liters (L)
1 hectoliter (hl)	= 100 L
1 decaliter (dal)	= 10 L
1 liter (L)	= 1 L
1 deciliter (dl)	= 0.1 L
1 centiliter (cl)	= 0.01 L
1 milliliter (ml)	= 0.001 L

Converting between units in the metric system involves moving the decimal point to the right or to the left. Listing the units in order from left to right and from largest to smallest will indicate how many places to move the decimal point and in which direction.

To convert 3 800 cm to meters, write the units of length in order from left to right and from largest to smallest.

km   hm   dam   m   dm   cm   mm

2 places

3 800 cm = 38.00 m

2 places

• Converting from cm to m requires moving 2 places to the left.

• Move the decimal point the same number of places and in the same direction.

ANSWER *The kilogram is equal to one thousand grams.*

In the metric system, all prefixes represent powers of 10. Therefore, when converting between units, we are multiplying or dividing by a power of 10.

Convert 2.7 kg to grams.

kg hg dag g dg cg mg

3 places

2.7 kg = 2 700 g

3 places

- Write the units of mass in order from left to right and from largest to smallest.
- Converting from kilogram to gram requires moving 3 places to the right.
- Move the decimal point the same number of places and in the same direction.

EXAMPLE 1 ■ Convert Units in the Metric System of Measurement

Convert.

a. 4.08 m to centimeters   **b.** 5.93 g to milligrams

c. 82 ml to liters   **d.** 9 kl to liters

Solution

a. Write the units of length from largest to smallest.

km hm dam (m) dm (cm) mm

Converting from meter to centimeter requires moving 2 places to the right.

4.08 m = 408 cm

b. Write the units of mass from largest to smallest.

kg hg dag (g) dg cg (mg)

Converting from gram to milligram requires moving 3 places to the right.

5.93 g = 5 930 mg

c. Write the units of capacity from largest to smallest.

kl hl dal (L) dl cl (ml)

Converting from milliliter to liter requires moving 3 places to the left.

82 ml = 0.082 L

d. Write the units of capacity from largest to smallest.

(kl) hl dal (L) dl cl ml

Converting from kiloliter to liter requires moving 3 places to the right.

9 kl = 9 000 L

CHECK YOUR PROGRESS 1   Convert.

a. 1 295 m to kilometers   **b.** 7 543 g to kilograms

c. 6.3 L to milliliters   **d.** 2 kl to liters

Solution   *See page S15.*

Other prefixes in the metric system are becoming more common as a result of technological advances in the computer industry. For example:

tera-   = 1 000 000 000 000
giga-   = 1 000 000 000
mega-   = 1 000 000
micro- = 0.000 001
nano-   = 0.000 000 001
pico-   = 0.000 000 000 001

A **bit** is the smallest unit of code that computers can read; it is a binary digit, either a 0 or a 1. Usually bits are grouped into bytes of 8 bits. Each **byte** stands for a letter, number, or any other symbol we might use in communicating information. For example, the letter W can be represented by 01010111. The amount of memory in a computer hard drive is measured in terabytes, gigabytes, and megabytes. The speed of a computer used to be measured in microseconds and then nanoseconds, but now the speed is measured in picoseconds.

Here are a few more examples of how these prefixes are used.

The mass of Earth gains 40 Gg (gigagrams) each year from captured meteorites and cosmic dust.

The average distance from Earth to the moon is 384.4 Mm (megameters), and the average distance from Earth to the sun is 149.5 Gm (gigameters).

The wavelength of yellow light is 590 nm (nanometers).

The diameter of a hydrogen atom is about 70 pm (picometers).

There are additional prefixes in the metric system, both larger and smaller. We may hear them more and more often as computer chips hold more and more information, as computers get faster and faster, and as we learn more and more about objects in our universe that are great distances away.

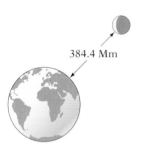

384.4 Mm

Math Matters   Correct Metric Usage

The U.S. Metric Association provides standards for usage of the metric system. Examples of correct and incorrect usage are listed below.

1. SI is the abbreviation for the Système International d'Unités, the modern version of the metric system. It is incorrect to abbreviate it as S.I.

2. The short forms for SI units, such as cm for centimeter, are called symbols, not abbreviations.

3. SI symbols do not end with a period unless they are the last word in a sentence. For example, 10 km is correct, and 10 km. is incorrect.

4. SI symbols should be preceded by digits (numbers), and a space must separate the digits from the symbol. For example, 5 kg is correct, and 5kg is incorrect.

5. SI symbols always are written in singular form. For example, 1 mg and 500 mg are correct; 500 mgs is incorrect. It is correct to pluralize metric unit names when they are written out, as in 500 milligrams.

6. In general, when pronouncing an SI unit involving a prefix, place the stress on the first syllable. Therefore, the preferred pronunciation of the word *kilometer* is KILL-oh-meet-ur rather than kill-AHM-it-ur.

EXAMPLE 2 ■ Solve an Application Involving Metric Units

The thickness of a single sheet of paper is 0.07 mm. Find the height in centimeters of a ream of paper. A ream is 500 sheets of paper.

Solution

$$0.07(500) = 35$$

 • Multiply the thickness of each sheet (0.07 mm) by the number of sheets in a ream (500). This is the height of a ream in millimeters.

$$35 \text{ mm} = 3.5 \text{ cm}$$

 • Convert millimeters to centimeters.

The height of a ream of this paper is 3.5 cm.

CHECK YOUR PROGRESS 2   One egg contains 274 mg of cholesterol. How many grams of cholesterol are in one dozen eggs?

Solution   *See page S15.*

Investigation

Electrical Energy

The **watt-hour** is used for measuring electrical energy. One watt-hour is the amount of energy required to lift 1 kg a distance of 370 m. A light bulb rated at 100 watts (W) will emit 100 watt-hours (Wh) of energy each hour.
   Recall that the prefix kilo- means 1 000.

1 kilowatt-hour (kWh) = 1 000 watt-hours (Wh)

EXAMPLE

A 150-watt bulb is on for 8 h. At 8¢ per kilowatt-hour, find the cost of the energy used.

Solution

$$150(8) = 1\ 200$$ • Find the number of watt-hours used.

$$1\ 200 \text{ Wh} = 1.2 \text{ kWh}$$ • Convert watt-hours to kilowatt-hours.

$$1.2(0.08) = 0.096$$ • Multiply the number of kilowatt-hours used by the cost, in dollars, per kilowatt-hour.

The cost of the energy used is $.096.

Investigation Exercises

1. Find the number of kilowatt-hours of energy used when a 150-watt light bulb burns for 200 h.

2. A fax machine is rated at 9 W when the machine is in standby mode and at 36 W when in operation. How many kilowatt-hours of energy are used during a week in which the fax machine is in standby mode for 39 h and in operation for 6 h?

(continued)

3. How much does it cost to run a 2 200-watt air conditioner for 8 h at 9¢ per kilowatt-hour? Round to the nearest cent.

4. A space heater is used for 3 h. The heater uses 1 400 W per hour. Electricity costs 11.1¢ per kilowatt-hour. Find the cost of using the electric heater. Round to the nearest cent.

5. A TV set rated at 1 800 W is on for an average of 3.5 h per day. At 7.2¢ per kilowatt-hour, find the cost of operating the set for 1 week. Round to the nearest cent.

6. A welder uses 6.5 kWh of energy each hour. Find the cost of using the welder for 6 h a day for 30 days. The cost is 9.4¢ per kilowatt-hour.

7. A microwave oven rated at 500 W is used an average of 20 minutes per day. At 8.7¢ per kilowatt-hour, find the cost of operating the oven for 30 days.

8. A 60-watt Sylvania Long Life Soft White Bulb has a light output of 835 lumens and an average life of 1 250 h. A 34-watt Sylvania Energy Saver Bulb has a light output of 400 lumens and an average life of 1 500 h.

   a. Is the light output of the Energy Saver Bulb more or less than half that of the Long Life Soft White Bulb?

   b. If electricity costs 10.8¢ per kilowatt-hour, what is the difference in cost between using the Long Life Soft White Bulb for 150 h and using the Energy Saver Bulb for 150 h? Round to the nearest cent.

Exercise Set 4.1

1. In the metric system, what is the basic unit of length? of liquid measure? of weight?

2. a. Explain how to convert meters to centimeters.

   b. Explain how to convert milliliters to liters.

In Exercises 3–26, name the unit in the metric system that most likely would be used to measure each of the following.

3. The distance from New York to London

4. The weight of a truck

5. A person's waist size

6. The amount of coffee in a mug

7. The weight of a thumbtack

8. The amount of water in a swimming pool

9. The distance a baseball player hits a baseball

10. A person's hat size

11. The amount of fat in a slice of cheddar cheese

12. A person's weight

13. The amount of maple syrup served with pancakes

14. The amount of water in a water cooler

15. The amount of vitamin C in a vitamin tablet

16. A serving of cereal

17. The width of a hair

18. A person's height

19. The amount of medication in an aspirin

20. The weight of a lawnmower

21. The weight of a slice of bread

22. The contents of a bottle of salad dressing

23. The amount of water a family uses monthly

24. The newspapers collected at a recycling center

25. The amount of liquid in a bowl of soup

26. The distance to the bank

27. **a.** Complete the table.

Metric System Prefix	Symbol	Magnitude	Means Multiply the Basic Unit By:
tera-	T	10^{12}	1 000 000 000 000
giga-	G	?	1 000 000 000
mega-	M	10^6	?
kilo-	?	?	1 000
hecto-	h	?	100
deca-	da	10^1	?
deci-	d	$\frac{1}{10}$	?
centi-	?	$\frac{1}{10^2}$	?
milli-	?	?	0.001
micro-	μ (mu)	$\frac{1}{10^6}$	?
nano-	n	$\frac{1}{10^9}$	?
pico-	p	?	0.000 000 000 001

b. How can the magnitude column in the table above be used to determine how many places to move the decimal point when converting to the basic unit in the metric system?

In Exercises 28–57, convert the given measure.

28. 42 cm = _____ mm

29. 91 cm = _____ mm

30. 360 g = _____ kg

31. 1 856 g = _____ kg

32. 5 194 ml = _____ L

33. 7 285 ml = _____ L

34. 2 m = _____ mm

35. 8 m = _____ mm

36. 217 mg = _____ g

37. 34 mg = _____ g

38. 4.52 L = _____ ml

39. 0.029 7 L = _____ ml

40. 8 406 m = _____ km

41. 7 530 m = _____ km

42. 2.4 kg = _____ g

43. 9.2 kg = _____ g

44. 6.18 kl = _____ L

45. 0.036 kl = _____ L

46. 9.612 km = _____ m

47. 2.35 km = _____ m

48. 0.24 g = _____ mg

49. 0.083 g = _____ mg

50. 298 cm = _____ m

51. 71.6 cm = _____ m

52. 2 431 L = _____ kl

53. 6 302 L = _____ kl

54. 0.66 m = _____ cm

55. 4.58 m = _____ cm

56. 243 mm = _____ cm

57. 92 mm = _____ cm

58. The Olympics

  a. One of the events in the summer Olympic Games is the 50 000-meter walk. How many kilometers do the entrants in this event walk?

  b. One of the events in the winter Olympic Games is 10 000-meter speed skating. How many kilometers do the entrants in this event skate?

59. Gemology A carat is a unit of weight equal to 200 mg. Find the weight in grams of a 10-carat precious stone.

60. Crafts How many pieces of material, each 75 cm long, can be cut from a bolt of fabric that is 6 m long?

61. Fundraising A walkathon had two checkpoints. One checkpoint was 1 400 m from the starting point. The second checkpoint was 1 200 m from the first checkpoint and was 1 800 m from the finish line. How long was the walk? Express the answer in kilometers.

62. Consumerism How many 240-milliliter servings are in a 2-liter bottle of cola? Round to the nearest whole number.

63. Business An athletic club uses 800 ml of chlorine each day for its swimming pool. How many liters of chlorine are used in a month of 30 days?

64. Carpentry Each of the four shelves in a bookcase measures 175 cm. Find the cost of the shelves when the price of lumber is $15.75 per meter.

65. Consumerism Find the total cost of three packages of ground meat weighing 540 g, 670 g, and 890 g if the price per kilogram is $9.89. Round to the nearest cent.

66. Orchards How many kilograms of fertilizer are necessary to fertilize 400 trees in an apple orchard if 300 g of fertilizer is used for each tree?

67. Consumerism The printed label from a container of milk is shown below. To the nearest whole number, how many 230-milliliter servings are in the container?

1 GAL. (3.78 L)

68. Consumerism A 1.19-kilogram container of Quaker Oats contains 30 servings. Find the number of grams in one serving of the oatmeal. Round to the nearest gram.

69. Health A patient is advised to supplement her diet with 2 g of calcium per day. The calcium tablets she purchases contain 500 mg of calcium per tablet. How many tablets per day should the patient take?

70. Chemistry A laboratory assistant is in charge of ordering acid for three chemistry classes of 30 students each. Each student requires 80 ml of acid. How many liters of acid should be ordered? The assistant must order by the whole liter.

71. Consumerism A case of 12 one-liter bottles of apple juice costs $19.80. A case of 24 cans, each can containing 340 ml of apple juice, costs $14.50. Which case of apple juice costs less per milliliter?

72. Construction A column assembly is being constructed in a building. The components are shown in the diagram below. What is the height of the column (the distance between the 1.25-centimeter plates)?

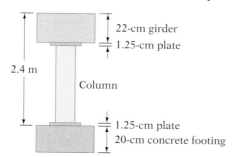

73. Light The distance between Earth and the sun is approximately 150 000 000 km. Light travels 300 000 000 m in 1 s. Approximately how long does it take for light to reach Earth from the sun?

74. Why is it advantageous to have internationally standardized units of measurement?

Extensions

CRITICAL THINKING

75. Business A service station operator bought 85 kl of gasoline for $38,500. The gasoline was sold for $.658 per liter. Find the profit on the 85 kl of gasoline.

76. Business For $149.50, a cosmetician buys 5 L of moisturizer and repackages it in 125-milliliter jars. Each jar costs the cosmetician $.55. Each jar of moisturizer is sold for $8.95. Find the profit on the 5 L of moisturizer.

77. Business A health food store buys nuts in 10-kilogram containers and repackages the nuts for resale. The store packages the nuts in 200-gram bags, costing $.06 each, and sells them for $2.89 per bag. Find the profit on a 10-kilogram container of nuts costing $75.

COOPERATIVE LEARNING

78. Form two debating teams. One team should argue in favor of changing to the metric system in the United States, and the other should argue against it.

SECTION 4.2 | # The U.S. Customary System

The U.S. Customary System of Measurement

In the first State of the Union Address, George Washington advocated "uniformity in the currency, weights, and measures of the United States." His recommendation led to the adoption of the U.S. Customary System of measurement.

The standard U.S. Customary System units of length are **inch, foot, yard,** and **mile.** The abbreviations for these units of length are in., ft, yd, and mi. Equivalences between units of length in the U.S. Customary System are:

$$1 \text{ ft} = 12 \text{ in.}$$
$$1 \text{ yd} = 3 \text{ ft}$$
$$1 \text{ yd} = 36 \text{ in.}$$
$$1 \text{ mi} = 5280 \text{ ft}$$

historical note

The ancient Greeks devised the foot measurement, which they usually divided into 16 fingers. It was the Romans who subdivided the foot into 12 units called *inches*. The word *inch* is derived from the Latin word *uncia*, meaning "a twelfth part."

The Romans also used a unit called *pace*, which equaled two steps. One thousand paces equaled one mile. The word for mile is derived from the Latin word *mille*, which means "thousand."

The word *quart* has its root in the Latin word *quartus*, which means "one-fourth"; a quart is one-fourth of a gallon. The same Latin word is the root of other English words such as *quarter*, *quadrilateral*, and *quartet*. ■

Weight is a measure of how strongly Earth is **pulling** on an object. The U.S. Customary System units of weight are **ounce, pound,** and **ton.** The abbreviation for ounce is oz, and the abbreviation for pound is lb. Equivalences between units of weight in the U.S. Customary System are:

$$1 \text{ lb} = 16 \text{ oz}$$
$$1 \text{ ton} = 2000 \text{ lb}$$

Liquids are measured in units of **capacity.** The standard U.S. Customary System units of capacity (and their abbreviations) are the **fluid ounce** (fl oz), **cup** (c), **pint** (pt), **quart** (qt), and **gallon** (gal). Equivalences between units of capacity in the U.S. Customary System are:

$$1 \text{ c} = 8 \text{ fl oz}$$
$$1 \text{ pt} = 2 \text{ c}$$
$$1 \text{ qt} = 4 \text{ c}$$
$$1 \text{ gal} = 4 \text{ qt}$$

Area is a measure of the amount of surface in a region. The standard U.S. Customary System units of area are **square inch** (in^2), **square foot** (ft^2), **square yard** (yd^2), **square mile** (mi^2), and **acre.** Equivalences between units of area in the U.S. Customary System are:

$$1 \text{ ft}^2 = 144 \text{ in}^2$$
$$1 \text{ yd}^2 = 9 \text{ ft}^2$$
$$1 \text{ acre} = 43,560 \text{ ft}^2$$
$$1 \text{ mi}^2 = 640 \text{ acres}$$

In solving application problems, scientists, engineers, and other professionals find it useful to include the units as they work through the solutions to problems so that they know the correct units associated with the answers. Using units to organize and check the correctness of an application is called **dimensional analysis.** Applying dimensional analysis to application problems requires converting units as well as multiplying and dividing units.

The equivalent measures listed above can be used to form *conversion rates* to change one unit of measurement to another. For example, the equivalent measures **1 mi** and **5280 ft** are used to form the following conversion rates:

$$\frac{1 \text{ mi}}{5280 \text{ ft}} \qquad \frac{5280 \text{ ft}}{1 \text{ mi}}$$

Because **1 mi = 5280 ft,** both of the conversion rates $\dfrac{1 \text{ mi}}{5280 \text{ ft}}$ and $\dfrac{5280 \text{ ft}}{1 \text{ mi}}$ are equal to 1.

To convert 3 mi to feet, multiply 3 mi by the conversion rate $\dfrac{5280 \text{ ft}}{1 \text{ mi}}$.

$$3 \text{ mi} = 3 \text{ mi} \cdot 1 = \frac{3 \text{ mi}}{1} \cdot \frac{5280 \text{ ft}}{1 \text{ mi}} = \frac{3 \cancel{\text{ mi}} \cdot 5280 \text{ ft}}{1 \cancel{\text{ mi}}} = 3 \cdot 5280 \text{ ft} = 15,840 \text{ ft}$$

There are important points to notice in the above example. First, you can think of dividing the numerator and denominator by the common unit "mile" just as you would divide the numerator and denominator of a fraction by a common factor. Second, the conversion rate $\dfrac{5280 \text{ ft}}{1 \text{ mi}}$ is equal to 1, and multiplying an expression by 1 does not change the value of the expression.

In the preceding example, we had the choice of two conversion rates, $\dfrac{1 \text{ mi}}{5280 \text{ ft}}$ or $\dfrac{5280 \text{ ft}}{1 \text{ mi}}$. In the conversion rate chosen, the unit in the numerator is the same as the unit desired in the answer (ft). The unit in the denominator is the same as the unit in the given measurement (mi).

QUESTION   *What conversion rate would you use to convert each of the following?*
a. *Feet to inches*   *b.* *Inches to feet*
c. *Pounds to ounces*   *d.* *Ounces to pounds*

EXAMPLE 1 ■ **Convert Units in the U.S. Customary System of Measurement**

Convert.

a. 36 fl oz to cups   **b.** $4\frac{1}{2}$ tons to pounds

Solution

a. The equivalence is 1 c = 8 fl oz. The conversion rate must have cups in the numerator and fluid ounces in the denominator: $\dfrac{1 \text{ c}}{8 \text{ fl oz}}$.

$$36 \text{ fl oz} = 36 \text{ fl oz} \cdot 1 = \frac{36 \text{ fl oz}}{1} \cdot \frac{1 \text{ c}}{8 \text{ fl oz}}$$

$$= \frac{36 \cancel{\text{ fl oz}} \cdot 1 \text{ c}}{8 \cancel{\text{ fl oz}}}$$

$$= \frac{36 \text{ c}}{8} = 4\frac{1}{2} \text{ c}$$

b. The equivalence is 1 ton = 2000 lb. The conversion rate must have pounds in the numerator and tons in the denominator: $\dfrac{2000 \text{ lb}}{1 \text{ ton}}$.

$$4\frac{1}{2} \text{ tons} = 4\frac{1}{2} \text{ tons} \cdot 1 = \frac{9}{2} \text{ tons} \cdot \frac{2000 \text{ lb}}{1 \text{ ton}}$$

$$= \frac{9 \text{ tons}}{2} \cdot \frac{2000 \text{ lb}}{1 \text{ ton}}$$

$$= \frac{9 \cancel{\text{ tons}} \cdot 2000 \text{ lb}}{2 \cdot 1 \cancel{\text{ ton}}}$$

$$= \frac{9 \cdot 2000 \text{ lb}}{2}$$

$$= \frac{18,000 \text{ lb}}{2}$$

$$= 9000 \text{ lb}$$

ANSWER   *a.* $\dfrac{12 \text{ in.}}{1 \text{ ft}}$ *b.* $\dfrac{1 \text{ ft}}{12 \text{ in.}}$ *c.* $\dfrac{16 \text{ oz}}{1 \text{ lb}}$ *d.* $\dfrac{1 \text{ lb}}{16 \text{ oz}}$

CHECK YOUR PROGRESS 1   Convert.

a. 40 in. to feet   **b.** 1 mi to yards

Solution   *See page S15.*

EXAMPLE 2 ■ Convert Units of Time

Convert 2.5 h to seconds.

Solution
We need to convert hours to minutes and minutes to seconds. The equivalences are
1 h = 60 min and 1 min = 60 s.
   Choose the conversion rates so that we can divide by the unit "hours" and by the unit "minutes."

$$2.5\,h = 2.5\,h \cdot 1 \cdot 1 = \frac{2.5\,h}{1} \cdot \frac{60\,min}{1\,h} \cdot \frac{60\,s}{1\,min}$$

$$= \frac{2.5\,\cancel{h} \cdot 60\,\cancel{min} \cdot 60\,s}{1 \cdot 1\,\cancel{h} \cdot 1\,\cancel{min}}$$

$$= \frac{2.5 \cdot 60 \cdot 60\,s}{1} = 9000\,s$$

CHECK YOUR PROGRESS 2   Convert 2880 min to days.

Solution   *See page S15.*

MathMatters   **Which is Heavier, a Pound of Feathers or a Pound of Gold?**

It would seem that a pound is a pound, whether it is feathers or gold. However, this is not the case.

   The weights of metals, such as gold and silver, are measured using the troy weight system. Originally the 12-ounce troy pound was the basis on which coins were minted and weighed. The weights of feathers, meat, people, and other non-metal quantities are measured using the avoirdupois weight system. In the Middle Ages, merchants preferred the 16-ounce avoirdupois pound because it is easily divided into halves, quarters, and eighths.

   In traditional English law, various pound weights were stated as multiples of the grain, which was originally the weight of a single barleycorn, or approximately 0.02 oz. A pound of feathers weighs 7000 grains, but a pound of gold weighs only 5760 grains. Therefore, a pound of feathers weighs more than a pound of gold.

   To complicate matters, since 1 avoirdupois pound contains 16 oz and 1 troy pound contains 12 oz, there are 437.5 grains in 1 avoirdupois ounce but 480 grains in a troy ounce. Thus an ounce of feathers weighs less than an ounce of gold.

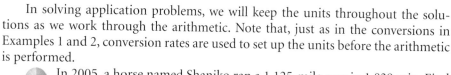

In solving application problems, we will keep the units throughout the solutions as we work through the arithmetic. Note that, just as in the conversions in Examples 1 and 2, conversion rates are used to set up the units before the arithmetic is performed.

 In 2005, a horse named Shaniko ran a 1.125-mile race in 1.829 min. Find Shaniko's average speed for that race in miles per hour. Round to the nearest tenth.

Shaniko's rate is $\dfrac{1.125 \text{ mi}}{1.829 \text{ min}}$.

- Write Shaniko's speed as a rate in fraction form. Speed is in *miles per minute*: the distance (1.125 mi) is in the numerator and the time (1.829 min) is in the denominator.

$$\frac{1.125 \text{ mi}}{1.829 \text{ min}} = \frac{1.125 \text{ mi}}{1.829 \text{ min}} \cdot \frac{60 \text{ min}}{1 \text{ h}}$$

- Multiply the fraction by the conversion rate $\dfrac{60 \text{ min}}{1 \text{ h}}$.

$$= \frac{67.5 \text{ mi}}{1.829 \text{ h}}$$

$$\approx 36.9 \text{ mph}$$

Shaniko's average speed was 36.9 mph.

EXAMPLE 3 ■ **Solve an Application Using Dimensional Analysis**

A carpet is to be installed in a room that is 20 ft wide and 30 ft long. At $28.50 per square yard, how much will it cost to carpet the room?

Solution

$A = LW$

- Use the formula for the area of a rectangle. $L = 30$ ft, $W = 20$ ft

$= 30 \text{ ft} \cdot 20 \text{ ft}$

$= 600 \text{ ft}^2$

- ft · ft can be written as ft².

$600 \text{ ft}^2 = \dfrac{600 \text{ ft}^2}{1} \cdot \dfrac{1 \text{ yd}^2}{9 \text{ ft}^2}$

- Use the conversion rate $\dfrac{1 \text{ yd}^2}{9 \text{ ft}^2}$ to convert the area to square yards.

$= \dfrac{600 \text{ yd}^2}{9}$

$= \dfrac{200 \text{ yd}^2}{3}$

$\text{Cost} = \dfrac{200 \text{ yd}^2}{3} \cdot \dfrac{\$28.50}{1 \text{ yd}^2}$

- Multiply the area by $\dfrac{\$28.50}{1 \text{ yd}^2}$ to find the cost.

$= \$1900$

The cost to carpet the room is $1900.

CHECK YOUR PROGRESS 3 Find the number of gallons of water in a fish tank that is 36 in. long and 23 in. wide and is filled to a depth of 16 in. Round to the nearest tenth of gallon. (*Note*: 1 gal = 231 in^3)

Solution   See page S15.

Conversion Between the U.S. Customary System and the Metric System

Because more than 90% of the world's population uses the metric system of measurement, converting U.S. Customary units to metric units is essential in trade and commerce—for example, in importing foreign goods and exporting domestic goods. Also, metric units are being used throughout the United States today. Cereal is packaged by the gram, 35-mm film is available, and soda is sold by the liter.

Approximate equivalences between the U.S. Customary System and the metric system are shown below.

Equivalences Between the U.S. Customary System and the Metric System

point of interest

The definition of one inch has been changed as a consequence of the wide acceptance of the metric system. One inch is now exactly 25.4 mm.

Units of Length	Units of Weight	Units of Capacity
1 in. = 2.54 cm	1 oz ≈ 28.35 g	1 L ≈ 1.06 qt
1 m ≈ 3.28 ft	1 lb ≈ 453.60 g	1 gal ≈ 3.79 L
1 m ≈ 1.09 yd	1 kg ≈ 2.20 lb	
1 mi ≈ 1.61 km		

These equivalences can be used to form conversion rates to change one measurement to another. For example, because 1 mi ≈ 1.61 km, the conversion rates $\dfrac{1 \text{ mi}}{1.61 \text{ km}}$ and $\dfrac{1.61 \text{ km}}{1 \text{ mi}}$ are each approximately equal to 1.

The procedure used to convert from one system to the other is identical to the conversion procedures performed in the U.S. Customary System in this section. Here is an example.

Convert 55 mi to kilometers.

$$55 \text{ mi} = \frac{55 \text{ mi}}{1}$$

$$\approx \frac{55 \text{ mi}}{1} \cdot \frac{1.61 \text{ km}}{1 \text{ mi}}$$

• The equivalence is **1 mi ≈ 1.61 km.** The conversion rate must have kilometers in the numerator and miles in the denominator: $\dfrac{1.61 \text{ km}}{1 \text{ mi}}$.

$$\approx \frac{55 \text{ m\!i}}{1} \cdot \frac{1.61 \text{ km}}{1 \text{ m\!i}}$$

• Divide the numerator and denominator by the common unit "mile."

$$\approx \frac{88.55 \text{ km}}{1}$$

• Multiply 55 times 1.61.

$$55 \text{ mi} \approx 88.55 \text{ km}$$

EXAMPLE 4 ■ Convert Units Between the U.S. Customary System and the Metric System

Convert.

a. 200 m to feet   **b.** 45 mph to kilometers per hour

Solution

a. $200 \text{ m} = \dfrac{200 \text{ m}}{1} \approx \dfrac{200 \text{ m}}{1} \cdot \dfrac{3.28 \text{ ft}}{1 \text{ m}} = 656 \text{ ft}$

b. $45 \text{ mph} = \dfrac{45 \text{ mi}}{1 \text{ h}} \approx \dfrac{45 \text{ mi}}{1 \text{ h}} \cdot \dfrac{1.61 \text{ km}}{1 \text{ mi}} = 72.45 \text{ km/h}$

CHECK YOUR PROGRESS 4   Convert. Round to the nearest hundredth.

a. 45 cm to inches   **b.** 75 km/h to miles per hour

Solution   *See page S15.*

EXAMPLE 5 ■ Solve an Application by Converting Units from the U.S. Customary System to the Metric System

The price of gasoline is $2.89 per gallon. Find the price per liter. Round to the nearest cent.

Solution

$\$2.89 \text{ per gallon} = \dfrac{\$2.89}{\text{gal}}$

$\approx \dfrac{\$2.89}{\text{gal}} \cdot \dfrac{1 \text{ gal}}{3.79 \text{ L}}$ • 1 gal ≈ 3.79 L. Use the conversion rate $\dfrac{1 \text{ gal}}{3.79 \text{ L}}$.

$\approx \$.76 \text{ per liter}$

The price is approximately $.76 per liter.

CHECK YOUR PROGRESS 5   The price of milk is $2.39 per gallon. Find the price per liter. Round to the nearest cent.

Solution   *See page S15.*

EXAMPLE 6 ■ Solve an Application by Converting Units from the Metric System to the U.S. Customary System

The price of gasoline is $.512 per liter. Find the price per gallon. Round to the nearest cent.

Solution

$$\$.512 \text{ per liter} = \frac{\$.512}{1 \text{ L}}$$

$$\approx \frac{\$.512}{1 \text{ L}} \cdot \frac{3.79 \text{ L}}{1 \text{ gal}}$$

• **1 gal** $\approx$ **3.79 L. Use the conversion rate** $\frac{3.79 \text{ L}}{1 \text{ gal}}$.

$$\approx \$1.94 \text{ per gallon}$$

The price is approximately $1.94 per gallon.

CHECK YOUR PROGRESS 6   The price of ice cream is $1.75 per liter. Find the price per gallon. Round to the nearest cent.

Solution   See page S15.

Investigation

Energy

Energy is defined as the ability to do work. Energy is stored in coal, in gasoline, in water behind a dam, and in one's own body.

One foot-pound (1 ft-lb) of energy is the amount of energy necessary to lift 1 pound a distance of 1 foot. One ft-lb is read as 1 foot pound.

To lift 500 lb a distance of 3 ft requires (3 ft)(500 lb) = 150 ft-lb of energy.

EXAMPLE

Find the energy required for a 150-pound person to climb a mile-high mountain.

Solution
In climbing the mountain, the person is lifting 150 lb a distance of 5280 ft.

Energy = (5280 ft)(150 lb) = 792,000 ft-lb

The energy required is 792,000 ft-lb.

Consumer items that use energy, such as furnaces, stoves, and air conditioners, are rated in terms of the **British thermal unit** (Btu). For example, a furnace might have a rating of 35,000 Btu per hour, which means that it releases 35,000 Btu of energy in 1 hour.

Because 1 Btu is approximately 778 ft-lb, the following conversion rate, equivalent to 1, is used:

$$\frac{778 \text{ ft-lb}}{1 \text{ Btu}}$$

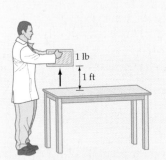
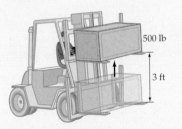

(continued)

EXAMPLE

A furnace is rated at 80,000 Btu per hour. How many foot-pounds of energy are released in 1 h?

Solution

$$80{,}000 \text{ Btu} = \frac{80{,}000 \text{ Btu}}{1} \cdot \frac{778 \text{ ft-lb}}{1 \text{ Btu}} \qquad \text{• Use the conversion rate } \frac{778 \text{ ft-lb}}{1 \text{ Btu}}.$$

$$= 62{,}240{,}000 \text{ ft-lb}$$

The furnace releases 62,240,000 ft-lb of energy in 1 h.

Investigation Exercises

1. A construction worker carries 3-pound blocks up a 12-foot flight of stairs. How many foot-pounds of energy are required to carry 600 blocks up the stairs?

2. A crane lifts a 1600-pound steel beam to the roof of a building 40 ft high. Find the amount of energy the crane requires in lifting the beam.

3. A furnance is rated at 56,000 Btu per hour. How many foot-pounds of energy does the furnace release in 1 h?

4. A furnace is rated at 48,000 Btu per hour. How many foot-pounds of energy are released in 1 h?

5. Find the amount of energy, in foot-pounds, given off when 1 lb of coal is burned. One pound of coal gives off 12,000 Btu of energy when burned.

6. Find the amount of energy, in foot-pounds, given off when 1 lb of gasoline is burned. One pound of gasoline gives off 21,000 Btu of energy when burned.

7. Mt. Washington is the highest mountain in the northeastern United States. How much more energy is required for a 180-pound person to climb Mt. Washington, a height of 6288 ft, than for a 120-pound person to climb the same mountain?

Exercise Set 4.2

1. Explain why we can multiply a distance measured in miles by the conversion rate $\dfrac{5280 \text{ ft}}{1 \text{ mi}}$.

2. Write two rates for the equivalence 1 gal = 4 qt.

3. Convert 64 in. to feet.

4. Convert 14 ft to yards.

5. Convert 42 oz to pounds.

6. Convert 4400 lb to tons.

7. Convert 7920 ft to miles.

8. Convert 42 c to quarts.

9. Convert 500 lb to tons.

10. Convert 90 oz to pounds.

11. Convert 10 qt to gallons.

12. How many pounds are in $1\frac{1}{4}$ tons?

13. How many fluid ounces are in $2\frac{1}{2}$ c?

14. How many ounces are in $2\frac{5}{8}$ lb?

15. Convert $2\frac{1}{4}$ mi to feet.

16. Convert 17 c to quarts.

17. Convert $7\frac{1}{2}$ in. to feet.

18. Convert $2\frac{1}{4}$ gal to quarts.

19. Convert 60 fl oz to cups.

20. Convert $1\frac{1}{2}$ qt to cups.

21. Convert $7\frac{1}{2}$ pt to quarts.

22. Convert 20 fl oz to pints.

23. How many yards are in $1\frac{1}{2}$ mi?

24. How many seconds are in 1 day?

25. **Time** When a person reaches the age of 35, for how many seconds has that person lived? Include 9 extra days for leap years.

26. **Interior Decorating** Fifty-eight feet of material is purchased for making pleated draperies. Find the total cost of the material if the price is $18 per yard.

27. **Catering** The Concord Theater serves punch during intermission. If each of 200 people drink 1 c of punch, how many gallons of punch should be prepared?

28. **Consumerism** A can of cranberry juice contains 25 fl oz. How many quarts of cranberry juice are in a case of 24 cans?

29. **Hiking** Five students are going backpacking in the desert. Each student requires 2 qt of water per day. How many gallons of water should the students take for a 3-day trip?

30. **Hiking** A hiker is carrying 5 qt of water. Water weighs $8\frac{1}{3}$ lb per gallon. Find the weight of the water carried by the hiker.

31. **Business** A garage mechanic purchases oil in a 50-gallon container for changing oil in customers' cars. After 35 oil changes requiring 5 qt each, how much oil is left in the 50-gallon container?

32. **Sound** The speed of sound is about 1100 ft/s. Find the speed of sound in miles per hour.

33. **Interior Decorating** Wall-to-wall carpeting is to be laid in the living room of the home shown in the floor plan below. At $24 per square yard, how much will the carpeting cost?

FIRST FLOOR

34. **Real Estate** A building lot with the dimensions shown in the diagram below is priced at $20,000 per acre. Find the price of the building lot.

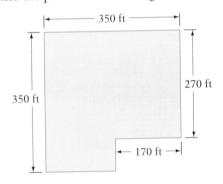

35. **Real Estate** A one-half acre commercial lot is selling for $3 per square foot. Find the price of the commercial lot.

In Exercises 36–55, convert. Round to the nearest hundredth or to the nearest cent.

36. Convert the 100-yard dash to meters.

37. Find the weight in kilograms of a 145-pound person.

38. Find the height in meters of a person 5 ft 8 in. tall.

39. Find the number of cups in 2 L of soda.

40. How many kilograms does a 15-pound turkey weigh?

41. Find the number of liters in 14.3 gal of gasoline.

42. Find the distance in feet of the 1 500-meter race.

43. Find the weight in pounds of an 86-kilogram person.

44. Find the number of gallons in 6 L of antifreeze.

45. Find the width in inches of 35-millimeter film.

46. Find the weight in ounces of 327 g of cereal.

47. How many gallons of water does a 24-liter aquarium hold?

48. Express 65 mph in kilometers per hour.

49. Convert 60 ft/s to meters per second.

50. Fat-free hot dogs cost $3.49 per pound. Find the price per kilogram.

51. Seedless watermelon costs $.59 per pound. Find the cost per kilogram.

52. Deck stain costs $24.99 per gallon. Find the cost per liter.

53. Express 80 km/h in miles per hour.

54. Express 30 m/s in feet per second.

55. Gasoline costs 83.5¢ per liter. Find the cost per gallon.

56. Fishing The largest trout ever caught in the state of Utah weighed 51 lb 8 oz. Find the trout's weight in kilograms. Round to the nearest tenth of a kilogram.

57. Earth Science The distance around Earth at the equator is 24,887 mi. What is this distance in kilometers?

58. Astronomy The distance from Earth to the sun is 93,000,000 mi. Calculate the distance from Earth to the sun in kilometers.

59. Is the statement true or false?

a. An ounce is less than a gram.

b. A liter is more than 4 c.

c. A meter is less than 3 ft.

d. A mile is more than 1 000 m.

e. A kilogram is greater than 16 oz.

f. 30 mph is less than 60 km/h.

Extensions
CRITICAL THINKING

60. a. Page 216 lists equivalences between the U.S. Customary System and the metric system of measurement. Explain why the word "rate" can be used in describing these equivalences.

b. Express each of the following as a unit rate:

1 in. = 2.54 cm; 1 kg ≈ 2.2 lb; 1 L ≈ 1.06 qt

61. For the following U.S. Customary System units, make an estimate of the metric equivalent. Then perform the conversion and see how close you came to the actual measurement.

60 mph	120 lb
6 ft	1 mi
1 gal	1 quarter-mile

EXPLORATIONS

62. Find examples of U.S. Customary System equivalences that were not included in this section. For example, 1 furlong = $\frac{1}{8}$ mi.

63. Write a paragraph describing the increased need for precision in our measurements as civilization progressed. Include a discussion of the need for precision in the aerospace industry.

| ## Basic Concepts of Euclidean Geometry

Lines and Angles

The word *geometry* comes from the Greek words for "earth" and "measure." Geometry was used by the ancient Egyptians to measure land and to build structures such as the pyramids.

Today geometry is used in many fields, such as physics, medicine, and geology. Geometry is also used in applied fields such as mechanical drawing and astronomy. Geometric forms are used in art and design.

historical note

Geometry is one of the oldest branches of mathematics. Around 350 B.C., Euclid (yo͞o′klĭd) of Alexandria wrote *Elements,* which contained all of the known concepts of geometry. Euclid's contribution to geometry was to unify various concepts into a single deductive system that was based on a set of postulates. ■

If you play a musical instrument, you know the meanings of the words *measure, rest, whole note,* and *time signature.* If you are a football fan, you have learned the terms *first down, sack, punt,* and *touchback.* Every field has its associated vocabulary. Geometry is no exception. We will begin by introducing two basic geometric concepts: point and line.

A **point** is symbolized by drawing a dot. A **line** is determined by two distinct points and extends indefinitely in both directions, as the arrows on the line at the right indicate. This line contains points A and B and is represented by $\overleftrightarrow{AB}$. A line can also be represented by a single letter, such as ℓ.

A **ray** starts at a point and extends indefinitely in *one* direction. The point at which a ray starts is called the **endpoint** of the ray. The ray shown at the right is denoted $\overrightarrow{AB}$. Point A is the endpoint of the ray.

A **line segment** is part of a line and has two endpoints. The line segment shown at the right is denoted by $\overline{AB}$.

QUESTION   *Classify each diagram as a line, a ray, or a line segment.*

a. E ———————————— F

b. C ———————————— D

c. J ———————————— K

The distance between the endpoints of $\overline{AC}$ is denoted by AC. If B is a point on $\overline{AC}$, then AC (the distance from A to C) is the sum of AB (the distance from A to B) and BC (the distance from B to C).

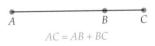

$AC = AB + BC$

Given $AB = 22$ cm and $BC = 13$ cm on $\overline{AC}$, find AC.

Make a drawing and write an equation to represent the distances between the points on the line segment.

Substitute the given distances for AB and BC into the equation. Solve for AC.

$$AB + BC = AC$$
$$22 + 13 = AC$$
$$35 = AC$$

$$AC = 35 \text{ cm}$$

ANSWER   *a. Ray*   *b. Line segment*   *c. Line*

EXAMPLE 1 ■ Find a Distance on a Line Segment

Given $MN = 14$ mm, $NO = 17$ mm, and $OP = 15$ mm on $\overline{MP}$, find MP.

Solution

$$MN + NO + OP = MP$$
• Write an equation to represent the distances on the line segment.

$$14 + 17 + 15 = MP$$
• Replace *MN* by 14, *NO* by 17, and *OP* by 15.

$$46 = MP$$
• Solve for *MP*.

$MP = 46$ mm

CHECK YOUR PROGRESS 1   Given $QR = 28$ cm, $RS = 16$ cm, and $ST = 10$ cm on $\overline{QT}$, find QT.

Solution   See page S16.

EXAMPLE 2 ■ Use an Equation to Find a Distance on a Line Segment

X, Y, and Z are all points on line ℓ. Given $XY = 9$ m and YZ is twice XY, find XZ.

Solution

$$XZ = XY + YZ$$
$$XZ = XY + 2(XY)$$ • *YZ* is twice *XY*.
$$XZ = 9 + 2(9)$$ • Replace *XY* by 9.
$$XZ = 9 + 18$$ • Solve for *XZ*.
$$XZ = 27$$

$XZ = 27$ m

CHECK YOUR PROGRESS 2 A, B, and C are all points on line ℓ. Given $BC = 16$ ft and $AB = \frac{1}{4}(BC)$, find AC.

Solution   See page S16.

In this section, we are discussing figures that lie in a plane. A **plane** is a flat surface with no thickness and no boundaries. It can be pictured as a desktop or blackboard that extends forever. Figures that lie in a plane are called **plane figures.**

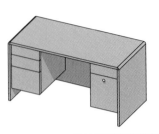

Lines in a plane can be intersecting or parallel. **Intersecting lines** cross at a point in the plane.

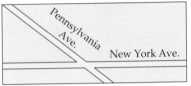

Parallel lines never intersect. The distance between them is always the same.

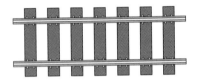

The symbol ∥ means "is parallel to." In the figure at the right, $j \parallel k$ and $\overline{AB} \parallel \overline{CD}$. Note that j contains $\overline{AB}$ and k contains $\overline{CD}$. Parallel lines contain parallel line segments.

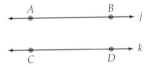

An **angle** is formed by two rays with the same endpoint. The **vertex** of the angle is the common endpoint of the two rays. The rays are called the **sides** of the angle.

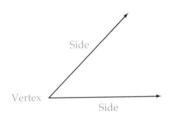

If A is a point on one ray of an angle, C is a point on the other ray, and B is the vertex, then the angle is called $\angle B$ or $\angle ABC$, where $\angle$ is the symbol for angle. Note that an angle can be named by the vertex or by giving three points, where the second point listed is the vertex. $\angle ABC$ could also be called $\angle CBA$.

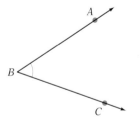

An angle can also be named by a variable written between the rays close to the vertex. In the figure at the right, $\angle x$ and $\angle QRS$ are two different names for the same angle, and $\angle y$ and $\angle SRT$ are two different names for the same angle. Note that in this figure, more than two rays share the endpoint R. In this case, the vertex alone cannot be used to name $\angle QRT$.

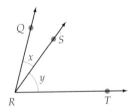

An angle is often measured in **degrees.** The symbol for degrees is a small raised circle, °. The angle formed by rotating a ray through a complete circle has a measure of 360°.

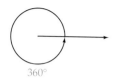

360°

A **protractor** is often used to measure an angle. Place the line segment near the bottom edge of the protractor on BC as shown in the figure below. Make sure the center of the line segment on the protractor is directly over the vertex. ∠ABC shown in the figure below measures 58°.

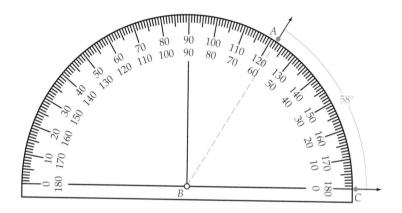

A 90° angle is called a **right angle.** The symbol ∟ represents a right angle.

90°

Perpendicular lines are intersecting lines that form right angles.

The symbol ⊥ means "is perpendicular to." In the figure at the right, $p \perp q$ and $\overline{AB} \perp \overline{CD}$. Note that line p contains $\overline{AB}$ and line q contains $\overline{CD}$. Perpendicular lines contain perpendicular line segments.

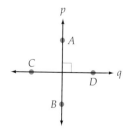

A **straight angle** is an angle whose measure is 180°. ∠AOB is a straight angle.

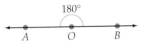

Straight angle

Complementary angles are two angles whose measures have the sum 90°.

∠A and ∠B at the right are complementary angles.

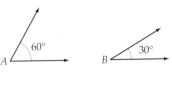

Supplementary angles are two angles whose measures have the sum 180°.

∠C and ∠D at the right are supplementary angles.

An **acute angle** is an angle whose measure is between 0° and 90°. ∠D above is an acute angle. An **obtuse angle** is an angle whose measure is between 90° and 180°. ∠C above is an obtuse angle.

The measure of ∠C is 110°. This is often written as $m\angle C = 110°$, where m is an abbreviation for "the measure of." This notation is used in Example 3.

EXAMPLE 3 ■ Determine If Two Angles are Complementary

Are angles E and F complementary angles?

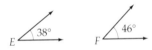

Solution

$$m\angle E + m\angle F = 38° + 46° = 84°$$

The sum of the measures of ∠E and ∠F is not 90°. Therefore, angles E and F are not complementary angles.

CHECK YOUR PROGRESS 3   Are angles G and H supplementary angles?

Solution   See page S16.

EXAMPLE 4 ■ Find the Measure of the Complement of an Angle

Find the measure of the complement of a 38° angle.

Solution

Complementary angles are two angles the sum of whose measures is 90°.
To find the measure of the complement, let *x* represent the complement of a 38° angle.
Write an equation and solve for *x*.

$$x + 38° = 90°$$
$$x = 52°$$

CHECK YOUR PROGRESS 4   Find the measure of the supplement of a 129° angle.

Solution   *See page S16.*

Adjacent angles are two angles that have a common vertex and a common side but have no interior points in common. In the figure at the right, $\angle DAC$ and $\angle CAB$ are adjacent angles.

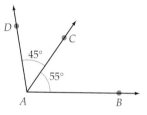

$m\angle DAC = 45°$ and $m\angle CAB = 55°$.

$$m\angle DAB = m\angle DAC + m\angle CAB$$
$$= 45° + 55° = 100°$$

In the figure at the right, $m\angle EDG = 80°$. The measure of $\angle FDG$ is three times the measure of $\angle EDF$. Find the measure of $\angle EDF$.

Let $x =$ the measure of $\angle EDF$.
Then $3x =$ the measure of $\angle FDG$.
Write an equation and solve for *x*.

$$m\angle EDF + m\angle FDG = m\angle EDG$$
$$x + 3x = 80°$$
$$4x = 80°$$
$$x = 20°$$

$$m\angle EDF = 20°$$

EXAMPLE 5 ■ **Solve a Problem Involving Adjacent Angles**

Find the measure of $\angle x$.

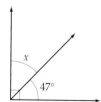

Solution

To find the measure of $\angle x$, write an equation using the fact that the sum of the measures of $\angle x$ and 47° is 90°. Solve for $m\angle x$.

$$m\angle x + 47° = 90°$$
$$m\angle x = 43°$$

CHECK YOUR PROGRESS 5   Find the measure of $\angle a$.

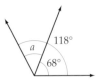

Solution   *See page S16.*

Angles Formed by Intersecting Lines

Four angles are formed by the intersection of two lines. If the two lines are perpendicular, each of the four angles is a right angle.

If the two lines are not perpendicular, then two of the angles formed are acute angles and two of the angles formed are obtuse angles. The two acute angles are always opposite each other, and the two obtuse angles are always opposite each other.

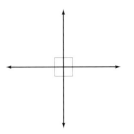

In the figure at the right, $\angle w$ and $\angle y$ are acute angles, and $\angle x$ and $\angle z$ are obtuse angles.

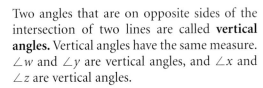

▼ **point of interest**

Many cities in the New World, unlike those in Europe, were designed using rectangular street grids. Washington, D.C. was designed this way, except that diagonal avenues were added, primarily for the purpose of enabling troop movement in the event the city required defense. As an added precaution, monuments were constructed at major intersections so that attackers would not have a straight shot down a boulevard.

Two angles that are on opposite sides of the intersection of two lines are called **vertical angles.** Vertical angles have the same measure. $\angle w$ and $\angle y$ are vertical angles, and $\angle x$ and $\angle z$ are vertical angles.

Recall that two angles that have a common vertex and a common side, but have no interior points in common, are called adjacent angles. For the figure shown above, $\angle x$ and $\angle y$ are adjacent angles, as are $\angle y$ and $\angle z$, $\angle z$ and $\angle w$, and $\angle w$ and $\angle x$. Adjacent angles of intersecting lines are supplementary angles.

Vertical angles have the same measure.

$$m\angle w = m\angle y$$
$$m\angle x = m\angle z$$

Adjacent angles formed by intersecting lines are supplementary angles.

$$m\angle x + m\angle y = 180°$$
$$m\angle y + m\angle z = 180°$$
$$m\angle z + m\angle w = 180°$$
$$m\angle w + m\angle x = 180°$$

EXAMPLE 6 ■ Solve a Problem Involving Intersecting Lines

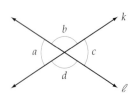

In the diagram at the right, $m\angle b = 115°$. Find the measures of angles a, c, and d.

Solution

$m\angle a + m\angle b = 180°$ • $\angle a$ is supplementary to $\angle b$ because $\angle a$ and $\angle b$ are
adjacent angles of intersecting lines.

$m\angle a + 115° = 180°$ • Replace $m\angle b$ with 115°.

$m\angle a = 65°$ • Subtract 115° from each side of the equation.

$m\angle c = 65°$ • $m\angle c = m\angle a$ because $\angle c$ and $\angle a$ are vertical angles.

$m\angle d = 115°$ • $m\angle d = m\angle b$ because $\angle d$ and $\angle b$ are vertical angles.

$m\angle a = 65°$, $m\angle c = 65°$, and $m\angle d = 115°$.

CHECK YOUR PROGRESS 6

In the diagram at the right, $m\angle a = 35°$.
Find the measures of angles *b*, *c*, and *d*.

Solution   *See page S16.*

A line that intersects two other lines at different points is called a **transversal.**

If the lines cut by a transversal *t* are parallel lines and the transversal is perpendicular to the parallel lines, all eight angles formed are right angles.

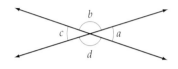

If the lines cut by a transversal *t* are parallel lines and the transversal is *not* perpendicular to the parallel lines, all four acute angles have the same measure and all four obtuse angles have the same measure. For the figure at the right:

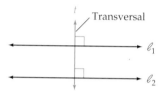

$$m\angle b = m\angle d = m\angle x = m\angle z$$
$$m\angle a = m\angle c = m\angle w = m\angle y$$

If two lines in a plane are cut by a transversal, then any two non-adjacent angles that are on opposite sides of the transversal and between the parallel lines are **alternate interior angles.** In the figure above, $\angle c$ and $\angle w$ are alternate interior angles, and $\angle d$ and $\angle x$ are alternate interior angles. Alternate interior angles formed by two parallel lines cut by a transversal have the same measure.

Two alternate interior angles formed by two parallel lines cut by a transversal have the same measure.

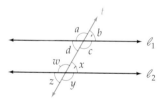

$$m\angle c = m\angle w$$
$$m\angle d = m\angle x$$

If two lines in a plane are cut by a transversal, then any two non-adjacent angles that are on opposite sides of the transversal and outside the parallel lines are **alternate exterior angles.** In the figure at the top of the following page,

Two alternate exterior angles formed by two parallel lines cut by a transversal have the same measure.

$$m\angle a = m\angle y$$
$$m\angle b = m\angle z$$

$\angle a$ and $\angle y$ are alternate exterior angles, and $\angle b$ and $\angle z$ are alternate exterior angles. Alternate exterior angles formed by two parallel lines cut by a transversal have the same measure.

If two lines in a plane are cut by a transversal, then any two angles that are on the same side of the transversal and are both acute angles or both obtuse angles are **corresponding angles.** For the figure at the right above, the following pairs of angles are corresponding angles: $\angle a$ and $\angle w$, $\angle d$ and $\angle z$, $\angle b$ and $\angle x$, $\angle c$ and $\angle y$. Corresponding angles formed by two parallel lines cut by a transversal have the same measure.

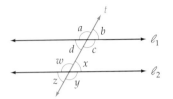

Two corresponding angles formed by two parallel lines cut by a transversal have the same measure.

$$m\angle a = m\angle w$$
$$m\angle d = m\angle z$$
$$m\angle b = m\angle x$$
$$m\angle c = m\angle y$$

QUESTION   *Which angles in the diagram at the right above have the same measure as angle a? Which angles have the same measure as angle b?*

EXAMPLE 7 ■ Solve a Problem Involving Parallel Lines Cut by a Transversal

In the diagram at the right, $\ell_1 \| \ell_2$ and $m\angle f = 58°$. Find the measures of $\angle a$, $\angle c$, and $\angle d$.

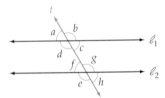

Solution

$m\angle a = m\angle f = 58°$ • $\angle a$ and $\angle f$ are corresponding angles.

$m\angle c = m\angle f = 58°$ • $\angle c$ and $\angle f$ are alternate interior angles.

$m\angle d + m\angle a = 180°$ • $\angle d$ is supplementary to $\angle a$.
$\quad\ m\angle d + 58° = 180°$ • Replace $m\angle a$ with 58°.
$\quad\quad\quad\ m\angle d = 122°$ • Subtract 58° from each side of the equation.

$m\angle a = 58°$, $m\angle c = 58°$, and $m\angle d = 122°$.

ANSWER   *Angles c, w, and y have the same measure as angle a. Angles d, x, and z have the same measure as angle b.*

CHECK YOUR PROGRESS 7

In the diagram at the right, $\ell_1 \| \ell_2$ and $m\angle g = 124°$. Find the measures of $\angle b$, $\angle c$, and $\angle d$.

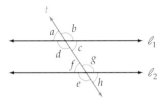

Solution   *See page S16.*

Math Matters   The Principle of Reflection

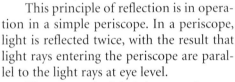

When a ray of light hits a flat surface, such as a mirror, the light is reflected at the same angle at which it hit the surface. For example, in the diagram at the left, $m\angle x = m\angle y$.

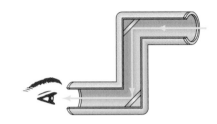

This principle of reflection is in operation in a simple periscope. In a periscope, light is reflected twice, with the result that light rays entering the periscope are parallel to the light rays at eye level.

The same principle is in operation on a billiard table. Assuming it has no "side spin," a ball bouncing off the side of the table will bounce off at the same angle at which it hit the side. In the figure below, $m\angle w = m\angle x$ and $m\angle y = m\angle z$.

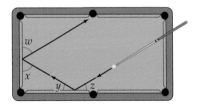

In the miniature golf shot illustrated below, $m\angle w = m\angle x$ and $m\angle y = m\angle z$.

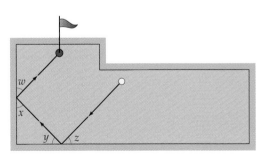

Angles of a Triangle

If the lines cut by a transversal are not parallel lines, the three lines will intersect at three points. In the figure at the right, the transversal t intersects lines p and q. The three lines intersect at points A, B, and C. These three points define three line segments, $\overline{AB}$, $\overline{BC}$, and $\overline{AC}$. The plane figure formed by these three line segments is called a **triangle.**

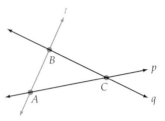

Each of the three points of intersection is the vertex of four angles. The angles within the region enclosed by the triangle are called **interior angles.** In the figure at the right, angles a, b, and c are interior angles. The sum of the measures of the interior angles of a triangle is 180°.

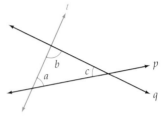

$$m\angle a + m\angle b + m\angle c = 180°$$

The Sum of the Measures of the Interior Angles of a Triangle

The sum of the measures of the interior angles of a triangle is 180°.

QUESTION   *Can the measures of the three interior angles of a triangle be 87°, 51°, and 43°?*

An **exterior angle of a triangle** is an angle that is adjacent to an interior angle of the triangle and is a supplement of the interior angle. In the figure at the right, angles m and n are exterior angles for angle a. **The sum of the measures of an interior angle and one of its exterior angles is 180°.**

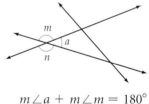

$$m\angle a + m\angle m = 180°$$
$$m\angle a + m\angle n = 180°$$

EXAMPLE 8 ■ Solve a Problem Involving the Angles of a Triangle

In the diagram at the right, $m\angle c = 40°$ and $m\angle e = 60°$. Find the measure of $\angle d$.

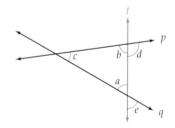

ANSWER   *No, because 87° + 51° + 43° = 181°, and the sum of the measures of the three interior angles of a triangle must be 180°.*

Solution

$m\angle a = m\angle e = 60°$ • $\angle a$ and $\angle e$ are vertical angles.

$m\angle c + m\angle a + m\angle b = 180°$ • The sum of the interior angles is **180°**.
$\quad 40° + 60° + m\angle b = 180°$ • Replace $m\angle c$ with **40°** and $m\angle a$ with **60°**.
$\quad\quad\quad 100° + m\angle b = 180°$ • Add $40° + 60°$.
$\quad\quad\quad\quad\quad m\angle b = 80°$ • Subtract 100° from each side of the equation.

$m\angle b + m\angle d = 180°$ • $\angle b$ and $\angle d$ are supplementary angles.
$\quad 80° + m\angle d = 180°$ • Replace $m\angle b$ with **80°**.
$\quad\quad\quad m\angle d = 100°$ • Subtract 80° from each side of the equation.

$m\angle d = 100°$

CHECK YOUR PROGRESS 8

In the diagram at the right, $m\angle c = 35°$ and $m\angle d = 105°$. Find the measure of $\angle e$.

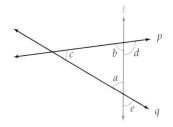

Solution   *See page S16.*

EXAMPLE 9 ■ Find the Measure of the Third Angle of a Triangle

Two angles of a triangle measure 43° and 86°. Find the measure of the third angle.

Solution

Use the fact that the sum of the measures of the interior angles of a triangle is 180°. Write an equation using x to represent the measure of the third angle. Solve the equation for x.

$\quad x + 43° + 86° = 180°$
$\quad\quad\quad x + 129° = 180°$ • Add 43° + 86°.
$\quad\quad\quad\quad\quad\quad x = 51°$ • Subtract 129° from each side of the equation.

The measure of the third angle is 51°.

> **TAKE NOTE**
>
> In this text, when we refer to the angles of a triangle, we mean the interior angles of the triangle unless specifically stated otherwise.

CHECK YOUR PROGRESS 9   One angle in a triangle is a right angle, and one angle measures 27°. Find the measure of the third angle.

Solution   *See page S16.*

Investigation

Preparing a Circle Graph

On page 225, a protractor was used to measure an angle. Preparing a circle graph requires the ability to use a protractor to draw angles.

To draw an angle of 142°, first draw a ray. Place a dot at the endpoint of the ray. This dot will be the vertex of the angle.

Place the line segment near the bottom edge of the protractor on the ray, as shown in the figure at the right. Make sure the center of the bottom edge of the protractor is located directly over the vertex point. Locate the position of the 142° mark. Place a dot next to the mark.

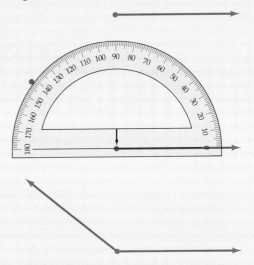

Remove the protractor and draw a ray from the vertex to the dot at the 142° mark.

Here is an example of how to prepare a circle graph.

The revenues (in thousands of dollars) from four departments of a car dealership for the first quarter of a recent year were

| New car sales: | $2100 | Used car/truck sales: | $1500 |
| New truck sales: | $1200 | Parts/service: | $700 |

To draw a circle graph to represent the percent that each department contributed to the total revenue from all four departments, proceed as follows:

Find the total revenue from all four departments.

$$2100 + 1200 + 1500 + 700 = 5500$$

Find what percent each department is of the total revenue of $5500 thousand.

New car sales: $\dfrac{2100}{5500} \approx 38.2\%$

New truck sales: $\dfrac{1200}{5500} \approx 21.8\%$

Used car/truck sales: $\dfrac{1500}{5500} \approx 27.3\%$

Parts/service: $\dfrac{700}{5500} \approx 12.7\%$

(continued)

Each percent represents the part of the circle for that sector. Because the circle contains 360°, multiply each percent by 360° to find the measure of the angle for each sector. Round to the nearest degree.

New car sales:	$0.382 \times 360° \approx 138°$
New truck sales:	$0.218 \times 360° \approx 78°$
Used car/truck sales:	$0.273 \times 360° \approx 98°$
Parts/service:	$0.127 \times 360° \approx 46°$

Draw a circle and use a protractor to draw the sectors representing the percents that each department contributed to the total revenue from all four departments.

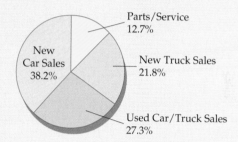

Parts/Service
12.7%

New Car Sales
38.2%

New Truck Sales
21.8%

Used Car/Truck Sales
27.3%

Investigation Exercises

Prepare a circle graph for the data provided in each exercise.

1. Shown below are American adults' favorite pizza toppings. (*Source:* Market Facts for Bolla wines)

Pepperoni:	43%
Sausage:	19%
Mushrooms:	14%
Vegetables:	13%
Other:	7%
Onions:	4%

2. According to a Pathfinder Research Group survey, more than 94% of American adults have heard of the Three Stooges. The choices of a favorite among those who have one are

Curly:	52%
Moe:	31%
Larry:	12%
Curly Joe:	3%
Shemp:	2%

(continued)

3. Only 18 million of the 67 million bicyclists in the United States own or have use of a helmet. The list below shows how often helmets are worn. (*Source:* U.S. Consumer Product Safety Commission)

Never or almost never:	13,680,000
Always or almost always:	2,412,000
Less than half the time:	1,080,000
More than half the time:	756,000
Unknown:	72,000

4. In a recent year, 10 million tons of glass containers were produced. Given below is a list of the numbers of tons used by various industries. The category "Other" includes containers used for items such as drugs and toiletries. (*Source:* Salomon Bros.)

Beer:	4,600,000 tons
Food:	3,500,000 tons
Wine and liquor:	900,000 tons
Soft drinks:	500,000 tons
Other:	500,000 tons

Exercise Set 4.3

1. Provide three names for the angle below.

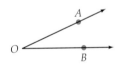

2. State the numbers of degrees in a full circle, a straight angle, and a right angle.

In Exercises 3–8, use a protractor to measure the angle to the nearest degree. State whether the angle is acute, obtuse, or right.

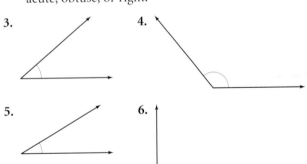

7.

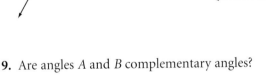

8.

9. Are angles *A* and *B* complementary angles?

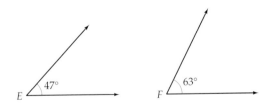

51°

39°

10. Are angles *E* and *F* complementary angles?

47°

63°

11. Are angles *C* and *D* supplementary angles?

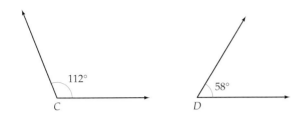

12. Are angles *G* and *H* supplementary angles?

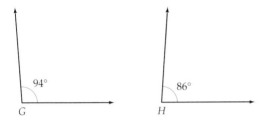

13. Find the complement of a 62° angle.

14. Find the complement of a 31° angle.

15. Find the supplement of a 162° angle.

16. Find the supplement of a 72° angle.

17. Given $AB = 12$ cm, $CD = 9$ cm, and $AD = 35$ cm, find the length of $\overline{BC}$.

18. Given $AB = 21$ mm, $BC = 14$ mm, and $AD = 54$ mm, find the length of $\overline{CD}$.

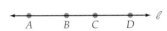

19. Given $QR = 7$ ft and RS is three times the length of $\overline{QR}$, find the length of $\overline{QS}$.

20. Given $QR = 15$ in. and RS is twice the length of $\overline{QR}$, find the length of $\overline{QS}$.

21. Given $EF = 20$ m and FG is one-half the length of $\overline{EF}$, find the length of $\overline{EG}$.

22. Given $EF = 18$ cm and FG is one-third the length of $\overline{EF}$, find the length of $\overline{EG}$.

23. Given $m \angle LOM = 53°$ and $m \angle LON = 139°$, find the measure of $\angle MON$.

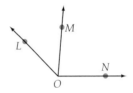

24. Given $m \angle MON = 38°$ and $m \angle LON = 85°$, find the measure of $\angle LOM$.

In Exercises 25 and 26, find the measure of $\angle x$.

25.   **26.**

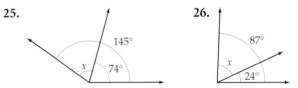

In Exercises 27–30, given that $\angle LON$ is a right angle, find the measure of $\angle x$.

27.   **28.**

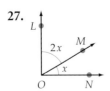

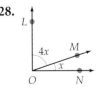

29.

30.

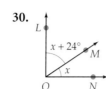

In Exercises 31–34, find the measure of $\angle a$.

31.

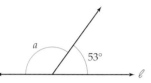

32.

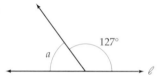

33.

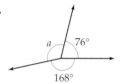

34.

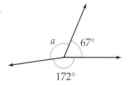

In Exercises 35–40, find the value of x.

35.

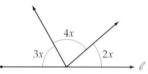

36.

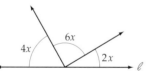

37.

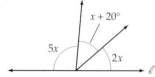

38.

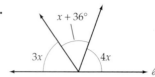

39.

40.

41. Given $m\angle a = 51°$, find the measure of $\angle b$.

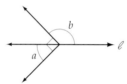

42. Given $m\angle a = 38°$, find the measure of $\angle b$.

In Exercises 43 and 44, find the measure of $\angle x$.

43.

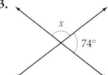

44.

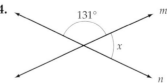

In Exercises 45 and 46, find the value of *x*.

45.

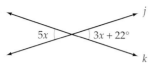

46.

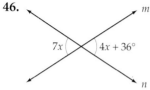

In Exercises 47–50, given that $\ell_1 \parallel \ell_2$, find the measures of angles *a* and *b*.

47.

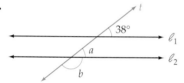

48.

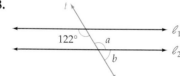

49.

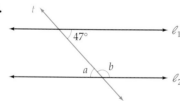

50.

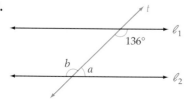

In Exercises 51–54, given that $\ell_1 \parallel \ell_2$, find *x*.

51.

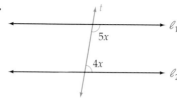

52.

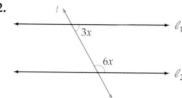

53.

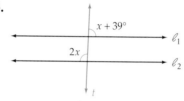

54.

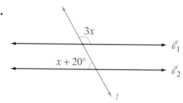

55. Given that $m\angle a = 95°$ and $m\angle b = 70°$, find the measures of angles *x* and *y*.

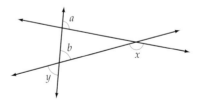

56. Given that $m\angle a = 35°$ and $m\angle b = 55°$, find the measures of angles *x* and *y*.

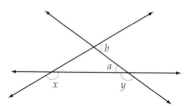

57. Given that $m \angle y = 45°$, find the measures of angles a and b.

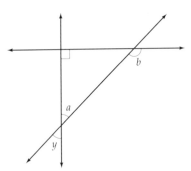

58. Given that $m \angle y = 130°$, find the measures of angles a and b.

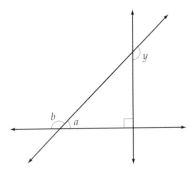

59. Given that $\overline{AO} \perp \overline{OB}$, express in terms of x the number of degrees in $\angle BOC$.

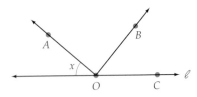

60. Given that $\overline{AO} \perp \overline{OB}$, express in terms of x the number of degrees in $\angle AOC$.

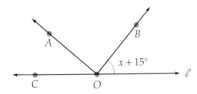

61. One angle in a triangle is a right angle, and one angle is equal to 30°. What is the measure of the third angle?

62. A triangle has a 45° angle and a right angle. Find the measure of the third angle.

63. Two angles of a triangle measure 42° and 103°. Find the measure of the third angle.

64. Two angles of a triangle measure 62° and 45°. Find the measure of the third angle.

65. A triangle has a 13° angle and a 65° angle. What is the measure of the third angle?

66. A triangle has a 105° angle and a 32° angle. What is the measure of the third angle?

67. Cut out a triangle and then tear off two of the angles, as shown below. Position the pieces you tore off so that angle a is adjacent to angle b and angle c is adjacent to angle b. Describe what you observe. What does this demonstrate?

68. Is the statement always true, sometimes true, or never true?

a. Two lines that are parallel to a third line are parallel to each other.

b. Vertical angles are complementary angles.

69. When a transversal intersects two parallel lines, which of the following are supplementary angles: vertical angles, adjacent angles, alternate interior angles, alternate exterior angles, corresponding angles?

70. What is the smallest possible whole number of degrees in an angle of a triangle? What is the largest possible whole number of degrees in an angle of a triangle?

Extensions

CRITICAL THINKING

71. The road mileage between San Francisco, California and New York City is 3036 mi. The air distance between these two cities is 2571 mi. Why do the distances differ?

72. How many dimensions does a point have? a line? a line segment? a ray? an angle?

73. Which line segment is longer, $\overline{AB}$ or $\overline{CD}$?

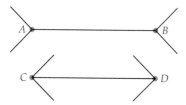

COOPERATIVE LEARNING

74. For the figure at the right, find the sum of the measures of angles *x*, *y*, and *z*.

75. For the figure at the right, explain why $m\angle a + m\angle b = m\angle x$. Write a statement that describes the relationship between the measure of an exterior angle of a triangle and the sum of the measures of its two opposite interior angles (the interior angles that are non-adjacent to the exterior angle). Use the statement to write an equation involving angles *a*, *c*, and *z*.

76. If $\overline{AB}$ and $\overline{CD}$ intersect at point *O*, and $m\angle AOC = m\angle BOC$, explain why $\overline{AB} \perp \overline{CD}$.

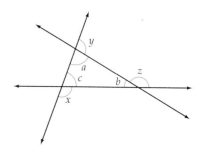

SECTION 4.4 | **Perimeter and Area of Plane Figures**

Perimeter of Plane Geometric Figures

A **polygon** is a closed figure determined by three or more line segments that lie in a plane. The line segments that form the polygon are called its **sides.** The figures below are examples of polygons.

▼ **point of interest**

Although a polygon is described in terms of the number of its sides, the word actually comes from the Latin word *polygonum,* meaning "many *angles.*"

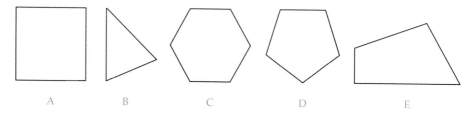

A **regular polygon** is one in which each side has the same length and each angle has the same measure. The polygons in Figures A, C, and D above are regular polygons.

The name of a polygon is based on the number of its sides. The table below lists the names of polygons that have from 3 to 10 sides.

Number of Sides	Name of Polygon
3	Triangle
4	Quadrilateral
5	Pentagon
6	Hexagon
7	Heptagon
8	Octagon
9	Nonagon
10	Decagon

The Pentagon in Arlington, Virginia

Triangles and quadrilaterals are two of the most common types of polygons. Triangles are distinguished by the numbers of equal sides and also by the measures of their angles.

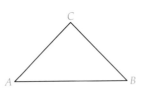

An **isosceles triangle** has exactly two sides of equal length. The angles opposite the equal sides are of equal measure.
$AC = BC$
$m \angle A = m \angle B$

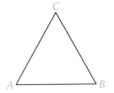

The three sides of an **equilateral triangle** are of equal length. The three angles are of equal measure.
$AB = BC = AC$
$m \angle A = m \angle B$
$\qquad = m \angle C$

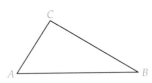

A **scalene triangle** has no two sides of equal length. No two angles are of equal measure.

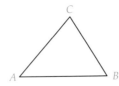

An **acute triangle** has three acute angles.

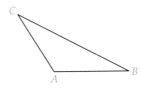

An **obtuse triangle** has one obtuse angle.

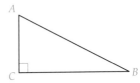

A **right triangle** has a right angle.

A **quadrilateral** is a four-sided polygon. Quadrilaterals are also distinguished by their sides and angles, as shown on the following page. Note that a rectangle, a square, and a rhombus are different forms of a parallelogram.

Parallelogram

Opposite sides parallel
Opposite sides equal in length
Opposite angles equal in measure

Rectangle

Opposite sides parallel
Opposite sides equal in length
All angles measure 90°
Diagonals equal in length

Square

Opposite sides parallel
All sides equal in length
All angles measure 90°
Diagonals equal in length

Quadrilateral

Four-sided polygon

Rhombus

Opposite sides parallel
All sides equal in length
Opposite angles equal in measure

Trapezoid

Two sides parallel

Isosceles Trapezoid

Two sides parallel
Nonparallel sides equal in length

QUESTION   **a.** *What distinguishes a rectangle from other parallelograms?*
                b. *What distinguishes a square from other rectangles?*

The **perimeter** of a plane geometric figure is a measure of the distance around the figure. Perimeter is used when buying fencing for a garden or determining how much baseboard is needed for a room.

The perimeter of a triangle is the sum of the lengths of the three sides.

Perimeter of a Triangle

Let a, b, and c be the lengths of the sides of a triangle. The perimeter, P, of the triangle is given by $P = a + b + c$.

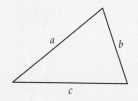

$$P = a + b + c$$

To find the perimeter of the triangle shown at the right, add the lengths of the three sides.

$P = 5 + 7 + 10 = 22$

The perimeter is 22 ft.

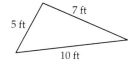

EXAMPLE 1 ■ Find the Perimeter of a Triangle

You want to sew bias binding along the edges of a cloth flag that has sides that measure $1\frac{1}{4}$ ft, $3\frac{1}{2}$ ft, and $3\frac{3}{4}$ ft. Find the length of bias binding needed.

ANSWER   *a. In a rectangle, all angles measure 90°. b. In a square, all sides are equal in length.*

Solution

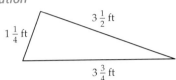

$P = a + b + c$

• Draw a diagram.

$P = 1\dfrac{1}{4} + 3\dfrac{1}{2} + 3\dfrac{3}{4}$

$P = 1\dfrac{1}{4} + 3\dfrac{2}{4} + 3\dfrac{3}{4}$

$P = 7\dfrac{6}{4}$

$P = 8\dfrac{1}{2}$

• Use the formula for the perimeter of a triangle.

• Replace a, b, and c with $1\frac{1}{4}$, $3\frac{1}{2}$, and $3\frac{3}{4}$. You can replace a with the length of any of the three sides, and then replace b and c with the lengths of the other two sides. The order does not matter. The result will be the same.

• $7\dfrac{6}{4} = 7\dfrac{3}{2} = 7 + \dfrac{3}{2} = 7 + 1\dfrac{1}{2} = 8\dfrac{1}{2}$

You need $8\frac{1}{2}$ ft of bias binding.

CHECK YOUR PROGRESS 1   A bicycle trail in the shape of a triangle has sides that measure $4\frac{3}{10}$ mi, $2\frac{1}{10}$ mi, and $6\frac{1}{2}$ mi. Find the total length of the bike trail.

Solution   See page S16.

The perimeter of a quadrilateral is the sum of the lengths of its four sides.

A **rectangle** is a quadrilateral with all right angles and opposite sides of equal length. Usually the length, L, of a rectangle refers to the length of one of the longer sides of the rectangle and the width, W, refers to the length of one of the shorter sides. The perimeter can then be represented as $P = L + W + L + W$.

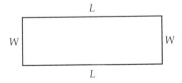

$P = L + W + L + W$

$P = 2L + 2W$

The formula for the perimeter of a rectangle is derived by combining like terms.

Perimeter of a Rectangle

Let L represent the length and W the width of a rectangle. The perimeter P of the rectangle is given by $P = 2L + 2W$.

EXAMPLE 2 ■ **Find the Perimeter of a Rectangle**

You want to trim a rectangular frame with a metal strip. The frame measures 30 in. by 20 in. Find the length of metal strip you will need to trim the frame.

Solution

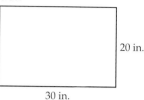

- Draw a diagram.

$P = 2L + 2W$

- Use the formula for the perimeter of a rectangle.

$P = 2(30) + 2(20)$

- The length is 30 in. Substitute 30 for *L*. The width is 20 in. Substitute 20 for *W*.

$P = 60 + 40$

$P = 100$

You will need 100 in. of the metal strip.

CHECK YOUR PROGRESS 2   Find the length of decorative molding needed to edge the tops of the walls in a rectangular room that is 12 ft long and 8 ft wide.

Solution   See page S16.

A **square** is a rectangle in which each side has the same length. Letting *s* represent the length of each side of a square, the perimeter of the square can be represented by $P = s + s + s + s$.

The formula for the perimeter of a square is derived by combining like terms.

$P = s + s + s + s$

$P = 4s$

Perimeter of a Square

Let *s* represent the length of a side of a square. The perimeter *P* of the square is given by $P = 4s$.

EXAMPLE 3 ■ Find the Perimeter of a Square

Find the length of fencing needed to surround a square corral that measures 60 ft on each side.

Solution

- Draw a diagram.

$P = 4s$

- Use the formula for the perimeter of a square.

$P = 4(60)$

- The length of a side is 60 ft. Substitute 60 for *s*.

$P = 240$

240 ft of fencing is needed.

CHECK YOUR PROGRESS 3 A homeowner plans to fence in the area around the swimming pool in the back yard. The area to be fenced in is a square measuring 24 ft on each side. How many feet of fencing should the homeowner purchase?

Solution See page S17.

historical note

Benjamin Banneker (băn′ĭ-kər) (1731–1806), a noted American scholar who was largely self-taught, was both a surveyor and an astronomer. As a surveyor, he was a member of the commission that defined the boundary lines and laid out the streets of the District of Columbia. (See the Point of Interest on page 228.) ∎

Figure $ABCD$ is a **parallelogram.** $\overline{BC}$ is the **base** of the parallelogram. Opposite sides of a parallelogram are equal in length, so $\overline{AD}$ is the same length as $\overline{BC}$, and $\overline{AB}$ is the same length as $\overline{CD}$.

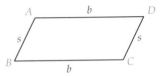

Let b represent the length of the base and s the length of an adjacent side. Then the perimeter of a parallelogram can be represented as $P = b + s + b + s$.

$$P = b + s + b + s$$

The formula for the perimeter of a parallelogram is derived by combining like terms.

$$P = 2b + 2s$$

Perimeter of a Parallelogram

Let b represent the length of the base of a parallelogram and s the length of a side adjacent to the base. The perimeter, P, of the parallelogram is given by $P = 2b + 2s$.

EXAMPLE 4 ▪ **Find the Perimeter of a Parallelogram**

You plan to trim the edge of a kite with a strip of red fabric. The kite is in the shape of a parallelogram with a base measuring 40 in. and a side measuring 28 in. Find the length of fabric needed to trim the kite.

Solution

• Draw a diagram.

$$P = 2b + 2s$$

• Use the formula for the perimeter of a parallelogram.

$$P = 2(40) + 2(28)$$

• The base is 40 in. Substitute 40 for b. The length of a side is 28 in. Substitute 28 for s.

$$P = 80 + 56$$

• Simplify using the Order of Operations Agreement.

$$P = 136$$

To trim the kite, 136 in. of fabric is needed.

CHECK YOUR PROGRESS 4 A flower bed is in the shape of a parallelogram that has a base of length 5 m and a side of length 7 m. Wooden planks are used to edge the garden. Find the length of wooden planks needed to surround the garden.

Solution See page S17.

▼ **point of interest**

A **glazier** is a person who cuts, fits, and installs glass, generally in doors and windows. Of particular challenge to a glazier are intricate stained glass window designs.

A **circle** is a plane figure in which all points are the same distance from point O, called the **center** of the circle.

A **diameter** of a circle is a line segment with endpoints on the circle and passing through the center. $\overline{AB}$ is a diameter of the circle at the right. The variable d is used to designate the length of a diameter of a circle.

A **radius** of a circle is a line segment from the center of the circle to a point on the circle. $\overline{OC}$ is a radius of the circle at the right above. The variable r is used to designate the length of a radius of a circle.

The length of the diameter is twice the length of the radius.

$$d = 2r \quad \text{or} \quad r = \frac{1}{2}d$$

The distance around a circle is called the **circumference**. The formula for the circumference, C, of a circle is:

$$C = \pi d$$

Because $d = 2r$, the formula for the circumference can also be written as:

$$C = 2\pi r$$

Circumference of a Circle

The circumference C of a circle with diameter d and radius r is given by $C = \pi d$ or $C = 2\pi r$.

✔ **TAKE NOTE**

Recall that an irrational number is a number whose decimal representation never terminates and does not have a pattern of numerals that keep repeating.

The formula for circumference uses the number π (pi), which is an irrational number. The value of π can be approximated by a fraction or by a decimal.

$$\pi \approx 3\frac{1}{7} \quad \text{or} \quad \pi \approx 3.14$$

The π key on a scientific calculator gives a closer approximation of π than 3.14. A scientific calculator is used in this section to find approximate values in calculations involving π.

Find the circumference of a circle with a diameter of 6 m.

CALCULATOR NOTE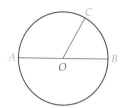

The π key on your calculator can be used to find decimal approximations of expressions that contain π. To perform the calculation at the right, enter

6 $\boxed{\times}\;\boxed{\pi}\;\boxed{=}$.

The diameter of the circle is given. Use the circumference formula that involves the diameter. $d = 6$.

$$C = \pi d$$
$$C = \pi(6)$$

The exact circumference of the circle is 6π m.

$$C = 6\pi$$

An approximate measure can be found by using the π key on a calculator.

An approximate circumference is 18.85 m.

$$C \approx 18.85$$

EXAMPLE 5 ■ Find the Circumference of a Circle

Find the circumference of a circle with a radius of 15 cm. Round to the nearest hundredth of a centimeter.

Solution

$C = 2\pi r$ • The radius is given. Use the circumference formula that involves the radius.

$C = 2\pi(15)$ • Replace r with 15.

$C = 30\pi$ • Multiply 2 times 15.

$C \approx 94.25$ • An approximation is asked for. Use the π key on a calculator.

The circumference of the circle is approximately 94.25 cm.

CHECK YOUR PROGRESS 5   Find the circumference of a circle with a diameter of 9 km. Give the exact measure.

Solution   *See page S17.*

EXAMPLE 6 ■ Application of Finding the Circumference of a Circle

A bicycle tire has a diameter of 24 in. How many feet does the bicycle travel when the wheel makes 8 revolutions? Round to the nearest hundredth of a foot.

24 in.

Solution

$24 \text{ in.} = 2 \text{ ft}$

• The diameter is given in inches, but the answer must be expressed in feet. Convert the diameter (24 in.) to feet. There are 12 in. in 1 ft. Divide 24 by 12.

$C = \pi d$

• The diameter is given. Use the circumference formula that involves the diameter.

$C = \pi(2)$

• Replace d with 2.

$C = 2\pi$

• This is the distance traveled in 1 revolution.

$8C = 8(2\pi) = 16\pi \approx 50.27$

• Find the distance traveled in 8 revolutions.

The bicycle will travel about 50.27 ft when the wheel makes 8 revolutions.

CHECK YOUR PROGRESS 6   A tricycle tire has a diameter of 12 in. How many feet does the tricycle travel when the wheel makes 12 revolutions? Round to the nearest hundredth of a foot.

Solution   *See page S17.*

Area of Plane Geometric Figures

Area is the amount of surface in a region. Area can be used to describe, for example, the size of a rug, a parking lot, a farm, or a national park. Area is measured in square units.

A square that measures 1 inch on each side has an area of 1 square inch, written 1 in^2.

A square that measures 1 centimeter on each side has an area of 1 square centimeter, written 1 cm^2.

1 in^2

1 cm^2

Larger areas are often measured in square feet (ft^2), square meters (m^2), square miles (mi^2), acres (43,560 ft^2), or any other square unit.

QUESTION   *a. What is the area of a square that measures 1 yard on each side?*

b. What is the area of a square that measures 1 kilometer on each side?

The area of a geometric figure is the number of squares (each of area 1 square unit) that are necessary to cover the figure. In the figures below, two rectangles have been drawn and covered with squares. In the figure on the left, 12 squares, each of area 1 cm^2, were used to cover the rectangle. The area of the rectangle is 12 cm^2. In the figure on the right, six squares, each of area 1 in^2, were used to cover the rectangle. The area of the rectangle is 6 in^2.

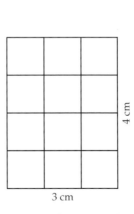

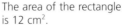
3 cm

4 cm

The area of the rectangle
is 12 cm^2.

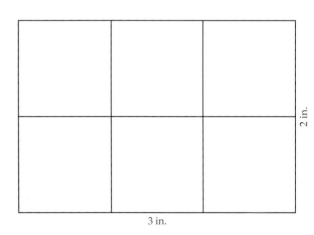
3 in.

2 in.

The area of the rectangle is 6 in^2.

Note from these figures that the area of a rectangle can be found by multiplying the length of the rectangle by its width.

ANSWER   *a. The area is 1 square yard, written 1 yd^2. b. The area is 1 square kilometer, written 1 km^2.*

Area of a Rectangle

Let L represent the length and W the width of a rectangle. The area A of the rectangle is given by $A = LW$.

QUESTION   *How many squares, each 1 inch on a side, are needed to cover a rectangle that has an area of 18 in^2?*

EXAMPLE 7 ▪ Find the Area of a Rectangle

How many square feet of sod are needed to cover a football field? A football field measures 360 ft by 160 ft.

Solution

160 ft

360 ft

• Draw a diagram.

$A = LW$

• Use the formula for the area of a rectangle.

$A = 360(160)$

• The length is **360** ft. Substitute **360** for *L*. The width is **160** ft. Substitute **160** for *W*. Remember that *LW* means "*L* times *W*."

$A = 57{,}600$

57,600 ft^2 of sod is needed.

• Area is measured in square units.

CHECK YOUR PROGRESS 7   Find the amount of fabric needed to make a rectangular flag that measures 308 cm by 192 cm.

Solution   *See page S17.*

✔ **TAKE NOTE**

Recall that the rules of exponents state that when multiplying variables with like bases, we add the exponents.

A square is a rectangle in which all sides are the same length. Therefore, both the length and the width of a square can be represented by s, and $A = LW = s \cdot s = s^2$.

s

$A = s \cdot s$

$A = s^2$

Area of a Square

Let s represent the length of a side of a square. The area A of the square is given by $A = s^2$.

ANSWER   *Eighteen squares, each 1 inch on a side, are needed to cover the rectangle.*

EXAMPLE 8 ■ Find the Area of a Square

A homeowner wants to carpet the family room. The floor is square and measures 6 m on each side. How much carpet should be purchased?

Solution

6 m

$A = s^2$

       • Use the formula for the area of a square.

$A = 6^2$

       • The length of a side is **6 m**. Substitute **6** for *s*.

$A = 36$

36 m² of carpet should be purchased.   • **Area is measured in square units.**

       • Draw a diagram.

CHECK YOUR PROGRESS 8   Find the area of the floor of a two-car garage that is in the shape of a square that measures 24 ft on a side.

Solution   *See page S17.*

Figure *ABCD* is a parallelogram. $\overline{BC}$ is the **base** of the parallelogram. $\overline{AE}$, perpendicular to the base, is the **height** of the parallelogram.

Any side of a parallelogram can be designated as the base. The corresponding height is found by drawing a line segment perpendicular to the base from the opposite side. In the figure at the right, $\overline{CD}$ is the base and $\overline{AE}$ is the height.

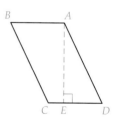

A rectangle can be formed from a parallelogram by cutting a right triangle from one end of the parallelogram and attaching it to the other end. The area of the resulting rectangle will equal the area of the original parallelogram.

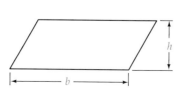

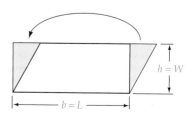

> **Area of a Parallelogram**
>
> Let *b* represent the length of the base and *h* the height of a parallelogram. The area *A* of the parallelogram is given by $A = bh$.

EXAMPLE 9 ■ **Find the Area of a Parallelogram**

A solar panel is in the shape of a parallelogram that has a base of 2 ft and a height of 3 ft. Find the area of the solar panel.

Solution

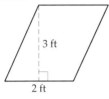

3 ft
2 ft

• Draw a diagram.

$A = bh$ • Use the formula for the area of a parallelogram.

$A = 2(3)$ • The base is 2 ft. Substitute 2 for *b*.
                    The height is 3 ft. Substitute 3 for *h*.
                    Remember that *bh* means "*b* times *h*."

$A = 6$

The area is 6 ft².   • Area is measured in square units.

CHECK YOUR PROGRESS 9   A fieldstone patio is in the shape of a parallelogram that has a base measuring 14 m and a height measuring 8 m. What is the area of the patio?

Solution   *See page S17.*

Figure *ABC* is a triangle. $\overline{AB}$ is the **base** of the triangle. $\overline{CD}$, perpendicular to the base, is the **height** of the triangle.

Any side of a triangle can be designated as the base. The corresponding height is found by drawing a line segment perpendicular to the base, or to an extension of the base, from the vertex opposite the base.

Consider triangle *ABC* with base *b* and height *h* shown at the right. By extending a line segment from *C* parallel to the base $\overline{AB}$ and equal in length to the base, a parallelogram is formed. The area of the parallelogram is *bh* and is twice the area of the original triangle. Therefore, the area of the triangle is one half the area of the parallelogram, or $\frac{1}{2}bh$.

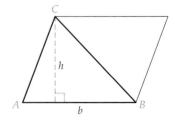

> **Area of a Triangle**
>
> Let b represent the length of the base and h the height of a triangle. The area A of the triangle is given by $A = \frac{1}{2}bh$.

EXAMPLE 10 ■ Find the Area of a Triangle

A riveter uses metal plates that are in the shape of a triangle with a base of 12 cm and a height of 6 cm. Find the area of one metal plate.

Solution

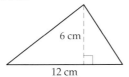

6 cm

12 cm

• Draw a diagram.

$A = \dfrac{1}{2}bh$ • Use the formula for the area of a triangle.

$A = \dfrac{1}{2}(12)(6)$ • The base is **12** cm. Substitute **12** for *b*. The height is **6** cm. Substitute **6** for *h*. Remember that *bh* means "*b* times *h*."

$A = 6(6)$

$A = 36$

The area is 36 cm².   • Area is measured in square units.

CHECK YOUR PROGRESS 10   Find the amount of felt needed to make a banner that is in the shape of a triangle with a base of 18 in. and a height of 9 in.

Solution   *See page S17.*

✔ **TAKE NOTE**

The bases of a trapezoid are the parallel sides of the figure.

Figure *ABCD* is a **trapezoid**. $\overline{AB}$, with length b_1, is one **base** of the trapezoid and $\overline{CD}$, with length b_2, is the other base. $\overline{AE}$, perpendicular to the two bases, is the **height.**

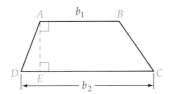

In the trapezoid at the right, the line segment $\overline{BD}$ divides the trapezoid into two triangles, *ABD* and *BCD*. In triangle *ABD*, b_1 is the base and h is the height. In triangle *BCD*, b_2 is the base and h is the height. The area of the trapezoid is the sum of the areas of the two triangles.

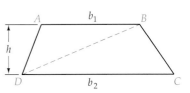

Area of trapezoid *ABCD* = Area of triangle *ABD* + area of triangle *BCD*

$$= \frac{1}{2}b_1h + \frac{1}{2}b_2h = \frac{1}{2}h(b_1 + b_2)$$

Area of a Trapezoid

Let b_1 and b_2 represent the lengths of the bases and h the height of a trapezoid. The area A of the trapezoid is given by $A = \frac{1}{2}h(b_1 + b_2)$.

EXAMPLE 11 ■ Find the Area of a Trapezoid

A boat dock is built in the shape of a trapezoid with bases measuring 14 ft and 6 ft and a height of 7 ft. Find the area of the dock.

Solution

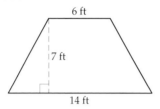

• Draw a diagram.

$$A = \frac{1}{2}h(b_1 + b_2)$$

• Use the formula for the area of a trapezoid.

$$A = \frac{1}{2} \cdot 7(14 + 6)$$

• The height is **7 ft.** Substitute **7** for h. The bases measure **14 ft** and **6 ft.** Substitute **14** and **6** for b_1 and b_2.

$$A = \frac{1}{2} \cdot 7(20)$$

$$A = 70$$

The area is 70 ft².

• Area is measured in square units.

CHECK YOUR PROGRESS 11 Find the area of a patio that has the shape of a trapezoid with a height of 9 ft and bases measuring 12 ft and 20 ft.

Solution *See page S17.*

The area of a circle is the product of π and the square of the radius.

$$A = \pi r^2$$

The Area of a Circle

The area A of a circle with radius of length r is given by $A = \pi r^2$.

Find the area of a circle that has a radius of 6 cm.

Use the formula for the area of a circle. $r = 6$.

$$A = \pi r^2$$
$$A = \pi(6)^2$$
$$A = \pi(36)$$

The exact area of the circle is 36π cm².

$$A = 36\pi$$

An approximate measure can be found by using the π key on a calculator.

$$A \approx 113.10$$

The approximate area of the circle is 113.10 cm².

EXAMPLE 12 ▪ **Find the Area of a Circle**

Find the area of a circle with a diameter of 10 m. Round to the nearest hundredth of a square meter.

Solution

$r = \dfrac{1}{2}d = \dfrac{1}{2}(10) = 5$ • Find the radius of the circle.

$A = \pi r^2$ • Use the formula for the area of a circle.

$A = \pi(5)^2$ • Replace *r* with **5**.

$A = \pi(25)$ • Square 5.

$A \approx 78.54$ • An approximation is asked for. Use the π key on a calculator.

The area of the circle is approximately 78.54 m².

CHECK YOUR PROGRESS 12   Find the area of a circle with a diameter of 12 km. Give the exact measure.

Solution   *See page S17.*

EXAMPLE 13 ▪ **Application of Finding the Area of a Circle**

How large a cover is needed for a circular hot tub that is 8 ft in diameter? Round to the nearest tenth of a square foot.

Solution

$r = \dfrac{1}{2}d = \dfrac{1}{2}(8) = 4$ • Find the radius of a circle with a diameter of 8 ft.

$A = \pi r^2$ • Use the formula for the area of a circle.

$A = \pi(4)^2$ • Replace *r* with **4**.

$A = \pi(16)$ • Square 4.

$A \approx 50.3$ • Use the π key on a calculator.

The cover for the hot tub must have an area of approximately 50.3 ft².

CHECK YOUR PROGRESS 13 How much material is needed to make a circular tablecloth that is to have a diameter of 4 ft? Round to the nearest hundredth of a square foot.

Solution *See page S17.*

MathMatters   **Möbius Bands**

Cut out a long, narrow, rectangular strip of paper.

Give the strip of paper a half-twist.

Put the ends together so that *A* meets *Z* and *B* meets *Y*. Tape the ends together. The result is a *Möbius band*.

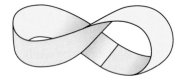

Make a Möbius band that is $1\frac{1}{2}$ in. wide. Use a pair of scissors to cut the Möbius band lengthwise down the middle, staying $\frac{3}{4}$ in. from each edge. Describe the result.

Make a Möbius band from plain white paper and then shade one side. Describe what remains unshaded on the Möbius band, and state the number of sides a Möbius band has.

Investigation

Slicing Polygons into Triangles[1]

Shown at the right is a triangle with three "slices" through it. Notice that the resulting pieces are six quadrilaterals and a triangle.

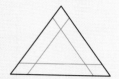

Investigation Exercises

1. Determine how you can slice a triangle so that all the resulting pieces are triangles. Use three slices. Each slice must cut all the way through the triangle.

2. Determine how you can slice each of the following polygons so that all the resulting pieces are triangles. Each slice must cut all the way through the polygon. Record the number of slices required for each polygon. *Note:* There may be more than one solution for a given polygon.

   a. Quadrilateral   **b.** Pentagon   **c.** Hexagon   **d.** Heptagon

Using Patterns in Experimentation

3. Try to cut a pie into the greatest number of pieces with only five straight cuts of a knife. An illustration showing how five cuts can produce 13 pieces is shown below. The correct answer, however, yields more than 13 pieces.

   A reasonable question is "How do I know when I have the maximum number of pieces?" To determine the answer, we suggest that you start with one cut, then two cuts, then three cuts, and so on. Try to discover a pattern for the greatest number of pieces that each successive cut can produce.

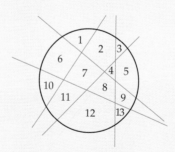

---

1. This activity is adapted from James Gary Propp's "The Slicing Game," *American Mathematical Monthly,* April 1996.

Exercise Set 4.4

1. Label the length of the rectangle L and the width of the rectangle W.

2. Label the base of the parallelogram b, the side s, and the height h.

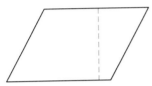

3. What is wrong with each statement?

   a. The perimeter is 40 m^2.

   b. The area is 120 ft.

In Exercises 4–7, name each polygon.

4. 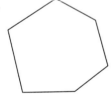   **5.**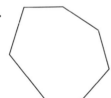

6. 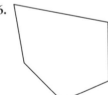   **7.**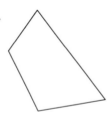

In Exercises 8–11, classify the triangle as isosceles, equilateral, or scalene.

8.    **9.**

10. 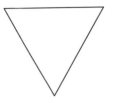   **11.**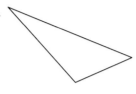

In Exercises 12–15, classify the triangle as acute, obtuse, or right.

12. 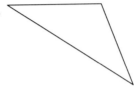   **13.**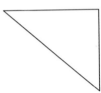

14. 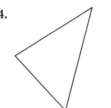   **15.**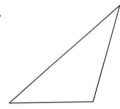

In Exercises 16–24, find (**a**) the perimeter and (**b**) the area of the figure.

16.
7 in.
11 in.

17.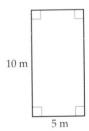
10 m
5 m

18.
8 ft
6 ft

19.
4 cm
4 cm

20.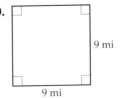
9 mi
9 mi

21.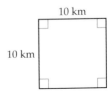
10 km
10 km

22.

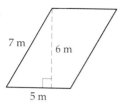

23.

24.

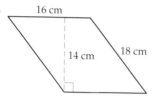

In Exercises 25–30, find (**a**) the circumference and (**b**) the area of the figure. Give both exact values and approximations to the nearest hundredth.

25.

26.

27.

28.

29.

30.

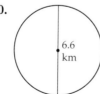

31. Fencing You need to fence in the triangular plot of land shown below. How many feet of fencing do you need?

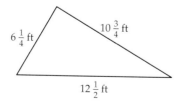

32. Gardens A flower garden in the yard of a historical home is in the shape of a triangle, as shown below. The wooden beams lining the edge of the garden need to be replaced. Find the total length of wood beams that must be purchased in order to replace the old beams.

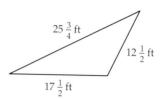

33. Yacht Racing The course of a yachting race is in the shape of a triangle with sides that measure $4\frac{3}{10}$ mi, $3\frac{7}{10}$ mi, and $2\frac{1}{2}$ mi. Find the total length of the course.

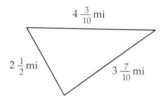

34. Physical Fitness An exercise course has stations set up along a path that is in the shape of a triangle with sides that measure $12\frac{1}{12}$ yd, $29\frac{1}{3}$ yd, and $26\frac{3}{4}$ yd. What is the entire length of the exercise course?

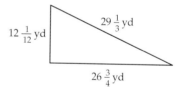

35. Fencing How many feet of fencing should be purchased to enclose a rectangular garden that is 20 ft long and 14 ft wide?

36. Sewing How many meters of binding are required to bind the edge of a rectangular quilt that measures 4 m by 6 m?

37. Perimeter Find the perimeter of a regular pentagon that measures 4 in. on each side.

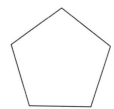

38. Interior Decorating Wall-to-wall carpeting is installed in a room that is 15 ft long and 10 ft wide. The edges of the carpet are held down by tack strips. How many feet of tack-strip material are needed?

39. Parks and Recreation The length of a rectangular park is 62 yd. The width is 45 yd. How many yards of fencing are needed to surround the park?

40. Perimeter What is the perimeter of a regular hexagon that measures 9 cm on each side?

41. Cross-Country A cross-country course is in the shape of a parallelogram with a base of length 3 mi and a side of length 2 mi. What is the total length of the cross-country course?

42. Parks and Recreation A rectangular playground has a length of 160 ft and a width of 120 ft. Find the length of hedge that surrounds the playground.

43. Sewing Bias binding is to be sewn around the edge of a rectangular tablecloth measuring 68 in. by 42 in. If the bias binding comes in packages containing 15 ft of binding, how many packages of bias binding are needed for the tablecloth?

44. Gardens Find the area of a rectangular flower garden that measures 24 ft by 18 ft.

45. Construction What is the area of a square patio that measures 12 m on each side?

46. Athletic Fields Artificial turf is being used to cover a playing field. If the field is rectangular with a length of 110 yd and a width of 80 yd, how much artificial turf must be purchased to cover the field?

47. Framing The perimeter of a square picture frame is 36 in. Find the length of each side of the frame.

48. Carpeting A square rug has a perimeter of 24 ft. Find the length of each edge of the rug.

49. Area The area of a rectangle is 400 in². If the length of the rectangle is 40 in., what is the width?

50. Area The width of a rectangle is 8 ft. If the area is 312 ft², what is the length of the rectangle?

51. Area The area of a parallelogram is 56 m². If the height of the parallelogram is 7 m, what is the length of the base?

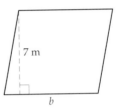

52. Storage Units You want to rent a storage unit. You estimate that you will need 175 ft² of floor space. In the Yellow Pages, you see the ad shown below. You want to rent the smallest possible unit that will hold everything you want to store. Which of the six units pictured in the ad should you select?

53. Sailing A sail is in the shape of a triangle with a base of 12 m and a height of 16 m. How much canvas was needed to make the sail?

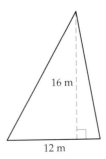

54. Gardens A vegetable garden is in the shape of a triangle with a base of 21 ft and a height of 13 ft. Find the area of the vegetable garden.

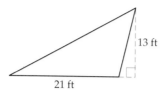

55. Athletic Fields How much artificial turf should be purchased to cover an athletic field that is in the shape of a trapezoid with a height of 15 m and bases that measure 45 m and 36 m?

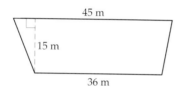

56. Land Area A township is in the shape of a trapezoid with a height of 10 km and bases measuring 9 km and 23 km. What is the land area of the township?

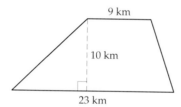

57. Parks and Recreation A city plans to plant grass seed in a public playground that has the shape of a triangle with a height of 24 m and a base of 20 m. Each bag of grass seed will seed 120 m². How many bags of seed should be purchased?

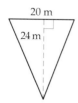

58. Interior Decorating The family room of your home is on the third floor of the house. Because of the pitch of the roof, two walls of the room are in the shape of

a trapezoid with bases that measure 16 ft and 24 ft and a height of 8 ft. You plan to wallpaper the two walls. One roll of wallpaper will cover 40 ft². How many rolls of wallpaper should you purchase?

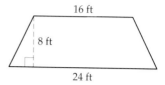

59. Home Maintenance You plan to stain the wooden deck at the back of your house. The deck is in the shape of a trapezoid with bases that measure 10 ft and 12 ft. The distance between the bases is 10 ft. A quart of stain will cover 55 ft². How many quarts of stain should you purchase?

60. Interior Decorating A fabric wall hanging is in the shape of a triangle that has a base of 4 ft and a height of 3 ft. An additional 1 ft² of fabric is needed for hemming the material. How much fabric should be purchased to make the wall hanging?

61. Interior Decorating You want to tile your kitchen floor. The floor measures 10 ft by 8 ft. How many square tiles that measure 2 ft along each side should you purchase for the job?

62. Interior Decorating You are wallpapering two walls of a den, one measuring 10 ft by 8 ft and the other measuring 12 ft by 8 ft. The wallpaper costs $24 per roll, and each roll will cover 40 ft². What is the cost of wallpapering the two walls?

63. Gardens An urban renewal project involves reseeding a garden that is in the shape of a square, 80 ft on each side. Each bag of grass seed costs $8 and will seed 1500 ft². How much money should be budgeted for buying grass seed for the garden?

64. Carpeting You want to install wall-to-wall carpeting in your family room. The floor plan is drawn below. If the cost of the carpet you would like to purchase is $19 per square yard, what is the cost of carpeting your family room? Assume there is no waste. *Hint:* 9 ft² = 1 yd²

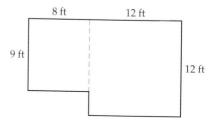

65. Interior Decorating You want to paint the walls of your bedroom. Two walls measure 16 ft by 8 ft, and the other two walls measure 12 ft by 8 ft. The paint you wish to purchase costs $17 per gallon, and each gallon will cover 400 ft² of wall. Find the total amount you will spend on paint.

66. Landscaping A walkway 2 m wide surrounds a rectangular plot of grass. The plot is 25 m long and 15 m wide. What is the area of the walkway?

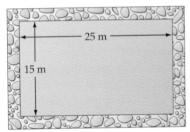

67. Draperies The material used to make pleated draperies for a window must be twice as wide as the width of the window. Draperies are being made for four windows, each 3 ft wide and 4 ft high. Because the drapes will fall slightly below the window sill and extra fabric is needed for hemming the drapes, 1 ft must be added to the height of each drape. How much material must be purchased to make the drapes?

68. Fencing How many feet of fencing should be purchased to enclose a circular flower garden that has a diameter of 18 ft? Round to the nearest tenth of a foot.

69. Carpentry Find the length of molding needed to surround a circular table that is 4.2 ft in diameter. Round to the nearest hundredth of a foot.

70. Sewing How much binding is needed to bind the edge of a circular rug that is 3 m in diameter? Round to the nearest hundredth of a meter.

71. Gardens Find the area of a circular flower garden that has a radius of 20 ft. Round to the nearest tenth of a square foot.

72. Pulleys A pulley system is diagrammed below. If pulley B has a diameter of 16 in. and is rotating at 240 revolutions per minute, how far does the belt travel each minute that the pulley system is in operation? Assume the belt does not slip as the pulley rotates. Round to the nearest inch.

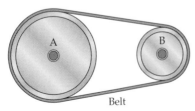

73. Bicycles A bicycle tire has a diameter of 18 in. How many feet does the bicycle travel when the wheel makes 20 revolutions? Round to the nearest hundredth of a foot.

74. Tricycles The front wheel of a tricycle has a diameter of 16 in. How many feet does the tricycle travel when the wheel makes 15 revolutions? Round to the nearest hundredth of a foot.

75. Telescopes The lens located on an astronomical telescope has a diameter of 24 in. Find the exact area of the lens.

76. Irrigation An irrigation system waters a circular field that has a 50-foot radius. Find the exact area watered by the irrigation system.

77. Pizza How much greater is the area of a pizza that has a radius of 10 in. than the area of a pizza that has a radius of 8 in.? Round to the nearest hundredth of a square inch.

78. Pizza A restaurant serves a small pizza that has a radius of 6 in. The restaurant's large pizza has a radius that is twice the radius of the small pizza. How much larger is the area of the large pizza? Round to the nearest hundredth of a square inch. Is the area of the large pizza more or less than twice the area of the small pizza?

79. Satellites There are two general types of satellite systems that orbit our Earth: geostationary Earth orbit (GEO) and non-geostationary, primarily low Earth orbit (LEO). Geostationary satellite systems orbit at a distance of 36,000 km above Earth. An orbit at this altitude allows the satellite to maintain a fixed position in relation to Earth. What is the distance traveled by a GEO satellite in one orbit around Earth? The radius of Earth at the equator is 6380 km. Round to the nearest kilometer.

80. Ball Fields How much farther is it around the bases of a baseball diamond than around the bases of a softball diamond? *Hint:* Baseball and softball diamonds are squares.

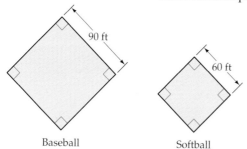

Baseball   Softball

81. Area Write an expression for the area of the shaded portion of the diagram. Leave the answer in terms of π and r.

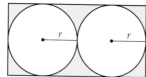

82. Area Write an expression for the area of the shaded portion of the diagram. Leave the answer in terms of π and r.

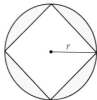

83. What is the name of a parallelogram in which all angles are the same measure?

84. **a.** Is every square a rectangle?

b. Is every rectangle a square?

85. Which has the greater perimeter, a square whose side measures 1 ft or a rectangle that has a length of 2 in. and a width of 1 in.?

86. If both the length and the width of a rectangle are doubled, how many times larger is the area of the resulting rectangle?

87. Is the statement always true, sometimes true, or never true?

a. Two triangles that have the same perimeter have the same area.

b. Two rectangles that have the same area have the same perimeter.

c. If two squares have the same area, then the sides of the squares have the same length.

d. An equilateral triangle is also an isosceles triangle.

e. All the radii (plural of radius) of a circle are equal.

f. All the diameters of a circle are equal.

Extensions

CRITICAL THINKING

88. Find the dimensions of a rectangle that has the same area as the shaded region in the diagram below. Write the dimensions in terms of the variable a.

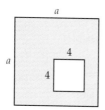

89. Find the dimensions of a rectangle that has the same area as the shaded region in the diagram below. Write the dimensions in terms of the variable x.

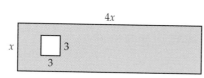

90. In the diagrams below, the length of each side of each square is 1 cm. Find the perimeter and the area of the eighth figure in the pattern.

a.

b.

c.

EXPLORATIONS

91. The perimeter of the square at the right is 4 units.

If two such squares are joined along one of the sides, the perimeter is 6 units. Note that it does not matter which sides are joined; the perimeter is still 6 units.

If three squares are joined, the perimeter of the resulting figure is 8 units for each possible placement of the squares.

Four squares can be joined in five different ways as shown. There are two possible perimeters: 10 units for A, B, C, and D, and 8 units for E.

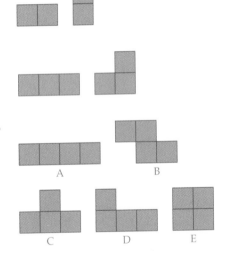

a. If five squares are joined, what is the maximum perimeter possible?

b. If five squares are joined, what is the minimum perimeter possible?

c. If six squares are joined, what is the maximum perimeter possible?

d. If six squares are joined, what is the minimum perimeter possible?

92. Suppose a circle is cut into 16 equal pieces, which are then arranged as shown at the right. The figure formed resembles a parallelogram. What variable expression could approximate the base of the parallelogram? What variable could approximate its height? Explain how the formula for the area of a circle is derived from this approach.

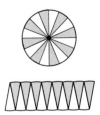

| SECTION 4.5 | **Properties of Triangles** |

Z Scale

N Scale

HO Scale

HO's ratio of 1:87 means that in every dimension, an HO scale model railroad car is $\frac{1}{87}$ the size of the real railroad car.

Similar Triangles

Similar objects have the same shape but not necessarily the same size. A tennis ball is similar to a basketball. A model ship is similar to an actual ship.

Similar objects have corresponding parts; for example, the rudder on the model ship corresponds to the rudder on the actual ship. The relationship between the sizes of each of the corresponding parts can be written as a ratio, and each ratio will be the same. If the rudder on the model ship is $\frac{1}{100}$ the size of the rudder on the actual ship, then the model wheelhouse is $\frac{1}{100}$ of the size of the actual wheelhouse, the width of the model is $\frac{1}{100}$ the width of the actual ship, and so on.

The two triangles ABC and DEF shown at the right are similar. Side $\overline{AB}$ corresponds to side $\overline{DE}$, side $\overline{BC}$ corresponds to side $\overline{EF}$, and side $\overline{AC}$ corresponds to side $\overline{DF}$. The ratios of the lengths of corresponding sides are equal.

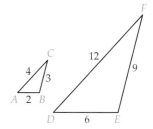

$$\frac{AB}{DE} = \frac{2}{6} = \frac{1}{3}, \quad \frac{BC}{EF} = \frac{3}{9} = \frac{1}{3}, \quad \text{and}$$

$$\frac{AC}{DF} = \frac{4}{12} = \frac{1}{3}$$

Because the ratios of corresponding sides are equal, several proportions can be formed.

$$\frac{AB}{DE} = \frac{BC}{EF}, \quad \frac{AB}{DE} = \frac{AC}{DF}, \quad \text{and} \quad \frac{BC}{EF} = \frac{AC}{DF}$$

The measures of corresponding angles in similar triangles are equal. Therefore,

$$m\angle A = m\angle D, \quad m\angle B = m\angle E, \quad \text{and}$$
$$m\angle C = m\angle F$$

Triangles ABC and DEF at the right are similar triangles. AH and DK are the heights of the triangles. The ratio of the heights of similar triangles equals the ratio of the lengths of corresponding sides.

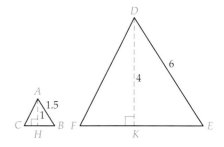

Ratio of corresponding sides $= \frac{1.5}{6} = \frac{1}{4}$

Ratio of heights $= \frac{1}{4}$

Properties of Similar Triangles

For **similar triangles,** the ratios of corresponding sides are equal. The ratio of corresponding heights is equal to the ratio of corresponding sides.

TAKE NOTE

The notation $\triangle ABC \sim \triangle DEF$ is used to indicate that triangle *ABC* is similar to triangle *DEF*.

The two triangles at the right are similar triangles. Find the length of side $\overline{EF}$. Round to the nearest tenth of a meter.

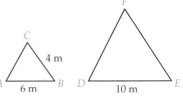

The triangles are similar, so the ratios of the lengths of corresponding sides are equal.

$$\frac{EF}{BC} = \frac{DE}{AB}$$

$$\frac{EF}{4} = \frac{10}{6}$$

$$6(EF) = 4(10)$$

$$6(EF) = 40$$

$$EF \approx 6.7$$

The length of side EF is approximately 6.7 m.

QUESTION   *What are two other proportions that can be written for the similar triangles shown above?*

EXAMPLE 1 ▪ Use Similar Triangles to Find the Unknown Height of a Triangle

Triangles *ABC* and *DEF* are similar. Find *FG*, the height of triangle *DEF*.

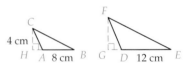

Solution

$$\frac{AB}{DE} = \frac{CH}{FG}$$ • For similar triangles, the ratio of corresponding sides equals the ratio of corresponding heights.

$$\frac{8}{12} = \frac{4}{FG}$$ • Replace *AB*, *DE*, and *CH* with their values.

$$8(FG) = 12(4)$$ • The cross products are equal.

$$8(FG) = 48$$

$$FG = 6$$ • Divide both sides of the equation by 8.

The height *FG* of triangle *DEF* is 6 cm.

CHECK YOUR PROGRESS 1

Triangles *ABC* and *DEF* are similar. Find *FG*, the height of triangle *DEF*.

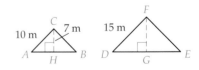

Solution   *See page S17.*

ANSWER   *In addition to $\frac{EF}{BC} = \frac{DE}{AB}$, we can write the proportions $\frac{DE}{AB} = \frac{DF}{AC}$ and $\frac{EF}{BC} = \frac{DF}{AC}$. These three proportions can also be written using the reciprocal of each fraction: $\frac{BC}{EF} = \frac{AB}{DE}$, $\frac{AB}{DE} = \frac{AC}{DF}$, and $\frac{BC}{EF} = \frac{AC}{DF}$. Also, the right and left sides of each proportion can be interchanged.*

Triangles *ABC* and *DEF* are similar triangles. Find the area of triangle *ABC*.

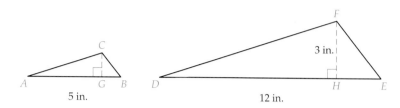

Solve a proportion to find the height of triangle *ABC*.

$$\frac{AB}{DE} = \frac{CG}{FH}$$

$$\frac{5}{12} = \frac{CG}{3}$$

$$12(CG) = 5(3)$$

$$12(CG) = 15$$

$$CG = 1.25$$

Use the formula for the area of a triangle.

$$A = \frac{1}{2}bh$$

The base is 5 in. The height is 1.25 in.

$$A = \frac{1}{2}(5)(1.25)$$

$$A = 3.125$$

The area of triangle *ABC* is 3.125 in².

If the three angles of one triangle are equal in measure to the three angles of another triangle, then the triangles are similar.

In triangle *ABC* at the right, line segment $\overline{DE}$ is drawn parallel to the base $\overline{AB}$. Because the measures of corresponding angles are equal, $m\angle x = m\angle r$ and $m\angle y = m\angle n$. We know that $m\angle C = m\angle C$. Thus the measures of the three angles of triangle *ABC* are equal, respectively, to the measures of the three angles of triangle *DEC*. Therefore, triangles *ABC* and *DEC* are similar triangles.

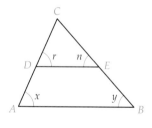

The sum of the measures of the three angles of a triangle is 180°. If two angles of one triangle are equal in measure to two angles of another triangle, then the third angles must be equal. Thus, we can say that if two angles of one triangle are equal in measure to two angles of another triangle, then the two triangles are similar.

In the figure at the right, $\overline{AB}$ intersects $\overline{CD}$ at point *O*. Angles *C* and *D* are right angles. Find the length of $\overline{DO}$.

First determine whether triangles *AOC* and *BOD* are similar.

$m\angle C = m\angle D$ because they are both right angles.

$m\angle x = m\angle y$ because vertical angles have the same measure.

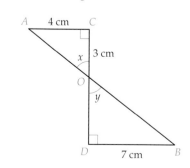

✔ **TAKE NOTE**

You can always create similar triangles by drawing a line segment inside the original triangle parallel to one side of the triangle. In the triangle below, $\overline{ST} \parallel \overline{QR}$ and triangle *PST* is similar to triangle *PQR*.

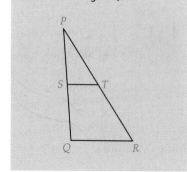

Because two angles of triangle *AOC* are equal in measure to two angles of triangle *BOD*, triangles *AOC* and *BOD* are similar.

Use a proportion to find the length of the unknown side.

$$\frac{AC}{BD} = \frac{CO}{DO}$$

$$\frac{4}{7} = \frac{3}{DO}$$

$$4(DO) = 7(3)$$

$$4(DO) = 21$$

$$DO = 5.25$$

The length of $\overline{DO}$ is 5.25 cm.

EXAMPLE 2 ■ Solve a Problem Involving Similar Triangles

In the figure at the right, $\overline{AB}$ is parallel to $\overline{DC}$, $\angle B$ and $\angle D$ are right angles, $AB = 12$ m, $DC = 4$ m, and $AC = 18$ m. Find the length of $\overline{CO}$.

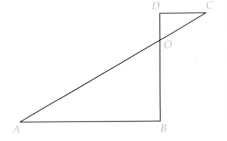

Solution

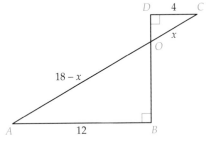

• Label the diagram using the given information. Let *x* represent *CO*. *AC* = *AO* + *CO*. Because *AC* = 18, *AO* = 18 − *x*.

$$\frac{DC}{BA} = \frac{CO}{AO}$$

$$\frac{4}{12} = \frac{x}{18 - x}$$

$$12x = 4(18 - x)$$

$$12x = 72 - 4x$$

$$16x = 72$$

$$x = 4.5$$

• Triangles *AOB* and *COD* are similar triangles. The ratios of corresponding sides are equal.

• The cross-products are equal.
• Use the Distributive Property.

The length of $\overline{CO}$ is 4.5 m.

CHECK YOUR PROGRESS 2

In the figure at the right, $\overline{AB}$ is parallel to $\overline{DC}$, $\angle A$ and $\angle D$ are right angles, $AB = 10$ cm, $CD = 4$ cm, and $DO = 3$ cm. Find the area of triangle AOB.

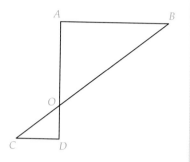

Solution   *See page S18.*

Math Matters   **Similar Polygons**

For similar triangles, the measures of corresponding angles are equal and the ratios of the lengths of corresponding sides are equal. The same is true for similar polygons: the measures of corresponding angles are equal and the lengths of corresponding sides are in proportion.

    Quadrilaterals *ABCD* and *LMNO* are similar.

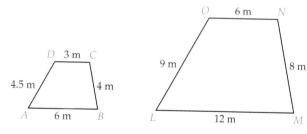

The ratio of the lengths of corresponding sides is $\dfrac{AB}{LM} = \dfrac{6}{12} = \dfrac{1}{2}$.

The ratio of the perimeter of *ABCD* to the perimeter of *LMNO* is:

$$\frac{\text{perimeter of } ABCD}{\text{perimeter of } LMNO} = \frac{17.5}{35} = \frac{1}{2}$$

Note that this ratio is the same as the ratio of corresponding sides. This is true for all similar polygons: If two polygons are similar, the ratio of their perimeters is equal to the ratio of the lengths of any pair of corresponding sides.

The Pythagorean Theorem

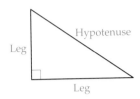

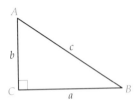

Recall that a right triangle contains one right angle. The side opposite the right angle is called the **hypotenuse.** The other two sides are called **legs.**

The vertices of the angles in a right triangle are usually labeled with the capital letters A, B, and C, with C reserved for the right angle. The side opposite angle A is side a, the side opposite angle B is side b, and c is the hypotenuse.

The figure at the right is a right triangle with legs measuring 3 units and 4 units and a hypotenuse measuring 5 units. Each side of the triangle is also the side of a square. The number of square units in the area of the largest square is equal to the sum of the numbers of square units in the areas of the smaller squares.

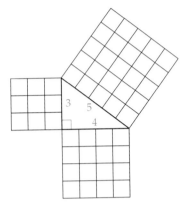

Square of the hypotenuse	$=$	sum of the squares of the two legs

$$5^2 = 3^2 + 4^2$$
$$25 = 9 + 16$$
$$25 = 25$$

The Greek mathematician Pythagoras is generally credited with the discovery that the square of the hypotenuse of a right triangle is equal to the sum of the squares of the two legs. This is called the **Pythagorean Theorem.**

point of interest

The first known proof of the Pythagorean Theorem is in a Chinese textbook that dates from 150 B.C. The book is called *Nine Chapters on the Mathematical Art*. The diagram below is from that book and was used in the proof of the theorem.

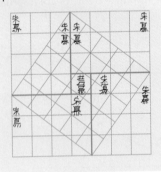

Pythagorean Theorem

If a and b are the lengths of the legs of a right triangle and c is the length of the hypotenuse, then $c^2 = a^2 + b^2$.

If the lengths of two sides of a right triangle are known, the Pythagorean Theorem can be used to find the length of the third side.

Consider a right triangle with legs that measure 5 cm and 12 cm. Use the Pythagorean Theorem, with $a = 5$ and $b = 12$, to find the length of the hypotenuse. (If you let $a = 12$ and $b = 5$, the result will be the same.) Take the square root of each side of the equation.

The length of the hypotenuse is 13 cm.

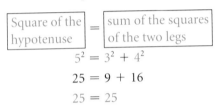

$$c^2 = a^2 + b^2$$
$$c^2 = 5^2 + 12^2$$
$$c^2 = 25 + 144$$
$$c^2 = 169$$
$$\sqrt{c^2} = \sqrt{169}$$
$$c = 13$$

✔ TAKE NOTE

The length of the side of a triangle cannot be negative. Therefore, we take only the principal, or positive, square root of 169.

CALCULATOR NOTE

The way in which you evaluate the square root of a number depends on the type of calculator you have. Here are two possible keystrokes to find $\sqrt{80}$:

80 $\boxed{\sqrt{}}$ $\boxed{=}$

or

$\boxed{\sqrt{}}$ 80 $\boxed{\text{ENTER}}$

The first method is used on many scientific calculators. The second method is used on many graphing calculators.

EXAMPLE 3 ■ Determine the Length of the Unknown Side of a Right Triangle

The length of one leg of a right triangle is 8 in. The length of the hypotenuse is 12 in. Find the length of the other leg. Round to the nearest hundredth of an inch.

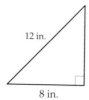

12 in.

8 in.

Solution

$$a^2 + b^2 = c^2$$ • Use the Pythagorean Theorem.

$$8^2 + b^2 = 12^2$$ • $a = 8$, $c = 12$

$$64 + b^2 = 144$$

$$b^2 = 80$$ • Solve for b^2. Subtract 64 from each side.

$$\sqrt{b^2} = \sqrt{80}$$ • Take the square root of each side of the equation.

$$b \approx 8.94$$ • Use a calculator to approximate $\sqrt{80}$.

The length of the other leg is approximately 8.94 in.

CHECK YOUR PROGRESS 3   The hypotenuse of a right triangle measures 6 m, and one leg measures 2 m. Find the measure of the other leg. Round to the nearest hundredth of a meter.

Solution   *See page S18.*

Investigation

Topology

In this section, we discussed similar figures—that is, figures with the same shape. The branch of geometry called *topology* is the study of even more basic properties of figures than their sizes and shapes. For example, look at the figures below. We could take a rubber band and stretch it into any one of these shapes.

   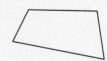   

All three of these figures are different shapes, but each can be turned into one of the others by stretching the rubber band.

(continued)

In topology, figures that can be stretched, molded, or bent into the same shape *without puncturing or cutting* belong to the same family. They are called **topologically equivalent.**

Rectangles, triangles, and circles are topologically equivalent.

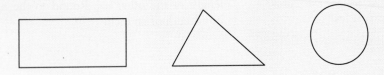

Line segments and wavy curves are topologically equivalent.

Note that the figures formed from a rubber band and those formed from a line segment are not topologically equivalent; to form a line segment from a rubber band, we would have to cut the rubber band.

In the following plane figures, the lines and curves are joined where they cross. The figures are topologically equivalent. They are not topologically equivalent to any of the figures shown above.

A **topologist** (a person who studies topology) is interested in identifying and describing different families of equivalent figures. Topology applies to solids as well as plane figures. For example, a topologist considers a brick, a potato, and a pool ball to be topologically equivalent to each other. Think of using modeling clay to form each of these shapes.

Investigation Exercises

Which of the figures listed is not topologically equivalent to the others?

1. a. Parallelogram   **b.** Square   **c.** Ray   **d.** Trapezoid

2. a. Wedding ring   **b.** Doughnut   **c.** Fork   **d.** Sewing needle

3. a. A   **b.** D   **c.** O   **d.** P   **e.** T

Exercise Set 4.5

In Exercises 1–4, find the ratio of the lengths of corresponding sides for the similar triangles.

1.

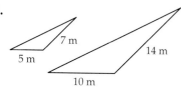

2.

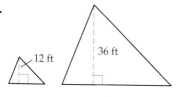

3.

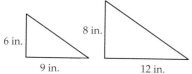

4.

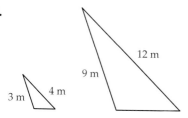

In Exercises 5–12, triangles *ABC* and *DEF* are similar triangles. Solve. Round to the nearest tenth.

5. Find side *DE*.

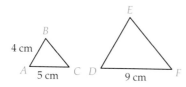

6. Find side *DE*.

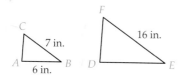

7. Find the height of triangle *DEF*.

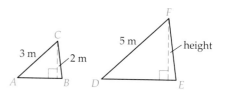

8. Find the height of triangle *ABC*.

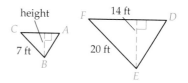

9. Find the perimeter of triangle *ABC*.

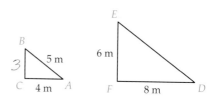

10. Find the perimeter of triangle *DEF*.

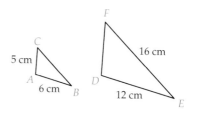

11. Find the perimeter of triangle *ABC*.

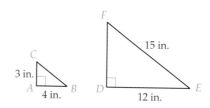

12. Find the area of triangle *DEF*.

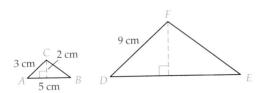

In Exercises 13 and 14, triangles *ABC* and *DEF* are similar triangles. Solve. Round to the nearest tenth.

13. Find the area of triangle *ABC*.

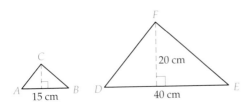

14. Find the area of triangle *DEF*.

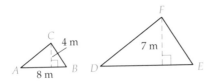

In Exercises 15–19, the given triangles are similar triangles. Use this fact to solve each exercise.

15. Find the height of the flagpole.

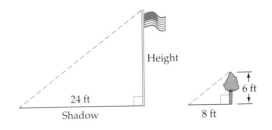

16. Find the height of the flagpole.

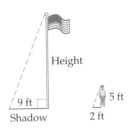

17. Find the height of the building.

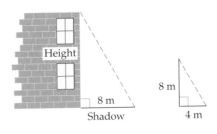

18. Find the height of the building.

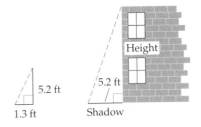

19. Find the height of the flagpole.

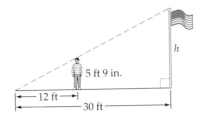

20. In the figure below, $\overline{BD} \parallel \overline{AE}$, *BD* measures 5 cm, *AE* measures 8 cm, and *AC* measures 10 cm. Find the length of $\overline{BC}$.

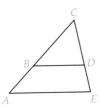

21. In the figure below, $\overline{AC} \parallel \overline{DE}$, *BD* measures 8 m, *AD* measures 12 m, and *BE* measures 6 m. Find the length of $\overline{BC}$.

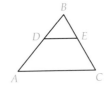

22. In the figure below, $\overline{DE} \parallel \overline{AC}$, DE measures 6 in., AC measures 10 in., and AB measures 15 in. Find the length of $\overline{DA}$.

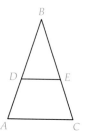

23. In the figure below, $\overline{AE} \parallel \overline{BD}$, $AB = 3$ ft, $ED = 4$ ft, and $BC = 3$ ft. Find the length of $\overline{CE}$.

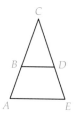

24. In the figure below, $\overline{MP}$ and $\overline{NQ}$ intersect at O, $NO = 25$ ft, $MO = 20$ ft, and $PO = 8$ ft. Find the length of $\overline{QO}$.

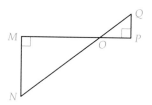

25. In the figure below, $\overline{MP}$ and $\overline{NQ}$ intersect at O, $NO = 24$ cm, $MN = 10$ cm, $MP = 39$ cm, and $QO = 12$ cm. Find the length of $\overline{OP}$.

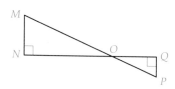

26. In the following figure, $\overline{MQ}$ and $\overline{NP}$ intersect at O, $NO = 12$ m, $MN = 9$ m, $PQ = 3$ m, and $MQ = 20$ m. Find the perimeter of triangle OPQ.

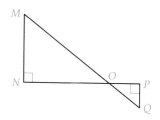

Surveying Surveyors often use similar triangles to measure distances that cannot be measured directly. This is illustrated in Exercises 27 and 28.

27. The diagram below represents a river of width CD. Triangles AOB and DOC are similar. The distances AB, BO, and OC were measured and found to have the lengths given in the diagram. Find CD, the width of the river.

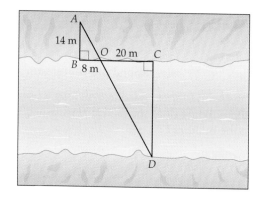

28. The diagram below shows how surveyors laid out similar triangles along the Winnepaugo River. Find the width d of the river.

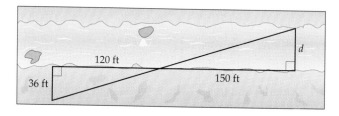

29. Determine whether the statement is always true, sometimes true, or never true.

   a. If two angles of one triangle are equal in measure to two angles of a second triangle, then the triangles are similar triangles.

   b. Two isosceles triangles are similar triangles.

   c. Two equilateral triangles are similar triangles.

   d. If an acute angle of a right triangle is equal to an acute angle of another right triangle, then the triangles are similar triangles.

In Exercises 30–38, find the unknown side of the triangle. Round to the nearest tenth.

30.

3 in.

4 in.

31.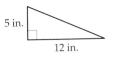

5 in.

12 in.

32.

5 cm

7 cm

33.

7 cm

9 cm

34.

15 ft

10 ft

35.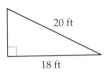

20 ft

18 ft

36.

4 cm   6 cm

37.

9 m   12 m

38.

9 yd

9 yd

In Exercises 39–43, solve. Round to the nearest tenth.

39. Home Maintenance A ladder 8 m long is leaning against a building. How high on the building does the ladder reach when the bottom of the ladder is 3 m from the building?

8 m

3 m

40. Mechanics Find the distance between the centers of the holes in the metal plate.

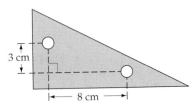

3 cm

8 cm

41. Travel If you travel 18 mi east and then 12 mi north, how far are you from your starting point?

42. Perimeter Find the perimeter of a right triangle with legs that measure 5 cm and 9 cm.

43. Perimeter Find the perimeter of a right triangle with legs that measure 6 in. and 8 in.

44. The lengths of two sides of a triangle are 10 in. and 14 in. What are the possible values for x, the length of the third side of the triangle?

Extensions

CRITICAL THINKING

45. **Home Maintenance** You need to clean the gutters of your home. The gutters are 24 ft above the ground. For safety, the distance a ladder reaches up a wall should be four times the distance from the bottom of the ladder to the base of the side of the house. Therefore, the ladder must be 6 ft from the base of the house. Will a 25-foot ladder be long enough to reach the gutters? Explain how you found your answer.

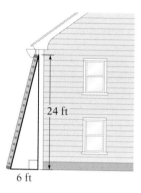

24 ft

6 ft

COOPERATIVE LEARNING

46. In the figure below, the height of a right triangle is drawn from the right angle perpendicular to the hypotenuse. (Recall that the hypotenuse of a right triangle is the side opposite the right angle.) Verify that the two smaller triangles formed are similar to the original triangle and similar to each other.

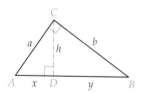

SECTION 4.6 | # Volume and Surface Area

Volume

In Section 4 of this chapter, we developed the geometric concepts of perimeter and area. Perimeter and area refer to plane figures (figures that lie in a plane). We are now ready to introduce the concept of *volume* of geometric solids.

Geometric solids are figures in space. Figures in space include baseballs, ice cubes, milk cartons, and trucks.

Volume is a measure of the amount of space occupied by a geometric solid. Volume can be used to describe, for example, the amount of trash in a land fill, the amount of concrete poured for the foundation of a house, or the amount of water in a town's reservoir.

Five common geometric solids are rectangular solids, spheres, cylinders, cones, and pyramids.

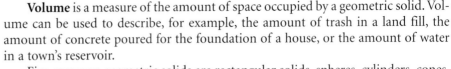

A **rectangular solid** is one in which all six sides, called **faces,** are rectangles. The variable L is used to represent the length of a rectangular solid, W is used to represent its width, and H is used to represent its height. A shoe box is an example of a rectangular solid.

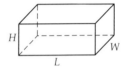

A **cube** is a special type of rectangular solid. Each of the six faces of a cube is a square. The variable s is used to represent the length of one side of a cube. A baby's block is an example of a cube.

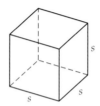

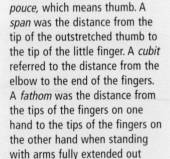

▼ **point of interest**

Originally, the human body was used as the standard of measure. A mouthful was used as a unit of measure in ancient Egypt; it was later referred to as a *half jigger.* In French, the word for inch is *pouce,* which means thumb. A *span* was the distance from the tip of the outstretched thumb to the tip of the little finger. A *cubit* referred to the distance from the elbow to the end of the fingers. A *fathom* was the distance from the tips of the fingers on one hand to the tips of the fingers on the other hand when standing with arms fully extended out from the sides. The *hand* is still used today to measure the heights of horses.

A cube that is 1 ft on each side has a volume of 1 cubic foot, which is written 1 ft³. A cube that measures 1 cm on each side has a volume of 1 cubic centimeter, written 1 cm³.

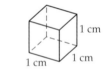

The volume of a solid is the number of cubes, each of volume 1 cubic unit, that are necessary to fill the solid. The volume of the rectangular solid at the right is 24 cm³ because it will hold exactly 24 cubes, each 1 cm on a side. Note that the volume can be found by multiplying the length times the width times the height.

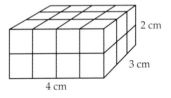

$$4 \cdot 3 \cdot 2 = 24$$

The volume of the solid is 24 cm³.

Volume of a Rectangular Solid

The volume V of a rectangular solid with length L, width W, and height H is given by $V = LWH$.

Volume of a Cube

The volume V of a cube with side of length s is given by $V = s^3$.

QUESTION *Which of the following are rectangular solids: a juice box, a milk carton, a can of soup, a compact disc, and a jewel case (plastic container in which a compact disc is packaged)?*

A **sphere** is a solid in which all points are the same distance from a point O, called the **center** of the sphere. A **diameter** of a sphere is a line segment with endpoints on the sphere and passing through the center. A **radius** is a line segment from the center to a point on the sphere. $\overline{AB}$ is a diameter and $\overline{OC}$ is a radius of the sphere shown at the right. A basketball is an example of a sphere.

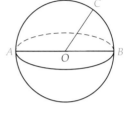

If we let d represent the length of a diameter and r represent the length of a radius, then $d = 2r$ or $r = \frac{1}{2}d$.

$$d = 2r \quad \text{or} \quad r = \frac{1}{2}d$$

Volume of a Sphere

The volume V of a sphere with radius of length r is given by $V = \frac{4}{3}\pi r^3$.

Find the volume of a rubber ball that has a diameter of 6 in.

First find the length of a radius of the sphere.	$r = \frac{1}{2}d = \frac{1}{2}(6) = 3$
Use the formula for the volume of a sphere.	$V = \frac{4}{3}\pi r^3$
Replace r with 3.	$V = \frac{4}{3}\pi(3)^3$
	$V = \frac{4}{3}\pi(27)$
The exact volume of the sphere is 36π in³.	$V = 36\pi$
An approximate measure can be found by using the π key on a calculator.	$V \approx 113.10$

The volume of the rubber ball is approximately 113.10 in³.

ANSWER *A juice box and a jewel case are rectangular solids.*

The most common cylinder, called a **right circular cylinder,** is one in which the bases are circles and are perpendicular to the height of the cylinder. The variable r is used to represent the length of the radius of a base of the cylinder, and h represents the height of the cylinder. In this text, only right circular cylinders are discussed.

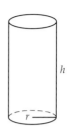

Volume of a Right Circular Cylinder

The volume V of a right circular cylinder is given by $V = \pi r^2 h$, where r is the radius of the base and h is the height of the cylinder.

A **right circular cone** is obtained when one base of a right circular cylinder is shrunk to a point, called the **vertex,** V. The variable r is used to represent the radius of the base of the cone, and h represents the height of the cone. The variable l is used to represent the **slant height,** which is the distance from a point on the circumference of the base to the vertex. In this text, only right circular cones are discussed. An ice cream cone is an example of a right circular cone.

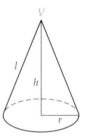

✔ TAKE NOTE

Note that πr^2 appears in the formula for the volume of a right circular cylinder and in the formula for the volume of a right circular cone. This is because, in each case, the base of the figure is a circle.

Volume of a Right Circular Cone

The volume V of a right circular cone is given by $V = \frac{1}{3}\pi r^2 h$, where r is the radius of the circular base and h is the height of the cone.

✔ TAKE NOTE

Recall that an isosceles triangle has two sides of equal length.

The base of a **regular pyramid** is a regular polygon and the sides are isosceles triangles (two sides of the triangle are the same length). The height, h, is the distance from the vertex, V, to the base and is perpendicular to the base. The variable l is used to represent the **slant height,** which is the height of one of the isosceles triangles on the face of the pyramid. The regular square pyramid at the right has a square base. This is the only type of pyramid discussed in this text. Many Egyptian pyramids are regular square pyramids.

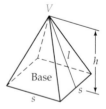

Pyramid at Giza

Volume of a Regular Square Pyramid

The volume V of a regular square pyramid is given by $V = \frac{1}{3}s^2 h$, where s is the length of a side of the base and h is the height of the pyramid.

QUESTION   *Which of the following units could not be used to measure the volume of a regular square pyramid?*

a. ft^3 *b.* m^3 *c.* yd^2 *d.* cm^3 *e.* mi

EXAMPLE 1 ■ Find the Volume of a Geometric Solid

Find the volume of a cube that measures 1.5 m on a side.

Solution

$V = s^3$ • Use the formula for the volume of a cube.

$V = 1.5^3$ • Replace *s* with **1.5**.

$V = 3.375$

The volume of the cube is 3.375 m³.

CHECK YOUR PROGRESS 1   The length of a rectangular solid is 5 m, the width is 3.2 m, and the height is 4 m. Find the volume of the solid.

Solution   *See page S18.*

EXAMPLE 2 ■ Find the Volume of a Geometric Solid

The radius of the base of a cone is 8 cm. The height of the cone is 12 cm. Find the volume of the cone. Round to the nearest hundredth of a cubic centimeter.

Solution

$V = \dfrac{1}{3}\pi r^2 h$ • Use the formula for the volume of a cone.

$V = \dfrac{1}{3}\pi(8)^2(12)$ • Replace *r* with **8** and *h* with **12**.

$V = \dfrac{1}{3}\pi(64)(12)$

$V = 256\pi$

$V \approx 804.25$ • Use the π key on a calculator.

The volume of the cone is approximately 804.25 cm³.

CHECK YOUR PROGRESS 2   The length of a side of the base of a regular square pyramid is 15 m and the height of the pyramid is 25 m. Find the volume of the pyramid.

Solution   *See page S18.*

ANSWER   *Volume is measured in cubic units. Therefore, the volume of a regular square pyramid could be measured in ft^3, m^3, or cm^3, but not in yd^2 or mi.*

EXAMPLE 3 ■ Find the Volume of a Geometric Solid

An oil storage tank in the shape of a cylinder is 4 m high and has a diameter of 6 m. The oil tank is two-thirds full. Find the number of cubic meters of oil in the tank. Round to the nearest hundredth of a cubic meter.

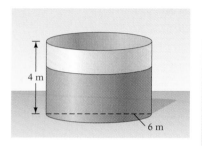

Solution

$$r = \frac{1}{2}d = \frac{1}{2}(6) = 3$$
• Find the radius of the base.

$$V = \pi r^2 h$$
• Use the formula for the volume of a cylinder.

$$V = \pi(3)^2(4)$$
• Replace *r* with **3** and *h* with **4**.

$$V = \pi(9)(4)$$

$$V = 36\pi$$

$$\frac{2}{3}(36\pi) = 24\pi$$
• Multiply the volume by $\frac{2}{3}$.

$$\approx 75.40$$
• Use the π key on a calculator.

There are approximately 75.40 m³ of oil in the storage tank.

CHECK YOUR PROGRESS 3 A silo in the shape of a cylinder is 16 ft in diameter and has a height of 30 ft. The silo is three-fourths full. Find the volume of the portion of the silo that is not being used for storage. Round to the nearest hundredth of a cubic foot.

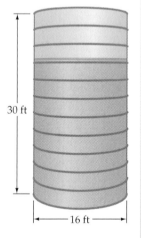

Solution *See page S18.*

Surface Area

The **surface area** of a solid is the total area on the surface of the solid. Suppose you want to cover a geometric solid with wallpaper. The amount of wallpaper needed is equal to the surface area of the figure.

When a rectangular solid is cut open and flattened out, each face is a rectangle. The surface area S of the rectangular solid is the sum of the areas of the six rectangles:

$$S = LW + LH + WH + LW + WH + LH$$

which simplifies to

$$S = 2LW + 2LH + 2WH$$

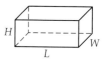
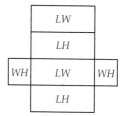

The surface area of a cube is the sum of the areas of the six faces of the cube. The area of each face is s^2. Therefore, the surface area S of a cube is given by the formula $S = 6s^2$.

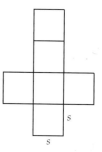

historical note

Pappus (păp′ŭs) of Alexandria (c. 290–350) was born in Egypt. We know the time period in which he lived because he wrote about his observation of the eclipse of the sun that took place in Alexandria on October 18, 320.

Pappus has been called the last of the great Greek geometers. His major work in geometry is *Synagoge,* or the *Mathematical Collection.* It consists of eight books. In Book V, Pappus proves that the sphere has a greater volume than any regular geometric solid with equal surface area. He also proves that if two regular solids have equal surface areas, the solid with the greater number of faces has the greater volume. ■

When a cylinder is cut open and flattened out, the top and bottom of the cylinder are circles. The side of the cylinder flattens out to a rectangle. The length of the rectangle is the circumference of the base, which is $2\pi r$; the width is h, the height of the cylinder. Therefore, the area of the rectangle is $2\pi rh$. The surface area S of the cylinder is

$$S = \pi r^2 + 2\pi rh + \pi r^2$$

which simplifies to

$$S = 2\pi r^2 + 2\pi rh$$

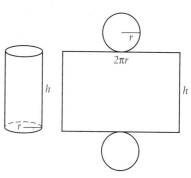

The surface area of a regular square pyramid is the area of the base plus the areas of the four isosceles triangles. The length of a side of the square base is s; therefore, the area of the base is s^2. The slant height l is the height of each triangle, and s is the length of the base of each triangle. The surface area S of a regular square pyramid is

$$S = s^2 + 4\left(\frac{1}{2}sl\right)$$

which simplifies to

$$S = s^2 + 2sl$$

Formulas for the surface areas of geometric solids are given below.

Surface Areas of Geometric Solids

The surface area, S, of a **rectangular solid** with length L, width W, and height H is given by $S = 2LW + 2LH + 2WH$.

The surface area, S, of a **cube** with sides of length s is given by $S = 6s^2$.

The surface area, S, of a **sphere** with radius r is given by $S = 4\pi r^2$.

The surface area, S, of a **right circular cylinder** is given by $S = 2\pi r^2 + 2\pi rh$, where r is the radius of the base and h is the height.

The surface area, S, of a **right circular cone** is given by $S = \pi r^2 + \pi rl$, where r is the radius of the circular base and l is the slant height.

The surface area, S, of a **regular square pyramid** is given by $S = s^2 + 2sl$, where s is the length of a side of the base and l is the slant height.

QUESTION   *Which of the following units could not be used to measure the surface area of a rectangular solid?*

    a. in^2 *b.* m^3 *c.* cm^2 *d.* ft^3 *e.* yd

Find the surface area of a sphere with a diameter of 18 cm.

First find the radius of the sphere. $r = \dfrac{1}{2}d = \dfrac{1}{2}(18) = 9$

Use the formula for the surface area of a sphere.

$$S = 4\pi r^2$$
$$S = 4\pi(9)^2$$
$$S = 4\pi(81)$$
$$S = 324\pi$$

The exact surface area of the sphere is 324π cm^2.

An approximate measure can be found by using the π key on a calculator. $S \approx 1017.88$

The approximate surface area is 1017.88 cm^2.

ANSWER   *Surface area is measured in square units. Therefore, the surface area of a rectangular solid could be measured in in^2 or cm^2, but not in m^3, ft^3, or yd.*

EXAMPLE 4 ■ **Find the Surface Area of a Geometric Solid**

The diameter of the base of a cone is 5 m and the slant height is 4 m. Find the surface area of the cone. Round to the nearest hundredth of a square meter.

Solution

$$r = \frac{1}{2}d = \frac{1}{2}(5) = 2.5$$ • Find the radius of the cone.

$$S = \pi r^2 + \pi r l$$ • Use the formula for the surface area of a cone.

$$S = \pi(2.5)^2 + \pi(2.5)(4)$$ • Replace *r* with 2.5 and *l* with 4.

$$S = \pi(6.25) + \pi(2.5)(4)$$

$$S = 6.25\pi + 10\pi$$

$$S = 16.25\pi$$

$$S \approx 51.05$$

The surface area of the cone is approximately 51.05 m².

CHECK YOUR PROGRESS 4   The diameter of the base of a cylinder is 6 ft and the height is 8 ft. Find the surface area of the cylinder. Round to the nearest hundredth of a square foot.

Solution   *See page S18.*

EXAMPLE 5 ■ **Find the Surface Area of a Geometric Solid**

Find the area of a label used to cover a soup can that has a radius of 4 cm and a height of 12 cm. Round to the nearest hundredth of a square centimeter.

Solution
The surface area of the side of a cylinder is given by $2\pi rh$.

$$\text{Area of the label} = 2\pi rh$$
$$= 2\pi(4)(12)$$
$$= 96\pi$$
$$\approx 301.59$$

The area of the label is approximately 301.59 cm².

CHECK YOUR PROGRESS 5   Which has a larger surface area, a cube with a side measuring 8 cm or a sphere with a diameter measuring 10 cm?

Solution   *See page S18.*

MathMatters   **Survival of the Fittest**

The ratio of an animal's surface area to the volume of its body is a crucial factor in its survival. The more square units of skin for every cubic unit of volume, the more rapidly the animal loses body heat. Therefore, animals living in a warm climate benefit from a higher ratio of surface area to volume, whereas those living in a cool climate benefit from a lower ratio.

Investigation

Water Displacement

A recipe for peanut butter cookies calls for 1 cup of peanut butter. Peanut butter is difficult to measure. If you have ever used a measuring cup to measure peanut butter, you know that there tend to be pockets of air at the bottom of the cup. And trying to scrape all of the peanut butter out of the cup and into the mixing bowl is a challenge.

A more convenient method of measuring 1 cup of peanut butter is to fill a 2-cup measuring cup with 1 cup of water. Then add peanut butter to the water until the water reaches the 2-cup mark. (Make sure all the peanut butter is below the top of the water.) Drain off the water, and the one cup of peanut butter drops easily into the mixing bowl.

This method of measuring peanut butter works because when an object sinks below the surface of the water, the object displaces an amount of water that is equal to the volume of the object.

(continued)

King Hiero of Syracuse commissioned a new crown from his goldsmith. When the crown was completed, Hiero suspected that the goldsmith stole some of the gold and replaced it with lead. Hiero asked Archimedes to determine, without defacing the surface of the crown, whether the crown was pure gold.

Archimedes knew that the first problem he had to solve was how to determine the volume of the crown. Legend has it that one day he was getting into a bathtub that was completely full of water, and the water splashed over the side. He surmised that the volume of water that poured over the side was equal to the volume of his submerged body. Supposedly, he was so excited that he jumped from the tub and went running through the streets yelling "Eureka! Eureka!", meaning "I found it! I found it!" — referring to the solution of the problem of determining the volume of the crown.

A sphere with a diameter of 4 in. is placed in a rectangular tank of water that is 6 in. long and 5 in. wide. How much does the water level rise? Round to the nearest hundredth of an inch.

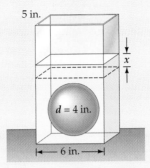

$V = \frac{4}{3}\pi r^3$ • Use the formula for the volume of a sphere.

$V = \frac{4}{3}\pi(2^3) = \frac{32}{3}\pi$ • $r = \frac{1}{2}d = \frac{1}{2}(4) = 2$

Let x represent the amount of the rise in water level. The volume of the sphere will equal the volume of the water displaced. As shown above, this volume is the rectangular solid with width 5 in., length 6 in., and height x in.

$V = LWH$ • Use the formula for the volume of a rectangular solid.

$\frac{32}{3}\pi = (6)(5)x$ • Substitute $\frac{32}{3}\pi$ for V, 6 for L, 5 for W, and x for H.

$\frac{32}{90}\pi = x$ • The exact height that the water will rise is $\frac{32}{90}\pi$.

$1.12 \approx x$ • Use a calculator to find an approximation.

The water will rise approximately 1.12 in.

Investigation Exercises

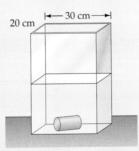

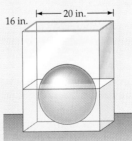

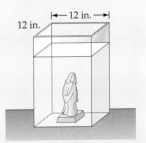

Figure 1   **Figure 2**   **Figure 3**

1. A cylinder with a 2-centimeter radius and a height of 10 cm is submerged in a tank of water that is 20 cm wide and 30 cm long (see Figure 1). How much does the water level rise? Round to the nearest hundredth of a centimeter.

(continued)

2. A sphere with a radius of 6 in. is placed in a rectangular tank of water that is 16 in. wide and 20 in. long (see Figure 2). The sphere displaces water until two-thirds of the sphere, with respect to its volume, is submerged. How much does the water level rise? Round to the nearest hundredth of an inch.

3. A chemist wants to know the density of a statue that weighs 15 lb. The statue is placed in a rectangular tank of water that is 12 in. long and 12 in. wide (see Figure 3). The water level rises 0.42 in. Find the density of the statue. Round to the nearest hundredth of a pound per cubic inch. *Hint:* Density = weight ÷ volume.

Exercise Set 4.6

In Exercises 1–6, find the volume of the figure. For calculations involving π, give both the exact value and an approximation to the nearest hundredth of a cubic unit.

1.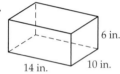
6 in.
14 in.   10 in.

2.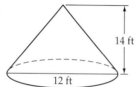
14 ft
12 ft

3.
5 ft
3 ft
3 ft

4.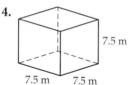
7.5 m
7.5 m   7.5 m

5.
3 cm

6.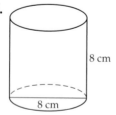
8 cm
8 cm

9.
5 m
4 m
4 m

10.
2 cm

11.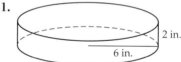
2 in.
6 in.

12.
9 ft
3 ft

13. Which of the following units could not be used to measure the volume of a cylinder: ft³, m³, yd², cm³, mi?

14. State whether feet, square feet, or cubic feet would be used to measure each of the following.
   a. The area of the surface of a lake
   b. The distance across a lake
   c. The volume of water in a lake

15. State whether feet, square feet, or cubic feet would be used to measure each of the following.
   a. The length of a driveway
   b. The volume of asphalt used to pave a driveway
   c. The area of a driveway that needs to be resealed

In Exercises 7–12, find the surface area of the figure. For calculations involving π, give both the exact value and an approximation to the nearest hundredth of a square unit.

7.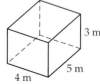
3 m
4 m   5 m

8.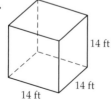
14 ft
14 ft
14 ft

In Exercises 16–49, solve.

16. **Volume** A rectangular solid has a length of 6.8 m, a width of 2.5 m, and a height of 2 m. Find the volume of the solid.

17. **Volume** Find the volume of a rectangular solid that has a length of 4.5 ft, a width of 3 ft, and a height of 1.5 ft.

18. **Volume** Find the volume of a cube whose side measures 2.5 in.

19. **Volume** The length of a side of a cube is 7 cm. Find the volume of the cube.

20. **Volume** The diameter of a sphere is 6 ft. Find the exact volume of the sphere.

21. **Volume** Find the volume of a sphere that has a radius of 1.2 m. Round to the nearest hundredth of a cubic meter.

22. **Volume** The diameter of the base of a cylinder is 24 cm. The height of the cylinder is 18 cm. Find the volume of the cylinder. Round to the nearest hundredth of a cubic centimeter.

23. **Volume** The height of a cylinder is 7.2 m. The radius of the base is 4 m. Find the exact volume of the cylinder.

24. **Volume** The radius of the base of a cone is 5 in. The height of the cone is 9 in. Find the exact volume of the cone.

25. **Volume** The height of a cone is 15 cm. The diameter of the cone is 10 cm. Find the volume of the cone. Round to the nearest hundredth of a cubic centimeter.

26. **Volume** The length of a side of the base of a regular square pyramid is 6 in. and the height of the pyramid is 10 in. Find the volume of the pyramid.

27. **Volume** The height of a regular square pyramid is 8 m and the length of a side of the base is 9 m. What is the volume of the pyramid?

28. **The Statue of Liberty** The index finger of the Statue of Liberty is 8 ft long. The circumference at the second joint is 3.5 ft. Use the formula for the volume of a cylinder to approximate the volume of the index finger on the Statue of Liberty. Round to the nearest hundredth of a cubic foot.

29. **Surface Area** The height of a rectangular solid is 5 ft, the length is 8 ft, and the width is 4 ft. Find the surface area of the solid.

30. **Surface Area** The width of a rectangular solid is 32 cm, the length is 60 cm, and the height is 14 cm. What is the surface area of the solid?

31. **Surface Area** The side of a cube measures 3.4 m. Find the surface area of the cube.

32. **Surface Area** Find the surface area of a cube with a side measuring 1.5 in.

33. **Surface Area** Find the exact surface area of a sphere with a diameter of 15 cm.

34. **Surface Area** The radius of a sphere is 2 in. Find the surface area of the sphere. Round to the nearest hundredth of a square inch.

35. **Surface Area** The radius of the base of a cylinder is 4 in. The height of the cylinder is 12 in. Find the surface area of the cylinder. Round to the nearest hundredth of a square inch.

36. **Surface Area** The diameter of the base of a cylinder is 1.8 m. The height of the cylinder is 0.7 m. Find the exact surface area of the cylinder.

37. **Surface Area** The slant height of a cone is 2.5 ft. The radius of the base is 1.5 ft. Find the exact surface area of the cone. The formula for the surface area of a cone is given on page 284.

38. **Surface Area** The diameter of the base of a cone is 21 in. The slant height is 16 in. What is the surface area of the cone? The formula for the surface area of a cone is given on page 284. Round to the nearest hundredth of a square inch.

39. **Surface Area** The length of a side of the base of a regular square pyramid is 9 in., and the pyramid's slant height is 12 in. Find the surface area of the pyramid.

40. **Surface Area** The slant height of a regular square pyramid is 18 m, and the length of a side of the base is 16 m. What is the surface area of the pyramid?

41. **Appliances** The volume of a freezer that is a rectangular solid with a length of 7 ft and a height of 3 ft is 52.5 ft³. Find the width of the freezer.

42. **Aquariums** The length of an aquarium is 18 in. and the width is 12 in. If the volume of the aquarium is 1836 in³, what is the height of the aquarium?

43. **Surface Area** The surface area of a rectangular solid is 108 cm². The height of the solid is 4 cm, and the length is 6 cm. Find the width of the rectangular solid.

44. **Surface Area** The length of a rectangular solid is 12 ft and the width is 3 ft. If the surface area is 162 ft², find the height of the rectangular solid.

45. Paint A can of paint will cover 300 ft² of surface. How many cans of paint should be purchased to paint a cylinder that has a height of 30 ft and a radius of 12 ft?

46. Ballooning A hot air balloon is in the shape of a sphere. Approximately how much fabric was used to construct the balloon if its diameter is 32 ft? Round to the nearest square foot.

47. Aquariums How much glass is needed to make a fish tank that is 12 in. long, 8 in. wide, and 9 in. high? The fish tank is open at the top.

48. Food Labels Find the area of a label used to completely cover the side of a cylindrical can of juice that has a diameter of 16.5 cm and a height of 17 cm. Round to the nearest hundredth of a square centimeter.

49. Surface Area The length of a side of the base of a regular square pyramid is 5 cm and the slant height of the pyramid is 8 cm. How much larger is the surface area of this pyramid than the surface area of a cone with a diameter of 5 cm and a slant height of 8 cm? Round to the nearest hundredth of a square centimeter.

In Exercises 50–55, find the volume of the figure. Round to the nearest hundredth of a cubic unit.

50.

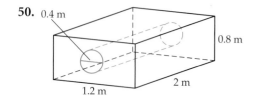

51.

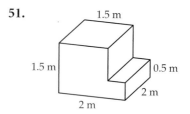

52.

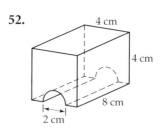

53.

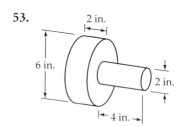

54.

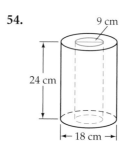

55.

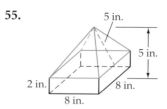

In Exercises 56–61, find the surface area of the figure. Round to the nearest hundredth of a square unit.

56.

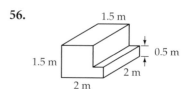

57.

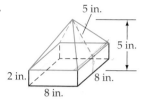

58.

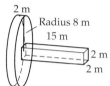

59.

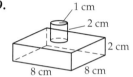

60.

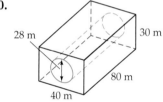

61.

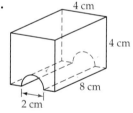

62. Oil Tanks A truck is carrying an oil tank, as shown in the figure below. If the tank is half full, how many cubic feet of oil is the truck carrying? Round to the nearest hundredth of a cubic foot.

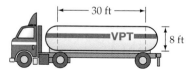

63. Concrete The concrete floor of a building is shown in the following figure. At a cost of $3.15 per cubic foot,

find the cost of having the floor poured. Round to the nearest cent.

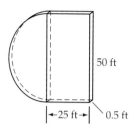

64. Swimming Pools How many liters of water are needed to fill the swimming pool shown below? (1 m^3 contains 1000 L.)

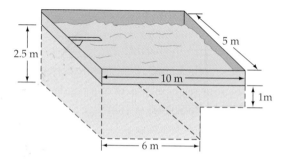

65. Paint A can of paint will cover 250 ft^2 of surface. Find the number of cans of paint that should be purchased to paint the exterior of the auditorium shown in the figure below.

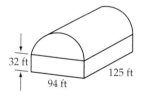

66. Metallurgy A piece of sheet metal is cut and formed into the shape shown below. Given that there is 0.24 g in 1 cm^2 of the metal, find the total number of grams of metal used. Round to the nearest hundredth of a gram.

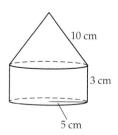

67. Plastering The walls of a room that is 25.5 ft long, 22 ft wide, and 8 ft high are being plastered. There are two doors in the room, each measuring 2.5 ft by 7 ft. Each of the six windows in the room measures 2.5 ft by 4 ft. At a cost of $.75 per square foot, find the cost of plastering the walls of the room.

68. Gold A solid sphere of gold with a radius of 0.5 cm has a value of $180. Find the value of a sphere of the same type of gold with a radius of 1.5 cm.

69. Swimming Pools A swimming pool is built in the shape of a rectangular solid. It holds 32,000 gal of water. If the length, width, and height of the pool are each doubled, how many gallons of water will be needed to fill the pool?

70. Determine whether the statement is always true, sometimes true, or never true.

   a. The slant height of a regular square pyramid is longer than the height.

   b. The slant height of a cone is shorter than the height.

   c. The four triangular faces of a regular square pyramid are equilateral triangles.

Extensions

CRITICAL THINKING

71. Half of a sphere is called a **hemisphere**. Derive formulas for the volume and surface area of a hemisphere.

72. a. Draw a two-dimensional figure that can be cut out and made into a right circular cone.

   b. Draw a two-dimensional figure that can be cut out and made into a regular square pyramid.

73. A sphere fits inside a cylinder as shown in the following figure. The height of the cylinder equals the diameter of the sphere. Show that the surface area of the sphere equals the surface area of the side of the cylinder.

74. a. What is the effect on the surface area of a rectangular solid of doubling the width and height?

   b. What is the effect on the volume of a rectangular solid of doubling the length and width?

   c. What is the effect on the volume of a cube of doubling the length of each side of the cube?

   d. What is the effect on the surface area of a cylinder of doubling the radius and height?

EXPLORATIONS

75. Explain how you could cut through a cube so that the face of the resulting solid is

   a. a square.   **b.** an equilateral triangle.

   c. a trapezoid.   **d.** a hexagon.

SECTION 4.7 | **Introduction to Trigonometry**

Trigonometric Functions of an Acute Angle

Given the lengths of two sides of a right triangle, it is possible to determine the length of the third side by using the Pythagorean Theorem. In some situations, however, it may not be practical or possible to know the lengths of two of the sides of a right triangle.

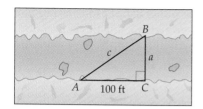

Consider, for example, the problem of engineers trying to determine the distance across a ravine so that they can design a bridge that can be built connecting the two sides. Look at the triangle to the left.

It is fairly easy to measure the length of the side of the triangle that is on the land (100 feet), but the lengths of sides a and c cannot be measured easily because of the ravine.

The study of *trigonometry*, a term that comes from two Greek words meaning "triangle measurement," began about 2000 years ago, partially as a means of solving surveying problems such as the one above. In this section, we will examine *right triangle* trigonometry—that is, trigonometry that applies only to right triangles.

When working with right triangles, it is convenient to refer to the side **opposite** an angle and to the side **adjacent** to (next to) an angle. The hypotenuse of a right triangle is not adjacent to or opposite either of the acute angles in the right triangle.

TAKE NOTE

In trigonometry, it is common practice to use Greek letters for the angles of a triangle. Here are some frequently used letters: α (alpha), β (beta), and θ (theta). The word *alphabet* is derived from the first two letters of the Greek alphabet, α and β.

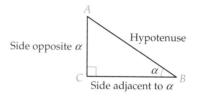

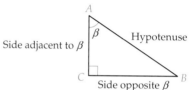

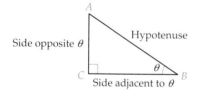

Consider the right triangle shown at the left. Six possible ratios can be formed using the lengths of the sides of the triangle.

$$\frac{\text{length of opposite side}}{\text{length of hypotenuse}} \qquad \frac{\text{length of hypotenuse}}{\text{length of opposite side}}$$

$$\frac{\text{length of adjacent side}}{\text{length of hypotenuse}} \qquad \frac{\text{length of hypotenuse}}{\text{length of adjacent side}}$$

$$\frac{\text{length of opposite side}}{\text{length of adjacent side}} \qquad \frac{\text{length of adjacent side}}{\text{length of opposite side}}$$

Each of these ratios defines a value of a trigonometric function of the acute angle θ. The functions are **sine** (sin), **cosine** (cos), **tangent** (tan), **cosecant** (csc), **secant** (sec), and **cotangent** (cot).

The Trigonometric Functions of an Acute Angle of a Right Triangle

If θ is an acute angle of a right triangle ABC, then

$$\sin \theta = \frac{\text{length of opposite side}}{\text{length of hypotenuse}} \qquad \csc \theta = \frac{\text{length of hypotenuse}}{\text{length of opposite side}}$$

$$\cos \theta = \frac{\text{length of adjacent side}}{\text{length of hypotenuse}} \qquad \sec \theta = \frac{\text{length of hypotenuse}}{\text{length of adjacent side}}$$

$$\tan \theta = \frac{\text{length of opposite side}}{\text{length of adjacent side}} \qquad \cot \theta = \frac{\text{length of adjacent side}}{\text{length of opposite side}}$$

As a convenience, we will write opp, adj, and hyp as abbreviations for *the length of the* opposite side, adjacent side, and hypotenuse, respectively. Using this convention, the definitions of the trigonometric functions are written

$$\sin \theta = \frac{\text{opp}}{\text{hyp}} \qquad \csc \theta = \frac{\text{hyp}}{\text{opp}}$$

$$\cos \theta = \frac{\text{adj}}{\text{hyp}} \qquad \sec \theta = \frac{\text{hyp}}{\text{adj}}$$

$$\tan \theta = \frac{\text{opp}}{\text{adj}} \qquad \cot \theta = \frac{\text{adj}}{\text{opp}}$$

All of the trigonometric functions have applications, but the sine, cosine, and tangent functions are used most frequently. For the remainder of this section, we will focus on those functions.

When working with trigonometric functions, be sure to draw a diagram and label the adjacent and opposite sides of an angle. For instance, in the definition above, if we had placed θ at angle *A*, then the triangle would have been labeled as shown at the left. The definitions of the functions remain the same.

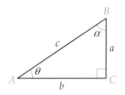

$$\sin \theta = \frac{\text{opp}}{\text{hyp}} \qquad \cos \theta = \frac{\text{adj}}{\text{hyp}} \qquad \tan \theta = \frac{\text{opp}}{\text{adj}}$$

QUESTION *For the right triangle shown at the left, indicate which side is*

    a. *adjacent to* ∠*A*   **b.** *opposite* θ
    c. *adjacent to* α   **d.** *opposite* ∠*B*

EXAMPLE 1 ■ Find the Values of Trigonometric Functions

For the right triangle at the right, find the values of sin θ, cos θ, and tan θ.

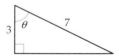

Solution
Use the Pythagorean Theorem to find the length of the side opposite θ.

$$a^2 + b^2 = c^2$$
$$3^2 + b^2 = 7^2 \qquad \bullet \; a = 3, c = 7$$
$$9 + b^2 = 49$$
$$b^2 = 40$$
$$b = \sqrt{40} = 2\sqrt{10}$$

Using the definitions of the trigonometric functions, we have

$$\sin \theta = \frac{\text{opp}}{\text{hyp}} = \frac{2\sqrt{10}}{7} \qquad \cos \theta = \frac{\text{adj}}{\text{hyp}} = \frac{3}{7} \qquad \tan \theta = \frac{\text{opp}}{\text{adj}} = \frac{2\sqrt{10}}{3}$$

CHECK YOUR PROGRESS 1 For the right triangle at the right, find the values of sin θ, cos θ, and tan θ.

Solution *See page S18.*

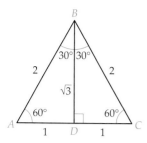

In Example 1, we gave the exact answers. In many cases, approximate values of trigonometric functions are given. The answers to Example 1, rounded to the nearest ten-thousandth, are

$$\sin \theta = \frac{2\sqrt{10}}{7} \approx 0.9035 \qquad \cos \theta = \frac{3}{7} \approx 0.4286 \qquad \tan \theta = \frac{2\sqrt{10}}{3} \approx 2.1082$$

There are many occasions when we will want to know the value of a trigonometric function for a given angle. Triangle ABC at the left is an equilateral triangle with sides of length 2 units and angle bisector $\overline{BD}$. Because $\overline{BD}$ bisects $\angle ABC$, the measures of $\angle ABD$ and $\angle CBD$ are both 30°. The angle bisector $\overline{BD}$ also bisects $\overline{AC}$. Therefore, $AD = 1$ and $DC = 1$. Using the Pythagorean Theorem, we can find the measure of BD.

$$(DC)^2 + (BD)^2 = (BC)^2$$
$$1^2 + (BD)^2 = 2^2$$
$$1 + (BD)^2 = 4$$
$$(BD)^2 = 3$$
$$BD = \sqrt{3}$$

Using the definitions of the trigonometric functions and triangle BCD, we can find the values of the sine, cosine, and tangent of 30° and 60°.

$$\sin 30° = \frac{\text{opp}}{\text{hyp}} = \frac{1}{2} = 0.5 \qquad\qquad \sin 60° = \frac{\text{opp}}{\text{hyp}} = \frac{\sqrt{3}}{2} \approx 0.8660$$

$$\cos 30° = \frac{\text{adj}}{\text{hyp}} = \frac{\sqrt{3}}{2} \approx 0.8660 \qquad \cos 60° = \frac{\text{adj}}{\text{hyp}} = \frac{1}{2} = 0.5$$

$$\tan 30° = \frac{\text{opp}}{\text{adj}} = \frac{1}{\sqrt{3}} \approx 0.5774 \qquad \tan 60° = \frac{\text{opp}}{\text{adj}} = \sqrt{3} \approx 1.7321$$

The properties of an equilateral triangle enabled us to calculate the values of the trigonometric functions for 30° and 60°. Calculating values of the trigonometric functions for most other angles, however, would be quite difficult. Fortunately, many calculators have been programmed to allow us to estimate these values.

To use a TI-83 or TI-84 Plus calculator to find tan 30°, confirm that your calculator is in "degree mode." Press the tan button and key in 30. Then press $\boxed{\text{ENTER}}$.

$$\tan 30° \approx 0.5774$$

Despite the fact that the values of many trigonometric functions are approximate, it is customary to use the equals sign rather than the approximately equals sign when writing these function values. Thus we write $\tan 30° = 0.5774$.

CALCULATOR NOTE

Just as distances can be measured in feet, miles, meters, and other units, angles can be measured in various units: degrees, radians, and grads. In this section, we use only degree measurements for angles, so be sure your calculator is in degree mode.

On a TI-83 or TI-84 Plus, press the $\boxed{\text{MODE}}$ key to determine whether the calculator is in degree mode.

```
Normal  Sci  Eng
Float 0123456789
Radian  Degree
Func  Par  Pol  Seq
Connected  Dot
Sequential  Simul
Real  a+bi  re^θi
Full  Horiz  G-T
```

EXAMPLE 2 ■ **Use a Calculator to Find the Value of a Trigonometric Function**

Use a calculator to find sin 43.8° to the nearest ten-thousandth.

Solution sin 43.8° = 0.6921

CHECK YOUR PROGRESS 2   Use a calculator to find tan 37.1° to the nearest ten-thousandth.

Solution   *See page S19.*

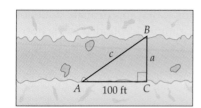

Using trigonometry, the engineers mentioned at the beginning of this section could determine the distance across the ravine after determining the measure of $\angle A$. Suppose the engineers measure the angle as $33.8°$. Now the engineers would ask, "Which trigonometric function, sine, cosine, or tangent, involves the side opposite an angle and the side adjacent to that angle?" Knowing that the tangent function is the required function, the engineers could write and solve the equation $\tan 33.8° = \dfrac{a}{100}$.

$$\tan 33.8° = \frac{a}{100}$$

$$100(\tan 33.8°) = a \qquad \bullet \text{ Multiply each side of the equation by 100.}$$

$$66.9 \approx a \qquad \bullet \text{ Use a calculator to find } \tan 33.8°. \text{ Multiply the result in the display by 100.}$$

The distance across the ravine is approximately 66.9 ft.

EXAMPLE 3 ■ **Find the Length of a Side of a Triangle**

For the right triangle shown at the left, find the length of side a. Round to the nearest hundredth of a meter.

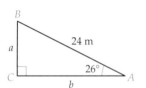

Solution

We are given the measure of $\angle A$ and the hypotenuse. We want to find the length of side a. Side a is opposite $\angle A$. The sine function involves the side opposite an angle and the hypotenuse.

$$\sin A = \frac{\text{opp}}{\text{hyp}}$$

$$\sin 26° = \frac{a}{24} \qquad \bullet \; A = 26°, \text{ hypotenuse} = 24 \text{ m}$$

$$24(\sin 26°) = a \qquad \bullet \text{ Multiply each side by 24.}$$

$$10.52 \approx a \qquad \bullet \text{ Use a calculator to find } \sin 26°. \text{ Multiply the result in the display by 24.}$$

The length of side a is approximately 10.52 m.

CHECK YOUR PROGRESS 3 For the right triangle shown at the right, find the length of side a. Round to the nearest hundredth of a foot.

Solution See page S19.

Inverse Trigonometric Functions

Sometimes it is necessary to find one of the acute angles in a right triangle. For instance, suppose it is necessary to find the measure of $\angle A$ in the figure at the left. Because the side adjacent to $\angle A$ is known and the hypotenuse is known, we can write

$$\cos A = \frac{\text{adj}}{\text{hyp}}$$

$$\cos A = \frac{25}{27}$$

The solution of this equation is the angle whose cosine is $\frac{25}{27}$. This angle can be found by using the $\cos^{-1}$ key on a calculator.

$$\cos^{-1}\left(\frac{25}{27}\right) \approx 22.19160657$$

To the nearest tenth, the measure of $\angle A$ is 22.2°.

The function $\cos^{-1}$ is called the *inverse cosine function*.

> **Definitions of the Inverse Sine, Inverse Cosine, and Inverse Tangent Functions**
>
> For $0° < x < 90°$:
>
> $y = \sin^{-1}(x)$ is notation for "y is the angle whose sine is x."
> $y = \cos^{-1}(x)$ is notation for "y is the angle whose cosine is x."
> $y = \tan^{-1}(x)$ is notation for "y is the angle whose tangent is x."

Note that $\sin^{-1}(x)$ is used to denote the inverse of the sine function. It is not the reciprocal of $\sin x$ but the notation used for its inverse. The same is true for $\cos^{-1}$ and $\tan^{-1}$.

EXAMPLE 4 ▪ **Evaluate an Inverse Trigonometric Function**

Use a calculator to find $\sin^{-1}(0.9171)$. Round to the nearest tenth of a degree.

Solution

$\sin^{-1}(0.9171) \approx 66.5°$ • The calculator must be in degree mode. Press the keys for the inverse sine function followed by .9171. Press $\boxed{\text{ENTER}}$.

CHECK YOUR PROGRESS 4   Use a calculator to find $\tan^{-1}(0.3165)$. Round to the nearest tenth of a degree.

Solution   *See page S19.*

EXAMPLE 5 ▪ **Find the Measure of an Angle Using the Inverse of a Trigonometric Function**

Given $\sin \theta = 0.7239$, find θ. Use a calculator. Round to the nearest tenth of a degree.

Solution
This is equivalent to finding $\sin^{-1}(0.7239)$. The calculator must be in degree mode.

$$\sin^{-1}(0.7239) \approx 46.4°$$
$$\theta \approx 46.4°$$

CHECK YOUR PROGRESS 5   Given $\tan \theta = 0.5681$, find θ. Use a calculator. Round to the nearest tenth of a degree.

Solution   *See page S19.*

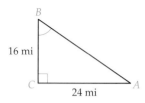

EXAMPLE 6 ■ Find the Measure of an Angle in a Right Triangle

For the right triangle shown at the left, find the measure of $\angle B$. Round to the nearest tenth of a degree.

Solution

We want to find the measure of $\angle B$, and we are given the lengths of the sides opposite $\angle B$ and adjacent to $\angle B$. The tangent function involves the side opposite an angle and the side adjacent to that angle.

$$\tan B = \frac{\text{opposite}}{\text{adjacent}}$$

$$\tan B = \frac{24}{16}$$

$$B = \tan^{-1}\left(\frac{24}{16}\right)$$

$$B \approx 56.3° \qquad \text{• Use the } \tan^{-1} \text{ key on a calculator.}$$

The measure of $\angle B$ is approximately 56.3°.

CHECK YOUR PROGRESS 6

For the right triangle shown at the left, find the measure of $\angle A$. Round to the nearest tenth of a degree.

Solution See page S19.

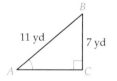

Angles of Elevation and Depression

The use of trigonometry is necessary in a variety of situations. One application, called **line-of-sight problems,** concerns an observer looking at an object.

Angles of elevation and depression are measured with respect to a horizontal line. If the object being sighted is above the observer, the acute angle formed by the line of sight and the horizontal line is an **angle of elevation.** If the object being sighted is below the observer, the acute angle formed by the line of sight and the horizontal line is an **angle of depression.**

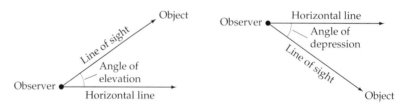

EXAMPLE 7 ■ Solve an Angle of Elevation Problem

The angle of elevation to the top of a flagpole 62 ft away is 34°. Find the height of the flagpole. Round to the nearest tenth of a foot.

Solution

Draw a diagram. To find the height h, write a trigonometric function that relates the given information and the unknown side of the triangle.

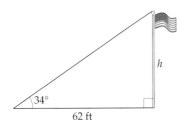

$$\tan 34° = \frac{h}{62}$$
$$62(\tan 34°) = h$$ • Multiply each side by 62.
$$41.8 \approx h$$ • Use a calculator to find $\tan 34°$. Multiply the result in the display by 62.

The height of the flagpole is approximately 41.8 ft.

CHECK YOUR PROGRESS 7   The angle of depression from the top of a lighthouse that is 20 m high to a boat on the water is 25°. How far is the boat from the center of the base of the lighthouse? Round to the nearest tenth of a meter.

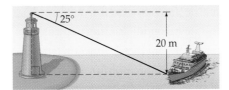

Solution *See page S19.*

Investigation

Approximating the Values of Trigonometric Functions

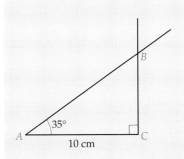

The value of a trigonometric function can be approximated by drawing a triangle with a given angle. To illustrate, we will choose an angle of 35°.

To find the tangent of 35° using the definitions given in this section, we can carefully construct a right triangle containing an angle of 35°. Measure the lengths of the sides opposite and adjacent to the 35° angle and compute the quotient of these values. Because any two right triangles containing an angle of 35° are similar, *the value of* tan 35° *is the same no matter what triangle we draw*.

Investigation Exercises

1. Draw a horizontal line segment 10 cm long with left endpoint *A* and right endpoint *C*. See the diagram at the left.

2. Using a protractor, draw a 35° angle at *A*.

3. Draw at *C* a vertical line that intersects the terminal side of angle *A* at *B*. Your drawing should be similar to the one at the left.

4. Measure line segment *BC*.

5. What is the approximate value of tan 35°?

6. Using your value for *BC* and the Pythagorean Theorem, estimate *AB*.

7. Estimate sin 35° and cos 35°.

8. What are the values of sin 35°, cos 35°, and tan 35° as produced by a calculator? Round to the nearest ten-thousandth.

Exercise Set 4.7

1. Use the right triangle at the right and sides *a*, *b*, and *c* to do the following:

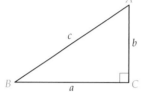

   a. Name the ratio for the trigonometric function sin *A*.

   b. Name the ratio for the trigonometric function sin *B*.

   c. Name the ratio for the trigonometric function cos *A*.

   d. Name the ratio for the trigonometric function cos *B*.

   e. Name the ratio for the trigonometric function tan *A*.

   f. Name the ratio for the trigonometric function tan *B*.

2. Explain the meaning of the notation $y = \sin^{-1}(x)$, $y = \cos^{-1}(x)$, and $y = \tan^{-1}(x)$.

In Exercises 3–10, find the values of sin θ, cos θ, and tan θ for the given right triangle. Give the exact values.

3.

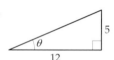

4.

5.

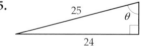

6.

7.

8.

9.

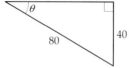

10.

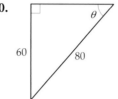

In Exercises 11–26, use a calculator to estimate the value of the trigonometric function. Round to the nearest tenthousandth.

11. cos 47°	**12.** sin 62°	**13.** tan 55°
14. cos 11°	**15.** sin 85.6°	**16.** cos 21.9°
17. tan 63.4°	**18.** sin 7.8°	**19.** tan 41.6°
20. cos 73°	**21.** sin 57.7°	**22.** tan 39.2°
23. sin 58.3°	**24.** tan 35.1°	**25.** cos 46.9°
26. sin 50°		

In Exercises 27–42, use a calculator. Round to the nearest tenth.

27. Given sin θ = 0.6239, find θ.

28. Given cos β = 0.9516, find β.

29. Find $\cos^{-1}(0.7536)$.

30. Find $\sin^{-1}(0.4478)$.

31. Given tan α = 0.3899, find α.

32. Given sin β = 0.7349, find β.

33. Find $\tan^{-1}(0.7815)$.

34. Find $\cos^{-1}(0.6032)$.

35. Given cos θ = 0.3007, find θ.

36. Given tan α = 1.588, find α.

37. Find $\sin^{-1}(0.0105)$.

38. Find $\tan^{-1}(0.2438)$.

39. Given sin β = 0.9143, find β.

40. Given cos θ = 0.4756, find θ.

41. Find $\cos^{-1}(0.8704)$.

42. Find $\sin^{-1}(0.2198)$.

43. You know the measure of $\angle A$ in a right triangle and the length of side a. Which trigonometric function, sine, cosine, or tangent, would you use to find the length of side c?

44. You know the measure of $\angle B$ in a right triangle and the length of side a. Which trigonometric function, sine, cosine, or tangent, would you use to find the length of side c?

45. You know the measure of $\angle A$ in a right triangle and the length of side b. Which trigonometric function, sine, cosine, or tangent, would you use to find the length of side a?

46. Which equation cannot be used to find x?

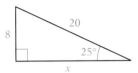

a. $\tan 25° = \dfrac{8}{x}$ **b.** $\cos 25° = \dfrac{x}{20}$ **c.** $\tan 65° = \dfrac{x}{8}$ **d.** $\tan 25° = \dfrac{x}{8}$

47. As an acute angle of a right triangle gets larger, the sine of the angle gets _____ and the cosine of the angle gets _____.

48. If $\sin \Phi = x$, then $\sin^{-1} x =$ _____.

In Exercises 49–62, draw a picture and label it. Set up an equation and solve it. Show all your work. Round an angle to the nearest tenth of a degree. Round the length of a side to the nearest hundredth of a unit.

49. Ballooning A balloon, tethered by a cable 997 feet long, was blown by a wind so that the cable made an angle of $57.6°$ with the ground. Find the height of the balloon.

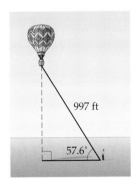

997 ft

57.6°

50. Roadways A road is inclined at an angle of $9.8°$ with the horizontal. Find the distance that one must drive on this road in order to be elevated 14.8 ft above the horizontal.

51. Home Maintenance A ladder 30.8 ft long leans against a building. If the foot of the ladder is 7.25 ft from the base of the building, find the angle the top of the ladder makes with the building.

52. **Aviation** A plane takes off from a field and climbs at an angle of 11.4° with the horizontal. Find the height of the plane after it has traveled a distance of 1250 ft.

53. **Guy Wires** A guy wire whose grounded end is 16 ft from the telephone pole it supports makes an angle of 56.7° with the ground. How long is the wire?

16 ft

54. **Angle of Depression** A lighthouse built at sea level is 169 ft tall. From its top, the angle of depression to a boat below measures 25.1°. Find the distance from the boat to the foot of the lighthouse.

55. **Angle of Elevation** At a point 39.3 ft from the base of a tree, the angle of elevation to its top measures 53.4°. Find the height of the tree.

53.4°

39.3 ft

56. **Angle of Depression** An artillery spotter in a plane that is at an altitude of 978 ft measures the angle of depression to an enemy tank as 28.5°. How far is the enemy tank from the point on the ground directly below the spotter?

57. **Home Maintenance** A 15-foot ladder leans against a house. The ladder makes an angle of 65° with the ground. How far up the side of the house does the ladder reach?

65°

58. **Angle of Elevation** Find the angle of elevation of the sun when a tree 40.5 ft high casts a shadow 28.3 ft long.

59. **Guy Wires** A television transmitter tower is 600 ft high. If the angle between the guy wire (attached at the top) and the tower is 55.4°, how long is the guy wire?

60. **Ramps** A ramp used to load a racing car onto a flatbed carrier is 5.25 m long, and its upper end is 1.74 m above the lower end. Find the angle between the ramp and the road.

61. **Angle of Elevation** The angle of elevation of the sun is 51.3° at a time when a tree casts a shadow 23.7 yd long. Find the height of the tree.

62. **Angle of Depression** From the top of a building 312 ft tall, the angle of depression to a flower bed on the ground below is 12.0°. What is the distance between the base of the building and the flower bed?

Extensions
CRITICAL THINKING

63. Can the value of sin θ or cos θ ever be greater than 1? Explain your answer.

64. Can the value of tan θ ever be greater than 1? Explain your answer.

65. Let sin $\theta = \dfrac{2}{3}$. Find the exact value of cos θ.

66. Let tan $\theta = \dfrac{5}{4}$. Find the exact value of sin θ.

67. Let cos $\theta = \dfrac{3}{4}$. Find the exact value of tan θ.

68. Let sin $\theta = \dfrac{\sqrt{5}}{4}$. Find the exact value of tan θ.

69. Let sin $\theta = a, a > 0$. Find cos θ.

70. Let tan $\theta = a, a > 0$. Find sin θ.

EXPLORATIONS

As we noted in this section, angles can also be measured in *radians*. To define a radian, first consider a circle of radius *r* and two radii $\overline{OA}$ and $\overline{OB}$. The angle θ formed by the two radii is a **central angle**. The portion of the circle between *A* and *B* is an **arc** of the circle and is written $\overset{\frown}{AB}$. We say that $\overset{\frown}{AB}$ subtends the angle θ. The length of the arc is *s*. (See Figure 1 below.)

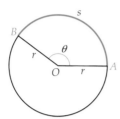

Figure 1

Definition of Radian

One **radian** is the measure of the central angle subtended by an arc of length *r*. The measure of θ in Figure 2 is 1 radian.

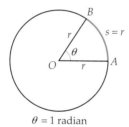

$\theta = 1$ radian

Figure 2

To find the radian measure of an angle subtended by an arc of length *s*, use the following formula.

Radian Measure

Given an arc of length *s* on a circle of radius *r*, the measure of the central angle subtended by the arc is $\theta = \dfrac{s}{r}$ radians.

For example, to find the measure in radians of the central angle subtended by an arc of 9 in. in a circle of radius 12 in., divide the length of the arc ($s = 9$ in.) by the length of the radius ($r = 12$ in.). See Figure 3.

$$\theta = \frac{9 \text{ in.}}{12 \text{ in.}} \text{ radian}$$
$$= \frac{3}{4} \text{ radian}$$

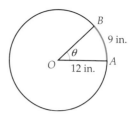

Figure 3

71. Find the measure in radians of the central angle subtended by an arc of 12 cm in a circle of radius 3 cm.

72. Find the measure in radians of the central angle subtended by an arc of 4 cm in a circle of radius 8 cm.

73. Find the measure in radians of the central angle subtended by an arc of 6 in. in a circle of radius 9 in.

74. Find the measure in radians of the central angle subtended by an arc of 12 ft in a circle of radius 10 ft.

Recall that the circumference of a circle is given by $C = 2\pi r$. Therefore, the radian measure of the central angle subtended by the circumference is $\theta = \dfrac{2\pi r}{r} = 2\pi$.

In degree measure, the central angle has a measure of 360°. Thus we have 2π radians = 360°. Dividing each side of the equation by 2 gives π radians = 180°. From the last equation, we can establish the conversion factors $\dfrac{\pi \text{ radians}}{180°}$ and $\dfrac{180°}{\pi \text{ radians}}$. These conversion factors are used to convert between radians and degrees.

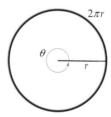

Conversion Between Radians and Degrees

• To convert from degrees to radians, multiply by $\dfrac{\pi \text{ radians}}{180°}$.

• To convert from radians to degrees, multiply by $\dfrac{180°}{\pi \text{ radians}}$.

For instance, to convert 30° to radians, multiply 30° by $\dfrac{\pi \text{ radians}}{180°}$.

$$30° = 30°\left(\frac{\pi \text{ radians}}{180°}\right)$$

$$= \frac{\pi}{6} \text{ radian} \qquad \bullet \text{ Exact answer}$$

$$\approx 0.5236 \text{ radian} \qquad \bullet \text{ Approximate answer}$$

To convert 2 radians to degrees, multiply 2 by $\dfrac{180°}{\pi \text{ radians}}$.

$$2 \text{ radians} = 2 \text{ radians}\left(\frac{180°}{\pi \text{ radians}}\right)$$

$$= \left(\frac{360}{\pi}\right)° \qquad \bullet \text{ Exact answer}$$

$$\approx 114.5916° \qquad \bullet \text{ Approximate answer}$$

75. What is the measure in degrees of 1 radian?

76. Is the measure of 1 radian larger or smaller than the measure of 1°?

In Exercises 77–82, convert degree measure to radian measure. Find an exact answer and an answer rounded to the nearest ten-thousandth.

77. 45°   **78.** 180°   **79.** 315°

80. 90°   **81.** 210°   **82.** 18°

In Exercises 83–88, convert radian measure to degree measure. For Exercises 86–88, find an exact answer and an answer rounded to the nearest ten-thousandth.

83. $\dfrac{\pi}{3}$ radians   **84.** $\dfrac{11\pi}{6}$ radians

85. $\dfrac{4\pi}{3}$ radians   **86.** 1.2 radians

87. 3 radians   **88.** 2.4 radians

CHAPTER 4   Summary

Key Terms

acute angle [p. 226]
acute triangle [p. 242]
adjacent angles [p. 227]
adjacent side [p. 293]
alternate exterior angles [p. 229]
alternate interior angles [p. 229]
angle [p. 224]
angle of depression [p. 298]
angle of elevation [p. 298]
area [p. 212]
base of a parallelogram [p. 246]
base of a triangle [p. 252]
base of a trapezoid [p. 253]
capacity [p. 204]
center of a circle [p. 247]
center of a sphere [p. 279]
circle [p. 247]
circumference [p. 247]
complementary angles [p. 226]

corresponding angles [p. 230]
cosecant [p. 293]
cosine [p. 293]
cotangent [p. 293]
cube [p. 278]
degree [p. 225]
diameter of a circle [p. 247]
diameter of a sphere [p. 279]
dimensional analysis [p. 212]
endpoint [p. 222]
equilateral triangle [p. 242]
exterior angle of a triangle [p. 232]
face of a rectangular solid [p. 278]
geometric solids [p. 277]
gram [p. 203]
height of a parallelogram [p. 251]
height of a triangle [p. 252]
height of a trapezoid [p. 253]
hypotenuse [p. 270]

interior angle of a triangle [p. 232]
intersecting lines [p. 224]
isosceles triangle [p. 242]
legs of a right triangle [p. 270]
line [p. 222]
line segment [p. 222]
line-of-sight problem [p. 298]
liter [p. 204]
mass [p. 203]
meter [p. 203]
obtuse angle [p. 226]
obtuse triangle [p. 242]
opposite side [p. 293]
parallel lines [p. 224]
parallelogram [p. 246]
perimeter [p. 243]
perpendicular lines [p. 225]
plane [p. 224]
plane figure [p. 224]
point [p. 222]
polygon [p. 241]
protractor [p. 225]
quadrilateral [p. 242]
radius of a circle [p. 247]
radius of a sphere [p. 279]
ray [p. 222]
rectangle [p. 244]
rectangular solid [p. 278]

regular polygon [p. 241]
regular pyramid [p. 280]
right angle [p. 225]
right circular cone [p. 280]
right circular cylinder [p. 280]
right triangle [p. 242]
scalene triangle [p. 242]
secant [p. 293]
sides of an angle [p. 224]
sides of a polygon [p. 241]
similar objects [p. 265]
similar triangles [p. 265]
sine [p. 293]
slant height [p. 280]
sphere [p. 279]
square [p. 245]
straight angle [p. 226]
supplementary angles [p. 226]
surface area [p. 282]
tangent [p. 293]
transversal [p. 229]
trapezoid [p. 253]
triangle [p. 232]
vertex [p. 224]
vertical angles [p. 228]
volume [p. 278]
weight [p. 203]

Essential Concepts

- **Basic Units in the Metric System**

Length:	meter
Mass:	gram
Capacity:	liter

- **Prefixes in the Metric System**

kilo-	=	1 000
hecto-	=	100
deca-	=	10
deci-	=	0.1
centi-	=	0.01
milli-	=	0.001

- **Units of Length in the U.S. Customary System**
  inch, foot, yard, mile

- **Units of Weight in the U.S. Customary System**
  ounce, pound, ton

- **Units of Capacity in the U.S. Customary System**
  fluid ounce, cup, pint, quart, gallon

- **Triangles**
  Sum of the measures of the interior angles $= 180°$

  Sum of the measures of an interior and a corresponding exterior angle $= 180°$

- **Perimeter Formulas**

Triangle:	$P = a + b + c$
Rectangle:	$P = 2L + 2W$
Square:	$P = 4s$
Parallelogram:	$P = 2b + 2s$
Circle:	$C = \pi d$ or $C = 2\pi r$

■ **Area Formulas**

Triangle: $A = \dfrac{1}{2}bh$

Rectangle: $A = LW$

Square: $A = s^2$

Circle: $A = \pi r^2$

Parallelogram: $A = bh$

Trapezoid: $A = \dfrac{1}{2}h(b_1 + b_2)$

■ **Volume Formulas**

Rectangular solid: $V = LWH$

Cube: $V = s^3$

Sphere: $V = \dfrac{4}{3}\pi r^3$

Right circular cylinder: $V = \pi r^2 h$

Right circular cone: $V = \dfrac{1}{3}\pi r^2 h$

Regular pyramid: $V = \dfrac{1}{3}s^2 h$

■ **Surface Area Formulas**

Rectangular solid: $S = 2LW + 2LH + 2WH$

Cube: $S = 6s^2$

Sphere: $S = 4\pi r^2$

Right circular cylinder: $S = 2\pi r^2 + 2\pi rh$

Right circular cone: $S = \pi r^2 + \pi rl$

Regular pyramid: $S = s^2 + 2sl$

■ **Pythagorean Theorem**

If a and b are the lengths of the legs of a right triangle and c is the length of the hypotenuse, then

$$c^2 = a^2 + b^2$$

■ **Similar Triangles**

The ratios of corresponding sides are equal. The ratio of corresponding heights is equal to the ratio of corresponding sides.

■ **The Trigonometric Functions of an Acute Angle of a Right Triangle**

If θ is an acute angle of a right triangle ABC, then

$$\sin \theta = \frac{\text{length of opposite side}}{\text{length of hypotenuse}}$$

$$\cos \theta = \frac{\text{length of adjacent side}}{\text{length of hypotenuse}}$$

$$\tan \theta = \frac{\text{length of opposite side}}{\text{length of adjacent side}}$$

$$\csc \theta = \frac{\text{length of hypotenuse}}{\text{length of opposite side}}$$

$$\sec \theta = \frac{\text{length of hypotenuse}}{\text{length of adjacent side}}$$

$$\cot \theta = \frac{\text{length of adjacent side}}{\text{length of opposite side}}$$

■ **Definitions of the Inverse Sine, Inverse Cosine, and Inverse Tangent Functions**

For $0° < x < 90°$:

$y = \sin^{-1}(x)$ is notation for "y is the angle whose sine is x."

$y = \cos^{-1}(x)$ is notation for "y is the angle whose cosine is x."

$y = \tan^{-1}(x)$ is notation for "y is the angle whose tangent is x."

CHAPTER 4   **Review Exercises**

1. Convert 1.24 km to meters.

2. Convert 0.45 g to milligrams.

3. Convert 72 oz to pounds.

4. Convert $1\dfrac{1}{4}$ mi to feet.

5. Convert 75 km/h to miles per hour. Round to the nearest hundredth.

6. Given that $m\angle a = 74°$ and $m\angle b = 52°$, find the measures of angles x and y.

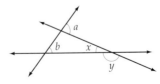

7. Triangles *ABC* and *DEF* are similar. Find the perimeter of triangle *ABC*.

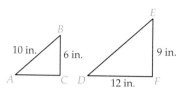

8. Find the volume of the geometric solid.

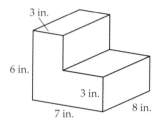

9. Find the measure of ∠*x*.

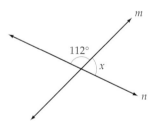

10. Find the surface area of the rectangular solid.

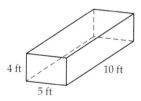

11. The length of a diameter of the base of a cylinder is 4 m, and the height of the cylinder is 8 m. Find the surface area of the cylinder. Give the exact value.

12. *A*, *B*, and *C* are all points on a line. Given that *BC* = 11 cm and that *AB* is three times the length of *BC*, find the length of *AC*.

13. Find *x*.

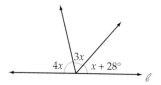

14. Find the area of a parallelogram that has a base of 6 in. and a height of 4.5 in.

15. Find the volume of the square pyramid.

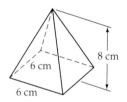

16. Find the circumference of a circle that has a diameter of 4.5 m. Round to the nearest hundredth of a meter.

17. Given that $\ell_1 \parallel \ell_2$, find the measures of angles *a* and *b*.

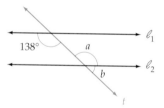

18. Find the supplement of a 32° angle.

19. Find the volume of a rectangular solid with a length of 6.5 ft, a width of 2 ft, and a height of 3 ft.

20. Two angles of a triangle measure 37° and 48°. Find the measure of the third angle.

21. The height of a triangle is 7 cm. The area of the triangle is 28 cm². Find the length of the base of the triangle.

22. Find the volume of a sphere that has a diameter of 12 mm. Give the exact value.

23. Framing The perimeter of a square picture frame is 86 cm. Find the length of each side of the frame.

24. **Paint** A can of paint will cover 200 ft^2 of surface. How many cans of paint should be purchased to paint a cylinder that has a height of 15 ft and a radius of 6 ft?

25. **Parks and Recreation** The length of a rectangular park is 56 yd. The width is 48 yd. How many yards of fencing are needed to surround the park?

26. **Patios** What is the area of a square patio that measures 9.5 m on each side?

27. **The Food Industry** In the butcher department of a supermarket, hamburger meat weighing 12 lb is equally divided and placed into 16 containers. How many ounces of hamburger meat is in each container?

28. **Baseball** Find the speed in feet per second of a baseball pitched at 87 mph.

29. **Catering** One hundred twenty-five people are expected to attend a reception. Assuming that each person drinks 400 ml of coffee, how many liters of coffee should be prepared?

30. **Landscaping** A walkway 2 m wide surrounds a rectangular plot of grass. The plot is 40 m long and 25 m wide. What is the area of the walkway?

31. Find the unknown side of the triangle. Round to the nearest hundredth of a foot.

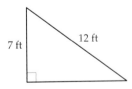

In Exercises 32 and 33, find the values of sin θ, cos θ, and tan θ for the given right triangle.

32.

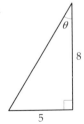

33.

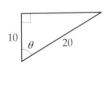

In Exercises 34–37, use a calculator. Round to the nearest tenth of a degree.

34. Find $\cos^{-1}(0.9013)$.

35. Find $\sin^{-1}(0.4871)$.

36. Given $\tan \beta = 1.364$, find β.

37. Given $\sin \theta = 0.0325$, find θ.

38. **Surveying** Find the distance across the marsh in the following figure. Round to the nearest tenth of a foot.

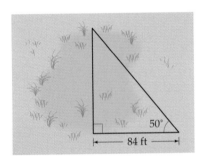

39. **Angle of Depression** The distance from a plane to a radar station is 200 mi, and the angle of depression is 40°. Find the number of ground miles from a point directly under the plane to the radar station. Round to the nearest tenth of a mile.

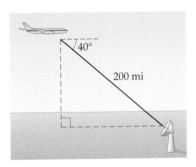

40. **Angle of Elevation** The angle of elevation from a point A on the ground to the top of a space shuttle is 27°. If point A is 110 ft from the base of the space shuttle, how tall is the space shuttle? Round to the nearest tenth of a foot.

Test

1. Convert 4 650 cm to meters.

2. Convert 4.1 L to milliliters.

3. Convert 42 yd to feet.

4. Convert $2\frac{1}{2}$ c to fluid ounces.

5. Convert 12 oz to grams.

6. Find the volume of a cylinder with a height of 6 m and a radius of 3 m. Round to the nearest hundredth of a cubic meter.

7. Find the perimeter of a rectangle that has a length of 2 m and a width of 1.4 m.

8. Find the complement of a 32° angle.

9. Find the area of a circle that has a diameter of 2 m. Round to the nearest hundredth of a square meter.

10. In the figure below, lines ℓ_1 and ℓ_2 are parallel. Angle x measures 30°. Find the measure of angle y.

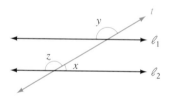

11. In the figure below, lines ℓ_1 and ℓ_2 are parallel. Angle x measures 45°. Find the measures of angles a and b.

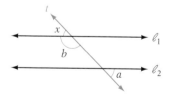

12. Find the area of a square that measures 2.25 ft on each side.

13. Find the volume of the figure. Give the exact value.

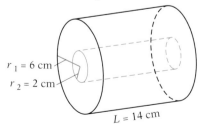

14. Triangles ABC and DEF are similar. Find side BC.

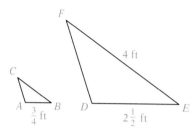

15. A right triangle has a 40° angle. Find the measures of the other two angles.

16. Find the measure of $\angle x$.

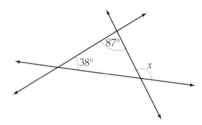

17. Find the area of the parallelogram shown below.

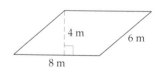

18. **Surveying** Find the width of the canal shown in the figure below.

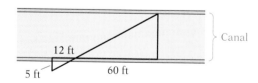

19. **Pizza** How much more area is in a pizza with radius 10 in. than in a pizza with radius 8 in.? Round to the nearest hundredth of a square inch.

20. For the right triangle shown below, determine the length of $\overline{BC}$. Round to the nearest hundredth of a centimeter.

21. Find the values of $\sin\theta$, $\cos\theta$, and $\tan\theta$ for the given right triangle.

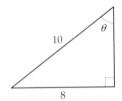

22. Angle of Elevation From a point 27 ft from the base of a Roman aqueduct, the angle of elevation to the top of the aqueduct is 78°. Find the height of the aqueduct. Round to the nearest foot.

23. Trees Find the cross-sectional area of a redwood tree that is 11 ft 6 in. in diameter. Round to the nearest hundredth of a square foot.

24. Toolbox A toolbox is 14 in. long, 9 in. wide, and 8 in. high. The sides and bottom of the toolbox are $\frac{1}{4}$ in. thick. The toolbox is open at the top. Find the volume of the interior of the toolbox in cubic inches.

25. The Food Industry A gourmet food store buys 24 lb of cheese for $126. The cheese is cut and packaged in 12-ounce packages that sell for $7.50 each. Find the difference between the store's purchase price of the cheese and the store's income from selling the packages of cheese.

CHAPTER 5

Linear Models

5.1 **Rectangular Coordinates and Functions**

5.2 **Properties of Linear Functions**

5.3 **Finding Linear Models**

5.4 **Linear Regression and Correlation**

All around us, we observe relationships in which some quantities are determined by others. The temperature outside changes with the time of day, different prices of gasoline can be matched to different dates in the past, and the value of a used car may depend on how long ago it was built. In mathematics, many of these relationships are considered functions. One of the most common types of functions is a linear function, in which consistent changes in one value cause consistent changes in the related value.

For example, a rule of thumb that is often used to predict the maximum heart rate for women who exercise is to subtract the age from 226. So if A is the age of a woman, her maximum predicted heart rate R is $R = 226 - A$. According to this rule, every time a woman gets one year older, her maximum heart rate decreases by one beat per minute. This consistent change is the defining characteristic of a linear function. We will see many additional examples of linear functions in this chapter.

Online Study Center

For online student resources, visit this textbook's Online Study Center at **college.hmco.com/pic/aufmannMTQR.**

| # Rectangular Coordinates and Functions

Introduction to Rectangular Coordinate Systems

When archaeologists excavate a site, a *coordinate grid* is laid over the site so that records can be kept not only of what was found but also of *where* it was found. The grid below is from an archaeological dig at Poggio Colla, a site about 20 miles northeast of Florence, Italy.

historical note

The concept of a coordinate system developed over time, culminating in 1637 with the publication of *Discourse on the Method for Rightly Directing One's Reason and Searching for Truth in the Sciences* by René Descartes (1596–1650) and *Introduction to Plane and Solid Loci* by Pierre de Fermat (1601–1665). Of the two mathematicians, Descartes is usually given more credit for developing the concept of a coordinate system. In fact, he became so famous in La Haye, the town in which he was born, that the town was renamed La Haye–Descartes. ■

In mathematics we encounter a similar problem, that of locating a point in a plane. One way to solve the problem is to use a *rectangular coordinate system.*

A **rectangular coordinate system** is formed by two number lines, one horizontal and one vertical, that intersect at the zero point of each line. The point of intersection is called the **origin.** The two number lines are called the **coordinate axes,** or simply the **axes.** Frequently, the horizontal axis is labeled the *x*-axis and the vertical axis is labeled the *y*-axis. In this case, the axes form what is called the **xy-plane.**

The two axes divide the plane into four regions called **quadrants,** which are numbered counterclockwise, using Roman numerals, from I to IV, starting at the upper right.

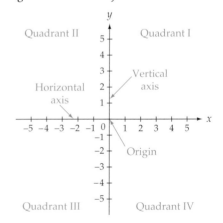

point of interest

The word *abscissa* has the same root as the word *scissors*. When open, a pair of scissors looks like an *x*.

Each point in the plane can be identified by a pair of numbers called an **ordered pair.** The first number of the ordered pair measures a horizontal change from the *y*-axis and is called the **abscissa,** or **x-coordinate.** The second number of the ordered pair measures a vertical change from the *x*-axis and is called the **ordinate,** or **y-coordinate.** The ordered pair (x, y) associated with a point is also called the **coordinates** of the point.

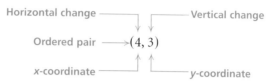

To **graph,** or **plot,** a point means to place a dot at the coordinates of the point. For example, to graph the ordered pair $(4, 3)$, start at the origin. Move 4 units to the right and then 3 units up. Draw a dot. To graph $(-3, -4)$, start at the origin. Move 3 units left and then 4 units down. Draw a dot.

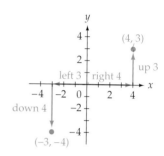

The **graph of an ordered pair** is the dot drawn at the coordinates of the point in the plane. The graphs of the ordered pairs $(4, 3)$ and $(-3, -4)$ are shown at the upper right.

The graphs of the points whose coordinates are $(2, 3)$ and $(3, 2)$ are shown at the right. Note that they are different points. The order in which the numbers in an ordered pair are listed is important.

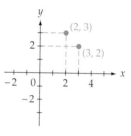

TAKE NOTE

This is *very* important. An *ordered pair* is a pair of coordinates, and the *order* in which the coordinates are listed matters.

If the axes are labeled with letters other than *x* and *y*, then we refer to the ordered pair using the given labels. For instance, if the horizontal axis is labeled *t* and the vertical axis is labeled *d*, then the ordered pairs are written as (t, d). We sometimes refer to the first number in an ordered pair as the **first coordinate** of the ordered pair and to the second number as the **second coordinate** of the ordered pair.

One purpose of a coordinate system is to draw a picture of the solutions of an **equation in two variables.** Examples of equations in two variables are shown at the right.

$$y = 3x - 2$$
$$x^2 + y^2 = 25$$
$$s = t^2 - 4t + 1$$

A **solution of an equation in two variables** is an ordered pair that makes the equation a true statement. For instance, as shown below, $(2, 4)$ is a solution of $y = 3x - 2$ but $(3, -1)$ is not a solution of the equation.

$y = 3x - 2$		
4	$3(2) - 2$	• *x* = 2, *y* = 4
4	$6 - 2$	
$4 = 4$		• Checks.

$y = 3x - 2$		
-1	$3(3) - 2$	• *x* = 3, *y* = −1
-1	$9 - 2$	
$-1 \neq 7$		• Does not check.

QUESTION   Is $(-2, 1)$ a solution of $y = 3x + 7$?

ANSWER   *Yes, because 1 = 3(−2) + 7.*

historical note

Maria Graëtana Agnesi
(an'yayzee)
(1718–1799) was probably the first woman to write a calculus text. A report on the text made by a committee of the Académie des Sciences in Paris stated: "It took much skill and sagacity to reduce, as the author has done, to almost uniform methods these discoveries scattered among the works of modern mathematicians and often presented by methods very different from each other. Order, clarity and precision reign in all parts of this work....We regard it as the most complete and best made treatise."

There is a graph named after Agnesi called the Witch of Agnesi. This graph came by its name because of an incorrect translation from Italian to English of a work by Agnesi. There is also a crater on Venus named after Agnesi called the Crater of Agnesi. ∎

The **graph of an equation in two variables** is a drawing of all the ordered-pair solutions of the equation. Many equations can be graphed by finding some ordered-pair solutions of the equation, plotting the corresponding points, and then connecting the points with a smooth curve.

EXAMPLE 1 ■ Graph an Equation in Two Variables

Graph $y = 3x - 2$.

Solution

To find ordered-pair solutions, select various values of x and calculate the corresponding values of y. Plot the ordered pairs. After the ordered pairs have been graphed, draw a smooth curve through the points. It is convenient to keep track of the solutions in a table.

When choosing values of x, we often choose integer values because the resulting ordered pairs are easier to graph.

x	$3x - 2 = y$	(x, y)
-2	$3(-2) - 2 = -8$	$(-2, -8)$
-1	$3(-1) - 2 = -5$	$(-1, -5)$
0	$3(0) - 2 = -2$	$(0, -2)$
1	$3(1) - 2 = 1$	$(1, 1)$
2	$3(2) - 2 = 4$	$(2, 4)$
3	$3(3) - 2 = 7$	$(3, 7)$

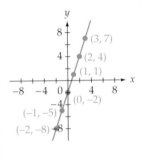

CHECK YOUR PROGRESS 1   Graph $y = -2x + 3$.

Solution   *See page S19.*

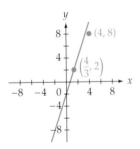

The graph of $y = 3x - 2$ is shown again at the left. Note that the ordered pair $\left(\frac{4}{3}, 2\right)$ is a solution of the equation and is a point on the graph. The ordered pair $(4, 8)$ is *not* a solution of the equation and is *not* a point on the graph. Every ordered-pair solution of the equation is a point on the graph, and every point on the graph is an ordered-pair solution of the equation.

EXAMPLE 2 ■ Graph an Equation in Two Variables

Graph $y = x^2 + 4x$.

Solution

Select various values of x and calculate the corresponding values of y. Plot the ordered pairs. After the ordered pairs have been graphed, draw a smooth curve through the points. The following table shows some ordered pair solutions.

x	$x^2 + 4x = y$	(x, y)
-5	$(-5)^2 + 4(-5) = 5$	$(-5, 5)$
-4	$(-4)^2 + 4(-4) = 0$	$(-4, 0)$
-3	$(-3)^2 + 4(-3) = -3$	$(-3, -3)$
-2	$(-2)^2 + 4(-2) = -4$	$(-2, -4)$
-1	$(-1)^2 + 4(-1) = -3$	$(-1, -3)$
0	$(0)^2 + 4(0) = 0$	$(0, 0)$
1	$(1)^2 + 4(1) = 5$	$(1, 5)$

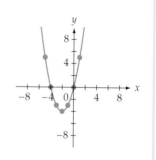

TAKE NOTE

As this example shows, it may be necessary to graph quite a number of points before a reasonably accurate graph can be drawn.

CHECK YOUR PROGRESS 2   Graph $y = -x^2 + 1$.

Solution   *See page S19.*

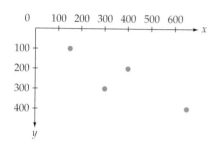

MathMatters   **Computer Software Program Coordinate Systems**

Some computer software programs use a coordinate system that is different from the *xy*-coordinate system we have discussed. For instance, in one particular software program, the origin $(0, 0)$ represents the top left point of a computer screen, as shown at the left. The points $(150, 100)$, $(300, 300)$, $(400, 200)$, and $(650, 400)$ are shown on the graph.

Introduction to Functions

An important part of mathematics is the study of the relationship between known quantities. Exploring relationships between known quantities frequently results in equations in two variables. For instance, as a car is driven, the fuel in the gas tank is burned. There is a correspondence between the number of gallons of fuel used and the number of miles traveled. If a car gets 25 miles per gallon, then the car consumes 0.04 gallon of fuel for each mile driven. For the sake of simplicity, we will assume that the car always consumes 0.04 gallon of gasoline for each mile driven. The equation $g = 0.04d$ defines how the number of gallons used, g, depends on the number of miles driven, d.

TAKE NOTE

A car that uses 1 gallon of gas to travel 25 miles uses $\frac{1}{25} = 0.04$ gallon to travel 1 mile.

Distance traveled (in miles), d	25	50	100	250	300
Fuel used (in gallons), g	1	2	4	10	12

The ordered pairs in this table are only some of the possible ordered pairs. Other possibilities are $(90, 3.6)$, $(125, 5)$, and $(235, 9.4)$. If all of the ordered pairs of the equation were drawn, the graph would appear as a portion of a line. The graph of the equation and the ordered pairs we have calculated are shown on the following page. Note that the graphs of all the ordered pairs are on the same line.

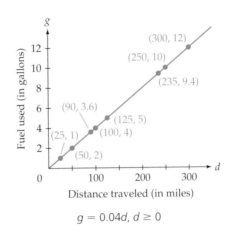

$$g = 0.04d, d \geq 0$$

QUESTION   *What is the meaning of the ordered pair (125, 5)?*

The ordered pairs, the graph, and the equation are all different ways of expressing the correspondence, or relationship, between the two variables. This correspondence, which pairs the number of gallons of fuel used with the number of miles driven, is called a *function.*

Here are some additional examples of functions, along with a specific example of each correspondence.

<table>
<tr><td>To each real number
5</td><td>there corresponds
⟶</td><td>its square
25</td></tr>
<tr><td>To each score on an exam
87</td><td>there corresponds
⟶</td><td>a grade
B</td></tr>
<tr><td>To each student
Alexander Sterling</td><td>there corresponds
⟶</td><td>a student identification number
S18723519</td></tr>
</table>

An important fact about each of these correspondences is that each result is *unique.* For instance, for the real number 5, there is *exactly one* square, 25. With this in mind, we now state the definition of a function.

> **TAKE NOTE**
>
> Because the square of any real number is a positive number or zero ($0^2 = 0$), the range of the function that pairs a number with its square contains the positive numbers and zero. Therefore, the range is the nonnegative real numbers. The difference between *nonnegative* numbers and *positive* numbers is that nonnegative numbers include zero; positive numbers do not include zero.

> **Function**
>
> A **function** is a correspondence, or relationship, between two sets called the **domain** and **range** such that for each element of the domain there corresponds *exactly one* element of the range.

ANSWER   *A car that gets 25 miles per gallon will require 5 gallons of fuel to travel a distance of 125 miles.*

Test Score	Grade
90–100	A
80–89	B
70–79	C
60–69	D
0–59	F

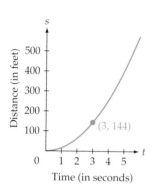

As an example of domain and range, consider the function that pairs a test score with a letter grade. The domain is the real numbers from 0 to 100. The range is the letters A, B, C, D, and F.

As shown at left, a function can be described in terms of ordered pairs or by a graph. Functions can also be defined by equations in two variables. For instance, when gravity is the only force acting on a falling body, a function that describes the distance s, in feet, an object will fall in t seconds is given by $s = 16t^2$.

Given a value of t (time), the value of s (the distance the object falls) can be found. For instance, given $t = 3$,

$$s = 16t^2$$
$$s = 16(3)^2 \quad \text{• Replace } t \text{ by 3.}$$
$$s = 16(9) \quad \text{• Simplify.}$$
$$s = 144$$

The object falls 144 feet in 3 seconds.

Because the distance the object falls *depends on* how long it has been falling, s is called the **dependent variable** and t is called the **independent variable.** Some of the ordered pairs of this function are (3, 144), (1, 16), (0, 0), and $\left(\frac{1}{4}, 1\right)$. The ordered pairs can be written as (t, s), where $s = 16t^2$. By substituting $16t^2$ for s, we can also write the ordered pairs as $(t, 16t^2)$. For the equation $s = 16t^2$, we say that "distance is a function of time." A graph of the function is shown at the left.

Not all equations in two variables define a function. For instance,

$$y^2 = x^2 + 9$$

is not an equation that defines a function because

$$5^2 = 4^2 + 9 \qquad \text{and} \qquad (-5)^2 = 4^2 + 9$$

The ordered pairs $(4, 5)$ and $(4, -5)$ both belong to the equation. Consequently, there are two ordered pairs with the same first coordinate, 4, but *different* second coordinates, 5 and -5. By definition, the equation does not define a function. The phrase "*y* is a function of *x*," or a similar phrase with different variables, is used to describe those equations in two variables that define functions.

Function notation is frequently used for equations that define functions. Just as the letter x is commonly used as a variable, the letter f is commonly used to name a function.

To describe the relationship between a number and its square using function notation, we can write $f(x) = x^2$. The symbol $f(x)$ is read "the value of f at x" or "f of x." The symbol $f(x)$ is the **value of the function** and represents the value of the dependent variable for a given value of the independent variable. We will often write $y = f(x)$ to emphasize the relationship between the independent variable x and the dependent variable y. Remember: y and $f(x)$ are different symbols for the same number. Also, the *name* of the function is f; the *value* of the function at x is $f(x)$.

The letters used to represent a function are somewhat arbitrary. All of the following equations represent the same function.

$$\left.\begin{array}{l} f(x) = x^2 \\ g(t) = t^2 \\ P(v) = v^2 \end{array}\right\} \quad \text{Each of these equations represents the squaring function.}$$

The process of finding $f(x)$ for a given value of x is called **evaluating the function.** For instance, to evaluate $f(x) = x^2$ when $x = 4$, replace x by 4 and simplify.

$$f(x) = x^2$$
$$f(4) = 4^2 = 16 \qquad \text{• Replace } x \text{ by 4. Then simplify.}$$

The *value* of the function is 16 when $x = 4$. This means that an ordered pair of the function is $(4, 16)$.

TAKE NOTE

To evaluate a function, you can use open parentheses in place of the variable. For instance,

$$s(t) = 2t^2 - 3t + 1$$
$$s(\ ) = 2(\ )^2 - 3(\ ) + 1$$

To evaluate the function, fill in each set of parentheses with the same number and then use the Order of Operations Agreement (see page 122) to evaluate the numerical expression on the right side of the equation.

EXAMPLE 3 ▪ Evaluate a Function

Evaluate $s(t) = 2t^2 - 3t + 1$ when $t = -2$.

Solution
$$s(t) = 2t^2 - 3t + 1$$
$$s(-2) = 2(-2)^2 - 3(-2) + 1 \qquad \text{• Replace } t \text{ by } -2. \text{ Then simplify.}$$
$$= 15$$

The value of the function is 15 when $t = -2$.

CHECK YOUR PROGRESS 3   Evaluate $f(z) = z^2 - z$ when $z = -3$.

Solution   See page S20.

Any letter or combination of letters can be used to name a function. In the next example, the letters *SA* are used to name a *Surface Area* function.

EXAMPLE 4 ▪ Application of Evaluating a Function

The surface area of a cube (the sum of the areas of each of the six faces) is given by $SA(s) = 6s^2$, where $SA(s)$ is the surface area of the cube and s is the length of one side of the cube. Find the surface area of a cube that has a side of length 10 centimeters.

Solution
$$SA(s) = 6s^2$$
$$SA(10) = 6(10)^2 \qquad \text{• Replace } s \text{ by 10.}$$
$$= 6(100) \qquad \text{• Simplify.}$$
$$= 600$$

The surface area of the cube is 600 square centimeters.

CHECK YOUR PROGRESS 4   A **diagonal** of a polygon is a line segment from one vertex to a nonadjacent vertex, as shown at the left. The total number of diagonals of a polygon is given by $N(s) = \dfrac{s^2 - 3s}{2}$, where $N(s)$ is the total number of diagonals and s is the number of sides of the polygon. Find the total number of diagonals of a polygon with 12 sides.

Solution   See page S20.

Diagonal

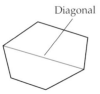

Math Matters   The Special Theory of Relativity

In 1905, Albert Einstein published a paper that set the framework for relativity theory. This theory, now called the Special Theory of Relativity, explains, among other
things, how mass changes for a body in motion. Essentially, the theory states that
the mass of a body is a function of its velocity, v. That is, the mass of a body increases
as its speed increases.

The function can be given by $M(v) = \dfrac{m_0}{\sqrt{1 - \dfrac{v^2}{c^2}}}$, where m_0 is the mass of the

body at rest (its mass when its speed is zero) and c is the velocity of light, which
Einstein showed was the same for all observers. The table below shows how a
5-kilogram mass increases as its speed becomes closer and closer to the speed of light.

Speed	Mass (kilograms)
30 meters/second—speed of a car on an expressway	5
240 meters/second—speed of a commercial jet	5
3.0×10^7 meters/second—10% of the speed of light	5.025
1.5×10^8 meters/second—50% of the speed of light	5.774
2.7×10^8 meters/second—90% of the speed of light	11.471

Note that for speeds of everyday objects, such as a car or plane, the increase in
mass is negligible. Physicists have verified these increases using particle accelerators
that can accelerate a particle such as an electron to more than 99.9% of the speed
of light.

Graphs of Functions

Often the graph of a function can be drawn by finding ordered pairs of the function, plotting the points corresponding to the ordered pairs, and then connecting
the points with a smooth curve.

For example, to graph $f(x) = x^3 + 1$, select several values of x and evaluate the
function at each value. Recall that $f(x)$ and y are different symbols for the same
quantity.

x	$f(x) = x^3 + 1$	(x, y)
-2	$f(-2) = (-2)^3 + 1 = -7$	$(-2, -7)$
-1	$f(-1) = (-1)^3 + 1 = 0$	$(-1, 0)$
0	$f(0) = (0)^3 + 1 = 1$	$(0, 1)$
1	$f(1) = (1)^3 + 1 = 2$	$(1, 2)$
2	$f(2) = (2)^3 + 1 = 9$	$(2, 9)$

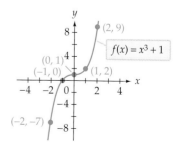

Plot the ordered pairs and draw a smooth curve through the points.

EXAMPLE 5 ■ **Graph a Function**

Graph $h(x) = x^2 - 3$.

Solution

x	$h(x) = x^2 - 3$	(x, y)
−3	$h(-3) = (-3)^2 - 3 = 6$	(−3, 6)
−2	$h(-2) = (-2)^2 - 3 = 1$	(−2, 1)
−1	$h(-1) = (-1)^2 - 3 = -2$	(−1, −2)
0	$h(0) = (0)^2 - 3 = -3$	(0, −3)
1	$h(1) = (1)^2 - 3 = -2$	(1, −2)
2	$h(2) = (2)^2 - 3 = 1$	(2, 1)
3	$h(3) = (3)^2 - 3 = 6$	(3, 6)

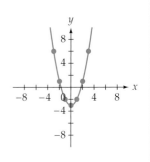

Plot the ordered pairs and draw a smooth curve through the points.

CHECK YOUR PROGRESS 5   Graph $f(x) = 2 - \dfrac{3}{4}x$.

Solution   *See page S20.*

Investigation

Dilations of a Geometric Figure

A **dilation** of a geometric figure changes the size of the figure by either enlarging it or reducing it. This is accomplished by multiplying the coordinates of the figure by a positive number called the **dilation constant.** Examples of enlarging (multiplying the coordinates by a number greater than 1) and reducing (multiplying the coordinates by a number between 0 and 1) a geometric figure are shown below.

▼ **point of interest**

Photocopy machines have reduction and enlargement features that function essentially as constants of dilation. The numbers are usually expressed as a percent. A copier selection of 50% reduces the size of the object being copied by 50%. A copier selection of 125% increases the size of the object being copied by 25%.

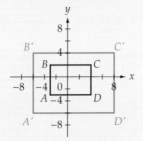

ABCD was enlarged by multiplying its coordinates by 2. The result is *A'B'C'D'*.

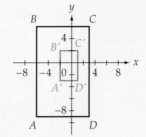

ABCD was reduced by multiplying its coordinates by $\frac{1}{3}$. The result is *A'B'C'D'*.

(continued)

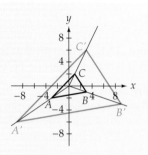

When each of the coordinates of a figure is multiplied by the same number in order to produce a dilation, the **center of dilation** will be the origin of the coordinate system. For triangle ABC at the left, a constant of dilation of 3 was used to produce triangle $A'B'C'$. Note that lines through the corresponding vertices of the two triangles intersect at the origin, the center of dilation. The center of dilation, however, can be any point in the plane.

Investigation Exercises

1. A dilation is performed on the figure with vertices $A(-2, 0), B(2, 0), C(4, -2),$ $D(2, -4),$ and $E(-2, -4)$.

   a. Draw the original figure and a new figure using 2 as the dilation constant.

   b. Draw the original figure and a new figure using $\frac{1}{2}$ as the dilation constant.

2. Because each of the coordinates of a geometric figure is multiplied by a number, the lengths of the sides of the figure will change. It is possible to show that the lengths change by a factor equal to the constant of dilation. In this exercise, you will examine the effect of a dilation on the angles of a geometric figure. Draw some figures and then draw a dilation of each figure using the origin as the center of dilation. Use a protractor to determine whether the measures of the angles of the dilated figure are different from the measures of the corresponding angles of the original figure.

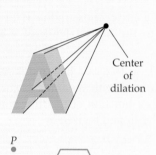

3. Graphic artists use centers of dilation to create three-dimensional effects. Consider the block letter A shown at the left. Draw another block letter A by changing the center of dilation to see how it affects the 3-D look of the letter. Programs such as PowerPoint use these methods to create various shading options for design elements in a presentation.

4. Draw an enlargement and a reduction of the figure at the left for the given center of dilation P.

5. On a blank piece of paper, draw a rectangle 4 inches by 6 inches with the center of the rectangle in the center of the paper. Make various photocopies of the rectangle using the reduction and enlargement settings on a copy machine. Where is the center of dilation for the copy machine?

Exercise Set 5.1

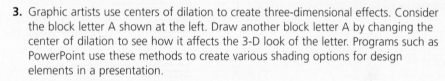
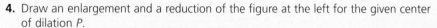

1. Graph the ordered pairs $(0, -1)$, $(2, 0)$, $(3, 2)$, and $(-1, 4)$.

2. Graph the ordered pairs $(-1, -3)$, $(0, -4)$, $(0, 4)$, and $(3, -2)$.

3. Draw a line through all points with an x-coordinate of 2.

4. Draw a line through all points with an x-coordinate of -3.

5. Draw a line through all points with a y-coordinate of -3.

6. Draw a line through all points with a y-coordinate of 4.

7. Graph the ordered-pair solutions of $y = x^2$ when $x = -2, -1, 0, 1,$ and 2.

8. Graph the ordered-pair solutions of $y = -x^2 + 1$ when $x = -2, -1, 0, 1,$ and 2.

9. Graph the ordered-pair solutions of $y = |x + 1|$ when $x = -5, -3, 0, 3$, and 5.

10. Graph the ordered-pair solutions of $y = -2|x|$ when $x = -3, -1, 0, 1$, and 3.

11. Graph the ordered-pair solutions of $y = -x^2 + 2$ when $x = -2, -1, 0$, and 1.

12. Graph the ordered-pair solutions of $y = -x^2 + 4$ when $x = -3, -1, 0, 1$, and 3.

13. Graph the ordered-pair solutions of $y = x^3 - 2$ when $x = -1, 0, 1$, and 2.

14. Graph the ordered-pair solutions of $y = -x^3 + 1$ when $x = -1, 0, 1$, and 2.

15. Without plotting the points, in which quadrant will each point lie?

    a. $(5, -2)$ **b.** $(-3, 7)$

16. Is the point $(-500, 500)$ on the graph of the equation $x + y = 0$?

In Exercises 17–26, graph each equation.

17. $y = 2x - 1$

18. $y = -3x + 2$

19. $y = \dfrac{2}{3}x + 1$

20. $y = -\dfrac{x}{2} - 3$

21. $y = \dfrac{1}{2}x^2$

22. $y = \dfrac{1}{3}x^2$

23. $y = 2x^2 - 1$

24. $y = -3x^2 + 2$

25. $y = |x - 1|$

26. $y = |x - 3|$

27. Does the equation $x^2 + y^2 = 4$ define a function?

28. Does the equation $p = 3t^2 + 2$ define p as a function of t?

In Exercises 29–36, evaluate the function for the given value.

29. $f(x) = 2x + 7; x = -2$

30. $y(x) = 1 - 3x; x = -4$

31. $f(t) = t^2 - t - 3; t = 3$

32. $P(n) = n^2 - 4n - 7; n = -3$

33. $v(s) = s^3 + 3s^2 - 4s - 2; s = -2$

34. $f(x) = 3x^3 - 4x^2 + 7; x = 2$

35. $T(p) = \dfrac{p^2}{p - 2}; p = 0$

36. $s(t) = \dfrac{4t}{t^2 + 2}; t = 2$

37. If $P(d)$ is the population of the world at the end of year d, which is larger, $P(1900)$ or $P(2000)$?

38. If $g(m)$ gives the amount (in gallons) of gasoline remaining in the tank of a particular car that has been driven m miles since the gas tank was filled, which is greater, $g(32)$ or $g(35)$?

39. **Geometry** The perimeter P of a square is a function of the length s of one of its sides and is given by $P(s) = 4s$.

    a. Find the perimeter of a square whose side is 4 meters.

    b. Find the perimeter of a square whose side is 5 feet.

40. **Geometry** The area of a circle is a function of its radius and is given by $A(r) = \pi r^2$.

    a. Find the area of a circle whose radius is 3 inches. Round to the nearest tenth of a square inch.

    b. Find the area of a circle whose radius is 12 centimeters. Round to the nearest tenth of a square centimeter.

41. **Sports** The height h, in feet, of a ball that is released 4 feet above the ground with an initial velocity of 80 feet per second is a function of the time t, in seconds, the ball is in the air and is given by

$$h(t) = -16t^2 + 80t + 4, \quad 0 \le t \le 5.04$$

    a. Find the height of the ball above the ground 2 seconds after it is released.

    b. Find the height of the ball above the ground 4 seconds after it is released.

42. **Forestry** The distance d, in miles, a forest fire ranger can see from an observation tower is a function of the height h, in feet, of the tower above level ground and is given by $d(h) = 1.5\sqrt{h}$.

    a. Find the distance a ranger can see whose eye level is 20 feet above level ground. Round to the nearest tenth of a mile.

    b. Find the distance a ranger can see whose eye level is 35 feet above level ground. Round to the nearest tenth of a mile.

43. **Sound** The speed s, in feet per second, of sound in air depends on the temperature t of the air in degrees Celsius and is given by $s(t) = \dfrac{1087\sqrt{t + 273}}{16.52}$.

a. What is the speed of sound in air when the temperature is 0°C (the temperature at which water freezes)? Round to the nearest foot per second.

b. What is the speed of sound in air when the temperature is 25°C? Round to the nearest foot per second.

44. Softball In a softball league in which each team plays every other team three times, the number of games N that must be scheduled depends on the number of teams n in the league and is given by $N(n) = \frac{3}{2}n^2 - \frac{3}{2}n$.

a. How many games must be scheduled for a league that has five teams?

b. How many games must be scheduled for a league that has six teams?

45. Mixtures The percent concentration P of salt in a particular salt water solution depends on the number of grams x of salt that are added to the solution and is given by $P(x) = \dfrac{100x + 100}{x + 10}$.

a. What is the original percent concentration of salt?

b. What is the percent concentration of salt after 5 more grams of salt are added?

46. Pendulums The time T, in seconds, it takes a pendulum to make one swing depends on the length of the pendulum and is given by $T(L) = 2\pi\sqrt{\dfrac{L}{32}}$, where L is the length of the pendulum in feet.

a. Find the time it takes the pendulum to make one swing if the length of the pendulum is 3 feet. Round to the nearest hundredth of a second.

b. Find the time it takes the pendulum to make one swing if the length of the pendulum is 9 inches. Round to the nearest tenth of a second.

In Exercises 47–62, graph the function.

47. $f(x) = 2x - 5$

48. $f(x) = -2x + 4$

49. $f(x) = -x + 4$

50. $f(x) = 3x - 1$

51. $g(x) = \dfrac{2}{3}x + 2$

52. $h(x) = \dfrac{5}{2}x - 1$

53. $F(x) = -\dfrac{1}{2}x + 3$

54. $F(x) = -\dfrac{3}{4}x - 1$

55. $f(x) = x^2 - 1$

56. $f(x) = x^2 + 2$

57. $f(x) = -x^2 + 4$

58. $f(x) = -2x^2 + 5$

59. $g(x) = x^2 - 4x$

60. $h(x) = x^2 + 4x$

61. $P(x) = x^2 - x - 6$

62. $P(x) = x^2 - 2x - 3$

Extensions

CRITICAL THINKING

63. Geometry Find the area of the rectangle.

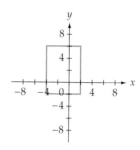

64. Geometry Find the area of the triangle.

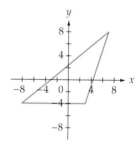

65. Suppose f is a function. Is it possible to have $f(2) = 4$ and $f(2) = 7$? Explain your answer.

66. Suppose f is a function and $f(a) = 4$ and $f(b) = 4$. Does this mean that $a = b$?

67. If $f(x) = 2x + 5$ and $f(a) = 9$, find a.

68. If $f(x) = x^2$ and $f(a) = 9$, find a.

69. Let $f(a, b) =$ the sum of a and b.

Let $g(a, b) =$ the product of a and b.

Find $f(2, 5) + g(2, 5)$.

70. Let $f(a, b) =$ the greatest common factor of a and b and let $g(a, b) =$ the least common multiple (see page 127) of a and b. Find $f(14, 35) + g(14, 35)$.

71. Given $f(x) = x^2 + 3$, for what value of x is $f(x)$ least?

72. Given $f(x) = -x^2 + 4x$, for what value of x is $f(x)$ greatest?

EXPLORATIONS

73. Consider the function given by

$$M(x, y) = \frac{x + y}{2} + \frac{|x - y|}{2}.$$

a. Complete the following table.

x	y	$M(x, y) = \dfrac{x + y}{2} + \dfrac{\|x - y\|}{2}$
−5	11	$M(-5, 11) = \dfrac{-5 + 11}{2} + \dfrac{\|-5 - 11\|}{2} = 11$
10	8	
−3	−1	
12	−13	
−11	15	

b. Extend the table by choosing some additional values of x and y.

c. How is the value of the function related to the values of x and y? *Hint:* For $x = -5$ and $y = 11$, the value of the function is 11, the value of y.

d. The function $M(x, y)$ is sometimes referred to as the *maximum function.* Why is this a good name for this function?

e. Create a *minimum function*—that is, a function that yields the minimum of two numbers x and y. *Hint:* The function is similar in form to the maximum function.

SECTION 5.2 | # Properties of Linear Functions

Intercepts

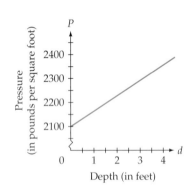

The graph at the left shows the pressure on a diver as the diver descends in the ocean. The equation of this graph can be represented by $P(d) = 64d + 2100$, where $P(d)$ is the pressure, in pounds per square foot, on a diver d feet below the surface of the ocean. By evaluating the function for various values of d, we can determine the pressure on the diver at those depths. For instance, when $d = 2$, we have

$$P(d) = 64d + 2100$$
$$P(2) = 64(2) + 2100$$
$$= 128 + 2100$$
$$= 2228$$

The pressure on a diver 2 feet below the ocean's surface is 2228 pounds per square foot.

The function $P(d) = 64d + 2100$ is an example of a *linear function.*

> **Linear Function**
>
> A **linear function** is one that can be written in the form $f(x) = mx + b$, where m is the coefficient of x and b is a constant.

For the linear function $P(d) = 64d + 2100$, $m = 64$ and $b = 2100$.

Here are some other examples of linear functions.

$$f(x) = 2x + 5 \quad \bullet\ m = 2,\ b = 5$$

$$g(t) = \frac{2}{3}t - 1 \quad \bullet\ m = \frac{2}{3},\ b = -1$$

$$v(s) = -2s \quad \bullet\ m = -2,\ b = 0$$

$$h(x) = 3 \quad \bullet\ m = 0,\ b = 3$$

$$f(x) = 2 - 4x \quad \bullet\ m = -4,\ b = 2$$

Note that different variables can be used to designate a linear function.

QUESTION   *Are the given functions linear functions?*

   a. $f(x) = 2x^2 + 5$ ***b.*** $g(x) = 1 - 3x$

Consider the linear function $f(x) = 2x + 4$. The graph of the function is shown below, along with a table listing some of its ordered pairs.

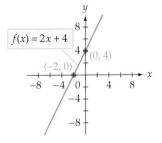

x	f(x) = 2x + 4	(x, y)
−3	$f(-3) = 2(-3) + 4 = -2$	$(-3, -2)$
−2	$f(-2) = 2(-2) + 4 = 0$	$(-2, 0)$
−1	$f(-1) = 2(-1) + 4 = 2$	$(-1, 2)$
0	$f(0) = 2(0) + 4 = 4$	$(0, 4)$
1	$f(1) = 2(1) + 4 = 6$	$(1, 6)$

✔ **TAKE NOTE**

Note that the graph of a *linear* function is a straight *line*. Observe that when the graph crosses the *x*-axis, the *y*-coordinate is 0. When the graph crosses the *y*-axis, the *x*-coordinate is 0. The table confirms these observations.

From the table and the graph, we can see that when $x = -2$, $y = 0$, and the graph crosses the *x*-axis at $(-2, 0)$. The point $(-2, 0)$ is called the ***x*-intercept** of the graph. When $x = 0$, $y = 4$, and the graph crosses the *y*-axis at $(0, 4)$. The point $(0, 4)$ is called the ***y*-intercept** of the graph.

ANSWER   *a. Because $f(x) = 2x^2 + 5$ has an x^2 term, f is not a linear function. **b.** Because $g(x) = 1 - 3x$ can be written in the form $f(x) = mx + b$ as $g(x) = -3x + 1$ ($m = -3$ and $b = 1$), g is a linear function.*

EXAMPLE 1 ■ Find the *x*- and *y*-Intercepts of a Graph

Find the *x*- and *y*-intercepts of the graph of $g(x) = -3x + 2$.

Solution

When a graph crosses the *x*-axis, the *y*-coordinate of the point is 0. Therefore, to find the *x*-intercept, replace $g(x)$ with 0 and solve the equation for *x*. [Recall that $g(x)$ is another name for *y*.]

$$g(x) = -3x + 2$$
$$0 = -3x + 2 \qquad \bullet \text{ Replace } g(x) \text{ with 0.}$$
$$-2 = -3x$$
$$\frac{2}{3} = x$$

The *x*-intercept is $\left(\frac{2}{3}, 0\right)$.

When a graph crosses the *y*-axis, the *x*-coordinate of the point is 0. Therefore, to find the *y*-intercept, evaluate the function when *x* is 0.

$$g(x) = -3x + 2$$
$$g(0) = -3(0) + 2 \qquad \bullet \text{ Evaluate } g(x) \text{ when } x = 0.$$
$$= 2$$

The *y*-intercept is $(0, 2)$.

CHECK YOUR PROGRESS 1 Find the *x*- and *y*-intercepts of the graph of $f(x) = \frac{1}{2}x + 3$.

Solution *See page S20.*

In Example 1, note that the *y*-coordinate of the *y*-intercept of $g(x) = -3x + 2$ has the same value as *b* in the equation $f(x) = mx + b$. This is always true.

> ### *y*-Intercept
>
> The *y*-intercept of the graph of $f(x) = mx + b$ is $(0, b)$.

If we evaluate the linear function that models pressure on a diver, $P(d) = 64d + 2100$, at 0, we have

$$P(d) = 64d + 2100$$
$$P(0) = 64(0) + 2100 = 2100$$

In this case, the *P*-intercept (the intercept on the vertical axis) is $(0, 2100)$. In the context of the application, this means that the pressure on a diver 0 feet below the ocean's surface is 2100 pounds per square foot. Another way of saying "zero feet below the ocean's surface" is "at sea level." Thus the pressure on the diver, or anyone else for that matter, at sea level is 2100 pounds per square foot.

Both the *x*- and *y*-intercepts can have meaning in the context of an application problem. This is demonstrated in the next example.

✔ **TAKE NOTE**

To find the *y*-intercept of $y = mx + b$ [we have replaced $f(x)$ by *y*], let $x = 0$. Then

$$y = mx + b$$
$$y = m(0) + b$$
$$= b$$

The *y*-intercept is $(0, b)$. This result is shown at the right.

✔ **TAKE NOTE**

We are working with the function $P(d) = 64d + 2100$. Therefore, the intercepts on the horizontal axis of a graph of the function are *d*-intercepts rather than *x*-intercepts, and the intercept on the vertical axis is a *P*-intercept rather than a *y*-intercept.

EXAMPLE 2 ■ Application of the Intercepts of a Linear Function

After a parachute is deployed, a function that models the height of the parachutist above the ground is $f(t) = -10t + 2800$, where $f(t)$ is the height, in feet, of the parachutist t seconds after the parachute is deployed. Find the intercepts on the vertical and horizontal axes and explain what they mean in the context of the problem.

Solution

To find the intercept on the vertical axis, evaluate the function when t is 0.

$$f(t) = -10t + 2800$$
$$f(0) = -10(0) + 2800 = 2800$$

The intercept on the vertical axis is $(0, 2800)$. This means that the parachutist is 2800 feet above the ground when the parachute is deployed.

To find the intercept on the horizontal axis, set $f(t)$ equal to 0 and solve for t.

$$f(t) = -10t + 2800$$
$$0 = -10t + 2800$$
$$-2800 = -10t$$
$$280 = t$$

The intercept on the horizontal axis is $(280, 0)$. This means that the parachutist reaches the ground 280 seconds after the parachute is deployed. **Note that the parachutist reaches the ground when $f(t) = 0$.**

CHECK YOUR PROGRESS 2

A function that models the descent of a certain small airplane is given by $g(t) = -20t + 8000$, where $g(t)$ is the height, in feet, of the airplane t seconds after it begins its descent. Find the intercepts on the vertical and horizontal axes, and explain what they mean in the context of the problem.

Solution   *See page S20.*

Slope of a Line

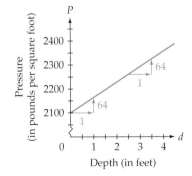

Consider again the linear function $P(d) = 64d + 2100$, which models the pressure on a diver as the diver descends below the ocean's surface. From the graph at the left, you can see that when the depth of the diver increases by 1 foot, the pressure on the diver increases by 64 pounds per square foot. This can be verified algebraically.

$$P(0) = 64(0) + 2100 = 2100$$ • Pressure at sea level
$$P(1) = 64(1) + 2100 = 2164$$ • Pressure after descending 1 foot
$$2164 - 2100 = 64$$ • Change in pressure

If we choose two other depths that differ by 1 foot, such as 2.5 feet and 3.5 feet (see the graph at the left), the change in pressure is the same.

$$P(2.5) = 64(2.5) + 2100 = 2260$$ • Pressure at **2.5** feet below the surface
$$P(3.5) = 64(3.5) + 2100 = 2324$$ • Pressure at **3.5** feet below the surface
$$2324 - 2260 = 64$$ • Change in pressure

The **slope** of a line is the change in the vertical direction caused by one unit of change in the horizontal direction. For $P(d) = 64d + 2100$, the slope is 64. In the context of the problem, the slope means that the pressure on a diver increases by 64 pounds per square foot for each additional foot the diver descends. Note that the slope (64) has the same value as the coefficient of d in $P(d) = 64d + 2100$. This connection between the slope and the coefficient of the variable in a linear function always holds.

> **Slope**
>
> For a linear function given by $f(x) = mx + b$, the slope of the graph of the function is m, the coefficient of the variable.

QUESTION *What is the slope of each of the following?*

a. $y = -2x + 3$ **b.** $f(x) = x + 4$ **c.** $g(x) = 3 - 4x$

d. $y = \dfrac{1}{2}x - 5$

The slope of a line can be calculated by using the coordinates of any two distinct points on the line and the following formula.

> **Slope of a Line**
>
> Let (x_1, y_1) and (x_2, y_2) be two points on a nonvertical line. Then the **slope** of the line through the two points is the ratio of the change in the y-coordinates to the change in the x-coordinates.
>
> $$m = \frac{\text{change in } y}{\text{change in } x} = \frac{y_2 - y_1}{x_2 - x_1}, x_1 \neq x_2$$

An easy way to remember the definition of slope is "rise over run." The change in y is the vertical distance (rise) between the points, and the change in x is the horizontal distance (run) between the points.

QUESTION *Why is the restriction $x_1 \neq x_2$ required in the definition of slope?*

EXAMPLE 3 ■ **Find the Slope of a Line Between Two Points**

Find the slope of the line between the two points.

a. $(-4, -3)$ and $(-1, 1)$ **b.** $(-2, 3)$ and $(1, -3)$
c. $(-1, -3)$ and $(4, -3)$ **d.** $(4, 3)$ and $(4, -1)$

ANSWER   *a.* -2 *b.* 1 *c.* -4 *d.* $\dfrac{1}{2}$

ANSWER   *If $x_1 = x_2$, then the difference $x_2 - x_1 = 0$. This would make the denominator 0, and division by 0 is not defined.*

Solution

a. $(x_1, y_1) = (-4, -3), (x_2, y_2) = (-1, 1)$

$$m = \frac{y_2 - y_1}{x_2 - x_1} = \frac{1 - (-3)}{-1 - (-4)} = \frac{4}{3}$$

The slope is $\frac{4}{3}$. A *positive* slope indicates that the line slopes *upward* to the right. For this particular line, the value of *y increases* by $\frac{4}{3}$ when *x* increases by 1.

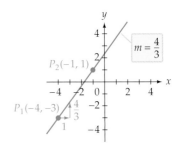

TAKE NOTE

When we talk about *y*-values increasing (as in part a) or decreasing (as in part b), we always mean as we move from left to right.

b. $(x_1, y_1) = (-2, 3), (x_2, y_2) = (1, -3)$

$$m = \frac{y_2 - y_1}{x_2 - x_1} = \frac{-3 - 3}{1 - (-2)} = \frac{-6}{3} = -2$$

The slope is -2. A *negative* slope indicates that the line slopes *downward* to the right. For this particular line, the value of *y decreases* by 2 when *x* increases by 1.

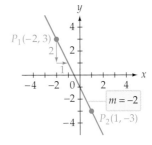

c. $(x_1, y_1) = (-1, -3), (x_2, y_2) = (4, -3)$

$$m = \frac{y_2 - y_1}{x_2 - x_1} = \frac{-3 - (-3)}{4 - (-1)} = \frac{0}{5} = 0$$

The slope is 0. A *zero* slope indicates that the line is *horizontal*. For a horizontal line, the value of *y stays the same* when *x* increases by any amount.

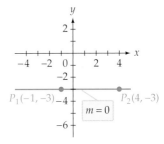

TAKE NOTE

A horizontal line has zero slope. A line that has no slope, or whose slope is undefined, is a vertical line. Note that when $y_1 = y_2$ in the formula for slope, the slope of the line through the two points is zero. When $x_1 = x_2$, the slope is undefined.

d. $(x_1, y_1) = (4, 3), (x_2, y_2) = (4, -1)$

$$m = \frac{y_2 - y_1}{x_2 - x_1} = \frac{-1 - 3}{4 - 4} = \frac{-4}{0} \qquad \text{Division by 0 is undefined.}$$

If the denominator of the slope formula is zero, the line has *no slope*. Sometimes we say that the slope of a vertical line is *undefined*.

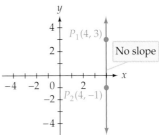

CHECK YOUR PROGRESS 3   Find the slope of the line between the two points.

a. $(-6, 5)$ and $(4, -5)$ **b.** $(-5, 0)$ and $(-5, 7)$

c. $(-7, -2)$ and $(8, 8)$ **d.** $(-6, 7)$ and $(1, 7)$

Solution   See page S20.

Suppose a jogger is running at a constant velocity of 6 miles per hour. Then the linear function $d = 6t$ relates the distance traveled d, in miles, to the time t, in hours, spent running. A table of values is shown below.

Time, t, in hours	0	0.5	1	1.5	2	2.5
Distance, d, in miles	0	3	6	9	12	15

✔ **TAKE NOTE**

Whether we write $f(t) = 6t$ or $d = 6t$, the equation represents a linear function. $f(t)$ and d are different symbols for the same quantity.

Because the equation $d = 6t$ represents a linear function, the slope of the graph of the equation is 6. This can be confirmed by choosing any two points on the graph shown below and finding the slope of the line between the two points. The points (0.5, 3) and (2, 12) are used here.

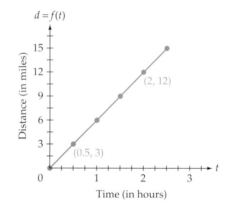

$$m = \frac{\text{change in } d}{\text{change in } t} = \frac{12 \text{ miles} - 3 \text{ miles}}{2 \text{ hours} - 0.5 \text{ hours}} = \frac{9 \text{ miles}}{1.5 \text{ hours}} = 6 \text{ miles per hour}$$

This example demonstrates that the slope of the graph of an object in uniform motion is the same as the velocity of the object. In a more general way, any time we discuss the constant velocity of an object, we are discussing the slope of the graph that describes the relationship between the distance the object travels and the time it travels.

EXAMPLE 4 ▪ **Application of the Slope of a Linear Function**

The function $T(x) = -6.5x + 20$ approximates the temperature $T(x)$, in degrees Celsius, at x kilometers above sea level. What is the slope of this function? Write a sentence that explains the meaning of the slope in the context of this application.

Solution
For the linear function $T(x) = -6.5x + 20$, the slope is the coefficient of x. Therefore, the slope is -6.5. The slope means that the temperature is decreasing (because the slope is negative) 6.5°C for each 1-kilometer increase in height above sea level.

CHECK YOUR PROGRESS 4 The distance that a homing pigeon can fly can be approximated by $d(t) = 50t$, where $d(t)$ is the distance, in miles, flown by the pigeon in t hours. Find the slope of this function. What is the meaning of the slope in the context of the problem?

Solution See page S21.

Math Matters   Galileo Galilei

Galileo Galilei (găl-ə′lā-ē) (1564–1642) was one of the most influential scientists of his time. In addition to inventing the telescope, with which he discovered the moons of Jupiter, Galileo successfully argued that Aristotle's assertion that heavy objects drop at a greater velocity than lighter ones was incorrect. According to legend, Galileo went to the top of the Leaning Tower of Pisa and dropped two balls at the same time, one weighing twice the other. His assistant, standing on the ground, observed that both balls reached the ground at the same time.

There is no historical evidence that Galileo actually performed this experiment, but he did do something similar. Galileo correctly reasoned that if Aristotle's assertion were true, then balls of different weights should roll down a ramp at different speeds. Galileo did carry out this experiment and was able to show that, in fact, balls of different weights reached the end of the ramp at the same time. Galileo was not able to determine why this happened, and it took Isaac Newton, born the same year that Galileo died, to formulate the first theory of gravity.

Leaning Tower of Pisa

Slope–Intercept Form of a Straight Line

The value of the slope of a line gives the change in y for a *1-unit* change in x. For instance, a slope of -3 means that y changes by -3 as x changes by 1; a slope of $\frac{4}{3}$ means that y changes by $\frac{4}{3}$ as x changes by 1. Because it is difficult to graph a change of $\frac{4}{3}$, it is easier to think of a fractional slope in terms of integer changes in x and y. As shown at the right, for a slope of $\frac{4}{3}$ we have

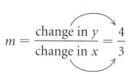

$$m = \frac{\text{change in } y}{\text{change in } x} = \frac{4}{3}$$

That is, for a slope of $\frac{4}{3}$, y changes by 4 as x changes by 3.

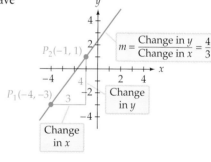

EXAMPLE 5 ■ Graph a Line Given a Point on the Line and the Slope

Draw the line that passes through $(-2, 4)$ and has slope $-\frac{3}{4}$.

Solution

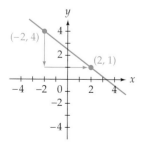

Place a dot at $(-2, 4)$ and then rewrite $-\frac{3}{4}$ as $\frac{-3}{4}$. Starting from $(-2, 4)$, move 3 units down (the change in y) and then 4 units to the right (the change in x). Place a dot at that location and then draw a line through the two points. See the graph at the left.

CHECK YOUR PROGRESS 5   Draw the line that passes through $(2, 4)$ and has slope -1.

Solution   See page S21.

Because the slope and y-intercept can be determined directly from the equation $f(x) = mx + b$, this equation is called the *slope–intercept form* of a straight line.

> **Slope–Intercept Form of the Equation of a Line**
>
> The graph of $f(x) = mx + b$ is a straight line with slope m and y-intercept $(0, b)$.

When a function is written in this form, it is possible to create a quick graph of the function.

EXAMPLE 6 ■ **Graph a Linear Function Using the Slope and y-Intercept**

Graph $f(x) = -\frac{2}{3}x + 4$ by using the slope and y-intercept.

Solution
From the equation, the slope is $-\frac{2}{3}$ and the y-intercept is $(0, 4)$. Place a dot at the y-intercept. We can write the slope as $m = -\frac{2}{3} = \frac{-2}{3}$. Starting from the y-intercept, move 2 units down and 3 units to the right and place another dot. Now draw a line through the two points.

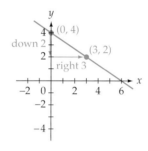

CHECK YOUR PROGRESS 6   Graph $y = \frac{3}{4}x - 5$ by using the slope and y-intercept.

Solution   *See page S21.*

Investigation

Negative Velocity

We can expand the concept of velocity to include negative velocity. Suppose a car travels in a straight line starting at a given point. If the car is moving to the right, then we say that its velocity is positive. If the car is moving to the left, then we say that its velocity is negative. For instance, a velocity of −45 miles per hour means the car is moving to the left at 45 miles per hour.

If we were to graph the motion of an object on a time–distance graph, a positive velocity would be indicated by a positive slope; a negative velocity would be indicated by a negative slope.

(continued)

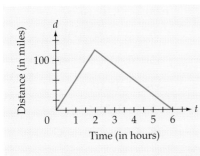

Investigation Exercises

The graph at the left represents a car traveling on a straight road. Answer the following questions on the basis of this graph.

   1. Between what two times is the car moving to the right?

   2. Between what two times does the car have a positive velocity?

   3. Between what two times is the car moving to the left?

   4. Between what two times does the car have a negative velocity?

   5. After 2 hours, how far is the car from its starting position?

   6. How long after the car leaves its starting position does it return to its starting position?

   7. What is the velocity of the car during its first 2 hours of travel?

   8. What is the velocity of the car during its last 4 hours of travel?

The graph below represents another car traveling on a straight road, but this car's motion is a little more complicated. Use this graph for the questions below.

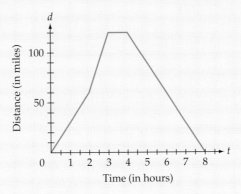

   9. What is the slope of the line between hours 3 and 4?

   10. What is the velocity of the car between hours 3 and 4? Is the car moving?

   11. During which of the following intervals of time is the absolute value of the velocity greatest: 0 to 2 hours, 2 to 3 hours, 3 to 4 hours, or 4 to 8 hours? (Recall that the absolute value of a real number a is the distance between a and 0 on the real number line.)

Exercise Set 5.2

In Exercises 1–14, find the x- and y-intercepts of the graph of the equation.

   1. $f(x) = 3x - 6$

   2. $f(x) = 2x + 8$

   3. $y = \dfrac{2}{3}x - 4$

   4. $y = -\dfrac{3}{4}x + 6$

   5. $y = -x - 4$

   6. $y = -\dfrac{x}{2} + 1$

   7. $3x + 4y = 12$

   8. $5x - 2y = 10$

   9. $2x - 3y = 9$

   10. $4x + 3y = 8$

   11. $\dfrac{x}{2} + \dfrac{y}{3} = 1$

   12. $\dfrac{x}{3} - \dfrac{y}{2} = 1$

   13. $x - \dfrac{y}{2} = 1$

   14. $-\dfrac{x}{4} + \dfrac{y}{3} = 1$

15. **Crickets** There is a relationship between the number of times a cricket chirps per minute and the air temperature. A linear model of this relationship is given by

$$f(x) = 7x - 30$$

where x is the temperature in degrees Celsius and $f(x)$ is the number of chirps per minute. Find and discuss the meaning of the x-intercept in the context of this application.

16. **Travel** An approximate linear model that gives the remaining distance, in miles, a plane must travel from Los Angeles to Paris is given by

$$s(t) = 6000 - 500t$$

where $s(t)$ is the remaining distance t hours after the flight begins. Find and discuss the meaning, in the context of this application, of the intercepts on the vertical and horizontal axes.

17. **Refrigeration** The temperature of an object taken from a freezer gradually rises and can be modeled by

$$T(x) = 3x - 15$$

where $T(x)$ is the Fahrenheit temperature of the object x minutes after being removed from the freezer. Find and discuss the meaning, in the context of this application, of the intercepts on the vertical and horizontal axes.

18. **Retirement Account** A retired biologist begins withdrawing money from a retirement account according to the linear model

$$A(t) = 100,000 - 2500t$$

where $A(t)$ is the amount, in dollars, remaining in the account t months after withdrawals begin. Find and discuss the meaning, in the context of this application, of the intercepts on the vertical and horizontal axes.

19. Which function has a graph that intersects the y-axis at a higher location, $f(x) = -3x + 7$ or $g(x) = 4x - 2$?

20. What do the graphs of $y = -\dfrac{2}{3}x + 4$ and $y = \dfrac{1}{5}x + 4$ have in common? Where do they intersect?

In Exercises 21–38, find the slope of the line containing the two points.

21. $(1, 3), (3, 1)$

22. $(2, 3), (5, 1)$

23. $(-1, 4), (2, 5)$

24. $(3, -2), (1, 4)$

25. $(-1, 3), (-4, 5)$

26. $(-1, -2), (-3, 2)$

27. $(0, 3), (4, 0)$

28. $(-2, 0), (0, 3)$

29. $(2, 4), (2, -2)$

30. $(4, 1), (4, -3)$

31. $(2, 5), (-3, -2)$

32. $(4, 1), (-1, -2)$

33. $(2, 3), (-1, 3)$

34. $(3, 4), (0, 4)$

35. $(0, 4), (-2, 5)$

36. $(-2, 3), (-2, 5)$

37. $(-3, -1), (-3, 4)$

38. $(-2, -5), (-4, -1)$

39. A linear function has x-intercept $(-5, 0)$ and y-intercept $(0, 2)$. Is the slope of the graph positive or negative?

40. The slope of a line is 0 and the line passes through the point $(4, 3)$. Find another point on the line.

41. **Travel** The graph below shows the relationship between the distance traveled by a motorist and the time of travel. Find the slope of the line between the two points shown on the graph. Write a sentence that states the meaning of the slope in the context of this application.

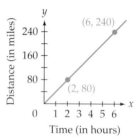

42. **Depreciation** The graph below shows the relationship between the value of a building and the depreciation allowed for income tax purposes. Find the slope of the line between the two points shown on the graph. Write a sentence that states the meaning of the slope in the context of this application.

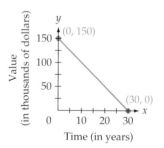

43. **Income Tax** The following graph shows the relationship between the amount of tax and the amount of taxable income between $29,050 and $70,350. Find the slope of the line between the two points shown

on the graph. Write a sentence that states the meaning of the slope in the context of this application.

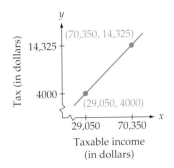

44. 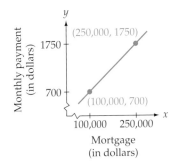 **Mortgages** The graph below shows the relationship between the monthly payment on a mortgage and the amount of the mortgage. Find the slope of the line between the two points shown on the graph. Write a sentence that states the meaning of the slope in the context of this application.

45. 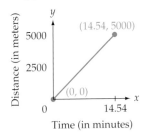 **Foot Races** The graph below shows the relationship between distance and time for the 5000-meter run for the world record by Deena Drossin in 2002. (Assume Drossin ran the race at a constant rate.) Find the slope of the line between the two points shown on the graph. Round to the nearest tenth. Write a sentence that states the meaning of the slope in the context of this application.

46. 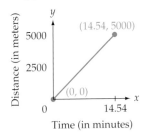 **Foot Races** The following graph shows the relationship between distance and time for the 10,000-meter run for the world record by Sammy Kipketer in 2002. (Assume Kipketer ran the race at a constant rate.) Find the slope of the line between

the two points shown on the graph. Round to the nearest tenth. Write a sentence that states the meaning of the slope in the context of this application.

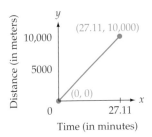

47. Graph the line that passes through the point $(-1, -3)$ and has slope $\frac{4}{3}$.

48. Graph the line that passes through the point $(-2, -3)$ and has slope $\frac{5}{4}$.

49. Graph the line that passes through the point $(-3, 0)$ and has slope -3.

50. Graph the line that passes through the point $(2, 0)$ and has slope -1.

In Exercises 51–56, graph each function using the slope and *y*-intercept.

51. $f(x) = \frac{1}{2}x + 2$

52. $f(x) = \frac{2}{3}x - 3$

53. $f(x) = -\frac{3}{2}x$

54. $f(x) = \frac{3}{4}x$

55. $f(x) = \frac{1}{3}x - 1$

56. $f(x) = -\frac{3}{2}x + 6$

57. Which equation has the steeper graph, $y = \frac{1}{4}x + 483$ or $y = \frac{2}{3}x - 216$?

58. If $g(r)$ is a linear function with negative slope, which is larger, $g(2)$ or $g(4)$?

59. If $f(x)$ is a linear function with slope $-\frac{1}{2}$, and $f(6) = 3$, what is the value of $f(7)$?

60. If $h(x)$ is a linear function with slope 2, and $h(7) = 26$, what is the value of $h(6)$?

Extensions

CRITICAL THINKING

61. **Jogging** Lois and Tanya start from the same place on a straight jogging course, at the same time, and jog in the same direction. Lois is jogging at 9 kilometers per

hour and Tanya is jogging at 6 kilometers per hour. The graph shows the distance each jogger has traveled in *x* hours and the distance between the joggers after *x* hours. Which line represents the distance Lois has traveled in *x* hours? Which line represents the distance Tanya has traveled in *x* hours? Which line represents the distance between Lois and Tanya after *x* hours?

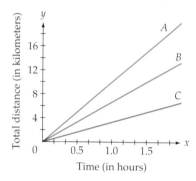

62. **Chemistry** A chemist is filling two cylindrical cans from a faucet that releases water at a constant rate. Can 1 has a diameter of 20 millimeters and can 2 has a diameter of 30 millimeters.

   a. In the following graph, which line represents the depth of the water in can 1 after *x* seconds?

   b. Use the graph to estimate the difference in the depths of the water in the two cans after 15 seconds.

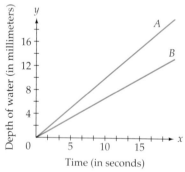

63. **ANSI** The American National Standards Institute (ANSI) states that the slope of a wheelchair ramp must not exceed $\frac{1}{12}$.

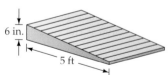

   Does the ramp pictured above meet the requirements of ANSI?

64. **ANSI** A ramp for a wheelchair must be 14 inches high. What is the minimum length of this ramp so that it meets the ANSI requirements stated in Exercise 63?

65. If $(2, 3)$ are the coordinates of a point on a line that has slope 2, what is the *y*-coordinate of the point on the line at which $x = 4$?

66. If $(-1, 2)$ are the coordinates of a point on a line that has slope -3, what is the *y*-coordinate of the point on the line at which $x = 1$?

67. If $(1, 4)$ are the coordinates of a point on a line that has slope $\frac{2}{3}$, what is the *y*-coordinate of the point on the line at which $x = -2$?

68. If $(-2, -1)$ are the coordinates of a point on a line that has slope $\frac{3}{2}$, what is the *y*-coordinate of the point on the line at which $x = -6$?

69. What effect does increasing the coefficient of *x* have on the graph of $y = mx + b$?

70. What effect does decreasing the coefficient of *x* have on the graph of $y = mx + b$?

71. What effect does increasing the constant term have on the graph of $y = mx + b$?

72. What effect does decreasing the constant term have on the graph of $y = mx + b$?

73. Do the graphs of all straight lines have *y*-intercepts? If not, give an example of one that does not.

74. If two lines have the same slope and the same *y*-intercept, must the graphs of the lines be the same? If not, give an example.

EXPLORATIONS

75. **Construction** When you climb a staircase, the flat part of a stair that you step on is called the *tread* of the stair. The *riser* is the vertical part of the stair. The slope of a staircase is the ratio of the length of the riser to the length of the tread. Because the design of a staircase may affect safety, most cities have building codes that give rules for the design of a staircase.

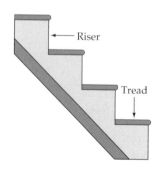

   a. The traditional design of a staircase calls for a 9-inch tread and an 8.25-inch riser. What is the slope of this staircase?

b. A newer design for a staircase uses an 11-inch tread and a 7-inch riser. What is the slope of this staircase?

c. ✏️ An architect is designing a house with a staircase that is 8 feet high and 12 feet long. Is the architect using the traditional design described in part a or the newer design described in part b? Explain your answer.

d. Staircases that have slopes between 0.5 and 0.7 are usually considered safer than those with slopes greater than 0.7. Design a safe staircase that goes from the first floor of a house to the second floor, which is 9 feet above the first floor.

e. Measure treads and risers for three staircases you encounter. Do these staircases match the traditional design described in part a or the newer design described in part b?

76. Geometry In the accompanying diagram, lines l_1 and l_2 are perpendicular with slopes m_1 and m_2, respectively, and $m_1 > 0$ and $m_2 < 0$. The horizontal line segment $\overline{AC}$ has length 1.

a. Show that the length of the vertical line segment $\overline{BC}$ is m_1.

b. Show that the length of $\overline{CD}$ is $-m_2$. Note that because m_2 is a negative number, $-m_2$ is a positive number.

c. Show that right triangles ACB and DCA are similar triangles. **Similar triangles** have the same shape; the corresponding angles are equal, and corresponding sides are in proportion. (*Suggestion:* Show that the measure of angle ADC equals the measure of angle BAC.)

d. Show that $\dfrac{m_1}{1} = \dfrac{1}{-m_2}$. Use the fact that the ratios of corresponding sides of similar triangles are equal.

e. Use the equation in part d to show that $m_1 m_2 = -1$.

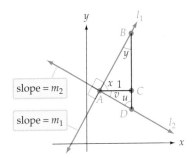

| # Finding Linear Models

Finding Linear Models

Suppose that a car uses 0.04 gallon of gas per mile driven and that the fuel tank, which holds 18 gallons of gas, is full. Using this information, we can determine a linear model for the amount of fuel remaining in the gas tank after driving x miles.

Recall that a linear function is one that can be written in the form $f(x) = mx + b$, where m is the slope of the line and b is the y-intercept. The slope is the rate at which the car is using fuel, 0.04 gallon per mile. Because the car is consuming the fuel, the amount of fuel in the tank is decreasing. Therefore, the slope is negative and we have $m = -0.04$.

The amount of fuel in the tank depends on the number of miles x the car has been driven. Before the car starts (that is, when $x = 0$), there are 18 gallons of gas in the tank. The y-intercept is $(0, 18)$.

Using this information, we can create the linear function.

> ✔️ **TAKE NOTE**
>
> When creating a linear model, the slope will be the quantity that is expressed using the word *per*. The car discussed at the right uses 0.04 gallon *per* mile. The slope is negative because the amount of fuel in the tank is decreasing.

$$f(x) = mx + b$$
$$f(x) = -0.04x + 18 \quad \text{• Replace } m \text{ by } -0.04 \text{ and } b \text{ by } 18.$$

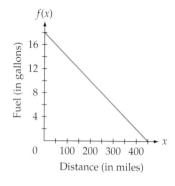

The linear function that models the amount of fuel remaining in the tank is given by $f(x) = -0.04x + 18$, where $f(x)$ is the amount of fuel, in gallons, remaining after driving x miles. The graph of the function is shown at the left.

The x-intercept of a graph is the point at which $f(x) = 0$. For this application, $f(x) = 0$ when there are 0 gallons of fuel remaining in the tank. Thus, replacing $f(x)$ by 0 in $f(x) = -0.04x + 18$ and solving for x will give the number of miles the car can be driven before running out of gas.

$$f(x) = -0.04x + 18$$
$$0 = -0.04x + 18 \qquad \text{• Replace } f(x) \text{ by 0.}$$
$$-18 = -0.04x$$
$$450 = x$$

The car can travel 450 miles before running out of gas.

Recall that the domain of a function is all possible values of x, and the range of a function is all possible values of $f(x)$. For the function $f(x) = -0.04x + 18$, which was used above to model the amount of fuel remaining in the gas tank of the car, the domain is only those values from 0 to 450, inclusive, because the fuel tank is empty when the car has traveled 450 miles (and the car cannot be driven a negative number of miles). The range consists of the values from 0 to 18, inclusive.

QUESTION   *Why does it not make sense for the domain of $f(x) = -0.04x + 18$ to exceed 450?*

EXAMPLE 1 ■ Application of Finding a Linear Model Given the Slope and *y*-Intercept

Suppose a 20-gallon gas tank contains 2 gallons when a motorist decides to fill up the tank. If the gas pump fills the tank at a rate of 0.1 gallon per second, find a linear function that models the amount of fuel in the tank t seconds after fueling begins.

Solution

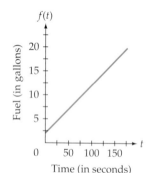

When fueling begins, at $t = 0$, there are 2 gallons of gas in the tank. Therefore, the y-intercept is $(0, 2)$. The slope is the rate at which fuel is being added to the tank. Because the amount of fuel in the tank is increasing, the slope is positive and we have $m = 0.1$. To find the linear function, replace m and b by their respective values.

$$f(t) = mt + b$$
$$f(t) = 0.1t + 2 \qquad \text{• Replace } m \text{ by 0.1 and } b \text{ by 2.}$$

The linear function is $f(t) = 0.1t + 2$, where $f(t)$ is the number of gallons of fuel in the tank t seconds after fueling begins.

ANSWER   *If $x > 450$, then $f(x) < 0$. This would mean that the tank has a negative amount of gas. For instance, $f(500) = -2$.*

CHECK YOUR PROGRESS 1   The boiling point of water at sea level is 100°C. The boiling point decreases 3.5°C per 1 kilometer increase in altitude. Find a linear function that gives the boiling point of water as a function of altitude.

Solution   *See page S21.*

For each of the previous linear models, the known point on the graph of the linear function was the y-intercept. This information enabled us to determine b for the linear function $f(x) = mx + b$. In some cases, a point other than the y-intercept is given. In such a case, the *point–slope formula* is used to find the equation of the line.

> **Point–Slope Formula of a Straight Line**
>
> Let (x_1, y_1) be a point on a line and let m be the slope of the line. Then the equation of the line can be found using the point–slope formula
>
> $$y - y_1 = m(x - x_1)$$

EXAMPLE 2 ■ **Find the Equation of a Line Given the Slope and a Point on the Line**

Find the equation of the line that passes through $(1, -3)$ and has slope -2.

Solution

$$y - y_1 = m(x - x_1)$$ • Use the point–slope formula.
$$y - (-3) = -2(x - 1)$$ • $m = -2$, $(x_1, y_1) = (1, -3)$
$$y + 3 = -2x + 2$$
$$y = -2x - 1$$

Note that we wrote the equation of the line as $y = -2x - 1$. We could also write the equation in function notation as $f(x) = -2x - 1$.

CHECK YOUR PROGRESS 2   Find the equation of the line that passes through $(-2, 2)$ and has slope $-\frac{1}{2}$.

Solution   *See page S21.*

EXAMPLE 3 ■ **Application of Finding a Linear Model Given a Point and the Slope**

Based on data from the *Kelley Blue Book*, the value of a certain car decreases approximately $250 per month. If the value of the car 2 years after it was purchased was $14,000, find a linear function that models the value of the car after x months of ownership. Use this function to find the value of the car after 3 years of ownership.

Solution

Let V represent the value of the car after x months. Then $V = 14{,}000$ when $x = 24$ (2 years is 24 months). A solution of the equation is $(24, 14{,}000)$. The car is decreasing in value at a rate of \$250 per month. Therefore, the slope is -250. Now use the point–slope formula to find the linear equation that models the function.

$$V - V_1 = m(x - x_1)$$
$$V - 14{,}000 = -250(x - 24) \qquad \bullet\ x_1 = 24,\ V_1 = 14{,}000,\ m = -250$$
$$V - 14{,}000 = -250x + 6000$$
$$V = -250x + 20{,}000$$

A linear function that models the value of the car after x months of ownership is $V(x) = -250x + 20{,}000$.

To find the value of the car after 3 years (36 months), evaluate the function when $x = 36$.

$$V(x) = -250x + 20{,}000$$
$$V(36) = -250(36) + 20{,}000 = 11{,}000$$

The value of the car is \$11,000 after 3 years of ownership.

CHECK YOUR PROGRESS 3 During a brisk walk, a person burns about 3.8 calories per minute. If a person has burned 191 calories in 50 minutes, determine a linear function that models the number of calories burned after t minutes.

Solution *See page S21.*

The next example shows how to find the equation of a line given two points on the line.

✔ **TAKE NOTE**

There are many ways to find the equation of a line. However, in every case, there must be enough information to determine a point on the line and to find the slope of the line. When you are doing problems of this type, look for different ways that information may be presented. For instance, in Example 4, even though the slope of the line is not given, knowing two points enables us to find the slope.

EXAMPLE 4 ■ **Find the Equation of a Line Given Two Points on the Line**

Find the equation of the line that passes through $P_1(6, -4)$ and $P_2(3, 2)$.

Solution

Find the slope of the line between the two points.

$$m = \frac{y_2 - y_1}{x_2 - x_1} = \frac{2 - (-4)}{3 - 6} = \frac{6}{-3} = -2$$

Use the point–slope formula to find the equation of the line.

$$y - y_1 = m(x - x_1)$$
$$y - (-4) = -2(x - 6) \qquad \bullet\ m = -2,\ x_1 = 6,\ y_1 = -4$$
$$y + 4 = -2x + 12$$
$$y = -2x + 8$$

CHECK YOUR PROGRESS 4 Find the equation of the line that passes through $P_1(-2, 3)$ and $P_2(4, 1)$.

Solution *See page S21.*

Math Matters   Perspective: Using Straight Lines in Art

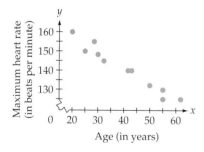

Many paintings we see today have a three-dimensional quality to them, even though they are painted on a flat surface. This was not always the case. It wasn't until the Renaissance that artists started to paint "in perspective." Using lines is one way to create this perspective. Here is a simple example.

Draw a dot, called the *vanishing point,* and a rectangle on a piece of paper. Draw windows as shown. To keep the perspective accurate, the lines through opposite corners of the windows should be parallel. A table in proper perspective is created in the same way.

This method of creating perspective was employed by Leonardo da Vinci. Use the Internet to find and print a copy of his painting *The Last Supper.* Using a ruler, see whether you can find the vanishing point by drawing two lines along the top edges of the tapestries on the sides of the painting.

Line Fitting

There are many situations in which a linear function can be used to approximate collected data. For instance, the table below shows the maximum exercise heart rates for specific individuals of various ages who exercise regularly.

Age, x, in years	20	25	30	32	43	55	28	42	50	55	62
Heart rate, y, in maximum beats per minute	160	150	148	145	140	130	155	140	132	125	125

The graph at the left, called a **scatter diagram,** is a graph of the ordered pairs of the table. These ordered pairs suggest that the maximum exercise heart rate for an individual decreases as the person's age increases.

Although these points do not lie on one line, it is possible to find a line that *approximately fits* the data. One way to do this is to select two data points and then find the equation of the line that passes through the two points. To do this, we first find the slope of the line between the two points and then use the point–slope formula to find the equation of the line. Suppose we choose $(20, 160)$ as P_1 and $(62, 125)$ as P_2. Then the slope of the line between P_1 and P_2 is

$$m = \frac{y_2 - y_1}{x_2 - x_1} = \frac{125 - 160}{62 - 20} = -\frac{35}{42} = -\frac{5}{6}$$

Now use the point–slope formula.

$$y - y_1 = m(x - x_1)$$

$$y - 160 = -\frac{5}{6}(x - 20) \qquad \bullet\ m = -\frac{5}{6}, x_1 = 20, y_1 = 160$$

$$y - 160 = -\frac{5}{6}x + \frac{50}{3} \qquad \bullet\ \text{Multiply by } -\frac{5}{6}.$$

$$y = -\frac{5}{6}x + \frac{530}{3} \qquad \bullet\ \text{Add 160 to each side of the equation.}$$

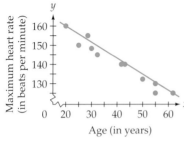

The graph of $y = -\frac{5}{6}x + \frac{530}{3}$

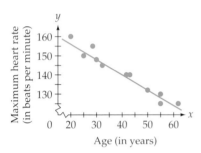

The graph of $y = -\frac{4}{5}x + 172$

The graph of $y = -\frac{5}{6}x + \frac{530}{3}$ is shown at the left. This line *approximates* the data and can be used to estimate maximum exercise heart rates for different ages. For example, an exercise physiologist could determine the recommended maximum exercise heart rate for a 28-year-old individual by replacing x in the equation by 28 and determining the value of y.

$$y = -\frac{5}{6}x + \frac{530}{3}$$

$$y = -\frac{5}{6}(28) + \frac{530}{3} \qquad \bullet \text{ Replace } x \text{ by 28.}$$

$$\approx 153.3$$

The maximum exercise heart rate recommended for a 28-year-old person is approximately 153 beats per minute.

If we had chosen different points to create the equation of a line, the result would have been a different equation. For instance, if we had used the points (30, 148) and (50, 132), we would have obtained the equation $y = -\frac{4}{5}x + 172$. Its graph is shown at the left.

Although this equation is different from the first equation we found, their graphs are similar, and they will return similar output values for the same input value. For instance, if we replace x by 28 in the second equation, we estimate that a 28-year-old person should have a maximum exercise heart rate of approximately 150 beats per minute.

If a set of data suggests a linear relationship, a general strategy for choosing a line that fits the data well is first to draw a scatter diagram. Imagine you are going to draw a line through the points. Choose two points from the scatter diagram that would be on or near this line. It is generally best not to choose two points that are close to each other. Use these two points to create the equation of the line. You may wish to graph your resulting equation to check the fit of the line visually.

EXAMPLE 5 ■ Fit a Linear Function to Data

The average prices per pound of white bread in U.S. cities, as recorded in August of various years, are listed in the table below.

Year	1997	1998	1999	2000	2001	2002	2003	2004	2005
Price per pound	0.872	0.869	0.884	0.923	0.991	1.012	0.996	0.996	1.060

(*Source:* Bureau of Labor Statistics)

a. Sketch a scatter diagram of the data.

b. Find a linear function that models the data.

c. Use this function to predict the price per pound of white bread in August of 2011.

Solution

a. We can simplify the data by using 0 for 1997, 1 for 1998, etc.

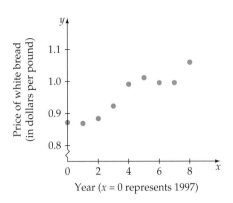

b. Looking at the scatter diagram, it appears that a line through the points $(3, 0.923)$ and $(6, 0.996)$ will fit the data points reasonably well.

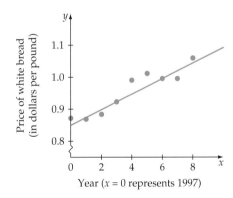

The slope of the line through these points is

$$m = \frac{y_2 - y_1}{x_2 - x_1} = \frac{0.996 - 0.923}{6 - 3} = \frac{0.073}{3}$$

and the equation of the line is

$$y - y_1 = m(x - x_1)$$

$$y - 0.923 = \frac{0.073}{3}(x - 3)$$

$$y - 0.923 = \frac{0.073}{3}x - 0.073$$

$$y = \frac{0.073}{3}x + 0.85$$

or approximately $y = 0.0243x - 0.85$. Thus the price per pound of white bread is approximately $f(x) = 0.0243x - 0.85$, where x is the number of years after August 1997.

c. The year 2011 corresponds to $x = 14$, so we evaluate the function when $x = 14$.

$$f(14) = 0.0243(14) + 0.85 = 1.1902$$

The estimated price of white bread in August 2011 is about $1.19 per pound.

CHECK YOUR PROGRESS 5   The populations of a city for various years are given in the table below.

Year	1992	1994	1996	1998	2000	2002	2004	2006
Population (thousands)	20.28	26.31	32.16	37.38	40.11	46.62	49.87	52.91

a. Sketch a scatter diagram of the data.

b. Write an equation for a linear function that models the data.

c. Use the function from part b to predict the city's population in the year 2025.

Solution   See page S21.

Investigation

A Linear Business Model

Two people decide to open a business reconditioning toner cartridges for copy machines. They rent a building for $7000 per year and estimate that building maintenance, taxes, and insurance will cost $6500 per year. Each person wants to make $12 per hour in the first year and will work 10 hours per day for 260 days of the year. Assume that it costs $28 to restore a cartridge and that the restored cartridge can be sold for $45.

Investigation Exercises

1. Write a linear function for the total cost C of operating the business and restoring n cartridges during the first year, not including the hourly wage the owners wish to earn.

2. Write a linear function for the total revenue R the business will earn during the first year by selling n cartridges.

3. How many cartridges must the business restore and sell annually to break even, not including the hourly wage the owners wish to earn?

4. How many cartridges must the business restore and sell annually for the owners to pay all expenses and earn the hourly wage they desire?

5. Suppose the entrepreneurs are successful in their business and are restoring and selling 25 cartridges each day of the 260 days they are open. What will be their hourly wage for the year if all the profit is shared equally?

6. As the company becomes successful and is selling and restoring 25 cartridges each day of the 260 days it is open, the entrepreneurs decide to hire a part-time employee 4 hours per day and to pay the employee $8 per hour. How many additional cartridges must be restored and sold each year just to cover the cost of the new employee? You can neglect employee costs such as social security, worker's compensation, and other benefits.

7. Suppose the company decides that it could increase its business by advertising. Answer Exercises 1, 2, 3, and 5 if the owners decide to spend $400 per month on advertising.

Exercise Set 5.3

In Exercises 1–8, find the equation of the line that passes through the given point and has the given slope.

1. $(0, 5)$, $m = 2$

2. $(2, 3)$, $m = \dfrac{1}{2}$

3. $(-1, 7)$, $m = -3$

4. $(0, 0)$, $m = \dfrac{1}{2}$

5. $(3, 5)$, $m = -\dfrac{2}{3}$

6. $(0, -3)$, $m = -1$

7. $(-2, -3)$, $m = 0$

8. $(4, -5)$, $m = -2$

In Exercises 9–16, find the equation of the line that passes through the given points.

9. $(0, 2)$, $(3, 5)$

10. $(0, -3)$, $(-4, 5)$

11. $(0, 3)$, $(2, 0)$

12. $(-2, -3)$, $(-1, -2)$

13. $(2, 0)$, $(0, -1)$

14. $(3, -4)$, $(-2, -4)$

15. $(-2, 5)$, $(2, -5)$

16. $(2, 1)$, $(-2, -3)$

17. A retired employee is paid a fixed monthly pension by her company. If $f(x) = mx + b$ is a linear function that gives the total amount of pension money the former employee has received x months after retiring, is m positive or negative?

18. A homeowner is draining his swimming pool. If $f(x) = mx + b$ is a linear function that models the amount of water remaining in the pool after x hours, is m positive or negative?

19. Hotel Industry The operator of a hotel estimates that 500 rooms per night will be rented if the room rate per night is $75. For each $10 increase in the price of a room, six fewer rooms per night will be rented. Determine a linear function that predicts the number of rooms that will be rented per night for a given price per room. Use this model to predict the number of rooms that will be rented if the room rate is $100 per night.

20. Construction A general building contractor estimates that the cost of building a new home is $30,000 plus $85 for each square foot of floor space in the house. Determine a linear function that gives the cost of building a house that contains x square feet of floor space. Use this model to determine the cost of building a house that contains 1800 square feet of floor space.

21. Travel A plane travels 830 miles in 2 hours. Determine a linear model that predicts the number of miles the plane can travel in a given interval of time. Use this model to predict the distance the plane will travel in $4\dfrac{1}{2}$ hours.

22. Compensation An account executive receives a base salary plus a commission. On $20,000 in monthly sales, an account executive would receive compensation of $1800. On $50,000 in monthly sales, an account executive would receive compensation of $3000. Determine a linear function that yields the compensation of an account executive for x dollars in monthly sales. Use this model to determine the compensation of an account executive who has $85,000 in monthly sales.

23. Car Sales A manufacturer of economy cars has determined that 50,000 cars per month can be sold at a price of $9000 per car. At a price of $8750, the number of cars sold per month would increase to 55,000. Determine a linear function that predicts the number of cars per month that will be sold at a price of x dollars. Use this model to predict the number of cars that will be sold per month at a price of $8500.

24. Calculator Sales A manufacturer of graphing calculators has determined that 10,000 calculators per week will be sold at a price of $95 per calculator. At a price of $90, it is estimated that 12,000 calculators will be sold. Determine a linear function that predicts the number of calculators that will be sold per week at a price of x dollars. Use this model to predict the number of calculators that will be sold per week at a price of $75.

25. A researcher would like to model the data shown in the scatter diagram below with a linear function. Which pair of points would be better to use to write a linear equation, the blue or the green?

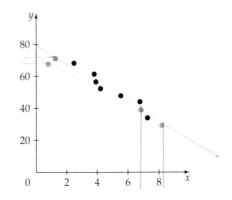

26. A biologist has recorded the growth of an animal population on the scatter diagram at the right. The horizontal axis represents the number of weeks she has been observing the animals. If she creates a linear model and predicts the population at the 80th week, which is a more likely result, 350 animals or 450 animals?

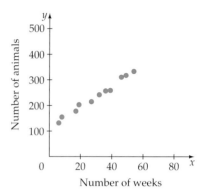

27. Stress A research hospital did a study on the relationship between stress and diastolic blood pressure. The results from eight patients in the study are given in the table below. Blood pressure values are measured in millimeters of mercury.

Stress test score, x	55	62	58	78	92	88	75	80
Blood pressure, y	70	85	72	85	96	90	82	85

a. Draw a scatter diagram of the data. Then use the points $(58, 72)$ and $(78, 85)$ to write the equation of a line that fits the points.

b. Use the equation to estimate the diastolic blood pressure of a person whose stress test score was 85. Round to the nearest tenth.

28. Hourly Wages The average hourly earnings, in dollars, of nonfarm workers in the United States for the years 1996 to 2003 are given in the table below. (*Source:* Bureau of Labor Statistics)

Year, x	1996	1997	1998	1999	2000	2001	2002	2003
Hourly wage, y	12.03	12.49	13.00	13.47	14.00	14.53	14.95	15.35

a. Draw a scatter diagram of the data. Then use the points corresponding to 1996 and 2000 to write the equation of a line that fits the points.

b. Use the equation to estimate the expected average hourly wage, rounded to to the nearest cent, in 2015.

29. High School Graduates The table below shows the numbers of students, in thousands, who have graduated from high school for the years 1996 to 2005. (*Source:* U.S. Bureau of Labor Statistics)

Year, x	1996	1997	1998	1999	2000	2001	2002	2003	2004	2005
Number of students (in thousands), y	2540	2633	2740	2786	2820	2837	2886	2929	2935	2944

a. Using $x = 0$ to correspond to 1996, draw a scatter diagram of the data and write the equation of a linear function that fits the data.

b. Use the model to estimate the expected number of high school graduates in 2010.

30. **Fuel Economy** An automotive engineer studied the relationship between the speed of a car and the number of miles traveled per gallon of fuel consumed at that speed. The results of the study are shown in the table below.

Speed (in miles per hour), x	40	25	30	50	60	80	55	35	45
Fuel Economy (in miles per gallon), y	26	27	28	24	22	21	23	27	25

   a. Draw a scatter diagram of the data and write the equation of a linear function that fits the data.

   b. Use the model to estimate the expected mileage, in miles per gallon, for a car traveling at 65 miles per hour. Round to the nearest mile per hour.

31. **Meteorology** A meteorologist studied the maximum temperatures at various latitudes for January of a certain year. The results of the study are shown in the table below.

Latitude (in °N), x	22	30	36	42	56	51	48
Maximum temperature (in °F), y	80	65	47	54	21	44	52

   a. Draw a scatter diagram of the data and write the equation of a linear function that fits the data.

   b. Use the model to estimate the expected maximum temperature in January at a latitude of 45°N. Round to the nearest degree.

32. **Zoology** A zoologist studied the running speeds of animals in terms of the animals' body lengths. The results of the study are shown in the table below.

Body length (in centimeters), x	1	9	15	16	24	25	60
Running speed (in meters per second), y	1	2.5	7.5	5	7.4	7.6	20

   a. Draw a scatter diagram of the data and write the equation of a linear function that fits the data.

   b. Use the model to estimate the expected running speed of a deer mouse, whose body length is 10 centimeters. Round to the nearest tenth of a meter per second.

Extensions

CRITICAL THINKING

33. A line passes through the points $(4, -1)$ and $(2, 1)$. Find the coordinates of three other points on the line.

34. If f is a linear function for which $f(1) = 3$ and $f(-1) = 5$, find $f(4)$.

35. The ordered pairs $(0, 1)$, $(4, 9)$, and $(3, n)$ are solutions of the same linear equation. Find n.

36. The ordered pairs $(2, 2)$, $(-1, 5)$, and $(3, n)$ are solutions of the same linear equation. Find n.

37. Is there a linear function whose graph passes through the points $(2, 4)$, $(-1, -5)$, and $(0, 2)$? If so, find the function and explain why there is such a function. If not, explain why there is no such function.

38. Is there a linear function whose graph passes through the points $(5, 1)$, $(4, 2)$, and $(0, 6)$? If so, find the function and explain why there is such a function. If not, explain why there is no such function.

39. Travel Assume that the maximum speed your car will travel varies linearly with the steepness of the hill it is climbing or descending. If the hill is 5° up, your car can travel 77 kilometers per hour. If the hill is 2° down ($-2°$), your car can travel 154 kilometers per hour. When your car's top speed is 99 kilometers per hour, how steep is the hill? State your answer in degrees, and note whether the car is climbing or descending.

EXPLORATIONS

40. Boating A person who can row at a rate of 3 miles per hour in calm water is trying to cross a river in which a current of 4 miles per hour runs perpendicular to the direction of rowing. See the figure at the right.

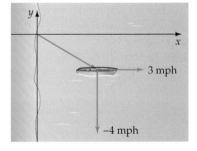

Because of the current, the boat is being pushed downstream at the same time that it is moving across the river. Because the boat is traveling at 3 miles per hour in the x direction, its horizontal position after t hours is given by $x = 3t$. The current is pushing the boat in the negative y direction at 4 miles per hour. Therefore, the boat's vertical position after t hours is given by $y = -4t$, where -4 indicates that the boat is moving downstream. The set of equations $x = 3t$ and $y = -4t$ are called **parametric equations,** and t is called the **parameter.**

a. What is the location of the boat after 15 minutes (0.25 hour)?

b. If the river is 1 mile wide, how far down the river will the boat be when it reaches the other shore? *Hint:* Find the time it takes the boat to cross the river by solving $x = 3t$ for t when $x = 1$. Then replace t by this value in $y = -4t$ and simplify.

c. For the parametric equations $x = 3t$ and $y = -4t$, write y in terms of x by solving $x = 3t$ for t and then substituting this expression into $y = -4t$.

41. Aviation In the diagram at the right, a plane flying at 5000 feet above sea level begins a gradual ascent.

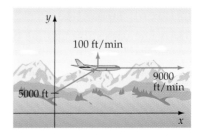

a. Determine parametric equations for the path of the plane. *Hint:* See Exercise 40.

b. What is the altitude of the plane 5 minutes after it begins its ascent?

c. What is the altitude of the plane after it has traveled 12,000 feet in the positive x direction?

| # Linear Regression and Correlation

Linear Regression

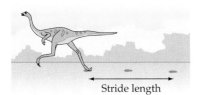

Stride length

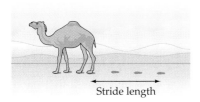

Stride length

In the previous section, we formed linear functions to model sets of data. The goal was to create the equation of a line that fit a scatter diagram of the data well, even though it is often not possible to fit the data exactly. We accomplished this task by carefully choosing two of the data points through which the line would pass. By computing slope and using the point–slope formula, we were able to write the desired linear equation.

This technique will often produce a very good model, but it has limitations. For instance, in many cases the best-fitting line is one that does not actually include any of the given data points. Of all the lines that could be chosen, statisticians and other scientists generally pick the *least-squares line*, which is also called the *regression line*. The **regression line** is the line for which the sum of the squares of the vertical distances between the data points and the line is a minimum. The process of finding the regression line is called **linear regression.**

As an illustration, the data in the table below show the results of an experiment comparing speed with length of stride for several dogs. (The stride length for an animal is the distance from a particular point on a footprint to that same point on the next footprint of the *same* foot.)

Stride length (meters)	1.5	1.7	2.0	2.4	2.7	3.0	3.2	3.5
Speed (meters per second)	3.7	4.4	4.8	7.1	7.7	9.1	8.8	9.9

A scatter diagram of the data, with stride length on the horizontal axis and speed on the vertical axis, is shown at the left. The scatter diagram suggests that a greater stride length produces a faster speed, and it appears that a linear model is appropriate. Imagine using a ruler to draw many different lines that fit the points in the scatter diagram. For each line drawn, measure the vertical distance between each point and the line. Square each of these distances, and find the total. If we were able to compare these totals for all possible lines, we would choose the one with the smallest sum. In this case, the line is approximately $y = 3.212x - 1.092$. This line is graphed on the scatter diagram below, and the directed vertical distances between each data point and the line are indicated.

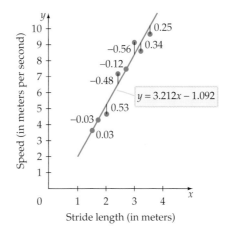

It would be a tedious process to find the equation of the regression line by hand. Fortunately, most graphing calculators have built-in capabilities to quickly determine the equation of the regression line once data have been entered. Here, using a TI-83/TI-84 Plus graphing calculator, we will show how to find the regression line for a set of data.

The table below shows the data collected by a chemistry student who is trying to determine a relationship between the temperature, in degrees Celsius, and volume, in liters, of 1 gram of oxygen at a constant pressure. Chemists refer to this relationship as Charles's Law.

Temperature, *T*, in degrees Celsius	−100	−75	−50	−25	0	25	50
Volume, *V*, in liters	0.43	0.5	0.57	0.62	0.7	0.75	0.81

To find the equation of the regression line for these data, we will first enter the data and create a scatter diagram. Press the $\boxed{\text{STAT}}$ key and then press $\boxed{\text{ENTER}}$ to select EDIT from the menu. This will bring up a table into which you can enter data. The first column should be labeled L1 and the second L2. (If your columns show different labels at the top, press $\boxed{\text{STAT}}$, select SetUpEditor from the menu, and press $\boxed{\text{ENTER}}$.) We will use L1 for the independent variable (temperature) and L2 for the dependent variable (volume). Enter the first number, −100, and press $\boxed{\text{ENTER}}$. The cursor will move down to the next entry in the column. When you have finished entering all the temperature values into the first column, arrow right and enter all the volume values into the second column. Screens from a TI-83/TI-84 Plus are shown at the left.

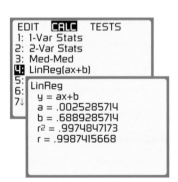

To view a scatter diagram on your calculator, press 〔STATPLOT〕 ($\boxed{\text{2ND}}$ $\boxed{\text{Y=}}$) and press $\boxed{\text{ENTER}}$ to select Plot1 from the menu. Your screen should look like the one at the left. Press $\boxed{\text{ENTER}}$ to turn the plot on. Be sure the other settings on your calculator match those shown here. You will usually need to go to the $\boxed{\text{WINDOW}}$ settings and enter appropriate values for Xmin, Xmax, Ymin, and Ymax for the entered data, or you can zoom to fit the scatter diagram in the screen automatically by pressing $\boxed{\text{ZOOM}}$ and selecting ZoomStat, which is option number 9.

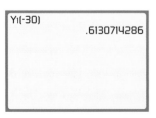

To find the linear regression equation, press the $\boxed{\text{STAT}}$ key again, highlight CALC, and arrow down to LinReg(ax+b). Press $\boxed{\text{ENTER}}$ to select this option, which will display LinReg(ax+b) on the home screen. Now you can just press $\boxed{\text{ENTER}}$ and the values for *a* and *b* will appear on the screen. Entering LinReg(ax+b)Y1 not only shows the results on the home screen, but also pastes the regression equation into Y1 in the Y= editor. This will enable you to graph or evaluate the regression equation easily.

For this set of data, the regression equation, with the coefficient and constant rounded to the nearest ten-millionth, is $V = 0.0025286T + 0.6889286$. To determine the volume of 1 gram of oxygen when the temperature is −30°C, replace *T* by −30 and evaluate the expression. This can be done using your calculator. Use the Y-VARS menu to place Y1 on the screen ($\boxed{\text{VARS}}$ ▷ $\boxed{\text{ENTER}}$ $\boxed{\text{ENTER}}$). Then enter the value −30, within parentheses, as shown at the left. After pressing $\boxed{\text{ENTER}}$, the volume will be displayed as approximately 0.61 liter.

✔ TAKE NOTE

If the regression equation from part a of Example 1 has been pasted into Y1, the answer to part b can be found by simply entering Y1(70).

EXAMPLE 1 ■ Find a Linear Regression Equation

Sodium thiosulfate is used by photographers to develop some types of film. The amount of this chemical that will dissolve in water depends on the temperature of the water. The table below gives the numbers of grams of sodium thiosulfate that will dissolve in 100 milliliters of water at various temperatures.

Temperature, x, in degrees Celsius	20	35	50	60	75	90	100
Sodium thiosulfate dissolved, y, in grams	50	80	120	145	175	205	230

a. Find the linear regression equation for these data.

b. How many grams of sodium thiosulfate does the model predict will dissolve in 100 milliliters of water when the temperature of the water is 70°C? Round to the nearest tenth of a gram.

Solution

a. Using a calculator, the regression equation is $y = 2.2517731x + 5.2482270$.

b. Evaluate the regression equation when $x = 70$.

$y = 2.2517731x + 5.2482270$

$= 2.2517731(70) + 5.2482270$ • Replace x by 70.

$= 162.872344$

Approximately 162.9 grams of sodium thiosulfate will dissolve when the temperature of the water is 70°C.

CHECK YOUR PROGRESS 1   The heights and weights of women swimmers on a college swim team are given in the table below.

Height, x, in inches	68	64	65	67	62	67	65
Weight, y, in pounds	132	108	108	125	102	130	105

a. Find the linear regression equation for these data.

b. Use your regression equation to estimate the weight of a woman swimmer who is 63 inches tall. Round to the nearest pound.

Solution   See page S21.

EXAMPLE 2 ■ Find a Linear Regression Equation

Earlier we compared speed and stride length for dogs. The table below lists data from a similar experiment performed with several adult men.

Stride length (meters)	2.5	3.0	3.3	3.5	3.8	4.0	4.2	4.5
Speed (meters per second)	3.4	4.9	5.5	6.6	7.0	7.7	8.3	8.7

a. Find the equation of the regression line for these data.

b. Use your regression equation to predict the average speeds of adult men with stride lengths of 2.8 meters and 4.8 meters. Round your results to the nearest tenth of a meter per second.

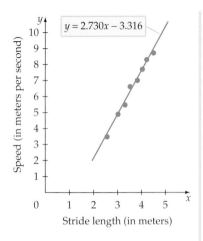

$y = 2.730x - 3.316$

Speed (in meters per second) vs. Stride length (in meters)

Solution

a. Using a calculator, the regression equation is approximately $y = 2.730x - 3.316$. The graph is shown at the left; you can see that the line fits the data well.

b. Evaluate the regression equation when $x = 2.8$.

$y = 2.730(2.8) - 3.316$ • Replace x by 2.8.

$\quad = 4.328$

Rounded to the nearest tenth, the predicted average speed for an adult man with a stride length of 2.8 meters is 4.3 meters per second.

Similarly, substituting 4.8 for x gives $y = 2.730(4.8) - 3.316 = 9.788$, so 9.8 meters per second is the predicted average speed for an adult man with a stride length of 4.8 meters.

CALCULATOR NOTE

Once you are finished analyzing a scatter diagram on the calculator, it is a good idea to turn the plot off so that it does not clutter the screen the next time you want to graph an equation. Access 〖STATPLOT〗 by pressing 2ND Y=. Press ENTER to select Plot1, arrow right to highlight Off, and press ENTER.

CHECK YOUR PROGRESS 2 The table below lists data from an experiment comparing speed and stride length for several camels.

Stride length (meters)	2.5	3.0	3.2	3.4	3.5	3.8	4.0	4.2
Speed (meters per second)	2.3	3.9	4.4	5.0	5.5	6.2	7.1	7.6

a. Find the equation of the regression line for these data.

b. Use your regression equation to predict the average speeds of camels with stride lengths of 2.7 meters and 4.5 meters. Round your results to the nearest tenth of a meter per second.

Solution   See page S22.

✔ TAKE NOTE

Sometimes values predicted by extrapolation are not reasonable. For instance, if we wish to predict the speed of a man with a stride length of $x = 20$ meters, the least-squares equation $y = 2.730x - 3.316$ gives us a speed of 50.7 meters per second. Because the maximum stride length of adult men is considerably less than 20 meters, we should not trust this prediction.

In Example 2, we first used the regression equation to estimate the speed corresponding to a stride length of 2.8 meters, a value that was between the given stride length values. This procedure is referred to as **interpolation.** When an equation is used to predict a value corresponding to a data point that is outside the given data values, as was the case with a stride length of 4.8 meters, the procedure is called **extrapolation.** See the figure below.

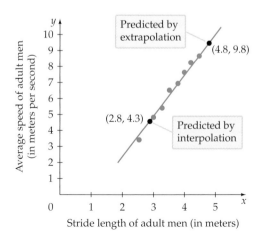

Math Matters   **Multiple Linear Regression**

Linear regression is a widely used tool in many fields. In some situations, however, more than one independent variable is needed for a thorough analysis. For example, a real estate appraiser weighs many factors when determining the value of a home, such as the size of the house and lot, the age and condition of the home, the location, and the quality of the neighborhood. An appraiser could collect data from a number of homes and assign a numeric value to each of several characteristics. These values can be considered as independent variables that determine the value of the home, the dependent variable. Because the data have more than one independent variable, we cannot draw a scatter diagram, but computer software can determine an equation that fits the data using a process called *multiple linear regression*. If four independent variables are used, x_1, x_2, x_3, x_4, the result is a linear equation (in several variables) of the form $y = ax_1 + bx_2 + cx_3 + dx_4 + e$. The appraiser could then use this equation in the future to determine the value of a property.

Linear Correlation Coefficient

A graphing calculator will find an equation of the regression line for almost any set of data, but that doesn't necessarily mean that a linear function is an accurate model for the data. How can we measure how good a "fit" the regression line is?

To determine the strength of a linear relationship between domain values and range values of a data set, statisticians use a statistic called the **linear correlation coefficient,** which is traditionally denoted by r.

If r is positive, the relationship between the domain and range values has a **positive correlation.** In this case, if the domain value increases, the range value also tends to increase. If r is negative, the linear relationship between the domain and range values has a **negative correlation.** In this case, if the domain value increases, the range value tends to decrease. The figure below shows some scatter diagrams along with the type of linear correlation that exists between the domain and range values. If r is positive, then the closer r is to 1, the stronger the linear relationship between the domain and range values and the better the fit of the regression line to the data. If r is negative, then the closer r is to -1, the stronger the linear relationship between the domain and range values and the better the fit of the regression line to the data.

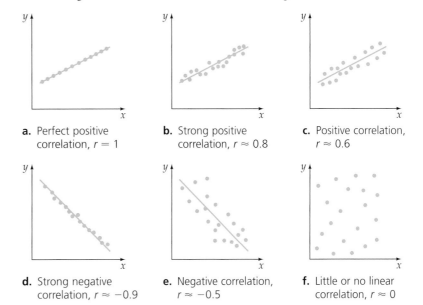

a. Perfect positive correlation, $r = 1$

b. Strong positive correlation, $r \approx 0.8$

c. Positive correlation, $r \approx 0.6$

d. Strong negative correlation, $r \approx -0.9$

e. Negative correlation, $r \approx -0.5$

f. Little or no linear correlation, $r \approx 0$

A TI-83 or TI-84 Plus graphing calculator will compute the value of the linear correlation coefficient r when it computes the equation of the regression line. You may have noticed the value displayed when the calculator displays the regression equation. If the value of r does not appear on the screen, press 2ND [CATALOG] (above the 0 key) and then scroll down to DiagnosticOn. Press ENTER twice. The next time you find a regression equation, the calculator will display the value of r along with the value of r^2.

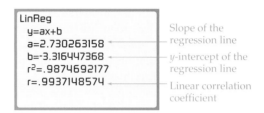

LinReg
  y=ax+b
  a=2.730263158 ← Slope of the regression line
  b=-3.316447368 ← y-intercept of the regression line
  r²=.9874692177
  r=.9937148574 ← Linear correlation coefficient

EXAMPLE 3 ■ Find a Linear Correlation Coefficient

Find the linear correlation coefficient for the data on stride length versus speed of an adult man in Example 2. Then find the linear correlation coefficient for the speed data for dogs on page 349. Which regression line is a better fit for the corresponding data?

Solution

After entering the data for adult men and finding the linear regression equation, the calculator gives the linear correlation coefficient as approximately $r = 0.9937$. Now clear the data for adult men and enter the data for dogs. The calculator gives the regression equation as approximately $y = 3.212x - 1.092$ and the correlation coefficient as approximately $r = 0.9864$. Both correlation coefficients are positive, but because the value of r for the adult men data is closer to 1 than the value of r for the dog data, the regression line for the adult men fits better than the one for the dogs.

CHECK YOUR PROGRESS 3   Find the linear correlation coefficient for stride length versus speed of a camel as given in Check Your Progress 2 on page 352. Round your result to the nearest hundredth.

Solution   See page S22.

QUESTION   For the data for adult men in Example 3, what is the significance of the fact that the linear correlation coefficient is positive?

The linear correlation coefficient indicates the strength of a linear relationship between two variables; however, it does *not* indicate the presence of a *cause-and-effect*

ANSWER   It indicates a positive correlation between a man's stride length and his speed. That is, as a man's stride length increases, his speed also increases.

relationship. For instance, the data in the table below show the hours per week that a student spent playing pool and the student's weekly algebra test scores for those same weeks.

Hours per week spent playing pool	4	5	7	8	10
Weekly algebra test score	52	60	72	79	83

The linear correlation coefficient for the ordered pairs in the table is $r \approx 0.98$. Thus there is a strong positive linear relationship between the student's algebra test scores and the time the student spent playing pool. This does not mean that the higher algebra test scores were caused by the increased time spent playing pool. The fact that the student's test scores increased with the increase in the time spent playing pool could be due to many other factors or it could just be a coincidence.

Investigation

An Application of Linear Regression

The zoology professor R. McNeill Alexander wanted to determine whether the stride length of a dinosaur, as determined by its fossilized footprints, could be used to estimate the speed of the dinosaur. Because no dinosaurs were available, he and fellow scientist A. S. Jayes carried out experiments with many types of animals, including adult men, dogs, camels, ostriches, and elephants. They collected data similar to that included in the examples in this section. The results of these experiments tended to support the idea that the speed of an animal is related to the animal's stride length. However, the data from each type of animal generated a different regression line.

Motivated by a strong desire to find a mathematical model that could be used to estimate the speed of any animal from its stride length, Alexander came up with the idea of using *relative stride lengths*. A **relative stride length** is the number obtained by dividing the stride length of an animal by the animal's leg length. That is,

$$\text{Relative stride length} = \frac{\text{stride length}}{\text{leg length}} \quad \text{(I)}$$

Thus a person with a leg length of 0.9 meter (distance from the hip to the ground) who runs steadily with a stride length of 4.5 meters has a relative stride length of (4.5 meters) ÷ (0.9 meters) = 5. Note that a relative stride length is a dimensionless quantity.

Because Alexander found it helpful to convert stride length to a dimensionless quantity (relative stride length), it was somewhat natural for him also to convert speed to a dimensionless quantity. His definition of *dimensionless speed* is

$$\text{Dimensionless speed} = \frac{\text{speed}}{\sqrt{\text{leg length} \times g}} \quad \text{(II)}$$

where g is the gravitational acceleration constant of 9.8 meters per second per second. At this point you may feel that things are getting a bit complicated and that

(continued)

you weren't really all that interested in the speed of a dinosaur anyway. However, once Alexander and Jayes converted stride lengths to *relative* stride lengths and speeds to *dimensionless* speeds, they discovered that many graphs of their data, even for different species, were nearly linear! To illustrate this concept, examine Table 5.1 below, in which the ordered pairs were formed by converting each ordered pair of speed data from the examples in this section from the form (stride length, speed) to the form (relative stride length, dimensionless speed). The conversions were calculated by using leg lengths of 0.8 meter for the adult men, 0.5 meter for the dogs, and 1.2 meters for the camels.

Table 5.1 *Dimensionless Speeds for Relative Stride Lengths*

a. Adult men

Relative stride length (*x*)	3.1	3.8	4.1	4.4	4.8	5.0	5.3	5.6
Dimensionless speed (*y*)	1.2	1.8	2.0	2.4	2.5	2.8	3.0	3.1

b. Dogs

Relative stride length (*x*)	3.0	3.4	4.0	4.8	5.4	6.0	6.4	7.0
Dimensionless speed (*y*)	1.7	2.0	2.2	3.2	3.5	4.1	4.0	4.5

c. Camels

Relative stride length (*x*)	2.1	2.5	2.7	2.8	2.9	3.2	3.3	3.5
Dimensionless speed (*y*)	0.7	1.1	1.3	1.5	1.6	1.8	2.1	2.2

A scatter diagram of the data in Table 5.1 is shown below. The scatter diagram shows a strong linear correlation. (You didn't expect a perfect linear correlation, did you? After all, we are working with camels, dogs, and adult men.) Although we have considered only three species, Alexander and Jayes were able to show a strong linear correlation for several species.

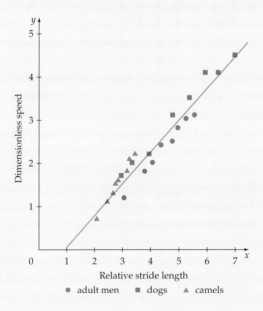

(continued)

Finally it is time to estimate the speed of a dinosaur. Consider a large theropod with a leg length of 2.5 meters. If the theropod's fossilized footprints show a stride length of 5 meters, then its relative stride length is $\frac{5\,m}{2.5\,m} = 2$. The least-squares regression line in the scatter diagram shows that a relative stride length of 2 has a dimensionless speed of about 0.8. If we use 2.5 meters for the leg length and 0.8 for the dimensionless speed and solve equation (II) (page 356) for speed, we get

$$\text{Speed} = (\text{dimensionless speed})\sqrt{\text{leg length} \times 9.8}$$
$$= (0.8)\sqrt{2.5 \times 9.8}$$
$$\approx 4.0 \text{ meters per second}$$

For more information about estimating the speeds of dinosaurs, consult the following article by Alexander and Jayes: "A dynamic similarity hypothesis for the gaits of quadrupedal mammals." *Journal of Zoology* 201:135–152, 1983.

Investigation Exercises

1. **a.** Use a calculator to find the equation of the regression line for *all* of the data in Table 5.1.

   b. Find the linear correlation coefficient for the regression line in part a.

2. The photograph at the left shows a set of sauropod tracks and a set of tracks made by a carnivore. These tracks were discovered by Roland Bird in 1938 in the Paluxy River bed, near the town of Glen Rose, Texas. Measurements of the sauropod tracks indicate an average stride length of about 4.0 meters. Assume that the sauropod that made the tracks had a leg length of 3.0 meters. Use the equation of the regression line from Investigation Exercise 1 to estimate the speed of the sauropod that produced the tracks.

sauropod   pachycephalosaur

3. A pachycephalosaur has an estimated leg length of 1.4 meters, and its footprints show a stride length of 3.1 meters. Use the equation of the regression line from Investigation Exercise 1 to estimate the speed of this pachycephalosaur.

Exercise Set 5.4

 In Exercises 1–6, find the equation of the regression line for the given data points. Round constants to the nearest hundredth.

1. $(2, 6), (3, 6), (4, 8), (6, 11), (8, 18)$

2. $(2, -3), (3, -4), (4, -9), (5, -10), (7, -12)$

3. $(-3, 11.8), (-1, 9.5), (0, 8.6), (2, 8.7), (5, 5.4)$

4. $(-7, -11.7), (-5, -9.8), (-3, -8.1), (1, -5.9), (2, -5.7)$

5. $(1, 4.1), (2, 6.0), (4, 8.2), (6, 11.5), (8, 16.2)$

6. $(2, 5), (3, 7), (4, 8), (6, 11), (8, 18), (9, 21)$

7. **Health** The U.S. Centers for Disease Control and Prevention (CDC) use a measure called body mass index (BMI) to determine whether a person is obese. A BMI between 25.0 and 29.9 is considered overweight, and a BMI of 30.0 or more is considered obese. The following table shows the percents of United States males 18 years old or older who were obese in the years indicated, judging on the basis of BMI. (*Source:* CDC, November 5, 2003)

Year	Percent Obese
1995	16.3
1996	16.3
1997	17.1
1998	18.4
1999	19.9
2000	20.6
2001	21.2
2002	23.1

a. Using 0 for 1995, 1 for 1996, and so on, find the linear regression equation for the data.

b. Use the equation from part a to predict the percent of overweight males in 2011.

8. **Health** The U.S. Centers for Disease Control and Prevention (CDC) use a measure called body mass index (BMI) to determine whether a person is overweight. A BMI between 25.0 and 29.9 is considered overweight, and a BMI of 30.0 or more is considered obese. The following table shows the percents of United States females 18 years old or older who were obese in the years indicated, judging on the basis of BMI. (*Source:* CDC, November 5, 2003)

a. Using 0 for 1995, 1 for 1996, and so on, find the equation of the regression line for the data.

b. Use the equation from part a to predict the percent of obese females in 2011.

9. 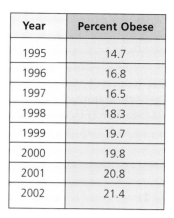 **Paleontology** The table below shows the length, in centimeters, of the humerus, and the total wingspan, in centimeters, of several pterosaurs, which are extinct flying reptiles. (*Source:* Southwest Educational Development Laboratory)

Year	Percent Obese
1995	14.7
1996	16.8
1997	16.5
1998	18.3
1999	19.7
2000	19.8
2001	20.8
2002	21.4

Pterosaur Data

Humerus	Wingspan	Humerus	Wingspan
24	600	20	500
32	750	27	570
22	430	15	300
17	370	15	310
13	270	9	240
4.4	68	4.4	55
3.2	53	2.9	50
1.5	24		

a. Find the equation of the regression line for the data. Round constants to the nearest hundredth.

b. Use the equation from part a to determine, to the nearest centimeter, the projected wingspan of a pterosaur if its humerus is 54 centimeters long.

10.  **Life Expectancy** The average remaining lifetimes for women in the United States are given in the following table. (*Source:* National Institute of Health)

Average Remaining Lifetimes for Women

Age	Years	Age	Years
0	79.4	65	19.2
15	65.1	75	12.1
35	45.7		

a. Find the equation of the regression line for the data.

b. Use the equation from part a to estimate the remaining lifetime of a woman of age 25.

11. Collected data are plotted in the scatter diagram below along with the regression line $y = ax + b$. If we use this equation to estimate a data point at $x = 22$, are we interpolating or extrapolating?

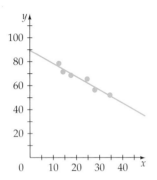

12. The data shown in the scatter diagram in Exercise 11 were modeled with the regression line $y = ax + b$. If we use the regression line equation to estimate a data point at $x = 48$, are we interpolating or extrapolating?

13. Which of the scatter diagrams below suggests the

a. strongest positive linear correlation between the x and y variables?

b. strongest negative linear correlation between the x and y variables?

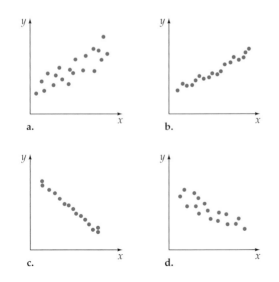

14. Which of the scatter diagrams below suggests

a. a near perfect positive linear correlation between the x and y variables?

b. little or no linear correlation between the x and y variables?

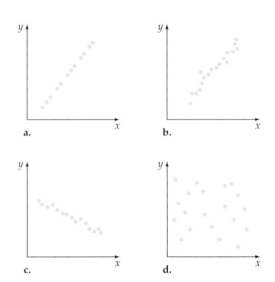

15. **Value of a Corvette** The following table gives the retail values of a 2004 Corvette Z06 for various odometer readings. (*Source:* Kelley Blue Book website, November 3, 2004)

Odometer Reading	Retail Value
13,000	$46,100
18,000	$44,600
20,000	$43,300
25,000	$41,975
29,000	$40,975
32,000	$39,750

a. Find the equation of the regression line for the data. Round constants to the nearest thousandth.

b. Use the equation from part a to predict the retail price of such a car with an odometer reading of 30,000.

c. Find the linear correlation coefficient for these data.

d. What is the significance of the fact that the linear correlation coefficient for these data is negative?

16. **Cellular Phone** The following table shows the approximate numbers of cellular telephone subscriptions in the United States for recent years.

U.S. Cellular Telephone Subscriptions

Year	1998	1999	2000	2001	2002	2003
Subscriptions, in millions	69	86	109	128	141	159

Source: CTIA Semiannual Wireless Survey. Data extracted from *The World Almanac and Book of Facts 2005*, page 398.

a. Using 0 for 1998, 1 for 1999, and so on, find the equation of the regression line for the data. Round constants to the nearest hundredth.

b. Find the linear correlation coefficient for the data.

c. On the basis of the value of the linear correlation coefficient, would you conclude that the data can be reasonably modeled by a linear function? Explain.

17. **Life Expectancy** The average remaining lifetimes for men in the United States are given in the following table. (*Source:* National Institute of Health)

Average Remaining Lifetimes for Men

Age	Years	Age	Years
0	73.6	65	15.9
15	59.4	75	9.9
35	40.8		

a. Find the regression line equation for the data. Round constants to the nearest hundredth.

b. Find the linear correlation coefficient for the data.

c. On the basis of the value of the linear correlation coefficient, would you conclude that the data can be reasonably modeled by a linear equation? Explain.

18. **Median Incomes** The following table lists the median weekly earnings, in dollars, for women and men for selected years from 1980 to 2002. (*Source:* U.S. Department of Labor, Bureau of Labor Statistics)

Median Weekly Earnings

	1980	1985	1990	1995	2000	2002
Women (x)	415	442	462	476	513	530
Men (y)	540	547	550	561	601	609

a. Find, to the nearest hundredth, the linear correlation coefficient for the data.

b. On the basis of your answer to part a, would you say that there is a strong linear relationship between the median weekly earnings of women and the median weekly earnings of men?

19. **Temperature** The data given in the following table show equivalent temperatures on the Celsius temperature scale and the Fahrenheit temperature scale.

Celsius ($x°$)	−40	0	100
Fahrenheit ($y°$)	−40	32	212

a. Find the linear correlation coefficient for the data.

b. What is the significance of the value found in part a?

c. Find the equation of the regression line.

d. Use the equation of the regression line from part c to predict the Fahrenheit temperature that corresponds to a Celsius temperature of 35°.

e. Is the procedure in part d an example of interpolation or extrapolation?

20. **Fitness** An aerobic exercise instructor uses the data given in the following table, which shows the recommended maximum exercise heart rates for individuals of the given ages, to design exercise programs.

Age (x years)	20	40	60
Maximum heart rate (y beats per minute)	170	153	136

a. Find the linear correlation coefficient for the data.

b. What is the significance of the value found in part a?

c. Find the equation of the regression line.

d. Use the equation of the regression line from part c to predict the maximum exercise heart rate for a person who is 72.

e. Is the procedure in part d an example of interpolation or extrapolation?

Extensions

CRITICAL THINKING

21. **Tuition** The following table shows the average annual tuition and fees of private and public 4-year colleges for the school years 1998–1999 through 2003–2004. (*Source:* The College Board. Extracted from *The New York Times Almanac, 2005*, page 354.)

Tuition and Fees

Year	Private	Public
1998–1999	$14,709	$3247
1999–2000	15,518	3362
2000–2001	16,233	3487
2001–2002	17,272	3725
2002–2003	18,273	4081
2003–2004	19,710	4694

Is there a strong linear relationship between average tuition at public 4-year colleges and that at private 4-year colleges? Use the linear correlation coefficient to support your answer.

22. **Fuel Efficiency** The following table shows the average fuel efficiency, in miles per gallon (mpg), of all cars sold in the United States during the years 1999 through 2004. (*Source:* Environmental Protection Agency. Extracted from *The World Almanac and Book of Facts, 2005*, page 237.)

Fuel Efficiency

Year	Miles per Gallon
1999	24.1
2000	24.1
2001	24.3
2002	24.5
2003	24.7
2004	24.6

Is there a strong linear relationship between year and average miles per gallon? Use the linear correlation coefficient to support your answer.

EXPLORATIONS

23. Another linear model that can be used to model data is called the *median-median line*. Use a statistics text or the Internet to read about the median-median line.

a. Find the equation of the median-median line for the data given in Exercise 7. You can do this by hand or by using a calculator.

b. Explain the type of situation in which it would be better to model the data using the median-median line than it would be to model the data using the regression line.

24. Search for data (in a magazine, in a newspaper, in an almanac, or on the Internet) that can be closely modeled by a linear equation.

a. Draw a scatter diagram of the data.

b. Find the equation of the regression line and the linear correlation coefficient for the data.

c. Graph the regression line on the scatter diagram in part a.

d. Use the equation of the regression line to predict a range value for a specific domain value.

Summary

Key Terms

abscissa [p. 313]
coordinate axes [p. 312]
coordinates of a point [p. 313]
dependent variable [p. 317]
domain [p. 316]
equation in two variables [p. 313]
evaluating a function [p. 317]
extrapolation [p. 352]
first coordinate [p. 313]
function [p. 316]
function notation [p. 317]
graph a point [p. 313]
graph of an equation in two variables [p. 314]
graph of an ordered pair [p. 313]
independent variable [p. 317]
interpolation [p. 352]
linear correlation coefficient [p. 353]
linear function [p. 324]
linear regression [p. 349]

negative correlation [p. 353]
ordered pair [p. 313]
ordinate [p. 313]
origin [p. 312]
plot a point [p. 313]
positive correlation [p. 353]
quadrants [p. 312]
range [p. 316]
rectangular coordinate system [p. 312]
regression line [p. 349]
scatter diagram [p. 341]
second coordinate [p. 313]
slope [p. 328]
solution of an equation in two variables [p. 313]
value of a function [p. 317]
x-coordinate [p. 313]
x-intercept [p. 325]
xy-plane [p. 312]
y-coordinate [p. 313]
y-intercept [p. 325]

Essential Concepts

■ **Function Notation**
The symbol $f(x)$ is the value of the function and represents the value of the dependent variable for a given value of the independent variable.

■ **Linear Function**
A linear function is one that can be written in the form $f(x) = mx + b$, where m is the coefficient of x and b is a constant. The slope of the graph of the function is m, the coefficient of x.

■ **y-Intercept of a Linear Function**
The y-intercept of the graph of $f(x) = mx + b$ is $(0, b)$.

■ **Slope of a Line**
Let (x_1, y_1) and (x_2, y_2) be two points on a nonvertical line. Then the slope of the line through the two points is the ratio of the change in the y-coordinates to the change in the x-coordinates.

$$m = \frac{\text{change in } y}{\text{change in } x} = \frac{y_2 - y_1}{x_2 - x_1}, \, x_1 \neq x_2$$

■ **Slope–Intercept Form of the Equation of a Line**
The graph of $f(x) = mx + b$ is a straight line with slope m and y-intercept $(0, b)$.

■ **Point–Slope Formula for a Straight Line**
Let (x_1, y_1) be a point on a line and let m be the slope of the line. Then the equation of the line can be found using the point–slope formula

$$y - y_1 = m(x - x_1).$$

■ The *regression line* or *least-squares line* is the line that minimizes the sum of the squares of the vertical distances between the data points and the line.

■ The *linear correlation coefficient r* is a measure of how well the regression line fits the given data. If the regression line has positive slope, then r is positive, and the closer r is to 1, the better the fit. If the slope of the regression line is negative, then r is negative, and the closer r is to -1, the better the fit.

Review Exercises

1. Draw a line through all points with an x-coordinate of 4.

2. Draw a line through all points with a y-coordinate of 3.

3. Graph the ordered-pair solutions of $y = 2x^2$ when $x = -2, -1, 0, 1,$ and 2.

4. Graph the ordered-pair solutions of $y = 2x^2 - 5$ when $x = -2, -1, 0, 1,$ and 2.

5. Graph the ordered-pair solutions of $y = -2x + 1$ when $x = -2, -1, 0, 1,$ and 2.

6. Graph the ordered-pair solutions of $y = |x + 1|$ when $x = -5, -3, 0, 3,$ and 5.

In Exercises 7–12, graph the function.

7. $y = -2x + 1$　　　　8. $y = 3x + 2$

9. $f(x) = x^2 + 2$　　　10. $f(x) = x^2 - 3x + 1$

11. $y = |x + 4|$　　　　12. $f(x) = 2|x| - 1$

In Exercises 13–16, evaluate the function for the given value.

13. $f(x) = 3x - 5; x = -3$

14. $g(x) = 2x - 6x^2; x = 2$

15. $p(t) = \dfrac{t}{2t + 1}; t = -1$

16. $r(s) = s^2 - 7s - 8; s = -1$

17. **Geometry** The surface area of a sphere is a function of its radius and is given by $A(r) = 4\pi r^2$, where r is the radius of the sphere.

　　a. Find the surface area of a sphere whose radius is 2.5 feet. Round to the nearest tenth of a square foot.

　　b. Find the surface area of a sphere whose radius is 8 centimeters. Round to the nearest tenth of a square centimeter.

18. **Physics** The distance s in feet that an object will fall in t seconds when gravity is the only force acting on the object is given by $s(t) = 16t^2$. If you drop a billiard ball from a bridge, how far will the ball drop in 3 seconds? (Assume that gravity is the only force acting on the ball.)

19. **Pendulums** The time T, in seconds, it takes a pendulum of length L, in feet, to make one swing is given by $T(L) = 2\pi\sqrt{\dfrac{L}{32}}$. Find the time it takes a pendulum to make one swing if the length of the pendulum is 2.5 feet. Round to the nearest hundredth of a second.

In Exercises 20–23, find the x- and y-intercepts of the graph of the function.

20. $f(x) = 2x + 10$　　　21. $f(x) = \dfrac{3}{4}x - 9$

22. $3x - 5y = 15$　　　　23. $4x + 3y = 24$

24. **Depreciation** The accountant for a small business uses the model $V(t) = 25,000 - 5000t$ to approximate the value $V(t)$ of a small truck t years after its purchase. Find and discuss the meaning of the intercepts, on the vertical and horizontal axes, in the context of this application.

In Exercises 25–28, find the slope of the line passing through the given points.

25. $(3, 2), (2, -3)$　　　26. $(-1, 4), (-3, -1)$

27. $(2, -5), (-4, -5)$　　28. $(5, 2), (5, 7)$

29. **The Film Industry** The following graph shows annual projections for revenue from home video rentals. Find the slope of the line and write a sentence that explains the meaning of the slope in the context of this application. (*Source:* Forrester Research)

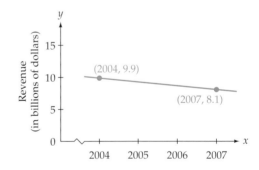

30. Graph the line that passes through the point $(3, -2)$ and has slope -2.

31. Graph the line that passes through the point $(-1, -3)$ and has slope $\dfrac{3}{4}$.

In Exercises 32–35, find the equation of the line that passing through the given point and has the given slope.

32. $(-2, 3), m = 2$　　　33. $(1, -4), m = 1$

34. $(-3, 1), m = \dfrac{2}{3}$　　35. $(4, 1), m = \dfrac{1}{4}$

36. Graph $f(x) = \dfrac{3}{2}x - 1$ using the slope and y-intercept.

37. Interior Decorating A dentist's office is being recarpeted. The cost of installing the new carpet is $100 plus $25 per square yard of carpeting.

   a. Determine a linear function for the cost of carpeting the office.

   b. Use this function to determine the cost of carpeting 32 square yards of floor space.

38. Gasoline Sales The manager of Valley Gas Mart has determined that 10,000 gallons of regular unleaded gasoline can be sold each week if the price is the same as that of Western QuickMart, a gas station across the street. If the manager increases the price $.02 above Western QuickMart's price, the manager will sell 500 fewer gallons per week. If the manager decreases the price $.02 below that of Western QuickMart, 500 more gallons of gasoline per week will be sold.

   a. Determine a linear function that predicts the number of gallons of gas per week that Valley Gas Mart can sell as a function of the price relative to that of Western QuickMart.

   b. Use the model to predict the number of gallons of gasoline Valley Gas Mart will sell if its price is $.03 below that of Western QuickMart.

39. **BMI** The body mass index (BMI) of a person is a measure of the person's ideal body weight. The data in the table below show the BMI values for different weights for a person 5'6″ tall.

| BMI Data for Person 5'6" Tall |||||
Weight (in pounds)	BMI	Weight (in pounds)	BMI
110	18	160	26
120	19	170	27
125	20	180	29
135	22	190	31
140	23	200	32
145	23	205	33
150	24	215	35

Source: Centers for Disease Control and Prevention

   a. Draw a scatter diagram of the data and write the equation of a linear function that fits the data.

   b. Use the model to estimate the BMI for a person 5'6″ tall whose weight is 158 pounds. Round to the nearest whole number.

40. **Cellular Phone Prevalence** The following table shows the percents of households that had a cellular phone in selected years. (*Source:* Consumer Electronics Association. Extracted from *The World Almanac and Book of Facts, 2005,* page 398.)

Percent of Cellular Phones in U.S. Households

Year	Percent
1990	5
1995	29
2000	60
2002	68
2003	70

   a. Using 0 for 1990, 5 for 1995, and so on, draw a scatter diagram of the data and write the equation of a linear function that fits the data.

   b. Use the model to estimate the percent of U.S. households that had a cellular phone in 1998. Round to the nearest percent.

41. You are given the following data.

x	10	12	14	15	16
y	8	7	5	4	1

   a. Find the equation of the regression line. Round constants to the nearest hundredth.

   b. Use the equation to predict the value of y for $x = 8$.

   c. What is the linear correlation coefficient? Round to the nearest thousandth.

42. **Physics** A student has recorded the data in the following table, which shows the distance a spring stretches in inches when a given weight, in pounds, is attached to the spring.

Weight, x	80	100	110	150	170
Distance, y	6.2	7.4	8.3	11.1	12.7

   a. Find the linear correlation coefficient.

   b. Find the equation of the regression line.

   c. Use the equation of the regression line from part b to predict the distance a weight of 195 pounds will stretch the spring.

43. **Internet** A test of an Internet service provider showed the following download times (in seconds) for files of various sizes (in kilobytes).

Download Times			
Size	Time	Size	Time
10.5	0.20	110	2.01
12.9	0.24	156	2.68
15	0.27	163	2.87
20	0.36	175	3.10
60	1.09	200	3.64
75	1.42	250	4.61

a. Find the equation of the regression line for these data. Round constants to the nearest thousandth.

b. On the basis of the value of the linear correlation coefficient, is a linear model of these data a reasonable model?

c. Use the equation of the regression line from part a to predict the expected download time of a file that is 100 kilobytes in size.

44. **Camera Sales** The following bar graph shows the sales, in millions of cameras, of analog and digital still cameras.

Worldwide Camera Sales

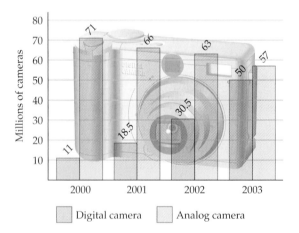

Source: Digital Photography Review, dpreview.com August 27, 2004.

a. Determine the linear regression equation of the number of sales of digital cameras as a function of the year. Use 0 for 2000, 1 for 2001, and so on.

b. Determine the linear regression equation of the number of sales of analog cameras as a function of the year.

c. Judging on the basis of the linear models found in parts a and b, were digital camera sales greater than or less than analog camera sales in 2005?

45. **Consumer Price Index** The U.S. Consumer Price Index (CPI) is a measure of the average change in prices over time. In the following table, the CPI for each year is based on a cost of $100 in 1967. For example, the CPI for the year 1985 is 322.2. This means that in 1985 it took $322.20 to purchase the same goods that cost $100 in the year 1967.

Consumer Price Index, 1975–2004 (1967 has a CPI of 100)

Year	1975	1980	1985	1990	1995	2000	2004*
CPI	161.2	248.8	322.2	391.4	456.5	515.8	562.0

*Average for first half of 2004
Source: Bureau of Labor Statistics, U.S. Dept. of Labor. Data taken from *The World Almanac and Book of Facts 2005*, page 109.

a. Find the equation of the regression line and the linear correlation coefficient for the data in the table. Let the year 1970 be represented by $x = 0$, 1975 by $x = 5$, and so on. Round constants to the nearest hundredth.

b. Use the equation of the regression line to predict the CPI for the year 2011. Round to the nearest tenth.

CHAPTER 5 **Test**

In Exercises 1 and 2, evaluate the function for the given value of the independent variable.

1. $s(t) = -3t^2 + 4t - 1;$ **2.** $f(x) = |x| + 2; x = -3$
$t = -2$

In Exercises 3 and 4, graph the function.

3. $f(x) = 2x - 3$

4. $f(x) = x^2 + 2x - 3$

5. Find the x- and y-intercepts of the line given by $5x - 3y = 8$.

6. Find the slope of the line that passes through $(3, -1)$ and $(-2, -4)$.

7. Find the equation of the line that passes through $(3, 5)$ and has slope $\frac{2}{3}$.

8. ✎ **Travel** The distance d, in miles, a small plane is from its final destination is given by $d(t) = 250 - 100t$, where t is the time, in hours, remaining for the flight. Find and discuss the meaning of the intercepts of the graph of the function.

9. **Farming** The manager of an orange grove has determined that when there are 320 trees per acre, the average yield per tree is 260 pounds of oranges. If the number of trees is increased to 330 trees per acre, the average yield per tree decreases to 245 pounds. Find a linear model for the average yield per tree as a function of the number of trees per acre.

10. **Farming** Giant pumpkin contests are popular at many state fairs. Suppose the data below show the weights, in pounds, of the winning pumpkins at recent fairs, where $x = 0$ represents 2000.

Year, x	0	1	2	3	4
Weight (in pounds), y	650	715	735	780	820

a. Draw a scatter diagram of the data and write the equation of a linear function that fits the data.

b. Use the model to predict the winning weight in 2008. Round to the nearest pound.

11. **Population** The following table shows the estimated population, in thousands, of England for various years.

Year	Population (thousands)
1994	48,229
2004	50,094
2011	51,967
2021	54,605
2031	56,832

Source: UK Office for National Statistics, www.statistics.gov.uk

Find the equation of the regression line and the linear regression coefficient for the data. According to the regression coefficient, how well does the regression line fit the data?

12. **Nutrition** The following table shows the percents of water and the numbers of calories in various canned soups to which 100 grams of water are added.

% Water	Calories
93.2	28
92.3	26
91.9	39
89.5	56
89.6	56
90.5	36
91.9	32
91.7	32

a. Find the equation of the regression line for the data. Round constants to the nearest hundredth.

b. Use the equation in part a to find the expected number of calories in a soup that is 89% water. Round to the nearest whole number.

CHAPTER 6

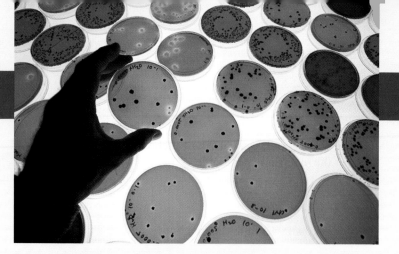

Nonlinear Models

6.1 **Introduction to Nonlinear Functions** **6.3** **Logarithmic Functions**
6.2 **Exponential Functions**

n Michael Crichton's novel *The Andromeda Strain*[1], he states

> A single cell of the bacterium *E. coli* would, under ideal circumstances, divide every twenty minutes. That is not particularly disturbing until you think about it, but the fact is that bacteria multiply geometrically: one becomes two, two become four, four become eight, and so on.

If a colony of *E. coli* originally contained 1000 bacteria, then Crichton's statement can be modeled by the exponential growth equation $A = 1000(2^{3t})$, where A is the number of *E. coli* present after t hours. This equation is an example of a nonlinear function called an *exponential function*. The table below shows the number of *E. coli* for various times t.

Time, t (in hours)	Number of *E. coli* Bacteria, $1000(2^{3t}) = A$
1 h	$1000(2^{3 \cdot 1}) = 1000(2^3) = 1000(8) = 8000$
2 h	$1000(2^{3 \cdot 2}) = 1000(2^6) = 1000(64) = 64{,}000$
3 h	$1000(2^{3 \cdot 3}) = 1000(2^9) = 1000(512) = 512{,}000$
4 h	$1000(2^{3 \cdot 4}) = 1000(2^{12}) = 1000(4096) = 4{,}096{,}000$

Crichton goes on to say,

> In this way it can be shown that in a single day, one cell of *E. coli* could produce a super-colony equal in size and weight to the entire planet Earth.

Of course, *ideal circumstances*, as stated by Crichton, mean that no bacteria die and there is a sustainable food supply. In practice, this does not happen and it would be necessary to modify the equation to reflect those conditions.

1. Michael Crichton (1969), *The Andromeda Strain*, Dell, N.Y., p. 247.

SECTION 6.1 | Introduction to Nonlinear Functions

Introduction to Polynomial Functions

In the previous chapter, we discussed *linear functions*. As we noted, these functions can be represented by $f(x) = mx + b$. The expression $mx + b$ is a linear polynomial or a first-degree polynomial, and we say that f is a linear function or a first-degree polynomial function.

A **polynomial function** in one variable is one in which the exponent on the variable is a nonnegative integer. The **degree of a polynomial function** in one variable is the largest exponent on the variable. Here are some examples of polynomial functions and their degrees.

Polynomial Function	Degree
$f(x) = 2x - 3$	1
$g(x) = 2x^2 - 4x - 7$ $f(t) = t^2$	2
$p(x) = 1 - 2x + 3x^2 + x^3$ $s(t) = 5t^3 - t + 8$	3
$z(x) = x^4 + 2x^3 - 13x^2 - 14x + 24$	4

Polynomial functions of degree greater than 1 are part of a class of functions that are called **nonlinear functions** because their graphs are not straight lines. Here are some graphs of polynomial functions of degree greater than 1.

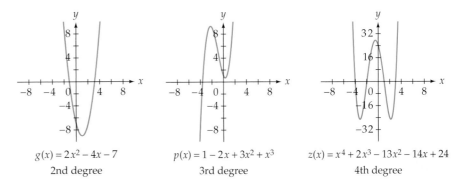

$g(x) = 2x^2 - 4x - 7$
2nd degree

$p(x) = 1 - 2x + 3x^2 + x^3$
3rd degree

$z(x) = x^4 + 2x^3 - 13x^2 - 14x + 24$
4th degree

TAKE NOTE

Fifth-degree polynomials are called quintic polynomials. Generally, however, polynomials of degree 4 and higher are referred to by their degrees. So, for instance, $f(x) = 3x^8 + 4x^5 - 3x^2 + 4$ represents an eighth-degree polynomial function.

Just as a first-degree polynomial function is also called a **linear function,** some of the other polynomial functions have special names.

Polynomial Function	Name
First-degree	Linear function
Second-degree	Quadratic function
Third-degree	Cubic function
Fourth-degree	Quartic function

The polynomials $2x + 4x^3 - 5 + 6x^2$ and $4x^3 + 6x^2 + 2x - 5$ are equal. However, it is customary to write polynomials in decreasing powers of the variable. This is called the **standard form of a polynomial.** Therefore, the standard form of $2x + 4x^3 - 5 + 6x^2$ is $4x^3 + 6x^2 + 2x - 5$.

To evaluate a polynomial function, replace the variable by its value and simplify the resulting numerical expression.

EXAMPLE 1 ■ Work with Polynomial Functions

Let $f(x) = 3x - 2x^2 - 4x^3 + 2$.

a. Write the polynomial in standard form.

b. Name the degree of the function.

c. Evaluate the function when $x = 3$.

Solution

a. Write the polynomial in decreasing powers of *x*.

$$f(x) = -4x^3 - 2x^2 + 3x + 2$$

b. The degree is 3, the largest exponent on the variable.

c. $f(x) = -4x^3 - 2x^2 + 3x + 2$

$f(3) = -4(3)^3 - 2(3)^2 + 3(3) + 2$ • Replace *x* by 3.

$= -4(27) - 2(9) + 3(3) + 2$ • Simplify the numerical expression.

$= -108 - 18 + 9 + 2 = -115$

CHECK YOUR PROGRESS 1   Let $g(t) = -4t + 3t^2 + 5$.

a. Write the polynomial in standard form.

b. Name the degree of the function.

c. Evaluate the function when $x = 3$.

Solution   *See page S22.*

Quadratic Functions

We will now focus special attention on quadratic, or second-degree, polynomial functions.

A **quadratic function** in a single variable x is a function of the form $f(x) = ax^2 + bx + c$, $a \neq 0$. Examples of quadratic functions are given below.

$f(x) = x^2 - 3x + 1$ • $a = 1, b = -3, c = 1$

$g(t) = -2t^2 - 4$ • $a = -2, b = 0, c = -4$

$h(p) = 4 - 2p - p^2$ • $a = -1, b = -2, c = 4$

$f(x) = 2x^2 + 6x$ • $a = 2, b = 6, c = 0$

▼ **point of interest**

The suspension cables of some bridges, such as the Golden Gate Bridge, have the shape of a parabola.

The photo above shows the roadway of the Golden Gate Bridge being assembled in sections and attached to suspender ropes. The bridge was opened to vehicles on May 28, 1937.

 TAKE NOTE

The axis of symmetry of the graph of a quadratic function is a vertical line. The vertex of the parabola lies on the axis of symmetry.

The graph of a quadratic function in a single variable x is a **parabola.** The graphs of two such quadratic functions are shown below.

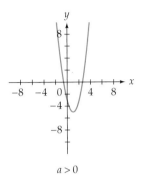

$a > 0$

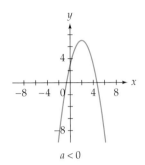

$a < 0$

The figure on the left is the graph of $f(x) = 2x^2 - 4x - 3$. The value of a is *positive* ($a = 2$) and the graph opens up. The figure on the right is the graph of $f(x) = -x^2 + 4x + 3$. The value of a is *negative* ($a = -1$) and the graph opens down. The point at which the graph of a parabola has a minimum or a maximum is called the *vertex* of the parabola. The **vertex** of a parabola is the point with the smallest y-coordinate when $a > 0$ and the point with the largest y-coordinate when $a < 0$.

The **axis of symmetry** of the graph of a quadratic function is a vertical line that passes through the vertex of the parabola. To understand the concept of the axis of symmetry of a graph, think of folding the graph along that line. The two portions of the graph will match up.

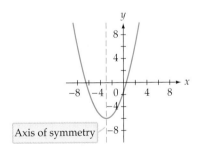

Axis of symmetry

The following formula enables us to determine the vertex of a parabola.

Vertex of a Parabola

Let $f(x) = ax^2 + bx + c$ be the equation of a parabola. The coordinates of the vertex are

$$\left(-\frac{b}{2a}, f\left(-\frac{b}{2a}\right)\right).$$

EXAMPLE 2 ■ **Find the Vertex of a Parabola**

Find the vertex of the parabola whose equation is $y = -3x^2 + 6x + 1$.

Solution

$$x = -\frac{b}{2a} = -\frac{6}{2(-3)} = 1$$ • Find the *x*-coordinate of the vertex.
                                               a = −3, b = 6

$$y = -3x^2 + 6x + 1$$
$$y = -3(1)^2 + 6(1) + 1$$ • Find the *y*-coordinate of the vertex by
$$y = 4$$ replacing *x* by **1** and solving for *y*.

The vertex is $(1, 4)$.

CHECK YOUR PROGRESS 2   Find the vertex of the parabola whose equation is $y = x^2 - 2$.

Solution   *See page S22.*

MathMatters   Paraboloids

The movie *Contact* was based on a novel by astronomer Carl Sagan. In the movie, Jodie Foster plays an astronomer who is searching for extraterrestrial intelligence. One scene from the movie takes place at the Very Large Array (VLA) in New Mexico. The VLA consists of 27 large radio telescopes whose dishes are paraboloids, the three-dimensional version of a parabola. A parabolic shape is used because of the following reflective property: When parallel rays of light, or radio waves, strike the surface of a parabolic mirror whose axis of symmetry is parallel to these rays, they are reflected to the same point.

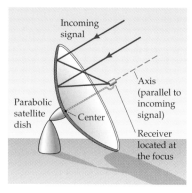

The photos above show the layout of the radio telescopes of the VLA and a more detailed picture of one of the telescopes. The figure at the far right shows the reflective property of a parabola. All the incoming rays are reflected to a point called the focus.

The reflective property of a parabola is also used in optical telescopes and headlights on a car. In the case of headlights, the bulb is placed at the focus and the light is reflected along parallel rays from the reflective surface of the headlight, thereby making a more concentrated beam of light.

Minimum and Maximum of a Quadratic Function

Note that for the graphs below, when $a > 0$, the vertex is the point with the minimum y-coordinate. When $a < 0$, the vertex is the point with the maximum y-coordinate.

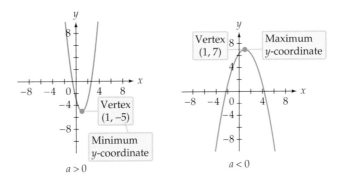

Finding the minimum or maximum value of a quadratic function is a matter of finding the vertex of the graph of the function.

EXAMPLE 3 ■ **Find the Minimum or Maximum Value of a Quadratic Function**

Find the maximum value of $f(x) = -2x^2 + 4x + 3$.

Solution

$$x = -\frac{b}{2a} = -\frac{4}{2(-2)} = 1$$

• Find the x-coordinate of the vertex. $a = -2$, $b = 4$

$$f(x) = -2x^2 + 4x + 3$$
$$f(1) = -2(1)^2 + 4(1) + 3$$

• Find the y-coordinate of the vertex by replacing x by **1** and solving for y.

$$f(1) = 5$$

• The vertex is $(1, 5)$.

The maximum value of the function is 5, the y-coordinate of the vertex.

CHECK YOUR PROGRESS 3   Find the minimum value of $f(x) = 2x^2 - 3x + 1$.

Solution   *See page S22.*

QUESTION   *The vertex of a parabola that opens up is $(-4, 7)$. What is the minimum value of the function?*

ANSWER   *The minimum value of the function is 7, the y-coordinate of the vertex.*

Applications of Quadratic Functions

EXAMPLE 4 ■ Application of Finding the Minimum Value of a Quadratic Function

A mining company has determined that the cost c, in dollars per ton, of mining a mineral is given by $c(x) = 0.2x^2 - 2x + 12$, where x is the number of tons of the mineral that is mined. Find the number of tons of the mineral that should be mined to minimize the cost. What is the minimum cost?

Solution

To find the number of tons of the mineral that should be mined to minimize the cost and to find the minimum cost, find the x- and y-coordinates of the vertex of the graph of $c(x) = 0.2x^2 - 2x + 12$.

$$x = -\frac{b}{2a} = -\frac{-2}{2(0.2)} = 5 \qquad \text{• Find the } x\text{-coordinate of the vertex.}$$
$$a = 0.2, b = -2$$

To minimize the cost, 5 tons of the mineral should be mined.

$$c(x) = 0.2x^2 - 2x + 12$$
$$c(5) = 0.2(5)^2 - 2(5) + 12 \qquad \text{• Find the } y\text{-coordinate of the vertex by}$$
$$c(5) = 7 \qquad\qquad\qquad \text{replacing } x \text{ by 5 and solving for } y.$$

The minimum cost per ton is $7.

CHECK YOUR PROGRESS 4   The height s, in feet, of a ball thrown straight up is given by $s(t) = -16t^2 + 64t + 4$, where t is the time in seconds after the ball is released. Find the time it takes the ball to reach its maximum height. What is the maximum height?

Solution   See page S22.

EXAMPLE 5 ■ Application of Finding the Maximum Value of a Quadratic Function

A lifeguard has 600 feet of rope with buoys attached to lay out a rectangular swimming area on a lake. If the beach forms one side of the rectangle, find the dimensions of the rectangle that will enclose the greatest swimming area.

Solution

Let l represent the length of the rectangle, let w represent the width of the rectangle, and let A (which is unknown) represent the area of the rectangle. See the figure at the left. Use these variables to write expressions for the perimeter and area of the rectangle.

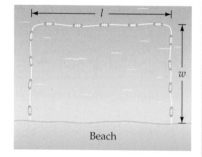

Beach

Perimeter: $w + l + w = 600$ • There is 600 feet of rope.
$$2w + l = 600$$

Area: $A = lw$

The goal is to maximize A. To do this, first write A in terms of a single variable. This can be accomplished by solving $2w + l = 600$ for l and then substituting into $A = lw$.

$$2w + l = 600 \qquad \text{• Solve for } l.$$
$$l = -2w + 600$$

$$A = lw$$
$$= (-2w + 600)w \qquad \text{• Substitute } -2w + 600 \text{ for } l.$$
$$A = -2w^2 + 600w \qquad \text{• Multiply. This is now a quadratic equation.}$$
$$a = -2, b = 600$$

Find the w-coordinate of the vertex.

$$w = -\frac{b}{2a} = -\frac{600}{2(-2)} = 150 \qquad a = -2, b = 600$$

The width is 150 feet. To find l, replace w by 150 in $l = -2w + 600$ and solve for l.

$$l = -2w + 600$$
$$l = -2(150) + 600 = -300 + 600 = 300 \qquad \text{• The length is 300 feet.}$$

The dimensions of the rectangle with maximum area are 150 feet by 300 feet.

CHECK YOUR PROGRESS 5   A mason is forming a rectangular floor for a storage shed. The perimeter of the rectangle is 44 feet. What dimensions will give the floor a maximum area?

Solution   *See page S22.*

Applications of Other Nonlinear Functions

We have just completed some applications of quadratic functions. However, polynomials of degree greater than 2 and other nonlinear functions are also important in mathematics and its applications.

EXAMPLE 6 ▪ Volume of a Box

An open box is made from a square piece of cardboard that measures 50 inches on a side. To construct the box, squares x inches on a side are cut from each corner of the cardboard. The remaining flaps are folded up to create a box.

a. Express the volume of the box as a polynomial function in x.

b. What is the volume of the box when squares 5 inches on a side are cut out?

c. Is it possible for the value of x to be 30? Explain your answer.

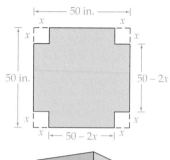

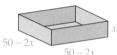

Solution

a. The volume of a box is a product of its length, width, and height. From the diagram on the preceding page, the length is $50 - 2x$, the width is $50 - 2x$, and the height is x. Therefore, the volume is given by

$$V = LWH$$

$$V(x) = (50 - 2x)(50 - 2x)x$$

$$= 4x^3 - 200x^2 + 2500x$$

The volume is given by $V(x) = 4x^3 - 200x^2 + 2500x$.

b. To find the volume when squares 5 inches on a side are cut out, evaluate the volume function when $x = 5$.

$$V(x) = 4x^3 - 200x^2 + 2500x$$

$$V(5) = 4(5)^3 - 200(5)^2 + 2500(5)$$

$$= 4(125) - 200(25) + 2500(5) = 8000$$

When squares 5 inches on a side are removed, the volume of the box is 8000 cubic inches.

c. If $x = 30$, then the value of $50 - 2x$ would be $50 - 2(30) = 50 - 60 = -10$. Because a length of -10 inches is not possible, the value of x cannot be 30.

CHECK YOUR PROGRESS 6   Express the surface area of the box in Example 6 as a function of x. The surface area is the sum of the areas of the four sides of the box and its bottom.

Solution   See page S22.

The examples to this point have involved polynomial functions, but there are other types of nonlinear functions. These include radical functions, such as $f(x) = \sqrt{x + 3}$, and rational functions in which the variable is in the denominator, as in $g(x) = \dfrac{3}{x + 2}$, or in both the numerator and denominator, as in $q(x) = \dfrac{x^2 + 3}{x - 4}$.

EXAMPLE 7 ■ **Apply the Pythagorean Theorem**

A lighthouse is 3 miles south of a port. A ship leaves the port and sails east at 15 mph.

a. Express the distance $d(t)$, in miles, between the ship and the lighthouse in terms of t, the number of hours the ship has been sailing.

b. Find the distance of the ship from the lighthouse after 3 hours. Round to the nearest tenth.

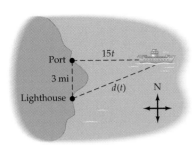

Solution

a. Because the ship is sailing at 15 mph, after t hours the ship has traveled $15t$ miles, as shown in the diagram. Using the Pythagorean Theorem $c^2 = a^2 + b^2$, where c is the length of the hypotenuse of a right triangle and a and b are the lengths of the legs, we have

$$[d(t)]^2 = (15t)^2 + 3^2$$ • The hypotenuse is $d(t)$; the legs are 15t and 3.

$$[d(t)]^2 = 225t^2 + 9$$

$$d(t) = \sqrt{225t^2 + 9}$$ • Take the square root of each side. Because distance is positive, we do not use the $\pm$ sign.

b. To find the distance after 3 hours, replace t by 3 and simplify.

$$d(t) = \sqrt{225t^2 + 9}$$

$$d(3) = \sqrt{225(3)^2 + 9} = \sqrt{2034}$$

$$\approx 45.1$$

After 3 hours, the ship is approximately 45.1 miles from the lighthouse.

CHECK YOUR PROGRESS 7   A plane flies directly over a radar station at an altitude of 2 miles and a speed of 400 miles per hour.

a. Express the distance $d(t)$, in miles, between the plane and the radar station in terms of t, the number of hours after the plane passes over the radar station.

b. Find the distance of the plane from the radar station after 3 hours. Round to the nearest tenth.

Solution   See page S22.

EXAMPLE 8 ■ Inventory Costs

One of the considerations for a retail company is the cost of maintaining its inventory. The annual inventory cost is the cost of storing the items plus the cost of reordering the items. A lighting store has determined that the annual cost, in dollars, of storing x 25-watt halogen bulbs is $0.10x$. The annual cost, in dollars, of reordering x 25-watt halogen bulbs is $\dfrac{0.15x + 2}{x}$.

a. Express the inventory cost, $C(x)$, for the halogen bulbs in terms of x.

b. Find the inventory cost if the company wants to maintain an inventory of 150 25-watt halogen bulbs. Round to the nearest cent.

Solution

a. $C(x) = 0.10x + \dfrac{0.15x + 2}{x}$ • Inventory cost is the sum of the storage and reordering costs.

b. $C(x) = 0.10x + \dfrac{0.15x + 2}{x}$

$$C(150) = 0.10(150) + \frac{0.15(150) + 2}{150}$$ • Replace x by 150.

$$\approx 15.16$$

The inventory cost is $15.16.

CHECK YOUR PROGRESS 8 A manufacturer has determined that the total cost, in dollars, of producing x straight-back wooden chairs is given by $C(x) = 35x + 500$. The average cost per chair, $A(x)$, is the quotient of the total cost and x.

a. Find the function for the average cost per chair.

b. What is the average cost per chair when the manufacturer produces 40 chairs?

Solution See page S23.

Investigation

Reflective Properties of a Parabola

The fact that the graph of $y = ax^2 + bx + c$ is a parabola is based on the following geometric definition of a parabola.

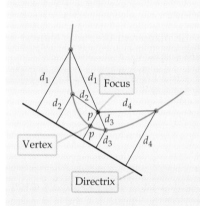

> **Parabola**
>
> A parabola is the set of points in the plane that are equidistant from a fixed line (the **directrix**) and a fixed point (the **focus**) not on the line.

This geometric definition of a parabola is illustrated in the figure at the left. Basically, for a point to be on a parabola, the distance from the point to the focus must equal the distance from the point to the directrix. Note also that the vertex is halfway between the focus and the directrix. This distance is traditionally labeled p.

The general form of the equation of a parabola that opens up with vertex at the origin can be written in terms of the distance p between the vertex and focus as $y = \frac{1}{4p}x^2$. For this equation, the coordinates of the focus are $(0, p)$. For instance, to find the coordinates of the focus for $y = \frac{1}{4}x^2$, let $\frac{1}{4p} = \frac{1}{4}$ and solve for p.

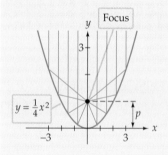

$$\frac{1}{4p} = \frac{1}{4}$$

$$1 = \frac{4p}{4} \qquad \text{• Multiply each side of the equation by } 4p.$$

$$1 = p \qquad \text{• Simplify the right side of the equation.}$$

The coordinates of the focus are $(0, 1)$.

Investigation Exercises

1. Find the coordinates of the focus for the parabola whose equation is $y = 0.4x^2$.

Optical telescopes work on the same principle as radio telescopes (see Math Matters, p. 372) except that light hits a mirror that has been shaped into a paraboloid. The light is reflected to the focus, where another mirror reflects it through a lens to the observer. See the diagram at the top of the following page.

(continued)

Palomar Observatory with the
shutters open

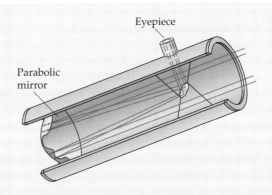

Eyepiece

Parabolic
mirror

2. The telescope at the Palomar Observatory in California has a parabolic mirror. The
circle at the top of the parabolic mirror has a 200-inch diameter. An equation that
approximates the parabolic cross-section of the surface of the mirror is
$y = \frac{1}{2639} x^2$, where x and y are measured in inches. How far is the focus from the
vertex of the mirror?

If a point on a parabola whose vertex is at the origin is known, then the equation of the
parabola can be found. For instance, if (4, 1) is a point on a parabola with vertex at the
origin, then we can find the equation as follows:

$$y = \frac{1}{4p} x^2$$ • **Begin with the general form of the equation of
a parabola.**

$$1 = \frac{1}{4p} (4)^2$$ • **The known point is (4, 1). Replace x by 4 and y by 1.**

$$1 = \frac{4}{p}$$ • **Solve for p.**

$$p = 4$$ • **$p = 4$ in the equation $y = \frac{1}{4p} x^2$.**

The equation of the parabola is $y = \frac{1}{16} x^2$.

Find a flashlight and measure the diameter of its lens cover and the depth of the
reflecting parabolic surface. See the diagram below.

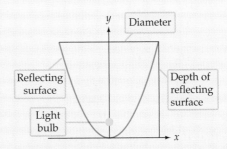

y Diameter

Reflecting
surface

Depth of
reflecting
surface

Light
bulb

x

3. Using coordinate axes as shown, find the equation of the parabola.

4. Find the location of the focus. Explain why the light bulb should be placed at
this point.

Exercise Set 6.1

In Exercises 1 to 6, **a.** write (if necessary) the polynomial function in standard form, **b.** give the degree of the polynomial function, and **c.** evaluate the function for the given values of the variable.

1. $f(x) = 2x^2 + 4x - 10, f(2)$

2. $f(x) = 1 + 2x^2, f(-2)$

3. $g(x) = x^2 + 2x^3 - 3x - 1, g(-1)$

4. $s(t) = 1 - t^2 - t^4, s(3)$

5. $y(z) = 2z^3 - 3z^2 + 4z - z^5 + 6, y(-2)$

6. $p(x) = -2x^3, p(-3)$

7. If $f(x) = x^2 + 1$ is evaluated for various values of x, will the result be **a.** positive, **b.** negative, or **c.** sometimes positive and sometimes negative?

8. Does the graph below appear to be the graph of a linear function or a nonlinear function?

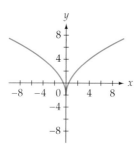

In Exercises 9 to 20, find the vertex of the graph of each equation.

9. $y = x^2 - 2$

10. $y = x^2 + 2$

11. $y = -x^2 - 1$

12. $y = -x^2 + 3$

13. $y = -\frac{1}{2}x^2 + 2$

14. $y = \frac{1}{2}x^2$

15. $y = 2x^2 - 1$

16. $y = x^2 - 2x$

17. $y = x^2 - x - 2$

18. $y = x^2 - 3x + 2$

19. $y = 2x^2 - x - 5$

20. $y = 2x^2 - x - 3$

In Exercises 21 to 28, find the minimum or maximum value of each quadratic function. State whether the value is a minimum or a maximum.

21. $f(x) = x^2 - 2x + 3$

22. $f(x) = 2x^2 + 4x$

23. $f(x) = -2x^2 + 4x - 5$

24. $f(x) = 3x^2 + 3x - 2$

25. $f(x) = x^2 + 3x - 1$

26. $f(x) = x^2 - 5x + 3$

27. $f(x) = -x^2 - x + 2$

28. $f(x) = -3x^2 + 4x - 2$

29. If two parabolas have the same vertex, are their graphs the same?

30. **Sports** The height s, in feet, of a rock thrown upward at an initial speed of 64 feet per second from a cliff 50 feet above an ocean beach is given by the function $s(t) = -16t^2 + 64t + 50$, where t is the time in seconds. Find the maximum height above the beach that the rock will attain.

31. **Sports** The height s, in feet, of a ball thrown upward at an initial speed of 80 feet per second from a platform 50 feet high is given by

$$s(t) = -16t^2 + 80t + 50$$

where t is the time in seconds. Find the maximum height above the ground that the ball will attain.

32. **Revenue** A manufacturer of microwave ovens believes that the revenue R, in dollars, the company receives is related to the price P, in dollars, of an oven by the function $R(P) = 125P - 0.25P^2$. What price will yield the maximum revenue?

33. **Manufacturing** A manufacturer of camera lenses estimates that the average monthly cost C of a lens is given by the function

$$C(x) = 0.1x^2 - 20x + 2000$$

where x is the number of lenses produced each month. Find the number of lenses the company should produce in order to minimize the average cost.

34. **Water Treatment** A pool is treated with a chemical to reduce the amount of algae. The amount of algae in the pool t days after the treatment can be approximated by the function $A(t) = 40t^2 - 400t + 500$. How many days after treatment will the pool have the least amount of algae?

35. **Civil Engineering** The suspension cable that supports a small footbridge hangs in the shape of a parabola. The height h, in feet, of the cable above the bridge is given by the function

$$h(x) = 0.25x^2 - 0.8x + 25$$

where x is the distance in feet from one end of the bridge. What is the minimum height of the cable above the bridge?

36. Annual Income The net annual income I, in dollars, of a family physician can be modeled by the equation $I(x) = -290(x - 48)^2 + 148,000$, where x is the age of the physician and $27 \le x \le 70$. Find **a.** the age at which the physician's income will be a maximum and **b.** the maximum income.

37. Recreation Karen is throwing an orange to her brother Saul, who is standing on the balcony of their home. The height h, in feet, of the orange above the ground t seconds after it is thrown is given by $h(t) = -16t^2 + 32t + 4$. If Saul's outstretched arms are 18 feet above the ground, will the orange ever be high enough so that he can catch it?

38. Football Some football fields are built in a parabolic-mound shape so that water will drain off the field. A model for the shape of the field is given by $h(x) = -0.00023475x^2 + 0.0375x$, where h is the height of the field in feet at a distance of x feet from the sideline. What is the maximum height of the field? Round to the nearest tenth.

39. Fountains The Buckingham Fountain in Chicago shoots water from a nozzle at the base of the fountain. The height h, in feet, of the water above the ground t seconds after it leaves the nozzle is given by $h(t) = -16t^2 + 90t + 15$. What is the maximum height of the water spout to the nearest tenth of a foot?

40. Stopping Distance On wet concrete, the stopping distance s, in feet, of a car traveling v miles per hour is given by $s(v) = -0.55v^2 + 1.1v$. At what speed could a car be traveling and still stop at a stop sign 44 feet away?

41. Fuel Efficiency The fuel efficiency of an average car is given by the equation

$$E(v) = -0.018v^2 + 1.476v + 3.4$$

where E is the fuel efficiency in miles per gallon and v is the speed of the car in miles per hour. What speed will yield the maximum fuel efficiency? What is the maximum fuel efficiency?

42. Ranching A rancher has 200 feet of fencing to build a rectangular corral alongside an existing fence. Determine the dimensions of the corral that will maximize the enclosed area.

43. Volume An open box is made from a square piece of cardboard that measures 60 inches on a side. To construct the box, squares x inches on a side are cut from each corner of the cardboard. The remaining flaps are folded up to create a box.

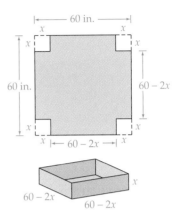

a. Express the volume of the box as a polynomial function in x.

b. What is the volume of the box when squares 7 inches on a side are cut from the cardboard?

44. Volume An open box is made from a square piece of cardboard that measures 2 meters on a side. To construct the box, squares x meter on a side are cut from each corner of the cardboard. The remaining flaps are folded up to create a box.

a. Express the volume of the box as a polynomial function in x.

b. What is the volume of the box when squares 25 centimeters (0.25 meter) on a side are cut from the cardboard?

45. Surface Area For the box in Exercise 43, express the surface area of the open box as a polynomial function in x.

46. Surface Area For the box in Exercise 44, express the surface area of the open box as a polynomial function in x.

47. Volume An open box is made from a rectangular piece of cardboard that measures 30 inches by 45 inches. To construct the box, squares x inches on a side are cut from each corner of the cardboard. The remaining flaps are folded up to create a box.

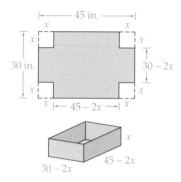

a. Express the volume of the box as a polynomial function in x.

b. What is the volume of the box when squares 6 inches on a side are cut from the cardboard?

48. Volume An open box is made from a rectangular piece of cardboard that measures 120 centimeters by 150 centimeters. To construct the box, squares x centimeters on a side are cut from each corner of the cardboard. The remaining flaps are folded up to create a box.

a. Express the volume of the box as a polynomial function in x.

b. What is the volume of the box when squares 30 centimeters on a side are cut from the cardboard?

49. Surface Area For the box in Exercise 47, express the surface area of the open box as a polynomial function in x.

50. Surface Area For the box in Exercise 48, express the surface area of the open box as a polynomial function in x.

51. Construction A piece of metal 8 feet long and 3 feet wide is used to make a trough to divert water by bending up the sides to equal heights, as shown in the figure.

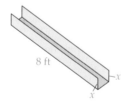

Express the volume of water that could run through the trough as a polynomial function in x.

52. Space Science The weight w, in pounds, of an astronaut h miles above Earth's surface can be approximated by $w(h) = M\left(\dfrac{3960}{h + 3960}\right)^2$, where M is the weight of the astronaut on Earth.

a. Suppose an astronaut weighs 125 pounds on Earth. Complete the table below to find the astronaut's weight at various heights above Earth.

h	10	20	50	100	500
$w(h)$					

b. On the basis of the values in the table, as h increases, $w(h)$ ____?____.

53. Bird Watching A bird flies directly over a bird watcher at a height of 100 feet and then flies in a straight line at 25 feet per second.

a. Express the distance $d(t)$, in feet, between the bird watcher and the bird in terms of t, the number of seconds after the bird passes over the bird watcher.

b. Find the distance the bird is from the bird watcher after 15 seconds. Round to the nearest tenth.

54. Hot-Air Balloon An observer is 50 feet from a hot-air balloon as the balloon starts to rise at a rate of 7 feet per second.

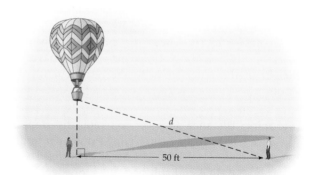

a. Express the distance $d(t)$, in feet, between the observer and the balloon in terms of t, the number of seconds after the balloon begins to rise.

b. Find the distance from the observer to the balloon after 1 minute. Round to the nearest tenth.

55. Inventory A car dealership has determined that the cost, in dollars, of storing x hybrid cars is $1000x$. The cost, in dollars, of reordering x hybrid cars is $\dfrac{15,000x + 20,000}{x}$.

a. Express the inventory cost, $C(x)$, for the cars in terms of x.

b. Find the inventory cost if the dealership wants to maintain an inventory of 25 hybrid cars.

56. Inventory A toy company has determined that the cost, in dollars, of storing x toys is $1.50x$. The cost, in dollars, of reordering x toys is $\dfrac{4x + 300}{x}$.

   a. Express the inventory cost, $C(x)$, for the toys in terms of x.

   b. Find the inventory cost if the toy company wants to maintain an inventory of 400 toys.

57. Speed of Sound If a person drops a rock into a deep well, the total elapsed time t from the time the rock is dropped until the person hears the sound of the rock hitting the water is the sum of the time it takes the rock to reach the bottom and the time it takes the sound of the rock hitting the water to reach the top of the well. The time, in seconds, it takes the rock to hit the water is given by $\dfrac{\sqrt{s}}{4}$, where s is the depth of the well in feet. The time, in seconds, it takes the sound to reach the top of the well is given by $\dfrac{s}{1130}$.

   a. Express the total elapsed time as a function of the depth of the well.

   b. What is the total elapsed time when the depth of the well is 100 feet? Round to the nearest tenth of a second.

58. Sustainable Yield Biologists use a function called *sustainable yield*, which is the difference between next year's population of an animal species and this year's population. A biologist estimates that next year's population of a certain salmon is given by $40\sqrt{p}$, where p is this year's population, in thousands.

   a. Express the sustainable yield of the salmon as a function of p.

   b. What is the value of the sustainable yield function when p is 2500 salmon?

Extensions
CRITICAL THINKING

In Exercises 59 and 60, find the value of k such that the graph of the equation contains the given point.

59. $f(x) = x^2 - 3x + k;\ (2, 5)$

60. $f(x) = 2x^2 + kx - 3;\ (4, -3)$

61. Does a quadratic function always have either a maximum or a minimum value?

62. If $f(x) = ax^2 + bx + c,\ a > 0$, does the graph of f become thinner or wider as a increases?

EXPLORATIONS

A real number x is called a **zero of a function** if the function evaluated at x is 0. That is, if $f(x) = 0$, then x is called a zero of the function. For instance, evaluating $f(x) = x^2 + x - 6$ when $x = -3$, we have

$$f(x) = x^2 + x - 6$$
$$f(-3) = (-3)^2 + (-3) - 6 \qquad \text{• Replace } x \text{ by } -3.$$
$$f(-3) = 9 - 3 - 6 = 0$$

For this function, $f(-3) = 0$, so -3 is a zero of the function.

63. Verify that 2 is a zero of $f(x) = x^2 + x - 6$ by showing that $f(2) = 0$.

The graph of $f(x) = x^2 + x - 6$ is shown below. Note that the graph crosses the x-axis at -3 and 2, the two zeros of the function. The points $(-3, 0)$ and $(2, 0)$ are x-intercepts of the graph.

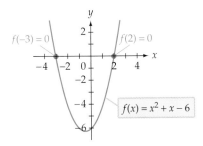

Consider the equation $0 = x^2 + x - 6$, which is $f(x) = x^2 + x - 6$ with $f(x)$ replaced by 0. The solutions of the equation are the zeros of the function. This important connection among the real zeros of a function, the x-intercepts of its graph, and the real solutions of an equation is the basis of using a graphing calculator to solve an equation.

The following method of solving a quadratic equation using a graphing calculator is based on a TI-83/TI-84 Plus calculator. Other calculators will necessitate a slightly different approach.

Approximate the solutions of $x^2 + 4x = 6$ by using a graphing calculator.

i. Write the equation in standard form.

$$x^2 + 4x = 6$$
$$x^2 + 4x - 6 = 0$$

Press $\boxed{Y=}$ and enter $x^2 + 4x - 6$ for Y1.

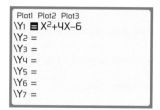

ii. Press $\boxed{GRAPH}$. If the graph does not appear on the screen, press $\boxed{ZOOM}$ 6.

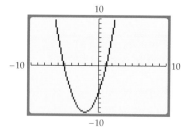

iii. Press $\boxed{2nd}$ CALC 2. Note that the selection for 2 says `zero`. This will begin the calculation of the zeros of the function, which are the solutions of the equation.

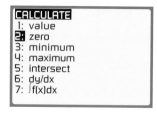

iv. At the bottom of the screen you will see `LeftBound?` This is asking you to move the blinking cursor so that it is to the *left* of the first x-intercept. Use the left arrow key to move the cursor to the left of the first x-intercept. The values of x and y that appear on your calculator may be different from the ones shown here.

Just be sure you are to the left of the x-intercept. When you are done, press $\boxed{ENTER}$.

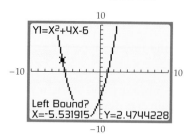

v. At the bottom of the screen you will see `Right Bound?` This is asking you to move the blinking cursor so that it is to the *right* of the x-intercept. Use the right arrow key to move the cursor to the right of the x-intercept. The values of x and y that appear on your calculator may be different from the ones shown here. Just be sure you are to the right of the x-intercept. When you are done, press $\boxed{ENTER}$.

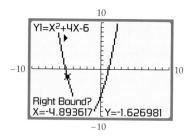

vi. At the bottom of the screen you will see `Guess?` This is asking you to move the blinking cursor so that it is close to the approximate x-intercept. Use the arrow keys to move the cursor to the approximate x-intercept. The values of x and y that appear on your calculator may be different from the ones shown here. When you are done, press $\boxed{ENTER}$.

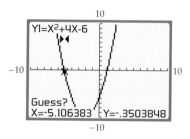

vii. The zero of the function is approximately -5.162278. Thus one solution of $x^2 + 4x = 6$ is approximately -5.162278. Also note that the value of y is given as Y1 $= ^-1E^-12$. This is the way the calculator writes a number in scientific notation. We would normally write Y1 $= -1.0 \times 10^{-12}$. This number is very close to zero.

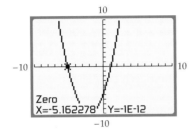

To find the other solution, repeat steps **iii** through **vi**. The screens are shown below.

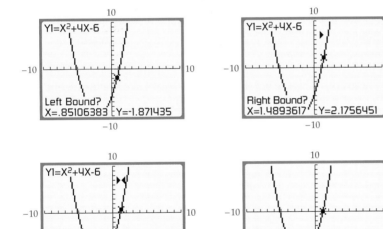

A second zero of the function is approximately 1.1622777. Thus, the two solutions of $x^2 + 4x = 6$ are approximately -5.162278 and 1.1622777.

 In Exercises 64–69, find the solution of each equation. Round to the nearest hundredth.

64. $x^2 + 3x - 4 = 0$

65. $x^2 - 4x - 5 = 0$

66. $x^2 + 3.4x = 4.15$

67. $2x^2 - \dfrac{5}{9}x = \dfrac{3}{8}$

68. $\pi x^2 - \sqrt{17}x - 2 = 0$

69. $\sqrt{2}x^2 + x - \sqrt{7} = 0$

The method described above can be extended to find the real zeros, and therefore the real roots, of any equation that can be graphed. In Exercises 70–72, solve each equation. The number after the equation states how many solutions exist for that equation.

70. $x^3 - 2x^2 - 5x + 6 = 0$; 3

71. $x^3 + 2x^2 - 11x - 12 = 0$; 3

72. $x^4 + 2x^3 - 13x^2 - 14x + 24 = 0$; 4

Exponential Functions

Introduction to Exponential Functions

In 1965, Gordon Moore, one of the cofounders of Intel Corporation, observed that the maximum number of transistors that could be placed on a microprocessor seemed to be doubling every 18 to 24 months. The table below shows how the maximum number of transistors on various Intel processors has changed over time. (*Source:* www.intel.com/technology/mooreslaw/index.htm)

Year, x	1971	1978	1982	1985	1989	1993	1997	1999	2000
Number of transistors per microprocessor (in thousands), y	2.3	29	120	275	1180	3100	7500	24,000	42,000

The curve in Figure 6.1 that approximately passes through the points associated with the data for the years 1971 to 2000 is the graph of a mathematical model of the data. The model is an *exponential function.* Because y is increasing (growing), it is an *exponential growth function.*

When light enters water, the intensity of the light decreases with the depth of the water. The graph in Figure 6.2 shows a model, for Lake Michigan, of the decrease in the percent of light available as the depth of the water increases. This model is also an exponential function. In this case, y is decreasing (decaying), and the model is an *exponential decay function.*

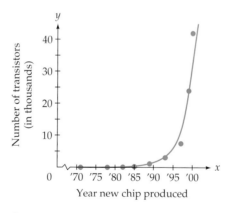

Figure 6.1

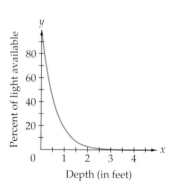

Figure 6.2

Exponential Function

The **exponential function** is defined by $f(x) = b^x$, where b is called the base, $b > 0$, $b \neq 1$, and x is any real number.

The base b of $f(x) = b^x$ must be positive. If the base were a negative number, the value of the function would not be a real number for some values of x. For instance, if $b = -4$ and $x = \frac{1}{2}$, then $f\left(\frac{1}{2}\right) = (-4)^{1/2} = \sqrt{-4}$, which is not a real number. Also, $b \neq 1$ because when $b = 1$, $b^x = 1^x = 1$, a constant function.

At the right we evaluate $f(x) = 2^x$ for $x = 3$ and $x = -2$.

$$f(x) = 2^x$$
$$f(3) = 2^3 = 8$$
$$f(-2) = 2^{-2} = \frac{1}{2^2} = \frac{1}{4}$$

To evaluate the exponential function $f(x) = 2^x$ for an irrational number such as $x = \sqrt{2}$, we use a rational approximation of $\sqrt{2}$ (for instance, 1.4142) and a calculator to obtain an approximation of the function.

$$f\left(\sqrt{2}\right) = 2^{\sqrt{2}} \approx 2^{1.4142} \approx 2.6651$$

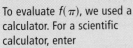

CALCULATOR NOTE

To evaluate $f(\pi)$, we used a calculator. For a scientific calculator, enter

3 $\boxed{y^x}$ π $\boxed{=}$

For the TI-83/84 Plus graphing calculator, enter

3 $\boxed{\wedge}$ $\boxed{2nd}$ π $\boxed{ENTER}$

EXAMPLE 1 ■ **Evaluate an Exponential Function**

Evaluate $f(x) = 3^x$ at $x = 2$, $x = -4$, and $x = \pi$. Round approximate results to the nearest hundred thousandth.

Solution
$$f(2) = 3^2 = 9$$
$$f(-4) = 3^{-4} = \frac{1}{3^4} = \frac{1}{81}$$
$$f(\pi) = 3^\pi \approx 3^{3.1415927} \approx 31.54428$$

CHECK YOUR PROGRESS 1 Evaluate $g(x) = \left(\frac{1}{2}\right)^x$ when $x = 3$, $x = -1$, and $x = \sqrt{3}$. Round approximate results to the nearest thousandth.

Solution *See page S23.*

Graphs of Exponential Functions

The graph of $f(x) = 2^x$ is shown in Figure 6.3. The coordinates of some of the points on the graph are given in the table.

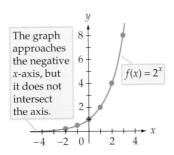

The graph approaches the negative x-axis, but it does not intersect the axis.

$f(x) = 2^x$

Figure 6.3

x	$f(x) = 2^x$	(x, y)
−2	$f(-2) = 2^{-2} = \frac{1}{4}$	$\left(-2, \frac{1}{4}\right)$
−1	$f(-1) = 2^{-1} = \frac{1}{2}$	$\left(-1, \frac{1}{2}\right)$
0	$f(0) = 2^0 = 1$	(0, 1)
1	$f(1) = 2^1 = 2$	(1, 2)
2	$f(2) = 2^2 = 4$	(2, 4)
3	$f(3) = 2^3 = 8$	(3, 8)

Observe that the values of *y increase* as *x* increases. This is an **exponential growth function.** This is typical of the graphs of all exponential functions for which the base is *greater than* 1. For the function $f(x) = 2^x$, $b = 2$, which is greater than 1.

Now consider the graph of an exponential function for which the base is between 0 and 1. The graph of $f(x) = \left(\frac{1}{2}\right)^x$ is shown in Figure 6.4. The coordinates of some of the points on the graph are given in the table.

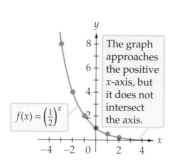

Figure 6.4

x	$f(x) = \left(\dfrac{1}{2}\right)^x$	(x, y)
−3	$f(-3) = \left(\dfrac{1}{2}\right)^{-3} = 8$	(−3, 8)
−2	$f(-2) = \left(\dfrac{1}{2}\right)^{-2} = 4$	(−2, 4)
−1	$f(-1) = \left(\dfrac{1}{2}\right)^{-1} = 2$	(−1, 2)
0	$f(0) = \left(\dfrac{1}{2}\right)^{0} = 1$	(0, 1)
1	$f(1) = \left(\dfrac{1}{2}\right)^{1} = \dfrac{1}{2}$	$\left(1, \dfrac{1}{2}\right)$
2	$f(2) = \left(\dfrac{1}{2}\right)^{2} = \dfrac{1}{4}$	$\left(2, \dfrac{1}{4}\right)$

Observe that the values of *y decrease* as *x* increases. This is an **exponential decay function.** This is typical of the graphs of all exponential functions for which the positive base is *less than* 1. For the function $f(x) = \left(\frac{1}{2}\right)^x$, $b = \frac{1}{2}$, which is less than 1.

QUESTION *Is $f(x) = 0.25^x$ an exponential growth function or an exponential decay function?*

EXAMPLE 2 ■ Graph an Exponential Function

State whether $g(x) = \left(\frac{3}{4}\right)^x$ is an exponential growth function or an exponential decay function. Then graph the function.

Solution
Because the base $\frac{3}{4}$ is less than 1, *g* is an exponential decay function. Because it is an exponential decay function, the *y*-values will decrease as *x* increases. The *y*-intercept of the graph is the point (0, 1), and the graph also passes through $\left(1, \frac{3}{4}\right)$. Plot a few additional points. Then draw a smooth curve through the points, as shown in the figure on the following page.

ANSWER *The base is 0.25, which is less than 1. The function is an exponential decay function.*

x	$g(x) = \left(\dfrac{3}{4}\right)^x$	(x, y)
-3	$g(-3) = \left(\dfrac{3}{4}\right)^{-3} = \dfrac{64}{27}$	$\left(-3, \dfrac{64}{27}\right)$
-2	$g(-2) = \left(\dfrac{3}{4}\right)^{-2} = \dfrac{16}{9}$	$\left(-2, \dfrac{16}{9}\right)$
-1	$g(-1) = \left(\dfrac{3}{4}\right)^{-1} = \dfrac{4}{3}$	$\left(-1, \dfrac{4}{3}\right)$
0	$g(0) = \left(\dfrac{3}{4}\right)^{0} = 1$	$(0, 1)$
1	$g(1) = \left(\dfrac{3}{4}\right)^{1} = \dfrac{3}{4}$	$\left(1, \dfrac{3}{4}\right)$
2	$g(2) = \left(\dfrac{3}{4}\right)^{2} = \dfrac{9}{16}$	$\left(2, \dfrac{9}{16}\right)$
3	$g(3) = \left(\dfrac{3}{4}\right)^{3} = \dfrac{27}{64}$	$\left(3, \dfrac{27}{64}\right)$

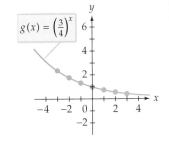

CHECK YOUR PROGRESS 2   State whether $f(x) = \left(\dfrac{3}{2}\right)^x$ is an exponential growth function or an exponential decay function. Then graph the function.

Solution   *See page S23.*

The Natural Exponential Function

The irrational number π is often used in applications that involve circles. Another irrational number, denoted by the letter e, is useful in applications that involve growth or decay.

n	$\left(1 + \dfrac{1}{n}\right)^n$
10	2.59374246
100	2.70481383
1000	2.71692393
10,000	2.71814593
100,000	2.71826824
1,000,000	2.71828047

The Number e

The number e is defined as the number that

$$\left(1 + \dfrac{1}{n}\right)^n$$

approaches as n increases without bound.

The letter e was chosen in honor of the Swiss mathematician Leonhard Euler. He was able to compute the value of e to several decimal places by evaluating $\left(1 + \dfrac{1}{n}\right)^n$ for large values of n, as shown at the left. The value of e accurate to eight decimal places is 2.71828183.

> **The Natural Exponential Function**
>
> For all real numbers x, the function defined by $f(x) = e^x$ is called the **natural exponential function.**

A calculator with an e^x key can be used to evaluate e^x for specific values of x. For instance,

$$e^2 \approx 7.389056, \qquad e^{3.5} \approx 33.115452, \qquad \text{and} \qquad e^{-1.4} \approx 0.246597$$

The graph of the natural exponential function can be constructed by plotting a few points or by using a graphing utility.

EXAMPLE 3 ■ **Graph a Natural Exponential Function**

  Graph $f(x) = e^x$.

Solution

Use a calculator to find range values for a few domain values. The range values in the table below have been rounded to the nearest tenth.

x	-2	-1	0	1	2
$f(x) = e^x$	0.1	0.4	1.0	2.7	7.4

Plot the points given in the table and then connect the points with a smooth curve. Because $e > 1$, as x increases, e^x increases. Thus the values of y increase as x increases. As x decreases, e^x becomes closer to zero. For instance, when $x = -5$, $e^{-5} \approx 0.0067$. Thus as x decreases, the graph gets closer and closer to the x-axis. The y-intercept is $(0, 1)$.

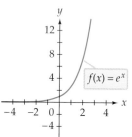

CHECK YOUR PROGRESS 3   Graph $f(x) = e^{-x} + 2$.

Solution   *See page S23.*

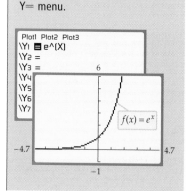

In the figure at the right, compare the graph of $f(x) = e^x$ with the graphs of $g(x) = 2^x$ and $h(x) = 3^x$. Because $2 < e < 3$, the graph of $f(x) = e^x$ is between the graphs of g and h.

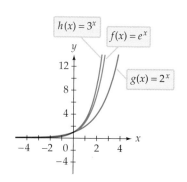

Applications of Exponential Functions

Many applications can be effectively modeled by exponential functions. For instance, when money is deposited in a compound interest account, the value of the money can be represented by an exponential growth function. When physicians use a test that involves a radioactive element, the amount of radioactivity remaining in the patient's body can be modeled by an exponential decay function.

✓ **TAKE NOTE**

Here is an application of exponential functions. Assume that you have a long strip of paper that is $\frac{1}{16}$ inch thick and can be folded in half infinitely many times. After how many folds would the paper be as high as the Empire State Building (1250 feet)?

EXAMPLE 4 ■ Application of an Exponential Function

 When an amount of money P is placed in an account that earns compound interest, the value A of the money after t years is given by the compound interest formula

$$A = P\left(1 + \frac{r}{n}\right)^{nt}$$

where r is the annual interest rate as a decimal and n is the number of compounding periods per year. Suppose $500 is placed in an account that earns 8% interest compounded daily. Find the value of the investment after 5 years.

Solution
Use the compound interest formula. Because interest is compounded daily, $n = 365$.

$$A = P\left(1 + \frac{r}{n}\right)^{nt}$$

$$= 500\left(1 + \frac{0.08}{365}\right)^{365(5)} \qquad • \ P = 500, r = 0.08, n = 365, t = 5$$

$$\approx 500(1.491759) \approx 745.88$$

After 5 years, there is $745.88 in the account.

CHECK YOUR PROGRESS 4 The radioactive isotope iodine-131 is used to monitor thyroid activity. The number of grams N of iodine-131 in the body t hours after an injection is given by $N(t) = 1.5\left(\frac{1}{2}\right)^{t/193.7}$. Find the number of grams of the isotope in the body 24 hours after an injection. Round to the nearest ten-thousandth.

Solution See page S23.

The next example is based on Newton's Law of Cooling. This exponential function can be used to model the temperature of something that is being cooled.

EXAMPLE 5 ■ Application of an Exponential Function

A cup of coffee is heated to 160°F and placed in a room that maintains a temperature of 70°F. The temperature T of the coffee after t minutes is given by $T(t) = 70 + 90e^{-0.0485t}$. Find the temperature of the coffee 20 minutes after it is placed in the room. Round to the nearest degree.

Solution

Evaluate the function $T(t) = 70 + 90e^{-0.0485t}$ for $t = 20$.

$$T(t) = 70 + 90e^{-0.0485t}$$

$$T(20) = 70 + 90e^{-0.0485(20)} \qquad \text{• Substitute 20 for } t.$$

$$\approx 70 + 34.1$$

$$\approx 104.1$$

After 20 minutes the temperature of the coffee is about 104°F.

CHECK YOUR PROGRESS 5 The function $A(t) = 200e^{-0.014t}$ gives the amount of aspirin, in milligrams, in a patient's bloodstream t minutes after the aspirin has been administered. Find the amount of aspirin in the patient's bloodstream after 45 minutes. Round to the nearest milligram.

Solution *See page S23.*

Math Matters Marie Curie

In 1903, Marie Sklodowska-Curie (1867–1934) became the first woman to receive a Nobel prize in physics. She was awarded the prize along with her husband, Pierre Curie, and Henri Becquerel for their discovery of radioactivity. In fact, Marie Curie coined the word *radioactivity*.

In 1911, Marie Curie became the first person to win a second Nobel prize, this time alone and in chemistry, for the isolation of radium. She also discovered the element polonium, which she named after Poland, her birthplace. The radioactive phenomena that she studied are modeled by exponential decay functions.

The stamp at the left was printed in 1938 to commemorate Pierre and Marie Curie. When the stamp was issued, there was a surcharge added to the regular price of the stamp. The additional revenue was donated to cancer research. Marie Curie died in 1934 from leukemia, which was caused by her constant exposure to radiation from the radioactive elements she studied.

In 1935, one of Marie Curie's daughters, Irene Joloit-Curie, won a Nobel prize in chemistry.

Exponential Regression

 We examined linear regression models in the last chapter. In some cases, an exponential function may more closely model a set of data.

TAKE NOTE

The value of a diamond is generally determined by its color, cut, clarity, and carat weight. These characteristics of a diamond are known as the four c's. In Example 6 we have assumed that the color, cut, and clarity of all of the diamonds are similar. This assumption enables us to model the value of each diamond as a function of just its carat weight.

▼ point of interest

The Hope Diamond, shown below, is the world's largest deep-blue diamond. It weighs 45.52 carats. We should not expect the function $y \approx 4067.6(1.3816)^x$ in Example 6 to yield an accurate value of the Hope Diamond because the Hope Diamond is not the same type of diamond as the diamonds in Example 6, and its weight is much larger.

The Hope Diamond is on display at the Smithsonian Museum of Natural History in Washington, D.C.

EXAMPLE 6 ■ Find an Exponential Regression Equation

A diamond merchant has determined the values of several white diamonds that have different weights, measured in carats, but are similar in quality. See the table below.

4.00 ct	3.00 ct	2.00 ct	1.75 ct	1.50 ct	1.25 ct	1.00 ct	0.75 ct	0.50 ct
$14,500	$10,700	$7900	$7300	$6700	$6200	$5800	$5000	$4600

Find an exponential growth function that models the values of the diamonds as a function of their weights and use the model to predict the value of a 3.5-carat diamond of similar quality.

Solution

Use a graphing calculator to find the regression equation. The calculator display below shows that the exponential regression equation is $y \approx 4067.6(1.3816)^x$, where x is the carat weight of the diamond and y is the value of the diamond.

```
ExpReg
 y = a*b^x
 a = 4067.641145
 b = 1.381644186
 r² = .994881215
 r = .9974373238
```

To use the regression equation to predict the value of a 3.5-carat diamond of similar quality, substitute 3.5 for x and evaluate.

$$y \approx 4067.6(1.3816)^x$$
$$y \approx 4067.6(1.3816)^{3.5}$$
$$y \approx 12{,}609$$

According to the modeling function, the value of a 3.5-carat diamond of similar quality is $12,609.

CHECK YOUR PROGRESS 6   The following table shows Earth's atmospheric pressure P at an altitude of a kilometers. Find an exponential function that models the atmospheric pressure as a function of altitude. Use the function to estimate, to the nearest tenth, the atmospheric pressure at an altitude of 24 kilometers.

Altitude, *a*, Above Sea Level, in kilometers	Atmospheric Pressure, *P*, in newtons per square centimeter
0	10.3
2	8.0
4	6.4
6	5.1
8	4.0
10	3.2
12	2.5
14	2.0
16	1.6
18	1.3

Solution   *See page S23.*

Investigation

Chess and Exponential Functions

According to legend, when Sissa Ben Dahir of India invented the game of chess, King Shirham was so impressed with the game that he summoned the game's inventor and offered him the reward of his choosing. The inventor pointed to the chessboard and requested, for his reward, one grain of wheat on the first square, two grains of wheat on the second square, four grains on the third square, eight grains on the fourth square, and so on for all 64 squares on the chessboard. The king considered this a very modest reward and said he would grant the inventor's wish.

Investigation Exercises

1. This portion of this Investigation will enable you to find a formula for the total amount of wheat on the first *n* squares of the chessboard. You may want to use the chart below as you answer the questions. It may help you to see a pattern.

Square number, *n*	1	2	3	4	5	6	7
Number of grains of wheat on square *n*	1	2	4				
Total number of grains of wheat on squares 1 through *n*	1	1 + 2 =	3 + 4 =				

a. How many grains of wheat are on each of the first seven squares?

(continued)

What is the total number (the sum) of grains of wheat on the first

    b. two squares?

    c. three squares?

    d. four squares?

    e. five squares?

    f. six squares?

    g. seven squares?

2. Use inductive reasoning to find a function that gives the total number (the sum) of grains of wheat on the first n squares of the chessboard. Test your function to ensure that it works for parts b through g.

3. If all 64 squares of the chessboard are piled with wheat as requested by Sissa Ben Dahir, how many grains of wheat are on the board?

4. A grain of wheat weighs approximately 0.000008 kilogram. Find the total weight of the wheat requested by Sissa Ben Dahir.

5. In a recent year, a total of 6.5×10^8 metric tons of wheat was produced in the world. At this level, how many years of wheat production would be required to fill the request of Sissa Ben Dahir? *Hint:* One metric ton equals 1000 kilograms.

Exercise Set 6.2

1. Given $f(x) = 3^x$, evaluate:

    a. $f(2)$ **b.** $f(0)$ **c.** $f(-2)$

2. Given $H(x) = 2^x$, evaluate:

    a. $H(-3)$ **b.** $H(0)$ **c.** $H(2)$

3. Given $g(x) = 2^{x+1}$, evaluate:

    a. $g(3)$ **b.** $g(1)$ **c.** $g(-3)$

4. Given $F(x) = 3^{x-2}$, evaluate:

    a. $F(-4)$ **b.** $F(-1)$ **c.** $F(0)$

5. Given $G(r) = \left(\frac{1}{2}\right)^{2r}$, evaluate:

    a. $G(0)$ **b.** $G\left(\frac{3}{2}\right)$ **c.** $G(-2)$

6. Given $R(t) = \left(\frac{1}{3}\right)^{3t}$, evaluate:

    a. $R\left(-\frac{1}{3}\right)$ **b.** $R(1)$ **c.** $R(-2)$

7. Given $f(x) = 2^{-x}$, is there a value of x for which $f(x)$ is a negative number? Explain.

8. Given $f(x) = \left(\frac{2}{3}\right)^x$, as the value of x increases, does $f(x)$ increase or decrease? Explain.

9. Does $f(x) = e^x$ represent exponential growth, exponential decay, or neither growth nor decay? Explain.

10. Does $f(x) = x^2$ represent exponential growth, exponential decay, or neither growth nor decay? Explain.

11. Given $h(x) = e^{x/2}$, evaluate the following. Round to the nearest ten-thousandth.

    a. $h(4)$ **b.** $h(-2)$ **c.** $h\left(\frac{1}{2}\right)$

12. Given $f(x) = e^{2x}$, evaluate the following. Round to the nearest ten-thousandth.

    a. $f(-2)$ **b.** $f\left(-\frac{2}{3}\right)$ **c.** $f(2)$

13. Given $H(x) = e^{-x+3}$, evaluate the following. Round to the nearest ten-thousandth.

    a. $H(-1)$ **b.** $H(3)$ **c.** $H(5)$

14. Given $g(x) = e^{-x/2}$, evaluate the following. Round to the nearest ten-thousandth.

    a. $g(-3)$ **b.** $g(4)$ **c.** $g\left(\frac{1}{2}\right)$

15. Given $F(x) = 2^{x^2}$, evaluate the following. Round to the nearest ten-thousandth.

    a. $F(2)$ **b.** $F(-2)$ **c.** $F\left(\frac{3}{4}\right)$

16. Given $Q(x) = 2^{-x^2}$, evaluate the following.

   a. $Q(3)$ **b.** $Q(-1)$ **c.** $Q(-2)$

17. Given $f(x) = e^{-x^2/2}$, evaluate the following. Round to the nearest ten-thousandth.

   a. $f(-2)$ **b.** $f(2)$ **c.** $f(-3)$

18. Given $h(x) = e^{-2x} + 1$, evaluate the following. Round to the nearest ten-thousandth.

   a. $h(-1)$ **b.** $h(3)$ **c.** $h(-2)$

In Exercises 19–28, graph the equation.

19. $f(x) = 2^x + 1$ **20.** $f(x) = 3^x - 2$

21. $g(x) = 3^{x/2}$ **22.** $h(x) = 2^{-x/2}$

23. $f(x) = 2^{x+3}$ **24.** $g(x) = 4^{-x} + 1$

25. $H(x) = 2^{2x}$ **26.** $F(x) = 2^{-x}$

27. $f(x) = e^{-x}$ **28.** $y(x) = e^{2x}$

Investments In Exercises 29 and 30, use the compound interest formula $A = P\left(1 + \dfrac{r}{n}\right)^{nt}$, where P is the amount deposited, A is the value of the money after t years, r is the annual interest rate in decimal form, and n is the number of compounding periods per year.

29. A computer network specialist deposits $2500 in a retirement account that earns 7.5% annual interest, compounded daily. What is the value of the investment after 20 years?

30. A $10,000 certificate of deposit (CD) earns 5% annual interest, compounded daily. What is the value of the investment after 20 years?

31. **Investments** Some banks now use continuous compounding of an amount invested. In this case, the equation that models the value of an initial investment of P dollars after t years at an annual interest rate of r is given by $A = Pe^{rt}$. Using this equation, find the value after 5 years of an investment of $2500 that earns 5% annual interest.

32. **Isotopes** An isotope of technetium is used to prepare images of internal body organs. This isotope has a half-life (time required for half the material to erode) of approximately 6 hours. If a patient is injected with 30 milligrams of this isotope, what will be the technetium level in the patient after 3 hours? Use the function $A(t) = 30\left(\frac{1}{2}\right)^{t/6}$, where A is the technetium level, in milligrams, in the patient after t hours. Round to the nearest tenth.

33. **Isotopes** Iodine-131 is an isotope that is used to study the functioning of the thyroid gland. This isotope has a half-life (time required for half the material to erode) of approximately 8 days. If a patient is given an injection that contains 8 micrograms of iodine-131, what will be the iodine level in the patient after 5 days? Use the function $A(t) = 8\left(\frac{1}{2}\right)^{t/8}$, where A is the amount of the isotope, in micrograms, in the patient after t days. Round to the nearest tenth.

34. **Welding** The percent of correct welds that a student can make will increase with practice and can be approximated by the function $P(t) = 100[1 - (0.75)^t]$, where P is the percent of correct welds and t is the number of weeks of practice. Find the percent of correct welds that a student will make after 4 weeks of practice. Round to the nearest percent.

35. **Music** The "concert A" note on a piano is the first A below middle C. When that key is struck, the string associated with the key vibrates 440 times per second. The next A above concert A vibrates twice as fast. An exponential function with a base of 2 is used to determine the frequency of the 11 notes between the two As. Find this function. *Hint:* The function is of the form $f(x) = k \cdot 2^{(cx)}$, where k and c are constants. Also, $f(0) = 440$ and $f(12) = 880$.

36. **Atmospheric Pressure** Atmospheric pressure changes as you rise above Earth's surface. At an altitude of h kilometers, where $0 < h < 80$, the pressure P in newtons per square centimeter is approximately modeled by the function

$$P(h) = 10.13e^{-0.116h}.$$

   a. What is the approximate pressure at 40 kilometers above Earth?

   b. What is the approximate pressure on Earth's surface?

   c. Does atmospheric pressure increase or decrease as you rise above Earth's surface?

37. **Automobiles** The number of automobiles in the United States in 1900 was around 8000. In the year 2000, the number of automobiles in the United States reached 200 million. Find an exponential model for the data and use the model to predict the number of automobiles in the United States in 2010. Use $t = 0$ to represent the year 1900. Round to the nearest hundred thousand automobiles.

38. **Panda Population** One estimate gives the world panda population as 3200 in 1980 and 590 in 2000. Find an exponential model for the data and use the model to predict the panda population in 2040. Use $t = 0$ to represent the year 1980. Round to the nearest whole number.

39. Polonium An initial amount of 100 micrograms of polonium decays to 75 micrograms in approximately 34.5 days. Find an exponential model for the amount of polonium in the sample after t days. Round to the nearest hundredth of a microgram.

40. **The Film Industry** The table below shows the number of multidisc DVDs, with three or more discs, released each year from 1999 to 2003. (*Source:* DVD Release Report)

Year, x	1999	2000	2001	2002	2003
Titles released, y	11	57	87	154	283

a. Find an exponential regression equation for these data, using $x = 0$ to represent 1995. Round to the nearest hundredth.

b. Use the equation to predict the number of multidisc DVDs that will be released in 2008.

41. **Meteorology** The table below shows the saturation of water in air at various air temperatures.

Temperature (in °C)	0	5	10	20	25	30
Saturation (in milliliters of water per cubic meter of air)	4.8	6.8	9.4	17.3	23.1	30.4

a. Find an exponential regression equation for these data. Round to the nearest thousandth.

b. Use the equation to predict the number of milliliters of water per cubic meter of air at a temperature of 15°C. Round to the nearest tenth.

42. **Snow Making** Artificial snow is made at a ski resort by combining air and water in a ratio that depends on the outside air temperature. The following table shows the rates of air flow needed for various temperatures.

Temperature (in °F)	0	5	10	15	20
Air flow (in cubic feet per minute)	3.0	3.6	4.7	6.1	9.9

a. Find an exponential regression equation for these data. Round to the nearest hundredth.

b. Use the equation to predict the air flow needed when the temperature is 25°F. Round to the nearest tenth of a cubic foot per minute.

Extensions

CRITICAL THINKING

An exponential model for population growth or decay can be accurate over a short period of time. However, this model begins to fail because it does not account for the natural resources necessary to support growth, nor does it account for death within the population. Another model, called the *logistic model,* can account for some of these effects. The logistic model is given by $P(t) = \dfrac{mP_0}{P_0 + (m - P_0)e^{-kt}}$, where $P(t)$ is the population at time t, m is the maximum population that can be supported, P_0 is the population when $t = 0$, and k is a positive constant that is related to the growth of the population.

43. Earth's Population One model of Earth's population is given by

$$P(t) = \frac{280}{4 + 66e^{-0.021t}}.$$

In this equation, $P(t)$ is the population in billions and t is the number of years after 1980. Round answers to the nearest hundred million people.

a. According to this model, what was Earth's population in the year 2000?

b. According to this model, what will be Earth's population in the year 2010?

c. If t is very large, say greater than 500, then $e^{-0.021t} \approx 0$. What does this suggest about the maximum population that Earth can support?

44. Wolf Population Game wardens have determined that the maximum wolf population in a certain preserve is 1000 wolves. Suppose the population of wolves in the preserve in the year 2000 was 500, and that k is estimated to be 0.025.

a. Find a logistic function for the number of wolves in the preserve in year t, where t is the number of years after 2000.

b. Find the estimated wolf population in 2015.

45. **Car Payments** The formula used to calculate a monthly lease payment or a monthly car payment (for a purchase rather than a lease) is given by

$$P = \frac{Ar(1 + r)^n - Vr}{(1 + r)^n - 1},$$

where P is the monthly payment, A is the amount of the loan, r is the *monthly* interest rate in decimal form, n is the number of months of the loan or lease, and V is the residual value of the car at the end of the lease. For a car purchase, $V = 0$.

a. If the annual interest rate for a loan is 9%, what is the monthly interest rate in decimal form?

b. Write the formula for a monthly car payment when the car is purchased rather than leased.

c. Suppose you lease a car for 5 years. Find the monthly lease payment if the lease amount is

$10,000, the residual value is $6000, and the annual interest rate is 6%.

d. Suppose you purchase a car and secure a 5-year loan for $10,000 at an annual interest rate of 6%. Find the monthly payment.

e. ✎ Why are the answers to parts c and d different?

The total amount C that has been repaid on a loan or lease is given by $C = \dfrac{(P - Ar)[(1 + r)^n - 1]}{r}$.

f. Using the lease payment in part c, find the total amount that will be repaid in 5 years. How much remains to be paid?

g. Using the monthly payment in part d, find the total amount that will be repaid in 5 years. How much remains to be paid?

h. ✎ Explain why the answers to parts f and g make sense.

| # Logarithmic Functions

Introduction to Logarithmic Functions

Time (in hours)	Number of Bacteria
0	1000
1	2000
2	4000
3	8000

Suppose a bacteria colony that originally contained 1000 bacteria doubles in size every hour. Then the table at the left would show the number of bacteria in that colony after 1, 2, and 3 hours.

The exponential function $A = 1000(2^t)$, where A is the number of bacteria in the colony at time t, is a model of the growth of the colony. For instance, when $t = 3$ hours, we have

$$A = 1000(2^t)$$
$$A = 1000(2^3) \qquad \bullet \text{ Replace } t \text{ by 3.}$$
$$A = 1000(8) = 8000$$

After 3 hours there are 8000 bacteria in the colony.

Now we ask, "How long will it take for there to be 32,000 bacteria in the colony?" To answer the question, we must solve the exponential equation $32,000 = 1000(2^t)$. By trial and error, we find that when $t = 5$,

$$A = 1000(2^t)$$
$$A = 1000(2^5) \qquad \bullet \text{ Replace } t \text{ by 5.}$$
$$A = 1000(32) = 32,000$$

After 5 hours there will be 32,000 bacteria in the colony.

Now suppose we want to know how long it will be before the colony reaches 50,000 bacteria. To answer that question, we must find t such that $50,000 = 1000(2^t)$. Using trial and error again, we find that

$$1000(2^5) = 32,000 \qquad \text{and} \qquad 1000(2^6) = 64,000$$

CALCULATOR NOTE

Using a calculator, we can verify that $2^{5.644} \approx 50$. On a graphing calculator, press 2 ^ 5.644 ENTER.

Because 50,000 is between 32,000 and 64,000, we conclude that t is between 5 and 6 hours. If we try $t = 5.5$ (halfway between 5 and 6), we get

$$A = 1000(2^t)$$
$$A = 1000(2^{5.5}) \qquad \text{• Replace } t \text{ by 5.5.}$$
$$A \approx 1000(45.25) = 45,250$$

After 5.5 hours there will be approximately 45,250 bacteria in the colony. Because this is less than 50,000, the value of t must be a little greater than 5.5.

We could continue to use trial and error to find the correct value of t, but it would be more efficient if we could just solve the exponential equation $50,000 = 1000(2^t)$ for t. If we follow the procedures for solving equations that were discussed earlier in the text, we have

$$50,000 = 1000(2^t)$$
$$50 = 2^t \qquad \text{• Divide each side of the equation by 1000.}$$

To proceed to the next step, it would be helpful to have a function that would find the power of 2 that produces 50.

Around the mid-sixteenth century, mathematicians created such a function, which we now call a *logarithmic function*. We write the solution of $50 = 2^t$ as $t = \log_2 50$. This is read "t equals the logarithm base 2 of 50" and it means "t equals the power of 2 that produces 50." When logarithms were first introduced, tables were used to find a numerical value of t. Today, a calculator is used. Using a calculator, we can approximate the value of t as 5.644. This means that $2^{5.644} \approx 50$.

The equivalence of the expressions $50 = 2^t$ and $t = \log_2 50$ are described in the following definition of logarithm.

✔ TAKE NOTE

Read $\log_b x$ as "the logarithm of x, base b" or "log base b of x."

> **Logarithm**
>
> For $b > 0$, $b \neq 1$, $y = \log_b x$ is equivalent to $x = b^y$.

For every exponential equation there is a corresponding logarithmic equation, and for every logarithmic equation there is a corresponding exponential equation. Here are some examples.

✔ TAKE NOTE

The idea of a function that performs the opposite of a given function occurs frequently. For instance, the opposite of a "doubling" function, one that doubles a given number, is a "halving" function, one that takes one-half of a given number. We could write $f(x) = 2x$ for the doubling function and $g(x) = \frac{1}{2}x$ for the halving function. We call these functions *inverse functions* of one another. In a similar manner, exponential and logarithmic functions are inverse functions of one another.

Exponential Equation	Logarithmic Equation
$2^5 = 32$	$\log_2 32 = 5$
$3^2 = 9$	$\log_3 9 = 2$
$5^{-2} = \dfrac{1}{25}$	$\log_5 \dfrac{1}{25} = -2$
$7^0 = 1$	$\log_7 1 = 0$

QUESTION   *Which of the following is the logarithmic form of $4^3 = 64$?*

    a. $\log_4 3 = 64$ ***b.*** $\log_3 4 = 64$ ***c.*** $\log_4 64 = 3$

ANSWER   **c.** $\log_4 64 = 3$ *is equivalent to* $4^3 = 64$.

The equation $y = \log_b x$ is the logarithmic form of $b^y = x$, and the equation $b^y = x$ is the exponential form of $y = \log_b x$. These two forms state exactly the same relationship between x and y.

historical note

Logarithms were developed independently by Jobst Burgi (1552–1632) and John Napier (1550–1617) as a means of simplifying the calculations of astronomers. The idea was to devise a method by which two numbers could be multiplied by performing additions. Napier is usually given credit for logarithms because he published his results first.

In Napier's original work, the logarithm of 10,000,000 was 0. After this work was published, Napier, in discussions with Henry Briggs (1561–1631), decided that tables of logarithms would be easier to use if the logarithm of 1 were 0. Napier died before new tables could be prepared, and Briggs took on the task. His table consisted of logarithms accurate to 30 decimal places, all accomplished without the use of a calculator! ∎

EXAMPLE 1 ▪ Write a Logarithmic Equation in Exponential Form and an Exponential Equation in Logarithmic Form

a. Write $2 = \log_{10}(x + 5)$ in exponential form.

b. Write $2^{3x} = 64$ in logarithmic form.

Solution

Use the definition of logarithm: $y = \log_b x$ if and only if $b^y = x$.

a. $2 = \log_{10}(x + 5)$ if and only if $10^2 = x + 5$.

b. $2^{3x} = 64$ if and only if $\log_2 64 = 3x$.

CHECK YOUR PROGRESS 1

a. Write $\log_2(4x) = 10$ in exponential form.

b. Write $10^3 = 2x$ in logarithmic form.

Solution   See page S23.

The relationship between the exponential and logarithmic forms can be used to evaluate some logarithms. The solutions to these types of problems are based on the Equality of Exponents Property.

Equality of Exponents Property

If $b > 0$ and $b^x = b^y$, then $x = y$.

EXAMPLE 2 ▪ Evaluate Logarithmic Expressions

Evaluate the logarithms.

a. $\log_8 64$ **b.** $\log_2\left(\dfrac{1}{8}\right)$

Solution

a. $\log_8 64 = x$ • Write an equation.

$\quad\quad 8^x = 64$ • Write the equation in its equivalent exponential form.

$\quad\quad 8^x = 8^2$ • Write 64 in exponential form using 8 as the base.

$\quad\quad\quad x = 2$ • Solve for x using the Equality of Exponents Property.

$\quad \log_8 64 = 2$

b. $\log_2\left(\dfrac{1}{8}\right) = x$ • Write an equation.

$\qquad 2^x = \dfrac{1}{8}$ • Write the equation in its equivalent exponential form.

$\qquad 2^x = 2^{-3}$ • Write $\dfrac{1}{8}$ in exponential form using 2 as the base.

$\qquad x = -3$ • Solve for x using the Equality of Exponents Property.

$\log_2\left(\dfrac{1}{8}\right) = -3$

CHECK YOUR PROGRESS 2   Evaluate the logarithms.

a. $\log_{10} 0.001$ **b.** $\log_5 125$

Solution   *See page S24.*

EXAMPLE 3 ■ Solve a Logarithmic Equation

Solve: $\log_3 x = 2$

Solution

$\qquad \log_3 x = 2$

$\qquad\quad 3^2 = x$ • Write the equation in its equivalent exponential form.

$\qquad\quad\; 9 = x$ • Simplify the exponential expression.

CHECK YOUR PROGRESS 3   Solve: $\log_2 x = 6$

Solution   *See page S24.*

Not all logarithms can be evaluated by rewriting the logarithm in its equivalent exponential form and using the Equality of Exponents Property. For instance, if we tried to evaluate $\log_{10} 18$, it would be necessary to solve the equivalent exponential equation $10^x = 18$. The difficulty here is trying to rewrite 18 in exponential form with 10 as a base.

Common and Natural Logarithms

Two of the most frequently used logarithmic functions are *common logarithms,* which have base 10, and *natural logarithms,* which have base e (the base of the natural exponential function).

> **Common and Natural Logarithms**
>
> The function defined by $f(x) = \log_{10} x$ is called the **common logarithmic function.** It is customarily written without the base as $f(x) = \log x$.
>
> The function defined by $f(x) = \log_e x$ is called the **natural logarithmic function.** It is customarily written as $f(x) = \ln x$.

Most scientific and graphing calculators have a log key for evaluating common logarithms and an ln key for evaluating natural logarithms. For instance,

$$\log 24 \approx 1.3802112 \quad \text{and} \quad \ln 81 \approx 4.3944492$$

EXAMPLE 4 ■ Solve Common and Natural Logarithmic Equations

Solve each of the following equations. Round to the nearest thousandth.

a. $\log x = -1.5$ **b.** $\ln x = 3$

Solution

a. $\log x = -1.5$
$10^{-1.5} = x$ • Write the equation in its equivalent exponential form.
$0.032 \approx x$ • Simplify the exponential expression.

b. $\ln x = 3$
$e^3 = x$ • Write the equation in its equivalent exponential form.
$20.086 \approx x$ • Simplify the exponential expression.

CHECK YOUR PROGRESS 4 Solve each of the following equations. Round to the nearest thousandth.

a. $\log x = -2.1$ **b.** $\ln x = 2$

Solution *See page S24.*

MathMatters Zipf's Law

George Zipf (1902–1950) was a lecturer in German and philology (the study of the change in a language over time) at Harvard University. Zipf's Law, as originally formulated, referred to the frequency of occurrence of a word in a book, magazine article, or other written material, and its rank. The word used most frequently had rank 1, the next most frequent word had rank 2, and so on. Zipf hypothesized that if the x-axis were the logarithm of a word's rank and the y-axis were the logarithm of the frequency of the word, then the graph of rank versus frequency would lie on approximately a straight line.

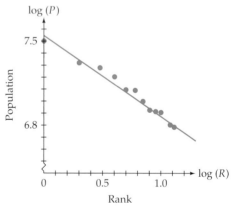

Zipf's Law has been extended to demographics. For instance, let the population of a U.S. state be P and its rank in population be R. The graphs of the points $(\log R, \log P)$ lie on approximately a straight line. The graph above shows Zipf's Law as applied to the populations and ranks of the 12 most populated states according to the 2000 census.

Recently, Zipf's Law has been applied to website traffic, where websites are ranked according to the numbers of hits they receive. Assumptions about web traffic are used by engineers and programmers who study ways to make the World Wide Web more efficient.

Graphs of Logarithmic Functions

The graph of a logarithmic function can be drawn by first rewriting the function in its exponential form. This procedure is illustrated in Example 5.

EXAMPLE 5 ■ **Graph a Logarithmic Function**

Graph $f(x) = \log_3 x$.

Solution

To graph $f(x) = \log_3 x$, first write the equation in the form $y = \log_3 x$. Then write the equivalent exponential equation $x = 3^y$. Because this equation is solved for x, choose values of y and calculate the corresponding values of x, as shown in the table below.

$x = 3^y$	$\frac{1}{9}$	$\frac{1}{3}$	1	3	9
y	-2	-1	0	1	2

Plot the ordered pairs and connect the points with a smooth curve, as shown at the left.

CHECK YOUR PROGRESS 5   Graph $f(x) = \log_5 x$.

Solution   *See page S24.*

The graphs of $y = \log x$ and $y = \ln x$ can be drawn on a graphing calculator by using the $\boxed{\texttt{log}}$ and $\boxed{\texttt{ln}}$ keys. The graphs are shown below for a TI-83/84 Plus calculator.

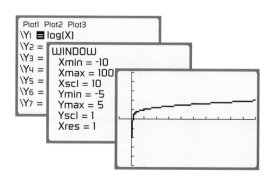

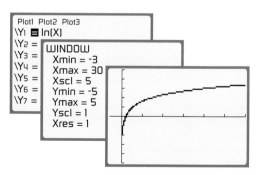

Applications of Logarithmic Functions

Many applications can be modeled by logarithmic functions.

EXAMPLE 6 ■ Application of a Logarithmic Function

During the 1980s and 1990s, the average time T of a major league baseball game tended to increase each year. If the year 1981 is represented by $x = 1$, then the function

$$T(x) = 149.57 + 7.63 \ln x$$

approximates the average time T, in minutes, of a major league baseball game for the years $x = 1$ to $x = 19$.

a. Use the function to determine the average time of a major league baseball game during the 1981 season and during the 1999 season. Round to the nearest hundredth of a minute.

b. By how much did the average time of a major league baseball game increase from 1981 to 1999?

Solution

a. The year 1981 is represented by $x = 1$ and the year 1999 by $x = 19$.

$$T(1) = 149.57 + 7.63 \ln(1) = 149.57$$

In 1981 the average time of a major league baseball game was about 149.57 minutes.

$$T(19) = 149.57 + 7.63 \ln(19) \approx 172.04$$

In 1999 the average time of a major league baseball game was about 172.04 minutes.

b. $T(19) - T(1) \approx 172.04 - 149.57 = 22.47$

From 1981 to 1999, the average time of a major league baseball game increased by about 22.47 minutes.

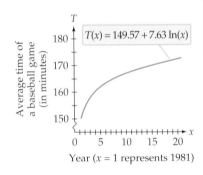

Average time of a baseball game (in minutes)

$T(x) = 149.57 + 7.63 \ln(x)$

Year ($x = 1$ represents 1981)

CHECK YOUR PROGRESS 6   The following function models the average typing speed S, in words per minute, of a student who has been typing for t months.

$$S(t) = 5 + 29 \ln(t + 1), \quad 0 \le t \le 9$$

a. Use the function to determine the student's average typing speed when the student first started to type and the student's average typing speed after 3 months. Round to the nearest whole word per minute.

b. By how much did the typing speed increase during the 3 months?

Solution   See page S24.

Logarithmic functions are often used to convert very large or very small numbers into numbers that are easier to comprehend. For instance, the *Richter scale,* which measures the magnitude of an earthquake, uses a logarithmic function to scale the intensity of an earthquake's shock waves I into a number M, which for most earthquakes is in the range of 0 to 10. The intensity I of an earthquake is often given in terms of the constant I_0, where I_0 is the intensity of the smallest earthquake (called a **zero-level earthquake**) that can be measured on a seismograph near the earthquake's epicenter. The following formula is used to compute the Richter scale magnitude of an earthquake.

The Richter Scale Magnitude of an Earthquake

An earthquake with an intensity of I has a Richter scale magnitude of

$$M = \log\left(\frac{I}{I_0}\right)$$

where I_0 is the measure of the intensity of a zero-level earthquake.

EXAMPLE 7 ▪ Find the Magnitude of an Earthquake

Find the Richter scale magnitude of the 2003 Amazonas, Brazil earthquake, which had an intensity of $I = 12{,}589{,}254 I_0$. Round to the nearest tenth.

Solution

$$M = \log\left(\frac{I}{I_0}\right) = \log\left(\frac{12{,}589{,}254 I_0}{I_0}\right) = \log(12{,}589{,}254) \approx 7.1$$

The 2003 Amazonas, Brazil earthquake had a Richter scale magnitude of 7.1.

CHECK YOUR PROGRESS 7   What is the Richter scale magnitude of an earthquake whose intensity is twice that of the Amazonas, Brazil earthquake in Example 7?

Solution   *See page S24.*

If you know the Richter scale magnitude of an earthquake, then you can determine the intensity of the earthquake.

EXAMPLE 8 ▪ Find the Intensity of an Earthquake

Find the intensity of the 2003 Colina, Mexico earthquake, which measured 7.6 on the Richter scale. Round to the nearest thousand.

Solution

$$\log\left(\frac{I}{I_0}\right) = 7.6$$

$$\frac{I}{I_0} = 10^{7.6} \qquad \text{• Write in exponential form.}$$

$$I = 10^{7.6} I_0 \qquad \text{• Solve for } I.$$

$$I \approx 39{,}810{,}717 I_0$$

The 2003 Colina, Mexico earthquake had an intensity that was approximately 39,811,000 times the intensity of a zero-level earthquake.

CHECK YOUR PROGRESS 8   On April 29, 2003, an earthquake measuring 4.6 on the Richter scale struck Fort Payne, Alabama. Find the intensity of the quake. Round to the nearest thousand.

Solution   *See page S24.*

✔ **TAKE NOTE**

Notice in Example 7 that we did not need to know the value of I_0 to determine the Richter scale magnitude of the quake.

historical note

The Richter scale was created by the seismologist Charles Francis Richter (rĭk´tər) (1900–1985) in 1935.

Richter was born in Ohio. At the age of 16, he and his mother moved to Los Angeles, where he enrolled at the University of Southern California. He went on to study physics at Stanford University. Richter was a professor of seismology at the Seismological Laboratory at California Institute of Technology (Caltech) from 1936 until he retired in 1970. ▪

point of interest

The pH scale was created by the Danish biochemist Søren Sørensen (sûr′ ən-s ən) in 1909 to measure the acidity of water used in the brewing of beer. pH is an abbreviation for *pondus hydrogenii,* which translates as "potential hydrogen."

✔ **TAKE NOTE**

One mole is equivalent to 6.022×10^{23} ions.

Logarithmic scales are also used in chemistry. Chemists use logarithms to determine the pH of a liquid, which is a measure of the liquid's **acidity** or **alkalinity.** (You may have tested the pH of a swimming pool or an aquarium.) Pure water, which is considered neutral, has a pH of 7.0. The pH scale ranges from 0 to 14, with 0 corresponding to the most acidic solutions and 14 to the most alkaline. Lemon juice has a pH of about 2, whereas household ammonia measures about 11.

Specifically, the acidity of a solution is a function of the hydronium-ion concentration of the solution. Because the hydronium-ion concentration of a solution can be very small (with values as low as 0.00000001), pH measures the acidity or alkalinity of a solution using a logarithmic scale.

The pH of a Solution

The pH of a solution with a hydronium-ion concentration of H^+ moles per liter is given by

$$pH = -\log[H^+]$$

EXAMPLE 9 ■ **Calculate the pH of a Liquid**

Find the pH of each liquid. Round to the nearest tenth.

a. Orange juice containing an H^+ concentration of 2.8×10^{-4} mole per liter

b. Milk containing an H^+ concentration of 3.97×10^{-7} mole per liter

c. A baking soda solution containing an H^+ concentration of 3.98×10^{-9} mole per liter

Solution

a. $pH = -\log[H^+]$
$pH = -\log(2.8 \times 10^{-4}) \approx 3.6$
The orange juice has a pH of about 3.6.

b. $pH = -\log[H^+]$
$pH = -\log(3.97 \times 10^{-7}) \approx 6.4$
The milk has a pH of about 6.4.

c. $pH = -\log[H^+]$
$pH = -\log(3.98 \times 10^{-9}) \approx 8.4$
The baking soda solution has a pH of about 8.4.

CHECK YOUR PROGRESS 9 Find the pH of each liquid. Round to the nearest tenth.

a. A cleaning solution containing an H^+ concentration of 2.41×10^{-13} mole per liter

b. A cola soft drink containing an H^+ concentration of 5.07×10^{-4} mole per liter

c. Rainwater containing an H^+ concentration of 6.31×10^{-5} mole per liter

Solution *See page S24.*

In Example 9, the hydronium-ion concentrations of orange juice and milk were given as 2.8×10^{-4} and 3.97×10^{-7}, respectively.

The following figure illustrates the pH scale, along with the corresponding hydronium-ion concentrations. A solution with a pH less than 7 is an **acid,** and a solution with a pH greater than 7 is an **alkaline solution,** or a **base.** Because the scale is logarithmic, a solution with a pH of 5 is 10 times more acidic than a solution with a pH of 6. From Example 9 we see that the orange juice and milk are acids, whereas the baking soda solution is a base.

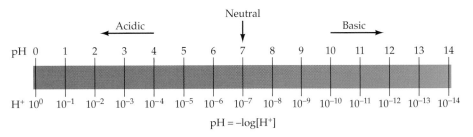

$$pH = -\log[H^+]$$

The figure above shows how the pH function scales small numbers on the H^+ axis into larger and more convenient numbers on the pH axis.

EXAMPLE 10 ▪ **Find the Hydronium-Ion Concentration of a Liquid**

A sample of blood has a pH of 7.3. Find the hydronium-ion concentration of the blood.

Solution

$$pH = -\log[H^+]$$

$$7.3 = -\log[H^+] \qquad \text{• Substitute 7.3 for pH.}$$

$$-7.3 = \log[H^+] \qquad \text{• Multiply both sides by } -1.$$

$$10^{-7.3} = H^+ \qquad \text{• Change to exponential form.}$$

$$5.0 \times 10^{-8} \approx H^+ \qquad \text{• Evaluate } 10^{-7.3} \text{ and write the answer in scientific notation.}$$

The hydronium-ion concentration of the blood is about 5.0×10^{-8} mole per liter.

CHECK YOUR PROGRESS 10   The water in the Great Salt Lake in Utah has a pH of 10.0. Find the hydronium-ion concentration of the water.

Solution   *See page S24.*

Investigation

Benford's Law

The authors of this text know some interesting details about your finances. For instance, of the last 100 checks you have written, about 30% are for amounts that start with a 1. Also, you have written about three times as many checks for amounts that start with a 2 as you have for amounts that start with a 7.

(continued)

▼ **point of interest**

Benford's Law has been used to identify fraudulent accountants. In most cases these accountants are unaware of Benford's Law and have replaced valid numbers with numbers selected at random. Their numbers do not conform to Benford's Law. Hence, an audit is warranted.

We are sure of these results because of a mathematical formula known as *Benford's Law*. This law was first discovered by the mathematician Simon Newcomb in 1881 and was then rediscovered by the physicist Frank Benford in 1938. Benford's Law states that the probability P that the first digit of a number selected from a wide range of numbers is a particular digit d is given by

$$P(d) = \log\left(1 + \frac{1}{d}\right)$$

Investigation Exercises

1. Use Benford's Law to complete the table and bar graph shown below.

d	$P(d) = \log(1 + 1/d)$
1	0.301
2	0.176
3	0.125
4	_____
5	_____
6	_____
7	_____
8	_____
9	_____

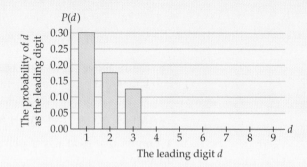

Benford's Law applies to most data with a wide range. For instance, it applies to

- the populations of the cities in the United States
- the numbers of dollars in the savings accounts at your local bank
- the numbers of miles driven during a month by each person in a state

2. Use Benford's Law to find the probability that, in a U.S. city selected at random, the number of telephones in that city will be a number starting with a 6.

3. Use Benford's Law to find how many times as many purchases you have made for dollar amounts that start with a 1 as for dollar amounts that start with a 9.

4. Explain why Benford's Law would not apply to the set of telephone numbers in a small city such as Le Mars, Iowa.

5. Explain why Benford's Law would not apply to the set of all the ages, in years, of the students at a local high school.

Exercise Set 6.3

In Exercises 1–8, write the exponential equation in logarithmic form.

1. $7^2 = 49$

2. $10^3 = 1000$

3. $5^4 = 625$

4. $2^{-3} = \dfrac{1}{8}$

5. $10^{-4} = 0.0001$

6. $3^5 = 243$

7. $10^y = x$

8. $e^y = x$

In Exercises 9–16, write the logarithmic equation in exponential form.

9. $\log_3 81 = 4$

10. $\log_2 16 = 4$

11. $\log_5 125 = 3$

12. $\log_4 64 = 3$

13. $\log_4 \dfrac{1}{16} = -2$

14. $\log_2 \dfrac{1}{16} = -4$

15. $\ln x = y$

16. $\log x = y$

In Exercises 17–24, evaluate the logarithm.

17. $\log_3 81$ **18.** $\log_7 49$

19. $\log 100$ **20.** $\log 0.001$

21. $\log_3 \dfrac{1}{9}$ **22.** $\log_7 \dfrac{1}{7}$

23. $\log_2 64$ **24.** $\log 0.01$

In Exercises 25–32, solve the equation for x.

25. $\log_3 x = 2$ **26.** $\log_5 x = 1$

27. $\log_7 x = -1$ **28.** $\log_8 x = -2$

29. $\log_3 x = -2$ **30.** $\log_5 x = 3$

31. $\log_4 x = 0$ **32.** $\log_4 x = -1$

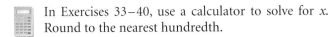 In Exercises 33–40, use a calculator to solve for x. Round to the nearest hundredth.

33. $\log x = 2.5$ **34.** $\log x = 3.2$

35. $\ln x = 2$ **36.** $\ln x = 4$

37. $\log x = 0.35$ **38.** $\log x = 0.127$

39. $\ln x = \dfrac{8}{3}$ **40.** $\ln x = \dfrac{1}{2}$

41. If b is a positive number not equal to 1, is the equation $\log_b 1 = 0$ always true, sometimes true, or never true?

42. If b is a positive number not equal to 1, is the equation $\log_b b = 1$ always true, sometimes true, or never true?

43. Suppose $c > 0$, $\log a = c$, and $\ln b = c$. Is $a < b$ or $a > b$?

44. Suppose $c < 0$, $\log a = c$, and $\ln b = c$. Is $a < b$ or $a > b$?

In Exercises 45–50, graph the function.

45. $g(x) = \log_2 x$ **46.** $g(x) = \log_4 x$

47. $f(x) = \log_3(2x - 1)$ **48.** $f(x) = -\log_2 x$

49. $f(x) = \log_2(x - 1)$ **50.** $f(x) = \log_3(x - 2)$

Light The percent of light that will pass through a material is given by the formula $\log P = -kd$, where P is the percent of light passing through the material, k is a constant that depends on the material, and d is the thickness of the material in centimeters. Use this formula for Exercises 51 and 52.

51. The constant k for a piece of opaque glass that is 0.5 centimeter thick is 0.2. Find the percent of light that will pass through the glass. Round to the nearest percent.

52. The constant k for a piece of tinted glass is 0.5. How thick is a piece of this glass that allows 60% of the light incident to the glass to pass through it? Round to the nearest hundredth of a centimeter.

Sound The number of decibels D of a sound can be given by the equation $D = 10(\log I + 16)$, where I is the power of the sound measured in watts. Use this formula for Exercises 53 and 54. Round to the nearest decibel.

53. Find the number of decibels of the sound of normal conversation. The power of the sound of normal conversation is approximately 3.2×10^{-10} watts.

54. The loudest sound made by an animal is made by the blue whale and can be heard over 500 miles away. The power of the sound is 630 watts. Find the number of decibels of the sound emitted by the blue whale.

pH of a Solution In Exercises 55 and 56, use the equation

$$pH = -\log(H^+)$$

where H^+ is the hydronium-ion concentration of a solution. Round to the nearest hundredth.

55. Find the pH of the digestive solution of the stomach, for which the hydronium-ion concentration is 0.045 mole per liter.

56. Find the pH of a morphine solution used to relieve pain for which the hydronium-ion concentration is 3.2×10^{-10} mole per liter.

Earthquakes In Exercises 57–61, use the Richter scale equation $M = \log \dfrac{I}{I_0}$, where M is the magnitude of an earthquake, I is the intensity of the shock waves, and I_0 is the measure of the intensity of a zero-level earthquake.

57. On May 21, 2003, an earthquake struck northern Algeria. The earthquake had an intensity of $I = 6,309,573 I_0$. Find the Richter scale magnitude of the earthquake. Round to the nearest tenth.

58. The earthquake on November 17, 2003 on the Aleutian Island of Moska had an intensity of $I = 63,095,734 I_0$. Find the Richter scale magnitude of the earthquake. Round to the nearest tenth.

59. An earthquake in Japan on March 2, 1933 measured 8.9 on the Richter scale. Find the intensity of the earthquake in terms of I_0. Round to the nearest whole number.

60. An earthquake that occurred in China in 1978 measured 8.2 on the Richter scale. Find the intensity of the earthquake in terms of I_0. Round to the nearest whole number.

61. How many times as strong is an earthquake whose magnitude is 8 as one whose magnitude is 6?

Astronomy Astronomers use the distance modulus function $M(r) = 5 \log r - 5$, where M is the distance modulus and r is the distance of a star from Earth in parsecs. (One parsec is approximately 1.92×10^{13} miles, or approximately 20 trillion miles.) Use this function for Exercises 62 and 63. Round to the nearest tenth.

62. The distance modulus of the star Betelgeuse is 5.89. How many parsecs from Earth is this star?

63. The distance modulus of Alpha Centauri is -1.11. How many parsecs from Earth is this star?

World's Oil Supply One model for the time it will take for the world's oil supply to be depleted is given by the function $T(r) = 14.29 \ln(0.00411r + 1)$, where r is the estimated oil reserves in billions of barrels and T is the time in years before that amount of oil is depleted. Use this function for Exercises 64 and 65. Round to the nearest tenth of a year.

64. How many barrels of oil are necessary to last 20 years?

65. How many barrels of oil are necessary to last 50 years?

Extensions

CRITICAL THINKING

66. As mentioned in this section, the main motivation for the development of logarithms was to aid astronomers and other scientists with arithmetic calculations. The idea was to allow scientists to multiply large numbers by adding the logarithms of the numbers. Dividing large numbers was accomplished by subtracting the logarithms of the numbers. Work through the following exercises to see how this was accomplished. For simplicity, we will use small numbers (2 and 3) to illustrate the procedure.

a. Write each of the equations $\log 2 = 0.30103$ and $\log 3 = 0.47712$ in exponential form.

b. Replace 2 and 3 in $x = 2 \cdot 3$ with the exponential expressions from part a. (In this exercise, we know that x is 6. However, if the two numbers were very large, the value of x would not be obvious.)

c. Simplify the exponential expression. Recall that to multiply two exponential expressions with the same base, you *add* the exponents.

d. If you completed part c correctly, you should have

$$x = 10^{0.77815}$$

Write this expression in logarithmic form.

e. Using a calculator, verify that the solution of the equation you created in part d is 6. *Note:* When tables of logarithms were used, a scientist would have looked through the table to find 0.77815 and would have observed that it was the logarithm of 6.

67. Replace the denominator 2 and the numerator 3 in $x = \frac{3}{2}$ by the exponential expressions from part a of Exercise 66. Simplify the expression by *subtracting* the exponents. Write the answer in logarithmic form and verify that $x = 1.5$.

68. Write a few sentences explaining how adding the logarithms of two numbers can be used to find the product of the two numbers.

69. Write a few sentences explaining how subtracting the logarithms of two numbers can be used to find the quotient of the two numbers.

EXPLORATIONS

Earthquakes Seismologists generally determine the Richter scale magnitude of an earthquake by examining a *seismogram*, an example of which is shown below.

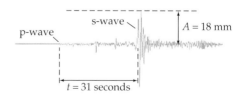

The magnitude of an earthquake cannot be determined just by examining the amplitude of a seismogram, because this amplitude decreases as the distance between the epicenter of the earthquake and the observation station increases. To account for the distance between the epicenter and the observation station, a seismologist examines a seismogram for both small waves, called *p-waves*, and larger waves, called *s-waves*. The Richter scale magnitude M of the earthquake is a function of both the amplitude A of the s-waves and the time t between the occurrence of the s-waves and the occurrence of the p-waves. In the 1950s, Charles Richter developed the formula at the right to determine the magnitude M of an earthquake from the data in a seismogram.

70. Find the Richter scale magnitude of the earthquake that produced the seismogram shown on page 410. Round to the nearest tenth.

71. Find the Richter scale magnitude of the earthquake that produced the seismogram shown at the right. Round to the nearest tenth.

The Amplitude-Time-Difference Formula

The Richter scale magnitude of an earthquake is given by

$$M = \log A + 3 \log 8t - 2.92$$

where A is the amplitude, in millimeters, of the s-waves on a seismogram and t is the time, in seconds, between the s-waves and the p-waves.

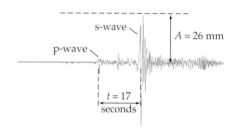

CHAPTER 6 **Summary**

Key Terms

acid [p. 407]
acidity and alkalinity [p. 406]
alkaline solution [p. 407]
axis of symmetry [p. 371]
base [p. 407]
common logarithm [p. 401]
common logarithmic function [p. 401]
degree of a polynomial function [p. 369]
e [p. 389]
exponential decay function [p. 388]
exponential function [p. 386]
exponential growth function [p. 388]
linear function [p. 369]

logarithm [p. 399]
maximum of a quadratic function [p. 373]
minimum of a quadratic function [p. 373]
natural exponential function [p. 390]
natural logarithm [p. 401]
natural logarithmic function [p. 401]
parabola [p. 371]
polynomial function [p. 369]
quadratic function [p. 370]
standard form of a polynomial [p. 370]
vertex of a parabola [p. 371]
zero-level earthquake [p. 404]

Essential Concepts

■ **Vertex of a Parabola**
Let $f(x) = ax^2 + bx + c$ be the equation of a parabola. The coordinates of the vertex are $\left(-\dfrac{b}{2a}, f\left(-\dfrac{b}{2a}\right)\right)$.

■ **Definition of Logarithm**
For $b > 0$, $b \neq 1$, $y = \log_b x$ is equivalent to $x = b^y$.

■ **Equality of Exponents Property**
If $b > 0$ and $b^x = b^y$, then $x = y$.

■ **Richter Scale**
An earthquake with an intensity of I has a Richter scale magnitude of

$$M = \log\left(\frac{I}{I_0}\right)$$

where I_0 is the measure of the intensity of a zero-level earthquake.

■ **The pH of a Solution**
The pH of a solution with a hydronium-ion concentration of H^+ moles per liter is given by $pH = -\log[H^+]$.

Review Exercises

In Exercises 1–8, evaluate the function for the given value.

1. $f(x) = 4x - 5;\ x = -2$

2. $g(x) = 2x^2 - x - 2;\ x = 3$

3. $s(t) = \dfrac{4}{3t - 5};\ t = -1$

4. $R(s) = s^3 - 2s^2 + s - 3;\ s = -2$

5. $f(x) = 2^{x-3};\ x = 5$

6. $g(x) = \left(\dfrac{2}{3}\right)^x;\ x = 2$

7. $T(r) = 2e^r + 1;\ r = 2.$ Round to the nearest hundredth.

8. $F(t) = e^{-t} - 3;\ t = 0.3.$ Round to the nearest hundredth.

In Exercises 9 to 12, graph the function.

9. $f(x) = 2^x - 3$ **10.** $f(x) = 3^{-x+2}$

11. $f(x) = \log_2(x + 2)$ **12.** $f(x) = x^2 - 3x + 1$

13. Geometry The volume of a sphere is a function of its radius and is given by $V(r) = \dfrac{4\pi r^3}{3}$, where r is the radius of the sphere.

   a. Find the volume of a sphere whose radius is 3 inches. Round to the nearest tenth.

   b. Find the volume of a sphere whose radius is 12 centimeters. Round to the nearest tenth.

14. Physics The height h of a ball that is released 5 feet above the ground with an initial velocity of 96 feet per second is a function of the time t the ball is in the air and is given by $h(t) = -16t^2 + 96t + 5$.

   a. Find the height of the ball above the ground 2 seconds after it is released.

   b. Find the height of the ball above the ground 4 seconds after it is released.

15. Chemistry The percent concentration P of sugar in a water solution depends on the number of grams x of sugar that is added to the solution and is given by $P(x) = \dfrac{100x + 100}{x + 20}$, where x is the number of grams of sugar that is added.

   a. What is the original percent concentration of sugar?

   b. What is the percent concentration of sugar after 10 more grams of sugar is added?

   c. Is P a linear or a nonlinear function?

16. Volume An open box is made from a rectangular piece of cardboard that measures 50 centimeters by 35 centimeters. To construct the box, squares x centimeters on a side are cut from each corner of the cardboard. The remaining flaps are folded up to create a box.

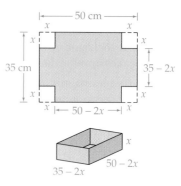

   a. Express the volume of the box as a polynomial function in x.

   b. What is the volume of the box when squares 8 centimeters on a side are cut from the cardboard?

17. Surface Area For the box in Exercise 16, express the surface area of the open box as a polynomial function in x.

18. Cycling A spectator is standing 10 feet from the path of a racing cyclist who is traveling at 40 feet per second in a straight path. See the diagram below.

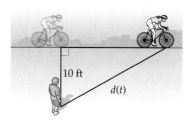

   a. Express the distance $d(t)$, in feet, between the spectator and the cyclist in terms of t, the number of seconds after the cyclist passes the spectator.

   b. Find the distance from the spectator to the cyclist after 1 minute. Round to the nearest tenth.

19. Inventory An electronics store has determined that the cost, in dollars, of storing x model ZX100 cell phones is $55x$. The cost, in dollars, of reordering x model ZX100 cell phones is $\dfrac{550x + 750}{x}$.

a. Express the inventory cost, $C(x)$, for the cell phones in terms of x.

b. Find the inventory cost if the store manager wants to maintain an inventory of 50 model ZX100 cell phones. Round to the nearest cent.

In Exercises 20–23, find the vertex of the graph of each function.

20. $y = x^2 + 2x + 4$ **21.** $y = -2x^2 - 6x + 1$

22. $f(x) = -3x^2 + 6x - 1$ **23.** $f(x) = x^2 + 5x - 1$

In Exercises 24–27, find the minimum or maximum value of each quadratic function.

24. $y = -x^2 + 4x + 1$ **25.** $y = 2x^2 + 6x - 3$

26. $f(x) = x^2 - 4x - 1$ **27.** $f(x) = -2x^2 + 3x - 1$

28. Physics The height s, in feet, of a rock thrown upward at an initial speed of 80 feet per second from a cliff 25 feet above an ocean beach is given by the function $s(t) = -16t^2 + 80t + 25$, where t is the time in seconds. Find the maximum height above the beach that the rock will attain.

29. Manufacturing A manufacturer of rewritable CDs (CD-RWs) estimates that the average daily cost C of producing a CD-RW is given by

$$C(x) = 0.01x^2 - 40x + 50{,}000$$

where x is the number of CD-RWs produced each day. Find the number of CD-RWs the company should produce in order to minimize the average daily cost.

30. Investments A \$5000 certificate of deposit (CD) earns 6% annual interest compounded daily. What is the value of the investment after 15 years?

31. Isotopes An isotope of technetium has a half-life of approximately 6 hours. If a patient is injected with 10 milligrams of this isotope, the number of milligrams, $N(t)$, of technetium in the patient after t hours is given by $N(t) = 10(2^{-t/6})$. What will be the technetium level in the patient after 2 hours?

32. Isotopes Iodine-131 has a half-life of approximately 8 days. If a patient is given an injection that contains 8 micrograms of iodine-131, the number of micrograms, $N(t)$, in the patient after t days is given by $N(t) = 8(2^{-t/8})$. What will be the iodine level in the patient after 10 days?

33. Golf A golf ball is dropped from a height of 6 feet. On each successive bounce, the ball rebounds to a height that is $\frac{2}{3}$ of the previous height. Find an exponential model for the height of the ball after the nth bounce. What is the height of the ball after the fifth bounce?

34. **Film Industry** When a new movie is released, there is initially a surge in the number of people who go to the movie. After 2 weeks, the attendance generally begins to drop off. The data in the following table show the number of people attending a certain movie, with $t = 0$ representing the time 2 weeks after the initial release.

t, number of weeks beyond 2 weeks after release	0	1	2	3	4
N, number of people (in thousands)	250	162	110	65	46

a. Find an exponential regression equation for these data.

b. Use the equation to predict the number of people who will attend the movie 8 weeks after it has been released.

In Exercises 35–38, evaluate the logarithm.

35. $\log_3 243$ **36.** $\log_2 \dfrac{1}{16}$

37. $\log_4 \dfrac{1}{4}$ **38.** $\log_2 64$

In Exercises 39–42, solve for x.

39. $\log_4 x = 3$ **40.** $\log_2 x = 5$

41. $\log_3 x = -1$ **42.** $\log_4 x = -2$

43. Astronomy Using the distance modulus formula $M = 5 \log r - 5$, where M is the distance modulus and r is the distance of a star from Earth in parsecs, find the distance from Earth to a star whose distance modulus is 3.2. Round to the nearest tenth.

44. Sound The number of decibels D of a sound can be given by the equation $D = 10(\log I + 16)$, where I is the power of the sound measured in watts. The pain threshold for a sound for most humans is approximately 0.01 watt. Find the number of decibels of this sound.

CHAPTER 6   **Test**

1. Evaluate each function for the given value.
   a. $s(t) = -3t^2 + 4t - 1; t = -2$
   b. $f(x) = 3^{x-4}; x = 2$

2. Write $2^a = b$ in logarithmic form.

3. Evaluate: $\log_5 125$

4. Solve for x: $\log_6 x = 2$

In Exercises 5 and 6, graph the function.

5. $f(x) = 2^x - 5$

6. $f(x) = \log_3(x - 1)$

7. Find the vertex of the graph of
   $f(x) = x^2 + 6x - 1$.

8. **Chemistry** The percent concentration P of salt in a water solution depends on the number of grams of salt that is added to the solution and is given by $P(x) = \dfrac{100x + 50}{x + 5}$, where x is the number of grams of salt that is added.
   a. What is the original percent concentration of salt?
   b. What is the percent concentration of salt after 10 grams of salt is added?

9. **Volume** An open box is made from a rectangular piece of cardboard that measures 38 inches by 28 inches. To construct the box, squares x inches on a side are cut from each corner of the cardboard. The remaining flaps are folded up to create a box.
   a. Express the volume of the box as a polynomial function in x.
   b. What is the volume of the box when squares 7 inches on a side are cut from the cardboard?

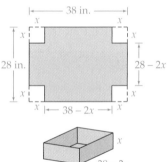

10. **Physics** The height h of a ball that is thrown straight up and released 4 feet above the ground is given by $h(t) = -16t^2 + 96t + 4$ where t is the time, in seconds, after the ball is released. Find the maximum height the ball attains.

11. **Newton's Law of Cooling** The temperature of a piece of metal is 70°F before it is placed in an oven that maintains a constant temperature of 500°F. The temperature of the metal after t minutes is given by $T = 500 - 430e^{-0.05t}$. Find the temperature of the metal after 30 minutes. Round to the nearest degree.

12. **Earthquakes** Two earthquakes struck Colombia, South America in the year 2000. One had a magnitude of 6.5 on the Richter scale and the second one had a magnitude of 5 on the Richter scale. How many times more intense was the first earthquake than the second one? Round to the nearest whole number.

CHAPTER 7

The Mathematics of Finance

7.1	Simple Interest
7.2	Compound Interest
7.3	Credit Cards and Consumer Loans

| 7.4 | Stocks, Bonds, and Mutual Funds |
| 7.5 | Home Ownership |

The table below shows the estimated annual expenses for a child born in 2003. The expenses are for the younger child in a two-child, two-parent home. From the table, we can see that the annual expense for a six-year-old child is $9730, and that $1670 of that amount is spent on food. Use the table to answer the following questions. (Answers are provided on the next page.)

1. Which age group has the highest annual expense? What might be some explanations for this age group having the highest annual expense?

2. For which ages is the annual cost of child care and education highest? Provide an explanation for this.

Estimated Annual Expenses for a Child Born in 2003								
Child's Age	Total	Housing	Food	Transportation	Clothing	Health Care	Child Care and Education	Miscellaneous
0–2	9510	3540	1130	1190	410	660	1570	1010
3–5	9780	3510	1310	1160	400	630	1740	1030
6–8	9730	3420	1670	1290	450	720	1110	1070
9–11	9600	3180	1960	1360	490	780	730	1100
12–14	10,350	3440	1980	1490	830	790	530	1290
15–17	10,560	2950	2200	1880	740	830	920	1040

Source: U.S. Department of Agriculture, Consumer Expenditure Survey Data

This table can also be used to estimate expenses for an only child by multiplying the total for any age group by 1.24.

3. Explain why the cost of raising an only child is greater than the cost of raising the second of two children.

4. What is the total estimated expense of raising an only child to the age of 18? And note that these data do not include the cost of college!

Given the cost of raising children, an understanding of the mathematics of finance is important for any parent. In this chapter, we will discuss the topics of simple interest, compound interest, consumer loans, investments, and home ownership.

Online Study Center

For online student resources, visit this textbook's Online Study Center at **college.hmco.com/pic/aufmannMTQR.**

| SECTION 7.1 | **Simple Interest** |

Simple Interest

historical note

The earliest loans date back to 3000 B.C., and interest on those loans may have extended over generations, not 4 or 5 years, as is the case for today's typical car loan. One of the first written records of an interest rate occurs in the Code of Hammurabi. Hammurabi ruled Babylon from 1795 to 1750 B.C. He is known for being the first ruler to write a set of laws that defined peoples' rights. In this Code, he allowed interest rates to be as high as $33\frac{1}{3}$%. ■

When you deposit money in a bank—for example, in a savings account—you are permitting the bank to use your money. The bank may lend the deposited money to customers to buy cars or make renovations on their homes. The bank pays you for the privilege of using your money. The amount paid to you is called **interest.** If you are the one borrowing money from a bank, the amount you pay for the privilege of using that money is also called interest.

The amount deposited in a bank or borrowed from a bank is called the **principal.** The amount of interest paid is usually given as a percent of the principal. The percent used to determine the amount of interest is called the **interest rate.** If you deposit $1000 in a savings account paying 5% interest, $1000 is the principal and the interest rate is 5%.

Interest paid on the original principal is called **simple interest.** The formula used to calculate simple interest is given below.

Simple Interest Formula

The simple interest formula is

$$I = Prt$$

where I is the interest, P is the principal, r is the interest rate, and t is the time period.

In the simple interest formula, the time t is expressed in the same period as the rate. For example, if the rate is given as an annual interest rate, then the time is measured in years; if the rate is given as a monthly interest rate, then the time must be expressed in months.

Interest rates are most commonly expressed as annual interest rates. Therefore, unless stated otherwise, we will assume the interest rate is an annual interest rate.

Interest rates are generally given as percents. Before performing calculations involving an interest rate, write the interest rate as a decimal.

ANSWERS TO QUESTIONS ON PAGE 415

1. *15–17 years. For example, car insurance for teenage drivers.*

2. *3–5 years. For example, child care for preschoolers is an expense, whereas public school is free.*

3. *For example, no hand-me-downs.*

4. *$209,659.20*

EXAMPLE 1 ■ Calculate Simple Interest

Calculate the simple interest earned in 1 year on a deposit of $1000 if the interest rate is 5%.

Solution

Use the simple interest formula. Substitute the following values into the formula: $P = 1000$, $r = 5\% = 0.05$, and $t = 1$.

$$I = Prt$$
$$I = 1000(0.05)(1)$$
$$I = 50$$

The simple interest earned is $50.

CHECK YOUR PROGRESS 1   Calculate the simple interest earned in 1 year on a deposit of $500 if the interest rate is 4%.

Solution   *See page S24.*

EXAMPLE 2 ■ Calculate Simple Interest

Calculate the simple interest due on a three-month loan of $2000 if the interest rate is 6.5%.

Solution
Use the simple interest formula. Substitute the values $P = 2000$ and $r = 6.5\% = 0.065$ into the formula. Because the interest rate is an annual rate, the time must be measured in years: $t = \dfrac{3 \text{ months}}{1 \text{ year}} = \dfrac{3 \text{ months}}{12 \text{ months}} = \dfrac{3}{12}$.

$$I = Prt$$

$$I = 2000(0.065)\left(\frac{3}{12}\right)$$

$$I = 32.5$$

The simple interest due is $32.50.

TAKE NOTE

If you perform the computation in Example 2 by hand, then we recommend that you reduce the fraction $\frac{3}{12}$ to $\frac{1}{4}$. If you use a calculator to perform the computation, it is not necessary to take the time to reduce $\frac{3}{12}$ to $\frac{1}{4}$.

CHECK YOUR PROGRESS 2   Calculate the simple interest due on a four-month loan of $1500 if the interest rate is 5.25%.

Solution   *See page S24.*

EXAMPLE 3 ■ Calculate Simple Interest

Calculate the simple interest due on a two-month loan of $500 if the interest rate is 1.5% per month.

Solution
Use the simple interest formula. Substitute the values $P = 500$ and $r = 1.5\% = 0.015$ into the formula. Because the interest rate is *per month*, the time period of the loan is expressed as the number of months: $t = 2$.

$$I = Prt$$

$$I = 500(0.015)(2)$$

$$I = 15$$

The simple interest due is $15.

point of interest

A PIRG (Public Interest Research Group) survey found that 29% of credit reports contained errors that could result in the denial of a loan. This is why financial advisors recommend that consumers check their credit ratings.

CHECK YOUR PROGRESS 3   Calculate the simple interest due on a five-month loan of $700 if the interest rate is 1.25% per month.

Solution   *See page S25.*

Remember that in the simple interest formula, time t is measured in the same period as the interest rate. Therefore, if the time period of a loan with an annual interest rate is given in days, it is necessary to convert the time period of the loan to a fractional part of a year. There are two methods for converting time from days to years: the exact method and the ordinary method. Using the exact method, the number of days of the loan is divided by 365, the number of days in a year.

Exact method: $t = \dfrac{\text{number of days}}{365}$

The ordinary method is based on there being an average of 30 days in a month and 12 months in a year ($30 \cdot 12 = 360$). Using this method, the number of days of the loan is divided by 360.

Ordinary method: $\quad t = \dfrac{\text{number of days}}{360}$

The ordinary method is used by most businesses. Therefore, unless otherwise stated, the ordinary method will be used in this text.

EXAMPLE 4 ■ Calculate Simple Interest

Calculate the simple interest due on a 45-day loan of $3500 if the annual interest rate is 8%.

Solution
Use the simple interest formula. Substitute the following values into the formula: $P = 3500$, $r = 8\% = 0.08$, and $t = \dfrac{\text{number of days}}{360} = \dfrac{45}{360}$.

$$I = Prt$$

$$I = 3500(0.08)\left(\dfrac{45}{360}\right)$$

$$I = 35$$

The simple interest due is $35.

CHECK YOUR PROGRESS 4   Calculate the simple interest due on a 120-day loan of $7000 if the annual interest rate is 5.25%.

Solution   *See page S25.*

The simple interest formula can be used to find the interest rate on a loan when the interest, principal, and time period of the loan are known. An example is given below.

EXAMPLE 5 ■ Calculate the Simple Interest Rate

The simple interest charged on a six-month loan of $3000 is $150. Find the simple interest rate.

Solution
Use the simple interest formula. Solve the equation for r.

$$I = Prt$$

$$150 = 3000(r)\left(\dfrac{6}{12}\right)$$

$$150 = 1500r \qquad \text{• } 3000\left(\dfrac{6}{12}\right) = 1500$$

$$0.10 = r \qquad \text{• Divide each side of the equation by 1500.}$$

$$r = 10\% \qquad \text{• Write the decimal as a percent.}$$

The simple interest rate on the loan is 10%.

CHECK YOUR PROGRESS 5   The simple interest charged on a six-month loan of $12,000 is $462. Find the simple interest rate.

Solution   See page S25.

Math Matters   Children Daydream About Being Rich

This chapter is about money, something everyone thinks about. But attitudes toward money differ. You might find it interesting to note that a survey of the world's six leading industrial nations revealed that about 62% of children aged 7 through 12 say that they think or daydream about "being rich." The breakdown by country is shown in the figure below.

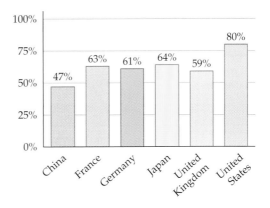

Percent of children aged 7–12 who say they think or daydream about "being rich"

Source: Roper Starch Worldwide for A. B. C. Research

Future Value and Maturity Value

point of interest

You may have heard the term *legal tender.* Section 102 of the Coinage Act of 1965 states, in part, that

All coins and currencies of the United States, regardless of when coins are issued, shall be legal tender for all debts, public and private, public charges, taxes, duties, and dues.

This means that when you offer U.S. currency to a creditor, you have made a valid and legal offer of payment of your debt.

When you borrow money, the total amount to be repaid to the lender is the sum of the principal and interest. This sum is calculated using the following future value or maturity value formula for simple interest.

> **Future Value or Maturity Value Formula for Simple Interest**
>
> The future value or maturity value formula for simple interest is
>
> $A = P + I$
>
> where A is the amount after the interest I has been added to the principal P.

This formula can be used for loans or investments. When used for a loan, A is the total amount to be repaid to the lender; this sum is called the **maturity value** of the loan. In Example 5, the simple interest charged on the loan of $3000 was $150. The maturity value of the loan is therefore $3000 + $150 = $3150.

For an investment, such as a deposit in a bank savings account, A is the total amount on deposit after the interest earned has been added to the principal. This sum is called the **future value** of the investment.

QUESTION   *Is the stated sum a maturity value or a future value?*

> **a.** *The sum of the principal and the interest on an investment*
>
> **b.** *The sum of the principal and the interest on a loan*

▼ **point of interest**

In the early 1970s, students borrowed an average of $2000 to cover the entire cost of their educations. (*Source:* Nellie Mae Student Loan Applications) Today, average student loan debt is more than $15,000 at public colleges and over $17,000 at private colleges. (*Source:* American Council of Education)

EXAMPLE 6 ■ **Calculate a Maturity Value**

Calculate the maturity value of a simple interest, eight-month loan of $8000 if the interest rate is 9.75%.

Solution

Step 1:   Find the interest. Use the simple interest formula. Substitute the values $P = 8000$, $r = 9.75\% = 0.0975$, and $t = \frac{8}{12}$ into the formula.

$$I = Prt$$

$$I = 8000(0.0975)\left(\frac{8}{12}\right)$$

$$I = 520$$

Step 2:   Find the maturity value. Use the maturity value formula for simple interest. Substitute the values $P = 8000$ and $I = 520$ into the formula.

$$A = P + I$$

$$A = 8000 + 520$$

$$A = 8520$$

The maturity value of the loan is $8520.

CHECK YOUR PROGRESS 6   Calculate the maturity value of a simple interest, nine-month loan of $4000 if the interest rate is 8.75%.

Solution   *See page S25.*

Recall that the simple interest formula states that $I = Prt$. We can substitute Prt for I in the future value or maturity value formula, as follows.

$$A = P + I$$

$$A = P + Prt$$

$$A = P(1 + rt) \qquad \text{• Use the Distributive Property.}$$

In the final equation, A is the future value of an investment or the maturity value of a loan, P is the principal, r is the interest rate, and t is the time period.

ANSWER   **a.** *Future value*   **b.** *Maturity value*

We used the formula $A = P + I$ in Example 6. The formula $A = P(1 + rt)$ is used in Examples 7 and 8. Note that two steps were required to find the solution in Example 6, but only one step is required in Examples 7 and 8.

EXAMPLE 7 ▪ **Calculate a Maturity Value Using $A = P(1 + rt)$**

Calculate the maturity value of a simple interest, three-month loan of $3800. The interest rate is 6%.

Solution

Substitute the following values into the formula $A = P(1 + rt)$: $P = 3800$, $r = 6\% = 0.06$, and $t = \frac{3}{12}$.

$$A = P(1 + rt)$$
$$A = 3800\left[1 + 0.06\left(\frac{3}{12}\right)\right]$$
$$A = 3800(1 + 0.015)$$
$$A = 3800(1.015)$$
$$A = 3857$$

The maturity value of the loan is $3857.

CHECK YOUR PROGRESS 7   Calculate the maturity value of a simple interest, one-year loan of $6700. The interest rate is 8.9%.

Solution   *See page S25.*

---

point of interest

The list below shows how adults aged 35 to 50 say they would distribute a $100,000 windfall among the given choices. (*Source:* Market Research Institute survey)

Put toward college education for their children: $36,200
Save for retirement: $26,100
Take care of parents: $24,800
Put toward a new home: $5800
Put toward a new car: $3800
Take a dream vacation: $3300

EXAMPLE 8 ▪ **Calculate a Future Value Using $A = P(1 + rt)$**

Find the future value after 1 year of $850 in an account earning 8.2% simple interest.

Solution

Because $t = 1$, $rt = r(1) = r$. Therefore, $1 + rt = 1 + r = 1 + 0.082 = 1.082$.

$$A = P(1 + rt)$$
$$A = 850(1.082)$$
$$A = 919.7$$

The future value of the account after 1 year is $919.70.

CHECK YOUR PROGRESS 8   Find the future value after 1 year of $680 in an account earning 6.4% simple interest.

Solution   *See page S25.*

Recall that the formula $A = P + I$ states that A is the amount after the interest has been added to the principal. Subtracting P from each side of this equation yields the following formula.

$$I = A - P$$

This formula states that the amount of interest paid is equal to the total amount minus the principal. This formula is used in Example 9.

EXAMPLE 9 ▪ Calculate the Simple Interest Rate

The maturity value of a three-month loan of $4000 is $4085. What is the simple interest rate?

Solution
First find the amount of interest paid. Subtract the principal from the maturity value.

$$I = A - P$$
$$I = 4085 - 4000$$
$$I = 85$$

Find the simple interest rate by solving the simple interest formula for r.

$$I = Prt$$
$$85 = 4000(r)\left(\frac{3}{12}\right)$$

$85 = 1000r$ • $4000\left(\dfrac{3}{12}\right) = 1000$

$0.085 = r$ • Divide each side of the equation by 1000.

$r = 8.5\%$ • Write the decimal as a percent.

The simple interest rate on the loan is 8.5%.

CHECK YOUR PROGRESS 9   The maturity value of a four-month loan of $9000 is $9240. What is the simple interest rate?

Solution   *See page S25.*

Investigation

Day-of-the-Year Table

Table 7.1 *Day-of-the-Year Table*

Day of the Month	Jan	Feb	Mar	Apr	May	Jun	Jul	Aug	Sep	Oct	Nov	Dec
1	1	32	60	91	121	152	182	213	244	274	305	335
2	2	33	61	92	122	153	183	214	245	275	306	336
3	3	34	62	93	123	154	184	215	246	276	307	337
4	4	35	63	94	124	155	185	216	247	277	308	338
5	5	36	64	95	125	156	186	217	248	278	309	339
6	6	37	65	96	126	157	187	218	249	279	310	340
7	7	38	66	97	127	158	188	219	250	280	311	341
8	8	39	67	98	128	159	189	220	251	281	312	342
9	9	40	68	99	129	160	190	221	252	282	313	343
10	10	41	69	100	130	161	191	222	253	283	314	344
11	11	42	70	101	131	162	192	223	254	284	315	345
12	12	43	71	102	132	163	193	224	255	285	316	346
13	13	44	72	103	133	164	194	225	256	286	317	347
14	14	45	73	104	134	165	195	226	257	287	318	348
15	15	46	74	105	135	166	196	227	258	288	319	349
16	16	47	75	106	136	167	197	228	259	289	320	350
17	17	48	76	107	137	168	198	229	260	290	321	351
18	18	49	77	108	138	169	199	230	261	291	322	352
19	19	50	78	109	139	170	200	231	262	292	323	353
20	20	51	79	110	140	171	201	232	263	293	324	354
21	21	52	80	111	141	172	202	233	264	294	325	355
22	22	53	81	112	142	173	203	234	265	295	326	356
23	23	54	82	113	143	174	204	235	266	296	327	357
24	24	55	83	114	144	175	205	236	267	297	328	358
25	25	56	84	115	145	176	206	237	268	298	329	359
26	26	57	85	116	146	177	207	238	269	299	330	360
27	27	58	86	117	147	178	208	239	270	300	331	361
28	28	59	87	118	148	179	209	240	271	301	332	362
29	29		88	119	149	180	210	241	272	302	333	363
30	30		89	120	150	181	211	242	273	303	334	364
31	31		90		151		212	243		304		365

During a leap year, add 1 day if February 29 falls between the two days under consideration.

(continued)

The Day-of-the-Year Table shown in Table 7.1 can be used to determine the number of days from one date to another date. For example, because May 15 is day 135 and August 23 is day 235, there are 235 − 135 = 100 days from May 15 to August 23.

The table can also be used to determine the due date of a loan. For example, a 120-day loan made on June 9, which is day 160, is due on day 160 + 120 = day 280, which is October 7.

EXAMPLE

Calculate the simple interest due on a $5000 loan made on September 20 and repaid on December 9 of the same year. The interest rate is 6%.

Solution
September 20 is day 263. December 9 is day 343.
343 − 263 = 80. The term of the loan is 80 days.

Use the simple interest formula. Substitute the following values into the formula: $P = 5000$, $r = 6\% = 0.06$, and $t = \dfrac{80}{360}$.

$$I = Prt$$

$$I = 5000(0.06)\left(\frac{80}{360}\right)$$

$$I \approx 66.67$$

The simple interest due is $66.67.

Investigation Exercises

1. Find the due date on a 180-day loan made on March 10.

2. Find the due date on a 90-day loan made on June 25.

3. Find the number of days from April 22 to November 8 of the same year.

4. Find the number of days from February 1 to July 1 during a leap year.

5. Find the number of days from December 18 to February 8 of the following year. Explain how you calculated the number.

6. Calculate the simple interest due on a $7500 loan made on January 30 and repaid on July 18 of the same year. The interest rate is 6.5%. The year is not a leap year.

7. Calculate the simple interest due on a $6000 loan made on May 18 and repaid on October 20 of the same year. The interest rate is 5.75%.

8. A $12,000 loan is made on February 21 of a leap year. The interest rate is 6.25%. The loan is repaid on November 15 of the same year. Calculate the simple interest paid on the loan.

9. A $15,000 loan is made on August 28. The interest rate is 7%. The loan is repaid on January 20 of the following year. Calculate the simple interest due on the loan.

Exercise Set 7.1

1. 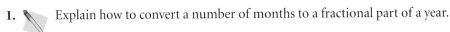 Explain how to convert a number of months to a fractional part of a year.

2. Explain how to convert a number of days to a fractional part of a year.

3. Explain what each variable in the simple interest formula represents.

In Exercises 4–17, calculate the simple interest earned. Round to the nearest cent.

4. $P = \$2000$, $r = 6\%$, $t = 1$ year
5. $P = \$8000$, $r = 7\%$, $t = 1$ year
6. $P = \$3000$, $r = 5.5\%$, $t = 6$ months
7. $P = \$7000$, $r = 6.5\%$, $t = 6$ months
8. $P = \$4200$, $r = 8.5\%$, $t = 3$ months
9. $P = \$9000$, $r = 6.75\%$, $t = 4$ months
10. $P = \$12{,}000$, $r = 7.8\%$, $t = 45$ days
11. $P = \$3000$, $r = 9.6\%$, $t = 21$ days
12. $P = \$4000$, $r = 8.4\%$, $t = 33$ days
13. $P = \$7000$, $r = 7.2\%$, $t = 114$ days
14. $P = \$800$, $r = 1.5\%$ monthly, $t = 3$ months
15. $P = \$2000$, $r = 1.25\%$ monthly, $t = 5$ months
16. $P = \$3500$, $r = 1.8\%$ monthly, $t = 4$ months
17. $P = \$1600$, $r = 1.75\%$ monthly, $t = 6$ months

In Exercises 18–23, use the formula $A = P(1 + rt)$ to calculate the maturity value of the simple interest loan.

18. $P = \$8500$, $r = 6.8\%$, $t = 6$ months
19. $P = \$15{,}000$, $r = 8.9\%$, $t = 6$ months
20. $P = \$4600$, $r = 9.75\%$, $t = 4$ months
21. $P = \$7200$, $r = 7.95\%$, $t = 4$ months
22. $P = \$13{,}000$, $r = 1.4\%$ monthly, $t = 3$ months
23. $P = \$2800$, $r = 9.2\%$, $t = 3$ months

In Exercises 24–29, calculate the simple interest rate.

24. $P = \$8000$, $I = \$500$, $t = 1$ year
25. $P = \$1600$, $I = \$120$, $t = 1$ year
26. $P = \$4000$, $I = \$190$, $t = 6$ months
27. $P = \$2000$, $I = \$105$, $t = 6$ months
28. $P = \$500$, $I = \$10.25$, $t = 3$ months
29. $P = \$1200$, $I = \$37.20$, $t = 4$ months

30. You open a savings account, deposit some money in it, and leave it there for a period of 6 months. Do you earn more interest on your money if you deposit $1000 in the account or if you deposit $1250?

31. You deposit money in a savings account. The annual simple interest rate on the account is 2.5%. Do you earn more interest on your money in a period of 5 months or a period of 8 months?

32. You deposit money in a savings account. You have a choice between an account that pays 2.185% annual simple interest and an account that pays 2.25% annual simple interest. If you leave your money in the account for 6 months, on which account would you earn more interest?

33. On January 1, you deposit equal amounts of money in two accounts. For account A, the simple interest rate is 12% per year. For account B, the simple interest rate is 1% per month. Assume you do not deposit any additional money in either account or withdraw any money from either account. At the end of 1 year, is the value of account A less than, equal to, or greater than the value of account B?

34. Suppose you deposit money in an account that earns 10% annual simple interest. Assuming you do not deposit additional money in or withdraw money from this account, how long would it take to earn as much interest as the original principal?

35. A worker deposits the proceeds from a bonus in an account that earns an annual simple interest rate of 5%. The worker leaves the money in the account for 10 years without adding or withdrawing money. Is the amount of interest earned in the eighth year less than, equal to, or more than the amount of interest earned in the fourth year?

36. **Simple Interest** Calculate the simple interest earned in 1 year on a deposit of $1900 if the interest rate is 8%.

37. **Simple Interest** Calculate the simple interest earned in 1 year on a deposit of $2300 if the interest rate is 7%.

38. **Simple Interest** Calculate the simple interest due on a three-month loan of $1400 if the interest rate is 7.5%.

39. **Simple Interest** You deposit $1500 in an account earning 10.4% interest. Calculate the simple interest earned in 6 months.

40. **Simple Interest** Calculate the simple interest due on a two-month loan of $800 if the interest rate is 1.5% per month.

41. **Simple Interest** Calculate the simple interest due on a 45-day loan of $1600 if the interest rate is 9%.

42. **Simple Interest** Calculate the simple interest due on a 150-day loan of $4800 if the interest rate is 7.25%.

43. **Maturity Value** Calculate the maturity value of a simple interest, eight-month loan of $7000 if the interest rate is 8.7%.

44. **Maturity Value** Calculate the maturity value of a simple interest, 10-month loan of $6600 if the interest rate is 9.75%.

45. **Maturity Value** Calculate the maturity value of a simple interest, one-year loan of $5200. The interest rate is 10.2%.

46. **Future Value** You deposit $880 in an account paying 9.2% simple interest. Find the future value of the investment after 1 year.

47. **Future Value** You deposit $750 in an account paying 7.3% simple interest. Find the future value of the investment after 1 year.

48. **Simple Interest Rate** The simple interest charged on a six-month loan of $6000 is $270. Find the simple interest rate.

49. **Simple Interest Rate** The simple interest charged on a six-month loan of $18,000 is $918. Find the simple interest rate.

50. **Simple Interest Rate** The maturity value of a four-month loan of $3000 is $3097. Find the simple interest rate.

51. **Simple Interest Rate** Find the simple interest rate on a three-month loan of $5000 if the maturity value of the loan is $5125.

52. **Late Payments** Your property tax bill is $1200. The county charges a penalty of 11% simple interest for late payments. How much do you owe if you pay the bill 2 months past the due date?

53. **Late Payments** Your electric bill is $132. You are charged 12% simple interest for late payments. How much do you owe if you pay the bill 1 month past the due date?

54. **Certificate of Deposit** If you withdraw part of your money from a certificate of deposit before the date of maturity, you must pay an interest penalty. Suppose you invested $5000 in a one-year certificate of deposit paying 8.5% interest. After 6 months, you decide to withdraw $2000. Your interest penalty is 3 months simple interest on the $2000. What interest penalty do you pay?

55. **Maturity Value** $10,000 is borrowed for 140 days at an 8% interest rate. Calculate the maturity value by the exact method and by the ordinary method. Which method yields the greater maturity value? Who benefits from using the ordinary method rather than the exact method, the borrower or the lender?

Extensions

CRITICAL THINKING

56. Interest has been described as a rental fee for money. Explain why this is an apt description of interest.

57. On July 31, at 4 P.M., you open a savings account that pays 5% interest, and you deposit $500 in the account. Your deposit is credited as of August 1. At the beginning of September, you receive a statement from the bank that shows that during the month of August, you received $2.15 in interest. The interest has been added to your account, bringing the total deposit to $502.15. At the beginning of October, you receive a statement from the bank that shows that during the month of September, you received $2.09 in interest on the $502.15 on deposit. Why did you receive less interest during the second month, when there was more money on deposit?

COOPERATIVE LEARNING

58. **Simple Interest a.** In the table below, the interest rate is an annual simple interest rate. Complete the table by calculating the simple interest due on the loan at the end of each month.

Loan Amount	Interest Rate	Period	Interest
$5000	6%	1 month	_____
$5000	6%	2 months	_____
$5000	6%	3 months	_____
$5000	6%	4 months	_____
$5000	6%	5 months	_____

Use the pattern of your answers in the table to find the simple interest due on a $5000 loan that has an annual simple interest rate of 6% for a period of:

b. 6 months   **c.** 7 months   **d.** 8 months   **e.** 9 months

Use your solutions to parts a through e to answer the following questions.

f. If you know the simple interest due on a one-month loan, explain how you can use that figure to calculate the simple interest due on a seven-month loan for the same principal and the same interest rate.

g. If the time period of a loan is doubled but the principal and interest rate remain the same, how many times as large is the simple interest due on the loan?

h. If the time period of a loan is tripled but the principal and interest rate remain the same, how many times as large is the simple interest due on the loan?

59. **Simple Interest a.** In the table below, the interest rate is an annual simple interest rate. Complete the table by calculating the simple interest due on the loan at the end of each month.

Loan Amount	Interest Rate	Period	Interest
$1000	6%	1 month	_____
$2000	6%	1 month	_____
$3000	6%	1 month	_____
$4000	6%	1 month	_____
$5000	6%	1 month	_____

Use the pattern of your answers in the table to determine the simple interest due on a one-month loan that has an annual simple interest rate of 6% when the principal is:

b. $6000   **c.** $7000   **d.** $8000   **e.** $9000

Use your solutions to parts a through e to answer the following questions.

f. If you know the simple interest due on a $1000 loan, explain how you can use that figure to calculate the simple interest due on an $8000 loan for the same time period and the same interest rate.

g. If the principal of a loan is doubled but the time period and interest rate remain the same, how many times as large is the simple interest due on the loan?

h. If the principal of a loan is tripled but the time period and interest rate remain the same, how many times as large is the simple interest due on the loan?

| # Compound Interest

Compound Interest

Simple interest is generally used for loans of 1 year or less. For loans of more than 1 year, the interest paid on the money borrowed is called *compound interest.* **Compound interest** is interest calculated not only on the original principal, but also on any interest that has already been earned.

To illustrate compound interest, suppose you deposit $1000 in a savings account earning 5% interest, compounded annually (once a year).

During the first year, the interest earned is calculated as follows.

$$I = Prt$$
$$I = \$1000(0.05)(1) = \$50$$

At the end of the first year, the total amount in the account is

$$A = P + I$$
$$A = \$1000 + \$50 = \$1050$$

During the second year, the interest earned is calculated using the amount in the account at the end of the first year.

$$I = Prt$$
$$I = \$1050(0.05)(1) = \$52.50$$

Note that the interest earned during the second year ($52.50) is greater than the interest earned during the first year ($50). This is because the interest earned during the first year was added to the original principal, and the interest for the second year was calculated using this sum. If the account earned simple interest rather than compound interest, the interest earned each year would be the same ($50).

At the end of the second year, the total amount in the account is the sum of the amount in the account at the end of the first year and the interest earned during the second year.

$$A = P + I$$
$$A = \$1050 + 52.50 = \$1102.50$$

The interest earned during the third year is calculated using the amount in the account at the end of the second year ($1102.50).

$$I = Prt$$
$$I = \$1102.50(0.05)(1) = \$55.125 \approx \$55.13$$

The interest earned each year keeps increasing. This is the effect of compound interest.

In this example, the interest is compounded annually. However, compound interest can be compounded semiannually (twice a year), quarterly (four times a year), monthly, or daily. The frequency with which the interest is compounded is called the **compounding period.**

If, in the preceding example, interest is compounded quarterly rather than annually, then the first interest payment on the $1000 in the account occurs after 3 months

$\left(t = \frac{3}{12} = \frac{1}{4}\right.$; 3 months is one-quarter of a year$\left.\right)$. That interest is then added to the account, and the interest earned for the second quarter is calculated using that sum.

End of 1st quarter: $I = Prt = \$1000(0.05)\left(\dfrac{3}{12}\right) = \$12.50$

$$A = P + I = \$1000 + \$12.50 = \$1012.50$$

End of 2nd quarter: $I = Prt = \$1012.50(0.05)\left(\dfrac{3}{12}\right) = \$12.65625 \approx \$12.66$

$$A = P + I = \$1012.50 + \$12.66 = \$1025.16$$

End of 3rd quarter: $I = Prt = \$1025.16(0.05)\left(\dfrac{3}{12}\right) = \$12.8145 \approx \$12.81$

$$A = P + I = \$1025.16 + \$12.81 = \$1037.97$$

End of 4th quarter: $I = Prt = \$1037.97(0.05)\left(\dfrac{3}{12}\right) = \$12.974625 \approx \$12.97$

$$A = P + I = \$1037.97 + \$12.97 = \$1050.94$$

The total amount in the account at the end of the first year is $1050.94.

When the interest is compounded quarterly, the account earns more interest ($50.94) than when the interest is compounded annually ($50). In general, an increase in the number of compounding periods results in an increase in the interest earned by an account.

In the example above, the formulas $I = Prt$ and $A = P + I$ were used to show the amount of interest added to the account each quarter. The formula $A = P(1 + rt)$ can be used to calculate A at the end of each quarter. For example, the amount in the account at the end of the first quarter is

$$A = P(1 + rt)$$
$$A = 1000\left[1 + 0.05\left(\dfrac{3}{12}\right)\right]$$
$$A = 1000(1.0125)$$
$$A = 1012.50$$

This amount, $1012.50, is the same as the amount calculated above using the formula $A = P + I$ to find the amount at the end of the first quarter.

The formula $A = P(1 + rt)$ is used in Example 1.

EXAMPLE 1 ■ Calculate Future Value

You deposit $500 in an account earning 6% interest, compounded semiannually. How much is in the account at the end of 1 year?

Solution

The interest is compounded every 6 months. Calculate the amount in the account after the first 6 months. $t = \dfrac{6}{12}$.

$$A = P(1 + rt)$$
$$A = 500\left[1 + 0.06\left(\dfrac{6}{12}\right)\right]$$
$$A = 515$$

Calculate the amount in the account after the second 6 months.

$$A = P(1 + rt)$$
$$A = 515\left[1 + 0.06\left(\frac{6}{12}\right)\right]$$
$$A = 530.45$$

The total amount in the account at the end of 1 year is $530.45.

CHECK YOUR PROGRESS 1   You deposit $2000 in an account earning 4% interest, compounded monthly. How much is in the account at the end of 6 months?

Solution   *See page S25.*

▼ **point of interest**

What do chief financial officers of corporations say is their main source of financial news? Their responses are shown below. (*Source:* Robert Half International)

Newspapers: 47%
Communications with
   colleagues: 15%
Television: 12%
Internet: 11%
Magazines: 9%
Radio: 5%
Don't know: 1%

In calculations that involve compound interest, the sum of the principal and the interest that has been added to it is called the **compound amount.** In Example 1, the compound amount is $530.45.

The calculations necessary to determine compound interest and compound amounts can be simplified by using a formula. Consider an amount P deposited in an account paying an annual interest rate r, compounded annually.

The interest earned during the first year is

$$I = Prt$$
$$I = Pr(1) \qquad \bullet\; t = 1$$
$$I = Pr$$

The compound amount A in the account after 1 year is the sum of the original principal and the interest earned during the first year:

$$A = P + I$$
$$A = P + Pr$$
$$A = P(1 + r) \qquad \bullet\; \text{Factor } P \text{ from each term.}$$

During the second year, the interest is calculated on the compound amount at the end of the first year, $P(1 + r)$.

$$I = Prt$$
$$I = P(1 + r)r(1) \qquad \bullet\; \text{Replace } P \text{ with } P(1 + r);\; t = 1.$$
$$I = P(1 + r)r$$

The compound amount A in the account after 2 years is the sum of the compound amount at the end of the first year and the interest earned during the second year:

$$A = P + I$$
$$A = P(1 + r) + P(1 + r)r \qquad \bullet\; \text{Replace } P \text{ with } P(1 + r) \text{ and } I \text{ with } P(1 + r)r.$$

$$A = 1[P(1 + r)] + [P(1 + r)]r$$
$$A = P(1 + r)(1 + r) \qquad \bullet\; \text{Factor } P(1 + r) \text{ from each term.}$$
$$A = P(1 + r)^2 \qquad \bullet\; \text{Write } (1 + r)(1 + r) \text{ as } (1 + r)^2.$$

During the third year, the interest is calculated on the compound amount at the end of the second year, $P(1 + r)^2$.

$$I = Prt$$
$$I = P(1 + r)^2 r(1) \qquad \text{• Replace } P \text{ with } P(1 + r)^2; t = 1.$$
$$I = P(1 + r)^2 r$$

The compound amount A in the account after 3 years is the sum of the compound amount at the end of the second year and the interest earned during the third year:

$$A = P + I$$
$$A = P(1 + r)^2 + P(1 + r)^2 r \qquad \text{• Replace } P \text{ with } P(1 + r)^2 \text{ and } I \text{ with } P(1 + r)^2 r.$$
$$A = 1[P(1 + r)^2] + [P(1 + r)^2]r$$
$$A = P(1 + r)^2(1 + r) \qquad \text{• Factor } P(1 + r)^2 \text{ from each term.}$$
$$A = P(1 + r)^3 \qquad \text{• Write } (1 + r)^2(1 + r) \text{ as } (1 + r)^3.$$

Note that the compound amount at the end of each year is the previous year's compound amount multiplied by $(1 + r)$. The exponent on $(1 + r)$ is equal to the number of compounding periods. Generalizing from this, we can state that the compound amount after n years is $A = P(1 + r)^n$.

In deriving this equation, interest was compounded annually; therefore, $t = 1$. Applying a similar argument for more frequent compounding periods, we derive the following compound amount formula. This formula enables us to calculate the compound amount for any number of compounding periods per year.

▼ **point of interest**

It is believed that U.S. currency is green because at the time of the introduction of the smaller-size bills in 1929, green pigment was readily available in large quantities. The color was resistant to chemical and physical changes, and green was psychologically associated with strong, stable government credit. (*Source:* www.moneyfactory.gov)

✔ **TAKE NOTE**

The compound amount formula can also be written as

$$A = P\left(1 + \frac{r}{n}\right)^{nt}.$$

Compound Amount Formula

The compound amount formula is

$$A = P(1 + i)^N$$

where A is the compound amount when P dollars is deposited at an interest rate of i per compounding period for N compounding periods.

Because i is the interest rate per compounding period and n is the number of compounding periods per year,

$$i = \frac{\text{annual interest rate}}{\text{number of compounding periods per year}} = \frac{r}{n}$$

$$N = (\text{number of compounding periods per year})(\text{number of years}) = nt$$

To illustrate how to calculate the values of i and N, consider an account earning 12% interest, compounded quarterly, for a period of 3 years.

The annual interest rate is 12%: $r = 12\%$.

When interest is compounded quarterly, there are four compounding periods per year: $n = 4$.

The time is 3 years: $t = 3$.

$$i = \frac{r}{n} = \frac{12\%}{4} = 3\%$$

$$N = nt = 4(3) = 12$$

The number of compounding periods in the 3 years is 12: $N = 12$. The account earns 3% interest per compounding period (per quarter): $i = 3\%$. Note that the interest rate per compounding period times the number of compounding periods per year equals the annual interest rate: $(3\%)(4) = 12\%$.

Recall that compound interest can be compounded annually (once a year), semiannually (twice a year), quarterly (four times a year), monthly, or daily. The possible values of n (the number of compounding periods per year) are recorded in the table below.

Values of n (number of compounding periods per year)

If interest is	then $n =$
compounded annually	1
compounded semiannually	2
compounded quarterly	4
compounded monthly	12
compounded daily	360

QUESTION *What is the value of i when the interest rate is 6%, compounded monthly?*

Recall that the future value of an investment is the value of the investment after the original principal has been invested for a period of time. In other words, it is the principal plus the interest earned. Therefore, it is the compound amount A in the compound amount formula.

EXAMPLE 2 ■ Calculate the Compound Amount

Calculate the compound amount when $10,000 is deposited in an account earning 8% interest, compounded semiannually, for 4 years.

Solution

Use the compound amount formula. $r = 8\% = 0.08$, $n = 2$, $t = 4$, $i = \dfrac{r}{n} = \dfrac{0.08}{2} = 0.04$, and $N = nt = 2(4) = 8$.

$$A = P(1 + i)^N$$
$$A = 10{,}000(1 + 0.04)^8$$
$$A = 10{,}000(1.04)^8$$
$$A \approx 10{,}000(1.368569)$$
$$A \approx 13{,}685.69$$

The compound amount after 4 years is approximately $13,685.69.

✔ **TAKE NOTE**

When using the compound amount formula, write the interest rate r as a decimal. Then calculate i.

ANSWER $i = \dfrac{r}{n} = \dfrac{6\%}{12} = 0.5\% = 0.005.$

CALCULATOR NOTE

When using a scientific calculator to solve the compound amount formula for *A*, use the keystroking sequence

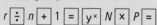

When using a graphing calculator, use the sequence

$P\ (\ 1\ +\ r\ \div\ n\ )\ \wedge\ N$

CHECK YOUR PROGRESS 2   Calculate the compound amount when $4000 is deposited in an account earning 6% interest, compounded monthly, for 2 years.

Solution   *See page S26.*

EXAMPLE 3 ■ Calculate Future Value

Calculate the future value of $5000 earning 9% interest, compounded daily, for 3 years.

Solution

Use the compound amount formula. $r = 9\% = 0.09$, $t = 3$. Interest is compounded daily. Use the ordinary method. $n = 360$, $i = \frac{r}{n} = \frac{0.09}{360} = 0.00025$, $N = nt = 360(3) = 1080$

$$A = P(1 + i)^N$$
$$A = 5000(1 + 0.00025)^{1080}$$
$$A = 5000(1.00025)^{1080}$$
$$A \approx 5000(1.3099202)$$
$$A \approx 6549.60$$

The future value after 3 years is approximately $6549.60.

CHECK YOUR PROGRESS 3   Calculate the future value of $2500 earning 9% interest, compounded daily, for 4 years.

Solution   *See page S26.*

The formula $I = A - P$ was used in Section 7.1 to find the interest earned on an investment or the interest paid on a loan. This same formula is used for compound interest. It is used in Example 4 to find the interest earned on an investment.

EXAMPLE 4 ■ Calculate Compound Interest

How much interest is earned in 2 years on $4000 deposited in an account paying 6% interest, compounded quarterly?

Solution

Calculate the compound amount. Use the compound amount formula. $r = 6\% = 0.06$, $n = 4$, $t = 2$, $i = \frac{r}{n} = \frac{0.06}{4} = 0.015$, $N = nt = 4(2) = 8$

$$A = P(1 + i)^N$$
$$A = 4000(1 + 0.015)^8$$
$$A = 4000(1.015)^8$$
$$A \approx 4000(1.1264926)$$
$$A \approx 4505.97$$

Calculate the interest earned. Use the formula $I = A - P$.

$$I = A - P$$
$$I = 4505.97 - 4000$$
$$I = 505.97$$

The amount of interest earned is approximately $505.97.

CHECK YOUR PROGRESS 4 How much interest is earned in 6 years on $8000 deposited in an account paying 9% interest, compounded monthly?

Solution   *See page S26.*

An alternative to using the compound amount formula is to use a calculator that has finance functions built into it. The TI-83/TI-84 Plus graphing calculator is used in Example 5.

EXAMPLE 5 ▦ **Calculate Compound Interest**

Use the finance feature of a calculator to determine the compound amount when $2000 is deposited in an account earning an interest rate of 12%, compounded quarterly, for 10 years.

Solution
On a TI-83 calculator, press | 2nd | [Finance] to display the FINANCE CALC menu.

For the TI-83 Plus or TI-84 Plus, press | APPS || ENTER |.

Press | ENTER | to select 1: TVM Solver.

> After N =, enter 40.
>
> After I% =, enter 12.
>
> After PV =, enter −2000. (See the note below.)
>
> After PMT =, enter 0.
>
> After P/Y =, enter 4.
>
> After C/Y =, enter 4.

Use the up arrow key to place the cursor at FV =.

Press | ALPHA | [Solve].

The solution is displayed to the right of FV =.

The compound amount is $6524.08.

Note: For most financial calculators and financial computer programs such as Excel, money that is paid out (such as the $2000 that is being deposited in this example) is entered as a negative number.

CHECK YOUR PROGRESS 5 Use the finance feature of a calculator to determine the compound amount when $3500 is deposited in an account earning an interest rate of 6%, compounded semiannually, for 5 years.

Solution   *See page S26.*

CALCULATOR NOTE

On the TI-83/84 Plus calculator screen below, the variable N is the number of compounding periods, I% is the interest rate, PV is the present value of the money, PMT is the payment, P/Y is the number of payments per year, C/Y is the number of compounding periods per year, and FV is the future value of the money. TVM represents Time Value of Money.

Present value is discussed a little later in this section. Enter the principal for the present value amount.

"Solve" is above the | ENTER | key on the calculator.

```
N=40
I%=12
PV=-2000
PMT=0
■FV=6524.075584
P/Y=4
C/Y=4
PMT: END BEGIN
```

Math Matters   Federal Funds Rate

The Federal Reserve Board meets a number of times each year to decide, among other things, the federal funds rate. The **federal funds rate** is the interest rate banks charge one another for overnight loans. The Federal Reserve Board can decide to lower, raise, or leave unchanged the federal funds rate. Its decision has a significant effect on the economy. For example, if the board increases the rate, then banks have to pay more to borrow money. If banks must pay higher rates, they are going to charge their customers a higher rate to borrow money.

The graph below shows the federal funds rates from March 2003 through December 2005.

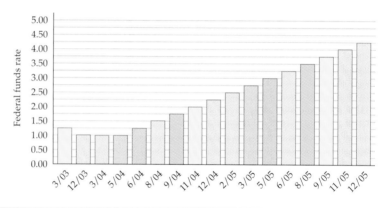

Present Value

The **present value** of an investment is the original principal invested, or the value of the investment before it earns any interest. Therefore, it is the principal (P) in the compound amount formula. Present value is used to determine how much money must be invested today in order for an investment to have a specific value at a future date.

The formula for the present value of an investment is found by solving the compound amount formula for P.

$$A = P(1 + i)^N$$

$$\frac{A}{(1 + i)^N} = \frac{P(1 + i)^N}{(1 + i)^N} \qquad \bullet \text{ Divide each side of the equation by } (1 + i)^N.$$

$$\frac{A}{(1 + i)^N} = P$$

Present Value Formula

The present value formula is

$$P = \frac{A}{(1 + i)^N}$$

where P is the original principal invested at an interest rate of i per compounding period for N compounding periods, and A is the compound amount.

The present value formula is used in Example 6.

EXAMPLE 6 ■ Calculate Present Value

How much money should be invested in an account that earns 8% interest, compounded quarterly, in order to have $30,000 in 5 years?

Solution
Use the present value formula. $r = 8\% = 0.08$, $n = 4$, $t = 5$, $i = \dfrac{r}{n} = \dfrac{0.08}{4} = 0.02$, $N = nt = 4(5) = 20$

$$P = \frac{A}{(1 + i)^N}$$

$$P = \frac{30,000}{(1 + 0.02)^{20}} = \frac{30,000}{1.02^{20}} \approx \frac{30,000}{1.485947396}$$

$$P \approx 20,189.14$$

$20,189.14 should be invested in the account in order to have $30,000 in 5 years.

CHECK YOUR PROGRESS 6 How much money should be invested in an account that earns 9% interest, compounded semiannually, in order to have $20,000 in 5 years?

Solution See page S26.

CALCULATOR NOTE

To calculate P in the present value formula using a calculator, you can first calculate $(1 + i)^N$. Next use the $\boxed{1/x}$ key or the $\boxed{x^{-1}}$ key. This will place the value of $(1 + i)^N$ in the denominator. Then multiply by the value of A.

EXAMPLE 7 ■ Calculate Present Value

Use the finance feature of a calculator to determine how much money should be invested in an account that earns 7% interest, compounded monthly, in order to have $50,000 in 10 years.

Solution
On a TI-83 calculator, press $\boxed{\text{2nd}}$ [FINANCE] to display the FINANCE CALC menu. For the TI-83 Plus or TI-84 Plus, press $\boxed{\text{APPS}}$ $\boxed{\text{ENTER}}$.

Press $\boxed{\text{ENTER}}$ to select 1 TVM Solver.

After N =, enter 120.

After I% =, enter 7.

After PMT =, enter 0.

After FV =, enter 50000.

After P/Y =, enter 12.

After C/Y =, enter 12.

```
N=120
I%=7
PV=█
PMT=0
FV=50000
P/Y=12
C/Y=12
PMT: END BEGIN
```

```
N=120
I%=7
■PV=-24879.81338
PMT=0
FV=50000
P/Y=12
C/Y=12
PMT: END BEGIN
```

Use the up arrow key to place the cursor at PV =.

Press ALPHA [Solve].

The solution is displayed to the right of PV =.

$24,879.81 should be invested in the account in order to have $50,000 in 10 years.

CHECK YOUR PROGRESS 7 Use the finance feature of a calculator to determine how much money should be invested in an account that earns 6% interest, compounded daily, in order to have $25,000 in 15 years.

Solution *See page S26.*

Inflation

▼ **point of interest**

The life of a bill depends on its denomination. The average lives of different denominations are shown below. (*Source:* Federal Reserve System)

$1: 22 months
$5: 2 years
$10: 3 years
$20: 4 years
$50: 9 years
$100: 9 years

We have discussed compound interest and its effect on the growth of an investment. After your money has been invested for a period of time in an account that pays interest, you will have more money than you originally deposited. But does that mean you will be able to buy more with the compound amount than you were able to buy with the original investment at the time you deposited the money? The answer is not necessarily, and the reason is the effect of inflation.

Suppose the price of a large-screen TV is $1500. You have enough money to purchase the TV, but decide to invest the $1500 in an account paying 6% interest, compounded monthly. After 1 year, the compound amount is $1592.52. But during that same year, the rate of inflation was 7%. The large-screen TV now costs

$$\$1500 \text{ plus } 7\% \text{ of } \$1500 = \$1500 + 0.07(\$1500)$$
$$= \$1500 + \$105$$
$$= \$1605$$

Because $1592.52 < $1605, you have actually lost purchasing power. At the beginning of the year, you had enough money to buy the large-screen TV; at the end of the year, the compound amount is not enough to pay for that same TV. Your money has actually lost value because it can buy less now than it could 1 year ago.

Inflation is an economic condition during which there are increases in the costs of goods and services. Inflation is expressed as a percent; for example, we speak of an annual inflation rate of 7%.

To calculate the effects of inflation, we use the same procedure we used to calculate compound amount. This process is illustrated in Example 8. Although inflation rates vary dramatically, in this section we will assume constant annual inflation rates, and we will use annual compounding in solving inflation problems. In other words, $n = 1$ for these exercises.

EXAMPLE 8 ■ **Calculate the Effect of Inflation on Salary**

Suppose your annual salary today is $35,000. You want to know what an equivalent salary will be in 20 years—that is, a salary that will have the same purchasing power. Assume a 6% inflation rate.

▼ **point of interest**

Inflation rates can change dramatically over time. The table below shows the five smallest and the five largest inflation rates since 1974. (*Source:* inflation.data.com)

The Five Smallest Inflation Rates	
Inflation Rate	Year
1.55%	1998
1.59%	2002
1.91%	1986
2.19%	1999
2.27%	2003

The Five Largest Inflation Rates	
Inflation Rate	Year
13.58%	1980
11.22%	1979
11.03%	1974
10.35%	1981
9.20%	1975

Solution

Use the compound amount formula. $r = 6\% = 0.06$. The inflation rate is an annual rate, so $n = 1$. $t = 20$, $i = \frac{r}{n} = \frac{0.06}{1} = 0.06$, $N = nt = 1(20) = 20$

$$A = P(1 + i)^N$$
$$A = 35,000(1 + 0.06)^{20}$$
$$A = 35,000(1.06)^{20}$$
$$A \approx 35,000(3.20713547)$$
$$A \approx 112,249.74$$

Twenty years from now, you need to earn an annual salary of approximately $112,249.74 in order to have the same purchasing power.

CHECK YOUR PROGRESS 8 Assume that the average new car sticker price in 2008 is $28,000. Use an annual inflation rate of 5% to estimate the average new car sticker price in 2025.

Solution See page S27.

The present value formula can be used to determine the effect of inflation on the future purchasing power of a given amount of money. Substitute the inflation rate for the interest rate in the present value formula. The compounding period is 1 year. Again we will assume a constant rate of inflation.

EXAMPLE 9 ■ **Calculate the Effect of Inflation on Future Purchasing Power**

Suppose you purchase an insurance policy in 2010 that will provide you with $250,000 when you retire in 2045. Assuming an annual inflation rate of 8%, what will be the purchasing power of the $250,000 in 2045?

Solution

Use the present value formula. $r = 8\%$ and $t = 35$. The inflation rate is an annual rate, so $n = 1$. $i = \frac{r}{n} = \frac{8\%}{1} = 8\% = 0.08$, $N = nt = 1(35) = 35$

$$P = \frac{A}{(1 + i)^N}$$
$$P = \frac{250,000}{(1 + 0.08)^{35}}$$
$$P \approx \frac{250,000}{14.785344}$$
$$P \approx 16,908.64$$

In 2045, the purchasing power of $250,000 will be approximately $16,908.64.

CHECK YOUR PROGRESS 9 Suppose you purchase an insurance policy in 2010 that will provide you with $500,000 when you retire in 40 years. Assuming an annual inflation rate of 7%, what will be the purchasing power of half a million dollars in 2050?

Solution See page S27.

Math Matters The Rule of 72

The **Rule of 72** states that the number of years for prices to double is approximately equal to 72 divided by the annual inflation rate.

$$\text{Years to double} = \frac{72}{\text{annual inflation rate}}$$

For example, at an annual inflation rate of 6%, prices will double in approximately 12 years.

$$\text{Years to double} = \frac{72}{\text{annual inflation rate}} = \frac{72}{6} = 12$$

Effective Interest Rate

When interest is compounded, the annual rate of interest is called the **nominal rate.** The **effective rate** is the simple interest rate that would yield the same amount of interest after 1 year. When a bank advertises a "7% annual interest rate compounded daily and yielding 7.25%," the nominal interest rate is 7% and the effective rate is 7.25%.

QUESTION   *A bank offers a savings account that pays 2.75% annual interest, compounded daily and yielding 2.79%. What is the effective rate on this account? What is the nominal rate?*

Consider $100 deposited at 6%, compounded monthly, for 1 year.

The future value after 1 year is $106.17.

$$A = P(1 + i)^N$$
$$A = 100(1 + 0.005)^{12}$$
$$A \approx 106.17$$

The interest earned in 1 year is $6.17.

$$I = A - P$$
$$I = 106.17 - 100$$
$$I = 6.17$$

Now consider $100 deposited at an annual simple interest rate of 6.17%.

The interest earned in 1 year is $6.17.

$$I = Prt$$
$$I = 100(0.0617)(1)$$
$$I = 6.17$$

The interest earned on $100 is the same when it is deposited at 6% compounded monthly as when it is deposited at an annual simple interest rate of 6.17%. 6.17% is the effective annual rate of a deposit that earns 6% compounded monthly.

ANSWER   *The effective rate is 2.79%. The nominal rate is 2.75%.*

In this example, $100 was used as the principal. When we use $100 for P, we multiply the interest rate by 100. Remember that the interest rate is written as a decimal in the equation $I = Prt$, and a decimal is written as a percent by multiplying by 100. Therefore, when $P = 100$, the interest earned on the investment ($6.17) is the same number as the effective annual rate (6.17%).

EXAMPLE 10 ■ Calculate the Effective Interest Rate

A credit union offers a certificate of deposit at an annual interest rate of 3%, compounded monthly. Find the effective rate. Round to the nearest hundredth of a percent.

Solution
Use the compound amount formula to find the future value of $100 after 1 year.
$r = 3\% = 0.03$, $n = 12$, $t = 1$, $i = \frac{r}{n} = \frac{0.03}{12} = 0.0025$, $N = nt = 12(1) = 12$

$$A = P(1 + i)^N$$
$$A = 100(1 + 0.0025)^{12}$$
$$A = 100(1.0025)^{12}$$
$$A \approx 100(1.030415957)$$
$$A \approx 103.04$$

Find the interest earned on the $100.

$$I = A - P$$
$$I = 103.04 - 100$$
$$I = 3.04$$

The effective interest rate is 3.04%.

CHECK YOUR PROGRESS 10   A bank offers a certificate of deposit at an annual interest rate of 4%, compounded quarterly. Find the effective rate. Round to the nearest hundredth of a percent.

Solution   *See page S27.*

To compare two investments or loan agreements, we could calculate the effective annual rate of each. However, a shorter method involves comparing the compound amounts of each. Because the value of $(1 + i)^N$ is the compound amount of $1, we can compare the value of $(1 + i)^N$ for each alternative.

EXAMPLE 11 ■ Compare Annual Yields

One bank advertises an interest rate of 5.5%, compounded quarterly, on a certificate of deposit. Another bank advertises an interest rate of 5.25%, compounded monthly. Which investment has the higher annual yield?

Solution

Calculate $(1 + i)^N$ for each investment.

$$i = \frac{r}{n} = \frac{0.055}{4}$$

$$N = nt = 4(1) = 4$$

$$(1 + i)^N = \left(1 + \frac{0.055}{4}\right)^4$$

$$\approx 1.0561448$$

$$i = \frac{r}{n} = \frac{0.0525}{12}$$

$$N = nt = 12(1) = 12$$

$$(1 + i)^N = \left(1 + \frac{0.0525}{12}\right)^{12}$$

$$\approx 1.0537819$$

Compare the two compound amounts.

$$1.0561448 > 1.0537819$$

An investment of 5.5% compounded quarterly has a higher annual yield than an investment that earns 5.25% compounded monthly.

CHECK YOUR PROGRESS 11 Which investment has the higher annual yield, one that earns 5% compounded quarterly or one that earns 5.25% compounded semiannually?

Solution *See page S27.*

Math Matters   Saving for Retirement

The tables below show results of surveys in which workers were asked questions about saving for retirement.

How many workers fear they've fallen behind in saving				
	56 and older	47 – 55	37 – 46	25 and younger
On track to save enough for retirement	39%	30%	32%	33%
Ahead of schedule	6%	5%	5%	7%
A little behind schedule	21%	27%	24%	28%
A lot behind schedule	30%	37%	37%	30%

Source: Annual Retirement Confidence Survey. Reprinted by permission of the Employee Benefits Research Institute.

Worker confidence in having enough money to live comfortably throughout their retirement years									
	1996	1997	1998	1999	2000	2001	2002	2003	2004
Very confident	19%	24%	22%	22%	25%	22%	23%	21%	24%
Somewhat confident	41%	41%	45%	47%	47%	41%	47%	45%	44%
Not too confident	23%	19%	18%	21%	18%	18%	19%	17%	18%
Not at all confident	16%	15%	13%	9%	10%	17%	10%	16%	13%

Source: Annual Retirement Confidence Survey. Reprinted by permission of the Employee Benefits Research Institute.

Investigation

Consumer Price Index

An **index number** measures the change in a quantity, such as cost, over a period of time. One of the most widely used indexes is the Consumer Price Index (CPI). The CPI, which includes the selling prices of about 400 key consumer goods and services, indicates the relative changes in the prices of these items over time. It measures the effect of inflation on the cost of goods and services.

The main components of the Consumer Price Index, shown below, are the costs of housing, food and beverages, transportation, medical care, clothing, recreation, and education.

> **✔ TAKE NOTE**
>
> The category "Recreation" includes television sets, cable TV, pets and pet products, sports equipment, and admission tickets.

Other 5.3%
Education and Communication 4.8%
Recreation 5.9%
Apparel and Upkeep 4.5%
Medical Care 5.8%
Housing 40.0%
17.6%
16.2%
Transportation
Food and Drinks

The Components of the Consumer Price Index

Source: U.S. Bureau of Labor Statistics

> **✔ TAKE NOTE**
>
> You can obtain current and historical data on the Consumer Price Index by visiting the website of the Bureau of Labor Statistics at www.bls.gov.

The CPI is a measure of the cost of living for consumers. The government publishes monthly and annual figures on the Consumer Price Index.

The Consumer Price Index has a base period, 1982–1984, from which to make comparisons. The CPI for the base period is 100. The CPI for December 2005 was 197.6. This means that $100 in the period 1982–1984 had the same purchasing power as $197.60 in December of 2005.

An index number is actually a percent written without a percent sign. The CPI of 197.6 for December of 2005 means that the average cost of consumer goods at that time was 197.6% of their cost in the 1982–1984 period.

The table below gives the CPI for various products in 2005.

Product	CPI
All items	193.2
Food and beverages	190.2
Housing	193.8
Apparel	120.5
Transportation	169.3
Medical care	320.6
Recreation	109.1
Education and communication	112.8

The Consumer Price Index, 2005

Source: U.S. Bureau of Labor Statistics

(continued)

Investigation Exercises

Solve the following.

1. The CPI for 1990 was 130.7. What percent of the base period prices were the consumer prices in 1990?

2. The CPI for 2001 was 177.1. The percent increase in consumer prices from 2001 to 2002 was 2.8%. Find the CPI for 2002. Round to the nearest tenth.

3. The CPI for 1988 was 118.3. The CPI for 1989 was 124. Find the percent increase in consumer prices for this time period.

4. Of the items listed in the table on the preceding page, are there any items that cost at least twice as much in 2005 as they cost during the base period? If so, which ones?

5. Of the items listed in the table on the preceding page, are there any items that cost more than one-and-one-half times as much in 2005 as they cost during the base years, but less than twice as much as they cost during the base period? If so, which ones?

6. If the cost of textbooks for one semester was $120 in the base years, how much did similar textbooks cost in 2005?

7. If a movie ticket cost $9 in 2005, what would a comparable movie ticket have cost during the base years?

8. The base year for the Consumer Confidence Index is 1985. The Consumer Confidence Index in November of 2005 was 98.9. Were consumers more confident in 1985 or in November of 2005?

Exercise Set 7.2

In Exercises 1–8, calculate the compound amount. Use the compound amount formula and a calculator.

1. $P = \$1200$, $r = 7\%$ compounded semiannually, $t = 12$ years

2. $P = \$3500$, $r = 8\%$ compounded semiannually, $t = 14$ years

3. $P = \$500$, $r = 9\%$ compounded quarterly, $t = 6$ years

4. $P = \$7000$, $r = 11\%$ compounded quarterly, $t = 9$ years

5. $P = \$8500$, $r = 9\%$ compounded monthly, $t = 10$ years

6. $P = \$6400$, $r = 6\%$ compounded monthly, $t = 3$ years

7. $P = \$9600$, $r = 9\%$ compounded daily, $t = 3$ years

8. $P = \$1700$, $r = 9\%$ compounded daily, $t = 5$ years

In Exercises 9–14, calculate the compound amount. Use a calculator with a financial mode.

9. $P = \$1600$, $r = 8\%$ compounded quarterly, $t = 10$ years

10. $P = \$4200$, $r = 6\%$ compounded semiannually, $t = 8$ years

11. $P = \$3000$, $r = 12\%$ compounded monthly, $t = 5$ years

12. $P = \$9800$, $r = 10\%$ compounded quarterly, $t = 4$ years

13. $P = \$1700$, $r = 9\%$ compounded semiannually, $t = 3$ years

14. $P = \$8600$, $r = 11\%$ compounded semiannually, $t = 5$ years

In Exercises 15–20, calculate the future value.

15. $P = \$7500$, $r = 12\%$ compounded monthly, $t = 5$ years

16. $P = \$1800$, $r = 9.5\%$ compounded annually, $t = 10$ years

17. $P = \$4600$, $r = 10\%$ compounded semiannually, $t = 12$ years

18. $P = \$9000$, $r = 11\%$ compounded quarterly, $t = 3$ years

19. $P = \$22,000$, $r = 9\%$ compounded monthly, $t = 7$ years

20. $P = \$5200$, $r = 8.1\%$ compounded daily, $t = 9$ years

In Exercises 21–26, calculate the present value.

21. $A = \$25,000$, $r = 10\%$ compounded quarterly, $t = 12$ years

22. $A = \$20,000$, $r = 12\%$ compounded monthly, $t = 5$ years

23. $A = \$40,000$, $r = 7.5\%$ compounded annually, $t = 35$ years

24. $A = \$10,000$, $r = 11\%$ compounded semiannually, $t = 30$ years

25. $A = \$15,000$, $r = 8\%$ compounded quarterly, $t = 5$ years

26. $A = \$50,000$, $r = 18\%$ compounded monthly, $t = 5$ years

27. Your are considering investing some money. One investment pays 7.25% annual interest compounded quarterly. A second investment pays 6.75% annual interest compounded quarterly. On which investment will you earn more interest?

28. You need to borrow some money. One bank offers a loan at 8.6% annual interest compounded monthly. A second institution offers a loan at 9.8% annual interest compounded monthly. On which loan will you pay less interest?

29. You need to borrow some money. One bank offers a loan at 7.5% annual interest compounded monthly. A second institution offers a loan at 7.5% annual interest compounded semiannually. On which loan will you pay less interest?

30. You are considering investing some money. One investment pays 4.25% annual interest compounded daily. A second investment pays 4.25% annual interest compounded annually. On which investment will you earn more interest?

31. On January 1, you deposit equal amounts of money in two accounts. For account A, the annual interest rate is 12% compounded annually. For account B, the monthly interest rate is 1% compounded monthly. Assume you do not deposit any additional money in either account or withdraw any money from either account. At the end of 1 year, would the value of account A be less than, equal to, or greater than the value of account B?

32. A worker deposits the proceeds from a bonus in an account that earns an annual interest rate of 5% compounded annually. If the worker leaves the money in the account for 10 years without adding or withdrawing money, would the amount of interest earned in the eighth year be less than, equal to, or more than the amount of interest earned in the fourth year?

33. Compound Amount Calculate the compound amount when $8000 is deposited in an account earning 8% interest, compounded quarterly, for 5 years.

34. Compound Amount Calculate the compound amount when $3000 is deposited in an account earning 10% interest, compounded semiannually, for 3 years.

35. Compound Amount If you leave $2500 in an account earning 9% interest, compounded daily, how much money will be in the account after 4 years?

36. Compound Amount What is the compound amount when $1500 is deposited in an account earning an interest rate of 6%, compounded monthly, for 2 years?

37. Future Value What is the future value of $4000 earning 12% interest, compounded monthly, for 6 years?

38. Future Value Calculate the future value of $8000 earning 8% interest, compounded quarterly, for 10 years.

39. Compound Interest How much interest is earned in 3 years on $2000 deposited in an account paying 6% interest, compounded quarterly?

40. Compound Interest How much interest is earned in 5 years on $8500 deposited in an account paying 9% interest, compounded semiannually?

41. Compound Interest Calculate the amount of interest earned in 8 years on $15,000 deposited in an account paying 10% interest, compounded quarterly.

42. Compound Interest Calculate the amount of interest earned in 6 years on $20,000 deposited in an account paying 12% interest, compounded monthly.

43. Present Value How much money should be invested in an account that earns 6% interest, compounded monthly, in order to have $15,000 in 5 years?

44. Present Value How much money should be invested in an account that earns 7% interest, compounded quarterly, in order to have $10,000 in 5 years?

45. Compound Interest $1000 is deposited for 5 years in an account that earns 9% interest.

   a. Calculate the simple interest earned.

   b. Calculate the interest earned if interest is compounded daily.

   c. How much more interest is earned on the account when the interest is compounded daily?

46. Compound Interest $10,000 is deposited for 2 years in an account that earns 12% interest.

   a. Calculate the simple interest earned.

   b. Calculate the interest earned if interest is compounded daily.

   c. How much more interest is earned on the account when the interest is compounded daily?

47. Future Value $15,000 is deposited for 4 years in an account earning 8% interest.

   a. Calculate the future value of the investment if interest is compounded semiannually.

   b. Calculate the future value if interest is compounded quarterly.

   c. How much greater is the future value of the investment when the interest is compounded quarterly?

48. Future Value $25,000 is deposited for 3 years in an account earning 6% interest.

   a. Calculate the future value of the investment if interest is compounded annually.

   b. Calculate the future value if interest is compounded semiannually.

   c. How much greater is the future value of the investment when the interest is compounded semiannually?

49. Future Value $10,000 is deposited for 2 years in an account earning 8% interest.

   a. Calculate the interest earned if interest is compounded semiannually.

   b. Calculate the interest earned if interest is compounded quarterly.

   c. How much more interest is earned on the account when the interest is compounded quarterly?

50. Future Value $20,000 is deposited for 5 years in an account earning 6% interest.

   a. Calculate the interest earned if interest is compounded annually.

   b. Calculate the interest earned if interest is compounded semiannually.

   c. How much more interest is earned on the account when the interest is compounded semiannually?

51. Loans To help pay your college expenses, you borrow $7000 and agree to repay the loan at the end of 5 years at 8% interest, compounded quarterly.

   a. What is the maturity value of the loan?

   b. How much interest are you paying on the loan?

52. Loans You borrow $6000 to help pay your college expenses. You agree to repay the loan at the end of 5 years at 10% interest, compounded quarterly.

   a. What is the maturity value of the loan?

   b. How much interest are you paying on the loan?

53. Present Value A couple plans to save for their child's college education. What principal must be deposited by the parents when their child is born in order to have $40,000 when the child reaches the age of 18? Assume the money earns 8% interest, compounded quarterly.

54. Present Value A couple plans to invest money for their child's college education. What principal must be deposited by the parents when their child turns 10 in order to have $30,000 when the child reaches the age of 18? Assume the money earns 8% interest, compounded quarterly.

55. Present Value You want to retire in 30 years with $1,000,000 in investments.

   a. How much money would you have to invest today at 9% interest, compounded daily, in order to have $1,000,000 in 30 years?

   b. How much will the $1,000,000 generate in interest each year if it is invested at 9% interest, compounded daily?

56. Present Value You want to retire in 40 years with $1,000,000 in investments.

   a. How much money must you invest today at 8.1% interest, compounded daily, in order to have $1,000,000 in 40 years?

   b. How much will the $1,000,000 generate in interest each year if it is invested at 8.1% interest, compounded daily?

57. Compound Amount You deposit $5000 in a two-year certificate of deposit (CD) earning 8.1% interest, compounded daily. At the end of the 2 years, you reinvest the compound amount in another two-year CD. The interest rate on the second CD is 7.2%, compounded daily. What is the compound amount when the second CD matures?

58. Compound Amount You deposit $7500 in a two-year certificate of deposit (CD) earning 9.9% interest, compounded daily. At the end of the 2 years, you reinvest the compound amount plus an additional $7500 in another two-year CD. The interest rate on the second CD is 10.8%, compounded daily. What is the compound amount when the second CD matures?

59. **Inflation** The average monthly rent for a three-bedroom apartment in San Luis Obispo, California is $1686. Using an annual inflation rate of 7%, find the average monthly rent in 15 years.

60. **Inflation** The average cost of housing in Greenville, North Carolina is $128,495. Using an annual inflation rate of 7%, find the average cost of housing in 10 years.

61. **Inflation** Suppose your salary in 2010 is $40,000. Assuming an annual inflation rate of 7%, what salary will you need to earn in 2015 in order to have the same purchasing power?

62. **Inflation** Suppose your salary in 2010 is $50,000. Assuming an annual inflation rate of 6%, what salary will you need to earn in 2020 in order to have the same purchasing power?

63. **Inflation** In 2009 you purchase an insurance policy that will provide you with $125,000 when you retire in 2049. Assuming an annual inflation rate of 6%, what will be the purchasing power of the $125,000 in 2049?

64. **Inflation** You purchase an insurance policy in the year 2010 that will provide you with $250,000 when you retire in 25 years. Assuming an annual inflation rate of 8%, what will be the purchasing power of the quarter million dollars in 2035?

65. **Inflation** A retired couple have a fixed income of $3500 per month. Assuming an annual inflation rate of 7%, what will be the purchasing power of their monthly income in 5 years?

66. **Inflation** A retired couple have a fixed income of $46,000 per year. Assuming an annual inflation rate of 6%, what will be the purchasing power of their annual income in 10 years?

In Exercises 67–74, calculate the effective annual rate for an investment that earns the given rate of return. Round to the nearest hundredth of a percent.

67. 7.2% interest compounded quarterly

68. 8.4% interest compounded quarterly

69. 7.5% interest compounded monthly

70. 6.9% interest compounded monthly

71. 8.1% interest compounded daily

72. 6.3% interest compounded daily

73. 5.94% interest compounded monthly

74. 6.27% interest compounded monthly

Inflation In Exercises 75–82, you are given the 2004 price of an item. Use an inflation rate of 6% to calculate its price in 2009, 2014, and 2024. Round to the nearest cent.

75. Gasoline: $2.00 per gallon

76. Milk: $3.35 per gallon

77. Loaf of bread: $2.69

78. Sunday newspaper: $2.25

79. Ticket to a movie: $9

80. Paperback novel: $12.00

81. House: $275,000

82. Car: $24,000

In Exercises 83–88, calculate the purchasing power using an annual inflation rate of 7%. Round to the nearest cent.

83. $50,000 in 10 years

84. $25,000 in 8 years

85. $100,000 in 20 years

86. $30,000 in 15 years

87. $75,000 in 5 years

88. $20,000 in 25 years

89. **a.** Complete the table.

Nominal Rate	Effective Rate
4% annual compounding	
4% semiannual compounding	
4% quarterly compounding	
4% monthly compounding	
4% daily compounding	

b. As the number of compounding periods increases, does the effective rate increase or decrease?

90. **Effective Interest Rate** Beth Chipman has money in a savings account that earns an annual interest rate of 3%, compounded quarterly. What is the effective rate of interest on Beth's account? Round to the nearest hundredth of a percent.

91. **Effective Interest Rate** Blake Hamilton has money in a savings account that earns an annual interest rate of 3%, compounded monthly. What is the effective rate of interest on Blake's savings? Round to the nearest hundredth of a percent.

92. **Annual Yield** One bank advertises an interest rate of 6.6%, compounded quarterly, on a certificate of deposit. Another bank advertises an interest rate of 6.25%, compounded monthly. Which investment has the higher annual yield?

93. **Annual Yield** Which investment has the higher annual yield, one earning 6% compounded quarterly or one earning 6.25% compounded semiannually?

94. **Annual Yield** Which investment has the higher annual yield, one earning 7.8% compounded monthly or one earning 7.5% compounded daily?

95. **Annual Yield** One bank advertises an interest rate of 5.8%, compounded quarterly, on a certificate of deposit. Another bank advertises an interest rate of 5.6%, compounded monthly. Which investment has a higher annual yield?

Extensions

CRITICAL THINKING

96. **Future Value a.** Using an 8% interest rate with interest compounded quarterly, calculate the future values in 2 years of $1000, $2000, and $4000.

    b. Based on your answers to part a, what is the future value if the investment is doubled again, to $8000?

97. **Interest Rates** The future value of $2000 deposited in an account for 25 years with interest compounded semiannually is $22,934.80. Find the interest rate.

98. **Present Value** You want to generate an interest income of $12,000 per month. How much principal must be invested at an interest rate of 8%, compounded monthly, to generate this amount of monthly income? Round to the nearest hundred thousand.

COOPERATIVE LEARNING

99. **Saving for a Purchase** You want to buy a motorcycle costing $20,000. You have two options:

    a. You can borrow $20,000 at an effective annual rate of 8% for 1 year.

    b. You can save the money you would have made in loan payments during 1 year and purchase the motorcycle.

    If you decide to save your money for 1 year, you will deposit the equivalent of 1 month's loan payment ($1800) in your savings account at the end of each month, and you will earn 5% interest on the account. (In determining the amount of interest earned each month, assume that each month is $\frac{1}{12}$ of 1 year.) You

will pay 28% of the interest earned on the savings account in income taxes. Also, an annual inflation rate of 7% will have increased the price of the motorcycle by 7%. If you choose the option of saving your money for 1 year, how much money will you have left after you pay the income tax and purchase the motorcycle?

EXPLORATIONS

100. **Continuous Compounding** In our discussion of compound interest, we used annual, semiannual, monthly, quarterly, and daily compounding periods. When interest is compounded daily, it is compounded 360 times a year. If interest were compounded twice daily, it would be compounded $360(2) = 720$ times a year. If interest were compounded four times a day, it would be compounded $360(4) = 1440$ times a year. Remember that the more frequent the compounding, the more interest earned on the account. Therefore, if interest is compounded more frequently than daily, an investment will earn even more interest than if interest is compounded daily.

Some banking institutions advertise **continuous compounding**, which means that the number of compounding periods per year gets very, very large. When compounding continuously, instead of using the compound amount formula $A = P(1 + i)^N$, the following formula is used.

$$A = Pe^{rt}$$

In this formula, A is the compound amount when P dollars is deposited at an annual interest rate of r percent compounded continuously for t years. The number e is approximately equal to 2.7182818.

The number e is found in many real-world applications. It is an irrational number, so its decimal representation never terminates or repeats. Scientific calculators have an $\boxed{e^x}$ key for evaluating exponential expressions in which e is the base.

To calculate the compound amount when $10,000 is invested for 5 years at an interest rate of 10%, compounded continuously, use the formula for continuous compounding. Substitute the following values into the formula: $P = 10,000$, $r = 10\% = 0.10$, and $t = 5$.

$$A = Pe^{rt}$$
$$A \approx 10,000(2.7182818)^{0.10(5)}$$
$$A \approx 16,487.21$$

The compound amount is $16,487.21.

In the following exercises, calculate the compound interest when interest is compounded continuously.

a. $P = \$5000$, $r = 8\%$, $t = 6$ years

b. $P = \$8000$, $r = 7\%$, $t = 15$ years

c. $P = \$12{,}000$, $r = 9\%$, $t = 10$ years

d. $P = \$7000$, $r = 6\%$, $t = 8$ years

e. $P = \$3000$, $r = 7.5\%$, $t = 4$ years

f. $P = \$9000$, $r = 8.6\%$, $t = 5$ years

Solve the following exercises.

g. Calculate the compound amount when $2500 is deposited in an account earning 11% interest, compounded continuously, for 12 years.

h. What is the future value of $15,000 earning 9.5% interest, compounded continuously, for 7 years?

i. How much interest is earned in 9 years on $6000 deposited in an account paying 10% interest, compounded continuously?

j. $25,000 is deposited for 10 years in an account that earns 8% interest. Calculate the future values of the investment if interest is compounded quarterly and if interest is compounded continuously. How much greater is the future value of the investment when interest is compounded continuously?

SECTION 7.3 | # Credit Cards and Consumer Loans

Credit Cards

When a customer uses a credit card to make a purchase, the customer is actually receiving a loan. Therefore, there is frequently an added cost to the consumer who purchases on credit. This added cost may be in the form of an annual fee or interest charges on purchases. A **finance charge** is an amount paid in excess of the cash price; it is the cost to the customer for the use of credit.

Most credit card companies issue monthly bills. The due date on the bill is usually 1 month after the billing date (the date the bill is prepared and sent to the customer). If the bill is paid in full by the due date, the customer pays no finance charge. If the bill is not paid in full by the due date, a finance charge is added to the next bill.

Suppose a credit card billing date is the 10th day of each month. If a credit card purchase is made on April 15, then May 10 is the billing date (the 10th day of the month following April). The due date is June 10 (one month from the billing date). If the bill is paid in full before June 10, no finance charge is added. However, if the bill is not paid in full, interest charges on the outstanding balance will start to accrue (be added) on June 10, and any purchase made after June 10 will immediately start accruing interest.

The most common method of determining finance charges is the **average daily balance method.** Interest charges are based on the credit card's average daily balance, which is calculated by dividing the sum of the total amounts owed each day of the month by the number of days in the billing period.

Average Daily Balance

$$\text{Average daily balance} = \frac{\text{sum of the total amounts owed each day of the month}}{\text{number of days in the billing period}}$$

An example of calculating the average daily balance follows.

Suppose an unpaid bill for $315 had a due date of April 10. A purchase of $28 was made on April 12, and $123 was charged on April 24. A payment of $50 was made on April 15. The next billing date is May 10. The interest on the average daily balance is 1.5% per month. Find the finance charge on the May 10 bill.

To find the finance charge, first prepare a table showing the unpaid balance for each purchase, the number of days the balance is owed, and the product of these numbers. A negative sign in the Payments or Purchases column of the table indicates that a payment was made on that date.

Date	Payments or Purchases	Balance Each Day	Number of Days Until Balance Changes	Unpaid Balance Times Number of Days
April 10 – 11		$315	2	$630
April 12 – 14	$28	$343	3	$1029
April 15 – 23	– $50	$293	9	$2637
April 24 – May 9	$123	$416	16	$6656
Total				$10,952

The sum of the total amounts owed each day of the month is $10,952.

Find the average daily balance.

$$\text{Average daily balance} = \frac{\text{sum of the total amounts owed each day of the month}}{\text{number of days in the billing period}}$$

$$= \frac{10,952}{30} \approx 365.07$$

Find the finance charge.

$$I = Prt$$
$$I = 365.07(0.015)(1)$$
$$I \approx 5.48$$

The finance charge on the May 10 bill is $5.48.

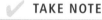

▼ **point of interest**

In 1950, Frank McNamara issued to 200 friends a card that could be used to pay for food at various restaurants in New York. The card, called a Diner's Card, spawned the credit card industry.

EXAMPLE 1 ■ **Calculate Interest on a Credit Card Bill**

An unpaid bill for $620 had a due date of March 10. A purchase of $214 was made on March 15, and $67 was charged on March 30. A payment of $200 was made on March 22. The interest on the average daily balance is 1.5% per month. Find the finance charge on the April 10 bill.

Solution

First calculate the sum of the total amounts owed each day of the month.

Date	Payments or Purchases	Balance Each Day	Number of Days Until Balance Changes	Unpaid Balance times Number of Days
March 10 – 14		$620	5	$3100
March 15 – 21	$214	$834	7	$5838
March 22 – 29	– $200	$634	8	$5072
March 30 – April 9	$67	$701	11	$7711
Total				$21,721

The sum of the total amounts owed each day of the month is $21,721.

Find the average daily balance.

$$\text{Average daily balance} = \frac{\text{sum of the total amounts owed each day of the month}}{\text{number of days in the billing period}}$$

$$= \frac{21,721}{31} \approx \$700.68$$

Find the finance charge.

$$I = Prt$$
$$I = 700.68(0.015)(1)$$
$$I \approx 10.51$$

The finance charge on the April 10 bill is $10.51.

CHECK YOUR PROGRESS 1 A bill for $1024 was due on July 1. Purchases totaling $315 were made on July 7, and $410 was charged on July 22. A payment of $400 was made on July 15. The interest on the average daily balance is 1.2% per month. Find the finance charge on the August 1 bill.

Solution   *See page S27.*

Math Matters   Credit Card Debt

The graph below shows how long it would take you to pay off a credit card debt of $3000, and the amount of interest you would pay, if you made the minimum monthly payment of 3% of the credit card balance each month.

If you have credit card debt and want to determine how long it will take you to pay off the debt, go to http://www.cardweb.com. There you will find a calculator that will calculate how long it will take you to pay off the debt and the amount of interest you will pay.

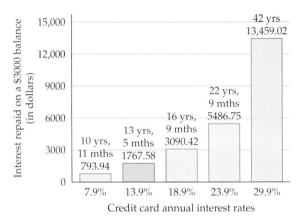

Source: CardWeb.com. Reprinted by permission of CardWeb.com.

Annual Percentage Rate

Federal law, in the form of the Truth in Lending Act, requires that credit customers be made aware of the cost of credit. This law, passed by Congress in 1969, requires that a business issuing credit inform the consumer of all details of a credit transaction, including the true annual interest rate. The **true annual interest rate,** also called the **annual percentage rate (APR)** or **annual percentage yield (APY),** is the effective annual interest rate on which credit payments are based.

The idea behind the APR is that interest is owed only on the *unpaid balance* of the loan. For instance, suppose you decide to borrow $2400 from a bank that advertises a 10% *simple* interest rate. You want a six-month loan and agree to repay the loan in six equal monthly payments. The simple interest due on the loan is

$$I = Prt$$

$$I = \$2400(0.10)\left(\frac{6}{12}\right)$$

$$I = \$120$$

The total amount to be repaid to the bank is

$$A = P + I$$

$$A = \$2400 + \$120$$

$$A = \$2520$$

The amount of each monthly payment is

$$\text{Monthly payment} = \frac{2520}{6} = \$420$$

During the first month you owe $2400. The interest on that amount is

$$I = Prt$$

$$I = \$2400(0.10)\left(\frac{1}{12}\right)$$

$$I = \$20$$

At the end of the first month, of the $420 payment you make, $20 is the interest payment and $400 is applied to reducing the loan. Therefore, during the second month you owe $2400 − $400 = $2000. The interest on that amount is

$$I = Prt$$

$$I = \$2000(0.10)\left(\frac{1}{12}\right) \approx 16.67$$

$$I = \$16.67$$

At the end of the second month, of the $420 payment you make, $16.67 is the interest payment and $403.33 is applied to reducing the loan. Therefore, during the third month you owe $2000 − $403.33 = $1596.67.

The point of these calculations is to demonstrate that each month the amount you owe is decreasing, and not by a constant amount. From our calculations, the loan decreased by $400 the first month and by $403.33 the second month.

The Truth in Lending Act stipulates that the interest rate for a loan be calculated only on the amount owed at a particular time, not on the original amount borrowed. All loans must be stated according to this standard, thereby making it possible for a consumer to compare different loans.

We can use the following formula to estimate the annual percentage rate (APR) on a simple interest rate installment loan.

Approximate Annual Percentage Rate (APR) Formula for a Simple Interest Rate Loan

The annual percentage rate (APR) of a simple interest rate loan can be approximated by

$$APR \approx \frac{2Nr}{N + 1}$$

where N is the number of payments and r is the simple interest rate.

For the loan described above, $N = 6$ and $r = 10\% = 0.10$.

$$APR \approx \frac{2Nr}{N + 1}$$

$$\approx \frac{2(6)(0.10)}{6 + 1} = \frac{1.2}{7} \approx 0.171$$

The annual percentage rate on the loan is approximately 17.1%. Recall that the simple interest rate was 10%, much less than the actual rate. The Truth in Lending Act provides the consumer with a standard interest rate, APR, so that it is possible to compare loans. The 10% simple interest loan described above is equivalent to an APR loan of about 17%.

EXAMPLE 2 ■ Calculate a Finance Charge and an APR

You purchase a refrigerator for $675. You pay 20% down and agree to repay the balance in 12 equal monthly payments. The finance charge on the balance is 9% simple interest.

a. Find the finance charge.

b. Estimate the annual percentage rate. Round to the nearest tenth of a percent.

Solution

a. To find the finance charge, first calculate the down payment.

Down payment = percent down × purchase price
$$= 0.20 \times 675 = 135$$

Amount financed = purchase price − down payment
$$= 675 - 135 = 540$$

Calculate the interest owed on the loan.

Interest owed = finance rate × amount financed
$$= 0.09 \times 540 = 48.60$$

The finance charge is $48.60.

b. Use the APR formula to estimate the annual percentage rate.

$$APR \approx \frac{2Nr}{N+1}$$

$$\approx \frac{2(12)(0.09)}{12+1} = \frac{2.16}{13} \approx 0.166$$

The annual percentage rate is approximately 16.6%.

CHECK YOUR PROGRESS 2 You purchase a washing machine and dryer for $750. You pay 20% down and agree to repay the balance in 12 equal monthly payments. The finance charge on the balance is 8% simple interest.

a. Find the finance charge.

b. Estimate the annual percentage rate. Round to the nearest tenth of a percent.

Solution See page S28.

Consumer Loans: Calculating Monthly Payments

The stated interest rate for a typical consumer loan, such as a car loan, is normally the annual percentage rate, APR, as required by the Truth in Lending Act. The payment amount for such a loan is given by the following formula.

Payment Formula for an APR Loan

The payment for a loan based on APR is given by

$$PMT = A\left(\frac{i}{1 - (1 + i)^{-n}}\right)$$

where PMT is the payment, A is the loan amount, i is the interest rate per payment period, and n is the total number of payments.

It is important to note that in this formula, i is the interest rate *per payment period*. For instance, if the annual interest rate is 9% and payments are made monthly, then

$$i = \frac{\text{annual interest rate}}{\text{number of payments per year}} = \frac{0.09}{12} = 0.0075$$

QUESTION *For a four-year loan repaid on a monthly basis, what is the value of n in the formula above?*

The payment formula given above is used to calculate monthly payments on most consumer loans. In Example 3 we calculate the monthly payment for a new television, and in Example 4 we calculate the monthly payment for a car loan.

CALCULATOR NOTE

You can calculate the monthly payment using the Finance option on a TI-83 or TI-84 Plus calculator. For a TI-83, press 2nd [FINANCE]; for a TI-83 Plus or TI-84 Plus, press APPS ENTER ENTER. Input the known values. Typically financial calculations use PV (present value) for the loan amount and FV (future value) for the amount owed at the end of the loan period, usually 0. P/Y = 12 and C/Y = 12 mean that payments and interest are calculated monthly (12 times a year). Now place the cursor at PMT= and press ALPHA [SOLVE].

```
N=48
I%=9.5
PV=5995
■PMT=-150.6132
FV=0
P/Y=12
C/Y=12
PMT: END BEGIN
```

The monthly payment is $150.61.

EXAMPLE 3 ■ Calculate a Monthly Payment

Integrated Visual Technologies is offering anyone who purchases a television an annual interest rate of 9.5% for 4 years. If Andrea Smyer purchases a 50-inch, flat screen television for $5995 from Integrated Visual Technologies, find her monthly payment.

Solution
To calculate the monthly payment, you will need a calculator. The following keystrokes will work on most scientific calculators.
First calculate i and store the result.

$$i = \frac{\text{annual interest rate}}{\text{number of payments per year}} = \frac{0.095}{12}$$

Keystrokes: 0.095 ÷ 12 = 0.00791667 STO

Calculate the monthly payment. For a four-year loan, $n = 4(12) = 48$.

$$PMT = A\left(\frac{i}{1 - (1 + i)^{-n}}\right)$$
$$= 5995\left(\frac{0.095/12}{1 - (1 + 0.095/12)^{-48}}\right) \approx 150.61$$

Keystrokes: 5995 × RCL = ÷ (1 − (1 + RCL)) y^x 48 +/−) =
The monthly payment is $150.61.

ANSWER $n = (\text{number of years}) \times (\text{number of payments per year}) = 4 \times 12 = 48$

CHECK YOUR PROGRESS 3   Carlos Menton purchases a new laptop computer from Knox Computer Solutions for $1499. If the sales tax is 4.25% of the purchase price and Carlos finances the total cost, including sales tax, for 3 years at an annual interest rate of 8.4%, find the monthly payment.

Solution   *See page S28.*

EXAMPLE 4 ■ **Calculate a Car Payment**

 A web page designer purchases a car for $18,395.

a. If the sales tax is 6.5% of the purchase price, find the amount of the sales tax.

b. If the car license fee is 1.2% of the purchase price, find the amount of the license fee.

c. If the designer makes a $2500 down payment, find the amount of the loan the designer needs.

d. Assuming the designer gets the loan in part c at an annual interest rate of 7.5% for 4 years, determine the monthly car payment.

Solution

a. Sales tax $= 0.065(18{,}395) = 1195.675$
   The sales tax is $1195.68.

b. License fee $= 0.012(18{,}395) = 220.74$
   The license fee is $220.74.

c. Loan amount $=$ purchase price $+$ sales tax $+$ license fee $-$ down payment
$$= 18{,}395 + 1195.68 + 220.74 - 2500$$
$$= 17{,}311.42$$
   The loan amount is $17,311.42.

d. To calculate the monthly payment, you will need a calculator. The following keystrokes will work on most scientific calculators.
   First calculate i and store the result.

$$i = \frac{APR}{12} = \frac{0.075}{12} = 0.00625$$

Keystrokes: 0.075 [÷] 12 [=] 0.00625 [STO]

Calculate the monthly payment. $n = 4(12) = 48$

$$PMT = A\left(\frac{i}{1 - (1 + i)^{-n}} \right)$$

$$= 17{,}311.42\left(\frac{0.00625}{1 - (1 + 0.00625)^{-48}} \right) \approx 418.57$$

Keystrokes:

17311.42 [×] [RCL] [=] [÷] [(] 1 [−] [(] 1 [+] [RCL] [)] [yˣ] 48 [+/−] [)] [=]

The monthly payment is $418.57.

CHECK YOUR PROGRESS 4   A school superintendent purchases a new sedan for $26,788.

a. If the sales tax is 5.25% of the purchase price, find the amount of the sales tax.

b. The superintendent makes a $2500 down payment and the license fee is $145. Find the amount the superintendent must finance.

c. Assuming the superintendent gets the loan in part b at an annual interest rate of 8.1% for 5 years, determine the superintendent's monthly car payment.

Solution   See page S28.

Math Matters   Payday Loans

An ad reads

> **Get cash until payday! Loans of $100 or more available.**

These ads refer to *payday* loans, which go by a variety of names, such as cash advance loans, check advance loans, post-dated check loans, or deferred deposit check loans. These types of loans are offered by finance companies and check-cashing companies.

Typically a borrower writes a personal check payable to the lender for the amount borrowed plus a *service fee.* The company gives the borrower the amount of the check minus the fee, which is normally a percent of the amount borrowed. The amount borrowed is usually repaid after payday, normally within a few weeks.

Under the Truth in Lending Act, the cost of a payday loan must be disclosed. Among other information, the borrower must receive, in writing, the APR for such a loan. To understand just how expensive these loans can be, suppose a borrower receives a loan for $100 for 2 weeks and pays a fee of $10. The APR for this loan can be calculated using the formula in this section or using a graphing calculator. Screens for a TI-83/84 Plus calculator are shown below.

```
N=1
I%=0
PV=100
PMT=-110
FV=0
P/Y=26
C/Y=26
PMT: END BEGIN
```

N = 1 (number of payments)
I% is unknown.
PV = 100 (amount borrowed)
PMT = −110 (the payment)
FV = 0 (no money is owed after all the payments)
P/Y = 26 (There are 26 two-week periods in 1 year.)
C/Y = 26

```
N=1
■I%=260
PV=100
PMT=-110
FV=0
P/Y=26
C/Y=26
PMT: END BEGIN
```

Place the cursor at I%=. Press ALPHA [SOLVE].

The annual interest rate is 260%.

To give you an idea of the enormity of a 260% APR, if the loan on the television set in Example 3, page 455, were based on a 260% interest rate, the monthly payment on the television set would be $1299.02!

The Federal Trade Commission (FTC) offers suggestions for consumers in need of credit. See www.ftc.gov/bcp/conline/pubs/alerts/pdayalrt.htm.

Consumer Loans: Calculating Loan Payoffs

Sometimes a consumer wants to pay off a loan before the end of the loan term. For instance, suppose you have a five-year car loan but would like to purchase a new car after owning your car for 4 years. Because there is still 1 year remaining on the loan, you must pay off the remaining loan before purchasing another car.

This is not as simple as just multiplying the monthly car payment by 12 to arrive at the payoff amount. The reason, as we mentioned earlier, is that each payment includes both interest and principal. By solving the payment formula for an APR loan for A, the amount of the loan, we can calculate the payoff amount, which is just the remaining principal.

APR Loan Payoff Formula

The payoff amount for a loan based on APR is given by

$$A = PMT\left(\frac{1 - (1 + i)^{-n}}{i}\right)$$

where A is the loan payoff, PMT is the payment, i is the interest rate per payment period, and n is the number of *remaining* payments.

EXAMPLE 5 ▪ Calculate a Payoff Amount

Allison Werke wants to pay off the loan on her jet ski that she has owned for 18 months. Allison's monthly payment is $284.67 on a two-year loan at an annual percentage rate of 8.7%. Find the payoff amount.

Solution

Because Allison has owned the jet ski for 18 months of a 24-month (two-year) loan, she has six payments remaining. Thus $n = 6$, the number of *remaining* payments. Here are the keystrokes to find the loan payoff.

Calculate i and store the result.

$$i = \frac{\text{annual interest rate}}{\text{number of payments per year}} = \frac{0.087}{12} = 0.00725$$

Keystrokes: 0.087 $\boxed{\div}$ 12 $\boxed{=}$ 0.00725 $\boxed{\text{STO}}$

Use the APR loan payoff formula.

$$A = PMT\left(\frac{1 - (1 + i)^{-n}}{i}\right)$$

$$= 284.67\left(\frac{1 - (1 + 0.00725)^{-6}}{0.00725}\right) \approx 1665.50$$

Keystrokes: 284.67 $\boxed{\times}$ $\boxed{(}$ 1 $\boxed{-}$ $\boxed{(}$ 1 $\boxed{+}$ $\boxed{\text{RCL}}$ $\boxed{)}$ $\boxed{y^x}$ 6 $\boxed{+/-}$ $\boxed{)}$ $\boxed{\div}$ $\boxed{\text{RCL}}$ $\boxed{=}$

The loan payoff is $1665.50.

CHECK YOUR PROGRESS 5 Aaron Jefferson has a five-year car loan based on an annual percentage rate of 8.4%. The monthly payment is $592.57. After 3 years, Aaron decides to purchase a new car and must pay off his car loan. Find the payoff amount.

Solution *See page S28.*

Car Leases

Leasing a car may result in lower monthly car payments. The primary reason for this is that at the end of the lease term, you do not own the car. Ownership of the car reverts to the dealer, who can then sell it as a used car and realize the profit from the sale.

The value of the car at the end of the lease term is called the **residual value** of the car. The residual value of a car is frequently based on a percent of the manufacturer's suggested retail price (MSRP) and normally varies between 40% and 60% of the MSRP, depending on the type of lease.

For instance, suppose the MSRP of a car is $18,500 and the residual value is 45% of the MSRP. Then

$$\text{Residual value} = 0.45 \cdot 18,500$$
$$= 8325$$

The residual value is $8325. This is the amount the dealer thinks the car will be worth at the end of the lease period. The person leasing the car, the lessee, usually has the option of purchasing the car at that price at the end of the lease.

In addition to the residual value of the car, the monthly lease payment for a car takes into consideration *net capitalized cost, the money factor, average monthly finance charge,* and *average monthly depreciation.* Each of these terms is defined below.

Net capitalized cost = negotiated price − down payment − trade-in value

$$\textbf{Money factor} = \frac{\text{annual interest rate as a percent}}{2400}$$

Average monthly finance charge
$$= (\text{net capitalized cost} + \text{residual value}) \times \text{money factor}$$

$$\textbf{Average monthly depreciation} = \frac{\text{net capitalized cost} - \text{residual value}}{\text{term of the lease in months}}$$

Using these definitions, we have the following formula for a monthly lease payment.

Monthly Lease Payment Formula

The monthly lease payment formula is given by $P = F + D$, where P is the monthly lease payment, F is the average monthly finance charge, and D is the average monthly depreciation of the car.

EXAMPLE 6 ■ Calculate a Monthly Lease Payment for a Car

The director of human resources for a company decides to lease a car for 30 months. Suppose the annual interest rate is 8.4%, the negotiated price is $29,500, there is no trade-in, and the down payment is $5000. Find the monthly lease payment. Assume that the residual value is 55% of the MSRP of $33,400.

Solution

Net capitalized cost = negotiated price − down payment − trade-in value
$$= 29,500 - 5000 - 0 = 24,500$$

Residual value = 0.55(33,400) = 18,370

$$\text{Money factor} = \frac{\text{annual interest rate as a percent}}{2400} = \frac{8.4}{2400} = 0.0035$$

✔ **TAKE NOTE**

When a person purchases a car, any state sales tax must be paid at the time of the purchase. However, with a lease, you make a state sales tax payment each month.

Suppose, for instance, that in Example 6 the state sales tax is 6% of the monthly lease payment. Then

Total monthly lease payment

= 354.38 + 0.06(354.38)

≈ 375.64

Average monthly finance charge

= (net capitalized cost + residual value) × money factor

= (24,500 + 18,370) × 0.0035

≈ 150.05

Average monthly depreciation = $\dfrac{\text{net capitalized cost} - \text{residual value}}{\text{term of the lease in months}}$

$= \dfrac{24,500 - 18,370}{30}$

≈ 204.33

Monthly lease payment

= average monthly finance charge + average monthly depreciation

= 150.05 + 204.33

= 354.38

The monthly lease payment is $354.38.

CHECK YOUR PROGRESS 6   Find the monthly lease payment for a car for which the negotiated price is $31,900, the annual interest rate is 8%, the length of the lease is 5 years, and the residual value is 40% of the MSRP of $33,395. There is no down payment or trade-in.

Solution   *See page S28.*

Investigation

Leasing Versus Buying a Car

In December 2005, the MSRP for a 2005 Dodge Dakota was $24,500. We will use this information to analyze the results of leasing versus buying the car. To ensure that we are making valid comparisons, we will assume:

There is no trade-in, the negotiated price is the MSRP, and a down payment of $2500 is made.

The license fee is 1.1% of the negotiated price.

If you purchase a car, then the state sales tax is 5.5% of the negotiated price. If you lease the car, then you make a state tax payment every month. The amount of each of the monthly state tax payments is equal to 5.5% of the monthly lease payment.

The annual interest rate for both a loan and a lease is 6%, and the term of the loan or the lease is 60 months.

The residual value when leasing the car is 45% of the MSRP.

(continued)

Investigation Exercises

1. Determine the loan amount to purchase the car.

2. Determine the monthly car payment to purchase the car.

3. How much will you have paid for the car (excluding maintenance) over the five-year term of the loan?

4. At the end of 5 years, you sell the car for 45% of the original MSRP (the residual lease value). Your net car ownership cost is the amount you paid over the 5 years minus the amount you realized from selling the car. What is your net ownership cost?

5. Determine the net capitalized cost for the car.

6. Determine the lease payment. Remember to include the sales tax that must be paid each month.

7. How much will you have paid to lease the car for the five-year term? Because the car reverts to the dealer after 5 years, the net ownership cost is the total of all the lease payments for the 5 years.

8. Which option, buying or leasing, results in the smaller net ownership cost?

9. List some advantages and disadvantages of buying and of leasing a car.

Exercise Set 7.3

In Exercises 1–4, calculate the finance charge for a credit card that has the given average daily balance and interest rate.

1. Average daily balance: $118.72; monthly interest rate: 1.25%

2. Average daily balance: $391.64; monthly interest rate: 1.75%

3. Average daily balance: $10,154.87; monthly interest rate: 1.5%

4. Average daily balance: $20,346.91; monthly interest rate: 1.25%

5. Average Daily Balance A credit card account had a $244 balance on March 5. A purchase of $152 was made on March 12, and a payment of $100 was made on March 28. Find the average daily balance if the next billing date is April 5.

6. Average Daily Balance A credit card account had a $768 balance on April 1. A purchase of $316 was made on April 5, and a payment of $200 was made on April 18. Find the average daily balance if the next billing date is May 1.

7. Finance Charges A charge account had a balance of $944 on May 5. A purchase of $255 was made on May 17, and a payment of $150 was made on May 20. The interest on the average daily balance is 1.5% per month. Find the finance charge on the June 5 bill.

8. Finance Charges A charge account had a balance of $655 on June 1. A purchase of $98 was made on June 17, and a payment of $250 was made on June 15. The interest on the average daily balance is 1.2% per month. Find the finance charge on the July 1 bill.

9. **Finance Charges** On August 10, a credit card account had a balance of $345. A purchase of $56 was made on August 15, and $157 was charged on August 27. A payment of $75 was made on August 15. The interest on the average daily balance is 1.25% per month. Find the finance charge on the September 10 bill.

10. **Finance Charges** On May 1, a credit card account had a balance of $189. Purchases of $213 were made on May 5, and $102 was charged on May 21. A payment of $150 was made on May 25. The interest on the average daily balance is 1.5% per month. Find the finance charge on the June 1 bill.

11. **Finance Charges** A charge of $213 is made on August 15 on a credit card. The billing date is September 1, and the due date is 30 days after the billing date. The bill is paid on September 24. Find the finance charge.

In Exercises 12 and 13, you may want to use the spreadsheet program available at the Online Study Center at **college.hmco.com/pic/aufmannMTQR**. This spreadsheet automates the finance charge procedure shown in this section.

12. **Finance Charges** The activity dates, companies, and amounts for a credit card bill are shown below. The due date of the bill is September 15. On August 15, there was an unpaid balance of $1236.43. Find the finance charge if the interest rate is 1.5% per month.

Activity Date	Company	Amount
August 15	Unpaid balance	1236.43
August 16	Veterinary clinic	125.00
August 17	Shell	23.56
August 18	Olive's restaurant	53.45
August 20	Seaside market	41.36
August 22	Monterey Hotel	223.65
August 25	Airline tickets	310.00
August 30	Bike 101	23.36
September 1	Trattoria Maria	36.45
September 9	Bookstore	21.39
September 12	Seaside Market	41.25
September 13	Credit card payment	−1345.00

13. **Finance Charges** The activity dates, companies, and amounts for a credit card bill are shown in the following table. The due date of the bill is July 10. On June 10, there was an unpaid balance of $987.81. Find the finance charge if the interest rate is 1.8% per month.

Activity Date	Company	Amount
June 10	Unpaid balance	987.81
June 11	Jan's Surf Shop	156.33
June 12	Albertson's	45.61
June 15	The Down Shoppe	59.84
June 16	NY Times Sales	18.54
June 20	Cardiff Delicatessen	23.09
June 22	The Olde Golf Mart	126.92
June 28	Lee's Hawaiian Restaurant	41.78
June 30	City Food Drive	100.00
July 2	Credit card payment	−1000.00
July 8	Safeway Stores	161.38

In Exercises 14–17, use the approximate annual percentage rate formula.

14. **APR** Chuong Ngo borrows $2500 from a bank that advertises a 9% simple interest rate and repays the loan in three equal monthly payments. Estimate the APR. Round to the nearest tenth of a percent.

15. **APR** Charles Ferrara borrows $4000 from a bank that advertises an 8% simple interest rate. If he repays the loan in six equal monthly payments, estimate the APR. Round to the nearest tenth of a percent.

16. **APR** Kelly Ang buys a computer system for $2400 and makes a 15% down payment. If Kelly agrees to repay the balance in 24 equal monthly payments at an annual simple interest rate of 10%, estimate the APR for Kelly's loan.

17. **APR** Jill Richards purchases a stereo system for $1500. She makes a 20% down payment and agrees to repay the balance in 12 equal payments. If the finance charge on the balance is 7% simple interest, estimate the APR. Round to the nearest tenth of a percent.

18. **Monthly Payments** Arrowood's Camera Store advertises a Canon Power Shot 6-megapixel camera for $400, including taxes. If you finance the purchase of this camera for 1 year at an annual percentage rate of 6.9%, find the monthly payment.

19. **Monthly Payments** Optics Mart offers a Meade ETX Astro telescope for $1249, including taxes. If you finance the purchase of this telescope for 2 years at an annual percentage rate of 7.2%, what is the monthly payment?

20. **APR** You purchase a big-screen TV and pay 20% down. You agree to repay the balance in 12 equal monthly payments. Would you prefer an annual percentage rate of 15.25% on the loan for the remaining balance, or an effective annual interest rate of 14.75% on the loan?

21. **Buying on Credit** Alicia's Surf Shop offers its 9′4″-long Noge Rider surfboard for $649. The sales tax is 7.25% of the purchase price.
    a. What is the total cost, including sales tax?
    b. If you make a down payment of 25% of the total cost, find the down payment.
    c. Assuming you finance the remaining cost at an annual interest rate of 5.7% for 6 months, find the monthly payment.

22. **Buying on Credit** Waterworld Marina offers a motorboat with a Mercury engine for $38,250. The sales tax is 6.5% of the purchase price.
    a. What is the total cost, including sales tax?
    b. If you make a down payment of 20% of the total cost, find the down payment.
    c. Assuming you finance the remaining cost at an annual interest rate of 5.7% for 3 years, find the monthly payment.

23. **Buying on Credit** After becoming a commercial pilot, Lorna Kao decides to purchase a Cessna 182 for $64,995. Assuming the sales tax is 5.5% of the purchase price, find each of the following.
    a. What is the total cost, including sales tax?
    b. If Lorna makes a down payment of 20% of the total cost, find the down payment.
    c. Assuming Lorna finances the remaining cost at an annual interest rate of 7.15% for 10 years, find the monthly payment.

24. **Buying on Credit** Donald Savchenko purchased new living room furniture for $2488. Assuming the sales tax is 7.75% of the purchase price, find each of the following.
    a. What is the total cost, including sales tax?
    b. If Donald makes a down payment of 15% of the total cost, find the down payment.
    c. Assuming Donald finances the remaining cost at an annual interest rate of 8.16% for 2 years, find the monthly payment.

25. **Car Payments** Luis Mahla purchases a Porsche Boxster for $42,600 and finances the entire amount at an annual

interest rate of 5.7% for 5 years. Find the monthly payment. Assume the sales tax is 6% of the purchase price and the license fee is 1% of the purchase price.

26. **Car Payments** Suppose you negotiate a selling price of $26,995 for a Ford Explorer. You make a down payment of 10% of the selling price and finance the remaining balance for 3 years at an annual interest rate of 7.5%. The sales tax is 7.5% of the selling price, and the license fee is 0.9% of the selling price. Find the monthly payment.

27. **Car Payments** Margaret Hsi purchases a late model Corvette for $24,500. She makes a down payment of $3000 and finances the remaining amount for 4 years at an annual interest rate of 8.5%. The sales tax is 5.5% of the selling price and the license fee is $331. Find the monthly payment.

28. **Car Payments** Chris Schmaltz purchases a Pontiac GTO for $34,119. Chris makes a down payment of $5000 and finances the remaining amount for 5 years at an annual interest rate of 7.6%. The sales tax is 6.25% of the selling price, and the license fee is $429. Find the monthly payment.

29. **Car Payments** Suppose you purchase a car for a total price of $25,445, including taxes and license fee, and finance that amount for 4 years at an annual interest rate of 8%.
    a. Find the monthly payment.
    b. What is the total amount of interest paid over the term of the loan?

30. **Car Payments** Adele Paolo purchased a Chevrolet Tracker for a total price of $21,425, including taxes and license fee, and financed that amount for 5 years at an annual interest rate of 7.8%.
    a. Find the monthly payment.
    b. What is the total amount of interest paid over the term of the loan?

31. **Loan Payoffs** Angela Montery has a five-year car loan for a Jeep Wrangler at an annual interest rate of 6.3% and a monthly payment of $603.50. After 3 years, Angela decides to purchase a new car. What is the payoff on Angela's loan?

32. **Loan Payoffs** Suppose you have a four-year car loan at an annual interest rate of 7.2% and a monthly payment of $587.21. After $2\frac{1}{2}$ years, you decide to purchase a new car. What is the payoff on your loan?

33. **Loan Payoffs** Suppose you have a four-year car loan at an annual interest rate of 8.9% and a monthly payment of $303.52. After 3 years, you decide to purchase a new car. What is the payoff on your loan?

34. **Loan Payoffs** Ming Li has a three-year car loan for a Mercury Sable at an annual interest rate of 9.3% and a monthly payment of $453.68. After 1 year, Ming decides to purchase a new car. What is the payoff on his loan?

35. **Car Leases** Suppose you decide to obtain a four-year lease for a car and negotiate a selling price of $28,990. The trade-in value of your old car is $3850. If you make a down payment of $2400, the money factor is 0.0027, and the residual value is $15,000, find each of the following.

    a. The net capitalized cost

    b. The average monthly finance charge

    c. The average monthly depreciation

    d. The monthly lease payment

36. **Car Leases** Marcia Scripps obtains a five-year lease for a Ford F-150 pickup and negotiates a selling price of $37,115. The trade-in value of her old car is $2950. Assuming she makes a down payment of $3000, the money factor is 0.0035, and the residual value is $16,500, find each of the following.

    a. The net capitalized cost

    b. The average monthly finance charge

    c. The average monthly depreciation

    d. The monthly lease payment

37. **Car Leases** Jorge Cruz obtains a three-year lease for a Dodge Stratus and negotiates a selling price of $22,100. The annual interest rate is 8.1%, the residual value is $15,000, and Jorge makes a down payment of $1000. Find each of the following.

    a. The net capitalized cost

    b. The money factor

    c. The average monthly finance charge

    d. The average monthly depreciation

    e. The monthly lease payment

38. **Car Leases** Suppose you obtain a five-year lease for a Porsche and negotiate a selling price of $165,000. The annual interest rate is 8.4%, the residual value is $85,000, and you make a down payment of $5000. Find each of the following.

    a. The net capitalized cost

    b. The money factor

    c. The average monthly finance charge

    d. The average monthly depreciation

    e. The monthly lease payment

Extensions

CRITICAL THINKING

39. Explain how the APR loan payoff formula can be used to determine the selling price of a car when the monthly payment, the annual interest rate, and the term of the loan on the car are known. Using your process, determine the selling price of a car that is offered for $235 per month for 4 years if the annual interest rate is 7.2%.

COOPERATIVE LEARNING

40. **Car Trade-Ins** You may have heard advertisements from car dealerships that say something like "Bring in your car, paid for or not, and we'll take it as a trade-in for a new car." The advertisement does not go on to say that you have to pay off the remaining loan balance or that the balance gets added to the price of the new car.

    a. Suppose you are making payments of $235.73 per month on a four-year car loan that has an annual interest rate of 8.4%. After making payments for 3 years, you decide to purchase a new car. What is the loan payoff?

    b. You negotiate a price, including taxes, of $18,234 for the new car. What is the actual amount you owe for the new car when the loan payoff is included?

    c. If you finance the amount in part b for 4 years at an annual interest rate of 8.4%, what is the new monthly payment?

41. **Car Leases** The residual value of a car is based on "average" usage. To protect a car dealership from abnormal usage, most car leases stipulate an average number of miles driven annually, that the tires be serviceable when the car is returned, and (in many cases) that all the manufacturer's recommended services be performed over the course of the lease. Basically, the dealership wants a car that can be put on the lot and sold as a used car without much effort.

    Suppose you decide to lease a car for 4 years. The net capitalized cost is $19,788, the residual value is 55% of the MSRP of $28,990, and the interest rate is 5.9%. In addition, you must pay a mileage penalty of $.20 for each mile over 48,000 miles that the car is driven during the 4 years of the lease.

a. Determine the monthly lease payment using the monthly lease payment formula given below. This formula is used by some financing agencies to calculate a monthly lease payment.

Monthly Lease Payment Formula

The monthly lease payment is given by $P = \dfrac{Ai(1+i)^n - Vi}{(1+i)^n - 1}$, where P is the monthly lease payment, A is the net capitalized cost, V is the residual value, i is the interest rate per payment period as a decimal, and n is the number of lease payments.

b. If the odometer reads 87 miles when you lease the car and you return the car with 61,432 miles, what mileage penalty will you have to pay?

c. If your car requires new tires at a cost of $635, what is the total cost, excluding maintenance, of the lease for the 4 years?

EXPLORATIONS

42. **Finance Charges** For most credit cards, no finance charge is added to the bill if the full amount owed is paid by the due date. However, if you do not pay the bill in full, the unpaid amount *and* all current charges are subject to a finance charge. This is a point that is missed by many credit card holders.

   To illustrate, suppose you have a credit card bill of $500 and you make a payment of $499, $1 less than the amount owed. Your credit card activity is as shown in the statement below.

Activity Date	Company	Amount
October 10	Unpaid balance	1.00
October 11	Rick's Tires	455.69
October 12	Costa's Internet Appliances	128.54
October 15	The Belgian Lion Restaurant	64.31
October 16	Verizon Wireless	33.57
October 20	Milton's Cake and Pie Shoppe	22.33
October 22	Fleming's Perfumes	65.00
October 24	Union 76	27.63
October 26	Amber's Books	42.31
November 2	Lakewood Meadows	423.88
November 8	Von's Grocery	55.64

a. Find the finance charge if the interest rate is 1.8% per month. Assume the due date of the bill is November 10. On October 10, the unpaid balance was the $1 you did not pay. You may want to use the spreadsheet program mentioned above Exercise 12.

b. Now assume that instead of paying $499, you paid $200, which is $299 less than the amount paid in part a. Calculate the finance charge.

c. How much more interest did you pay in part b than in part a?

d. If you took the $299 difference in payment and deposited it in an account that earned 1.8% simple interest (the credit card rate) for 1 month, how much interest would you earn?

e. 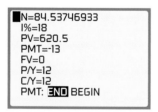 Explain why the answers to parts c and d are the same.

43. a. Car Leases Use the formula in Exercise 41 to find the monthly lease payment, excluding sales tax, for a car for which the net capitalized cost is $23,488, the residual value is $12,500, the annual interest rate is 7%, and the term of the lease is 4 years.

b. Suppose you negotiate a net capitalized cost of $26,445 for a car and a residual value of $14,000. If the annual interest rate is 6.5% and the term of the lease is 5 years, find the monthly lease payment. The sales tax is 6.25% of the lease payment.

c. Use the formula in Exercise 41 to find the lease payment for Exercise 37. What is the difference between the payment calculated using this method and the payment calculated using the method in Exercise 37?

d. Use the formula in Exercise 41 to find the lease payment for Exercise 38. What is the difference between the payment calculated using this method and the payment calculated using the method in Exercise 38?

44. Credit Card Debt The APR loan payoff formula can be used to determine how many months it will take to pay off a credit card debt if the minimum monthly payment is made each month. For instance, suppose you have a credit card bill of $620.50, the minimum payment is $13, and the interest rate is 18% per year. Using a graphing calculator, we can determine the number of months n it will take to pay off the debt. Enter the values shown on the calculator screen at the right. Move the cursor to N = and press ALPHA [SOLVE]. It will take over 84 months (or approximately 7 years) to pay off the credit card debt, assuming you do not make additional purchases.

```
N=84.53746933
I%=18
PV=620.5
PMT=-13
FV=0
P/Y=12
C/Y=12
PMT: END BEGIN
```

a. Find the number of months it will take to pay off a credit card debt of $1283.34 if the minimum payment is $27 and the annual interest rate is 19.6%. Round to the nearest month.

b. How much interest will be paid on the credit card debt in part a?

c. If you have credit card debt, determine how many months it would take to pay off your debt by making the minimum monthly payments. How much interest would you pay? You may want to go to the website at http://www.cardweb.com, which was mentioned in the Math Matters on page 452, to determine the answers.

| SECTION 7.4 | **Stocks, Bonds, and Mutual Funds** |

Stocks

historical note

Have you heard the tongue-in-cheek advice for making a fortune in the stock market that goes something like "Buy low, sell high?" The phrase is adapted from a comment made by Hetty Green (1834–1916), who was also known as the Witch of Wall Street. She turned a $1 million inheritance into $100 million and became the richest woman in the world. In today's terms, that $100 million would be worth over $2 billion. When asked how she became so successful, she replied, "There is no secret in fortune making. All you have to do is buy cheap and sell dear, act with thrift and shrewdness, and be persistent." ■

Stocks, bonds, and mutual funds are investment vehicles, but they differ in nature.

When owners of a company want to raise money, generally to expand their business, they may decide to sell part of the company to investors. An investor who purchases a part of the company is said to own *stock* in the company. Stock is measured in shares; a **share of stock** in a company is a certificate that indicates partial ownership in the company. The owners of the certificates are called **stockholders** or **shareholders.** As owners, the stockholders share in the profits or losses of the corporation.

A company may distribute profits to its shareholders in the form of **dividends.** A dividend is usually expressed as a per-share amount—for example, $.07 per share.

EXAMPLE 1 ■ Calculate Dividends Paid to a Stockholder

A stock pays an annual dividend of $.84 per share. Calculate the dividends paid to a shareholder who has 200 shares of the company's stock.

Solution

$$(\$.84 \text{ per share}) \times (200 \text{ shares}) = \$168$$

The shareholder receives $168 in dividends.

CHECK YOUR PROGRESS 1   A stock pays an annual dividend of $.72 per share. Calculate the dividends paid to a shareholder who has 550 shares of the company's stock.

Solution   See page S29.

The **dividend yield,** which is used to compare companies' dividends, is the amount of the dividend divided by the stock price and is expressed as a percent. Determining a dividend yield is similar to calculating the simple interest rate earned on an investment. You can think of the dividend as the interest earned, the stock price as the principal, and the yield as the interest rate.

▼ **point of interest**

Usually the dividends earned on shares of stock are sent to shareholders. However, in a **dividend reinvestment plan,** or DRIP, the dividends earned are applied toward the purchase of more shares.

EXAMPLE 2 ■ Calculate a Dividend Yield

A stock pays an annual dividend of $1.75 per share. The stock is trading at $70. Find the dividend yield.

Solution

$$I = Prt$$
$$1.75 = 70r(1)$$

• Let I = the annual dividend and P = the stock price. The time is **1** year.

$$1.75 = 70r$$

$$0.025 = r \qquad \bullet \text{ Divide each side of the equation by 70.}$$

The dividend yield is 2.5%.

CHECK YOUR PROGRESS 2 A stock pays an annual dividend of $.82 per share. The stock is trading at $51.25. Find the dividend yield.

Solution See page S29.

New York Stock Exchange

The **market value** of a share of stock is the price for which a stockholder is willing to sell a share of the stock and a buyer is willing to purchase it. Shares are always sold to the highest bidder. A **brokerage firm** is a dealer of stocks that acts as your agent when you want to buy or sell shares of stock. The **brokers** in the firm charge commissions for their service. Most trading of stocks happens on a stock exchange. **Stock exchanges** are businesses whose purpose it is to bring together buyers and sellers of stock. The largest stock exchange in the United States is the New York Stock Exchange. Shares of stock are also bought and sold through the National Association of Securities Dealers Automated Quotation System, which is commonly referred to as the NASDAQ. Every working day, each stock exchange provides financial institutions, Internet website hosts, newspapers, and other publications with data on the trading activity of all the stocks traded on that exchange. Table 7.2 is a portion of a stock table printed in the *Wall Street Journal*.

Table 7.2

YTD % CHG	52-WEEK HI	52-WEEK LO	STOCK (SYM)	DIV	YLD %	PE	VOL 100s	CLOSE	NET CHG	YTD % CHG	52-WEEK HI	52-WEEK LO	STOCK (SYM)	DIV	YLD %	PE	VOL 100s	CLOSE	NET CHG
8.0	53.54	40.25 ♣	PanPacProp **PNP**	2.17	4.2	21	1407	51.45	−0.28	25.1	19.99	6.26	PeriniCp **PCR**		...	6	649	11.45	−0.12
−40.3	75.44	32.10	PanPharm **PRX**		...	10	7652	38.88	0.18	0.4	22.59	13.96	PerkinElmer **PKI**	.28	1.6	29	4475	17.13	−0.27
16.1	12.35	6.08	ParTch **PTC**		...	22	7	9.28	−0.10	▲ 35.7	11.12	7	PermRltyTr **PBT**	.83e	7.4		820	11.21	0.10
−13.0	30.70	19	ParkElchm **PKE**	.24	1.0	43	1344	23.05	−0.56	−11.1	14.76	9.67	PerotSys A **PER**		...	26	5341	11.99	−0.69
36.5	4.49	1.65	ParkerDri **PKD**		...	dd	8091	3.48	−0.32	−11.1	14.32	7.37	Petrobrs ADS **PZE**		...	...	306	9.77	−0.08
−3.4	61	43.90	ParkerHan **PH**	.76	1.3	20	4114	57.45	−0.63	−4.0	53.17	37.56	PetroCnda ADS **PCZ**	.60g	...	...	641	47.35	0.55
5.4	48.70	36.80	ParkwyProp **PKY**	2.60	5.9	18	210	43.85	−0.45	−12.0	63.70	29.25	PtroChna ADS **PTR**	2.16e	4.3	...	2327	50.20	−0.20
−8.9	60.15	46.74 ♣	PartnerRe **PRE**	1.36	2.6	6	2982	52.86	0.26	▲ 41.7	32.88	12.80	Ptrokzkhstn A **PKZ**	.30eg	.9	7	5930	31.90	−0.83
21.0	31.68	14.65	PatinaOil **POG s**	.20	.7	18	4810	29.65	−0.30	−4.2	35.64	17.76	PetrlBra ADS **PBR**	1.76e	6.3	5	10196	28.02	0.03
45.4	20	11.46	PaxarCp **PXR**		...	25	4144	19.48	0.34	−4.3	31.94	16.56	PtrlBras ADS A **PBRA**	1.76e	6.9	...	4861	25.52	−0.04
−4.3	17.72	11.96	PaylessShoe **PSS**		...	...	9130	12.82	−0.25	15.1	43.50	27	PfeiffrVac **PV**	.85e	2.1	...	9	40.32	1.42
33.9	58.34	28.61	Peabdy Egy **BTU**	.50	.9	31	3750	55.85	−0.22	−8.2	38.89	29.43	Pfizer **PFE**	.68	2.1	31	169915	32.45	0.25
0.5	12.80	9.17	Pearson ADS **PSO**	.45e	4.0	...	287	11.27	−0.08	1.3	90.52	41.11	PhelpDodg **PD**	.25e	.3	16	14948	77.08	−1.29
15.7	71.62	37.50	PediatrixMed **PDX**		...	18	3743	63.75	0.50	−0.1	25.57	23.20	PhilAuthInd **POB**	1.64	6.5	...	72	25.08	0.08
▲ 3.2	15.05	14.40	Pengrowth **n**	.17p	...	...	5904	15.29	0.31	▲ 32.5	23.41	8.78	PhlpLngDst **PHI**		...	...	2246	23.09	0.48
−2.0	16.10	11.46	PennAmGp **PNG**	.24	1.8	10	387	13	−0.02	−18.9	33.38	20.27	PhlpsEl **PHG**	.44e	1.9	...	14433	23.59	−0.54
1.2	21.58	12.85	PennEngrg **PNN**	.28	1.5	24	137	19.25	−0.54	3.4	19.95	13.72	PhillipsVanH **PVH**	.15	.8	dd	976	18.35	−0.38
−1.1	2.67	1.47	PennTreaty **PTA**		...	dd	609	1.82	0.08	−15.8	14.53	8.67	PhoenixCos **PNX**	.16e	1.6	14	3784	10.14	−0.32
34.8	40	19.83 ♣	PennVirginia **PVA s**	.45	1.2	24	554	37.50	−0.05	−14.9	43.80	28.50	PhoenixCos **un**	1.81	5.6	...	2	32.11	−0.68
9.6	39	27.25 ♣	PennVARes **PVR**	2.16f	5.8	...	603	37.56	0.06	48.9	3.20	0.95	PhosphtRes **PLP**		...	dd	212	2.83	0.04
51.0	41.50	17.25	PenneyJC **JCP**	.50	1.3	dd	66959	39.69	−0.51	−4.1	43.95	37.23	PidmntNG **PNY**	1.72	4.1	...	1308	41.67	0.07
−2.7	37.87	30	♣ PA Reit **PEI**	2.16	6.1	13	1027	35.31	−0.31	−19.9	26.44	16.82	Pier 1 **PIR**	.40	2.3	14	7163	17.52	−0.32
37.0	34.75	18.38	Pentair **PNR s**	.44f	1.4	19	7705	31.30	0.21	77.2	32.09	11.25	PilgrmPr **PPC**	.06	.2	20	3660	28.94	0.59
−6.3	46.03	38.50	PeopEngy **PGL**	2.16	5.5	16	2262	39.38	0.22	15.7	14.93	5.80	PinnacleEnt **PNK**		...	dd	1197	10.78	−0.51
−13.8	29.38	14.05	PepBoys **PBY**	.27	1.4	dd	11159	19.71	−0.68	1.9	41.50	32.87 ♣	PinaclWCap **PNW**	1.80	4.4	14	2792	40.80	0.24
−5.8	21.71	16.70	PepcoHldg **POM**	1.00	5.4	17	3317	18.40	0.07	−23.9	31.25	20.85	PioneerCp **PIO**	.23e	1.1	...	263	21.44	−0.28
16.0	31.40	20.39	PepsiBttlng **PBG**	.20f	.7	17	11900	28.04	−0.15	12.1	37.50	22.76	PionrNtrlRes **PXD**	.10e	.3	11	11662	35.78	0.25
10.1	21.67	13.03	PepsiAm **PAS**	.08	.4	15	3066	18.85	−0.05	−2.9	59.59	38.70	PiperJaffray **PJC n**		...	...	1259	40.35	−0.82
9.8	55.71	43.35	PepsiCo **PEP**	.92	1.8	23	33068	51.20	0.12	4.7	45.21	37.25	PitneyBws **PBI**	1.22	2.9	19	8263	42.51	−0.13
44.3	25.26	8.34	Perdigao **ADS PDA**	.55e	2.2	...	32	25.04	0.29	−8.8	19.23	12.23	PlacrDome **PDG**	.10g	.6	35	15195	16.33	0.20

The headings at the tops of the columns are repeated below, along with the information for Pier 1 (with stock symbol PIR).

YTD %CHG	52-WEEK HI	52-WEEK LO	STOCK (SYM)	DIV	YLD %	PE	VOL 100s	CLOSE	NET CHG
−19.9	26.44	16.82	Pier 1 (PIR)	.40	2.3	14	7163	17.52	−0.32

YTD % CHG The number −19.9 in the first column indicates that the price of a share of Pier 1 stock has decreased 19.9% so far this calendar year.

52-WEEK HI/LO The next two numbers show that in the last 52 weeks, the highest price a share of the stock sold for was $26.44, and the lowest price was $16.82.

DIV The number .40 under the column headed DIV means that the company is currently paying an annual dividend of $.40 per share of stock.

YLD% The current dividend yield on the company's stock is 2.3%. Thus the dividend of $.40 is 2.3% of the current purchase price of a share of the stock.

PE The heading PE refers to the price–earnings ratio, the purchase price per share divided by the earnings per share.

VOL 100s 7163 is the number of shares sold that day, in hundreds. 7163 × 100 = 716,300. 716,300 shares of Pier 1 stock were sold that day.

CLOSE The next number, 17.52, indicates that the closing price of the stock was $17.52. This means that in the final trade of the day, before the market closed, the price of a share of Pier 1 stock was $17.52.

NET CHG The number −0.32 under the heading NET CHG indicates that the day's closing price was $.32 lower than that of the previous trading day.

▼ **point of interest**

Young adults age 18 to 34 make up 19% of the stock investors in this country. According to a survey of investors in this age group, 93% plan to use investment money for retirement, 52% are saving for a child's college education, 23% report that they will use proceeds to buy a home, and 19% have plans to start a business. (*Source:* NASDAQ Stock Market)

EXAMPLE 3 ■ Calculate Profits or Losses and Expenses in Selling Stock

Suppose you owned 500 shares of stock in J.C. Penney (which is listed as PenneyJC in Table 7.2). You purchased the shares at a price of $23.90 per share and sold them at the closing price of the stock given in Table 7.2.

a. Ignoring dividends, what was your profit or loss on the sale of the stock?

b. If your broker charges 2.4% of the total sale price, what was the broker's commission?

Solution

a. From Table 7.2, the selling price per share was $39.69.
The selling price per share is greater than the purchase price per share.
You made a profit on the sale of the stock.

$$\text{Profit} = \text{selling price} - \text{purchase price}$$
$$= 500(\$39.69) - 500(\$23.90)$$
$$= \$19,845 - \$11,950$$
$$= \$7895$$

The profit on the sale of the stock was $7895.

b. Commission = 2.4%(selling price)
$$= 0.024(\$19,845)$$
$$= \$476.28$$

The broker's commission was $476.28.

CHECK YOUR PROGRESS 3   Use Table 7.2. Suppose you bought 300 shares of Pfizer at the 52-week low and sold the shares at the 52-week high.

a. Ignoring dividends, what was your profit or loss on the sale of the stock?

b. If your broker charges 2.1% of the total sale price, what was the broker's commission? Round to the nearest cent.

Solution   *See page S29.*

Bonds

When a corporation issues stock, it is *selling* part of the company to the stockholders. When it issues a **bond,** the corporation is *borrowing* money from the bondholders; a **bondholder** lends money to a corporation. Corporations, the federal government, government agencies, states, and cities all issue bonds. These entities need money to operate—for example, to fund the federal deficit, repair roads, or build a new factory—so they borrow money from the public by issuing bonds.

Bonds are usually issued in units of $1000. The price paid for the bond is the **face value.** The issuer promises to repay the bondholder on a particular day, called the **maturity date,** at a given rate of interest, called the **coupon.**

Assume that a bond with a $1000 face value has a 5% coupon and a 10-year maturity date. The bondholder collects interest payments of $50 in each of those 10 years. The payments are calculated using the simple interest formula, as shown below.

$$I = Prt$$
$$I = 1000(0.05)(1)$$
$$I = 50$$

At the end of the 10-year period, the bondholder receives from the issuer the $1000 face value of the bond.

▼ **point of interest**

Municipal bonds are issued by states, cities, counties, and other governments to raise money to build schools, highways, sewer systems, hospitals, and other projects for the public good. The income from many municipal bonds is exempt from federal and/or state taxes.

EXAMPLE 4 ■ Calculate Interest Payments on a Bond

A bond with a $10,000 face value has a 3% coupon and a five-year maturity date. Calculate the total of the interest payments paid to the bondholder.

Solution
Use the simple interest formula to find the annual interest payments. Substitute the following values into the formula: $P = 10,000$, $r = 3\% = 0.03$, and $t = 1$.

$$I = Prt$$
$$I = 10,000(0.03)(1)$$
$$I = 300$$

Multiply the annual interest payments by the term of the bond.

$$300(5) = 1500$$

The total of the interest payments paid to the bondholder is $1500.

CHECK YOUR PROGRESS 4　A bond has a $15,000 face value, a four-year maturity, and a 3.5% coupon. What is the total of the interest payments paid to the bondholder?

Solution　*See page S29.*

A key difference between stocks and bonds is that stocks make no promises about dividends or returns, whereas the issuer of a bond guarantees that, provided the issuer remains solvent, it will pay back the face value of the bond plus interest.

Mutual Funds

An **investment trust** is a company whose assets are stocks and bonds. The purpose of such a company is not to manufacture a product but to purchase stocks and bonds with the hope that their value will increase. A **mutual fund** is an example of an investment trust.

When investors purchase shares in a mutual fund, they are adding their money to a pool along with many other investors. The investments within a mutual fund are called the fund's portfolio. The investors in a mutual fund share the fund's profits or losses from the investments in the portfolio.

An advantage of owning shares of a mutual fund is that your money is managed by full-time professionals whose job it is to research and evaluate stocks; you own stocks without having to choose which individual stocks to buy or decide when to sell them. Another advantage is that by owning shares in the fund, you have purchased shares of stock in many different companies. This diversification may reduce some of the risks of investing.

Because a mutual fund owns many different stocks, each share of the fund owns a fractional interest in each of the companies. Each day, the value of a share in the fund, called the **net asset value of the fund,** or **NAV,** depends on the performance of the stocks in the fund. It is calculated by the following formula.

Net Asset Value of a Mutual Fund

The net asset value of a mutual fund is given by

$$NAV = \frac{A - L}{N}$$

where A is the total fund assets, L is the total fund liabilities, and N is the number of shares outstanding.

EXAMPLE 5 ■ Calculate the Net Asset Value of, and the Number of, Shares Purchased in a Mutual Fund

A mutual fund has $600 million worth of stock, $5 million worth of bonds, and $1 million in cash. The fund's total liabilities amount to $2 million. There are 25 million shares outstanding. You invest $15,000 in this fund.

a. Calculate the *NAV*.

b. How many shares will you purchase?

Solution

a. $NAV = \dfrac{A - L}{N}$

$= \dfrac{606 \text{ million} - 2 \text{ million}}{25 \text{ million}}$ • A = 600 million + 5 million + 1 million = 606 million, L = 2 million, N = 25 million

$= 24.16$

The *NAV* of the fund is $24.16.

b. $\dfrac{15{,}000}{24.16} \approx 620$ • Divide the amount invested by the *NAV* of the fund. Round down to the nearest whole number.

You will purchase 620 shares of the mutual fund.

CHECK YOUR PROGRESS 5   A mutual fund has $750 million worth of stock, $750,000 in cash, and $1,500,000 in other assets. The fund's total liabilities amount to $1,500,000. There are 20 million shares outstanding. You invest $10,000 in this fund.

a. Calculate the *NAV*.

b. How many shares will you purchase?

Solution   See page S29.

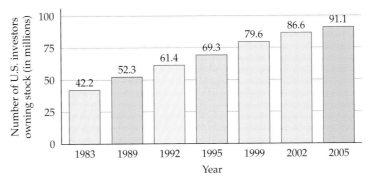

Math Matters   Growth of Number of Shareholders

Where do Americans invest their money? You might correctly assume that the largest number of Americans invest in real estate, as every homeowner is considered to have an investment in real estate. But more and more Americans are investing their money in the stock market. The graph below shows the growth in the number of U.S. investors owning stock in selected years from 1983 to 2005.

(*Source*: ICI/SIA Equity Ownership Surveys, New York Stock Exchange, and U.S. Census Bureau)

Investigation

Treasury Bills

✔ **TAKE NOTE**

You can obtain more information about the various Treasury securities at http://www.publicdebt. treas.gov. Using this site, investors can buy securities directly from the government, thereby avoiding the service fee charged by banks and brokerage firms.

The bonds issued by the United States government are called Treasuries. Some investors prefer to invest in Treasury bills, rather than the stock market, because their investments are backed by the federal government. As such they are considered the safest of all investments. They are grouped into three categories.

  U.S. Treasury bills have maturities of less than 1 year.
  U.S. Treasury notes have maturities ranging from 2 to 10 years.
  U.S. Treasury bonds have maturities ranging from 10 to 30 years.

This Investigation will focus on Treasury bills.

The **face value** of a Treasury bill is the amount of money received on the maturity date of the bill. Treasury bills are sold on a **discount basis;** that is, the interest on the bill is computed and subtracted from the face value to determine its cost.

Suppose a company invests in a $50,000 United States Treasury bill at 3.35% interest for 28 days. The bank through which the bill is purchased charges a service fee of $15. What is the cost of the Treasury bill?

To find the cost, first find the interest. Use the simple interest formula.

$$I = Prt$$

$$= 50,000(0.0335)\left(\frac{28}{360}\right)$$

$$\approx 130.28$$

The interest earned is $130.28.

(*continued*)

Find the cost of the Treasury bill.

Cost = (face value − interest) + service fee
= (50,000 − 130.28) + 15
= 49,869.72 + 15
= 49,884.72

The cost of the Treasury bill is $49,884.72.

Investigation Exercises

1. The face value of a Treasury bill is $30,000. The interest rate is 2.32% and the bill matures in 182 days. The bank through which the bill is purchased charges a service fee of $15. What is the cost of the Treasury bill?

2. The face value of a 91-day Treasury bill is $20,000. The interest rate is 2.96%. The purchaser buys the bill through Treasury Direct and pays no service fee. Calculate the cost of the Treasury bill.

3. A company invests in a 29-day, $60,000 United States Treasury bill at 2.28% interest. The bank charges a service fee of $35. Calculate the cost of the Treasury bill.

4. A $40,000 United States Treasury bill, purchased at 1.96% interest, matures in 92 days. The purchaser is charged a service fee of $20. What is the cost of the Treasury bill?

Exercise Set 7.4

1. **Annual Dividends** A stock pays an annual dividend of $1.02 per share. Calculate the dividends paid to a shareholder who has 375 shares of the company's stock.

2. **Annual Dividends** A stock pays an annual dividend of $.58 per share. Calculate the dividends paid to a shareholder who has 1500 shares of the company's stock.

3. **Annual Dividends** Calculate the dividends paid to a shareholder who has 850 shares of a stock that is paying an annual dividend of $.63 per share.

4. **Annual Dividends** Calculate the dividends paid to a shareholder who has 400 shares of a stock that is paying an annual dividend of $.91 per share.

5. **Dividend Yield** Find the dividend yield for a stock that pays an annual dividend of $1.24 per share and has a current price of $49.375. Round to the nearest hundredth of a percent.

6. **Dividend Yield** The Blackburn Computer Company has declared an annual dividend of $.50 per share. The stock is trading at $40 per share. Find the dividend yield.

7. **Dividend Yield** A stock that pays an annual dividend of $.58 per share has a current price of $31.75. Find the dividend yield. Round to the nearest hundredth of a percent.

8. **Dividend Yield** The Moreau Corporation is paying an annual dividend of $.65 per share. If the price of a share of the stock is $81.25, what is the dividend yield on the stock?

Use the portion of the stock table shown below for Exercises 9 to 16. Round dollar amounts to the nearest cent when necessary.

YTD % CHG	52-WEEK HI	52-WEEK LO	STOCK (SYM)	DIV	YLD %	PE	VOL 100s	CLOSE	NET CHG
12.7	51.52	34.45	GallaherGp **GLH**	2.12e	4.4	...	415	47.95	–0.15
–0.7	19.05	12.52	GameStop A **GME**		...	14	3149	15.30	–0.38
–5.7	91.38	75.59♣	Gannett **GCI**	1.00	1.2	18	10658	84.06	–0.07
–4.2	25.72	16.99	Gap Inc **GPS**	.09	.4	19	37506	22.24	–0.35
15.9	30.30	19.95♣	GardnrDenvr **GDI**		...	18	530	27.66	0.08
8.5	13.75	8.86	Gartner **IT**		...	61	7366	12.27	–0.33
12.0	13.06	8.52	Gartner B **ITB**		...	61	2521	12.19	–0.33
–10.0	6.85	3.64	Gateway **GTW**		...	dd	12852	4.14	–0.25
–4.0	32.70	17.70	GaylEnt **GET**		...	dd	1248	28.66	–0.25
9.3	13.53	8.75	GenCorp **GY**	.12	1.0	dd	1987	11.77	–0.14
1.5	68.25	36.82	Genentech **DNA** s		...	74	28074	47.50	0.54
15.3	10.23	6.26	GenlCbl **BGC**		...	dd	2002	9.40	–0.26
9.1	101.94	75.59	GenDynam **GD**	1.44	1.5	18	4787	98.59	–1.44
6.1	34.57	27.18	GenElec **GE**	.80	2.4	21	157197	32.87	–0.39
12.7	35.30	21.82	GenGrthProp **GGP** s	1.20	3.8	25	6392	31.27	0.02
63.6	30.42	11.13	GenMaritime **GMR**		...	8	3176	28.80	–0.20
–0.8	49.17	43.75	GenMills **GIS**	1.24f	2.8	16	7247	44.92	–0.44
–19.3	55.55	36.11	GenMotor **GM**	2.00	4.6	6	57115	43.07	–0.23
36.9	25.67	14.30	Genesco **GCO**		...	15	1688	20.71	–0.71
6.6	26.10	14	♣GeneseWY A **GWR** s		...	18	1085	22.38	–0.32
13.9	40.20	29.83	GenuinePart **GPC**	1.20	3.2	18	3739	37.80	–0.38
16.7	23.50	18.75	GnwrthFnl A **GNW** n	.68p	...	...	8890	22.76	0.04
10.8	29.18	25	GnwrthFnl un n	.33p	...	...	3122	28.15	...
–18.9	24.69	16.40♣	GeoGrp **GGI**		...	5	328	18.50	–0.25
▲ 23.9	35.99	20.71	GA Gulf **GGC**	.32	.9	22	3543	35.79	0.04
9.3	38.60	20.68♣	GA Pac **GP**	.50	1.5	15	8039	33.53	–0.32

9. **Stock Tables** For Gap, Inc. (GPS):

   a. What is the difference between the highest and lowest prices paid for this stock during the last 52 weeks?

   b. Suppose you own 750 shares of this stock. What dividend do you receive this year?

   c. How many shares of this stock were sold yesterday?

   d. Did the price of a share of this stock increase or decrease during the day shown in the table?

   e. What was the price of a share of this stock at the start of the trading day yesterday?

10. **Stock Tables** For General Motors (GM):

   a. What is the difference between the highest and lowest prices paid for this stock during the last 52 weeks?

   b. Suppose you own 750 shares of this stock. What dividend do you receive this year?

   c. How many shares of this stock were sold yesterday?

   d. Did the price of a share of this stock increase or decrease during the day shown in the table?

   e. What was the price of a share of this stock at the start of the trading day yesterday?

11. **Stock Purchases** At the closing price per share of General Dynamics (GD), how many shares of the stock can you purchase for $5000?

12. **Stock Purchases** At the closing price per share of GenCorp (GY), how many shares of the stock can you purchase for $2500?

13. **Stock Sale** Suppose you owned 1000 shares of stock in Gateway (GTW). You purchased the shares at a price of $3.85 per share and sold them at the closing price of the stock given in the table.

   a. Ignoring dividends, what was your profit or loss on the sale of the stock?

   b. If your broker charges 1.9% of the total sale price, what was the broker's commission?

14. **Stock Sale** Gary Walters owned 400 shares of stock in General Electric (GE). He purchased the shares at a price of $27.80 per share and sold them at the closing price of the stock given in the table.

   a. Ignoring dividends, what was Gary's profit or loss on the sale of the stock?

   b. If his broker charges 2.5% of the total sale price, what was the broker's commission?

15. **Stock Sale** Michelle Desjardins bought 800 shares of Genesco (GCO) at the 52-week low and sold the shares at the 52-week high shown in the table.

   a. Ignoring dividends, what was Michelle's profit or loss on the sale of the stock?

   b. If her broker charges 2.3% of the total sale price, what was the broker's commission?

16. **Stock Sale** Suppose you bought 1200 shares of General Mills (GIS) at the 52-week low and sold the shares at the 52-week high shown in the table.

   a. Ignoring dividends, what was your profit or loss on the sale of the stock?

   b. If your broker charges 2.25% of the total sale price, what was the broker's commission?

17. I own shares of Byplex stock that paid $20 in dividends this year. My brother owns twice as many shares of Byplex stock as I do. What dividend did his shares of Byplex stock earn this year?

18. This year the market value of the shares of Trilynx stock is twice what it was last year but the dollar amount of the dividend per share has stayed the same. What is the current dividend yield this year as compared with the current dividend yield last year?

19. This year the price per share of Dromax stock is twice what it was last year. The earnings per share is unchanged from last year. What is the PE ratio this year as compared with the PE ratio last year?

20. The PE of a certain stock was 15.6 on January 1, 2006 and 17.2 on January 1, 2007. The price of the stock on January 1, 2006 was the same as the price of the stock on January 1, 2007. Did the earnings per share of the stock decrease, remain the same, or increase during that one-year period?

21. Suppose a company decides to split its stock 2 for 1 on June 1, which means there are twice as many shares as previously and the price of each share is half the previous price. Does the PE of the stock also change on June 1?

22. Bonds A bond with a face value of $6000 and a 4.2% coupon has a five-year maturity. Find the annual interest paid to the bondholder.

23. Bonds The face value of a bond is $15,000. It has a 10-year maturity and a 3.75% coupon. What is the annual interest paid to the bondholder?

24. Bonds A bond with an $8000 face value has a 3.5% coupon and a three-year maturity. What is the total of the interest payments paid to the bondholder?

25. Bonds A bond has a $12,000 face value, an eight-year maturity, and a 2.95% coupon. Find the total of the interest payments paid to the bondholder.

26. Mutual Funds A mutual fund has total assets of $50,000,000 and total liabilities of $5,000,000. There are 2,000,000 shares outstanding. Find the net asset value of the mutual fund.

27. Mutual Funds A mutual fund has total assets of $25,000,000 and total liabilities of $250,000. There are 1,500,000 shares outstanding. Find the net asset value of the mutual fund.

28. Mutual Funds A mutual fund has total assets of $15 million and total liabilities of $1 million. There are 2 million shares outstanding. You invest $5000 in this fund. How many shares will you purchase?

29. Mutual Funds A mutual fund has total assets of $12 million and total liabilities of $2 million. There are 1 million shares outstanding. You invest $2500 in this fund. How many shares will you purchase?

30. Mutual Funds A mutual fund has $500 million worth of stock, $500,000 in cash, and $1 million in other assets. The fund's total liabilities amount to $2 million. There are 10 million shares outstanding.

You invest $12,000 in this fund. How many shares will you purchase?

31. Mutual Funds A mutual fund has $250 million worth of stock, $10 million worth of bonds, and $1 million in cash. The fund's total liabilities amount to $1 million. There are 13 million shares outstanding. You invest $10,000 in this fund. How many shares will you purchase?

Extensions

CRITICAL THINKING

32. Load and No-Load Funds All mutual funds carry fees. One type of fee is called a "load." This is an additional fee that generally is paid at the time you invest your money in the mutual fund. A no-load mutual fund does not charge this up-front fee.

Suppose you invested $2500 in a 4% load mutual fund 2 years ago. The 4% fee was paid out of the $2500 invested. The fund has earned 8% during each of the past 2 years. There was a management fee of 0.015% charged at the end of each year. A friend of yours invested $2500 2 years ago in a no-load fund that has earned 6% during each of the past 2 years. This fund charged a management fee of 0.15% at the end of each year. Find the difference between the values of the two investments now.

COOPERATIVE LEARNING

33. Investing in the Stock Market (This activity assumes that the instructor has assigned each student in the class to a group of three or four students.) Imagine that your group has $10,000 to invest in each of 10 stocks. Use the stock table in today's paper or use the Internet to choose your 10 stocks and to determine the price per share you would pay for each stock. Determine the number of shares of each stock you will purchase. Check the value of each stock every business day for the next 4 weeks. Assume that you sell your shares at the end of the fourth week. Calculate the group's profit or loss over the four-week period. Compare your profits or losses with those of the other groups in your class.

EXPLORATIONS

34. Find a mutual fund table in a daily newspaper. You can find one in the same section in which the stock tables are printed. Explain the meaning of the heading of each column in the table.

| **Home Ownership**

Initial Expenses

When you purchase a home, you generally make a down payment and finance the remainder of the purchase price with a loan obtained through a bank or savings and loan association. The amount of the down payment can vary, but it is normally between 10% and 30% of the selling price. The **mortgage** is the amount that is borrowed to buy the real estate. The amount of the mortgage is the difference between the selling price and the down payment.

Mortgage = selling price − down payment

This formula is used to find the amount of the mortgage. For example, suppose you buy a $240,000 home with a down payment of 25%. First find the down payment by computing 25% of the purchase price.

Down payment = 25% of 240,000 = 0.25(240,000)
= 60,000

Then find the mortgage by subtracting the down payment from the selling price.

Mortgage = selling price − down payment
= 240,000 − 60,000
= 180,000

The mortgage is $180,000.

The down payment is generally the largest initial expense in purchasing a home, but there are other expenses associated with the purchase. These payments are due at the closing, when the sale of the house is finalized, and are called **closing costs.** The bank may charge fees for attorneys, credit reports, loan processing, and title searches. There may also be a **loan origination fee.** This fee is usually expressed in **points.** One point is equal to 1% of the mortgage.

Suppose you purchase a home and obtain a loan for $180,000. The bank charges a fee of 1.5 points. To find the charge for points, multiply the loan amount by 1.5%.

Points = 1.5% of 180,000 = 0.015(180,000)
= 2700

The charge for points is $2700.

EXAMPLE 1 ■ Calculate a Down Payment and Closing Costs

The purchase price of a home is $392,000. A down payment of 20% is made. The bank charges $450 in fees plus $2\frac{1}{2}$ points. Find the total of the down payment and the closing costs.

Solution
First find the down payment.

Down payment = 20% of 392,000 = 0.20(392,000)
= 78,400

The down payment is $78,400.

Next find the mortgage.

$$\text{Mortgage} = \text{selling price} - \text{down payment}$$
$$= 392{,}000 - 78{,}400$$
$$= 313{,}600$$

The mortgage is $313,600.

Then, calculate the charge for points.

$$\text{Points} = 2\frac{1}{2}\% \text{ of } 313{,}600 = 0.025(313{,}600) \qquad \bullet \ 2\frac{1}{2}\% = 2.5\% = 0.025$$
$$= 7840$$

The charge for points is $7840.

Finally, find the sum of the down payment and the closing costs.

$$78{,}400 + 450 + 7840 = 86{,}690$$

The total of the down payment and the closing costs is $86,690.

CHECK YOUR PROGRESS 1   The purchase price of a home is $410,000. A down payment of 25% is made. The bank charges $375 in fees plus 1.75 points. Find the total of the down payment and the closing costs.

Solution   *See page S29.*

Mortgages

When a bank agrees to provide you with a mortgage, you agree to pay off that loan in monthly payments. If you fail to make the payments, the bank has the right to **foreclose,** which means that the bank takes possession of the property and has the right to sell it.

There are many types of mortgages available to home buyers today, so the terms of mortgages differ considerably. Some mortgages are **adjustable rate mortgages (ARMs).** The interest rate charged on an ARM is adjusted periodically to more closely reflect current interest rates. The mortgage agreement specifies exactly how often and by how much the interest rate can change.

A **fixed rate mortgage,** or **conventional mortgage,** is one in which the interest rate charged on the loan remains the same throughout the life of the mortgage. For a fixed rate mortgage, the amount of the monthly payment also remains unchanged throughout the term of the loan.

The term of a mortgage can vary. Terms of 15, 20, 25, and 30 years are most common.

The monthly payment on a mortgage is the **mortgage payment.** The amount of the mortgage payment depends on the amount of the mortgage, the interest rate on the loan, and the term of the loan. This payment is calculated by using the payment formula for an APR loan given in Section 7.3. We will restate the formula here.

> **Mortgage Payment Formula**
>
> The mortgage payment for a mortgage is given by
>
> $$PMT = A\left(\frac{i}{1 - (1 + i)^{-n}}\right)$$
>
> where PMT is the monthly mortgage payment, A is the amount of the mortgage, i is the interest rate per payment period, and n is the total number of payments.

EXAMPLE 2 ■ Calculate a Mortgage Payment

Suppose Allison Sommerset purchases a condominium and secures a loan of $134,000 for 30 years at an annual interest rate of 6.5%.

a. Find the monthly mortgage payment.
b. What is the total of the payments over the life of the loan?
c. Find the amount of interest paid on the loan over the 30 years.

Solution

a. First calculate i and store the result.

$$i = \frac{\text{annual interest rate}}{\text{number of payments per year}} = \frac{0.065}{12}$$

Keystrokes: 0.065 ÷ 12 = 0.00541667 STO

Calculate the monthly payment. For a 30-year loan, $n = 30(12) = 360$.

$$PMT = A\left(\frac{i}{1 - (1 + i)^{-n}}\right)$$

$$= 134{,}000\left(\frac{0.065/12}{1 - (1 + 0.065/12)^{-360}}\right) \approx 846.97$$

Keystrokes:

134000 × RCL = ÷ (1 − (1 + RCL) y^x 360 +/−) =

The monthly mortgage payment is $846.97.

b. To determine the total of the payments, multiply the number of payments (360) by the monthly payment ($846.97).

846.97(360) = 304,909.20

The total of the payments over the life of the loan is $304,909.20.

c. To determine the amount of interest paid, subtract the mortgage from the total of the payments.

304,909.20 − 134,000 = 170,909.20

The amount of interest paid over the life of the loan is $170,909.20.

CHECK YOUR PROGRESS 2 Suppose Antonio Scarletti purchases a home and secures a loan of $223,000 for 25 years at an annual interest rate of 7%.

a. Find the monthly mortgage payment.
b. What is the total of the payments over the life of the loan?
c. Find the amount of interest paid on the loan over the 25 years.

Solution See page S29.

▼ **point of interest**

Home buyers rated the following characteristics as "extremely important" in their purchase decisions. (*Source:* Copyright © 1997 Dow Jones & Co., Inc. Republished with permission of Dow Jones & Co., Inc. Permission conveyed through Copyright Clearance Center, Inc.)

Natural, open space: 77%
Walking and biking paths: 74%
Gardens with native plants: 56%
Clustered retail stores: 55%
Wilderness area: 52%
Outdoor pool: 52%
Community recreation center: 52%
Interesting little parks: 50%

A portion of a mortgage payment pays the current interest owed on the loan, and the remaining portion of the mortgage payment is used to reduce the principal owed on the loan. This process of paying off the principal and the interest, which is similar to paying a car loan, is called **amortizing the loan.**

In Example 2, the mortgage payment on a $134,000 mortgage at 6.5% for 30 years was $846.97. The portion of the first payment that is interest and the portion that is applied to the principal can be calculated using the simple interest formula.

$$I = Prt$$

$$= 134{,}000(0.065)\left(\frac{1}{12}\right) \qquad \bullet\ P = 134{,}000, \text{ the current loan amount;}$$
$$r = 0.065;\ t = \frac{1}{12}$$

$$\approx 725.83 \qquad\qquad\qquad \bullet\ \text{Round to the nearest cent.}$$

Of the $846.97 mortgage payment, $725.83 is an interest payment. The remainder is applied toward reducing the principal.

$$\text{Principal reduction} = \$846.97 - \$725.83 = \$121.14$$

After the first month's mortgage payment, the balance on the loan (the amount that remains to be paid) is calculated by subtracting the principal paid on the mortgage from the original mortgage amount.

$$\text{Loan balance after first month} = \$134{,}000 - \$121.14 = \$133{,}878.86$$

The portion of the second mortgage payment that is applied to interest and the portion that is applied to the principal can be calculated in the same manner. In the calculation, the figure used for the principal P is the current balance on the loan, $133,878.86.

$$I = Prt$$

$$= 133{,}878.86(0.065)\left(\frac{1}{12}\right) \qquad \bullet\ P = 133{,}878.86, \text{ the current loan amount;}$$
$$r = 0.065;\ t = \frac{1}{12}$$

$$\approx 725.18 \qquad\qquad\qquad \bullet\ \text{Round to the nearest cent.}$$
$$\text{Principal reduction} = \$846.97 - \$725.18 = \$121.79$$

Of the second mortgage payment, $725.18 is an interest payment and $121.79 is a payment toward the principal.

$$\text{Loan balance after second month} = \$133{,}878.86 - \$121.79 = \$133{,}757.07$$

The interest payment, principal payment, and balance on the loan can be calculated in this manner for all of the mortgage payments throughout the life of the loan—all 360 of them! Or a computer can be programmed to make these calculations and print out the information. The printout is called an **amortization schedule.** It lists, for each mortgage payment, the payment number, the interest payment, the amount applied toward the principal, and the resulting balance to be paid.

Each month, the amount of the mortgage payment that is an interest payment decreases and the amount applied toward the principal increases. This is because you are paying interest on a decreasing balance each month. Mortgage payments early in the life of a mortgage are largely interest payments; mortgage payments late in the life of a mortgage are largely payments toward the principal.

The partial amortization schedule below shows the breakdown for the first 12 months of the loan in Example 2.

Amortization Schedule

Loan Amount	$134,000.00
Interest Rate	6.50%
Term of Loan	30
Monthly Payment	$846.97

Month	Amount of Interest	Amount of Principal	New Loan Amount
1	$725.83	$121.14	$133,878.86
2	$725.18	$121.79	$133,757.07
3	$724.52	$122.45	$133,634.61
4	$723.85	$123.12	$133,511.50
5	$723.19	$123.78	$133,387.71
6	$722.52	$124.45	$133,263.26
7	$721.84	$125.13	$133,138.13
8	$721.16	$125.81	$133,012.32
9	$720.48	$126.49	$132,885.84
10	$719.80	$127.17	$132,758.66
11	$719.11	$127.86	$132,630.80
12	$718.42	$128.55	$132,502.25

QUESTION   *Using the amortization schedule above, how much of the loan has been paid off after 1 year?*

EXAMPLE 3 ■ Calculate Principal and Interest for a Mortgage Payment

You purchase a condominium for $98,750 and obtain a 30-year, fixed rate mortgage at 7.25%. After paying a down payment of 20%, how much of the second payment is interest and how much is applied toward the principal?

Solution

First find the down payment by calculating 20% of the purchase price.

$$0.20(98,750) = 19,750$$

The down payment is $19,750.

Find the mortgage by subtracting the down payment from the purchase price.

$$98,750 - 19,750 = 79,000$$

The mortgage is $79,000.

ANSWER   *After 1 year (12 months), the loan amount is $132,502.25. The original loan was $134,000. The amount that has been paid off is*
$134,000 - $132,502.25 = $1497.75.

Calculate the mortgage payment.

$$i = \frac{\text{annual interest rate}}{\text{number of payments per year}} = \frac{0.0725}{12}$$

Keystrokes: 0.0725 $\boxed{\div}$ 12 $\boxed{=}$ 0.00604167 $\boxed{\text{STO}}$

Calculate the monthly payment. For a 30-year loan, $n = 30(12) = 360$.

$$PMT = A\left(\frac{i}{1 - (1 + i)^{-n}}\right)$$

$$= 79{,}000\left(\frac{0.0725/12}{1 - (1 + 0.0725/12)^{-360}}\right) \approx 538.92$$

Keystrokes:

79000 $\boxed{\times}$ $\boxed{\text{RCL}}$ $\boxed{=}$ $\boxed{\div}$ $\boxed{(}$ 1 $\boxed{-}$ $\boxed{(}$ 1 $\boxed{+}$ $\boxed{\text{RCL}}$ $\boxed{)}$ $\boxed{y^x}$ 360 $\boxed{+/-}$ $\boxed{)}$ $\boxed{=}$

The monthly payment is $538.92.

Find the amount of interest paid on the first mortgage payment by using the simple interest formula.

$$I = Prt$$

$$= 79{,}000(0.0725)\left(\frac{1}{12}\right)$$ • $P = 79{,}000$, the current loan amount;
$r = 0.0725$; $t = \dfrac{1}{12}$

$$\approx 477.29$$ • Round to the nearest cent.

Find the principal paid on the first mortgage payment by subtracting the interest paid from the monthly mortgage payment.

$$538.92 - 477.29 = 61.63$$

Calculate the balance on the loan after the first mortgage payment by subtracting the principal paid from the mortgage.

$$79{,}000 - 61.63 = 78{,}938.37$$

Find the amount of interest paid on the second mortgage payment.

$$I = Prt$$

$$= 78{,}938.37(0.0725)\left(\frac{1}{12}\right)$$ • $P = 78{,}938.37$, the current loan amount;
$r = 0.0725$; $t = \dfrac{1}{12}$

$$\approx 476.92$$ • Round to the nearest cent.

The interest paid on the second payment was $476.92.

Find the principal paid on the second mortgage payment.

$$538.92 - 476.92 = 62.00$$

The principal paid on the second payment was $62.

CHECK YOUR PROGRESS 3   You purchase a home for $295,000. You obtain a 30-year conventional mortgage at 6.75% after paying a down payment of 25% of the purchase price. Of the first month's payment, how much is interest and how much is applied toward the principal?

Solution   *See page S30.*

When a home is sold before the term of the loan has expired, the homeowner must pay the lender the remaining balance on the loan. To calculate that balance, we can use the APR loan payoff formula from Section 7.3.

APR Loan Payoff Formula

The payoff amount for a mortgage is given by

$$A = PMT\left(\frac{1 - (1 + i)^{-n}}{i}\right)$$

where A is the loan payoff, PMT is the mortgage payment, i is the interest rate per payment period, and n is the number of *remaining* payments.

EXAMPLE 4 ■ Calculate a Mortgage Payoff

A homeowner has a monthly mortgage payment of $645.32 on a 30-year loan at an annual interest rate of 7.2%. After making payments for 5 years, the homeowner decides to sell the house. What is the payoff for the mortgage?

Solution

Use the APR loan payoff formula. The homeowner has been making payments for 5 years, or 60 months. There are 360 months in a 30-year loan, so there are $360 - 60 = 300$ remaining payments.

$$A = PMT\left(\frac{1 - (1 + i)^{-n}}{i}\right)$$

$$= 645.32\left(\frac{1 - (1 + 0.006)^{-300}}{0.006}\right) \qquad \bullet \; PMT = 645.32; \; i = \frac{0.072}{12} = 0.006;$$

$$\approx 89,679.01 \qquad\qquad\qquad n = 300, \text{ the number of remaining payments}$$

Here are the keystrokes to compute the payoff on a scientific calculator. The same calculation using a graphing calculator is shown at the left.

```
N=300
I%=7.2
■PV=89679.0079
PMT=-645.32
FV=0
P/Y=12
C/Y=12
PMT: END BEGIN
```

Calculate i: 0.072 $\boxed{\div}$ 12 $\boxed{=}$ 0.006 $\boxed{\text{STO}}$
Calculate the payoff: 645.32 $\boxed{\times}$ $\boxed{(}$ 1 $\boxed{-}$ $\boxed{(}$ 1 $\boxed{+}$ $\boxed{\text{RCL}}$ $\boxed{)}$ $\boxed{y^x}$ 300 $\boxed{+/-}$ $\boxed{)}$ $\boxed{\div}$ $\boxed{\text{RCL}}$ $\boxed{=}$

The loan payoff is $89,679.01.

CHECK YOUR PROGRESS 4 Ava Rivera has a monthly mortgage payment of $846.82 on her condo. After making payments for 4 years, she decides to sell the condo. If she has a 25-year loan at an annual interest rate of 6.9%, what is the payoff for the mortgage?

Solution See page S30.

Math Matters   **Biweekly and Two-Step Mortgages**

A variation of the fixed rate mortgage is the *biweekly mortgage*. Borrowers make payments on a 30-year loan, but they pay half of a monthly payment every 2 weeks, which adds up to 26 half-payments a year, or 13 monthly payments. The extra monthly payment each year can result in the loan being paid off in about $17\frac{1}{2}$ years.

Another type of mortgage is the *two-step mortgage*. Its name is derived from the fact that the life of the loan has two steps, or stages. The first stage is a low fixed rate for the first 7 years of the loan, and the second stage is a different, and probably higher, fixed rate for the remaining 23 years of the loan. This loan is appealing to those homeowners who do not anticipate owning the home beyond the initial low-interest-rate period; they do not need to worry about the increased interest rate during the second stage.

Ongoing Expenses

In addition to a monthly mortgage payment, there are other ongoing expenses associated with home ownership. Among these expenses are the costs of insurance, property tax, and utilities such as heat, electricity, and water.

Services such as schools, police and fire protection, road maintenance, and recreational services, which are provided by cities and counties, are financed by the revenue received from taxes levied on real property, or property taxes. Property tax is normally an annual expense that can be paid on a monthly, quarterly, semiannual, or annual basis.

Homeowners who obtain a mortgage must carry fire insurance. This insurance guarantees that the lender will be repaid in the event of a fire.

▼ **point of interest**

The home ownership rate in the United States for the first quarter of 2004 was 68.6%. The following list gives home owner-ship rates by region during the same quarter.

　　Northeast: 65.1%
　　Midwest: 73.5%
　　South: 70.3%
　　West: 63.7%

It is interesting to note that the home ownership rate in the United States in 1950 was 55.0%, significantly lower than it is today. (*Source:* U.S. Bureau of the Census)

EXAMPLE 5 ■ **Calculate a Total Monthly Payment**

A homeowner has a monthly mortgage payment of $1145.60 and an annual prop-erty tax bill of $1074. The annual fire insurance premium is $600. Find the total monthly payment for the mortgage, property tax, and fire insurance.

Solution

Find the monthly property tax bill by dividing the annual property tax by 12.

　　$1074 \div 12 = 89.50$

The monthly property tax bill is $89.50.

Find the monthly fire insurance bill by dividing the annual fire insurance premium by 12.

　　$600 \div 12 = 50$

The monthly fire insurance bill is $50.

Find the sum of the mortgage payment, the monthly property tax bill, and the monthly fire insurance bill.

　　$1145.60 + 89.50 + 50.00 = 1285.10$

The monthly payment for the mortgage, property tax, and fire insurance is $1285.10.

CHECK YOUR PROGRESS 5   A homeowner has a monthly mortgage payment of $1492.89, an annual property tax bill of $2332.80, and an annual fire insurance premium of $450. Find the total monthly payment for the mortgage, property tax, and fire insurance.

Solution   *See page S30.*

Investigation

Home Ownership Issues

There are a number of issues that a person must think about when purchasing a home. One such issue is the difference between the interest rate on which the loan payment is based and the APR. For instance, a bank may offer a loan at an annual interest rate of 6.5%, but then go on to say that the APR is 7.1%.

The discrepancy is a result of the Truth in Lending Act. This act requires that the APR be based on *all* loan fees. This includes points and other fees associated with the purchase. To calculate the APR, a computer or financial calculator is necessary.

Suppose you decide to purchase a home and you secure a 30-year, $285,000 loan at an annual interest rate of 6.5%.

Investigation Exercises

1. Calculate the monthly payment for the loan.

2. If points are 1.5% of the loan amount, find the fee for points.

3. Add the fee for points to the loan amount. This is the modified mortgage on which the APR is calculated.

4. Using the result from Investigation Exercise 3 as the mortgage, and the monthly payment from Investigation Exercise 1, determine the interest rate. (This is where the financial or graphing calculator is necessary. See page 454 for details.) The result is the APR required by the Truth in Lending Act. For this example we have included only points. In most situations, other fees would be included as well.

Another issue to research when purchasing a home is that of points and mortgage interest rates. Usually paying higher points results in a lower mortgage interest rate. The question for the homebuyer is: Should I pay higher points for a lower mortgage interest rate, or pay lower points for a higher mortgage interest rate? The answer to that question depends on many factors, one of which is the amount of time the homeowner plans on staying in the home.

Consider two typical situations for a 30-year, $100,000 mortgage. Option 1 offers an annual mortgage interest rate of 7.25% and a loan origination fee of 1.5 points. Option 2 offers an annual interest rate of 7% and a loan origination fee of 2 points.

5. Calculate the monthly payments for Option 1 and Option 2.

6. Calculate the loan origination fees for Option 1 and Option 2.

7. What is the total amount paid, including points, after 2 years for each option?

8. What is the total amount paid, including points, after 3 years for each option?

9. Which option is more cost effective if you stay in the home for 2 years or less? Which option is more cost effective if you stay in the home for 3 years or more? Explain your answer.

Exercise Set 7.5

1. **Mortgages** You buy a $258,000 home with a down payment of 25%. Find the amount of the down payment and the mortgage amount.

2. **Mortgages** Greg Walz purchases a home for $325,000 with a down payment of 10%. Find the amount of the down payment and the mortgage amount.

3. **Points** Clarrisa Madison purchases a home and secures a loan of $250,000. The bank charges a fee of 2.25 points. Find the charge for points.

4. **Points** Jerome Thurber purchases a home and secures a loan of $170,000. The bank charges a fee of $2\frac{3}{4}$ points. Find the charge for points.

5. **Closing Costs** The purchase price of a home is $309,000. A down payment of 30% is made. The bank charges $350 in fees plus 3 points. Find the total of the down payment and the closing costs.

6. **Closing Costs** The purchase price of a home is $243,000. A down payment of 20% is made. The bank charges $425 in fees plus 4 points. Find the total of the down payment and the closing costs.

7. **Closing Costs** The purchase price of a condominium is $121,500. A down payment of 25% is made. The bank charges $725 in fees plus $3\frac{1}{2}$ points. Find the total of the down payment and the closing costs.

8. **Closing Costs** The purchase price of a manufactured home is $159,000. A down payment of 20% is made. The bank charges $815 in fees plus 1.75 points. Find the total of the down payment and the closing costs.

9. **Mortgage Payments** Find the mortgage payment for a 25-year loan of $129,000 at an annual interest rate of 7.75%.

10. **Mortgage Payments** Find the mortgage payment for a 30-year loan of $245,000 at an annual interest rate of 6.5%.

11. **Mortgage Payments** Find the mortgage payment for a 15-year loan of $223,500 at an annual interest rate of 8.15%.

12. **Mortgage Payments** Find the mortgage payment for a 20-year loan of $149,900 at an annual interest rate of 8.5%.

13. Suppose a house is purchased. Which results in a lower mortgage, a 20% down payment or a 25% down payment?

14. Suppose a house is purchased and a 25% down payment is made. Which mortgage carries the higher monthly payment, a 30-year fixed rate mortgage at an annual interest rate of 6.75% or a 20-year fixed rate mortgage at an annual interest rate of 6.75%?

15. Suppose a house is purchased. Which combination results in lower initial expenses, paying a 20% down payment and a fee of 2 points or paying a 21% down payment and a fee of 1 point?

16. Suppose a homeowner obtains a traditional 30-year mortgage. If $458 of the first month's payment is interest, will the amount of interest paid in the second month be more than, equal to, or less than $458?

17. Suppose a homeowner obtains a traditional 30-year mortgage at an annual interest rate of 5%. For the first month of the loan, is the amount of interest owed less than 5% of the loan amount, equal to 5% of the loan amount, or more than 5% of the loan amount?

18. **Mortgage Payments** Leigh King purchased a townhouse and obtained a 30-year loan of $152,000 at an annual interest rate of 7.75%.

    a. What is the mortgage payment?

    b. What is the total of the payments over the life of the loan?

    c. Find the amount of interest paid on the mortgage loan over the 30 years.

19. **Mortgage Payments** Richard Miyashiro purchased a condominium and obtained a 25-year loan of $199,000 at an annual interest rate of 8.25%.

    a. What is the mortgage payment?

    b. What is the total of the payments over the life of the loan?

    c. Find the amount of interest paid on the mortgage loan over the 25 years.

20. **Interest Paid** Ira Patton purchased a home and obtained a 15-year loan of $219,990 at an annual interest rate of 8.7%. Find the amount of interest paid on the loan over the 15 years.

21. **Interest Paid** Leona Jefferson purchased a home and obtained a 30-year loan of $437,750 at an annual interest rate of 7.5%. Find the amount of interest paid on the loan over the 30 years.

22. **Principal and Interest** Marcel Thiessen purchased a home for $208,500 and obtained a 15-year, fixed rate mortgage at 9% after paying a down payment of 10%.

Of the first month's mortgage payment, how much is interest and how much is applied to the principal?

23. **Principal and Interest** You purchase a condominium for $173,000. You obtain a 30-year, fixed rate mortgage loan at 12% after paying a down payment of 25%. Of the second month's mortgage payment, how much is interest and how much is applied to the principal?

24. **Principal and Interest** You purchase a cottage for $185,000. You obtain a 20-year, fixed rate mortgage loan at 12.5% after paying a down payment of 30%. Of the second month's mortgage payment, how much is interest and how much is applied to the principal?

25. **Principal and Interest** Fay Nguyen purchased a second home for $183,000 and obtained a 25-year, fixed rate mortgage loan at 9.25% after paying a down payment of 30%. Of the second month's mortgage payment, how much is interest and how much is applied to the principal?

26. **Loan Payoffs** After making payments of $913.10 for 6 years on your 30-year loan at 8.5%, you decide to sell your home. What is the loan payoff?

27. **Loan Payoffs** Christopher Chamberlain has a 25-year mortgage loan at an annual interest rate of 7.75%. After making payments of $1011.56 for $3\frac{1}{2}$ years, Christopher decides to sell his home. What is the loan payoff?

28. **Loan Payoffs** Iris Chung has a 15-year mortgage loan at an annual interest rate of 7.25%. After making payments of $672.39 for 4 years, Iris decides to sell her home. What is the loan payoff?

29. **Loan Payoffs** After making payments of $736.98 for 10 years on your 30-year loan at 6.75%, you decide to sell your home. What is the loan payoff?

30. **Total Monthly Payment** A homeowner has a mortgage payment of $996.60, an annual property tax bill of $594, and an annual fire insurance premium of $300. Find the total monthly payment for the mortgage, property tax, and fire insurance.

31. **Total Monthly Payment** Malcolm Rothschild has a mortgage payment of $1753.46, an annual property tax bill of $1023, and an annual fire insurance premium of $780. Find the total monthly payment for the mortgage, property tax, and fire insurance.

32. **Total Monthly Payment** Baka Onegin obtains a 25-year mortgage loan of $259,500 at an annual interest rate of 7.15%. Her annual property tax bill is $1320 and her annual fire insurance premium is $642. Find the total monthly payment for the mortgage, property tax, and fire insurance.

33. **Total Monthly Payment** Suppose you obtain a 20-year mortgage loan of $198,000 at an annual interest rate of 8.4%. The annual property tax bill is $972 and the annual fire insurance premium is $486. Find the total monthly payment for the mortgage, property tax, and fire insurance.

34. **Mortgage Loans** Consider a mortgage loan of $150,000 at an annual interest rate of 8.125%.
    a. How much greater is the monthly mortgage payment if the term is 15 years rather than 30 years?
    b. How much less is the amount of interest paid over the life of the 15-year loan than over the life of the 30-year loan?

35. **Mortgage Loans** Consider a mortgage loan of $359,960 at an annual interest rate of 7.875%.
    a. How much greater is the monthly mortgage payment if the term is 15 years rather than 30 years?
    b. How much less is the amount of interest paid over the life of the 15-year loan than over the life of the 30-year loan?

36. **Mortgage Loans** The Mendez family is considering a mortgage loan of $349,500 at an annual interest rate of 6.75%.
    a. How much greater is their monthly mortgage payment if the term is 20 years rather than 30 years?
    b. How much less is the amount of interest paid over the life of the 20-year loan than over the life of the 30-year loan?

37. **Mortgage Loans** Herbert Bloom is considering a mortgage loan of $322,495 at an annual interest rate of 7.5%.
    a. How much greater is his monthly mortgage payment if the term is 20 years rather than 30 years?
    b. How much less is the amount of interest paid over the life of the 20-year loan than over the life of the 30-year loan?

38. **Affordability** A couple has saved $25,000 for a down payment on a home. Their bank requires a minimum down payment of 20%. What is the maximum price they can offer for a house in order to have enough money for the down payment?

39. **Affordability** You have saved $18,000 for a down payment on a house. Your bank requires a minimum down payment of 15%. What is the maximum price you can offer for a home in order to have enough money for the down payment?

40. **Affordability** You have saved $39,400 to make a down payment and pay the closing costs on your future

home. Your bank informs you that a 15% down payment is required and that the closing costs should be $380 plus 4 points. What is the maximum price you can offer for a home in order to have enough money for the down payment and the closing costs?

Extensions

CRITICAL THINKING

41. **Amortization Schedules** Suppose you have a 30-year mortgage loan for $119,500 at an annual interest rate of 8.25%. For which monthly payment does the amount of principal paid first exceed the amount of interest paid? For this exercise, you will need a spreadsheet program for producing amortization schedules. You can find one at the Online Study Center at **college.hmco.com/pic/aufmannMTQR.**

42. **Amortization Schedules** Does changing the amount of the loan in Exercise 41 change the number of the monthly payment for which the amount of principal paid first exceeds the amount of interest paid? For this exercise, you will need a spreadsheet program for producing amortization schedules. You can find one at the Online Study Center at **college.hmco.com/pic/aufmannMTQR.**

43. **Amortization Schedules** Does changing the interest rate of the loan in Exercise 41 change the number of the monthly payment for which the amount of principal paid first exceeds the amount of interest paid? For this exercise, you will need a spreadsheet program for producing amortization schedules. You can find one at the Online Study Center at **college.hmco.com/pic/aufmannMTQR.**

44. Suppose you are considering a mortgage loan of $250,000 at an annual interest rate of 8%. Explain why the monthly payment for a 15-year mortgage loan is not twice the monthly payment for a 30-year mortgage loan.

COOPERATIVE LEARNING

45. **Buying and Selling a Home** Suppose you buy a house for $208,750, make a down payment that is 30% of the purchase price, and secure a 30-year loan for the balance at an annual interest rate of 7.75%. The points on the loan are 1.5% and there are additional lender fees of $825.

   a. How much is due at closing? Note that the down payment is due at closing.

   b. After 5 years, you decide to sell your house. What is the loan payoff?

   c. Because of inflation, you were able to sell your house for $248,000. Assuming the selling fees are 6% of the selling price, what are the proceeds from the sale after deducting selling fees? Do not include the interest paid on the mortgage. Remember to consider the loan payoff.

   d. The percent return on an investment $=$ $$\frac{\text{proceeds from sale}}{\text{total closing costs}} \times 100.$$ Find the percent return on your investment. Round to the nearest percent.

46. **Affordability** Assume that you are a computer programmer for a company and that your gross pay is $72,000 per year. You are paid monthly and take home $4479.38 per month. You are considering purchasing a condominium and have looked at one you would like to buy. The purchase price is $134,000. The condominium development company will arrange financing for the purchase. A down payment of 20% of the selling price is required. The interest rate on the 30-year, fixed rate mortgage loan is 9%. The lender charges 2.5 points. The appraisal fee, title fee, and recording fee total $325. The property taxes on the condominium are currently $1152 per year.

The condominium management company charges a monthly maintenance fee for landscaping, trash removal, snow removal, maintenance on the buildings, management costs, and insurance on the property. The monthly fee is currently $120. This fee includes insurance on the buildings and land only, so you will need to purchase separate insurance coverage for your furniture and other personal possessions. You have talked to the insurance agent through whom you buy tenant homeowner's insurance (to cover the personal possessions in your apartment), and you were told that such coverage will cost you approximately $192 per year.

The condominium uses electricity for cooking, heating, and lighting. You have been informed that the unit will use approximately 8750 kilowatt-hours per year. The cost of electricity is currently $.07 per kilowatt-hour.

You have saved $36,000 in anticipation of buying a home, but you do not want your savings account balance to fall below $6000. (You want to have some funds to fall back on in case of an emergency.) You have $900 in your checking account, but you do not want your checking account balance to fall below $600.

   a. Calculate the amount to be paid for the down payment and closing costs. Are you willing to take this much out of your accounts?

b. Determine the mortgage payment.

c. Calculate the total of the monthly payments related to ownership of the condominium (mortgage, maintenance fee, property tax, utilities, insurance).

d. Find the difference between your monthly take-home pay and the monthly expenses related to condominium ownership (part c). This figure indicates how much money is available to pay for all other expenses, including food, transportation, clothing, entertainment, dental bills, telephone bills, car bills, etc.

e. What percent of your monthly take-home pay are the condominium-related expenses (part c)? Round to the nearest tenth of a percent.

f. Do you think that, financially, you can handle the purchase of this condominium? Why?

CHAPTER 7   **Summary**

Key Terms

adjustable rate mortgage (ARM) [p. 478]
amortization schedule [p. 480]
amortize a loan [p. 480]
annual percentage rate (APR) [p. 452]
annual percentage yield (APY) [p. 452]
average daily balance method [p. 450]
bond [p. 470]
bondholder [p. 470]
broker [p. 468]
brokerage firm [p. 468]
closing costs [p. 477]
compound amount [p. 431]
compound interest [p. 429]
compounding period [p. 429]
conventional mortgage [p. 478]
coupon [p. 470]
dividend [p. 467]
dividend yield [p. 467]
effective interest rate [p. 440]
face value [p. 470]
finance charge [p. 449]
fixed rate mortgage [p. 478]
future value [p. 420]
inflation [p. 438]
interest [p. 416]
interest rate [p. 416]
investment trust [p. 471]
loan origination fee [p. 477]
market value [p. 468]

maturity date [p. 470]
maturity value [p. 419]
mortgage [p. 477]
mortgage payment [p. 478]
mutual fund [p. 471]
nominal interest rate [p. 440]
points [p. 477]
present value [p. 436]
principal [p. 416]
residual value [p. 459]
share of stock [p. 467]
simple interest [p. 416]
stock exchange [p. 468]
stockholders or shareholders [p. 467]

Essential Concepts

- **Approximate Annual Percentage Rate (APR) Formula**

$$APR = \frac{2Nr}{N + 1}$$

  N is the number of payments and r is the simple interest rate.

- **APR Loan Payoff Formula**

$$A = PMT\left(\frac{1 - (1 + i)^{-n}}{i}\right)$$

  A is the loan payoff, PMT is the payment, i is the interest rate per payment period, and n is the number of *remaining* payments.

■ **Average Daily Balance Method**

Average daily balance

$$= \frac{\text{sum of the total amounts owed each day of the month}}{\text{number of days in the billing period}}$$

■ **Compound Amount Formula**

$A = P(1 + i)^N$

A is the compound amount when P dollars is deposited at an interest rate of i per compounding period for N compounding periods.

$$i = \frac{\text{annual interest rate}}{\text{number of compounding periods per year}} = \frac{r}{n}$$

$N = $ (number of compounding periods per year)
$\times$ (number of years) $= nt$

■ **Future Value/Maturity Value Formulas for Simple Interest**

$A = P + I$
$A = P(1 + rt)$

■ **Monthly Lease Payment Formula**

$P = F + D$

P is the monthly lease payment, F is the average monthly finance charge, and D is the average monthly depreciation of the car, where:

Net capitalized cost
    $= $ negotiated price $-$ down payment
        $-$ trade-in value

$$\text{Money factor} = \frac{\text{annual interest rate as a percent}}{2400}$$

Average monthly finance charge
    $= $ (net capitalized cost $+$ residual value)
        $\times$ money factor

Average monthly depreciation

$$= \frac{\text{net capitalized cost} - \text{residual value}}{\text{term of the lease in months}}$$

■ **Mortgage Amount**

Mortgage $=$ selling price $-$ down payment

■ **Net Asset Value of a Mutual Fund**

$$NAV = \frac{A - L}{N}$$

NAV is the net asset value of the market fund, A is the total fund assets, L is the total fund liabilities, and N is the number of shares outstanding.

■ **Payment Formula for an APR Loan**

$$PMT = A\left(\frac{i}{1 - (1 + i)^{-n}} \right)$$

PMT is the payment, A is the loan amount, i is the interest rate per payment period, and n is the total number of payments.

■ **Present Value Formula**

$$P = \frac{A}{(1 + i)^N}$$

P is the original principal invested at an interest rate of i per compounding period for N compounding periods, and A is the compound amount.

■ **Simple Interest Formula**

$I = Prt$

I is the interest, P is the principal, r is the interest rate, and t is the time period.

■ **Time Period of a Loan: Converting from Days to Years**

Exact method: $t = \dfrac{\text{number of days}}{365}$

Ordinary method: $t = \dfrac{\text{number of days}}{360}$

CHAPTER 7 Review Exercises

1. **Simple Interest** Calculate the simple interest due on a four-month loan of $2750 if the interest rate is 6.75%.

2. **Simple Interest** Find the simple interest due on an eight-month loan of $8500 if the interest rate is 1.15% per month.

3. **Simple Interest** What is the simple interest earned in 120 days on a deposit of $4000 if the interest rate is 6.75%?

4. **Maturity Value** Calculate the maturity value of a simple interest, 108-day loan of $7000 if the interest rate is 10.4%.

5. **Simple Interest Rate** The simple interest charged on a three-month loan of $6800 is $127.50. Find the simple interest rate.

6. **Compound Amount** Calculate the compound amount when $3000 is deposited in an account earning 6.6% interest, compounded monthly, for 3 years.

7. **Compound Amount** What is the compound amount when $6400 is deposited in an account earning an interest rate of 6%, compounded quarterly, for 10 years?

8. **Future Value** Find the future value of $6000 earning 9% interest, compounded daily, for 3 years.

9. **Compound Interest** Calculate the amount of interest earned in 4 years on $600 deposited in an account paying 7.2% interest, compounded daily.

10. **Present Value** How much money should be invested in an account that earns 8% interest, compounded semiannually, in order to have $18,500 in 7 years?

11. **Loans** To help pay your college expenses, you borrow $8000 and agree to repay the loan at the end of 5 years at 7% interest, compounded quarterly.

    a. What is the maturity value of the loan?

    b. How much interest are you paying on the loan?

12. **Present Value** A couple plans to save for their child's college education. What principal must be deposited by the parents when their child is born in order to have $80,000 when the child reaches the age of 18? Assume the money earns 8% interest, compounded quarterly.

13. **Dividend Yield** A stock pays an annual dividend of $.66 per share. The stock is trading at $60. Find the dividend yield.

14. **Bonds** A bond with a $20,000 face value has a 4.5% coupon and a 10-year maturity. Calculate the total of the interest payments paid to the bondholder.

15. **Inflation** In 2001, the price of 1 pound of red delicious apples was $1.29. Use an annual inflation rate of 6% to calculate the price of 1 pound of red delicious apples in 2011. Round to the nearest cent.

16. **Inflation** You purchase a bond that will provide you with $75,000 in 8 years. Assuming an annual inflation rate of 7%, what will be the purchasing power of the $75,000 in 8 years?

17. **Effective Interest Rate** Calculate the effective interest rate of 5.90% compounded monthly. Round to the nearest hundredth of a percent.

18. **Annual Yield** Which has the higher annual yield, 5.2% compounded quarterly or 5.4% compounded semiannually?

19. **Average Daily Balance** A credit card account had a $423.35 balance on March 11. A purchase of $145.50 was made on March 18, and a payment of $250 was made on March 29. Find the average daily balance if the billing date is April 11.

20. **Finance Charges** On September 10, a credit card account had a balance of $450. A purchase of $47 was made on September 20, and $157 was charged on September 25. A payment of $175 was made on September 28. The interest on the average daily balance is 1.25% per month. Find the finance charge on the October 10 bill.

21. **APR** Arlene McDonald borrows $1500 from a bank that advertises a 7.5% simple interest rate and repays the loan in six equal monthly payments.

    a. Find the monthly payment.

    b. Estimate the APR. Round to the nearest tenth of a percent.

22. **APR** Suppose you purchase a DVD player for $449, make a 10% down payment, and agree to repay the balance in 12 equal monthly payments. The finance charge on the balance is 7% simple interest.

    a. Find the monthly payment.

    b. Estimate the APR. Round to the nearest tenth.

23. **Monthly Payments** Photo Experts offers a Nikon camera for $999, including taxes. If you finance the purchase of this camera for 2 years at an annual interest rate of 8.5%, find the monthly payment.

24. **Monthly Payments** Abeni Silver purchases a plasma high-definition television for $9499. The sales tax is 6.25% of the purchase price.

    a. What is the total cost, including sales tax?

    b. If Abeni makes a down payment of 20% of the total cost, find the down payment.

    c. Assuming Abeni finances the remaining cost at an annual interest rate of 8% for 3 years, find the monthly payment.

25. **Car Payments** Suppose you decide to purchase a new car. You go to a credit union to get pre-approval for your loan. The credit union offers you an annual interest rate of 7.2% for 3 years. The purchase price of the car you select is $28,450, including taxes, and you make a 20% down payment. What is your monthly payment?

26. **Loan Payoffs** Dasan Houston obtains a $28,000, five-year car loan for a Windstar at an annual interest rate of 5.9%.

    a. Find the monthly payment.

    b. After 3 years, Dasan decides to purchase a new car. What is the payoff on his loan?

27. Car Leases Nami Coffey obtains a five-year lease for a Chrysler 300 and negotiates a selling price of $32,450. Assuming she makes a down payment of $3000, the money factor is 0.004, and the residual value is $16,000, find each of the following.

   a. The net capitalized cost

   b. The average monthly finance charge

   c. The average monthly depreciation

   d. The monthly lease payment using a sales tax rate of 7.5%

28. Stock Sale Suppose you purchased 500 shares of stock at a price of $28.75 per share and sold them for $39.40 per share.

   a. Ignoring dividends, what was your profit or loss on the sale of the stock?

   b. If your broker charges 1.3% of the total sale price, what was the broker's commission?

29. Mutual Funds A mutual fund has total assets of $34 million and total liabilities of $4 million. There are 2 million shares outstanding. You invest $3000 in this fund. How many shares will you purchase?

30. Closing Costs The purchase price of a seaside cottage is $459,000. A down payment of 20% is made. The bank charges $815 in fees plus 1.75 points. Find the total of the down payment and the closing costs.

31. Mortgage Payments Suppose you purchase a condominium and obtain a 30-year loan of $255,800 at an annual interest rate of 6.75%.

   a. What is the mortgage payment?

   b. What is the total of the payments over the life of the loan?

   c. Find the amount of interest paid on the mortgage loan over the 30 years.

32. Mortgage Payments and Loan Payoffs Garth Santacruz purchased a condominium and obtained a 25-year loan of $189,000 at an annual interest rate of 7.5%.

   a. What is the mortgage payment?

   b. After making payments for 10 years, Garth decides to sell his home. What is the loan payoff?

33. Total Monthly Payments Geneva Goldberg obtains a 15-year loan of $278,950 at an annual interest rate of 7%. Her annual property tax bill is $1134 and her annual fire insurance premium is $681. Find the total monthly payment for the mortgage, property tax, and fire insurance.

CHAPTER 7   **Test**

---

1. Simple Interest Calculate the simple interest due on a three-month loan of $5250 if the interest rate is 8.25%.

2. Simple Interest Find the simple interest earned in 180 days on a deposit of $6000 if the interest rate is 6.75%.

3. Maturity Value Calculate the maturity value of a simple interest, 200-day loan of $8000 if the interest rate is 9.2%.

4. Simple Interest Rate The simple interest charged on a two-month loan of $7600 is $114. Find the simple interest rate.

5. Compound Amount What is the compound amount when $4200 is deposited in an account earning an interest rate of 7%, compounded monthly, for 8 years?

6. Compound Interest Calculate the amount of interest earned in 3 years on $1500 deposited in an account paying 6.3% interest, compounded daily.

7. Maturity Value To help pay for a new truck, you borrow $10,500 and agree to repay the loan in 4 years at 9.5% interest, compounded monthly.

   a. What is the maturity value of the loan?

   b. How much interest are you paying on the loan?

8. Present Value A young couple wants to save money to buy a house. What principal must be deposited by the couple in order to have $30,000 in 5 years? Assume the money earns 6.25% interest, compounded daily.

9. Dividend Yield A stock that has a market value of $40 pays an annual dividend of $.48 per share. Find the dividend yield.

10. Bonds Suppose you purchase a $5000 bond that has a 3.8% coupon and a 10-year maturity. Calculate the total of the interest payments you will receive.

11. **Inflation** In 2002 the median value of a single-family house was $158,200. Use an annual inflation rate of 7% to calculate the median value of a single family house in 2022. (*Source:* moneycentral. msn.com)

12. **Effective Interest Rate** Calculate the effective interest rate of 6.25% compounded quarterly. Round to the nearest hundredth of a percent.

13. **Annual Yield** Which has the higher annual yield, 4.4% compounded monthly or 4.6% compounded semiannually?

14. **Finance Charges** On October 15, a credit card account had a balance of $515. A purchase of $75 was made on October 20, and a payment of $250 was made on October 28. The interest on the average daily balance is 1.8% per month. Find the finance charge on the November 15 bill.

15. **APR** Suppose you purchase a Joranda hand-held computer for $629, make a 15% down payment, and agree to repay the balance in 12 equal monthly payments. The finance charge on the balance is 9% simple interest.

   a. Find the monthly payment.

   b. Estimate the APR. Round to the nearest tenth of a percent.

16. **Monthly Payments** Technology Pro offers a new computer for $1899, including taxes. If you finance the purchase of this computer for 3 years at an annual percentage rate of 9.25%, find your monthly payment.

17. **Stock Sale** Suppose you purchased 800 shares of stock at a price of $31.82 per share and sold them for $25.70 per share.

   a. Ignoring dividends, what was your profit or loss on the sale of the stock?

   b. If your broker charges 1.1% of the total sale price, what was the broker's commission?

18. **Mutual Funds** A mutual fund has total assets of $42 million and total liabilities of $6 million. There are 3 million shares outstanding. You invest $2500 in this fund. How many shares will you purchase?

19. **Monthly Payments** Kalani Canfield purchases a high-speed color laser printer for $6575. The sales tax is 6.25% of the purchase price.

   a. What is the total cost, including sales tax?

   b. If Kalani makes a down payment of 20% of the total cost, find the down payment.

   c. Assuming Kalani finances the remaining cost at an annual interest rate of 7.8% for 3 years, find the monthly payment.

20. **Closing Costs** The purchase price of a house is $262,250. A down payment of 20% is made. The bank charges $815 in fees plus 3.25 points. Find the total of the down payment and the closing costs.

21. **Mortgage Payments and Loan Payoffs** Bernard Mason purchased a house and obtained a 30-year loan of $236,000 at an annual interest rate of 6.75%.

   a. What is the mortgage payment?

   b. After making payments for 5 years, Bernard decides to sell his home. What is the loan payoff?

22. **Total Monthly Payment** Zelda MacPherson obtains a 20-year loan of $312,000 at an annual interest rate of 7.25%. Her annual property tax bill is $1044 and her annual fire insurance premium is $516. Find the total monthly payment for the mortgage, property tax, and fire insurance.

CHAPTER 8

Probability and Statistics

The U.S. Census Bureau collects data on the population of the United States. It then issues *statistical* reports that indicate changes and trends in the U.S. population. For instance, in 2002 there were approximately 105 males for every 100 females between the ages of 5 and 17 years. However, in the category of people 85 years and older, there were approximately 46 men for every 100 women. See the graph below.

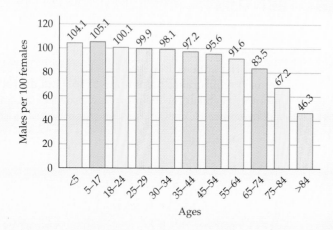

Source: U.S. Census Bureau, Current Population Survey, March 2002

Here are some other statistics from the Census Bureau.

- The *mean* (average) commuting time to work in the United States is approximately 25.5 minutes.

- In 1910, the mean annual family income in the United States was $687. In 2003, the mean annual family income was approximately $59,100.

However, the *median* annual family income in 2003 was approximately $43,500. The difference between the mean and the median is one of the topics of this chapter.

Online Study Center

For online student resources, visit this textbook's Online Study Center at **college.hmco.com/pic/aufmannMTQR.**

| ## Counting Methods

The Counting Principle

Counting may sound like a simple topic, but as we will see it can be much more complex than one might at first guess. The study of counting the different results of a task is a branch of mathematics called **combinatorics.**

In combinatorics, an activity with an observable outcome is called an **experiment.** For example, if a coin is flipped, the side facing upward will be a head or a tail. The two possible outcomes can be listed as {H,T}. If a regular six-sided die is rolled, the possible outcomes are ▫, ▫, ▫, ▫, ▫, ▫ . The outcomes can be listed as {1, 2, 3, 4, 5, 6}.

If there are not too many different outcomes, we may be able simply to list all of the possibilities.

EXAMPLE 1 ■ Outcomes of the Roll of Two Dice

Suppose we roll two standard six-sided dice, first one colored green followed by a second one colored red. How many different results are possible?

Solution
We can organize all the possibilities in a grid arrangement, where the first die in each pair is green and the second is red, as shown below.

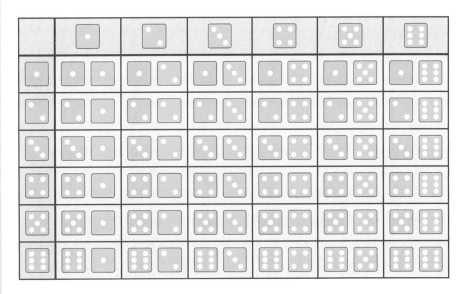

Outcomes of Rolling Two Dice

We can see that there are 36 different possibilities.

CHECK YOUR PROGRESS 1   Four coins are tossed, first a quarter, then a dime, then a nickel, and finally a penny. List all of the possible outcomes and determine the total number of ways these four coins can be tossed.

Solution   *See page S30.*

Rolling two dice, as shown in Example 1, is an example of a **multistage** experiment. (This experiment involved two stages; first one die was rolled, then a second die was rolled.)

In multistage experiments, simply drawing or listing all the possible outcomes can be tedious and difficult to organize. To help, we can draw what is called a **tree diagram.** The technique is demonstrated in the next example.

EXAMPLE 2 ▪ Use a Tree Diagram

A computer store offers a three-component computer system. The system consists of a central processing unit (CPU), a hard drive, and a monitor. The vendor offers two different CPUs, three different hard drives, and two monitors from which to choose. How many distinct computer systems are possible?

Solution

We can organize the information by letting C_1 and C_2 represent the two CPUs; H_1, H_2, and H_3 represent the three hard drives; and M_1 and M_2 represent the two monitors.

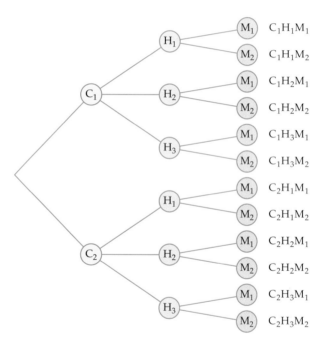

Each "branch" of the tree diagram represents a possible option; by following a path from left to right we assemble one of the outcomes of the experiment. There are 12 possible computer systems.

CHECK YOUR PROGRESS 2   A true-false test consists of three questions. Draw a tree diagram that shows the number of ways to answer the three questions.

Solution   See page S30.

For each of the previous problems, the possible outcomes were listed and then counted to determine the number of different outcomes. However, it is not always possible or practical to list and count outcomes. For example, the number of different five-card poker hands that can be drawn from a regular deck of 52 playing cards is 2,598,960. (See Example 8.) Trying to create a list of these hands would be quite time consuming.

In Example 1, we found that after tossing two dice, there were 36 possible outcomes. We can arrive at this result without listing the outcomes by finding the product of the number of outcomes of the green die and the number of outcomes of the red die.

$$\begin{bmatrix} \text{Outcomes} \\ \text{of green die} \end{bmatrix} \times \begin{bmatrix} \text{Outcomes} \\ \text{of red die} \end{bmatrix} = \begin{bmatrix} \text{Number of} \\ \text{outcomes} \end{bmatrix}$$
$$\qquad 6 \qquad\quad \times \qquad 6 \qquad = \qquad 36$$

Consider again the problem of selecting a computer system consisting of three components. By using a tree diagram, we listed the 12 possible computer systems. Another way to arrive at this result is to find the product of the number of options available for each type of component.

$$\begin{bmatrix} \text{Number of} \\ \text{CPUs} \end{bmatrix} \times \begin{bmatrix} \text{Number of} \\ \text{hard drives} \end{bmatrix} \times \begin{bmatrix} \text{Number of} \\ \text{monitors} \end{bmatrix} = \begin{bmatrix} \text{Number of} \\ \text{systems} \end{bmatrix}$$
$$\quad 2 \qquad\quad \times \qquad 3 \qquad\quad \times \qquad 2 \qquad = \qquad 12$$

The basis of counting all of the outcomes of a multistage experiment without listing them is the *counting principle.*

Counting Principle

Let E be a multistage experiment. If $n_1, n_2, n_3, \ldots, n_k$ are the numbers of possible outcomes of each of the k stages of E, then there are $n_1 \cdot n_2 \cdot n_3 \cdot \cdots \cdot n_k$ possible outcomes for E.

EXAMPLE 3 ■ Use the Counting Principle

In horse racing, a *trifecta* consists of choosing the exact order of the first three horses across the finish line. If there are eight horses in the race, how many trifectas are possible, assuming there are no ties?

Solution
Any one of the eight horses can be first, so $n_1 = 8$. Since a horse cannot finish both first and second, there are seven horses that can finish second; thus $n_2 = 7$. Similarly, there are six horses that can finish third; $n_3 = 6$. By the counting principle, there are $8 \cdot 7 \cdot 6 = 336$ possible trifectas.

CHECK YOUR PROGRESS 3   Nine runners are entered in a 100-meter dash for which a gold medal, a silver medal, and a bronze medal will be awarded for first, second, and third place finishes, respectively. How many possible ways can the medals be awarded?

Solution   See page S30.

Permutations

A **permutation** is an arrangement of distinct objects in a definite order. The counting principle can be used to determine the total number of possible permutations of a group of objects.

For instance, the possible permutations of the three symbols μ, ν, and θ are

Permutation 1	$\mu \, \nu \, \theta$
Permutation 2	$\mu \, \theta \, \nu$
Permutation 3	$\theta \, \mu \, \nu$
Permutation 4	$\theta \, \nu \, \mu$
Permutation 5	$\nu \, \mu \, \theta$
Permutation 6	$\nu \, \theta \, \mu$

Notice that there are three choices for which symbol to place first. Once the first symbol has been picked, there are two choices for the second symbol, and finally only one choice for the third symbol. By the counting principle, there are $3 \cdot 2 \cdot 1 = 6$ permutations of the three symbols.

Similarly, if we have four distinct objects, there are four choices for the first, three choices for the second, two choices for the third, and only one choice for the fourth position. Thus there are $4 \cdot 3 \cdot 2 \cdot 1 = 24$ permutations of four objects.

In general, n objects can be arranged in $n \cdot (n - 1) \cdot (n - 2) \cdot \cdots \cdot 3 \cdot 2 \cdot 1$ different permutations. This product is given a special name.

> **n factorial**
>
> n factorial is the product of the natural numbers 1 through n and is symbolized by $n!$.

For instance, $8! = 8 \cdot 7 \cdot 6 \cdot 5 \cdot 4 \cdot 3 \cdot 2 \cdot 1 = 40{,}320$. By convention, $0! = 1$.

> ✔ **TAKE NOTE**
>
> We discussed factorials earlier in the text. It is a key concept for counting, so we have restated it here.
>
> Many calculators can directly compute factorials. (On the TI-83/83 Plus/84 Plus graphing calculators, for example, $n!$ is accessible in the probability menu after pressing the MATH key.)

EXAMPLE 4 ■ Evaluate Factorial Expressions

Evaluate: **a.** $5! - 3!$ **b.** $\dfrac{9!}{6!}$

Solution

a. $5! - 3! = (5 \cdot 4 \cdot 3 \cdot 2 \cdot 1) - (3 \cdot 2 \cdot 1) = 120 - 6 = 114$

b. A factorial can be written in terms of smaller factorials; this concept is useful for simplifying ratios of factorial expressions.

$$\frac{9!}{6!} = \frac{9 \cdot 8 \cdot 7 \cdot 6 \cdot 5 \cdot 4 \cdot 3 \cdot 2 \cdot 1}{6!} = \frac{9 \cdot 8 \cdot 7 \cdot \cancel{6!}}{\cancel{6!}} = 9 \cdot 8 \cdot 7 = 504$$

CHECK YOUR PROGRESS 4   Evaluate: **a.** $7! + 4!$ **b.** $\dfrac{8!}{4!}$

Solution   *See page S30.*

> ✔ **TAKE NOTE**
>
> When computing with factorials, it is sometimes convenient not to write the entire product. For instance, we can write
>
> $10! = 10 \cdot 9 \cdot 8!$
>
> $7! = 7 \cdot 6 \cdot 5 \cdot 4!$
>
> $5! = 5 \cdot 4!$
>
> In Example 4b, we used this convention to write $9! = 9 \cdot 8 \cdot 7 \cdot 6!$.

There are $n!$ different ways of arranging n objects if we are arranging *all* of the objects. In many cases, however, we may use only *some* of the objects. For example, consider the five symbols θ, ω, ε, ∂, and τ. When only three symbols are selected from these five, there are (by the counting principle) $5 \cdot 4 \cdot 3 = 60$ possible permutations.

The number of permutations of n distinct objects of which k are selected can be generalized by the following formula.

Permutation Formula

The number of permutations, $P(n, k)$, of n distinct objects selected k at a time is

$$P(n, k) = \frac{n!}{(n-k)!}$$

EXAMPLE 5 ■ Choose a Tennis Team

A university tennis team consists of six players who are ranked from 1 through 6. If a tennis coach has 10 players from which to choose, how many different tennis teams can the coach select?

Solution

Because the players on the tennis team are ranked from 1 through 6, a team with player A in position 1 would be different from a team with player A in position 2. Therefore, the number of different teams is the number of permutations of 10 players selected six at a time.

$$P(10, 6) = \frac{10!}{(10-6)!} = \frac{10!}{4!} = \frac{10 \cdot 9 \cdot 8 \cdot 7 \cdot 6 \cdot 5 \cdot 4!}{4!} = 151{,}200$$

There are 151,200 possible tennis teams.

CHECK YOUR PROGRESS 5   A college golf team consists of five players who are ranked from 1 through 5. If a golf coach has eight players from which to choose, how many different golf teams can the coach select?

Solution   *See page S31.*

Math Matters   **Shuffling Cards**

A standard deck of playing cards consists of 52 different cards. Each shuffle of the deck results in a new arrangement of the cards. Another way of stating this would be to say that each shuffle results in a new permutation of the cards. There are $P(52, 52) = 52! \approx 8 \times 10^{67}$ (that's 8 with 67 zeros after it) possible arrangements.

Suppose a deck has each of the four suits arranged in order from ace through king. How many shuffles are necessary to have a randomly ordered deck in the sense that any card will be equally likely to occur in any position in the deck?

Two mathematicians, Dave Bayer of Columbia University and Persi Diaconis of Harvard University, have shown that seven shuffles are enough. Their proof has many applications to complicated counting problems. One problem in particular is analyzing speech patterns. Solving this problem is critical if we are going to be able to "talk" to a computer.

Combinations

In the preceding examples, the order of the arrangement of objects was important. These are permutations. If a telephone extension is 2537, then the digits must be dialed in exactly that order. On the other hand, if you were to receive a one dollar bill, a five dollar bill, and a ten dollar bill, you would have $16 regardless of the order in which you received the bills. A **combination** is a collection of objects for which the order is not important. The three-letter sequences *acb* and *bca* are *different* permutations but the *same* combination.

Consider the problem of finding the number of combinations possible when choosing three letters from the letters *a*, *b*, *c*, *d*, and *e* (without any repetitions). For each choice of three letters, there are $3! = 6$ permutations. For instance, choosing the letters *a*, *d*, and *e* gives the following six permutations.

| *ade* | *aed* | *dea* | *dae* | *ead* | *eda* |

Because there are six permutations and each permutation is the *same* combination, the number of permutations is six times the number of combinations. This is true each time three letters are selected. Therefore, to find the number of combinations of five objects chosen three at a time, divide the number of permutations by $3! = 6$. The number of combinations of five objects chosen three at a time is

$$\frac{P(5, 3)}{3!} = \frac{5!/(5 - 3)!}{3!} = \frac{5!}{3! \cdot (5 - 3)!} = \frac{5!}{3! \cdot 2!} = \frac{5 \cdot 4 \cdot 3!}{3! \cdot 2!} = \frac{5 \cdot 4}{2 \cdot 1} = 10$$

✓ **TAKE NOTE**

Some graphing calculators use nPr to represent $P(n, r)$ and nCr for $C(n, r)$. To find $C(11, 5)$, for instance, enter 11 nCr 5, and then press ENTER.

Combination Formula

The number of combinations, $C(n, k)$, of n distinct objects chosen k at a time is

$$C(n, k) = \frac{P(n, k)}{k!} = \frac{n!}{k! \cdot (n - k)!}$$

EXAMPLE 6 ▦ **Choosing a Basketball Team**

A basketball team consists of 11 players.

a. How many different ways can a coach choose the starting five players, assuming the position of a player is not considered?

b. How many different ways can a coach choose the starting five players if the positions of the players are considered?

Solution

a. This is a combination problem (rather than a permutation problem) because the order in which the coach chooses the players is not important. The starting five players P_1, P_2, P_3, P_4, P_5 are the same starting five players as P_3, P_5, P_1, P_2, P_4.

$$C(11, 5) = \frac{11!}{5! \cdot (11 - 5)!} = \frac{11!}{5! \cdot 6!} = \frac{11 \cdot 10 \cdot 9 \cdot 8 \cdot 7 \cdot 6!}{5! \cdot 6!}$$

$$= \frac{11 \cdot 10 \cdot 9 \cdot 8 \cdot 7}{5 \cdot 4 \cdot 3 \cdot 2 \cdot 1} = 462$$

There are 462 possible five-player starting teams.

b. This time the same five players could be chosen but for different positions, and we would consider these different arrangements. So we want to count permutations rather than combinations.

$$P(11, 5) = \frac{11!}{(11 - 5)!} = \frac{11!}{6!} = \frac{11 \cdot 10 \cdot 9 \cdot 8 \cdot 7 \cdot 6!}{6!}$$
$$= 11 \cdot 10 \cdot 9 \cdot 8 \cdot 7 = 55{,}440$$

There are 55,440 permutations.

CHECK YOUR PROGRESS 6 A softball team consists of 16 players. How many ways can a coach choose the starting 10 players, assuming position is not considered?

Solution *See page S31.*

Some counting problems require the use of more than one counting technique to determine the total number of possible outcomes.

EXAMPLE 7 ■ Choose a Committee

A committee of five professors is to be chosen from five mathematicians and six economists. How many different committees are possible if the committee must include two mathematicians and three economists?

Solution

Because a committee of professors *A*, *B*, *C*, *D*, and *E* is exactly the same committee as professors *B*, *D*, *E*, *A*, and *C*, choosing a committee is an example of a combination. There are five mathematicians from which two are chosen, which is $C(5, 2)$ combinations. There are six economists from which three are chosen, which is $C(6, 3)$ combinations. Therefore, by the counting principle, there are $C(5, 2) \cdot C(6, 3)$ ways to choose both two mathematicians and three economists.

$$C(5, 2) \cdot C(6, 3) = \frac{5!}{2! \cdot 3!} \cdot \frac{6!}{3! \cdot 3!} = 10 \cdot 20 = 200$$

There are 200 possible committees consisting of two mathematicians and three economists.

CHECK YOUR PROGRESS 7 An IRS auditor randomly chooses five tax returns to audit from a stack of 10 tax returns, four of which are from corporations and six of which are from individuals. How many different ways can the auditor choose the five tax returns if the auditor wants to include three corporate and two individual returns?

Solution *See page S31.*

✔ **TAKE NOTE**

A standard deck of playing cards consists of 52 cards in four suits: spades (♠), hearts (♥), diamonds (♦), and clubs (♣). Each suit has 13 cards: 2 through 10, jack, queen, king, and ace. (Here we will not include the two jokers.)

A Standard Deck of
Playing Cards

EXAMPLE 8 ■ Choose Cards from a Deck

From a standard deck of playing cards, a hand of five cards is chosen.

a. How many different five-card hands are possible?

b. How many different five-card hands consist of two kings and three queens? (This hand is an example of a "full house" in the game of poker.)

c. How many different hands consist of five cards of the same suit? (This is a "flush" in poker.)

Solution

a. Since the order in which the cards are chosen is not important, we need to determine the number of combinations of 52 cards taken five at a time.

$$C(52, 5) = \frac{52!}{5! \cdot 47!} = \frac{52 \cdot 51 \cdot 50 \cdot 49 \cdot 48}{5!} = 2{,}598{,}960$$

There are 2,598,960 different five-card hands.

b. There are $C(4, 2)$ ways of choosing two kings from the four kings in the deck and $C(4, 3)$ ways of choosing three queens from the four queens. By the counting principle, there are $C(4, 2) \cdot C(4, 3)$ ways of choosing two kings and three queens.

$$C(4, 2) \cdot C(4, 3) = \frac{4!}{2! \cdot 2!} \cdot \frac{4!}{3! \cdot 1!} = 6 \cdot 4 = 24$$

There are 24 ways of choosing two kings and three queens.

c. First, there are four different suits from which to choose. For each suit, there are $C(13, 5)$ ways of choosing five cards from the 13 cards of that suit. By the counting principle, there are $4 \cdot C(13, 5)$ total outcomes.

$$4 \cdot C(13, 5) = 4 \cdot 1287 = 5148$$

There are 5148 ways of choosing five cards of the same suit from a standard deck of playing cards.

CHECK YOUR PROGRESS 8   From a standard deck of playing cards, a hand of five cards is chosen. How many different hands contain exactly four cards of the same suit?

Solution   *See page S31.*

Investigation

Choosing Numbers in Keno

A popular gambling game called keno is played in many casinos. In keno, there are 80 balls numbered from 1 to 80. The casino chooses 20 balls randomly from the 80 balls. These are "lucky balls" because if a gambler also chooses some of the numbers on these balls, there is a possibility of winning money. The amount that is won depends on the

(continued)

number of lucky numbers the gambler has selected. The number of ways in which a casino can choose 20 balls from 80 is

$$C(80, 20) = \frac{80!}{20! \cdot 60!} \approx 3{,}535{,}000{,}000{,}000{,}000{,}000$$

Once the casino chooses the 20 lucky balls, the remaining 60 balls are unlucky for the gambler. A gambler who chooses five numbers will have from zero to five lucky numbers.

Let's consider the case in which two of the five numbers chosen by the gambler are lucky numbers. Because five numbers were chosen, there must also be three unlucky numbers among the five numbers. The number of ways of choosing two lucky numbers from 20 lucky numbers is $C(20, 2)$. The number of ways of choosing three unlucky numbers from 60 is $C(60, 3)$. By the counting principle, there are $C(20, 2) \cdot C(60, 3) = 190 \cdot 34{,}220 = 6{,}501{,}800$ ways to choose two lucky and three unlucky numbers.

Investigation Exercises

For each of the following exercises, assume that a gambler playing keno has randomly chosen four numbers.

1. How many ways can the gambler choose no lucky numbers?

2. How many ways can the gambler choose exactly one lucky number?

3. How many ways can the gambler choose exactly two lucky numbers?

4. How many ways can the gambler choose exactly three lucky numbers?

5. How many ways can the gambler choose all four lucky numbers?

Exercise Set 8.1

In Exercises 1–10, list all the possible outcomes of each experiment.

1. Select an even single-digit whole number.

2. Select an odd single-digit whole number.

3. Select one day from the days of the week.

4. Select one month from the months of the year.

5. Toss a coin twice.

6. Toss a coin three times.

7. Roll a single die and then toss a coin.

8. Toss a coin and then choose a digit from the digits 1 through 4.

9. Andy, Bob, and Cassidy need to be seated in seats A1, A2, and A3 at a concert.

10. First, second, and third prizes must be awarded to the paintings of Angela, Hector, and Quinh.

In Exercises 11–16, draw a tree diagram that shows all the possible outcomes of each experiment.

11. A dinner menu allows a customer to choose from two salads, three dinner entrees, and two desserts.

12. A new car promotion allows a buyer to choose from three body styles, two radios, and two interior color schemes.

13. Roll a four-sided die twice.

14. Toss a coin and then roll a six-sided die.

15. A two-character category code is created for employees of a company, where the first character is one of the letters A–D, and the second character is one of the numbers 1–5.

16. A three-character category code is used for customers of a store, where the first character is N, S, E, or W (for North, South, East, or West), the second character is either a 5 or a 6 (for 2005 or 2006), and the third character is either Y or N (for Yes or No).

In Exercises 17–22, use the counting principle to determine the number of possible outcomes of each experiment.

17. Two digits are selected from the digits 1, 2, 3, 4, and 5, where the same digit can be used twice.

18. Two digits are selected from the digits 1, 2, 3, 4, and 5, where the same digit cannot be used twice.

19. A multiple-choice test consisting of 15 questions is completed, where each question has four possible answers.

20. A true-false quiz consisting of 20 questions is completed.

21. Four-digit telephone extensions are generated, where the first digit cannot be a 0, 1, 8, or 9.

22. Three-letter codes are generated using only vowels (a, e, i, o, and u). The same letter cannot be used twice in one code.

In Exercises 23–26, use the following experiment. Four cards labeled A, B, C, and D are randomly placed in four boxes labeled A, B, C, and D, one card to each box.

23. In how many different ways can the cards be placed in the boxes?

24. In how many different ways can the cards be placed in the boxes if no card can be in a box with the same letter?

25. In how many different ways can the cards be placed in the boxes if *at least* one card is placed in the box with the same letter?

26. If you add the answers of Exercises 24 and 25, is the sum the same as the answer to Exercise 23? Why or why not?

In Exercises 27–48, evaluate each expression.

27. $8!$

28. $5!$

29. $9! - 5!$

30. $8! + 3!$

31. $\dfrac{5!}{2!}$

32. $\dfrac{7!}{5!}$

33. $\dfrac{8!}{3!}$

34. $\dfrac{12!}{6!}$

35. $P(7, 3)$

36. $P(8, 6)$

37. $P(9, 6)$

38. $P(10, 5)$

39. $P(6, 0)$

40. $P(4, 4)$

41. $C(9, 2)$

42. $C(8, 6)$

43. $C(12, 0)$

44. $C(11, 11)$

45. $C(6, 2) \cdot C(7, 3)$

46. $C(7, 5) \cdot C(9, 4)$

47. $3! \cdot C(8, 5)$

48. $\dfrac{C(10, 4) \cdot C(5, 2)}{C(15, 6)}$

49. If $7! = 5040$, find $8!$.

50. If $9! = 362{,}880$, find $10!$.

51. A selection is made from 10 distinct objects. Will the number of combinations ever exceed the number of permutations?

52. When the counting principle is used to count the arrangements of distinct objects, is the result the number of permutations of the objects or the number of combinations of the objects?

53. Geometry Seven points are drawn in a plane, no three of which are on the same straight line. How many different lines can be drawn that pass through two of the seven points?

54. Geometry A pentagon is a five-sided figure. A diagonal is a line segment connecting any two nonadjacent vertices. How many diagonals are possible?

diagonal

55. Corporate Leadership The board of directors of a corporation must select a president, a secretary, and a treasurer. How many possible ways can this be accomplished if there are 20 members on the board of directors?

56. PINs The personal identification numbers (PINs) used by a certain automatic teller machine (ATM) are sequences of four letters.

a. How many different PINs are possible?

b. If no two letters in a PIN can be the same, how many different PINs are possible?

57. **House Painting** A house painter offers six base colors for the exterior of a home, eight different colors for the trim, and a choice of five colors for accents. How many different color combinations are possible?

58. **Vehicle Purchase** A car manufacturer offers three different body styles on its new SUV, two different engine sizes, 12 different exterior colors, and eight different interior options. How many different versions of the SUV can be purchased?

59. **License Plates** One state's automobile license plates consist of a single digit from 1 through 9, followed by three letters, followed by three digits from 0 through 9. How many unique license plates are possible?

60. **Restaurant Choices** A restaurant offers a special pizza with any five toppings. If the restaurant has 12 toppings available, how many different special pizzas are possible?

61. **Interview Scheduling** Seven people are interviewed for a possible promotion. In how many ways can the seven candidates be scheduled for the interviews?

62. **Shift Scheduling** One shift at a manufacturing plant requires the operation of four possible machines. If eight people are qualified to operate any of the four machines, how many different shifts are possible?

63. **Shift Scheduling** Five women and four men have volunteered to serve for one hour each to answer phones during a nine-hour telethon.

    a. How many schedules are possible if there are no restrictions on the order of the schedule?

    b. How many schedules are possible if two people came in the same car and would like to have consecutive hours?

64. **Car Rental** A rental car agency has 12 identical cars and seven identical vans.

    a. A group taking a field trip needs to rent six vehicles. In how many different ways is this possible?

    b. If the group needs to rent four cars and two vans, in how many different ways can they select their vehicles?

65. **Human Resources** The personnel director of a company must select four finalists from a group of 10 candidates for a job opening. How many different groups of four can the director select as finalists?

66. **Quality Control** A quality control inspector receives a shipment of 15 DVD players, from which the inspector must choose three for testing. How many different groups of three players can the inspector choose?

67. **Panel Discussion** Twelve executives, six women and six men, are to be seated in chairs for a panel discussion.

    a. How many arrangements are possible if the men and women are to alternate seats, beginning with a woman?

    b. How many arrangements are possible if the men must all be seated together?

68. **Student Committees** A committee of 16 students must select a president, a vice president, a secretary, and a treasurer. How many possible ways can this be accomplished?

69. **Library Science** Ten volumes of an encyclopedia are arranged on a bookshelf. How many different arrangements are possible?

70. **Committee Selection** A committee of six students is chosen from eight juniors and eight seniors.

    a. How many different committees are possible?

    b. How many different committees are possible that include three juniors and three seniors?

71. **Reading List** A student must read three of seven books for an English class. How many different selections can the student make?

72. **True-False Quiz** A quiz consists of 10 true-false questions.

    a. In how many distinct ways can the quiz be completed if no answers are left blank?

    b. In how many ways can the quiz be completed if five questions must be marked true and the other five must be marked false?

73. **Testing** A professor gives students 15 possible essay questions for an upcoming test. If the professor chooses three of these questions for an exam, how many different exams are possible?

74. **Olympic Events** A gold medal, a silver medal, and a bronze medal are awarded in an Olympic event. In how many possible ways can the medals be awarded in a 200-meter sprint in which there are nine runners?

75. Swim Team Twelve athletes are on a college swim team.

  a. In how many ways can the coach select eight swimmers for an upcoming race?

  b. In how many ways can the coach choose four of the swimmers to compete in a relay race in which order is important?

76. Softball Teams Eighteen people decide to play softball. In how many ways can the 18 people be divided into two teams of nine people?

77. Bowling Teams Fifteen people decide to join a bowling league. In how many ways can the 15 people be divided into three teams of five people each?

78. Computer Password The password for a computer system consists of three letters followed by three numbers (0 through 9). If a letter or number cannot be repeated, how many different passwords are possible?

Extensions

CRITICAL THINKING

79. Computer Security WPA-PSK, a wireless network security system, suggests that users choose a password, at least eight characters long, that is a random selection of the letters a through z and the numbers 0 through 9, where repetition of a letter or number is allowed.

  a. If a person chooses an eight-character password, how many different passwords are possible?

  b. If a person chooses a nine-character password, how many different passwords are possible?

  c. Computer hackers use programs to try to gain access to computer networks by finding users' passwords. Suppose such a program could check 1000 possible passwords each second (current personal computers are not that fast). How long, to the nearest year, would it take such a computer to try all possible passwords that are eight characters long? Assume that there are 31,536,000 seconds in 1 year.

  d. How long, to the nearest year, would it take the computer in part c to try all possible passwords that are nine characters long?

80. Concert Seating Six friends, three women and three men, go to a concert and have tickets for six consecutive seats.

  a. How many seating arrangements are possible if each person can choose any of the six seats?

  b. How many seating arrangements are possible if women and men alternate in the seats?

  c. How many seating arrangements are possible if the women sit together and the men sit together?

  d. Suppose Jocelyn will not sit next to Jonathan. How many seating arrangements are possible?

81. DNA Structure One strand of a DNA (deoxyribonucleic acid) molecule consists of sequences of elements among which are four bases: adenine, cytosine, guanine, and thymine. Suppose a portion of a DNA strand contains 10 of these bases.

  a. How many different sequences of these bases are possible for this strand? Assume that there are no restrictions on the bases.

  b. How many different sequences of these bases are possible for this strand if the strand contains three adenine, two cytosine, four guanine, and one thymine?

EXPLORATIONS

Internet Addresses Computers connected to the Internet communicate using unique "IP addresses." For the current address scheme, called IPv4, each address is of the form X.X.X.X, where X represents any one-, two-, or three-digit number from 0 through 255. (For instance, the website for the White House, www.whitehouse.gov, corresponds to the IP address 198.137.240.91.) Because of an increased demand for Internet addresses, a new address system called IPv6 is being gradually introduced.

82. How many unique IPv4 Internet addresses are possible?

83. Research why IPv6 is necessary and how it differs from IPv4. What are some of the benefits of this new system?

84. How many unique IPv6 Internet addresses are possible?

85. How many times more IPv6 addresses are there than IPv4 addresses?

86. The surface area of Earth is approximately 5.5×10^{15} square feet. How many IPv6 addresses could be allocated to each square foot of Earth?

| # Introduction to Probability

Sample Spaces and Events

In California, the likelihood of someone selecting the winning lottery numbers in the "Super Lotto Plus" game is approximately 1 in 41,000,000. In contrast, the likelihood of being struck by lightning is about 1 chance in 1,000,000. Comparing the likelihood of winning the Super Lotto Plus with that of being struck by lightning indicates that you are 41 times more likely to be struck by lightning than to pick the winning California lottery numbers.

The likelihood of a certain event occuring is described by a number between 0 and 1. (The number can also be expressed as a percent.) This number is called the **probability** of the event. An event that is not very likely has a probability close to 0; an event that is very likely has a probability close to 1 (100%). For instance, the probability of being struck by lightning is close to 0. However, if you randomly choose a basketball player from the National Basketball Association, the probability that the player is over 6 feet tall is very likely and therefore close to 1. Because any event has between a 0% and a 100% chance of occuring, probabilities are always between 0 and 1, inclusive.

Probabilities are calculated by considering **experiments,** which are activities with observable outcomes. Here are some examples of experiments:

- Flip a coin and observe the outcome as heads or tails.

- Select a company and observe its annual profit.

- Record the time a person spends at the checkout line in a supermarket.

The **sample space** of an experiment is the set of all possible outcomes of the experiment. For example, consider tossing a coin three times and observing the outcome as heads or tails. Using H for heads and T for tails, the sample space is

$$S = \{HHH, HHT, HTH, HTT, THH, THT, TTH, TTT\}$$

Note that the sample space consists of *every* possible outcome of tossing the coin three times.

EXAMPLE 1 ■ **Find a Sample Space**

A single die is rolled once. What is the sample space for this experiment?

Solution
The sample space is all possible outcomes of the experiment.

$$S = \{\boxdot, \boxdot, \boxdot, \boxdot, \boxdot, \boxdot\}$$

CHECK YOUR PROGRESS 1   A coin is tossed twice. What is the sample space for this experiment?

Solution   *See page S31.*

An **event** is a subset of a sample space. Using the sample space of Example 1, here are some possible events:

- There are an even number of pips (dots) on the upward face. The event is

  $E_1 = \{\boxdot, \boxdot, \boxdot\}.$

- The number of pips on the upward face is greater than 4. The event is

  $E_2 = \{\boxdot, \boxdot\}.$

- The number of pips on the upward face is less than 20. The event is

  $E_3 = \{\boxdot, \boxdot, \boxdot, \boxdot, \boxdot, \boxdot\}.$

  Because the number of pips on the upward face is always less than 20, this event will always occur. The event and the sample space are the same.

- The number of pips on the upward face is greater than 15. The event is $E_4 = \varnothing$, the empty set. This is an impossible event, because it is not possible for the number facing up to be greater than 15.

Outcomes of some experiments are *equally likely*, which means that the chance of any one outcome is just as likely as the chance of any other. For instance, if four balls of the same size but different colors, red, blue, green, and white, are placed in a box and a blindfolded person chooses one ball, the chance of choosing a blue ball is the same as the chance of choosing a ball of any other color.

In the case of equally likely outcomes, the probability of an event is based on the number of elements in the event and the number of elements in the sample space.

> ✔ **TAKE NOTE**
>
> As an example of an experiment whose outcomes are not equally likely, consider tossing a thumbtack and recording whether it lands with the point up or on its side. There are only two possible outcomes for this experiment, but the outcomes are not equally likely.

Basic Probability Formula

For an experiment with sample space S of *equally likely outcomes*, the probability $P(E)$ of an event E is given by

$$P(E) = \frac{n(E)}{n(S)}$$

where $n(E)$ is the number of elements in the event and $n(S)$ is the number of elements in the sample space.

Because each outcome from rolling a fair die is equally likely, the probability of the events E_1 through E_4 described above can be determined from the basic probability formula.

$$P(E_1) = \frac{3}{6} \longleftarrow \text{Number of elements in } E_1 \\ \frac{}{} \longleftarrow \text{Number of elements in the sample space}$$

$$= \frac{1}{2}$$

The probability of rolling an even number of pips on a single roll of one die is $\frac{1}{2}$ (or 50%).

$$P(E_2) = \frac{2 \longleftarrow \text{Number of elements in } E_2}{6 \longleftarrow \text{Number of elements in the sample space}}$$

$$= \frac{1}{3}$$

The probability of rolling a number greater than 4 on a single roll of one die is $\frac{1}{3}$.

$$P(E_3) = \frac{6 \longleftarrow \text{Number of elements in } E_3}{6 \longleftarrow \text{Number of elements in the sample space}}$$

$$= 1$$

The probability of rolling a number less than 20 on a single roll of one die is 1 (or 100%). The probability of any event that is *certain* to occur is 1.

$$P(E_4) = \frac{0 \longleftarrow \text{Number of elements in } E_4}{6 \longleftarrow \text{Number of elements in the sample space}}$$

$$= 0$$

The probability of rolling a number greater than 15 on a single roll of one die is 0.

EXAMPLE 2 ■ Find a Probability Involving Tossing Coins

A fair coin, one for which it is equally likely that heads or tails will result from a single toss of the coin, is tossed three times. What is the probability that two heads and one tail are tossed?

Solution

Determine the number of elements in the sample space. This must include every possible toss of heads or tails (in order) in three tosses of a coin.

$$S = \{\text{HHH, HHT, HTH, HTT, THH, THT, TTH, TTT}\}$$

The elements in the event are: $E = \{\text{HHT, HTH, THH}\}$.

$$P(E) = \frac{n(E)}{n(S)} = \frac{3}{8}$$

The probability is $\frac{3}{8}$.

CHECK YOUR PROGRESS 2   If a fair die is rolled once, what is the probability that an odd number will appear on the upward face?

Solution   See page S31.

QUESTION   *Is it possible that the probability of some event could be 1.23?*

ANSWER   *No. All probabilities must be between 0 and 1, inclusive.*

Figure 8.1 *Outcomes from Rolling Two Dice*

EXAMPLE 3 ■ Find a Probability Involving Rolling Dice

Two fair dice are tossed once. What is the probability that the sum of the pips on the upward faces of the two dice equals 8?

Solution

The 36 possible outcomes of rolling two dice are shown in Figure 8.1. Therefore, $n(S) = 36$. Let E represent the event that the sum of the pips on the upward faces is 8. These outcomes are circled in Figure 8.1. By counting the number of circled pairs, $n(E) = 5$.

$$P(E) = \frac{n(E)}{n(S)} = \frac{5}{36}$$

The probability that the sum of the pips is 8 on the roll of two dice is $\frac{5}{36}$.

CHECK YOUR PROGRESS 3   Two fair dice are tossed once. What is the probability that the sum of the pips on the upward faces of the two dice equals 7?

Solution   *See page S31.*

EXAMPLE 4 ■ Find a Probability Involving Playing Cards

A five-card hand is dealt from a standard 52-card deck of playing cards. What is the probability that the hand has five cards of the same suit?

Solution

Let E be the event of drawing five cards of the same suit. In Example 8 in Section 8.1, we determined that the total number of outcomes for E is $4 \cdot C(13, 5) = 5148$, so $n(E) = 5148$. The sample space is the total number of different five-card hands, which is given by $C(52, 5) = 2{,}598{,}960$. Thus $n(S) = 2{,}598{,}960$. The probability of E is

$$P(E) = \frac{n(E)}{n(S)} = \frac{5148}{2{,}598{,}960} \approx 0.00198$$

There is only about a 0.2% chance of being dealt five cards of the same suit.

CHECK YOUR PROGRESS 4   A committee of five professors is chosen from five mathematicians and six economists. If the members are chosen randomly, find the probability that the committee will consist of two mathematicians and three economists.

Solution   *See page S31.*

A probability such as those calculated in the preceding examples is sometimes referred to as a *theoretical* probability. We assume that, in theory, we have a perfectly balanced coin in Example 2 and we calculate the probability based on the fact that each event is equally likely. Similarly, we assume that the dice in Example 3 are equally likely to land with any of the six faces upward, and that the deck of cards in Example 4 is well shuffled so that every card has an equal chance of being dealt.

When a probability is based on data gathered from an experiment, it is called an **experimental probability** or an **empirical probability.** For instance, if we tossed a thumbtack 100 times and recorded the number of times it landed "point up," the results might be as shown in the table at the left. From this, the empirical probability of "point up" is

Point up	15
Side	85
Total	100

$$P(\text{point up}) = \frac{15}{100} = 0.15$$

EXAMPLE 5 ■ Empirical Probability

A survey of the registrar of voters office in a city showed the following information on the ages and party affiliations of registered voters. If one voter is chosen from this survey, what is the probability that the voter is a Republican? Round to the nearest hundredth.

Age	Republican	Democrat	Independent	Other	Total
18–28	205	432	98	112	847
29–38	311	301	109	83	804
39–49	250	251	150	122	773
50+	272	283	142	107	804
Total	1038	1267	499	424	3228

Solution

Let R be the event that a Republican is selected. Then

$$P(R) = \frac{1038}{3228} \quad \longleftarrow \quad \text{Number of Republicans in the survey}$$
$$\phantom{P(R) = \frac{1038}{3228}} \quad \longleftarrow \quad \text{Total number of people surveyed}$$
$$\approx 0.32$$

The probability that the selected person is a Republican is 0.32.

CHECK YOUR PROGRESS 5   Using the data from Example 5, what is the probability that a randomly selected person is between the ages of 39 and 49? Round to the nearest hundredth.

Solution   *See page S31.*

Sometimes the chances of an event occurring are given in terms of *odds.*

Odds in Favor

The **odds in favor** of an event is the ratio of the number of favorable outcomes of an experiment to the number of unfavorable outcomes.

For instance, in Example 2, to find the odds in favor of two heads and one tail when a coin is tossed three times, determine the favorable events, $E = \{\text{HHT, HTH, THH}\}$ and the unfavorable events, $U = \{\text{HHH, HTT, THT, TTH, TTT}\}$. Then

$$\text{Odds in favor of two heads and one tail} = \frac{n(E)}{n(U)} = \frac{3}{5}$$

This is read "the odds in favor of two heads and one tail is 3 to 5." It is important to note that $\frac{3}{5}$ is *not* a probability. It is the ratio of the number of favorable events to the number of unfavorable events.

Given the odds in favor of an event, the probability of the event can be calculated as

$$P(E) = \frac{n(E)}{n(E) + n(U)}$$

where $n(E)$ is the number of favorable events and $n(F)$ is the number of unfavorable events.

The *odds against* an event is the ratio of the number of unfavorable outcomes of an experiment to the number of favorable outcomes.

> **Odds Against**
>
> The **odds against** an event is the ratio of the number of unfavorable outcomes of an experiment to the number of favorable outcomes.

For instance, in horse racing, the odds against a horse winning a race are posted. Odds posted as 7 to 2 mean that in $9(7 + 2)$ races, the horse is estimated to lose 7 times and win 2 times.

EXAMPLE 6 ■ Calculate Odds

A charity sells 100 raffle tickets. There is one grand prize and four smaller prizes. What are the odds in favor of winning a prize?

Solution
There are five favorable outcomes and 95 unfavorable outcomes. The odds in favor of winning a prize are $\frac{5}{95} = \frac{1}{19}$. The odds in favor of winning a prize are 1 to 19.

CHECK YOUR PROGRESS 6   What are the odds against rolling a sum of 7 when two dice are tossed once?

Solution   *See page S32.*

The Addition Rule for Probabilities

Suppose we draw a single card from a standard deck of playing cards. The sample space S is the 52 cards of the deck. Therefore, $n(S) = 52$. Now consider the events

E_1 = A 4 is drawn = {♠4, ♥4, ♦4, ♣4}

E_2 = A spade is drawn

= {♠A, ♠2, ♠3, ♠4, ♠5, ♠6, ♠7, ♠8, ♠9, ♠10, ♠J, ♠Q, ♠K}

It is possible, on one draw, to satisfy the conditions of both events: The ♠4 could be drawn. This card is an element of both E_1 and E_2.

Now compare the events

$E_3 =$ A 5 is drawn $= \{\spadesuit 5,\ \heartsuit 5,\ \diamondsuit 5,\ \clubsuit 5\}$

$E_4 =$ A king is drawn $= \{\spadesuit K,\ \heartsuit K,\ \diamondsuit K,\ \clubsuit K\}$

In this case, it is not possible to draw one card that satisfies the conditions of both events. Two events that cannot both occur at the same time are called **mutually exclusive events.** Thus, the events E_3 and E_4 are mutually exclusive events, whereas E_1 and E_2 are not.

Mutually Exclusive Events

Two events A and B are **mutually exclusive** if they cannot occur at the same time. That is, A and B are mutually exclusive when $A \cap B = \varnothing$.

The probability of either of two mutually exclusive events occurring can be determined by adding the probabilities of the individual events.

Probability of Mutually Exclusive Events

If A and B are two mutually exclusive events, then the probability of A or B written $P(A \cup B)$, occurring is given by $P(A \cup B) = P(A) + P(B)$.

EXAMPLE 7 ▪ Find a Probability Involving Playing Cards

Suppose a single card is drawn from a standard deck of playing cards. Find the probability of drawing a 5 or a king.

Solution
Let $A = \{\clubsuit 5, \spadesuit 5, \heartsuit 5, \diamondsuit 5\}$ and $B = \{\clubsuit K, \spadesuit K, \heartsuit K, \diamondsuit K\}$. There are 52 cards in a standard deck of playing cards; thus $n(S) = 52$. Because the events are mutually exclusive, we can use the formula for the probability of mutually exclusive events.

$P(A \cup B) = P(A) + P(B)$ • Addition formula for the probability of mutually exclusive events

$= \dfrac{1}{13} + \dfrac{1}{13} = \dfrac{2}{13}$ • $P(A) = \dfrac{4}{52} = \dfrac{1}{13},\ P(B) = \dfrac{4}{52} = \dfrac{1}{13}$

The probability of drawing a 5 or a king is $\frac{2}{13}$.

CHECK YOUR PROGRESS 7   Two fair dice are tossed once. What is the probability of rolling a 7 or an 11? The sample space for this experiment is given on page 510.

Solution   *See page S32.*

Consider the experiment of rolling two dice. Let A be the event of rolling a sum of 8 and let B be the event of rolling a double (the same number on both dice).

$A = \{\boxed{}\boxed{},\ \boxed{}\boxed{},\ \boxed{}\boxed{},\ \boxed{}\boxed{},\ \boxed{}\boxed{}\}$

$B = \{\boxed{}\boxed{},\ \boxed{}\boxed{},\ \boxed{}\boxed{},\ \boxed{}\boxed{},\ \boxed{}\boxed{},\ \boxed{}\boxed{}\}$

✓ **TAKE NOTE**

The $P(A \cap B)$ term in the addition rule for probabilities is subtracted to compensate for the overcounting of the first two terms of the formula. If two events are mutually exclusive, then $A \cap B = \varnothing$. Therefore, $n(A \cap B) = 0$ and $P(A \cap B) = \frac{n(A \cap B)}{n(S)} = 0$. Thus, for mutually exclusive events, the addition rule for probabilities is the same as the formula for the probability of mutually exclusive events.

These events are *not* mutually exclusive because it is possible to satisfy the conditions of each event on one toss of the dice—a 🎲 🎲 could be rolled. Therefore, $P(A \cup B)$, the probability of a sum of 8 or a double, cannot be calculated using the formula for the probability of mutually exclusive events given above. However, a modification of that formula can be used.

Addition Rule for Probabilities

Let A and B be two events in a sample space S. Then

$$P(A \cup B) = P(A) + P(B) - P(A \cap B)$$

Using this formula with

$$A = \{\text{⚃⚁}, \text{⚂⚃}, \text{⚃⚃}, \text{⚄⚂}, \text{⚅⚁}\}$$
$$B = \{\text{⚀⚀}, \text{⚁⚁}, \text{⚂⚂}, \text{⚃⚃}, \text{⚄⚄}, \text{⚅⚅}\}$$
$$A \cap B = \{\text{⚃⚃}\}$$

the probability of A or B can be calculated.

$$P(A \cup B) = P(A) + P(B) - P(A \cap B)$$
$$= \frac{5}{36} + \frac{6}{36} - \frac{1}{36} \qquad \bullet\ P(A) = \frac{5}{36},\ P(B) = \frac{6}{36},\ P(A \cap B) = \frac{1}{36}$$
$$= \frac{10}{36} = \frac{5}{18}$$

On a single roll of two dice, the probability of rolling a sum of 8 or a double is $\frac{5}{18}$.

Conditional Probability

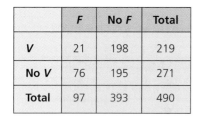

V: Vaccinated
F: Contracted the flu

	F	No F	Total
V	21	198	219
No V	76	195	271
Total	97	393	490

The table at the left shows data from an experiment for testing the effectiveness of a flu vaccine. From the table, we can calculate the probability that one person randomly selected from this population will have the flu.

$$P(F) = \frac{n(F)}{n(S)} = \frac{97}{490} \approx 0.198 \qquad \bullet\ n(F) = 97,\ n(S) = 490$$

Now consider a slightly different situation. We can ask, "What is the probability that a person randomly chosen from this population will contract the flu *given* that the person has received the flu vaccination?"

In this case we know that the person has received a flu vaccination and we want to determine the probability that someone from that group will contract the flu. Therefore, the only part of the table that is of concern is the top row. In this case, we have

$$P(F \text{ given } V) = \frac{21}{219} \approx 0.096$$

Thus the probability that someone will contract the flu given that the person has been vaccinated is 0.096.

The probability of an event B occurring based on knowing that event A has already occurred is called a *conditional probability*, and is denoted by $P(B\,|\,A)$.

Conditional Probability Formula

Let A and B be two events in a sample space S. Then the conditional probability of B given that A has occurred is

$$P(B \mid A) = \frac{P(A \cap B)}{P(A)}$$

The symbol $P(B \mid A)$ is read "the probability of B given A."

To see how this formula applies to the flu data given previously, let $S = \{$all people participating in the test$\}$, let $F = \{$people contracting the flu$\}$, and let $V = \{$people who were vaccinated$\}$.

Then $F \cap V = \{$people who contracted the flu *and* were vaccinated$\}$.

$$P(F \mid V) = \frac{P(F \cap V)}{P(V)} = \frac{\dfrac{21}{490}}{\dfrac{219}{490}}$$

- $P(F \cap V) = \dfrac{n(F \cap V)}{n(S)} = \dfrac{21}{490}$

- $P(V) = \dfrac{n(V)}{n(S)} = \dfrac{219}{490}$

$$= \frac{21}{219} \approx 0.096$$

The probability that a person selected from this population will contract the flu given that the person has received the vaccination, $P(F \mid V)$, is 0.096. Our answer agrees with the calculation we performed directly from the table, but the conditional probability formula allows us to calculate conditional probabilities even when we cannot compute them directly.

EXAMPLE 8 ■ Find a Conditional Probability

The data in the table below show the results of a survey for determining the numbers of adults who have had financial help from their parents for certain purchases.

Age	College Tuition	Buy a Car	Buy a House	Total
18–28	405	253	261	919
29–39	389	219	392	1000
40–49	291	146	245	682
50–59	150	71	112	333
60+	62	15	98	175
Total	1297	704	1108	3109

If one person is selected from this survey, what is the probability that the person received financial help for purchasing a home given that the person is between the ages of 29 and 39?

Solution

Let $B = \{$adults receiving financial help for a home purchase$\}$ and
$A = \{$adults between 29 and 39$\}$. From the table, $n(A \cap B) = 392$, $n(A) = 1000$,
and $n(S) = 3109$. Using the conditional probability formula, we have

$$P(B\,|\,A) = \frac{P(A \cap B)}{P(A)} = \frac{\dfrac{392}{3109}}{\dfrac{1000}{3109}}$$

- $P(A \cap B) = \dfrac{n(A \cap B)}{n(S)} = \dfrac{392}{3109}$

- $P(A) = \dfrac{n(A)}{n(S)} = \dfrac{1000}{3109}$

$$= \frac{392}{1000} = 0.392$$

The conditional probability of B given A is 0.392.

CHECK YOUR PROGRESS 8   A pair of dice are tossed once. What is the probability that the result is a sum of 6 given that the result is not a sum of 7?

Solution   See page S32.

Suppose two cards are drawn, without replacement, from a standard deck of playing cards. Let A be the event that an ace is drawn on the first draw and let B be the event that an ace is drawn on the second draw. Then the probability that an ace is drawn on the first *and* second draws is $P(A$ and $B)$ and is written $P(A \cap B)$. To find this probability, we can solve the conditional probability formula for $P(A \cap B)$. The result is called the product rule for probabilities.

Product Rule for Probabilities

If A and B are two events from the sample space S, then

$$P(A \cap B) = P(A) \cdot P(B\,|\,A)$$

For the problem of drawing an ace from a standard deck of playing cards on the first and second draws, $P(A \cap B)$ is the product of $P(A)$, the probability that the first drawn card is an ace, and $P(A\,|\,B)$, the probability of an ace on the second draw *given* that the first card drawn was an ace.

The tree diagram at the left shows the possible outcomes of drawing two cards from a deck. On the first draw, there are four aces in the deck of 52 cards. Therefore, $P(A) = \frac{4}{52} = \frac{1}{13}$. On the second draw, there are only 51 cards remaining and only three aces (an ace was drawn on the first draw). Therefore, $P(B\,|\,A) = \frac{3}{51} = \frac{1}{17}$. Putting these calculations together, we have

$$P(A \cap B) = P(A) \cdot P(B\,|\,A)$$

$$= \frac{1}{13} \cdot \frac{1}{17} = \frac{1}{221}$$

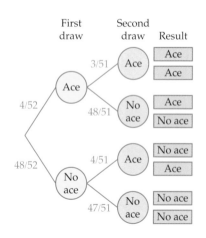

The probability of drawing an ace on both the first and second draws is $\frac{1}{221}$.

The product rule for probabilities can be extended to more than two events. Basically the idea is that the probability that a certain sequence of events will occur in succession is the product of the probabilities of each of the events *given* that the preceding events have occurred.

> **Probability of Successive Events**
>
> The probability of two or more events occurring in succession is the product of the conditional probabilities of the individual events.

EXAMPLE 9 ■ Find the Probability of Successive Events

A box contains four red, three white, and five green balls. Suppose three balls are randomly selected from the box in succession without replacement.

a. What is the probability that first a red, then a white, and then a green ball are selected?

b. What is the probability that two white balls followed by one green ball are selected?

Solution

a. Let $A = \{$a red ball is selected first$\}$, let $B = \{$a white ball is selected second$\}$, and let $C = \{$a green ball is selected third$\}$. Then

$$P(A \cap B \cap C) = P(A) \cdot P(B|A) \cdot P(C|A \cap B)$$
$$= \frac{4}{12} \cdot \frac{3}{11} \cdot \frac{5}{10} = \frac{1}{22}$$

The probability of choosing a red, then a white, then a green ball is $\frac{1}{22}$.

b. Let $A = \{$a white ball is selected first$\}$, let $B = \{$a white ball is selected second$\}$, and let $C = \{$a green ball is selected third$\}$. Then

$$P(A \cap B \cap C) = P(A) \cdot P(B|A) \cdot P(C|A \cap B)$$
$$= \frac{3}{12} \cdot \frac{2}{11} \cdot \frac{5}{10} = \frac{1}{44}$$

The probability of choosing two white balls followed by one green ball is $\frac{1}{44}$.

CHECK YOUR PROGRESS 9 A standard deck of playing cards is shuffled, and three cards are dealt. Find the probability that the cards dealt are a spade followed by a heart followed by another spade.

Solution See page S32.

TAKE NOTE

In part (a), there are originally 12 balls in the box. After a red ball is selected, there are only 11 balls remaining, of which three are white. Thus $P(B|A) = \frac{3}{11}$. After a red and a white ball are selected, there are 10 balls left, of which five are green. Thus $P(C|A \cap B) = \frac{5}{10}$.

In part (b), we have a similar situation. However, after a white ball is selected, there are 11 balls remaining, of which only two are white. Therefore, $P(B|A) = \frac{2}{11}$.

Independent Events

Earlier in this section we considered the probability of drawing two aces from a standard deck of playing cards. Because the cards were drawn without replacement, the probability of an ace on the second draw *depended* on the result of the first draw.

Now consider the case of tossing a coin twice. The outcome of the first coin toss has no effect on the outcome of the second toss. So the probability of the coin flipping to a head or tail on the second toss is not affected by the result of the first toss. When the outcome of a first event does not affect the outcome of a second event, the events are called *independent*. In symbols, if A is the first event and B is the second, this can be written $P(B \mid A) = P(B)$.

Independent Events

If A and B are two events in a sample space and $P(B \mid A) = P(B)$, then A and B are called **independent events.**

As a mathematical verification, consider tossing a coin twice. We can compute the probability that the second toss comes up heads, given that the first coin toss came up heads. If A is the event of a head on the first toss, then $A = \{HH, HT\}$. Let B be the event of a head on the second toss. Then $B = \{HH, TH\}$. The sample space is $S = \{HH, HT, TH, TT\}$. The conditional probability $P(B \mid A)$ (the probability of a head on the second toss given a head on the first toss) is

$$P(B \mid A) = \frac{P(A \cap B)}{P(A)} = \frac{\dfrac{1}{4}}{\dfrac{1}{2}} = \frac{1}{2} \qquad \bullet\ P(A \cap B) = \frac{n(A \cap B)}{n(S)} = \frac{1}{4}.$$

$$\bullet\ P(A) = \frac{n(A)}{n(S)} = \frac{2}{4} = \frac{1}{2}$$

Thus $P(B \mid A) = \frac{1}{2}$. Note, however, that $P(B) = \frac{n(B)}{n(S)} = \frac{2}{4} = \frac{1}{2}$. Therefore, $P(B \mid A) = P(B)$, and the events are independent.

In general, this allows us to simplify the product rule when two events are independent; the probability of two independent events occurring in succession is simply the product of the probabilities of the individual events.

✔ **TAKE NOTE**

The formula for the probability of independent events can be extended to more than two events. If E_1, E_2, E_3, and E_4 are independent events, then the probability that all four events will occur is $P(E_1) \cdot P(E_2) \cdot P(E_3) \cdot P(E_4)$.

Product Rule for Independent Events

If A and B are two independent events from the sample space S, then $P(A \cap B) = P(A) \cdot P(B)$.

✔ **TAKE NOTE**

See page 495 for all the possible outcomes of rolling two dice.

EXAMPLE 10 ■ **Probability of Independent Events**

A pair of dice are tossed twice. What is the probability that the first roll is a sum of 7 and the second roll is a sum of 11?

Solution

The rolls of a pair of dice are independent; the probability of a sum of 11 on the second roll does not depend on the outcome of the first roll. Let $A = \{$sum of 7 on the first roll$\}$ and let $B = \{$sum of 11 on the second roll$\}$. Then

$$P(A \cap B) = P(A) \cdot P(B) = \frac{6}{36} \cdot \frac{2}{36} = \frac{1}{108}$$

CHECK YOUR PROGRESS 10   A coin is tossed three times. What is the probability that heads appears on all three tosses?

Solution   *See page S32.*

Conditional probability is used in many real-world situations, such as determining the efficacy of a drug test, verifying the accuracy of genetic testing, and analyzing evidence in legal proceedings.

EXAMPLE 11 ■ Drug Testing and Conditional Probability

Suppose that a company claims that it has a test that is 95% effective in determining whether an athlete is using a steroid. That is, if an athlete is using a steroid, the test will be positive 95% of the time. In the case of a negative result, the company says its test is 97% accurate. That is, even if an athlete is not using steroids, it is possible that the test will be positive in 3% of the cases. Suppose this test is given to a group of athletes in which 10% of the athletes are using steroids. What is the probability that a randomly chosen athlete actually uses steroids given that the athlete's test is positive?

Solution

Let S be the event that an athlete uses steroids and let T be the event that the test is positive. Then the probability we wish to determine is $P(S \mid T)$. Using the conditional probability formula, we have

$$P(S \mid T) = \frac{P(S \cap T)}{P(T)}$$

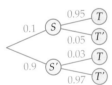

A tree diagram, shown at the left, can be used to calculate this probability. A positive test result can occur in two ways: either an athlete using steroids correctly tests positive, or an athlete not using steroids incorrectly tests positive. The probability of a positive test result, $P(T)$, corresponds to an athlete following path ST or path $S'T$ in the tree diagram. (S' symbolizes no steroid use.) $P(S \cap T)$, the probability of using steroids and a positive test, is path ST. Thus,

$$P(S \mid T) = \frac{P(S \cap T)}{P(T)}$$

$$= \frac{0.1 \cdot 0.95}{0.1 \cdot 0.95 + 0.9 \cdot 0.03} \approx 0.779$$

Given that an athlete tests positive, the probability that the athlete actually uses steroids is approximately 77.9%.

CHECK YOUR PROGRESS 11   A pharmaceutical company has a test that is 95% effective in determining whether a person has a certain genetic defect. It is possible, however, that the test may give false positives in 4% of the cases. Suppose this particular genetic defect occurs in 2% of the population. Given that a person tests positive, what is the probability that the person actually has the defect?

Solution   *See page S32.*

Applications to Genetics

The Human Genome Project (a genome is an organism's complete set of DNA, deoxyribonucleic acid) was completed, to the limits of current technology, in April 2003. Researchers hope to use this information to treat and prevent certain hereditary diseases.

The concept behind this project began with Gregor Mendel and his work on flower color and how it was transmitted from generation to generation. From his studies, Mendel concluded that flower color seems to be predictable in future generations by making some assumptions of a plant's color "determiner." He concluded that red was a *dominant* determiner of color and that white was a *recessive* determiner. Today, geneticists talk about the *gene* for flower color and the *allele* of the gene, which determines whether the gene consists of two dominant alleles (two red), a dominant and a recessive allele (a red and a white), or two recessive alleles (two white). Since red is the dominant allele, the flower will be white only if no dominant allele is present.

Later work by Reginald Punnett (1875–1967) showed how to determine the probability of certain flower colors by using a **Punnett square.** Using a capital letter for a dominant allele (say, R for red) and the corresponding lowercase letter for the recessive allele (r for white), Punnett arranged the genes of the parent in a square. A parent could be RR, Rr, or rr.

Suppose that the genotypes (genetic compositions) of two parents are as shown below (one parent's genotype shown in the top row and the other parent's genotype shown in the left column). The genotypes of the offspring are shown in the body of the table and are the result of combining one allele from each parent.

Parents	R	r
r	Rr	rr
r	Rr	rr

Because each of the genotypes of the offspring is equally likely, the probability that a flower is red is 0.5 (two Rr genotypes of the four possible) and the probability that the flower is white is 0.5 (two rr genotypes of the four possible).

EXAMPLE 12 ■ Find a Probability Involving Genetics

A child will have cystic fibrosis if the child inherits the recessive gene from both parents. Using F for the normal allele and f for the mutant allele, suppose a parent who is Ff (said to be a *carrier*) and a parent who is FF (does not have the mutant allele) decide to have a child.

a. What is the probability that the child will have cystic fibrosis? To have the disease, the child must be ff.

b. What is the probability that the child will be a carrier?

Solution

Make a Punnett square.

Parents	F	F
F	FF	FF
f	Ff	Ff

a. To have the disease, the child must be ff. From the table, there is no combination of the alleles that will produce ff. Therefore, the child cannot have the disease, and the probability is 0.

b. To be a carrier, one allele must be f. From the table, there are two cases out of four in which the child will have one f. The probability that the child will be a carrier is $\frac{2}{4} = \frac{1}{2}$.

CHECK YOUR PROGRESS 12   For a certain type of hamster, the color cinnamon, C, is dominant, and white, c, is recessive. If both parents are Cc, what is the probability that an offspring will be white?

Solution   *See page S33.*

Investigation

Complement of an Event and Sharing Birthdays

Have you, at a party or some other social gathering, discovered that two or more people share the same birthday? It seems to be an amazing coincidence when it occurs, but how rare is it?

To analyze this question, we will compute the probability of the *complement*, E^c, of an event. $E^c = \{$elements of the sample space not in $E\}$. Basically, E^c is *not E*.

$P(E^c)$ and $P(E)$ are related by

$$P(E) + P(E^c) = 1 \quad \text{or} \quad P(E) = 1 - P(E^c)$$

For instance, if E is the event of rolling any number from 1 to 5 on a single roll of a die, then $E = \{$ ⚀ ⚁ ⚂ ⚃ ⚄ $\}$ and E^c (not E) is $E^c = \{$ ⚅ $\}$. By the equation above,

$$P(E) = 1 - P(E^c)$$
$$= 1 - \frac{1}{6} = \frac{5}{6}$$

(continued)

✔ TAKE NOTE

For the example at the right, it would have been just as easy to calculate $P(E)$ as

$$P(E) = \frac{n(E)}{n(S)} = \frac{5}{6}$$

The advantage of using the complement occurs when determining $n(E)$ is difficult, as it is in the birthday problem.

To begin the analysis of the birthday problem, suppose four people have gathered for a dinner party. (For simplicity, we will ignore February 29th in leap years.) Let E be the event that two or more people share a birthday. Then E^c (E complement) is the event that no one shares the same birthday.

To find $P(E^c)$, we start with one of the guests, whose birthday can be any day of the year. The second person cannot have that same birthday, so that person has 364 possible days for his or her birthday from the total of 365. Thus the conditional probability that the second guest has a different birthday, given that we know the first person's birthday, is $\frac{364}{365}$. Similarly, the third person has 363 possible birthdays that do not coincide with those of the first two guests, so the conditional probability that the third person does not share a birthday with either of the first two guests, given that we know the birthdays of the first two people, is $\frac{363}{365}$. The probability of the fourth guest having a distinct birthday is similarly $\frac{362}{365}$. We can use the product rule for probabilities to find the probability that all these conditions are met—that is, that none of the four share a birthday:

$$P(E^c) = \frac{364}{365} \cdot \frac{363}{365} \cdot \frac{362}{365} \approx 0.984$$

Then $P(E) = 1 - P(E^c) \approx 0.016$, so there is about a 1.6% chance that in a group of four people, two or more will have the same birthday.

Investigation Exercises

1. If eight people are present at a meeting, find the probability that two or more have a common birthday. Round to the nearest thousandth.

2. Compute the probability that two or more people in a group of 15 will have the same birthday. Round to the nearest thousandth.

3. If 23 people are in attendance at a party, what is the probability that two or more will share a birthday? Round to the nearest thousandth.

4. In a group of 40 people, estimate the probability that at least two people share a birthday. Compute the probability to check your guess. Round to the nearest thousandth.

Exercise Set 8.2

In Exercises 1–4, list the elements of the sample space of each experiment.

1. A coin is flipped three times.

2. An even number between 1 and 11 is selected at random.

3. An ace is pulled from a deck of playing cards.

4. A diamond is pulled from a deck of playing cards.

In Exercises 5–8, a fair coin is tossed four times. Compute the probability of each event occurring.

5. Two heads and two tails

6. One head and three tails

7. All tails

8. All four coin tosses show the same face.

In Exercises 9–14, two dice are rolled. Determine the probability of each of the following.

9. A sum of 11

10. A sum of 4

11. A sum of 6 or doubles

12. A sum of 7 or doubles

13. An odd-number sum or a sum less than 4

14. An even-number sum or a sum greater than 10

In Exercises 15–18, a single card is drawn from a deck. Find the probability of each of the following events.

15. An 8 or a spade

16. An ace or a red card

17. A jack or a face card

18. A spade or a red card

Salary Survey In Exercises 19–22, a random survey asked respondents their current annual salaries. The results are given in the table below.

Salary Range	Number Responding
Less than $18,000	24
$18,000–27,999	41
$28,000–35,999	52
$36,000–45,999	58
$46,000–59,999	43
$60,000–79,999	39
$80,000–99,999	22
$100,000 or more	14

If a respondent of the survey is selected at random, compute the probability of each of the following. (Round to two decimal places.)

19. The respondent earns between $36,000 and $46,000 annually.

20. The respondent earns between $60,000 and $80,000 per year.

21. The respondent earns at least $80,000 per year.

22. The respondent earns less than $36,000 annually.

23. Births in Alaska During a recent year in Alaska, 5238 boys and 4984 girls were born. If a newborn is selected at random from that year, what is the probability that the baby is a girl?

24. If a pair of dice are rolled once, what are the odds in favor of rolling a sum of 9?

25. If a card is randomly pulled from a standard deck of playing cards, what are the odds in favor of pulling a heart?

26. The odds against a certain football team winning its next game are 8 to 3. What is the probability that the team will win the game?

27. A contest is advertising that the odds against winning the first prize are 100 to 1. What is the probability of winning first prize?

Education Level In Exercises 28–31, a random survey asked 850 respondents their highest levels of complete education. The results are given in the table below.

Education Completed	Number Responding
No high school diploma	52
High school diploma	234
Associate's degree	274
Bachelor's degree	187
Master's degree	67
Ph.D. or professional degree	36

If a respondent of the survey is selected at random, compute the probability of each of the following.

28. The respondent did not complete high school.

29. The respondent has an associate's degree or two years of college (but not more).

30. The respondent has a Ph.D. or professional degree.

31. The respondent has a degree beyond a bachelor's degree.

32. Suppose A and B are two events with $P(A) = 0.6$ and $P(B) = 0.5$. Based on these probabilities, a person calculated that the probability of A or B is $P(A \cup B) = P(A) + P(B) = 0.6 + 0.5 = 1.1$. Is this correct?

33. Suppose A and B are two events with $P(A) = 0.6$ and $P(B) = 0.5$. Based on these probabilities, a person calculated that the probability of A and B is $P(A \cap B) = P(A) \cdot P(B) = 0.6 \cdot 0.5 = 0.3$. Is this correct?

34. Monopoly In the game of Monopoly, a player rolls two dice. If the player rolls doubles three times in a row, the player goes to jail. What is the probability of rolling doubles three times in a row?

35. Monopoly In the game of Monopoly, a player rolls two dice. Noel is currently on a square five squares from Park Place. What is the probability that Noel will land on Park Place on either of her next two rolls?

Playing Cards In Exercises 36–39, a hand of five cards is dealt from a standard deck of playing cards.

36. Find the probability that the hand will contain all four aces.

37. Find the probability that the hand will contain three jacks and two queens.

38. Find the probability that the hand will contain exactly two 7s.

39. Find the probability that the hand will consist of the ace, king, queen, jack, and 10 of the same suit.

Roulette In Exercises 40–45 use the casino game roulette. Roulette is played by spinning a wheel with 38 numbered slots. The numbers 1 through 36 appear on the wheel, half of them colored black and half red, and two slots numbered 0 and 00 are colored green. A ball is placed on the spinning wheel and allowed to come to rest in one of the slots. Bets are placed on where the ball will land.

40. What is the probability the ball will stop in a black slot?

41. You can bet that the ball will land on an odd number. What is the probability of this occuring?

42. You can bet that the ball will land on any number from 1 to 12. What is the probability of this occuring?

43. You can bet that the ball will land on any particular number. What is the probability of this occuring?

44. You can bet that the ball will land on 0 or 00. What is the probability of this occuring?

45. You can bet that the ball will land on certain groups of six numbers (such as 1–6). What is the probability of this occuring?

Employment Status In Exercises 46–49, use the data in the table below, which show the employment status of individuals in a particular town by age group.

	Full-Time	Part-Time	Unemployed	
0–17	24	164	371	559
18–25	185	203	148	536
26–34	348	67	27	442
35–49	581	179	104	864
50+	443	162	173	778
	1581	775	823	3179

46. If a person in this town is selected at random, find the probability that the individual is employed part-time, given that the person selected is between the ages of 35 and 49.

47. If a person in the town is randomly selected, what is the probability that the individual is unemployed, given that the person selected is over 50 years old?

48. A person from the town is randomly selected; what is the probability that the individual is employed full-time, given that the person selected is between 18 and 49 years of age?

49. A person from the town is randomly selected; what is the probability that the individual is employed part-time, given that the person is at least 35 years old?

50. A pair of dice are tossed. Find the probability that the sum on the two dice is 8, given that the sum is even.

51. A pair of dice are tossed. Find the probability that the sum on the two dice is 12, given that doubles are rolled.

52. What is the probability of drawing two cards in succession (without replacement) from a deck and having them both be face cards?

53. Two cards are drawn from a deck without replacement, one after the other. Find the probability that both cards are hearts.

54. Two cards are drawn from a deck without replacement. What is the probability that the first card is a spade and the second card is red?

In Exercises 55–58, a snack-size bag of M&Ms candies is opened. Inside, there are 12 red candies, 12 blue, 7 green, 13 brown, 3 orange, and 10 yellow. Three candies are pulled from the bag in succession, without replacement.

55. Determine the probability that the first candy is blue, the second is red, and the third is green.

56. Determine the probability that the first candy is brown, the second is orange, and the third is yellow.

57. What is the probability that the first two candies are green and the third is red?

58. What is the probability that the first candy is orange, the second is blue, and the third is orange?

Playing Cards In Exercises 59–62, three cards are dealt from a shuffled deck of playing cards.

59. Find the probability that the first card is red, the second is black, and the third is red.

60. Find the probability that the first two cards are clubs and the third is a spade.

61. What is the probability that the three cards are, in order, an ace, a face card, and an 8?

62. What is the probability that the three cards are, in order, a red card, a club, and another red card?

School Attendance In Exercises 63–66, the probability that a student enrolled at a local high school will be absent today is 0.04, assuming that the student was in attendance the previous school day. However, if a student was absent the previous day, the probability that he or she will be absent today is 0.11.

63. What is the probability that a student will be absent exactly three days in a row?

64. What is the probability that a student will be absent two days in a row, but then show up the third day?

65. Find the probability that a student will be absent one day, will attend the next day, but then will be absent the third day.

66. Find the probability that a student will be absent precisely four days in a row.

In Exercises 67–70, determine whether or not the events are independent.

67. A single die is rolled and then rolled a second time.

68. Numbered balls are pulled from a bin one by one to determine the winning lottery numbers.

69. Numbers are written on slips of paper in a hat; one person pulls out a slip of paper without replacing it, then a second person pulls out a slip of paper.

70. In order to determine who goes first in a game, one person picks a number between 1 and 10, then a second person picks a number between 1 and 10.

71. Genetics The Punnett square for flower color below shows two parents, each of genotype Rr, where R corresponds to the dominant red flower allele and r represents the recessive white flower allele. (See Example 12.)

Parents	R	r
R	RR	Rr
r	Rr	rr

What is the probability that the offspring of these parents will have white flowers?

72. Genetics One parent plant, with red flowers, has genotype RR, and the other, with white flowers, has genotype rr, where R is the dominant allele for a red flower and r is the recessive allele for a white flower. Draw a Punnett square for these parents, and compute the probability of the offspring having white flowers.

73. Genetics The eye color of mice is determined by a dominant allele E, corresponding to black eyes, and a recessive allele e, corresponding to red eyes. If two mice have offspring, one parent of genotype EE and the other of type ee, draw a Punnett square for the offspring and compute the probability of a child having red eyes.

74. Genetics The height of a certain plant is determined by a dominant allele T, corresponding to tall plants, and a recessive allele t, for short (or dwarf) plants. If the parent plants both have genotype Tt, draw a Punnett square and compute the probability that the offspring plants will be tall.

75. Drug Testing A company that performs drug testing guarantees that its test determines a positive result with 97% accuracy. However, the test also gives 6% false positives. If 5% of those being tested actually have the drug present in their bloodstreams, find the probability that a person testing positive has actually been using drugs.

76. Gene Testing A test for a genetic disorder can detect the disorder with 94% accuracy. However, the test will incorrectly report a positive result for 3% of those without the disorder. If 12% of the population has the disorder, find the probability that a person testing positive actually does have the genetic disorder.

77. Disease Testing A pharmaceutical company has developed a test for a rare disease that is present in 0.5% of the population. The test is 98% accurate in determining a positive result, and the chance of a false positive is 4%. What is the probability that someone who tests positive actually has the disease?

78. Disease Testing When used together, the ELISA and Western Blot tests for HIV are more than 99% accurate in determining a positive result. If we assume a 99% accuracy rate for correctly identifying positive results and 0.00005 false positives, find the probability that someone who tests positive has HIV. (It is estimated that 0.6% of U.S. residents are infected with HIV.)

Extensions
CRITICAL THINKING

79. Suppose you are standing at a street corner and decide that you will flip a coin to decide whether you will go north or south from your current position. When you reach the next intersection, you repeat the procedure. (This problem is a simplified version of what is called a *random walk* problem. Problems of this type are important in economics, physics, chemistry, biology, and other disciplines.)

a. After performing this experiment three times, what is the probability that you will be three blocks north of your original position?

b. After performing this experiment four times, what is the probability that you will be two blocks north of your original position?

c. After performing this experiment four times, what is the probability that you will be back to your original position?

Powerball Lottery Exercises 80 and 81 use the Powerball Lottery game, in which a player selects five regular numbers from 1 through 55 and an additional red number from 1 to 42. The jackpot is awarded to anyone matching all six of these numbers to the numbers drawn.

80. Find the probability of matching all five regular numbers.

81. Find the probability of matching all five regular numbers and the red number.

California Lottery Exercises 82–85 use the California Lottery's game "Super Lotto Plus," in which a player selects five numbers from 1 through 47 and an additional "MEGA number" from 1 to 27. The jackpot, normally several million dollars, is awarded to anyone matching all six of these numbers to the numbers drawn.

82. Prize money (typically $10,000–$20,000) is also awarded if a player matches the first five numbers, but not the MEGA number. Compute the probability that this will occur.

83. Find the probability of matching all five numbers plus the MEGA number and winning the jackpot.

84. A prize of $10 is given for matching three of the five numbers, but not the MEGA number. What is the probability of winning the $10 prize?

85. Approximately $100 is awarded if three of the five numbers are matched along with the MEGA number. Compute the probability that this will occur.

COOPERATIVE LEARNING

86. From a standard deck of playing cards, choose four red cards and four black cards. Deal the eight cards, face up, in two rows of four cards each. A good event is that each column contains a red card and a black card. (The column can be red/black or black/red.) A bad event is any other situation. [Although this problem is stated in terms of cards, there is a very practical application. If several proteins in a cell break (say, from radiation therapy) and then reattach, it is possible that the new protein is harmful to the cell.]

a. Perform this experiment 50 times and keep a record of the numbers of good events and bad events.

b. From your data, what is the probability of a good event?

c. Repeat parts a and b again.

d. Combining the 100 trials you have run, what is the probability of a good event?

e. Calculate the theoretical probability of a good event.

EXPLORATIONS

87. Sicherman Dice George Sicherman proposed an alternative pair of dice that give the same sums, with the same probabilities, as a regular pair of dice. The pair is unusual in that the dice are not identical and the same number of pips can appear on more than one face. Sicherman dice have the following structure.

$$\text{Die 1:} \quad \{1, 3, 4, 5, 6, 8\}$$
$$\text{Die 2:} \quad \{1, 2, 2, 3, 3, 4\}$$

It has been shown that this pair of dice is the only possible pair that gives the same probabilities as a regular pair of dice. (See Martin Gardner, *Mathematical Games*, Scientific American, Vol. 238 No. 2, pp. 19–32.)

a. If the two Sicherman dice are rolled, list the sample space for the experiment. (*Note:* Two numbers appear twice on the second die, but they are considered separate results.)

b. Compute the probability of obtaining each possible sum from rolling the two dice, and verify that these are the same as the probabilities resulting from rolling a regular pair of dice.

c. Suppose the following pair of dice are rolled.

$$\text{Die 1:} \quad \{1, 2, 3, 4, 5, 6\}$$
$$\text{Die 2:} \quad \{0, 0, 0, 6, 6, 6\}$$

Show that the sum of the two faces can be any of the numbers 1 through 12, and that each occurs with equal probability.

88. **The Monty Hall Problem** A famous probability puzzle has its history in the game show "Let's Make a Deal," of which Monty Hall was the host, and goes something like the following. Suppose you are appearing on the show, and you are shown three closed doors. Behind one of the doors is a new car; behind the other two are goats. If you select the door hiding the car, you win the car. Clearly the probability of randomly choosing the correct door is 1/3. After you make a selection, the show's host, who knows where the car is, opens one of the other two doors and reveals a goat. He then asks if you would like to switch to another door. Should you stay with your original choice, or switch?

Most people at first would say that it makes no difference. However, computer simulations that play the game over and over have shown that you *should* switch. In fact, you will double your chances of winning the car if you give up your first choice. It can be proven mathematically that the probability of winning if you switch is 2/3. Search the Internet for the "Monty Hall problem" and write a short essay describing how to determine this probability, and why it is advantageous to switch doors.

SECTION 8.3 | Measures of Central Tendency

The Arithmetic Mean

Statistics involves the collection, organization, summarization, presentation, and interpretation of data. The branch of statistics that involves the collection, organization, summarization, and presentation of data is called **descriptive statistics.** The branch that interprets and draws conclusions from the data is called **inferential statistics.**

Statisticians often collect data from small portions of a large group in order to determine information about the group. For instance, to determine who will be elected as the next president of the United States, an organization may poll a small group of voters. From the information it obtains from the small group, it will make conjectures about the voting preferences of the entire group of voters. In such situations the entire group under consideration is known as the **population,** and any subset of the population is called a **sample.**

Because of practical restraints such as time and money, it is common to apply descriptive statistical procedures to a sample of a population and then to make use of inferential statistics to deduce conclusions about the population. Obviously, some samples are more representative of the population than others.

One of the most basic statistical concepts involves finding *measures of central tendency* of a set of numerical data. Here is a scenario in which it would be helpful to find numerical values that locate, in some sense, the *center* of a set of data. Elle is a senior at a university. In a few months she plans to graduate and start a career as a graphic artist. A sample of five graphic artists from her class shows that they have received job offers with the following yearly salaries.

$34,000 $30,500 $41,000 $32,500 $39,500

Before Elle interviews for a job, she wishes to determine an *average* of these five salaries. This average should be a "central" number around which the salaries cluster. We will consider three types of averages, known as the *arithmetic mean,* the *median,* and the *mode.* Each of these averages is a **measure of central tendency** for numerical data.

The *arithmetic mean* is the most commonly used measure of central tendency. The arithmetic mean of a set of numbers is often referred to as simply the *mean.* To find the mean for a set of data, find the sum of the data values and divide by the number of data values. For instance, to find the mean of the five salaries listed above, Elle would divide the sum of the salaries by 5.

$$\text{Mean} = \frac{\$34{,}000 + \$30{,}500 + \$41{,}000 + \$32{,}500 + \$39{,}500}{5}$$

$$= \frac{\$177{,}500}{5} = \$35{,}500$$

The mean suggests that Elle can expect a job offer at a salary of $35,500.

In statistics it is often necessary to find the sum of a set of numbers. The traditional symbol used to indicate a summation is the Greek letter *sigma,* Σ. Thus the notation Σx, called **summation notation,** denotes the sum of all the numbers in a given set. The use of summation notation enables us to define the mean as follows.

> **Mean**
>
> The **mean** of n numbers is the sum of the numbers divided by n.
>
> $$\text{Mean} = \frac{\Sigma x}{n}$$

It is traditional to denote the mean of a *sample* by $\bar{x}$ (which is read as "x bar") and to denote the mean of a *population* by the Greek letter μ (lower case *mu*).

EXAMPLE 1 ▪ Find a Mean

Six friends in a biology class received test grades of

　　92,　84,　65,　76,　88,　and　90

Find the mean of these test scores.

Solution

$$\bar{x} = \frac{\Sigma x}{n} = \frac{92 + 84 + 65 + 76 + 88 + 90}{6} = \frac{495}{6} = 82.5$$

The mean of these test scores is 82.5.

CHECK YOUR PROGRESS 1　Four separate blood tests revealed that a patient had total blood cholesterol levels of

　　245,　235,　220,　and　210

Find the mean of the blood cholesterol levels.

Solution　*See page S33.*

From a physical perspective, numerical data can be represented by weights on a seesaw. The mean of the data is represented by the balance point of the seesaw. For instance, the mean of 1, 3, 5, 5, 5, and 8 is 4.5. If equal weights are placed at the locations 1, 3, and 8 and three weights are placed at 5, then the seesaw will balance at 4.5.

A Physical Interpretation of the Mean

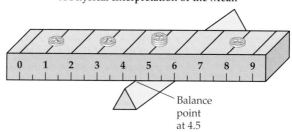

Balance point at 4.5

The Median

Another type of average is the *median*. Essentially, the median is the *middle number* or the *mean of the two middle numbers* in a list of numbers that have been arranged in numerical order from smallest to largest or from largest to smallest. Any list of numbers that is arranged in numerical order from smallest to largest or from largest to smallest is a **ranked list.**

▼ **point of interest**

The average price of the homes in a neighborhood is often stated in terms of the median price of the homes that have been sold over a given time period. The median price, rather than the mean, is used because it is easy to calculate and is less sensitive to extreme prices. The median prices of homes can vary dramatically, even for areas that are relatively close to each other. For instance, in March of 2004, the median sales price of a home in one ZIP code in Honolulu, Hawaii was $625,000. In an adjacent ZIP code, the median sales price was $472,100. (*Source:* DQNews.com, August 22, 2004)

> **Median**
>
> The **median** of a *ranked list* of n numbers is:
>
> ■ the middle number if n is odd.
>
> ■ the mean of the two middle numbers if n is even.

EXAMPLE 2 ■ **Find a Median**

Find the median for the data in each of the following lists.

a. 4, 8, 1, 14, 9, 21, 12   **b.** 46, 23, 92, 89, 77, 108

Solution

a. The list 4, 8, 1, 14, 9, 21, 12 contains seven numbers. The median of a list with an odd number of numbers is found by ranking the numbers and finding the middle number. Ranking the numbers from smallest to largest gives 1, 4, 8, 9, 12, 14, 21. The middle number is 9. Thus 9 is the median.

b. The list 46, 23, 92, 89, 77, 108 contains six numbers. The median of a list of data with an even number of numbers is found by ranking the numbers and computing the mean of the two middle numbers. Ranking the numbers from smallest to largest gives 23, 46, 77, 89, 92, 108. The two middle numbers are 77 and 89. The mean of 77 and 89 is 83. Thus 83 is the median of the data.

CHECK YOUR PROGRESS 2   Find the median for the data in each of the following lists.

a. 14, 27, 3, 82, 64, 34, 8, 51   **b.** 21.3, 37.4, 11.6, 82.5, 17.2

Solution   *See page S33.*

QUESTION   *The median of the ranked list 3, 4, 7, 11, 17, 29, 37 is 11. If the maximum value 37 is increased to 55, what effect will this have on the median?*

The Mode

A third type of average is the *mode.*

> **Mode**
>
> The **mode** of a list of numbers is the number that occurs most frequently.

ANSWER   *The median will remain the same because 11 will still be the middle number in the ranked list.*

Some lists of numbers do not have a mode. For instance, in the list 1, 6, 8, 10, 32, 15, 49, each number occurs exactly once. Because no number occurs more often than the other numbers, there is no mode.

A list of numerical data can have more than one mode. For instance, in the list 4, 2, 6, 2, 7, 9, 2, 4, 9, 8, 9, 7, the number 2 occurs three times and the number 9 occurs three times. Each of the other numbers occurs less than three times. Thus 2 and 9 are both modes of the data.

EXAMPLE 3 ▪ Find a Mode

Find the mode of the data in each of the following lists.

a. 18, 15, 21, 16, 15, 14, 15, 21   **b.** 2, 5, 8, 9, 11, 4, 7, 23

Solution

a. In the list 18, 15, 21, 16, 15, 14, 15, 21, the number 15 occurs more often than the other numbers. Thus 15 is the mode.

b. Each number in the list 2, 5, 8, 9, 11, 4, 7, 23 occurs only once. Because no number occurs more often than the others, there is no mode.

CHECK YOUR PROGRESS 3   Find the mode of the data in each of the following lists.

a. 3, 3, 3, 3, 3, 4, 4, 5, 5, 5, 8   **b.** 12, 34, 12, 71, 48, 93, 71

Solution   *See page S33.*

The mean, the median, and the mode are all averages; however, they are generally not equal and they have different properties. The following summary illustrates some of the properties of each type of average.

Comparative Properties of the Mean, the Median, and the Mode

The *mean* of a set of data:

- is the most sensitive of the averages. A change in any of the numbers changes the mean.
- can be different from each of the numbers in the set.
- can be changed drastically by changing an extreme value.

The *median* of a set of data:

- is usually not changed by changing an extreme value.
- is generally easy to compute.

The *mode* of a set of data:

- may not exist, and when it does exist it may not be unique.
- is one of the numbers in the set, provided a mode exists.
- is generally not changed by changing an extreme value.
- is generally easy to compute.

In the following example, we compare the mean, the median, and the mode of the salaries of five employees of a small company.

Salaries: $370,000 $60,000 $32,000 $16,000 $16,000

The sum of the five salaries is $494,000. Hence the mean is

$$\frac{\$494,000}{5} = \$98,800$$

The median is the middle number, $32,000. Because the $16,000 salary occurs the most, the mode is $16,000. The data contain one extreme value that is much larger than the other values. This extreme value makes the mean considerably larger than the median. Most of the employees of this company would probably agree that the median of $32,000 better represents the average of the salaries than does either the mean or the mode.

Math Matters   Average Rate for a Round Trip

Suppose you average 60 miles per hour on a one-way trip of 60 miles. On the return trip you average 30 miles per hour. You might be tempted to think that the average of 60 miles per hour and 30 miles per hour, which is 45 miles per hour, is the average rate for the entire trip. However, this is not the case. Because you were traveling more slowly on the return trip, the return trip took longer than the time spent going to your destination. More time was spent traveling at the lower speed. Thus the average rate for the round trip is less than the average (mean) of 60 miles per hour and 30 miles per hour.

To find the actual average rate for the round trip, use the formula

$$\text{Average rate} = \frac{\text{total distance}}{\text{total time}}$$

The total round-trip distance is 120 miles. The time spent going to your destination was 1 hour and the time spent on the return trip was 2 hours. The total time for the round trip was 3 hours. Thus

$$\text{Average rate} = \frac{\text{total distance}}{\text{total time}} = \frac{120}{3} = 40 \text{ miles per hour}$$

The Weighted Mean

A value called the *weighted mean* is often used when some data values are more important than others. For instance, many professors determine a student's course grade from the student's tests and the final examination. Consider the situation in which a professor counts the final examination score as two test scores. To find the weighted mean of the student's scores, the professor first assigns a weight to each score. In this case the professor could assign each of the test scores a weight of 1 and the final exam score a weight of 2. A student with test scores of 65, 70, and 75 and a final examination score of 90 has a weighted mean of

$$\frac{(65 \times 1) + (70 \times 1) + (75 \times 1) + (90 \times 2)}{5} = \frac{390}{5} = 78$$

Note that the numerator of the above weighted mean is the sum of the products of each test score and its corresponding weight. The number 5 in the denominator is the sum of all the weights ($1 + 1 + 1 + 2 = 5$). The above procedure can be generalized as follows.

The Weighted Mean

The **weighted mean** of the n numbers $x_1, x_2, x_3, \ldots, x_n$ with the respective assigned weights $w_1, w_2, w_3, \ldots, w_n$ is

$$\text{Weighted mean} = \frac{\Sigma(x \cdot w)}{\Sigma w}$$

where $\Sigma(x \cdot w)$ is the sum of the products formed by multiplying each number by its assigned weight, and Σw is the sum of all the weights.

▼ **point of interest**

Grade-point averages (GPAs) can be determined by using the weighted mean formula. See Exercises 23 and 24 on page 537.

EXAMPLE 4 ■ Find a Weighted Mean

An instructor determines a student's weighted mean from quizzes, tests, and a project. Each test counts as four quizzes and the project counts as eight quizzes. Larry has quiz scores of 70 and 55. His test scores are 90, 72, and 68. His project score is 85. Find Larry's weighted mean for the course.

Solution

If we assign each quiz score a weight of 1, then each test score will have a weight of 4, and the project will have a weight of 8. The sum of all the weights is 22.

Weighted mean

$$= \frac{(70 \times 1) + (55 \times 1) + (90 \times 4) + (72 \times 4) + (68 \times 4) + (85 \times 8)}{22}$$

$$= \frac{1725}{22} \approx 78.4$$

Larry's weighted mean is approximately 78.4.

CHECK YOUR PROGRESS 4   Find Larry's weighted mean in Example 4 if a test score counts as two quiz scores and the project counts as six quiz scores.

Solution   See page S33.

Data that have not been organized or manipulated in any manner are called **raw data.** A large collection of raw data may not provide much pertinent information that can be readily observed. A **frequency distribution,** which is a table that lists observed events and the frequency of occurrence of each observed event, is often used to organize raw data. For instance, consider Table 8.1, which lists the number of cable television connections for each of 40 homes in a subdivision.

Table 8.1 *Numbers of Cable Television Connections per Household*

2	0	3	1	2	1	0	4
2	1	1	7	2	0	1	1
0	2	2	1	3	2	2	1
1	4	2	5	2	3	1	2
2	1	2	1	5	0	2	5

The frequency distribution in Table 8.2 below was constructed using the data in Table 8.1. The first column of the frequency distribution consists of the numbers 0, 1, 2, 3, 4, 5, 6, and 7. The corresponding frequency of occurrence of each of the numbers in the first column is listed in the second column.

Table 8.2 *A Frequency Distribution for Table 8.1*

Observed Event Number of Cable Television Connections, x	Frequency Number of Households, f, with x Cable Television Connections
0	5
1	12
2	14
3	3
4	2
5	3
6	0
7	1
	40 total

This row indicates that there are 14 households with two cable television connections.

The formula for a weighted mean can be used to find the mean of the data in a frequency distribution. The only change is that the weights $w_1, w_2, w_3, \ldots, w_n$ are replaced with the frequencies $f_1, f_2, f_3, \ldots, f_n$. This procedure is illustrated in the next example.

EXAMPLE 5 ■ Find the Mean of Data Displayed in a Frequency Distribution

Find the mean of the data in Table 8.2.

Solution

The numbers in the right-hand column of Table 8.2 are the frequencies f of the numbers in the first column. The sum of all the frequencies is 40.

Mean

$$= \frac{\Sigma(x \cdot f)}{\Sigma f}$$

$$= \frac{(0 \cdot 5) + (1 \cdot 12) + (2 \cdot 14) + (3 \cdot 3) + (4 \cdot 2) + (5 \cdot 3) + (6 \cdot 0) + (7 \cdot 1)}{40}$$

$$= \frac{79}{40}$$

$$= 1.975$$

The mean number of cable connections per household for the homes in the subdivision is 1.975.

CHECK YOUR PROGRESS 5   A housing subdivision consists of 45 homes. The following frequency distribution shows the number of homes in the subdivision that are two-bedroom homes, the number that are three-bedroom homes, the number that are four-bedroom homes, and the number that are five-bedroom homes. Find the mean number of bedrooms for the 45 homes.

Observed Event Number of Bedrooms, *x*	*Frequency* Number of Homes with *x* Bedrooms
2	5
3	25
4	10
5	5
	45 total

Solution   *See page S33.*

Investigation

Linear Interpolation and Animation

Linear interpolation is a method used to find a particular number between two given numbers. For instance, if a table lists the two entries 0.3156 and 0.8248, then the value exactly halfway between the numbers is the mean of the numbers, which is 0.5702. To find the number that is 20% of the way from 0.3156 to 0.8248, compute 0.2 times the difference between the numbers and, because the first number is smaller than the second number, add this result to the smaller number.

$$0.8248 - 0.3156 = 0.5092 \quad \longleftarrow \quad \text{Difference between the table entries}$$

$$0.2 \cdot (0.5092) = 0.10184 \quad \longleftarrow \quad \text{0.2 times the above difference}$$

$$0.3156 + 0.10184 = 0.41744 \quad \longleftarrow \quad \text{Interpolated result, which is 20\% of the way between the two table entries}$$

The above linear interpolation process can be used to find an intermediate number that is any specified fraction of the difference between two given numbers.

Investigation Exercises

1. Use linear interpolation to find the number that is 70% of the way from 1.856 to 1.972.

2. Use linear interpolation to find the number that is 30% of the way from 0.8765 to 0.8652. Note that because 0.8765 is larger than 0.8652, 30% of the difference between 0.8765 and 0.8652 must be subtracted from 0.8765 to find the desired number.

3. A calculator shows that $\sqrt{2} \approx 1.414$ and $\sqrt{3} \approx 1.732$. Use linear interpolation to estimate $\sqrt{2.4}$. *Hint:* Find the number that is 40% of the difference between 1.414 and 1.732 and add this number to the smaller number, 1.414. Round your estimate to the nearest thousandth.

4. We know that $2^1 = 2$ and $2^2 = 4$. Use linear interpolation to estimate $2^{1.2}$.

5. At the present time, a football player weighs 325 pounds. There are 90 days until the player needs to report to spring training at a weight of 290 pounds. The player wants to lose weight at a constant rate. That is, the player wants to lose the same amount of weight each day of the 90 days. What weight, to the nearest tenth of a pound, should the player attain in 25 days?

Graphic artists use computer drawing programs, such as Adobe Illustrator, to draw the intermediate frames of an animation. For instance, in the following figure, the artist drew the small green apple on the left and the large ripe apple on the right. The drawing program used interpolation procedures to draw the five apples between the two apples drawn by the artist.

(continued)

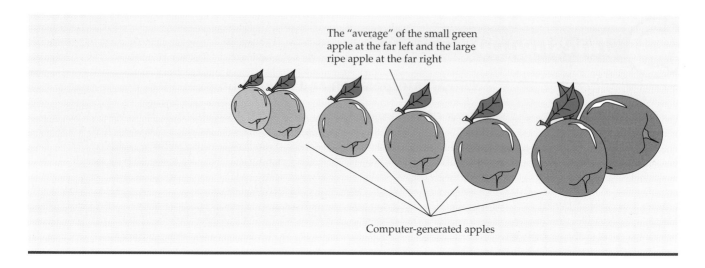

The "average" of the small green apple at the far left and the large ripe apple at the far right

Computer-generated apples

Exercise Set 8.3

In Exercises 1–10, find the mean, the median, and the mode(s), if any, of the given data. Round noninteger means to the nearest tenth.

1. 2, 7, 5, 7, 14

2. 8, 3, 3, 17, 9, 22, 19

3. 11, 8, 2, 5, 17, 39, 52, 42

4. 101, 88, 74, 60, 12, 94, 74, 85

5. 2.1, 4.6, 8.2, 3.4, 5.6, 8.0, 9.4, 12.2, 56.1, 78.2

6. 5, 5, 5, 5, 5, 5, 5, 5, 5, 5, 5, 5, 5

7. 255, 178, 192, 145, 202, 188, 178, 201

8. 118, 105, 110, 118, 134, 155, 166, 166, 118

9. −12, −8, −5, −5, −3, 0, 4, 9, 21

10. −8.5, −2.2, 4.1, 4.1, 6.4, 8.3, 9.7

11. **a.** If one number in a set of data is changed, will this necessarily change the mean of the set? Explain.

   b. If one number in a set of data is changed, will this necessarily change the median of the set? Explain.

12. If a set of data has a mode, then *must* the mode be one of the numbers in the set? Explain.

13. Suppose that the mean of a data set is M and the median is N. If the largest value in the data set is replaced with an even larger value, is the mean of the new data set less than M, equal to M, or greater than M? Is the median of the new data set less than N, equal to N, or greater than N?

14. Suppose that the mean of a data set is M and the median is N. If the smallest value in the data set is replaced with an even smaller value, is the mean of the new data set less than M, equal to M, or greater than M? Is the median of the new data set less than N, equal to N, or greater than N?

15. **Academy Awards** The following table displays the ages of female actors when they starred in their Oscar-winning Best Actor performances.

Ages of Best Female Actor Award Recipients, Academy Awards, 1971–2004

34	26	37	42	41	35	31	41	33	31	74	33
49	38	61	21	41	26	80	42	29	33	35	
45	49	39	34	26	25	33	35	35	28	30	

Find the mean, the median, and the mode(s) of the data in the table.

16. Academy Awards The following table displays the ages of male actors when they starred in their Oscar-winning Best Actor performances.

Ages of Best Male Actor Award Recipients, Academy Awards, 1971–2004

41	48	48	56	38	60	30	40	42	37	76	39
52	45	35	61	43	51	32	42	54	52	37	
38	31	45	60	46	40	36	47	29	43	37	

a. Find the mean, the median, and the mode(s) of the data in the table.

b. How do the results of part a compare with the results of Exercise 15?

17. Dental Schools Dental schools provide urban statistics to their students.

a. Use the following data to decide which of the two cities you would pick to set up your practice in.

Cloverdale: Population, 18,250;

median price of a home, $167,000;

dentists, 12; median age, 49;

mean number of patients, 1294.5

Barnbridge: Population, 27,840;

median price of a home, $204,400;

dentists, 17.5; median age, 53;

mean number of patients, 1148.7

b. Explain how you made your decision.

18. Expense Reports A salesperson records the following daily expenditures during a 10-day trip.

$185.34, $234.55, $211.86, $147.65, $205.60
$216.74, $1345.75, $184.16, $320.45, $88.12

In your opinion, does the mean or the median of the expenditures best represent the salesperson's average daily expenditure? Explain your reasoning.

19. Course Grades A professor grades students on three tests, four quizzes, and a final examination. Each test counts as two quizzes and the final examination counts as two tests. Sara has test scores of 85, 75, and 90. Sara's quiz scores are 96, 88, 60, and 76. Her final examination score is 85. Use the weighted mean formula to find Sara's average for the course.

20. Course Grades A professor grades students on three tests, three quizzes, and a final examination. Each quiz counts as one-half a test and the final examination counts as three tests. Dan has test scores of 80, 65, and 86. Dan's quiz scores are 86, 80, and 60. His final examination score is 84. Use the weighted mean formula to find Dan's average for the course.

21. Calculate a Course Grade A professor grades students on five tests, a project, and a final examination. Each test counts as 10% of the course grade. The project counts as 20% of the course grade. The final examination counts as 30% of the course grade. Samantha has test scores of 70, 65, 82, 94, and 85. Samantha's project score is 92. Her final examination score is 80. Use the weighted mean formula to find Samantha's average for the course. *Hint:* The sum of all the weights is 100% = 1.

22. Calculate a Course Grade A professor grades students on four tests, a term paper, and a final examination. Each test counts as 15% of the course grade. The term paper counts as 20% of the course grade. The final examination counts as 20% of the course grade. Alan has test scores of 80, 78, 92, and 84. Alan received an 84 on his term paper. His final examination score was 88. Use the weighted mean formula to find Alan's average for the course. *Hint:* The sum of all the weights is 100% = 1.

Grade-Point Average Many colleges use the four-point grading system: A = 4, B = 3, C = 2, D = 1, and F = 0. In Exercises 23 and 24, use the weighted mean formula to compute the grade-point average for each student. *Hint:* The units represent the "weights" of the letter grades that the student received in a course.

23. *Dillon's Grades, Fall Semester*

Course	Course Grade	Course Units
English	B	4
History	A	3
Chemistry	D	3
Algebra	C	4

24. *Janet's Grades, Spring Semester*

Course	Course Grade	Course Units
Biology	A	4
Statistics	B	3
Business Law	C	3
Psychology	F	2
CAD	B	2

In Exercises 25–28, find the mean, the median, and all modes of the data in the given frequency distribution.

25. *Points Scored by Lynn*

Points Scored in a Basketball Game	Frequency
2	6
4	5
5	6
9	3
10	1
14	2
19	1

26. *Mystic Pizza Company*

Hourly Pay Rates for Employees	Frequency
$8.00	14
$11.50	9
$14.00	8
$16.00	5
$19.00	2
$22.50	1
$35.00	1

27. *Quiz Scores*

Scores on a Biology Quiz	Frequency
2	1
4	2
6	7
7	12
8	10
9	4
10	3

28. *Ages of Science Fair Contestants*

Age	Frequency
7	3
8	4
9	6
10	15
11	11
12	7
13	1

Another measure of central tendency for a set of data is called the *midrange*. The **midrange** is defined as the value that is halfway between the minimum data value and the maximum data value. That is,

$$\text{Midrange} = \frac{\text{minimum value} + \text{maximum value}}{2}$$

The midrange is often stated as the *average* of a set of data in situations in which there is a large amount of data and the data are constantly changing. Many weather reports state the average daily temperature of a city as the midrange of the temperatures achieved during that day. For instance, if the minimum daily temperature of a city was 60° and the maximum daily temperature was 90°, then the midrange of the temperatures is $\frac{60° + 90°}{2} = 75°$.

29. Meteorology Find the midrange of the following daily temperatures, which were recorded at three-hour intervals.

52°, 65°, 71°, 74°, 76°, 75°, 68°, 57°, 54°

30. Meteorology Find the midrange of the following daily temperatures, which were recorded at three-hour intervals.

−6°, 4°, 14°, 21°, 25°, 26°, 18°, 12°, 2°

31. Meteorology During a 24-hour period on January 23–24, 1916, the temperature in Browning, Montana decreased from a high of 44°F to a low of −56°F. Find the midrange of the temperatures during this 24-hour period. (*Source: Time Almanac 2002*, page 609)

32. Meteorology During a two-minute period on January 22, 1943, the temperature in Spearfish, South Dakota increased from a low of −4°F to a high of 45°F. Find the midrange of the temperatures during this two-minute period. (*Source: Time Almanac 2002*, page 609)

33. Testing After six biology tests, Ruben has a mean score of 78. What score does Ruben need on the next test to raise his average (mean) to 80?

34. **Testing** After four algebra tests, Alisa has a mean score of 82. One more 100-point test is to be given in this class. All of the test scores are of equal importance. Is it possible for Alisa to raise her average (mean) to 90? Explain.

35. **Baseball** For the first half of a baseball season, a player had 92 hits out of 274 times at bat. The player's batting average was $\frac{92}{274} \approx 0.336$. During the second half of the season, the player had 60 hits out of 282 times at bat. The player's batting average was $\frac{60}{282} \approx 0.213$.

 a. What is the average (mean) of 0.336 and 0.213?

 b. What is the player's batting average for the complete season?

 c. Does the answer in part a equal the average in part b?

36. **Commuting Times** Mark averaged 60 miles per hour during the 30-mile trip to college. Because of heavy traffic he was able to average only 40 miles per hour during the return trip. What was Mark's average speed for the round trip?

Extensions

CRITICAL THINKING

37. The mean of 12 numbers is 48. Removing one of the numbers causes the mean to decrease to 45. What number was removed?

38. Find eight numbers such that the mean, the median, and the mode of the numbers are all 45, and no more than two of the numbers are the same.

39. The average rate for a trip is given by

$$\text{Average rate} = \frac{\text{total distance}}{\text{total time}}$$

If a person travels to a destination at an average rate of r_1 miles per hour and returns over the same route to the original starting point at an average rate of r_2 miles per hour, show that the average rate for the round trip is

$$r = \frac{2r_1 r_2}{r_1 + r_2}$$

40. Pick six numbers and compute the mean and the median of the numbers.

 a. Now add 12 to each of your original numbers and compute the mean and the median for this new set of numbers.

 b. How does the mean of the new set of data compare with the mean of the original set of data?

 c. How does the median of the new set of data compare with the median of the original set of data?

COOPERATIVE LEARNING

Consider the data in the following table.

Summary of Yards Gained in Two Football Games

	Game 1	Game 2	Combined Statistics for Both Games
Warren	12 yards on 4 carries Average: 3 yards/carry	78 yards on 16 carries Average: 4.875 yards/carry	90 yards on 20 carries Average: 4.5 yards/carry
Barry	120 yards on 30 carries Average: 4 yards/carry	100 yards on 20 carries Average: 5 yards/carry	220 yards on 50 carries Average: 4.4 yards/carry

- In the first game, Barry has the best average.
- In the second game, Barry has the best average.
- If the statistics for the games are combined, Warren has the best average.

You may be surprised by the above results. After all, how can it be that Barry has the best average in game 1 and game 2, but he does not have the best average for both games? In statistics, an example such as this is known as a **Simpson's paradox.**

Form groups of three or four students to work Exercises 41–43.

41. Baseball Consider the following data.

Batting Statistics for Two Baseball Players

	First Month	**Second Month**	**Both Months**
Dawn	2 hits; 5 at-bats Average: ?	19 hits; 49 at-bats Average: ?	? hits; ? at-bats Average: ?
Joanne	29 hits; 73 at-bats Average: ?	31 hits; 80 at-bats Average: ?	? hits; ? at-bats Average: ?

Is this an example of a Simpson's paradox? Explain.

42. Testing Consider the following data.

Test Scores for Two Students

	English	**History**	**English and History Combined**
Wendy	84, 65, 72, 91, 99, 84 Average: ?	66, 84, 75, 77, 94, 96, 81 Average: ?	Average: ?
Sarah	90, 74 Average: ?	68, 78, 98, 76, 68, 92, 88, 86 Average: ?	Average: ?

Is this an example of a Simpson's paradox? Explain.

EXPLORATIONS

43. Create your own example of a Simpson's paradox.

SECTION 8.4 | **Measures of Dispersion**

Table 8.3 *Test Scores*

Alan	**Tara**
55	80
80	76
97	77
80	83
68	84
100	80
Mean: 80	Mean: 80
Median: 80	Median: 80
Mode: 80	Mode: 80

The Range

In the preceding section we introduced the mean, the median, and the mode. Each of these statistics is a type of average that is designed to measure central tendencies of the data from which it was derived. Some characteristics of a set of data may not be evident from an examination of averages. For instance, consider the test scores for Alan and Tara, as shown in Table 8.3.

The mean, the median, and the mode of Alan's test scores and Tara's test scores are identical; however, an inspection of the test scores shows that Alan's scores are widely scattered, whereas all of Tara's scores are within a few units of the mean. This example shows that average values do not reflect the *spread* or *dispersion* of data. To measure the spread or dispersion of data, we must introduce statistical values known as the *range* and the *standard deviation*.

> ### Range
>
> The **range** of a set of data values is the difference between the largest data value and the smallest data value.

EXAMPLE 1 ■ Find a Range

Find the range of Alan's test scores in Table 8.3.

Solution
Alan's highest test score is 100 and his lowest test score is 55. The range of Alan's test scores is $100 - 55 = 45$.

CHECK YOUR PROGRESS 1   Find the range of Tara's test scores in Table 8.3.

Solution   *See page S33.*

Math Matters   A World Record Range

The tallest man for whom there is irrefutable evidence was Robert Pershing Wadlow. On June 27, 1940, Wadlow was 8 feet 11.1 inches tall. The shortest man for whom there is reliable evidence is Gul Mohammad. On July 19, 1990, he was 22.5 inches tall. (*Source:* Guinness World Records 2001) The range of the heights of these men is $107.1 - 22.5 = 84.6$ inches.

Robert Wadlow

The Standard Deviation

220-Yard Dash Times (in seconds)

Race	Sprinter 1	Sprinter 2
1	23.8	24.1
2	24.0	24.2
3	24.1	24.1
4	24.4	24.2
5	23.9	24.1
6	24.5	25.8
Range	0.7	1.7

The range of a set of data is easy to compute, but it can be deceiving. The range is a measure that depends only on the two most extreme values, and as such it is very sensitive. For instance, the table at the left shows the times for two sprinters in six track meets. The range of times for the first sprinter is 0.7 second, and the range for the second sprinter is 1.7 seconds. If you consider only range values, then you might conclude that the first sprinter's times are more consistent than those of the second sprinter. However, a closer examination shows that if you exclude the time of 25.8 seconds by the second sprinter in the sixth race, then the second sprinter has a range of 0.1 second. On this basis one could argue that the second sprinter has a more consistent performance record.

The next measure of dispersion that we will consider is called the *standard deviation*. It is less sensitive to a change in an extreme value than is the range. The standard deviation of a set of numerical data makes use of the individual amount that each data value deviates from the mean. These deviations, represented by $(x - \bar{x})$, are positive when the data value x is greater than the mean $\bar{x}$, and are negative when x is less than the mean $\bar{x}$. The sum of all the deviations $(x - \bar{x})$ is 0 for all sets of data. For instance, consider the sample data 2, 6, 11, 12, 14. For these data, $\bar{x} = 9$. The individual deviations of the data values from the mean are shown in the table on the following page. Note that the sum of the deviations is 0.

Because the sum of all the deviations of the data values from the mean is *always* 0, we cannot use the sum of the deviations as a measure of dispersion for a set of data.

Deviations from the Mean

x	$x - \bar{x}$
2	$2 - 9 = -7$
6	$6 - 9 = -3$
11	$11 - 9 = 2$
12	$12 - 9 = 3$
14	$14 - 9 = 5$
Sum of the deviations $\longrightarrow$	0

✔ **TAKE NOTE**

You may question why a denominator of $n - 1$ is used instead of n when we compute a sample standard deviation. The reason is that a sample standard deviation is often used to estimate the population standard deviation, and it can be shown mathematically that the use of $n - 1$ tends to yield better estimates.

What is needed is a procedure that can be applied to the deviations so that the sum of the numbers that are derived by adjusting the deviations is not always 0. The procedure that we will use *squares* each of the deviations $(x - \bar{x})$ to make each of them nonnegative. The sum of the squares of the deviations is then divided by a constant that depends on the number of data values. We then compute the square root of this result. The following definitions show that the formula for calculating the standard deviation of a population differs slightly from the formula used to calculate the standard deviation of a sample.

Standard Deviations of Samples and Populations

If $x_1, x_2, x_3, \ldots, x_n$ is a *population* of n numbers with a mean of μ, then the **standard deviation** of the population is $\sigma = \sqrt{\dfrac{\Sigma(x - \mu)^2}{n}}$.

If $x_1, x_2, x_3, \ldots, x_n$ is a *sample* of n numbers with a mean of $\bar{x}$, then the **standard deviation** of the sample is $s = \sqrt{\dfrac{\Sigma(x - \bar{x})^2}{n - 1}}$.

Most statistical applications involve a sample rather than a population, which is the complete set of data values. Sample standard deviations are designated by the lowercase letter *s*. In those cases in which we *do* work with a population, we designate the standard deviation of the population by σ, which is the lowercase Greek letter sigma. To calculate the standard deviation of n numbers, it is helpful to use the following procedure.

Procedure for Computing a Standard Deviation

1. Determine the mean of the n numbers.

2. For each number, calculate the deviation (difference) between the number and the mean of the numbers.

3. Calculate the square of each of the deviations and find the sum of these squared deviations.

4. If the data is a *population,* then divide the sum by n. If the data is a *sample,* then divide the sum by $n - 1$.

5. Find the square root of the quotient in Step 4.

EXAMPLE 2 ■ **Find the Standard Deviation**

The following numbers were obtained by sampling a population.

2, 4, 7, 12, 15

Find the standard deviation of the sample.

Solution

Step 1: The mean of the numbers is

$$\bar{x} = \frac{2 + 4 + 7 + 12 + 15}{5} = \frac{40}{5} = 8$$

Step 2: For each number, calculate the deviation between the number and the mean.

x	$x - \bar{x}$
2	$2 - 8 = -6$
4	$4 - 8 = -4$
7	$7 - 8 = -1$
12	$12 - 8 = 4$
15	$15 - 8 = 7$

TAKE NOTE

Because the sum of the deviations is always 0, you can use this to check your arithmetic. That is, if your deviations from the mean do not have a sum of 0, then you know you have made an error.

Step 3: Calculate the square of each of the deviations in Step 2, and find the sum of these squared deviations.

x	$x - \bar{x}$	$(x - \bar{x})^2$
2	$2 - 8 = -6$	$(-6)^2 = 36$
4	$4 - 8 = -4$	$(-4)^2 = 16$
7	$7 - 8 = -1$	$(-1)^2 = 1$
12	$12 - 8 = 4$	$4^2 = 16$
15	$15 - 8 = 7$	$7^2 = \underline{49}$
		118

The sum of the squared deviations

Step 4: Because we have a sample of $n = 5$ values, divide the sum 118 by $n - 1$, which is 4.

$$\frac{118}{4} = 29.5$$

Step 5: The standard deviation of the sample is $s = \sqrt{29.5}$. To the nearest hundredth, the sample standard deviation is $s = 5.43$.

CHECK YOUR PROGRESS 2 A student has the following quiz scores: 5, 8, 16, 17, 18, 20. Find the standard deviation of this population of quiz scores. Round to the nearest hundredth.

Solution *See page S33.*

In the next example we use standard deviations to determine which company produces batteries that are most consistent with regard to life expectancy.

EXAMPLE 3 ▪ Use Standard Deviations

A consumer group has tested a sample of eight size D batteries from each of three companies. The results of the tests are shown in the following table. According to these tests, which company produces batteries for which the values representing hours of constant use have the smallest standard deviation?

Company	Hours of Constant Use per Battery
EverSoBright	6.2, 6.4, 7.1, 5.9, 8.3, 5.3, 7.5, 9.3
Dependable	6.8, 6.2, 7.2, 5.9, 7.0, 7.4, 7.3, 8.2
Beacon	6.1, 6.6, 7.3, 5.7, 7.1, 7.6, 7.1, 8.5

Solution

The mean for each sample of batteries is 7 hours.

The batteries from EverSoBright have a standard deviation of

$$s_1 = \sqrt{\frac{(6.2 - 7)^2 + (6.4 - 7)^2 + \cdots + (9.3 - 7)^2}{7}}$$

$$= \sqrt{\frac{12.34}{7}} \approx 1.328 \text{ hours}$$

The batteries from Dependable have a standard deviation of

$$s_2 = \sqrt{\frac{(6.8 - 7)^2 + (6.2 - 7)^2 + \cdots + (8.2 - 7)^2}{7}}$$

$$= \sqrt{\frac{3.62}{7}} \approx 0.719 \text{ hour}$$

The batteries from Beacon have a standard deviation of

$$s_3 = \sqrt{\frac{(6.1 - 7)^2 + (6.6 - 7)^2 + \cdots + (8.5 - 7)^2}{7}}$$

$$= \sqrt{\frac{5.38}{7}} \approx 0.877 \text{ hour}$$

The batteries from Dependable have the *smallest* standard deviation. According to these results, the Dependable company produces the most consistent batteries with regard to life expectancy under constant use.

CHECK YOUR PROGRESS 3   A consumer testing agency has tested the strengths of three brands of $\frac{1}{8}$-inch rope. The results of the tests are shown in the following table. According to the sample test results, which company produces $\frac{1}{8}$-inch rope for which the breaking point has the smallest standard deviation?

Company	Breaking Point of $\frac{1}{8}$-Inch Rope, in Pounds
Trustworthy	122, 141, 151, 114, 108, 149, 125
Brand X	128, 127, 148, 164, 97, 109, 137
NeverSnap	112, 121, 138, 131, 134, 139, 135

Solution   *See page S34.*

Many calculators have built-in features for calculating the mean and standard deviation of a set of numbers. The next example illustrates these features on a TI-83/84 Plus graphing calculator.

EXAMPLE 4 ■ Use a Calculator to Find the Mean and Standard Deviation

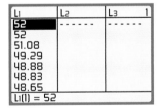 Use a graphing calculator to find the mean and standard deviation of the times in the following table. Because the table contains all the winning times for this race (up to the year 2004), the data set is a population.

Olympic Women's 400-Meter Dash Results (in seconds), 1964–2004

| 52.0 | 52.0 | 51.08 | 49.29 | 48.88 | 48.83 | 48.65 | 48.83 | 48.25 | 49.11 | 49.41 |

Solution

On a TI-83/84 Plus calculator, press [STAT] [ENTER] and then enter the above times into list **[L1]**. See the calculator display below. Press [STAT] [▶] [ENTER] [ENTER]. The calculator displays the mean and standard deviations shown below. Because we are working with a population, we are interested in the population standard deviation. From the calculator screen, $\bar{x} \approx 49.666$ and $\sigma x \approx 1.296$ seconds.

TAKE NOTE

Because the calculation of the population mean and the sample mean are the same, a graphing calculator uses the same symbol $\boxed{\bar{x}}$ for both. The symbols for the population standard deviation, $\boxed{\sigma x}$, and the sample standard deviation, $\boxed{s x}$, are different.

TI-83/84 Plus Display of List 1

L1	L2	L3	1
52	- - - - - -	- - - - - -	
52			
51.08			
49.29			
48.88			
48.83			
48.65			

L1(1) = 52

TI-83/84 Plus Display of $\bar{x}$, s, and σ

```
1-Var Stats
  x̄=49.66636364      ←——— Mean
  x=546.33
  x²=27152.6879
  Sx=1.358802949    ←——— Sample standard
                            deviation
  σx=1.295567778    ←——— Population standard
↓n=11                       deviation
```

CHECK YOUR PROGRESS 4 Use a calculator to find the mean and the population standard deviation of the race times in the following table.

Olympic Men's 400-Meter Dash Results (in seconds), 1896–2004

54.2	49.4	49.2	53.2	50.0	48.2	49.6	47.6	47.8
46.2	46.5	46.2	45.9	46.7	44.9	45.1	43.8	44.66
44.26	44.60	44.27	43.87	43.50	43.49	43.84	44.00	

Solution *See page S34.*

The Variance

A statistic known as the *variance* is also used as a measure of dispersion. The **variance** of a given set of data is the square of the standard deviation of the data. The following chart shows the mathematical notations that are used to denote standard deviations and variances.

Notations for Standard Deviation and Variance

σ is the standard deviation of a population.

σ^2 is the variance of a population.

s is the standard deviation of a sample.

s^2 is the variance of a sample.

EXAMPLE 5 ■ Find the Variance

Find the variance of the sample given in Example 2.

Solution
In Example 2, we found $s = \sqrt{29.5}$. Variance is the square of the standard deviation. Thus the variance is $s^2 = \left(\sqrt{29.5}\right)^2 = 29.5$.

CHECK YOUR PROGRESS 5   Find the variance of the population given in Check Your Progress 2.

Solution   See page S34.

QUESTION   *Can the variance of a data set be smaller than the standard deviation of the data set?*

Although the variance of a set of data is an important measure of dispersion, it has a disadvantage that is not shared by the standard deviation: the variance does not have the same unit of measure as the original data. For instance, if a set of data consists of times measured in hours, then the variance of the data will be measured in *square* hours. The standard deviation of this data set is the square root of the variance, and as such it is measured in hours, which is a more intuitive unit of measure.

ANSWER   *Yes. The variance is smaller than the standard deviation whenever the standard deviation is less than 1.*

Investigation

A Geometric View of Variance and Standard Deviation[1]

The following geometric explanation of the variance and standard deviation of a set of data is designed to provide you with a deeper understanding of these important concepts.

Consider the data $x_1, x_2, \ldots, x_n$, which are arranged in ascending order. The average, or mean, of these data is

$$\mu = \frac{\Sigma x_i}{n}$$

and the variance is

$$\sigma^2 = \frac{\Sigma(x_i - \mu)^2}{n}$$

In the second formula, each term $(x_i - \mu)^2$ can be pictured as the area of a square whose sides are of length $|x_i - \mu|$, the distance between the ith data value and the mean. We will refer to these squares as *tiles*, denoting by T_i the area of the tile associated with the data value x_i. Thus $\sigma^2 = \frac{\Sigma T_i}{n}$, which means that the variance may be thought of as the *area of the averaged-sized tile* and the standard deviation σ as the length of a side of this averaged-sized tile. By drawing the tiles associated with a data set, as shown below, you can visually estimate an averaged-size tile and thus you can roughly approximate the variance and standard deviation.

✔ **TAKE NOTE**

Up to this point we have used $\mu = \dfrac{\Sigma x}{n}$ as the formula for the mean. However, many statistics texts use the formula $\mu = \dfrac{\Sigma x_i}{n}$ for the mean. Letting the subscript i vary from 1 to n helps us to remember that we are finding the sum of all the numbers $x_1, x_2, \ldots, x_n$.

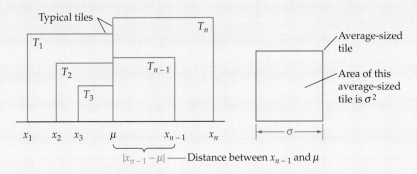

A Typical Data Set, with Its Associated Tiles and Average-Sized Tile

These geometric representations of variance and standard deviation enable us to see visually how these values are used as measures of the dispersion of a set of data. If all of the data are bunched up near the mean, it is clear that the average-sized tile will be small and, consequently, so will its side length, which represents the standard deviation. But if even a small number of data values lie far from the mean, the average-sized tile may be rather large, and thus its side length will also be large.

(continued)

1. Adapted with permission from "Chebyshev's Theorem: A Geometric Approach," *The College Mathematics Journal*, Vol. 26, No. 2, March 1995. Article by Pat Touhey, College Misericordia, Dallas, PA 18612.

Investigation Exercises

1. This exercise makes use of the geometric procedure just explained to calculate the variance and standard deviation of the population 2, 5, 7, 11, 15. The following figure shows the given set of data labeled on a number line, along with its mean, which is 8.

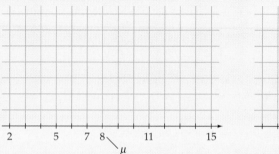 

Average tile

    a. Draw the tile associated with each of the five data values 2, 5, 7, 11, and 15.

    b. Label each tile with its area.

    c. Find the sum of the areas of all the tiles.

    d. Find the average (mean) of the areas of all five tiles.

    e. To the right of the above number line, draw a tile whose area is the average found in part d.

    f. What is the variance of the data? What geometric figure represents the variance?

    g. What is the standard deviation of the data? What geometric figure represents the standard deviation?

2. **a. to g.** Repeat all of the steps described in Investigation Exercise 1 for the data set

$$6, 8, 9, 11, 16$$

    h. Which of the data sets in these two exercises has the larger mean? Which data set has the larger standard deviation?

Exercise Set 8.4

1. **Meteorology** During a 12-hour period on December 24, 1924, the temperature in Fairfield, Montana dropped from a high of 63°F to a low of −21°F. What was the range of the temperatures during this period? (*Source: Time Almanac 2002*, page 609)

2. **Meteorology** During a two-hour period on January 12, 1911, the temperature in Rapid City, South Dakota dropped from a high of 49°F to a low of −13°F. What was the range of the temperatures during this period? (*Source: Time Almanac 2002*, page 609)

In Exercises 3–12, find the range, the standard deviation, and the variance of each given *sample*. Round noninteger results to the nearest tenth.

3. 1, 2, 5, 7, 8, 19, 22

4. 3, 4, 7, 11, 12, 12, 15, 16

5. 2.1, 3.0, 1.9, 1.5, 4.8

6. 5.2, 11.7, 19.1, 3.7, 8.2, 16.3

7. 48, 91, 87, 93, 59, 68, 92, 100, 81

8. 93, 67, 49, 55, 92, 87, 77, 66, 73, 96, 54

9. 4, 4, 4, 4, 4, 4, 4, 4, 4, 4, 4, 4, 4, 4, 4, 4, 4

10. 8, 6, 8, 6, 8, 6, 8, 6, 8, 6, 8, 6, 8

11. −8, −5, −12, −1, 4, 7, 11

12. −23, −17, −19, −5, −4, −11, −31

13. If the standard deviations of two data sets are equal, are the means of the data sets necessarily equal?

14. A vending machine is supposed to fill a soda cup with 8.5 ounces of soda. On one day, a quality control manager found that the mean fill level was 8.5 ounces with a standard deviation of 0.005 ounce. On a different day, the mean fill level was 8.5 ounces and the standard deviation was 0.1 ounce. Because the mean fill levels were the same, the manager concluded that no adjustments were needed. Is this correct?

15. **Mountain Climbing** A mountain climber plans to buy some rope to use as a lifeline. Which of the following would be the better choice? Explain why you think your choice is the better choice.

Rope A: Mean breaking strength: 500 pounds; standard deviation of 300 pounds

Rope B: Mean breaking strength: 400 pounds; standard deviation of 40 pounds

16. **Lotteries** Which would you expect to be the larger: the standard deviation of five random numbers picked from 1 to 47 in the California Super Lotto, or the standard deviation of five random numbers picked from 1 to 51 in the multistate PowerBall lottery?

17. **Weights of Students** Which would you expect to be the larger standard deviation: the standard deviation of the weights of 25 students in a first-grade class, or the standard deviation of the weights of 25 students in a college statistics course?

18. Evaluate the accuracy of the following statement: When the mean of a data set is large, the standard deviation will be large.

19. **Super Bowl** The following table lists the winning and losing scores for the Super Bowl games from 1972 through 2005.

Super Bowl Results, 1972–2005

33–14	16–6	26–21	20–16	27–17
16–7	21–17	27–17	55–10	35–21
23–7	32–14	38–9	20–19	31–24
16–13	27–10	38–16	37–24	34–19
24–3	35–31	46–10	52–17	23–16
14–7	31–19	39–20	30–13	34–7
20–17	48–21	32–29	24–21	

a. Find the mean and the *population* standard deviation of the winning scores. Round each result to the nearest tenth.

b. Find the mean and the *population* standard deviation of the losing scores. Round each result to the nearest tenth.

c. Which of the two data sets has the larger mean? Which of the two data sets has the larger standard deviation?

20. **Academy Awards** The following tables list the ages of female and male actors when they starred in their Oscar-winning Best Actor performances.

Ages of Best Female Actor Award Recipients, Academy Awards, 1971–2004

34	26	37	42	41	35	31	41	33	31	74	33
49	38	61	21	41	26	80	42	29	33	35	
45	49	39	34	26	25	33	35	35	28	30	

Ages of Best Male Actor Award Recipients,
Academy Awards, 1971–2004

41	48	48	56	38	60	30	40	42	37	76	39
52	45	35	61	43	51	32	42	54	52	37	
38	31	45	60	46	40	36	47	29	43	37	

a. Find the mean and the *sample* standard deviation of the ages of the female recipients. Round each result to the nearest tenth.

b. Find the mean and the *sample* standard deviation of the ages of the male recipients. Round each result to the nearest tenth.

c. Which of the two data sets has the larger mean? Which of the two data sets has the larger standard deviation?

21. **Baseball** The following tables list the numbers of home runs hit by the leaders in the National and American Leagues from 1971 to 2004.

Home Run Leaders, 1971–2004

National League											
48	40	44	36	38	38	52	40	48	48	31	37
40	36	37	37	49	39	47	40	38	35	46	
43	40	47	49	70	65	50	73	49	47	48	

American League											
33	37	32	32	36	32	39	46	45	41	22	39
39	43	40	40	49	42	36	51	44	43	46	
40	50	52	56	56	48	47	52	57	47	43	

a. Find the mean and the *population* standard deviation of the numbers of home runs hit by the leaders in the National League. Round each result to the nearest tenth.

b. Find the mean and the *population* standard deviation of the numbers of home runs hit by the leaders in the American League. Round each result to the nearest tenth.

c. Which of the two data sets has the larger mean? Which of the two data sets has the larger standard deviation?

22. **Triathlon** The following table lists the winning times for the men's and women's Ironman Triathlon World Championships, held in Kailua-Kona, Hawaii.

Ironman Triathlon World Championships
(Winning times rounded to the nearest minute)

Men	**(1980–2003)**		**Women**	**(1980–2003)**	
9:25	8:31	8:04	11:21	9:01	9:07
9:38	8:09	8:33	12:01	9:01	9:32
9:08	8:28	8:24	10:54	9:14	9:24
9:06	8:19	8:17	10:44	9:08	9:13
8:54	8:09	8:21	10:25	8:55	9:26
8:51	8:08	8:31	10:25	8:58	9:29
8:29	8:20	8:30	9:49	9:20	9:08
8:34	8:21	8:23	9:35	9:17	9:12

a. Find the mean and the *population* standard deviation of the winning times of the female athletes. *Note:* Convert each time to hours. For instance, a time of 12:55 (12 hours 55 minutes) is equal to $12 + \frac{55}{60} = 12.91\overline{6}$ hours. Round each result to the nearest tenth.

b. Find the mean and the *population* standard deviation of the winning times of the male athletes. Round each result to the nearest tenth.

c. Which of the two data sets has the larger mean? Which of the two data sets has the larger standard deviation?

23. **Political Science** The table on the following page lists the U.S. presidents and their ages at inauguration. President Cleveland has two entries because he served two nonconsecutive terms.

Washington	57	J. Adams	61
Jefferson	57	Madison	57
Monroe	58	J. Q. Adams	57
Jackson	61	Van Buren	54
W. H. Harrison	68	Tyler	51
Polk	49	Taylor	64
Fillmore	50	Pierce	48
Buchanan	65	Lincoln	52
A. Johnson	56	Grant	46
Hayes	54	Garfield	49
Arthur	50	Cleveland	47, 55
B. Harrison	55	McKinley	54
T. Roosevelt	42	Taft	51
Wilson	56	Harding	55
Coolidge	51	Hoover	54
F. D. Roosevelt	51	Truman	60
Eisenhower	62	Kennedy	43
L. B. Johnson	55	Nixon	56
Ford	61	Carter	52
Reagan	69	G. H. W. Bush	64
Clinton	46	G. W. Bush	54

Source: The World Almanac and Book of Facts, 2005

Find the mean and the *population* standard deviation of the ages. Round each result to the nearest tenth.

24. **Political Science** The following table lists the deceased U.S. presidents as of November 2004, and their ages at death.

Washington	67	J. Adams	90
Jefferson	83	Madison	85
Monroe	73	J. Q. Adams	80
Jackson	78	Van Buren	79
W. H. Harrison	68	Tyler	71
Polk	53	Taylor	65
Fillmore	74	Pierce	64
Buchanan	77	Lincoln	56
A. Johnson	66	Grant	63
Hayes	70	Garfield	49
Arthur	56	Cleveland	71
B. Harrison	67	McKinley	58
T. Roosevelt	60	Taft	72
Wilson	67	Harding	57
Coolidge	60	Hoover	90
F. D. Roosevelt	63	Truman	88
Eisenhower	78	Kennedy	46
L. B. Johnson	64	Nixon	81
Reagan	93		

Source: The World Almanac and Book of Facts, 2005

Find the mean and the *population* standard deviation of the ages. Round each result to the nearest tenth.

Extensions

CRITICAL THINKING

25. Pick five numbers and compute the *population* standard deviation of the numbers.

  a. Add a nonzero constant c to each of your original numbers and compute the standard deviation of this new population.

  b. Use the results of part a and inductive reasoning to state what happens to the standard deviation of a population when a nonzero constant c is added to each data item.

26. Pick six numbers and compute the *population* standard deviation of the numbers.

  a. Double each of your original numbers and compute the standard deviation of this new population.

  b. Use the results of part a and inductive reasoning to state what happens to the standard deviation of a population when each data item is multiplied by a positive constant k.

27. **a.** All of the numbers in a sample are the same number. What is the standard deviation of the sample?

  b. If the standard deviation of a sample is 0, must all of the numbers in the sample be the same number?

  c. If two samples both have the same standard deviation, are the samples necessarily identical?

28. Under what condition would the variance of a sample be equal to the standard deviation of the sample?

EXPLORATIONS

29. **a.** Use a calculator to compare the standard deviation of the population 1, 2, 3, 4, 5 and the value $\sqrt{\dfrac{5^2 - 1}{12}}$.

b. Use a calculator to compare the standard deviation of the population 1, 2, 3, ..., 10 and the value $\sqrt{\dfrac{10^2 - 1}{12}}$.

c. Use a calculator to compare the standard deviation of the population 1, 2, 3, ..., 15 and the value $\sqrt{\dfrac{15^2 - 1}{12}}$.

d. Make a conjecture about the standard deviation of the population 1, 2, 3, ..., n and the value $\sqrt{\dfrac{n^2 - 1}{12}}$.

30. Find, without using a calculator, the standard deviation of the population

$$3001, 3002, 3003, 3004, \ldots, 3010$$

Hint: Use your answer to part b of Exercise 25 and your conjecture from part d of Exercise 29.

SECTION 8.5 | Measures of Relative Position

z-Scores

When you take a course in college, it is natural to wonder how you will do compared with the other students. Will you finish in the top 10% or will you be closer to the middle? One statistic that is often used to measure the position of a data value with respect to other values is known as the *z-score* or the *standard score.*

> **z-Score**
>
> The **z-score** for a given data value x is the number of standard deviations that x is above or below the mean of the data. The following formulas show how to calculate the z-score for a data value x in a population and in a sample.
>
> Population: $z_x = \dfrac{x - \mu}{\sigma}$ Sample: $z_x = \dfrac{x - \bar{x}}{s}$

QUESTION   *Must the z-score for a data value be a positive number?*

In the next example, we use a student's z-scores for two tests to determine how well the student did on each test in comparison with the other students.

ANSWER   *No. The z-score for a data value x is positive if x is greater than the mean, it is 0 if x is equal to the mean, and it is negative if x is less than the mean.*

EXAMPLE 1 ■ Compare *z*-Scores

Raul has taken two tests in his chemistry class. He scored 72 on the first test, for which the mean of all scores was 65 and the standard deviation was 8. He received a 60 on the second test, for which the mean of all scores was 45 and the standard deviation was 12. In comparison with the other students, did Raul do better on the first test or the second test?

Solution

Find the *z*-score for each test.

$$z_{72} = \frac{72 - 65}{8} = 0.875 \qquad z_{60} = \frac{60 - 45}{12} = 1.25$$

Raul scored 0.875 standard deviation above the mean on the first test and 1.25 standard deviations above the mean on the second test. These *z*-scores indicate that in comparison with his classmates, Raul scored better on the second test than he did on the first test.

CHECK YOUR PROGRESS 1   Cheryl has taken two quizzes in her history class. She scored 15 on the first quiz, for which the mean of all scores was 12 and the standard deviation was 2.4. Her score on the second quiz, for which the mean of all scores was 11 and the standard deviation was 2.0, was 14. In comparison with her classmates, did Cheryl do better on the first quiz or the second quiz?

Solution   See page S34.

The *z*-score equation $z_x = \dfrac{x - \bar{x}}{s}$ involves four variables. If the values of any three of the four variables are known, you can solve for the unknown variable. This procedure is illustrated in the next example.

EXAMPLE 2 ■ Use *z*-Scores

A consumer group tested a sample of 100 light bulbs. It found that the mean life expectancy of the bulbs was 842 hours, with a standard deviation of 90. One particular light bulb from the DuraBright Company had a *z*-score of 1.2. What was the life span of this light bulb?

Solution

Substitute the given values into the *z*-score equation and solve for *x*.

$$z_x = \frac{x - \bar{x}}{s}$$

$$1.2 = \frac{x - 842}{90}$$

$$108 = x - 842$$

$$950 = x$$

The light bulb had a life span of 950 hours.

CHECK YOUR PROGRESS 2   Roland received a score of 70 on a test for which the mean score was 65.5. Roland has learned that the *z*-score for his test is 0.6. What is the standard deviation for this set of test scores?

Solution   See page S34.

Percentiles

Most standardized examinations provide scores in terms of *percentiles*, which are defined as follows.

---

pth Percentile

A value x is called the **pth percentile** of a data set provided p% of the data values are less than x.

EXAMPLE 3 ■ Using Percentiles

According to the U.S. Department of Labor, the median annual salary in 2003 for a physical therapist was $57,720. If the 85th percentile for the annual salary of a physical therapist was $71,500, find the percent of physical therapists whose annual salaries were

a. more than $57,720.

b. less than $71,500.

c. between $57,720 and $71,500.

Solution

a. By definition, the median is the 50th percentile. Therefore, 50% of the physical therapists earned more than $57,720 per year.

b. Because $71,500 is the 85th percentile, 85% of all physical therapists made less than $71,500.

c. From parts a and b, 85% − 50% = 35% of the physical therapists earned between $57,720 and $71,500.

CHECK YOUR PROGRESS 3 According to the U.S. Department of Labor, the median annual salary in 2003 for a police dispatcher was $28,288. If the 30th percentile for the annual salary of a police dispatcher was $25,640, find the percent of police dispatchers whose annual salaries were

a. less than $28,288.

b. more than $25,640.

c. between $25,640 and $28,288.

Solution   *See page S34.*

The following formula can be used to find the percentile that corresponds to a particular data value in a set of data.

Percentile for a Given Data Value

Given a set of data and a data value x,

$$\text{Percentile of score } x = \frac{\text{number of data values less than } x}{\text{total number of data values}} \cdot 100$$

EXAMPLE 4 ■ Find a Percentile

On a reading examination given to 900 students, Elaine's score of 602 was higher than the scores of 576 of the students who took the examination. What is the percentile for Elaine's score?

Solution

$$\text{Percentile} = \frac{\text{number of data values less than 602}}{\text{total number of data values}} \cdot 100$$

$$= \frac{576}{900} \cdot 100$$

$$= 64$$

Elaine's score of 602 places her at the 64th percentile.

CHECK YOUR PROGRESS 4   On an examination given to 8600 students, Hal's score of 405 was higher than the scores of 3952 of the students who took the examination. What is the percentile for Hal's score?

Solution   *See page S34.*

Math Matters   **Standardized Tests and Percentiles**

Standardized tests, such as the Scholastic Assessment Test (SAT), are designed to measure all students by a *single* standard. The SAT is used by many colleges as part of their admissions criteria. The SAT I is a three-hour examination that measures verbal and mathematical reasoning skills. Scores on each portion of the test range from 200 to 800 points. SAT scores are generally reported in points and percentiles. Sometimes students are confused by the percentile scores. For instance, if a student scores 650 points on the mathematics portion of the SAT and is told that this score is in the 85th percentile, this *does not* indicate that the student answered 85% of the questions correctly. An 85th percentile score means that the student scored higher than 85% of the students who took the test. Consequently, the student scored lower than 15% (100% − 85%) of the students who took the test.

Quartiles

The three numbers Q_1, Q_2, and Q_3 that partition a data set into four (approximately) equal portions are called the **quartiles** of the data. For instance, for the data set below, the values $Q_1 = 11$, $Q_2 = 29$, and $Q_3 = 104$ are the quartiles of the data.

2, 5, 5, 8, 11, 12, 19, 22, 23, 29, 31, 45, 83, 91, 104, 159, 181, 312, 354

$$Q_1 \qquad\qquad Q_2 \qquad\qquad Q_3$$

The quartile Q_1 is called the *first quartile*. The quartile Q_2 is called the *second quartile*. It is the median of the data. The quartile Q_3 is called the *third quartile*. The following method of finding quartiles makes use of medians.

> **The Median Procedure for Finding Quartiles**
>
> **1.** Rank the data.
>
> **2.** Find the median of the data. This is the second quartile, Q_2.
>
> **3.** The first quartile, Q_1, is the median of the data values smaller than Q_2. The third quartile, Q_3, is the median of the data values larger than Q_2.

EXAMPLE 5 ▪ Use Medians to Find the Quartiles of a Data Set

The following table lists the calories per 100 milliliters of 25 popular beers. Find the quartiles of the data.

Calories, per 100 Milliliters, of Selected Beers

43	37	42	40	53	62	36	32	50	49
26	53	73	48	45	39	45	48	40	56
41	36	58	42	39					

Solution

Step 1: Rank the data, as shown in the following table.

1) 26	**2)** 32	**3)** 36	**4)** 36	**5)** 37	**6)** 39	**7)** 39	**8)** 40	**9)** 40
10) 41	**11)** 42	**12)** 42	**13)** 43	**14)** 45	**15)** 45	**16)** 48	**17)** 48	**18)** 49
19) 50	**20)** 53	**21)** 53	**22)** 56	**23)** 58	**24)** 62	**25)** 73		

Step 2: The median of these 25 data values has a rank of 13. Thus the median is **43**. The second quartile Q_2 is the median of the data, so $Q_2 = 43$.

Step 3: There are 12 data values less than the median and 12 data values greater than the median. The first quartile is the median of the data values less than the median. Thus Q_1 is the mean of the data values with ranks of 6 and 7.

$$Q_1 = \frac{39 + 39}{2} = 39$$

The third quartile is the median of the data values greater than the median. Thus Q_3 is the mean of the data values with ranks of 19 and 20.

$$Q_3 = \frac{50 + 53}{2} = 51.5$$

CHECK YOUR PROGRESS 5 The following table lists the weights, in ounces, of 15 avocados in a random sample. Find the quartiles of the data.

Weights (in ounces) of Avocados

12.4	10.8	14.2	7.5	10.2	11.4	12.6	12.8	13.1	15.6
9.8	11.4	12.2	16.4	14.5					

Solution See page S34.

Box-and-Whisker Plots

A **box-and-whisker plot** (sometimes called a **box plot**) is often used to provide a visual summary of a set of data. A box-and-whisker plot shows the median, the first and third quartiles, and the minimum and maximum values of a data set. See the figure below.

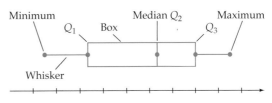

A Box-and-Whisker Plot

> **✔ TAKE NOTE**
>
> If a set of data is not ranked, then you need to rank the data so that you can easily determine the median, the first and third quartiles, and the minimum and maximum data values.

Construction of a Box-and-Whisker Plot

1. Draw a horizontal scale that extends from the minimum data value to the maximum data value.

2. Above the scale, draw a rectangle (box) with its left side at Q_1 and its right side at Q_3.

3. Draw a vertical line segment across the rectangle at the median, Q_2.

4. Draw a horizontal line segment, called a whisker, that extends from Q_1 to the minimum and another whisker that extends from Q_3 to the maximum.

EXAMPLE 6 ■ Construct a Box-and-Whisker Plot

Construct a box-and-whisker plot for the data set in Example 5.

Solution
For the data set in Example 5, we determined that $Q_1 = 39$, $Q_2 = 43$, and $Q_3 = 51.5$. The minimum data value for the data set is 26 and the maximum data value is 73. Thus the box-and-whisker plot is as shown in Figure 8.2.

> *historical note*
>
> **John W. Tukey** (1915–2000) made many important contributions to the field of statistics during his career at Princeton University and Bell Labs. He invented box plots when he was working on exploratory data analysis. ■

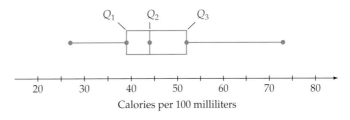

Figure 8.2 *A Box-and-Whisker Plot of the Data in Example 5*

CHECK YOUR PROGRESS 6   Construct a box-and-whisker plot for the following data.

The Numbers of Occupied Rooms in a Resort During an 18-Day Period

86	77	58	45	94	96	83	76	75
65	68	72	78	85	87	92	55	61

Solution   *See page S34.*

Box plots have become popular because they are easy to construct and they illustrate several important features of a data set in a simple diagram. Note from Figure 8.2 that we can easily estimate

- the quartiles of the data.
- the range of the data.
- the position of the middle half of the data as shown by the length of the box.

Some graphing calculators, such as the TI-83/84 Plus, can be used to produce box-and-whisker plots. For instance, on a TI-83/84 Plus, you enter the data into a list, as shown in Figure 8.3. The `WINDOW` menu is used to enter appropriate boundaries that contain all the data. Use the key sequence `2nd` `[STAT PLOT]` `ENTER` and choose from the `Type` menu the box-and-whisker plot icon. The `GRAPH` key is then used to display the box-and-whisker plot. After the calculator displays a box-and-whisker plot, the `TRACE` key and the `▶` key enable you to view Q_1, Q_2, Q_3, and the minimum and maximum values of your data set.

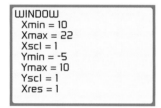

TAKE NOTE

The following data were used to produce the box plot shown at the right.

21.2, 20.5, 17.0, 16.8, 16.8, 16.5, 16.2, 14.0, 13.7, 13.3, 13.1, 13.0, 12.4, 12.1, 12.0

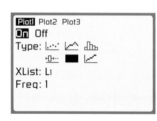

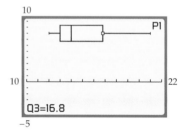

Figure 8.3 *TI-83/84 Plus Screen Displays*

Investigation

Stem-and-Leaf Diagrams

The relative position of each data value in a small set of data can be graphically displayed by using a *stem-and-leaf diagram*. For instance, consider the following history test scores:

65, 72, 96, 86, 43, 61, 75, 86, 49, 68, 98, 74, 84, 78, 85, 75, 86, 73

In the stem-and-leaf diagram on the following page, we have organized the history test scores by placing all of the scores that are in the 40s in the top row, the scores that are

(continued)

A Stem-and-Leaf Diagram of a Set of History Test Scores

Stems	Leaves
4	3 9
5	
6	1 5 8
7	2 3 4 5 5 8
8	4 5 6 6 6
9	6 8

Legend: 8│6 represents 86

in the 50s in the second row, the scores that are in the 60s in the third row, and so on. The tens digits of the scores have been placed to the left of the vertical line. In this diagram they are referred to as *stems*. The ones digits of the test scores have been placed in the proper row to the right of the vertical line. In this diagram they are the *leaves*. It is now easy to make observations about the distribution of the scores. Only two of the scores are in the 90s. Six of the scores are in the 70s, and none of the scores are in the 50s. The lowest score is 43 and the highest is 98.

Steps in the Construction of a Stem-and-Leaf Diagram

1. Determine the stems and list them in a column from smallest to largest or largest to smallest.

2. List the remaining digit of each stem as a leaf to the right of the stem.

3. Include a *legend* that explains the meaning of the stems and the leaves. Include a *title* for the diagram.

The choice of how many leading digits to use as the stem will depend on the particular data set. For instance, consider the following data set, in which a travel agent has recorded the amounts spent by customers for a cruise.

Amounts Spent for a Cruise, Summer of 2005

$3600	$4700	$7200	$2100	$5700	$4400	$9400
$6200	$5900	$2100	$4100	$5200	$7300	$6200
$3800	$4900	$5400	$5400	$3100	$3100	$4500
$4500	$2900	$3700	$3700	$4800	$4800	$2400

One method of choosing the stems is to let each thousands digit be a stem and each hundreds digit be a leaf. If the stems and leaves are assigned in this manner, then the notation 2│1, with a stem of 2 and a leaf of 1, represents a cost of $2100, and 5│4 represents a cost of $5400. A stem-and-leaf diagram can now be constructed by writing all of the stems in a column from smallest to largest to the left of a vertical line, and writing the corresponding leaves to the right of the line.

Amounts Spent for a Cruise

Stems	Leaves
2	1 1 4 9
3	1 1 6 7 7 8
4	1 4 5 5 7 8 8 9
5	2 4 4 7 9
6	2 2
7	2 3
8	
9	4

Legend:
7│3 represents $7300

(continued)

Sometimes two sets of data can be compared by using a *back-to-back stem-and-leaf diagram,* in which common stems are listed in the middle column of the diagram. Leaves from one data set are displayed to the right of the stems, and leaves from the other data set are displayed to the left. For instance, the back-to-back stem-and-leaf diagram below shows the test scores for two classes that took the same test. It is easy to see that the 8 A.M. class did better on the test because it had more scores in the 80s and 90s and fewer scores in the 40s, 50s, and 60s. The number of scores in the 70s was the same for both classes.

Biology Test Scores

8 A.M. Class		10 A.M. Class
2	4	5 8
7	5	6 7 9 9
5 8	6	2 3 4 8
1 2 3 3 3 7 8	7	1 3 3 5 5 6 8
4 4 5 5 6 8 8 9	8	2 3 6 6 6
2 4 5 5 8	9	4 5

Legend: 3 | 7
represents 73

Legend: 8 | 2
represents 82

Investigation Exercises

1. The following table lists the ages of customers who purchased a cruise. Construct a stem-and-leaf diagram for the data.

Ages of Customers Who Purchased a Cruise

32	45	66	21	62	68	72
61	55	23	38	44	77	64
46	50	33	35	42	45	51
51	28	40	41	52	52	33

2. Construct a back-to-back stem-and-leaf diagram for the winning and losing scores given in Exercise 19 in Section 8.4 (page 549). What information is revealed by your diagram?

3. Construct a back-to-back stem-and-leaf diagram for the data given in the two tables in Exercise 20 in Section 8.4 (pages 549–550). What information is revealed by your diagram?

4. Construct a back-to-back stem-and-leaf diagram for the data given in the two tables in Exercise 21 in Section 8.4 (page 550). What information is revealed by your diagram?

Exercise Set 8.5

In Exercises 1–4, round each z-score to the nearest hundredth.

1. A data set has a mean of $\bar{x} = 75$ and a standard deviation of 11.5. Find the z-score for each of the following.
   - **a.** $x = 85$ **b.** $x = 95$
   - **c.** $x = 50$ **d.** $x = 75$

2. A data set has a mean of $\bar{x} = 212$ and a standard deviation of 40. Find the z-score for each of the following.
   - **a.** $x = 200$ **b.** $x = 224$
   - **c.** $x = 300$ **d.** $x = 100$

3. A data set has a mean of $\bar{x} = 6.8$ and a standard deviation of 1.9. Find the z-score for each of the following.
   - **a.** $x = 6.2$ **b.** $x = 7.2$
   - **c.** $x = 9.0$ **d.** $x = 5.0$

4. A data set has a mean of $\bar{x} = 4010$ and a standard deviation of 115. Find the z-score for each of the following.
   - **a.** $x = 3840$ **b.** $x = 4200$
   - **c.** $x = 4300$ **d.** $x = 4030$

5. If the z-score for a certain data item is negative, is the data item negative, is it positive, or is there insufficient information to answer the question?

6. Suppose the z-score for a certain data item x_1 is less than the z-score for a second data item x_2 from the same data set. Is $x_1 < x_2$, is $x_1 = x_2$, is $x_1 > x_2$, or is there insufficient information to answer the question?

7. **Blood Pressure** A blood pressure test was given to 450 women ages 20 to 36. It showed that their mean systolic blood pressure was 119.4 mm Hg, with a standard deviation of 13.2 mm Hg.
   - **a.** Determine the z-score, to the nearest hundredth, for a woman who had a systolic blood pressure reading of 110.5 mm Hg.
   - **b.** The z-score for one woman was 2.15. What was her systolic blood pressure reading?

8. **Fruit Juice** A random sample of 1000 oranges showed that the mean amount of juice per orange was 7.4 fluid ounces, with a standard deviation of 1.1 fluid ounces.

   - **a.** Determine the z-score, to the nearest hundredth, of an orange that produced 6.6 fluid ounces of juice.
   - **b.** The z-score for one orange was 3.15. How much juice was produced by this orange? Round to the nearest tenth of a fluid ounce.

9. **Cholesterol** A test involving 380 men ages 20 to 24 found that their blood cholesterol levels had a mean of 182 mg/dl and a standard deviation of 44.2 mg/dl.
   - **a.** Determine the z-score, to the nearest hundredth, for one of the men who had a blood cholesterol level of 214 mg/dl.
   - **b.** The z-score for one man was −1.58. What was his blood cholesterol level? Round to the nearest hundredth.

10. **Tire Wear** A random sample of 80 tires showed that the mean mileage per tire was 41,700 miles, with a standard deviation of 4300 miles.
   - **a.** Determine the z-score, to the nearest hundredth, for a tire with a life of 46,300 miles.
   - **b.** The z-score for one tire was −2.44. What mileage did this tire provide? Round your result to the nearest hundred miles.

11. **Test Scores** Which of the following three test scores is the highest relative score?
   - **a.** A score of 65 on a test with a mean of 72 and a standard deviation of 8.2
   - **b.** A score of 102 on a test with a mean of 130 and a standard deviation of 18.5
   - **c.** A score of 605 on a test with a mean of 720 and a standard deviation of 116.4

12. **Physical Fitness** Which of the following fitness scores is the highest relative score?
   - **a.** A score of 42 on a test with a mean of 31 and a standard deviation of 6.5
   - **b.** A score of 1140 on a test with a mean of 1080 and a standard deviation of 68.2
   - **c.** A score of 4710 on a test with a mean of 3960 and a standard deviation of 560.4

13. **Reading Test** On a reading test, Shaylen's score of 455 was higher than the scores of 4256 of the 7210 students who took the test. Find the percentile, rounded to the nearest percent, for Shaylen's score.

14. Placement Exam On a placement examination, Rick scored lower than 1210 of the 12,860 students who took the exam. Find the percentile, rounded to the nearest percent, for Rick's score.

15. Test Scores Kevin scored at the 65th percentile on a test given to 9840 students. How many students scored lower than Kevin?

16. Test Scores Rene scored at the 84th percentile on a test given to 12,600 students. How many students scored higher than Rene?

17. Median Income In 2004, the median family income in the United States was $57,500. (*Source:* U.S. Department of Housing and Urban Development) If the 88th percentile for the 2004 median four-person family income was $70,400, find the percents of families whose 2004 incomes were

   a. more than $57,500.   **b.** more than $70,400.

   c. between $57,500 and $70,400.

18. What percentile is represented by Q_1, the first quartile? What percentile is represented by the median? What percentile is represented by Q_3, the third quartile?

19. When creating a box-and-whisker plot, which of the following are used: median, mode, standard deviation, range, or z-score?

20. Monthly Rents A recent survey by the U.S. Census Bureau determined that the median monthly housing rent was $632. If the first quartile for monthly housing rent was $497, find the percents of monthly housing rents that were

   a. more than $497.   **b.** less than $632.

   c. between $497 and $632.

21. Commute to School A survey was given to 18 students. One question asked about the one-way distances the students had to travel to attend college. The results, in miles, are shown in the following table. Use the median procedure for finding quartiles to find the first, second, and third quartiles of the data.

Distances Traveled to Attend College								
12	18	4	5	26	41	1	8	10
10	3	28	32	10	85	7	5	15

22. Prescriptions The following table shows the numbers of prescriptions a doctor wrote each day for a 36-day period. Use the median procedure for finding quartiles to find the first, second, and third quartiles of the data.

Numbers of Prescriptions Written per Day					
8	12	14	10	9	16
7	14	10	7	11	16
11	12	8	14	13	10
9	14	15	12	10	8
10	14	8	7	12	15
14	10	9	15	10	12

23. **Earned Run Averages** The following table shows the earned run averages (ERAs) for the baseball teams in the American League and the National League during the 2004 season. Draw box-and-whisker plots for the teams' ERAs in each league. Place one plot above the other so that you can compare the data. Write a few sentences that explain any differences you found by using the box-and-whisker plots.

Team Pitching Statistics, 2004 Regular Season Earned Run Averages (ERAs)

American League		National League	
Minnesota	4.04	Atlanta	3.75
Oakland	4.17	St. Louis	3.75
Boston	4.19	Chicago	3.83
Anaheim	4.28	Los Angeles	4.01
Texas	4.54	San Diego	4.03
New York	4.69	Houston	4.05
Baltimore	4.71	Florida	4.10
Seattle	4.76	New York	4.10
Tampa Bay	4.82	Milwaukee	4.26
Cleveland	4.82	Pittsburgh	4.31
Chicago	4.91	Montréal	4.33
Toronto	4.93	San Francisco	4.34
Detroit	4.93	Philadelphia	4.47
Kansas City	5.16	Arizona	4.98
		Cincinnati	5.21
		Colorado	5.54

Source: STATS, Inc.

24. **Super Bowl** The following data are the points scored during the regular season by the two teams that played in the 2004 Super Bowl.

Points Scored by New England Patriots							
0	31	23	17	38	17	19	9
30	12	23	38	12	27	21	31

Points Scored by Carolina Panthers							
24	12	23	19	23	17	23	10
27	20	20	16	14	20	20	37

Draw a box-and-whisker plot of each set of data, placing one plot above the other. Write a few sentences explaining any differences you found between the two plots.

25. **Salary Comparison** The table below shows the median weekly salaries for women and men for selected occupations. Draw box-and-whisker plots for these data. Place one plot above the other so that you can compare the data. Write a few sentences that explain any differences you found.

Occupation	Women's Median Weekly Earnings	Men's Median Weekly Earnings
Chief executive	$1243	$1736
Operations manager	$966	$1170
Systems manager	$904	$1271
Financial manager	$1280	$1437
Human resources dir.	$823	$1314
Purchasing manager	$844	$1297
Education administrator	$878	$1172
Medical services mgr.	$954	$1149
Property manager	$638	$849
Claims adjuster	$648	$868
Management analyst	$977	$1267
Auditor	$756	$1041

Source: U.S. Department of Labor, September 2004.

26. **Home Sales** The following table shows the median sales prices of existing single-family homes in the United States in the four regions of the country for the years 1993 through 2003. Prices have been rounded to the nearest hundred.

a. Draw a box-and-whisker plot of the data for each of the four regions. Write a few sentences that explain any differences you found.

b. Use the Internet to find the median price of a single-family home in your region of the United States for the current year. How does this median price compare with the 2003 median price shown in the following table?

Median Prices of Homes Sold in the United States, 1993–2003

Year	Northeast	Midwest	South	West
1993	162,600	125,000	115,000	135,000
1994	169,000	132,900	116,900	140,400
1995	180,000	134,000	124,500	141,000
1996	186,900	137,500	125,000	153,900
1997	190,000	149,900	129,600	160,000
1998	200,000	157,500	135,800	163,500
1999	210,500	164,000	145,900	173,700
2000	227,400	169,700	148,000	196,400
2001	246,400	172,600	155,400	213,600
2002	264,300	178,000	163,400	238,500
2003	264,500	184,300	168,100	260,900

Source: National Association of REALTORS.

Extensions

CRITICAL THINKING

27. a. The population 3, 4, 9, 14, and 20 has a mean of 10 and a standard deviation of 6.356. The z-scores for each of the five data values are $z_3 \approx -1.101$, $z_4 \approx -0.944$, $z_9 \approx -0.157$, $z_{14} \approx 0.629$, and $z_{20} \approx 1.573$. Find the mean and the standard deviation of these z-scores.

b. The population 2, 6, 12, 17, 22, and 25 has a mean of 14 and a standard deviation of 8.226. The z-scores for each of the six data values are $z_2 \approx -1.459$, $z_6 \approx -0.973$, $z_{12} \approx -0.243$, $z_{17} \approx 0.365$, $z_{22} \approx 0.973$, and $z_{25} \approx 1.337$. Find the mean and the standard deviation of these z-scores.

c. Use the results of part a and part b to make a conjecture about the mean and standard deviation of the z-scores for any set of data.

28. For each of the following, determine whether the statement is true or false.

a. For any given set of data, the median of the data equals the mean of Q_1 and Q_3.

b. For any given set of data, $Q_3 - Q_2 = Q_2 - Q_1$.

c. A *z*-score for a given data value *x* in a set of data can be a negative number.

d. If a student answers 75% of the questions on a test correctly, then the student's score on the test will place the student at the 75th percentile.

EXPLORATIONS

29. Some data sets include values so large or so small that they differ significantly from the rest of the data. Such data values are referred to as **outliers.** An outlier may be the result of an error, such as an incorrect measurement or a recording error, or it may be a legitimate data value. Consult a statistics text to find the mathematical formula that is used to determine whether a given data value in a set of data is an outlier. Use the formula to find all outliers for the data set in Exercise 21 on page 562.

SECTION 8.6 | **Normal Distributions**

Frequency Distributions and Histograms

Large sets of data are often displayed using *grouped frequency distributions* or *histograms.* For instance, consider the following situation. An Internet service provider (ISP) has installed new computers. To estimate the new download times its subscribers will experience, the ISP surveyed 1000 of its subscribers to determine the time required for each subscriber to download a particular file from the Internet site music.net. The results of that survey are summarized in Table 8.4.

Table 8.4 *A Grouped Frequency Distribution with 12 Classes*

Download Time (in seconds)	Number of Subscribers
0–5	6
5–10	17
10–15	43
15–20	92
20–25	151
25–30	192
30–35	190
35–40	149
40–45	90
45–50	45
50–55	15
55–60	10

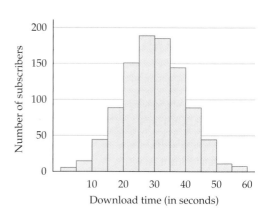

Figure 8.4 *A Histogram for the Frequency Distribution in Table 8.4*

Table 8.4 is called a **grouped frequency distribution.** It shows how often (frequently) certain events occurred. Each interval, 0–5, 5–10, and so on, is called a **class.** This distribution has 12 classes. For the 10–15 class, 10 is the **lower class**

boundary and 15 is the **upper class boundary.** Any data value that lies on a common boundary is assigned to the higher class. The *graph* of a frequency distribution is called a **histogram.** A histogram provides a pictorial view of how the data are distributed. In Figure 8.4 on the preceding page, the height of each bar of the histogram indicates how many subscribers experienced the download times shown by the class at the base of the bar.

Examine the distribution in Table 8.5. It shows the *percent* of subscribers that are in each class, as opposed to the frequency distribution in Table 8.4 on the preceding page, which shows the *number* of customers in each class. The type of frequency distribution that lists the *percent* of data in each class is called a **relative frequency distribution.** The **relative frequency histogram** in Figure 8.5 was drawn by using the data in the relative frequency distribution. It shows the *percent* of subscribers along its vertical axis.

Table 8.5 *A Relative Frequency Distribution*

Download Time (in seconds)	Percent of Subscribers
0–5	0.6
5–10	1.7
10–15	4.3
15–20	9.2
20–25	15.1
25–30	19.2
30–35	19.0
35–40	14.9
40–45	9.0
45–50	4.5
50–55	1.5
55–60	1.0

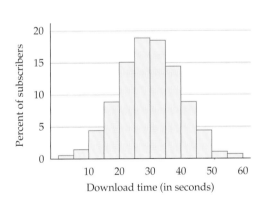

Figure 8.5 *A Relative Frequency Histogram*

One advantage of using a relative frequency distribution instead of a grouped frequency distribution is that there is a direct correspondence between the percent values of the relative frequency distribution and probabilities. For instance, in the relative frequency distribution in Table 8.5, the percent of the data that lie between 35 and 40 seconds is 14.9%. Thus, if a subscriber is chosen at random, the probability that the subscriber will require at least 35 seconds but less than 40 seconds to download the music file is 0.149.

EXAMPLE 1 ■ **Use a Relative Frequency Distribution**

Use the relative frequency distribution in Table 8.6 to determine

a. the *percent* of subscribers who required at least 25 seconds to download the file.

b. the *probability* that a subscriber chosen at random will require at least 5 but less than 20 seconds to download the file.

Solution

a. The percent of data in all the classes with a lower boundary of 25 seconds or more is the sum of the percents for all of the classes highlighted in red in the distribution below. Thus the percent of subscribers who required at least 25 seconds to download the file is 69.1%. See Table 8.6.

Table 8.6

Download Time (in seconds)	Percent of Subscribers	
0–5	0.6	
5–10	1.7	
10–15	4.3	Sum is 15.2%
15–20	9.2	
20–25	15.1	
25–30	19.2	
30–35	19.0	
35–40	14.9	
40–45	9.0	Sum is 69.1%
45–50	4.5	
50–55	1.5	
55–60	1.0	

b. The percent of data in all the classes with a lower boundary of at least 5 seconds and an upper boundary of 20 seconds or less is the sum of the percents in all of the classes highlighted in blue in the distribution above. Thus the percent of subscribers who required at least 5 but less than 20 seconds to download the file is 15.2%. The probability that a subscriber chosen at random will require at least 5 but less than 20 seconds to download the file is 0.152. See Table 8.6.

CHECK YOUR PROGRESS 1 Use the relative frequency distribution in Table 8.5 to determine

a. the *percent* of subscribers who required less than 25 seconds to download the file.

b. the *probability* that a subscriber chosen at random will require at least 10 seconds but less than 30 seconds to download the file.

Solution See page S35.

There is a geometric analogy between percent of data and probabilities and the relative frequency histogram for the data. For instance, the percent of data described in part a of Example 1 corresponds to the *area* represented by the red bars in the histogram in Figure 8.6. The percent of data described in part b of Example 1 corresponds to the area represented by the blue bars in the histogram in Figure 8.7.

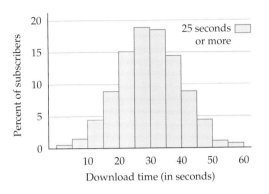

Figure 8.6

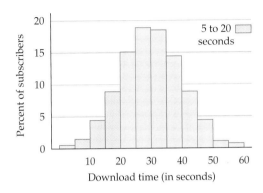

Figure 8.7

Normal Distributions and The Empirical Rule

A histogram of a set of data provides us with a tool that can indicate patterns or trends in the distribution of the data. The terms *uniform, bimodal, symmetrical, skewed,* and *normal* are used to describe the distributions of sets of data.

A **uniform distribution,** shown in the figure below, occurs when all of the observed events occur with the same frequency. The graph of a uniform distribution remains at the same height over the range of the data. Some random processes produce distributions that are uniform or nearly uniform. For example, if the spinner below is used to generate numbers, then in the *long run* each of the numbers 1, 2, 3, ..., 8 will be generated with approximately the same frequency.

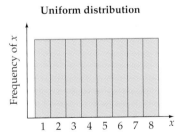

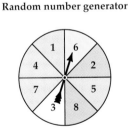

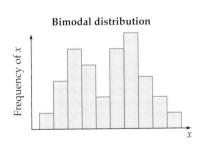

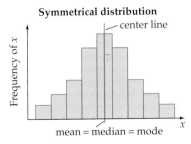

A **bimodal distribution,** shown in the histogram at the left, is produced when two *nonadjacent* classes occur more frequently than any of the other classes. Bimodal distributions are often produced by samples that contain data from two very different populations.

A **symmetrical distribution,** shown at the left, is symmetrical about a vertical center line. If you fold a symmetrical distribution along the center line, the right side of the distribution will match the left side. The following sets of data are examples of distributions that are nearly symmetrical: the weights of all male students, the heights of all teenage females, the prices of a gallon of regular gasoline in a large city, the mileages for a particular type of automobile tire, and the amounts of soda dispensed by a vending machine per day. In a symmetrical distribution, the mean, the median, and the mode are all equal and are located at the center of the distribution.

Skewed distributions, shown in the figures below, can be identified by the fact that their distributions have a longer *tail* on one side of the distribution and a shorter tail on the other side. A distribution is skewed to the *left* if it has a longer tail on the left and is skewed to the *right* if it has a longer tail on the right. In a distribution that is skewed to the left, the mean is less than the median, which is less than the mode. In a distribution that is skewed to the right, the mode is less than the median, which is less than the mean.

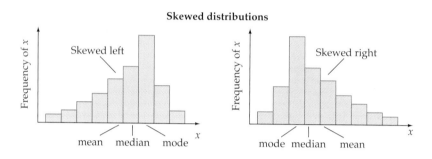

Skewed distributions

Many examinations yield test scores that have skewed distributions. For instance, if a test designed for students in the sixth grade is given to students in a ninth-grade class, most of the scores will be high. The distribution of the test scores will be skewed to the left, as shown above.

Discrete data are separated from each other by an increment, or "space." For instance, only whole numbers are used to record the number of points that a basketball player scores in a game. The possible numbers of points that the player can score, which we will represent by s, are restricted to 0, 1, 2, 3, 4, …. The variable s is a **discrete variable.** Different scores are separated from each other by at least 1. A variable that is based on counting procedures is a discrete variable. Histograms are generally used to show the distributions of discrete variables.

Continuous data can take on the values of all real numbers in some interval. For example, the possible times that it takes to drive to the grocery store are continuous data. The times are not restricted to natural numbers such as 4 minutes or 5 minutes. In fact, the time may be any part of a minute, or even of a second, if we care to measure that precisely. A variable such as the time t, that is based on *measuring* with smaller and smaller units, is called a **continuous variable.** Continuous curves, rather than histograms, are used to show the distributions of continuous variables.

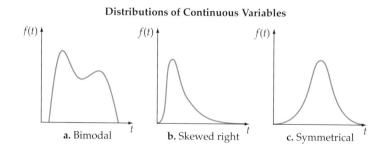

Distributions of Continuous Variables

In some cases we use a continuous curve to display the distribution of a set of discrete data. For instance, in those cases in which we have a large set of data and very small class intervals, the shape of the top of the histogram approaches a smooth curve. See the two figures below. When graphing the distributions of very large sets of data with very small class intervals, it is common practice to replace the histogram with a smooth continuous curve.

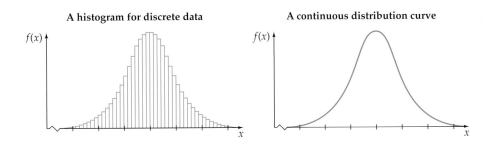

One of the most important statistical distributions is known as a *normal distribution*. The precise mathematical definition of a normal distribution is given by the equation in the Take Note at the left. However, for many applied problems, it is sufficient to know that all normal distributions have the following properties.

TAKE NOTE

If *x* is a continuous variable with mean μ and standard deviation σ, then its **normal distribution** is given by

$$f(x) = \frac{e^{-(1/2)(x-\mu/\sigma)^2}}{\sigma\sqrt{2\pi}}$$

Recall from Section 6.2 that the irrational number *e* is about 2.7182818.

Properties of a Normal Distribution

The distribution of data in a normal distribution has a bell shape that is symmetrical about a vertical line through its center. The mean, the median, and the mode of a normal distribution are all equal and they are located at the center of the distribution.

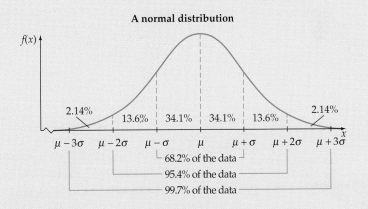

A normal distribution

The Empirical Rule: In a normal distribution, about

68.2% of the data lie within one standard deviation of the mean.
95.4% of the data lie within two standard deviations of the mean.
99.7% of the data lie within three standard deviations of the mean.

The Empirical Rule can be used to solve many applied problems.

EXAMPLE 2 ■ Use the Empirical Rule to Solve an Application

A survey in 2004 of 1000 U.S. gas stations found that the price charged for a gallon of regular gas could be closely approximated by a normal distribution with a mean of $1.88 and a standard deviation of $0.20. How many of the stations charged

a. between $1.48 and $2.28 for a gallon of regular gas?

b. less than $2.08 for a gallon of regular gas?

c. more than $2.28 for a gallon of regular gas?

Solution

a. The $1.48 per gallon price is two standard deviations below the mean. The $2.28 price is two standard deviations above the mean. In a normal distribution, 95.4% of all data lie within two standard deviations of the mean. See Figure 8.8. Therefore, approximately

$$(95.4\%)(1000) = (0.954)(1000) = 954$$

of the stations charged between $1.48 and $2.28 for a gallon of regular gas.

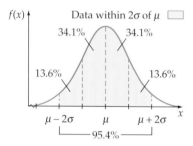

Figure 8.8

b. The $2.08 price is one standard deviation above the mean. See Figure 8.9. In a normal distribution, 34.1% of all data lie between the mean and one standard deviation above the mean. Thus, approximately

$$(34.1\%)(1000) = (0.341)(1000) = 341$$

of the stations charged between $1.88 and $2.08 for a gallon of regular gasoline. Half of the stations charged less than the mean. Therefore, about $341 + 500 = 841$ of the stations charged less than $2.08 for a gallon of regular gas.

c. The $2.28 price is two standard deviations above the mean. In a normal distribution, 95.4% of all data are within two standard deviations of the mean. This means that the other 4.6% of the data will lie either above two standard deviations above the mean or below two standard deviations below the mean. We are interested only in the data that are more than two standard deviations above the mean, which is $\frac{1}{2}$ of 4.6%, or 2.3%, of the data. See Figure 8.10. Thus, about $(2.3\%)(1000) = (0.023)(1000) = 23$ of the stations charged more than $2.28 for a gallon of regular gas.

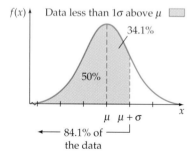

Figure 8.9

CHECK YOUR PROGRESS 2 A vegetable distributor knows that during the month of August, the weights of its tomatoes are normally distributed with a mean of 0.61 pound and a standard deviation of 0.15 pound.

a. What percent of the tomatoes weigh less than 0.76 pound?

b. In a shipment of 6000 tomatoes, how many tomatoes can be expected to weigh more than 0.31 pound?

c. In a shipment of 4500 tomatoes, how many tomatoes can be expected to weigh from 0.31 pound to 0.91 pound?

Solution See page S35.

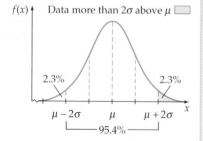

Figure 8.10

QUESTION   *Can the Empirical Rule be applied to all data sets?*

The Standard Normal Distribution

It is often helpful to convert the values of a continuous variable x to z-scores, as we did in the preceding section by using the z-score formulas

$$z_x = \frac{x - \bar{x}}{s} \quad \text{or} \quad z_x = \frac{x - \mu}{\sigma}$$

If the original distribution of x values is a normal distribution, then the corresponding distribution of z-scores will also be a normal distribution. This normal distribution of z-scores is called the *standard normal distribution*. See Figure 8.11. It has a mean of 0 and a standard deviation of 1, and it was first used by the French mathematician Abraham De Moivre (1667–1754).

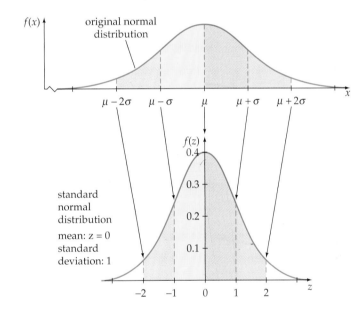

Figure 8.11 *Conversion of a Normal Distribution to the Standard Normal Distribution*

The Standard Normal Distribution

The **standard normal distribution** is the normal distribution for the continuous variable z that has a mean of 0 and a standard deviation of 1.

Tables and calculators are often used to determine the area of a portion of the standard normal distribution. For example, Table 8.7 gives the approximate areas of the standard normal distribution between the mean 0 and z standard deviations from the mean. Table 8.7 indicates that the area A of the standard normal distribution from the mean 0 up to $z = 1.34$ is 0.410 square unit.

ANSWER   *No. The Empirical Rule can only be applied to normal distributions.*

Table 8.7 *Areas Under the Standard Normal Curve*

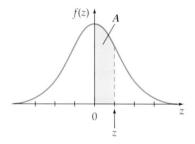

Figure 8.12 *A is the area of the shaded region*

z	A	z	A	z	A	z	A	z	A	z	A
.00	.000	.56	.212	1.12	.369	1.68	.454	2.24	.487	2.80	.497
.01	.004	.57	.216	1.13	.371	1.69	.454	2.25	.488	2.81	.498
.02	.008	.58	.219	1.14	.373	1.70	.455	2.26	.488	2.82	.498
.03	.012	.59	.222	1.15	.375	1.71	.456	2.27	.488	2.83	.498
.04	.016	.60	.226	1.16	.377	1.72	.457	2.28	.489	2.84	.498
.05	.020	.61	.229	1.17	.379	1.73	.458	2.29	.489	2.85	.498
.06	.024	.62	.232	1.18	.381	1.74	.459	2.30	.489	2.86	.498
.07	.028	.63	.236	1.19	.383	1.75	.460	2.31	.490	2.87	.498
.08	.032	.64	.239	1.20	.385	1.76	.461	2.32	.490	2.88	.498
.09	.036	.65	.242	1.21	.387	1.77	.462	2.33	.490	2.89	.498
.10	.040	.66	.245	1.22	.389	1.78	.462	2.34	.490	2.90	.498
.11	.044	.67	.249	1.23	.391	1.79	.463	2.35	.491	2.91	.498
.12	.048	.68	.252	1.24	.393	1.80	.464	2.36	.491	2.92	.498
.13	.052	.69	.255	1.25	.394	1.81	.465	2.37	.491	2.93	.498
.14	.056	.70	.258	1.26	.396	1.82	.466	2.38	.491	2.94	.498
.15	.060	.71	.261	1.27	.398	1.83	.466	2.39	.492	2.95	.498
.16	.064	.72	.264	1.28	.400	1.84	.467	2.40	.492	2.96	.498
.17	.067	.73	.267	1.29	.401	1.85	.468	2.41	.492	2.97	.499
.18	.071	.74	.270	1.30	.403	1.86	.469	2.42	.492	2.98	.499
.19	.075	.75	.273	1.31	.405	1.87	.469	2.43	.492	2.99	.499
.20	.079	.76	.276	1.32	.407	1.88	.470	2.44	.493	3.00	.499
.21	.083	.77	.279	1.33	.408	1.89	.471	2.45	.493	3.01	.499
.22	.087	.78	.282	1.34	.410	1.90	.471	2.46	.493	3.02	.499
.23	.091	.79	.285	1.35	.411	1.91	.472	2.47	.493	3.03	.499
.24	.095	.80	.288	1.36	.413	1.92	.473	2.48	.493	3.04	.499
.25	.099	.81	.291	1.37	.415	1.93	.473	2.49	.494	3.05	.499
.26	.103	.82	.294	1.38	.416	1.94	.474	2.50	.494	3.06	.499
.27	.106	.83	.297	1.39	.418	1.95	.474	2.51	.494	3.07	.499
.28	.110	.84	.300	1.40	.419	1.96	.475	2.52	.494	3.08	.499
.29	.114	.85	.302	1.41	.421	1.97	.476	2.53	.494	3.09	.499
.30	.118	.86	.305	1.42	.422	1.98	.476	2.54	.494	3.10	.499
.31	.122	.87	.308	1.43	.424	1.99	.477	2.55	.495	3.11	.499
.32	.126	.88	.311	1.44	.425	2.00	.477	2.56	.495	3.12	.499
.33	.129	.89	.313	1.45	.426	2.01	.478	2.57	.495	3.13	.499
.34	.133	.90	.316	1.46	.428	2.02	.478	2.58	.495	3.14	.499
.35	.137	.91	.319	1.47	.429	2.03	.479	2.59	.495	3.15	.499
.36	.141	.92	.321	1.48	.431	2.04	.479	2.60	.495	3.16	.499
.37	.144	.93	.324	1.49	.432	2.05	.480	2.61	.495	3.17	.499
.38	.148	.94	.326	1.50	.433	2.06	.480	2.62	.496	3.18	.499
.39	.152	.95	.329	1.51	.434	2.07	.481	2.63	.496	3.19	.499
.40	.155	.96	.331	1.52	.436	2.08	.481	2.64	.496	3.20	.499
.41	.159	.97	.334	1.53	.437	2.09	.482	2.65	.496	3.21	.499
.42	.163	.98	.336	1.54	.438	2.10	.482	2.66	.496	3.22	.499
.43	.166	.99	.339	1.55	.439	2.11	.483	2.67	.496	3.23	.499
.44	.170	1.00	.341	1.56	.441	2.12	.483	2.68	.496	3.24	.499
.45	.174	1.01	.344	1.57	.442	2.13	.483	2.69	.496	3.25	.499
.46	.177	1.02	.346	1.58	.443	2.14	.484	2.70	.497	3.26	.499
.47	.181	1.03	.348	1.59	.444	2.15	.484	2.71	.497	3.27	.499
.48	.184	1.04	.351	1.60	.445	2.16	.485	2.72	.497	3.28	.499
.49	.188	1.05	.353	1.61	.446	2.17	.485	2.73	.497	3.29	.499
.50	.191	1.06	.355	1.62	.447	2.18	.485	2.74	.497	3.30	.500
.51	.195	1.07	.358	1.63	.448	2.19	.486	2.75	.497	3.31	.500
.52	.198	1.08	.360	1.64	.449	2.20	.486	2.76	.497	3.32	.500
.53	.202	1.09	.362	1.65	.451	2.21	.486	2.77	.497	3.33	.500
.54	.205	1.10	.364	1.66	.452	2.22	.487	2.78	.497		
.55	.209	1.11	.367	1.67	.453	2.23	.487	2.79	.497		

Because the standard normal distribution is symmetrical about the mean of 0, we can also use Table 8.7 to find the area of a region that is located to the left of the mean. This process is explained in Example 3.

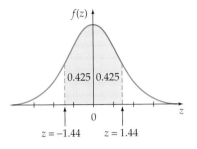

Figure 8.13 *Symmetrical Region*

EXAMPLE 3 ■ **Use Symmetry to Determine an Area**

Find the area of the standard normal distribution between $z = -1.44$ and $z = 0$.

Solution
Because the standard normal distribution is symmetrical about the center line $z = 0$, the area of the standard normal distribution between $z = -1.44$ and $z = 0$ is equal to the area between $z = 0$ and $z = 1.44$. See Figure 8.13. The entry in Table 8.7 associated with $z = 1.44$ is 0.425. Thus the area of the standard normal distribution between $z = -1.44$ and $z = 0$ is 0.425 square unit.

CHECK YOUR PROGRESS 3 Find the area of the standard normal distribution between $z = -0.67$ and $z = 0$.

Solution *See page S35.*

In Figure 8.14, the region to the right of $z = 0.82$ is called a *tail region*. A **tail region** is a region of the standard normal distribution to the right of a positive z-value or to the left of a negative z-value. To find the area of a tail region, we subtract the entry in Table 8.7 from 0.500. This procedure is illustrated in the next example.

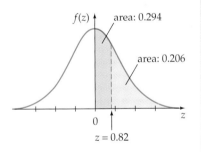

Figure 8.14 *Area of a Tail Region*

EXAMPLE 4 ■ **Find the Area of a Tail Region**

Find the area of the standard normal distribution to the right of $z = 0.82$.

Solution
Table 8.7 indicates that the area from $z = 0$ to $z = 0.82$ is 0.294 square unit. The area to the right of $z = 0$ is 0.500 square unit. Thus the area to the right of $z = 0.82$ is $0.500 - 0.294 = 0.206$ square unit. See Figure 8.14.

CHECK YOUR PROGRESS 4 Find the area of the standard normal distribution to the left of $z = -1.47$.

Solution *See page S35.*

> **The Standard Normal Distribution, Areas, Percents, and Probabilities**
>
> In the standard normal distribution, the area under the graph of the distribution over an interval is the same as
>
> ■ the percent of all the data in the population that have z-scores that lie in the interval.
>
> ■ the probability that the z-score of a data value selected at random will lie in the interval.

Because the area of a portion of the standard normal distribution can be interpreted as a percent of the data or as a probability that the variable lies in an interval, we can use the standard normal distribution to solve many application problems.

EXAMPLE 5 ■ Solve an Application

A soda machine dispenses soda into 14-ounce cups. Tests show that the actual amount of soda dispensed is normally distributed, with a mean of 12 ounces and a standard deviation of 0.8 ounce.

a. What percent of cups will receive less than 11 ounces of soda?

b. What percent of cups will receive between 10.8 ounces and 12.2 ounces of soda?

c. If a cup is chosen at random, what is the probability that the machine will overflow the cup?

Solution

a. The z-score for 11 ounces is

$$z_{11} = \frac{11 - 12}{0.8} = -1.25$$

Table 8.7 indicates that 0.394 (39.4%) of the data in a normal distribution are between $z = 0$ and $z = 1.25$. Because the data are normally distributed, 39.4% of the data are also between $z = 0$ and $z = -1.25$. The percent of data to the left of $z = -1.25$ is 50% − 39.4% = 10.6%. See Figure 8.15. Thus 10.6% of the cups filled by the soda machine will receive less than 11 ounces of soda.

b. The z-score for 12.2 ounces is

$$z_{12.2} = \frac{12.2 - 12}{0.8} = 0.25$$

Table 8.7 indicates that 0.099 (9.9%) of the data in a normal distribution are between $z = 0$ and $z = 0.25$.

The z-score for 10.8 ounces is

$$z_{10.8} = \frac{10.8 - 12}{0.8} = -1.5$$

Table 8.7 indicates that 0.433 (43.3%) of the data in a normal distribution are between $z = 0$ and $z = 1.5$. Because the data are normally distributed, 43.3% of the data are also between $z = 0$ and $z = -1.5$. See Figure 8.16. Thus the percent of the cups that the vending machine will fill with between 10.8 ounces and 12.2 ounces of soda is

$$43.3\% + 9.9\% = 53.2\%$$

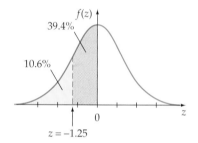

Figure 8.15 *Portion of Data to the Left of $z = -1.25$*

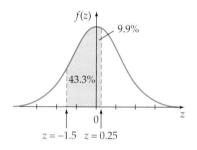

Figure 8.16 *Portion of Data Between Two z-Scores*

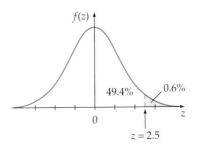

Figure 8.17 *Portion of Data to the Right of z = 2.5*

c. A cup will overflow if it receives more than 14 ounces of soda. The *z*-score for 14 ounces is

$$z_{14} = \frac{14 - 12}{0.8} = 2.5$$

Table 8.7 indicates that 0.494 (49.4%) of the data in the standard normal distribution are between $z = 0$ and $z = 2.5$. The percent of data to the right of $z = 2.5$ is determined by subtracting 49.4% from 50%. See Figure 8.17. Thus 0.6% of the time the machine produces an overflow, and the probability that a cup chosen at random will overflow is 0.006.

CHECK YOUR PROGRESS 5 A study of the careers of professional football players shows that the lengths of their careers are nearly normally distributed, with a mean of 6.1 years and a standard deviation of 1.8 years.

a. What percent of professional football players have a career of more than 9 years?

b. If a professional football player is chosen at random, what is the probability that the player will have a career of between 3 and 4 years?

Solution *See page S35.*

MathMatters **Find the Area of a Portion of the Standard Normal Distribution by Using a Calculator**

Some calculators can be used to find the area of a portion of the standard normal distribution. For instance, the TI-83/84 Plus screen displays in Figure 8.18 both indicate that the area of the standard normal distribution from a lower bound of $z = 0$ to an upper bound of $z = 1.34$ is about 0.409877 square unit. This is a more precise value than the entry given in Table 8.7, which is 0.410.

Select the `normalcdf(` function from the `DISTR` menu. Enter your lower bound, followed by your upper bound. Press [ENTER].

The `ShadeNorm` instruction in the `DISTR, DRAW` menu draws the standard normal distribution and shades the area between your lower bound and your upper bound.

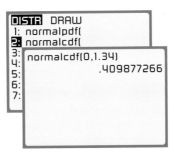

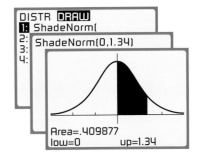

Figure 8.18 *TI-83/84 Plus Screen Displays*

Investigation

Cut-Off Scores

There are many applications for which it is necessary to find a *cut-off score,* which is a score that separates data into two groups such that the data in one group satisfy a certain requirement and the data in the other group do not satisfy the requirement. If the data are normally distributed, then we can find a cut-off score by the method shown in the following example.

EXAMPLE

The OnTheGo Company manufactures laptop computers. A study indicates that the life spans of its computers are normally distributed, with a mean of 4.0 years and a standard deviation of 1.2 years. For how long should the company warrant its computers if the company wishes less than 4% of its computers to fail during the warranty period?

Solution

Figure 8.19 shows a standard normal distribution with 4% of the data to the left of some unknown z-score and 46% of the data to the right of the z-score but to the left of the mean of 0. Using Table 8.7, we find that the z-score associated with an area of $A = 0.46$ is 1.75. Our unknown z-score is to the left of 0, so it must be negative. Thus $z_x = -1.75$. If we let x represent the time in years that a computer is in use, then x is related to the z-scores by the formula

$$z_x = \frac{x - \bar{x}}{s}$$

Solving for x with $\bar{x} = 4.0$, $s = 1.2$, and $z = -1.75$ gives us

$$-1.75 = \frac{x - 4.0}{1.2}$$

$$(-1.75)(1.2) = x - 4.0$$

$$x = 4.0 - 2.1$$

$$x = 1.9 \qquad \text{The cut-off score}$$

Hence the company can provide a 1.9-year warranty and expect less than 4% of its computers to fail during the warranty period.

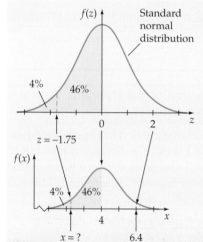

Figure 8.19 *Finding a Cut-Off Score*

Investigation Exercises

1. A professor finds that the grades in a large class are normally distributed. The mean of the grades is 64 and the standard deviation is 10. If the professor decides to give an A grade to the students in the top 9% of the class, what is the cut-off score for an A?

(continued)

2. The results of a statewide examination of the reading skills of sixth-grade students were normally distributed, with a mean score of 104 and a standard deviation of 16. The students in the top 10% are to receive an award, and those in the bottom 14% will be required to take a special reading class.

    a. What score does a student need in order to receive an award?

    b. What is the cut-off score that will be used to determine whether a student will be required to take the special reading class?

3. A secondary school system finds that the 440-yard dash times of its female students are normally distributed, with an average time of 72 seconds and a standard deviation of 5.5 seconds. What time does a runner need in order to be in the 9% of the runners with the best times? Round to the nearest hundredth of a second.

Exercise Set 8.6

1. Physician Incomes The 2003 median income for family practice physicians was $130,000. (*Source:* American Academy of Family Physicians, © 2004) The distribution of the physicians' incomes is skewed to the right. Is the mean of these incomes greater than or less than $130,000?

2. Testing At a university, 500 law students took an examination. One student completed the exam in 24 minutes; however, the mode for the times is 50 minutes. The distribution of the times the students took to complete the exam is skewed to the left. Is the mean of these times greater than or less than 50 minutes?

3. Select the best answer. In a normal distribution,

    a. the mean, median, and mode are equal.

    b. the mean and median are equal, but the mode can be different.

    c. the mean, median, and mode can all be different.

    d. none of the above.

4. If A_1 is the area under a normal distribution to the left of $z_1 = -1.43$ and A_2 is the area under a normal distribution to the right of $z_2 = 1.43$, then is $A_1 < A_2$, is $A_1 = A_2$, or is $A_1 > A_2$?

5. Sketch two normal distributions whose means are equal but whose standard deviations are different.

6. Sketch two normal distributions whose standard deviations are equal but whose means are different.

In Exercises 7–12, use the Empirical Rule on page 569 to answer each question.

7. What percent of the data lie

    a. within two standard deviations of the mean?

    b. above one standard deviation above the mean?

    c. between one standard deviation below the mean and two standard deviations above the mean?

8. What percent of the data lie

    a. within three standard deviations of the mean?

    b. below two standard deviations below the mean?

    c. between two standard deviations below the mean and three standard deviations above the mean?

9. Shipping During a certain week an overnight delivery company found that the weights of its parcels had a mean of 24 ounces and a standard deviation of 6 ounces.

    a. What percent of the parcels weighed between 12 ounces and 30 ounces?

    b. What percent of the parcels weighed more than 42 ounces?

10. **Baseball** A baseball franchise finds that the attendance at its home games has a mean of 16,000 and a standard deviation of 4000.

    a. What percent of the home games have attendances between 12,000 and 20,000 people?

    b. What percent of the home games have attendances of less than 8000 people?

11. **Traffic** A highway study of 8000 vehicles that passed by a checkpoint found that their speeds had a mean of 61 miles per hour and a standard deviation of 7 miles per hour.

    a. How many of the vehicles had speeds of more than 68 miles per hour?

    b. How many of the vehicles had speeds of less than 40 miles per hour?

12. **Women's Heights** A survey of 1000 women aged 20 to 30 found that their heights had a mean of 65 inches and a standard deviation of 2.5 inches.

    a. How many of the women had heights that were within one standard deviation of the mean?

    b. How many of the women had heights that were between 60 inches and 70 inches?

In Exercises 13–20, find the area, to the nearest thousandth, of the standard normal distribution between the given z-scores.

13. $z = 0$ and $z = 1.5$ 14. $z = 0$ and $z = 1.9$

15. $z = 0$ and $z = -1.85$ 16. $z = 0$ and $z = -2.3$

17. $z = 1$ and $z = 1.9$ 18. $z = 0.7$ and $z = 1.92$

19. $z = -1.47$ and $z = 1.64$

20. $z = -0.44$ and $z = 1.82$

In Exercises 21–28, find the area, to the nearest thousandth, of the indicated region of the standard normal distribution.

21. The region where $z > 1.3$

22. The region where $z > 1.92$

23. The region where $z < -2.22$

24. The region where $z < -0.38$

25. The region where $z > -1.45$

26. The region where $z < 1.82$

27. The region where $z < 2.71$

28. The region where $z < 1.92$

In Exercises 29–34, find the z-score, to the nearest hundredth, that satisfies the given condition.

29. 0.200 square unit of the standard normal distribution is to the right of z.

30. 0.227 square unit of the standard normal distribution is to the right of z.

31. 0.184 square unit of the standard normal distribution is to the left of z.

32. 0.330 square unit of the standard normal distribution is to the left of z.

33. 0.363 square unit of the standard normal distribution is to the right of z.

34. 0.440 square unit of the standard normal distribution is to the left of z.

In Exercises 35–44, use Table 8.7 to answer each question. *Note:* Round z-scores to the nearest hundredth and then find the required A-values using Table 8.7 on page 572.

35. A population is normally distributed with a mean of 44.8 and a standard deviation of 12.4.

    a. What percent of the data are greater than 51.0?

    b. What percent of the data are between 47.9 and 63.4?

36. A population is normally distributed with a mean of 6.8 and a standard deviation of 1.2.

    a. What percent of the data are less than 7.2?

    b. What percent of the data are between 7.1 and 9.5?

37. A population is normally distributed with a mean of 580 and a standard deviation of 160.

    a. What percent of the data are less than 404?

    b. What percent of the data are between 460 and 612?

38. A population is normally distributed with a mean of 3010 and a standard deviation of 640.

    a. What percent of the data are greater than 2818?

    b. What percent of the data are between 2562 and 4162?

39. **Cereal Weight** The weights of all the boxes of corn flakes filled by a machine are normally distributed, with a mean weight of 14.5 ounces and a standard deviation of 0.4 ounce. What percent of the boxes will

    a. weigh less than 14 ounces?

    b. weigh between 13.5 ounces and 15.5 ounces?

40. **Telephone Calls** A telephone company has found that the lengths of its long-distance telephone calls are normally distributed, with a mean of 225 seconds and

a standard deviation of 55 seconds. What percent of its long distance calls are

a. longer than 340 seconds?

b. between 200 and 300 seconds?

41. Rope Strength The breaking point of a particular type of rope is normally distributed, with a mean of 350 pounds and a standard deviation of 24 pounds. What is the probability that a piece of this rope chosen at random will have a breaking point of

a. less than 320 pounds?

b. between 340 and 370 pounds?

42. Tire Wear The mileages for WearEver tires are normally distributed, with a mean of 48,000 miles and a standard deviation of 7400 miles. What is the probability that the WearEver tires you purchase will provide a mileage of

a. more than 60,000 miles?

b. between 40,000 and 50,000 miles?

43. Bank Lines The amounts of time customers spend waiting in line at a bank are normally distributed, with a mean of 2.5 minutes and a standard deviation of 0.75 minute. Find the probability that the time a customer spends waiting is

a. less than 3 minutes.

b. less than 1 minute.

44. IQ Tests A psychologist finds that the intelligence quotients of a group of patients are normally distributed, with a mean of 102 and a standard deviation of 16. Find the percent of the patients with IQs

a. above 114.

b. between 90 and 118.

45. **Heights** Consider the data set of the heights of all babies born in the United States during a particular year. Do you think this data set is nearly normally distributed? Explain.

46. **Weights** Consider the data set of the weights of all Valencia oranges grown in California during a particular year. Do you think this data set is nearly normally distributed? Explain.

Extensions

CRITICAL THINKING

In Exercises 47–55, determine whether the given statement is true or false.

47. The standard normal distribution has a mean of 0.

48. Every normal distribution is a bell-shaped distribution.

49. If a distribution is symmetrical about a vertical line down its center, then it is a normal distribution.

50. The mean of a normal distribution is always larger than the standard deviation of the distribution.

51. The standard deviation of the standard normal distribution is 1.

52. If a data value x from a normal distribution is positive, then its z-score must also be positive.

53. All normal distributions have a mean of 0.

54. Let x be the number of people who attended a baseball game today. The variable x is a discrete variable.

55. The time of day d in the lobby of a bank is measured with a digital clock. The variable d is a continuous variable.

56. a. How does the area of the standard normal distribution for $0 \leq z < 1$ compare with the area of the standard normal distribution for $0 \leq z \leq 1$?

b. Explain the reasoning you used to answer part a.

57. Determine the two z-scores that bound the middle 60% of the data in a normal distribution.

58. Determine the approximate z-scores for the first quartile and the third quartile of the standard normal distribution.

EXPLORATIONS

59. The mathematician Pafnuty Chebyshev (cha-bǐ´shôf) (1821–1894) is well known for a theorem that concerns the distribution of data in any data set. This theorem is known as Chebyshev's Theorem. Consult a statistics text or the Internet to find information on Chebyshev's Theorem. Write a statement of Chebyshev's Theorem and use the theorem to find the minimum percent of data in any data set that must be within two standard deviations of the mean.

| **Inferential Statistics**

Hypothesis Testing

Suppose you are going to play a game in which a fair coin is required. Before the game begins, you decide to toss a coin 100 times and record the numbers of heads and tails. The table below shows some possible results from tossing the coin 100 times. For each row, ask yourself whether you think the coin, based on the numbers of heads and tails in that row, is fair.

Row	Heads	Tails	Coin is Fair, Yes or No?
1	50	50	?
2	53	47	?
3	56	44	?
4	59	41	?
5	62	38	?
6	65	35	?

If we asked several people to answer the question of whether the results in various rows indicate a fair coin, there would be general agreement that the results in rows 1 and 2 probably indicate a fair coin, and the results in rows 5 and 6 probably indicate a coin that is not fair. There would be disagreement as to whether the results in rows 3 and 4 indicate a fair coin.

A goal of *inferential statistics* is to provide standard guidelines to determine how much variability from a stated or assumed fact is allowed before it is concluded that the stated fact may not be true. In the case of tossing a coin 100 times, inferential statistics would be used to determine how many heads and tails would have to occur before it would be concluded that the coin is not fair.

An event is said to be **statistically significant** at the $p\%$ level if the probability of the event is less than $p\%$. The values of $p\%$ are typically set at 1% and 5%. For instance, the probability of tossing a fair coin 100 times and having 59 or more heads is approximately 4.6%. Because 4.6% < 5%, the event of 59 or more heads in 100 tosses of a fair coin is *statistically significant at the 5% level*. However, because 4.6% > 1%, the event is not statistically significant at the 1% level.

To determine whether a particular event is statistically significant, it is necessary to calculate the probabilities of certain events. Typically, we can use the normal distribution to calculate these probabilities based on what is called the *Central Limit Theorem*.

A precise statement of the Central Limit Theorem can be found in a statistics text. For our purposes, we will present a less formal statement.

> ✔ **TAKE NOTE**
>
> To discuss statistical significance, there must be an assumed fact that acts as a standard with which a sample is compared. For tossing a coin, the assumed fact is that the coin is fair. If an assumed fair coin is tossed 100 times and 80 heads occur, the event of 80 heads suggests that the coin is not fair and the event of 80 heads is statistically significant.

> ✔ **TAKE NOTE**
>
> The basic idea of this theorem is that the distribution of the *means of samples* from any distribution (think of a histogram) can be approximated by a normal distribution.

Central Limit Theorem

The distribution of the means of a large number of samples of size n taken from a population is approximately a normal distribution.

The distribution of the sample means taken from a population is usually referred to as the *sampling distribution of means*.

For instance, consider the population $U = \{1, 2, 3, 4, 5, 6, 7, 8, 9, 10, 11, 12, 13, 14, 15\}$. A histogram of this population is shown at the right. This is a very nonnormal-looking distribution with $\mu = 8$.

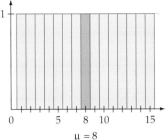
$\mu = 8$

If we now take every possible sample of five elements (we could have chosen a different number) from U, find the mean of each of those samples, and then create a histogram, we get a graph similar to the one at the right. The horizontal axis shows the mean of each sample and the vertical axis shows the number of times that mean occurs. For instance, there were 72 samples whose mean was 6.

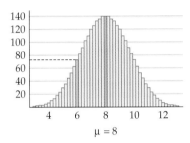
$\mu = 8$

The distribution of the means of the samples is similar to a normal distribution. The mean of this distribution is $\mu = 8$, the same as the mean of the original population.

Hypothesis testing is the formal method of determining whether an event is statistically significant. Hypothesis testing requires stating two hypotheses: the **null hypothesis,** symbolized by H_0, and the **alternative hypothesis,** symbolized by H_1, or sometimes by H_A. The null hypothesis is the assumed fact that acts as a standard with which a sample is compared. It is usually stated in terms of equality. For tossing a coin, the null hypothesis would be that the coin is fair or, stated in terms of proportions, that the proportion of heads is 0.5. Symbolically, we write H_0: $p = 0.5$, where p is the proportion of heads that one would expect from a fair coin.

The alternative hypothesis, sometimes called the research hypothesis, is that the accepted standard is not true. The alternative hypothesis is usually stated in terms of an inequality. For tossing a coin, the alternative hypothesis would be that the coin is not fair in that there are too many heads. Symbolically, we write H_1: $p > 0.5$, meaning that the proportion of heads is greater than 0.5 and the coin is not fair.

EXAMPLE 1 ■ Determine Null and Alternative Hypotheses

The manufacturer of a laser printer claims that the mean print speed in black-and-white mode is 12 pages per minute (ppm). You have purchased this printer and have timed how long the printer takes to print a document. On the basis of your sample printing, you think the mean print speed is lower than suggested by the manufacturer. If you were to set up a hypothesis test to determine whether the manufacturer's claim is correct, what would be the null hypothesis and what would be the alternative hypothesis?

Solution
The null hypothesis is the accepted standard, which in this case is the manufacturer's stated mean print speed of 12 ppm. The alternative hypothesis is that the printer is slower than the stated speed of 12 ppm. Thus,

H_0: $\mu = 12$ ppm   and H_1: $\mu < 12$ ppm

CHECK YOUR PROGRESS 1 After repeatedly rolling a pair of dice and keeping track of the number of times a sum of 7 occurs, you think there is a possibility that the dice are "loaded," producing more 7s than the $\frac{1}{6}$ proportion that one would expect from fair dice. If you were to set up a hypothesis test to determine whether the dice are fair, what would be the null hypothesis and what would be the alternative hypothesis?

Solution See page S35.

In Example 1, we just stated the null and alternative hypotheses. Actual testing of the hypotheses requires several steps. One of those steps is to calculate a *test statistic*. There are many different test statistics used by statisticians. We will use only two of them, both of which are based on the Central Limit Theorem and the normal distribution.

Test Statistics

1. To test a hypothesis about a proportion, use

$$z = \frac{p - \hat{p}}{\sqrt{\dfrac{\hat{p}(1 - \hat{p})}{n}}}$$

where z is the test statistic, p is the proportion found from an experiment, $\hat{p}$ is the proportion from the null hypothesis, and n is the number of times the experiment was performed.

2. To test a hypothesis about a mean, use

$$z = \frac{\bar{x} - \mu}{\dfrac{\sigma}{\sqrt{n}}}$$

where z is the test statistic, $\bar{x}$ is the mean of the sample taken from the population, μ is the mean of the population, σ is the standard deviation of the population, and n is the size of the sample.

✔ **TAKE NOTE**

There are cases in which the standard deviation of the population, σ, is unknown. In these cases, if the sample size is greater than 30, the sample standard deviation is used as an approximation of the population standard deviation. If the sample size is less than 30 and σ is unknown, a different test statistic is used.

We now have all the pieces needed to test a hypothesis. The test is carried out by completing several steps.

Steps for Testing a Hypothesis

1. Determine the null and alternative hypotheses.
2. Calculate a test statistic.
3. Compare the test statistic value with a corresponding *critical z-value* for the significance level of the test.

Critical z-Values for 1% and 5% Significance Levels

The critical z-values at the 1% significance level are $z_{0.01} = \pm 2.33$.

The critical z-values at the 5% significance level are $z_{0.05} = \pm 1.65$.

In each case, the plus sign is used when the alternative hypothesis uses a greater than symbol; the minus sign is used when the alternative hypothesis uses a less than symbol.

4. Based on the result of Step 3, use the following table to choose whether to reject H_0 in favor of H_1 or fail to reject H_0 in favor of H_1.

H_1 Stated Using Less Than ($<$)	H_1 Stated Using Greater Than ($>$)
Test statistic $<$ critical value, reject H_0 in favor of H_1.	Test statistic $>$ critical value, reject H_0 in favor of H_1.
Test statistic $>$ critical value, fail to reject H_0.	Test statistic $<$ critical value, fail to reject H_0.

EXAMPLE 2 ■ Hypothesis Testing with a Proportion

An economist believes that the proportion of people living at or below the poverty level in a certain city is less than the national proportion of 12%. A random survey of 300 people in the city found that 21 people were living at or below the poverty level. On the basis of this sample, should the economist conclude, at the 1% level of significance, that there is a smaller proportion of people living at or below the poverty level in this city?

Solution

1. The null hypothesis is that the proportion of people living at or below the poverty level in this city is the same as the national proportion: H_0: $\hat{p} = 0.12$. The alternative hypothesis is that the proportion of people living at or below the poverty level in this city is less than the national proportion: H_1: $\hat{p} < 0.12$.

2. This is a proportion hypothesis test. $p = \frac{21}{300} = 0.07$, $\hat{p} = 0.12$, and $n = 300$.

$$z = \frac{p - \hat{p}}{\sqrt{\frac{\hat{p}(1-\hat{p})}{n}}}$$

$$z = \frac{0.07 - 0.12}{\sqrt{\frac{0.12(1-0.12)}{300}}}$$

$$z = \frac{-0.05}{\sqrt{\frac{0.12(0.88)}{300}}} \approx -2.67$$

3. The significance level is 1% and the alternative hypothesis uses a less than symbol. The critical z-value is $z_{0.01} = -2.33$. Compare the test statistic value with the critical z-value: $z \approx -2.67 < -2.33$.

4. Because the value of the test statistic is less than the critical z-value, we reject the null hypothesis. This means that on the basis of the sample, the economist would conclude that the proportion of people at or below the poverty level in this city is less than the national proportion.

CHECK YOUR PROGRESS 2   Cancer clusters are regions of the country in which there are more people with a certain type of cancer than in the general population. Suppose the nationwide prevalence of a certain cancer is 1.5%. To investigate whether a certain region of the country should be classified as a cancer cluster, researchers tested 200 people in the region and found that four people had the particular cancer. On the basis of this sample, should the region be designated a cancer cluster at the 5% level of significance?

Solution   *See page S35.*

In Example 2, we rejected the null hypothesis on the basis of the sample drawn from that population. Suppose, however, that the sample of 300 people showed that there were 30 people at or below the poverty level. In this case,

$$z = \frac{p - \hat{p}}{\sqrt{\dfrac{\hat{p}(1 - \hat{p})}{n}}}$$

$$z = \frac{0.10 - 0.12}{\sqrt{\dfrac{0.12(1 - 0.12)}{300}}} \qquad \bullet\, p = \frac{30}{300} = 0.10$$

$$z = \frac{-0.02}{\sqrt{\dfrac{0.12(0.88)}{300}}} \approx -1.07$$

Comparing this value with the critical z-value, we have $z \approx -1.07 > -2.33$. In this case, we would fail to reject H_0. Note that we do not say *accept* H_0. All that is known is that, based on this sample, there is no reason to believe that the alternative hypothesis is true.

EXAMPLE 3 ■ **Hypothesis Testing About a Mean**

An agronomist believes that increasing the nitrogen fertilizer in soil planted with corn will increase the mean number of bushels per acre of corn. The current level of corn production is 120 bushels per acre, with a standard deviation of 9.8 bushels per acre. To test this hypothesis, the agronomist adds additional nitrogen to 40 different one-acre plots of land. The corn harvested from those 40 acres yields a mean of 125 bushels per acre. On the basis of this sample, should the agronomist, at the 5% significance level, conclude that the increase in nitrogen fertilizer resulted in a greater number of bushels per acre of corn?

Solution

1. The null hypothesis is that the yield is 120 bushels of corn per acre: H_0: $\mu = 120$. The alternative hypothesis is that, after adding more nitrogen fertilizer, the yield is more than 120 bushels of corn per acre: H_1: $\mu > 120$.

2. This is a hypothesis about a mean. $\bar{x} = 125$, $\mu = 120$, $\sigma = 9.8$, and $n = 40$.

$$z = \frac{\bar{x} - \mu}{\dfrac{\sigma}{\sqrt{n}}}$$

$$z = \frac{125 - 120}{\dfrac{9.8}{\sqrt{40}}}$$

$$z \approx 3.23$$

3. The significance level is 5% and the alternative hypothesis uses a greater than symbol. The critical z-value is $z_{0.05} = 1.65$. Compare the test statistic value with the critical z-value: $z \approx 3.23 > 1.65$.

4. Because the value of the test statistic is greater than the critical z-value, we reject the null hypothesis at the 5% level of significance. This means that, on the basis of the sample, the agronomist could conclude that adding more nitrogen to the fertilizer increases yield.

CHECK YOUR PROGRESS 3   A consumer advocacy group believes that an automobile manufacturer's claim that its hybrid car gets 42 miles per gallon (mpg) is too high. To test the claim, the group tests 30 of the manufacturer's hybrid cars and finds the mean gasoline consumption of the cars to be 40 mpg with a standard deviation of 2.5 mpg. Do these data suggest that the manufacturer's claim is incorrect at the 1% level of significance?

Solution   *See page S36.*

Confidence Intervals

For hypothesis testing, there is a known standard that can be tested. For instance, if we are trying to determine whether a coin is fair, the known standard is that the proportion of heads in a sample should be close to 0.5. There are other instances, however, for which a standard is unknown.

Consider, for example, a pollster for a television producer who is trying to determine what proportion of a population likes a new television sitcom. Because the proportion of the population that likes the show is unknown, the pollster tries to determine a possible range of values for the proportion of the population that likes the show. The interval the pollster finds is called a **confidence interval.**

Although there are many different confidence intervals that may be found, two of the most frequently used confidence intervals are the 95% confidence interval and the 99% confidence interval. These intervals are based on the normal distribution, as shown in the figures below.

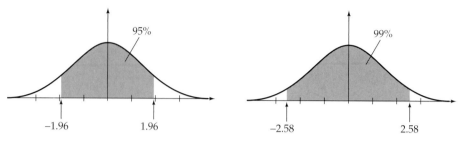

The 95% confidence interval is a range of values in which we would expect, on the basis of a sample, some characteristic of a population to lie 95% of the time. A similar statement is true for the 99% confidence interval.

Confidence intervals are based on the Central Limit Theorem and the hypothesis testing equations.

Confidence Interval for a Proportion	Confidence Interval for a Mean
95% confidence interval $p - 1.96\sqrt{\dfrac{p(1-p)}{n}} < \hat{p} < p + 1.96\sqrt{\dfrac{p(1-p)}{n}}$	95% confidence interval $\bar{x} - 1.96\dfrac{\sigma}{\sqrt{n}} < \mu < \bar{x} + 1.96\dfrac{\sigma}{\sqrt{n}}$
99% confidence interval $p - 2.58\sqrt{\dfrac{p(1-p)}{n}} < \hat{p} < p + 2.58\sqrt{\dfrac{p(1-p)}{n}}$	99% confidence interval $\bar{x} - 2.58\dfrac{\sigma}{\sqrt{n}} < \mu < \bar{x} + 2.58\dfrac{\sigma}{\sqrt{n}}$

EXAMPLE 4 ■ Find a Confidence Interval for a Proportion

A newspaper conducted a survey of voters in an upcoming mayoral election to determine the popularity of a certain candidate. In a survey of 500 voters, 240 people indicated that they would vote for the candidate. What is the 95% confidence interval for the true proportion of voters who would vote for the candidate? Round to the nearest hundredth.

Solution
Use the formula for the 95% confidence interval for a proportion. From the survey, $n = 500$ and $p = \frac{240}{500} = 0.48$.

$$p - 1.96\sqrt{\frac{p(1-p)}{n}} < \hat{p} < p + 1.96\sqrt{\frac{p(1-p)}{n}}$$

$$0.48 - 1.96\sqrt{\frac{0.48(1-0.48)}{500}} < \hat{p} < 0.48 + 1.96\sqrt{\frac{0.48(1-0.48)}{500}}$$

$$0.48 - 0.04 < \hat{p} < 0.48 + 0.04$$

$$0.44 < \hat{p} < 0.52$$

The 95% confidence interval for the true proportion of voters who would vote for the candidate is $0.44 < \hat{p} < 0.52$.

CHECK YOUR PROGRESS 4   A chemist collected 200 samples of water from different locations in a certain reservoir and found that 15 of them contained unusually high concentrations of zinc. What is the 99% confidence interval for the true proportion of zinc in the reservoir?

Solution   See page S36.

In Example 4, we determined a confidence interval for a proportion. In Example 5, we determine a confidence interval for the mean of a population.

TAKE NOTE

In Example 5, suppose that the *population* mean of weights is 6.4 pounds. A 99% confidence interval means that if repeated random samples of 40 children were taken from hospital records, the population mean would be contained in 99% of the random samples.

The graph below shows some of the possible confidence intervals. The one for Example 5 is shown in blue. Note that one of the intervals, shown in red, does not contain the population mean.

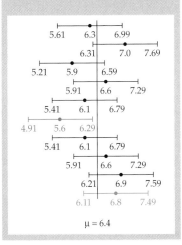

EXAMPLE 5 ■ Find a Confidence Interval for a Population Mean

The records from births of 40 children at a hospital showed that the mean weight was 6.8 pounds with a standard deviation of 1.7 pounds. What is the 99% confidence interval for the mean weight of all children born at this hospital?

Solution
The sample mean is $\bar{x} = 6.8$ pounds and the standard deviation is $\sigma = 1.7$ pounds. The sample size is $n = 40$. Use the formula for the 99% confidence interval for the mean. Round to the nearest hundredth.

$$\bar{x} - 2.58 \frac{\sigma}{\sqrt{n}} < \mu < \bar{x} + 2.58 \frac{\sigma}{\sqrt{n}}$$

$$6.8 - 2.58 \frac{1.7}{\sqrt{40}} < \mu < 6.8 + 2.58 \frac{1.7}{\sqrt{40}}$$

$$6.8 - 0.69 < \mu < 6.8 + 0.69$$

$$6.11 < \mu < 7.49$$

The 99% confidence interval for the population mean is $6.11 < \mu < 7.49$.

CHECK YOUR PROGRESS 5   Golf club manufacturers test drivers for what is called the coefficient of restitution (COR), which is a measure of how much a golf ball will bounce off the clubhead. The mean COR of 60 drivers tested by the manufacturer was 0.81, with a standard deviation of 0.04. What is the 95% confidence interval for the COR of all drivers from this manufacturer?

Solution   See page S36.

Margin of Error

If you have listened to a news story or read a newspaper account of the results of a survey, you may have seen a notation about the *margin of error* in the poll. This error is related to the confidence interval, the proportion, and the size of the sample that is taken from the population.

A percent margin of error at the 95% confidence level means that if the same poll were repeated many times, the results would lie in a range given by the percent margin of error 95% of the time. A similar statement can be made for the 99% confidence level.

Margin of Error

95% confidence level: Margin of error $= 1.96 \sqrt{\dfrac{p(1-p)}{n}}$.

99% confidence level: Margin of error $= 2.58 \sqrt{\dfrac{p(1-p)}{n}}$.

In each case, n is the size of a random sample taken from the population and p is the sample proportion.

EXAMPLE 6 ■ Find a Margin of Error for a Proportion

A survey of 1500 baseball fans found that 630 fans thought that a computer generated strike zone should be used to call balls and strikes instead of using an umpire. Find the margin of error, to the nearest tenth of a percent, at the 95% confidence level for this survey.

Solution

Use the formula for the margin of error at the 95% confidence level with $p = \dfrac{630}{1500} = 0.42$ and $n = 1500$.

$$\text{Margin of error} = 1.96 \sqrt{\dfrac{p(1-p)}{n}}$$

$$= 1.96 \sqrt{\dfrac{0.42(0.58)}{1500}} \approx 0.0249$$

The margin of error, to the nearest tenth of a percent, is 2.5%.

CHECK YOUR PROGRESS 6 A survey of 500 physicians found that 350 of them would like to spend more time with their patients. Find the margin of error, to the nearest tenth of a percent, at the 99% confidence level for this survey.

Solution See page S37.

Sometimes a pollster is interested only in the *greatest* margin of error. In this case, the pollster would assume that $p = 0.5$ and, based on that value for p, would calculate the margin of error.

Greatest Margin of Error

95% confidence level: Greatest margin of error $= \dfrac{0.98}{\sqrt{n}}$.

99% confidence level: Greatest margin of error $= \dfrac{1.29}{\sqrt{n}}$.

In each case, n is the size of a random sample taken from the population.

EXAMPLE 7 ■ **Find the Greatest Margin of Error**

A poll of 1000 homeowners in a community asked whether they were in favor of installing speed bumps at intersections where school children cross. Find the greatest margin of error at the 95% confidence level. Round to the nearest tenth of a percent.

Solution

Use the formula for the greatest margin of error at the 95% confidence level.

$$\text{Greatest margin of error} = \frac{0.98}{\sqrt{n}}$$

$$= \frac{0.98}{\sqrt{1000}} \approx 0.031$$

The greatest margin of error at the 95% confidence level is 3.1%.

CHECK YOUR PROGRESS 7   A poll of 500 people asked whether they liked the taste of a new soft drink. Find the greatest margin of error, to the nearest tenth of a percent, at the 99% confidence level.

Solution   See page S37.

Note in Example 7 that the percent of homeowners who were in favor of the speed bumps is not given. We found only the greatest margin of error. If additionally we knew that the sample indicated that 30% of the homeowners were in favor of the speed bumps, then we could use the margin of error formula to conclude that the true percent of the homeowners who preferred the speed bumps at the 95% confidence level was

$$\text{Greatest margin of error} = 1.96 \sqrt{\frac{p(1-p)}{n}}$$

$$= 1.96 \sqrt{\frac{0.30(0.70)}{1000}} \approx 0.0284$$

The greatest percent margin of error, to the nearest tenth of a percent, is 2.8%.

Investigation

A Fair Die?

Obtain a standard, six-sided die. We normally assume a die is a fair die, meaning that each of the six numbers on the die has an equal chance of appearing on the top face when the die is rolled. However, we would probably be surprised if, after many rolls, each of the numbers on the die appeared exactly the same number of times. How do we know if the die is biased in some way? Does it make a difference which of the six numbers we focus on?

(continued)

In this Investigation, you will analyze the fairness of the die using several hypothesis tests. The Investigation will work best with several people contributing, ideally six. Take turns rolling the die for a total of 60 rolls, and have a member of the group record the number of ones that appear, the number of twos, and so on.

Investigation Exercises

1. Compute the proportion of each number rolled. For instance, if the number 5 comes up in 12 of the 60 rolls, then the proportion of fives is $\frac{12}{60} = \frac{1}{5}$. What is the proportion for each number that you would expect by random chance?

2. Have one member of the group design a hypothesis test for the number of ones that appear, to test the fairness of the die. State the null and alternative hypotheses. Similarly, the next member should design a hypothesis test for the number of twos that appear, and so on.

3. Compute the test statistic for each of the six hypothesis tests.

4. Based on a 5% level of significance, determine the result of each hypothesis test. Do all six tests give the same result? If not, suggest some possible explanations for why the results might differ.

5. What are the results of the hypothesis tests if you use a 1% level of significance? Do any of your results change?

Exercise Set 8.7

1. **Tossing Coins** A referee, trying to determine whether a coin is fair, tosses the coin 20 times, resulting in six heads and 14 tails. Is this result statistically significant at the 5% level? The probability of tossing a coin 20 times and getting 14 or more tails is approximately 5.8%.

2. **Dice** A Monopoly player suspects that a six-sided die is biased. She rolls the die 100 times and sees the die come up 3 only eight times. Is this result statistically significant at the 1% level if the probability of rolling eight or fewer 3s in 100 rolls is approximately 0.95%?

3. **Lottery** A state sells lottery tickets at thousands of locations. In the last 50 drawings, three of the winning tickets were purchased at the same grocery store. Compared with what you would expect by random chance, do you think this event is statistically significant?

4. **Bacterial Infections** Of the 340 poultry-related salmonella infections reported in a particular state over the last year, 215 were traced to poultry purchased from the same grocery chain. Compared with what you would expect by random chance, do you think this event is statistically significant? Explain your answer.

5. **Fuel Consumption** A car manufacturer claims that its new hybrid model gets 48 miles per gallon (mpg) under normal highway conditions. A new owner of this model has run mileage tests on her car and suspects that her car gets less than the mileage claimed by the manufacturer. If she were to set up a hypothesis test to determine whether her car meets the manufacturer's mileage claim, what would be the null hypothesis and what would be the alternative hypothesis?

6. **Pizza Delivery Times** The owner of a pizza parlor in a small community advertises that his pizzas have a mean delivery time of 45 minutes. The owner suspects that the actual mean delivery time is greater than 45 minutes. If he were to set up a hypothesis test to determine whether his advertised claim is correct,

what would be the null hypothesis and what would be the alternative hypothesis?

7. **Roulette** In the game of roulette, a ball is rolled onto a wheel and stops in one of 38 numbered slots. The slots are equally spaced, which should mean that the ball is equally likely to land on any number. However, a casino manager observed that in the last 400 plays, the ball has landed on number 33 a total of 17 times. The manager is suspicious that the wheel is not a fair wheel.

   a. What is the proportion of 33s that we would expect by random chance?

   b. The manager would like to set up a hypothesis test to determine whether the wheel is fair. State an appropriate null hypothesis and the corresponding alternative hypothesis.

   c. Compute the test statistic for the observed number of 33s.

   d. According to the hypothesis test, should the manager conclude that the wheel is not fair? Use a 5% level of significance.

   e. Should the manager conclude, at the 1% significance level, that the wheel is not fair?

8. **Quality Control** A manufacturer of computer chips typically makes thousands of memory chips from a single silicon wafer. It is the manufacturer's policy to discard the entire batch from a wafer if more than 2% of the chips are defective. A quality control engineer tests 125 chips made from the same wafer and finds four defective chips.

   a. State an appropriate null hypothesis and an alternative hypothesis that the engineer could use to test this batch of chips.

   b. Compute the test statistic for the batch of chips.

   c. Based on a 5% significance level, should the engineer discard the entire batch of memory chips?

9. **Restless Legs Syndrome** A team of medical researchers has developed a new drug to treat restless legs syndrome (RLS). Before the development of this new drug, 22% of the patients with RLS were relieved of their symptoms by taking conventional medications. A recent study of 325 patients with RLS showed that 105 patients were relieved of their symptoms by taking the new drug. Can this study be used to justify the claim that the new drug benefits a greater percent of RLS patients than the previously prescribed conventional medications? Use a 1% level of significance.

10. **Teenage Smoking** In 2001, the Centers for Disease Control published a report claiming that 28.5% of high school students smoke (meaning that a student had smoked a cigarette within the past 30 days). A high school principal claimed her student population had a lower smoking rate. A random survey was conducted at the school, and 36 of the 180 students interviewed said they smoked. At the 1% confidence level, is the principal's claim justified?

11. **Ticket Prices** A national organization of theatre owners reported that movie theatre ticket prices for the year 2007 were normally distributed with a mean of $6.91 and a standard deviation of $1.15. During the year 2007, a group of students sampled 38 theatres in San Diego County. Their sample had a mean of $7.45.

   a. The students decided to use a hypothesis test to determine whether the mean 2007 movie theatre ticket price in San Diego County was greater than the national mean. State the null hypothesis and the alternative hypothesis for this test.

   b. Compute the test statistic.

   c. Should the students, at the 1% significance level, conclude that the mean 2007 movie theatre ticket price in San Diego County was greater than the national mean?

12. **Golf** A consumer testing company is testing a new golf ball called the BestFlite 2007. The company uses a mechanical driving machine to test how far the golf balls travel when hit by the machine. The previous model, BestFlite 2006, had a mean driving distance of 274.0 yards with a standard deviation of 4.2 yards. The BestFlite company claims that the new 2007 ball has a longer mean driving distance than the 2006 model. The testing company has performed a driving distance test of the BestFlite 2007 golf balls. The test consisted of 35 driving distance trials with a mean of 275.5 yards.

   a. The testing company has decided to use a hypothesis test to determine whether the 2007 golf balls have a mean driving distance that is greater than the mean driving distance of the 2006 golf balls. State the null hypothesis and the alternative hypothesis for this test.

   b. Compute the test statistic.

   c. Should the testing company, at the 5% significance level, conclude that the 2007 golf balls have a mean driving distance that is greater than the mean driving distance of the 2006 golf balls?

   d. Should the testing company, at the 1% significance level, conclude that the 2007 golf balls have a mean driving distance that is greater than the mean driving distance of the 2006 golf balls?

13. **Soda Machine** A deluxe model soda machine dispenses soda into large cups. The amount of soda the machine is designed to dispense is 16.0 ounces, with a standard deviation of 0.5 ounce. An accuracy test of one of the machines showed that for 36 samples it dispensed an average of 15.9 ounces per cup. On the basis of this test, should the tester, at the 5% significance level, conclude that this machine has a mean dispensing amount that is less than 16.0 ounces per cup?

14. **Life Span of Batteries** A company manufactures flashlight batteries. It claims that during constant use its batteries have a mean life span of 12.5 hours with a standard deviation of 0.45 hour. A recent test of 32 of these batteries yielded a mean life span of 12.2 hours. On the basis of this test, should the tester, at the 1% significance level, conclude that these batteries have a mean life span that is less than 12.5 hours?

15. **Election Survey** A television station conducted a survey of voters in an upcoming mayoral election to determine the popularity of a certain candidate. In the survey of 350 voters, 195 people indicated that they would vote for the candidate. What is the 99% confidence interval for the true proportion of voters who would vote for the candidate? Round to the nearest hundredth.

16. **Environmental Issues** An Environmental Protection Agency (EPA) agent collected 65 samples of water from different locations in a small creek and found that five of them contained unusually high levels of ammonia-nitrogen. The agent suspected that the high ammonia-nitrogen levels were a result of the increased use of lawn fertilizers in the surrounding area. What is the 95% confidence interval for the true proportion of high ammonia-nitrogen levels in the creek? Round to the nearest hundredth.

17. **Gasoline Conservation** A recent survey of 750 U.S. drivers found that 326 were driving less to save gas. What is the 95% confidence interval for the true proportion of drivers who are driving less to save gas? Round to the nearest hundredth.

18. **Work Week** In a recent survey of 625 workers, 402 reported that they worked more than 40 hours per week. What is the 99% confidence interval for the true proportion of workers who work more than 40 hours per week? Round to the nearest hundredth.

19. **Avocados** During a harvest, a farmer finds that in a random sample of 220 of his Fuerte avocados, the mean weight is 15.4 ounces with a standard deviation of 1.1 ounces. What is the 95% confidence interval for the mean weight of this avocado crop? Round to the nearest hundredth of an ounce.

20. **Oranges** During a recent harvest, a citrus farmer found that in a random sample of 85 of her navel oranges, the mean amount of juice produced per orange was 3.84 ounces with a standard deviation of 0.35 ounce. What is the 99% confidence interval for the mean amount of juice produced per orange for this orange crop? Round to the nearest hundredth of an ounce.

21. **Rental Prices** A consumer rights organization found that in a random sample of 180 two-bedroom apartment rentals in its metropolitan area, the mean monthly apartment rental price was $765 with a standard deviation of $92. What is the 99% confidence interval for the mean of all the monthly two-bedroom apartment rental prices in this metropolitan area? Round to the nearest dollar.

22. **Camera Prices** Selena is interested in buying a Canon Rebel XT 8.0MP digital camera with an 18-55 mm lens. On the Internet she found 32 stores from which she can purchase this camera. The mean price of the camera at these stores is $735.00, with a standard deviation of $29.44. What is the 95% confidence interval for the mean of all the Canon Rebel XT 8.0MP camera prices?

23. **Instant Replay System** A poll of 450 football fans found that 348 fans were in favor of keeping the instant replay system, which allows coaches to challenge on-field calls of plays. Find the margin of error, to the nearest tenth of a percent, at the 95% confidence level for this poll.

24. **Laser Eye Surgery** A survey of 285 patients who have recently had laser eye surgery found that 209 were satisfied with the results. Find the margin of error, to the nearest tenth of a percent, at the 99% confidence level for this survey.

25. **Global Warming** In March of 2006, a Gallup poll of 1000 adults asked if they thought that global warming will pose a serious threat in their lifetimes. About one-third of the adults gave positive responses. Find the greatest margin of error for this poll, at the 99% confidence level. Round to the nearest tenth of a percent.

26. **Election Survey** A poll of 855 voters asked whether they would vote for the incumbent candidate for governor in the coming election. Find the greatest margin of error for this poll at the 99% confidence level. Round to the nearest tenth of a percent.

Extensions

CRITICAL THINKING

27. How is the greatest margin of error affected if we quadruple the sample size?

28. Explain what you need to do with the sample size to decrease the greatest margin of error by a factor of 3—for example, from 6% to 2%.

29. A company needs to determine a confidence interval for the mean of some data. Which confidence interval will have the smallest range, the 95% confidence interval or the 99% confidence interval?

30. Explain how to derive the formula

$$\text{Greatest margin of error} = \frac{0.98}{\sqrt{n}}$$

from the formula

$$\text{Margin of error} = 1.96 \sqrt{\frac{p(1-p)}{n}}.$$

COOPERATIVE LEARNING

31. Tossing Coins Flip a coin 100 times and record how many heads and how many tails appear. Design a hypothesis test to determine whether the coin is fair. State the null and alternative hypotheses, and compute the test statistic for the 100 coin flips. Based on a 5% level of significance, what is the conclusion of the hypothesis test?

EXPLORATIONS

32. **Election Survey** The upset win of Harry S. Truman over Thomas E. Dewey in the 1948 presidential election came after many newspapers and polls had predicted that Dewey would win. Use the Internet to find information about this election and write a few sentences that explain some of the reasons why the election results contrasted with the preelection polls.

Photo Source: Eagleton Institute of Politics
http://www.eagleton.rutgers.edu/e-gov/
e-politicalarchive-1948election.htm

CHAPTER 8 **Summary**

Key Terms

alternative hypothesis [p. 581]
bimodal distribution [p. 567]
box-and-whisker plot [p. 557]
Central Limit Theorem [p. 580]
class [p. 564]
combination [p. 500]
combinatorics [p. 495]
confidence interval [p. 585]
continuous data [p. 568]
continuous variable [p. 568]
descriptive statistics [p. 527]
discrete data [p. 568]
discrete variable [p. 568]
empirical (experimental) probability [p. 511]
event [p. 508]
experiment [p. 495]

experimental probability [p. 511]
frequency distribution [p. 533]
grouped frequency distribution [p. 564]
histogram [p. 565]
independent events [p. 518]
inferential statistics [p. 527]
interpolation [p. 535]
lower (upper) class boundary [pp. 564–565]
measures of central tendency [p. 527]
mutually exclusive events [p. 513]
normal distribution [p. 569]
null hypothesis [p. 581]
odds [p. 511]
outlier [p. 564]
percentile [p. 554]
population [p. 527]

Essential Concepts

- **Counting Principle**
  Let E be a multistage experiment. If $n_1, n_2, n_3, \ldots, n_k$ are the numbers of possible outcomes of each of the k stages of E, then there are $n_1 \cdot n_2 \cdot n_3 \cdot \cdots \cdot n_k$ possible outcomes for E.

- **n factorial**
  n factorial is the product of the natural numbers 1 through n and is symbolized by $n!$.

- **Permutation Formula**
  The number of permutations, $P(n, k)$, of n distinct objects selected k at a time is $P(n, k) = \dfrac{n!}{(n - k)!}$.

- **Combination Formula**
  The number of combinations of n distinct objects chosen k at a time is $C(n, k) = \dfrac{P(n, k)}{k!} = \dfrac{n!}{k! \cdot (n - k)!}$.

- **Basic Probability Formula**
  For an experiment with sample space S of *equally likely outcomes*, the probability $P(E)$ of an event E is given by $P(E) = \dfrac{n(E)}{n(S)}$, where $n(E)$ is the number of elements in the event and $n(S)$ is the number of elements in the sample space.

- **Probability of Mutually Exclusive Events**
  If A and B are two mutually exclusive events, then the probability of A or B occuring is $P(A \cup B) = P(A) + P(B)$.

- **Addition Rule for Probabilities**
  Let A and B be two events in a sample space S. Then $P(A \cup B) = P(A) + P(B) - P(A \cap B)$.

- **Conditional Probability Formula**
  Let A and B be two events in a sample space S. Then the conditional probability of B given that A has occurred is $P(B|A) = \dfrac{P(A \cap B)}{P(A)}$.

- **Product Rule for Probabilities**
  If A and B are two events from the sample space S, then $P(A \cap B) = P(A) \cdot P(B|A)$.

- **Probability of Successive Events**
  The probability of two or more events occurring in succession is the product of the conditional probabilities of the individual events.

- The mean of n numbers is the sum of the numbers divided by n. The symbol for the sample mean is $\bar{x}$. The symbol for the population mean is μ.

- The median of a *ranked list* of n numbers is the middle number if n is odd and is the mean of the two middle numbers if n is even.

- The mode of a list of numbers is the number that occurs most frequently.

- The weighted mean of the n numbers $x_1, x_2, x_3, \ldots, x_n$ with the respective assigned weights $w_1, w_2, w_3, \ldots, w_n$ is $\dfrac{\Sigma(x \cdot w)}{\Sigma w}$, where $\Sigma(x \cdot w)$ is the sum of the products formed by multiplying each number by its assigned weight, and Σw is the sum of all the weights.

- The range of a set of data values is the difference between the largest data value and the smallest data value.

- The standard deviation of the population $x_1, x_2, x_3, \ldots, x_n$ with a mean of μ is
  $$\sigma = \sqrt{\frac{\Sigma(x - \mu)^2}{n}}.$$

- The standard deviation of the sample $x_1, x_2, x_3, \ldots, x_n$ with a mean of $\bar{x}$ is $s = \sqrt{\dfrac{\Sigma(x - \bar{x})^2}{n - 1}}$.

■ The z-score for a given data value x is the number of standard deviations that x is above or below the mean of the data. Population: $z_x = \dfrac{x - \mu}{\sigma}$.

Sample: $z_x = \dfrac{x - \bar{x}}{s}$.

■ **The Empirical Rule:** In a normal distribution about 68.2% of the data lie within one standard deviation of the mean.

95.4% of the data lie within two standard deviations of the mean.

99.7% of the data lie within three standard deviations of the mean.

■ In the standard normal distribution, the *area* of the distribution from $z = a$ to $z = b$ represents the probability that z lies in the interval from a to b.

■ **Test Statistics**

1. To test a hypothesis about a proportion, use

$$z = \frac{p - \hat{p}}{\sqrt{\dfrac{\hat{p}(1 - \hat{p})}{n}}}$$

where z is the test statistic, p is the proportion found from an experiment, $\hat{p}$ is the proportion from the null hypothesis, and n is the number of times the experiment was performed.

2. To test a hypothesis about a mean, use

$$z = \frac{\bar{x} - \mu}{\dfrac{\sigma}{\sqrt{n}}}$$

where z is the test statistic, $\bar{x}$ is the mean of the sample taken from the population, μ is the mean of the population, σ is the standard deviation of the population, and n is the size of the sample.

■ The critical z-values for a greater than or less than hypothesis test at the 1% significance level are $z_{0.01} = \pm 2.33$.

■ The critical z-values for a greater than or less than hypothesis test at the 5% significance level are $z_{0.05} = \pm 1.65$.

■ 95% confidence interval for a proportion p:

$$p - 1.96\sqrt{\frac{p(1 - p)}{n}} < \hat{p} < p + 1.96\sqrt{\frac{p(1 - p)}{n}}$$

■ 99% confidence interval for a proportion p:

$$p - 2.58\sqrt{\frac{p(1 - p)}{n}} < \hat{p} < p + 2.58\sqrt{\frac{p(1 - p)}{n}}$$

■ 95% confidence interval for a mean μ:

$$\bar{x} - 1.96\frac{\sigma}{\sqrt{n}} < \mu < \bar{x} + 1.96\frac{\sigma}{\sqrt{n}}$$

■ 99% confidence interval for a mean μ:

$$\bar{x} - 2.58\frac{\sigma}{\sqrt{n}} < \mu < \bar{x} + 2.58\frac{\sigma}{\sqrt{n}}$$

■ 95% confidence level for margin of error:

$$1.96\sqrt{\frac{p(1 - p)}{n}}$$

■ 99% confidence level for margin of error:

$$2.58\sqrt{\frac{p(1 - p)}{n}}$$

■ 95% confidence level for greatest margin of error:

$$\frac{0.98}{\sqrt{n}}$$

■ 99% confidence level for greatest margin of error:

$$\frac{1.29}{\sqrt{n}}$$

CHAPTER 8   **Review Exercises**

In Exercises 1–4, list the elements of the sample space for the given experiment.

1. Two-digit numbers are formed, with replacement, from the digits 1, 2, and 3.

2. Two-digit numbers are formed, without replacement, from the digits 2, 6, and 8.

3. Use a tree diagram to display the outcomes that can result from tossing four coins.

4. A *biquinary code* is a code that consists of two binary digits (0 or 1), both of which cannot be the same, followed by five binary digits on which there are no restrictions. How many biquinary codes are possible?

In Exercises 5–10, evaluate each expression.

5. $7!$

6. $8! - 4!$

7. $\dfrac{9!}{2!3!4!}$

8. $P(10, 6)$

9. $P(8, 3)$

10. $\dfrac{C(6, 2) \cdot C(8, 3)}{C(14, 5)}$

11. A matching test has seven definitions, which are to be paired with seven words. Assuming each word corresponds to exactly one definition, how many different matches are possible by just random matching?

12. A matching test has seven definitions to be matched with five words. Assuming each word corresponds to exactly one definition, how many different matches are possible by just random matching?

13. How many distinct arrangements are possible using the letters of *letter*?

14. Twelve identical coins are tossed. How many distinct arrangements are possible consisting of four heads and eight tails?

15. A professor assigns 25 homework problems, of which 10 will be graded. How many different sets of 10 problems can the professor choose to grade?

16. A stockbroker recommends 11 stocks to a client. If the client will invest in three of the stocks, how many different three-stock portfolios can be selected?

17. If it is equally likely that a child will be a boy or a girl, compute the probability that a family of three children will have one boy and two girls.

18. If a coin is tossed three times, what is the probability of getting no heads?

19. A large company currently employs 5739 men and 7290 women. If an employee is selected at random, what is the probability that the employee is a woman?

In Exercises 20 and 21, use the table below, which shows the number of students at a university who are currently in each class level.

Class Level	Number of Students
First-year	642
Sophomore	549
Junior	483
Senior	445
Graduate student	376

20. If a student is selected at random, what is the probability that the student is an upper division undergraduate student (junior or senior)?

21. If a student is selected at random, what is the probability that the student is not a graduate student?

In Exercises 22–25, a pair of dice is tossed.

22. Find the probability that the sum of the pips on the two upward faces is 9.

23. Find the probability that the sum showing on the two dice is an even number or a number less than 5.

24. What is the probability that the sum showing on the two dice is 9, given that the sum is odd?

25. What is the probability that the sum showing on the two dice is 8, given that doubles were rolled?

In Exercises 26–29, a single card is selected from a standard deck of playing cards.

26. What is the probability that the card is a heart or a black card?

27. What is the probability that the card is a heart or a jack?

28. What is the probability that the card is not a 3?

29. What is the probability that the card is red, given that it is not a club?

30. The length of hair of a particular rodent is determined by a dominant allele H, corresponding to long hair, and a recessive allele h, which is associated with short hair. Draw a Punnett square for parents of genotypes Hh and hh, and compute the probability that offspring of these parents will have short hair.

31. Two cards are drawn from a standard deck of playing cards. The probability that exactly one card is an ace is 0.145 and the probability that exactly one card is a face card (jack, queen, or king) is 0.362. The probability that the two cards drawn from the deck are an ace or a face card is 0.489. Find the probability that the two cards are an ace and a face card.

32. A recent survey asked 1000 people whether they liked cheese-flavored corn chips (642 people), jalapeño-flavored chips (487 people), or both (302 people). If one person is chosen from this survey, what is the probability that the person does not like either of the two flavors?

In Exercises 33–37, a box contains 24 different colored chips that are identical in size. Five are black, four are red, eight are white, and seven are yellow.

33. If a chip is selected at random, what is the probability that the chip will be yellow or white?

34. If a chip is selected at random, what are the odds in favor of getting a red chip?

35. If a chip is selected at random, find the probability that the chip is yellow given that it is not white.

36. If five chips are randomly chosen (without replacement), what is the probability that none of them is red?

37. If three chips are chosen (without replacement), find the probability that the first one is yellow, the second is white, and the third is yellow.

In Exercises 38–42, use the table below, which shows the numbers of voters in a city who voted for or against a proposition, or abstained from voting, according to their party affiliations.

	For	Against	Abstained
Democrat	8452	2527	894
Republican	2593	5370	1041
Independent	1225	712	686

38. If a voter is chosen at random, compute the probability that the person voted against the proposition.

39. If a voter is chosen at random, compute the probability that the person is a Democrat or an Independent.

40. If a voter is randomly chosen, what is the probability that the person abstained from voting on the proposition and is not a Republican?

41. A voter is randomly selected; what is the probability that the individual voted for the proposition, given that the voter is a registered Independent?

42. A voter is randomly selected; what is the probability that the individual is registered as a Democrat, given that the person voted against the proposition?

43. A veterinarian uses a test to determine whether or not a dog has a disease that affects 7% of the dog population. The test correctly gives positive results for 98% of the dogs that have the disease, but gives false positives for 4% of the dogs that do not have the disease. If a dog tests positive, what is the probability that the dog has the disease?

44. Suppose that in a certain region of the country, if it rains one day there is a 65% chance it will rain the next day. If it does not rain on a given day, there is only a 15% chance it will rain the following day. What is the probability that, if it is not raining today, it will rain tomorrow but will not rain the next day?

45. The numbers of runs scored by the Chicago White Sox in their 2005 postseason games were:

$$1, 2, 2, 5, 5, 5, 5, 6, 7, 7, 8, 14$$

Find the mean, median, and mode of these data.

46. A set of data has a mean of 16, a median of 15, and a mode of 14. Which of these numbers must be one of the data values?

47. Write a set of data with five different data values for which the mean and median are 55.

48. State whether the mean, the median, or the mode is being used.

  a. In 2006, there were as many people aged 25 and younger in the world as there were people aged 25 and older.

  b. The majority of full-time students carry a load of 15 credit hours per semester.

  c. The average annual rainfall is 27.3 inches.

49. The lengths of cantilever bridges in the United States are shown below. Find the mean, the median, and the range of the data.

Bridge	Length (in feet)
Baton Rouge (Louisiana)	235
Commodore John Barry (Pennsylvania)	644
Greater New Orleans (Louisiana)	576
Longview (Washington)	200
Patapsco River (Maryland)	200
Queensboro (New York)	182
Tappan Zee (New York)	212
Transbay Bridge (California)	400

50. In a 4.0-point grading system, the letter grades have the following numerical values.

A = 4.00	B− = 2.67	D+ = 1.33
A− = 3.67	C+ = 2.33	D = 1.00
B+ = 3.33	C = 2.00	D− = 0.67
B = 3.00	C− = 1.67	F = 0.00

Use the weighted mean formula to find the grade-point average for a student with the following grades.

Course	Credits	Grade
Mathematics	3	A
English	3	C+
Computer	2	B−
Biology	4	B
Art	1	A

51. A teacher finds that the test scores for a group of 40 students have a mean of 72 and a standard deviation of 8.

a. If Anne has a test score of 82, then what is Anne's *z*-score?

b. What is Anne's percentile score?

52. An airline recorded the times for a ground crew to unload the baggage from an airplane. The recorded times, in minutes, were 12, 18, 20, 14, and 16. Find the *sample* standard deviation and the variance of these times. Round your results to the nearest hundredth.

53. The following table gives the average annual admission prices for U.S. movie theaters for the years from 1995 to 2004.

Year	Ticket Price
2004	$6.21
2003	6.03
2002	5.80
2001	5.65
2000	5.39
1999	5.06
1998	4.69
1997	4.59
1996	4.42
1995	4.35

Source: http://www.natoonline.org/
statisticstickets.htm

Find the mean, the median, and the standard deviation of this *sample* of admission prices.

54. One student received test scores of 85, 92, 86, and 89. A second student received scores of 90, 97, 91, and 94 (exactly 5 points more on each test). What is the relationship between

a. the means of their test scores?

b. the standard deviations of their test scores?

55. A *population* data set has a mean of $\mu = 81$ and a standard deviation of 5.2. Find the *z*-scores for each of the following.

a. $x = 72$ **b.** $x = 84$

56. The cholesterol levels for 10 adults are as follows.

310, 185, 254, 221, 170, 214, 172, 208, 164, 182

Draw a box-and-whisker plot of the data.

57. The following histogram shows the distribution of the test scores for a history test.

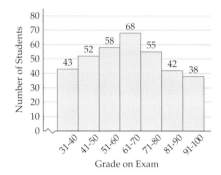

a. How many students scored more than 70 on the test?

b. How many students took the test?

c. What is the uniform class width?

d. What percent of the students scored 60 or less on the test? Round to the nearest tenth.

58. Use the relative frequency distribution below to determine

a. the *percent* of the states that paid an average teacher salary of at least $42,000.

b. the *probability* that a state, selected at random, paid an average teacher salary of at least $50,000 but less than $58,000.

Salaries($)	Percent of States
34,000 – 37,999	7.8
38,000 – 41,999	33.3
42,000 – 45,999	23.5
46,000 – 49,999	7.8
50,000 – 53,999	11.8
54,000 – 57,999	11.8
58,000 – 61,999	4.0

Source: Rankings and Estimates, NEA Research, June 2005

59. A professor gave a final examination to 110 students. Eighteen students had examination scores that were more than one standard deviation above the mean. With this information, can you conclude that 18 of the students had examination scores that were less than one standard deviation below the mean? Explain.

60. The amounts of time customers spend waiting in line at the ticket counter at an amusement park are normally distributed with a mean of 6.5 minutes and a standard deviation of 1 minute. Find the probability that the time a customer will spend waiting is

   a. at least 8 minutes.

   b. less than 6 minutes.

61. The weights of all the sacks of dog food filled by a machine are normally distributed with a mean weight of 50 pounds and a standard deviation of 0.02 pound. What percent of the sacks will

   a. weigh less than 49.97 pounds?

   b. weigh between 49.98 and 50.06 pounds?

62. A telephone company finds that the life spans of its telephones are normally distributed with a mean of 6.5 years and a standard deviation of 0.5 year.

   a. What percent of its telephones will last at least 7.25 years?

   b. What percent of its telephones will last between 5.8 years and 6.8 years?

   c. What percent of its telephones will last less than 6.9 years?

63. A gambler playing blackjack is doubting that the dealer is using a full, standard deck of cards. The probability of being dealt blackjack (a 10 or a face card in addition to an ace) is $\frac{32}{663}$, but of the last 200 deals, only five were blackjack.

   a. The player wants to use a hypothesis test to determine whether the deck of cards is complete. State an appropriate null hypothesis and the corresponding alternative hypothesis.

   b. Compute the test statistic for the observed 200 deals.

   c. Based on a 5% significance level, should the player conclude that the deck of cards is not complete?

64. A manufacturer of tires claims that its tires will last 40,000 miles. A survey of 325 owners of the tires resulted in a mean life span of 38,500 miles with a standard deviation of 4200 miles.

   a. State a null hypothesis and an alternative hypothesis that could be used to perform a hypothesis test on these data.

   b. Compute the test statistic for the survey results.

   c. Based on this survey, should we conclude, at the 1% significance level, that the manufacturer's claim is false?

65. A computer chip manufacturer claims that its EC-17 chips run at a mean speed of 2.8 gigahertz with a standard deviation of 0.15 gigahertz. To test this claim, an engineering firm measured the speed of 45 EC-17 chips and found that the mean speed of the sample was 2.7 gigahertz. On the basis of this sample, should the engineers conclude, at the 1% level of significance, that the speed of the EC-17 chips is lower than the 2.8 gigahertz claimed by the company?

66. The proportion of students who graduate from high school in the United States is approximately 78%. A recent survey of 600 students in a large city found that 456 graduated from high school. At the 5% level of significance, is there an indication that there is a smaller proportion of students graduating from high school in this city?

67. A company cooked 125 bags of its microwave popcorn in various ovens in an attempt to determine the cooking time. The mean cooking time was 3.7 minutes with a standard deviation of 1.6 minutes. Based on this sample, what is the 95% confidence interval that the company can use for the cooking time of its microwave popcorn? Round to the nearest tenth.

68. A sample of 40 identically equipped cars was tested for highway fuel consumption. The mean fuel consumption was 28 miles per gallon (mpg), with a standard deviation of 0.78 mpg. Based on this sample, what is the 99% confidence interval for the mean highway fuel consumption of this car?

69. A ratings firm is responsible for estimating the number of listeners of a local radio station. The company conducted random phone surveys of 800 residents. Find the greatest margin of error at the 99% confidence level. Round to the nearest tenth of a percent.

1. A computer system can be configured using one of three processors of different speeds, one of four different disk drives of different sizes, one of three different monitors, and one of two different graphics cards. How many different computer systems are possible?

2. The math center at a college has 20 computers. If 12 students walk into the math center to use a computer, in how many different ways can the students be assigned to a computer?

3. A coin and a regular six-sided die are tossed together once. What is the probability that the coin shows a head or the die has a 5 on the upward face?

4. What is the probability of drawing two cards in succession (without replacement) from a standard deck of playing cards and having them both be hearts?

5. Three coins are tossed once. What are the odds in favor of getting three heads?

6. A new medical test can determine whether or not a human has a disease that affects 5% of the population. The test correctly gives positive results for 99% of the people who have the disease, but gives false positives for 3% of the people who do not have the disease. If a person tests positive, what is the probability that the person has the disease? Round to the nearest hundreth.

7. The table below shows the numbers of men and the numbers of women who responded either positively or negatively to a new commercial. If one person is chosen from this group, find the probability that the person is a woman given that the person responded negatively.

	Positive	Negative	Total
Men	684	736	1420
Women	753	642	1395
Total	1437	1378	2815

8. Straight or curly hair for a hamster is determined by a dominant allele S, corresponding to straight hair, and a recessive allele s, which produces curly hair. Draw a Punnett square for parents of genotypes Ss and ss, and compute the probability that offspring of these parents will have curly hair.

9. The total points scored by the top 10 players during the 2005 WNBA finals were 17, 23, 33, 37, 41, 51, 51, 58, 69, and 74. What are the mean, median, and mode of these data?

10. A student's final grade in a statistics class is based on the following schedule: mean of tests, 50%; mean of quizzes, 20%; final exam, 30%. Find the numeric grade for a student whose mean test score was 86, whose mean quiz score was 88, and whose final exam score was 92.

11. A set of data has a mean of $\mu = 34$ and a standard deviation of $\sigma = 2.4$. Find the z-score for each of the following. Round to the nearest hundredth.

   a. 37   b. 29

12. The mean weight of babies born at a certain hospital was 6.4 pounds, with a standard deviation of 1.5 pounds. If the weights are normally distributed, find

   a. the probability that a baby born at the hospital weighs more than 7 pounds.

   b. the probability that a baby born at the hospital weighs between 5 and 8 pounds.

   c. the probability that a baby born at the hospital weighs less than 7.9 pounds.

13. The Mega Numbers selected from 20 consecutive drawings of the California SuperLotto Plus lottery were 2, 20, 13, 24, 17, 6, 13, 22, 5, 19, 8, 11, 7, 17, 8, 5, 2, 18, 23, and 20. Draw a box-and-whisker plot for these data.

14. The ages of the participants in a backgammon tournament are given in the following frequency distribution.

Age Group	Number of Participants
18 – 27	7
28 – 37	12
38 – 47	21
48 – 57	27
58 – 67	22
68 – 77	8
78 – 87	3

  a. What was the total number of participants?

  b. How many participants were at least 48 years old?

  c. What percent of the participants were less than 38 years old?

15. A software company advertises that users who call for support should not have to wait on hold longer than 15 minutes. The customer service manager reviewed the records for one business day, which showed 62 calls with a mean hold time of 17.4 minutes and a standard deviation of 4.3 minutes.

  a. The manager plans to use a hypothesis test to determine whether the advertised average hold time is valid. State an appropriate null hypothesis and the corresponding alternative hypothesis for the test.

  b. Compute the test statistic for the hold times.

  c. Based on the hypothesis test at the 5% significance level, should the manager conclude that the advertised 15-minute hold time is no longer valid?

16. A new movie was shown at several film festivals to try to determine its box office appeal. The studio surveyed 175 people who watched the film, of whom 55 said they liked the film.

  a. Determine the 95% confidence interval for the true proportion of moviegoers who would like the film.

  b. At the 99% level, what is the greatest margin of error? Round to the nearest tenth.

CHAPTER 9

Apportionment and Voting

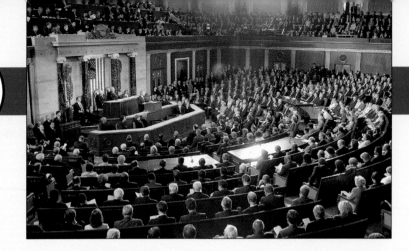

| 9.1 | **Introduction to Apportionment** | 9.3 | **Weighted Voting Systems** |
| 9.2 | **Introduction to Voting** | | |

In this chapter, we discuss two of the most fundamental principles of democracy: the right to vote and the value of that vote. The U.S. Constitution, in Article I, Section 2, states in part that

> The House of Representatives shall be composed of members chosen every second year by the people of the several states, and the electors in each state shall have the qualifications requisite for electors of the most numerous branch of the state legislature. . . . Representatives and direct taxes shall be apportioned among the several states which may be included within this union, according to their respective numbers. . . . The actual Enumeration shall be made within three years after the first meeting of the Congress of the United States, and within every subsequent term of ten years, in such manner as they shall by law direct. The number of Representatives shall not exceed one for every thirty thousand, but each state shall have at least one Representative; . . .

This article of the Constitution requires that "Representatives . . . be *apportioned* [our italics] among the several states . . . according to their respective numbers. . . ." That is, the number of representatives each state sends to Congress should be based on its population. Because populations change over time, this article also requires that the number of people within a state should be counted "within every subsequent term of ten years." This is why we have a census every 10 years.

The way representatives are *apportioned* has been a contentious issue since the founding of the United States. The first presidential veto was issued by George Washington in 1792 because he did not approve of the way the House of Representatives decided to apportion the number of representatives each state would have. Ever since that first veto, the issue of how to apportion membership in the House of Representatives among the states has been revisited many times.

Online Study Center

For online student resources, visit this textbook's Online Study Center at **college.hmco.com/pic/aufmannMTQR.**

| # Introduction to Apportionment

historical note

The original Article I, Section 2 of the U.S. Constitution included wording as to how to count citizens. According to this article, the numbers "shall be determined by adding to the whole number of free persons, including those bound to service for a term of years, and excluding Indians not taxed, three fifths of all other Persons." The "three fifths of all other Persons" meant that a slave was counted as only $\frac{3}{5}$ of a person. This article was modified by Section 1 of the 14th Amendment to the Constitution, which states, in part, that "All persons born or naturalized in the United States, and subject to the jurisdiction thereof, are citizens of the United States and of the state wherein they reside...." ∎

The mathematical investigation into **apportionment,** which is a method of dividing a whole into various parts, has its roots in the U.S. Constitution. (See the chapter opener.) Since 1790, when the House of Representatives first attempted to apportion itself, various methods have been used to decide how many voters would be represented by each member of the House. The two competing plans in 1790 were put forward by Alexander Hamilton and Thomas Jefferson.

To illustrate how the Hamilton and Jefferson plans were used to calculate the number of representatives each state should have, we will consider the fictitious country of Andromeda, with a population of 20,000 and five states. The population of each state is given in the table at the right.

Andromeda's constitution calls for 25 representatives to be chosen from these states. The number of representatives is to be apportioned according to the states' respective populations.

Andromeda

State	Population
Apus	11,123
Libra	879
Draco	3518
Cephus	1563
Orion	2917
Total	20,000

The Hamilton Plan

Under the Hamilton plan, the total population of the country (20,000) is divided by the number of representatives (25). This gives the number of citizens represented by each representative. This number is called the *standard divisor.*

> **Standard Divisor**
>
> $$\text{Standard divisor} = \frac{\text{total population}}{\text{number of people to apportion}}$$

For Andromeda, we have

$$\text{Standard divisor} = \frac{\text{total population}}{\text{number of people to apportion}} = \frac{20,000}{25} = 800$$

✔ **TAKE NOTE**

Today apportionment is applied to situations other than a population of people. For instance, *population* could refer to the number of math classes offered at a college or the number of fire stations in a city. Nonetheless, the definition of *standard divisor* is still given as though people were involved.

QUESTION *What is the meaning of the number 800 calculated above?*

Now divide the population of each state by the standard divisor and round the quotient *down* to a whole number. For example, both 15.1 and 15.9 would be rounded to 15. Each whole number quotient is called a *standard quota.*

ANSWER *It is the number of citizens represented by each representative.*

> **Standard Quota**
>
> The **standard quota** is the whole number part of the quotient of a population divided by the standard divisor.

State	Population	Quotient	Standard Quota
Apus	11,123	$\dfrac{11,123}{800} \approx 13.904$	13
Libra	879	$\dfrac{879}{800} \approx 1.099$	1
Draco	3518	$\dfrac{3518}{800} \approx 4.398$	4
Cephus	1563	$\dfrac{1563}{800} \approx 1.954$	1
Orion	2917	$\dfrac{2917}{800} \approx 3.646$	3
		Total	22

From the calculations in the above table, the total number of representatives is 22, not 25 as required by Andromeda's constitution. When this happens, the Hamilton plan calls for revisiting the calculation of the quotients and assigning an additional representative to the state with the largest decimal remainder. This process is continued until the number of representatives equals the number required by the constitution. For Andromeda, we have

✔ **TAKE NOTE**

Additional representatives are assigned according to the largest decimal remainders. Because the sum of the standard quotas came to only 22 representatives, we must add three more representatives. The states with the three highest decimal remainders are Cephus (1.<u>954</u>), Apus (13.<u>904</u>), and Orion (3.<u>646</u>). Thus each of these states gets an additional representative.

State	Population	Quotient	Standard Quota	Number of Representatives
Apus	11,123	$\dfrac{11,123}{800} \approx 13.904$	13	14
Libra	879	$\dfrac{879}{800} \approx 1.099$	1	1
Draco	3518	$\dfrac{3518}{800} \approx 4.398$	4	4
Cephus	1563	$\dfrac{1563}{800} \approx 1.954$	1	2
Orion	2917	$\dfrac{2917}{800} \approx 3.646$	3	4
		Total	22	25

The Jefferson Plan

As we saw with the Hamilton plan, dividing by the standard divisor and then rounding down does not always yield the correct number of representatives. In the previous example, we were three representatives short. The Jefferson plan attempts to overcome this difficulty by using a *modified standard divisor*. This number is chosen, by trial and error, so that the sum of the standard quotas is equal to the total number of representatives. In a specific apportionment calculation, there may be more than one number that can serve as the modified standard divisor. In the following apportionment calculation, we used 740 as our modified standard divisor. However, 741 also can be used as the modified standard divisor.

✔ **TAKE NOTE**

The modified standard divisor will always be smaller than the standard divisor.

State	Population	Quotient	Standard Quota	Number of Representatives
Apus	11,123	$\frac{11{,}123}{740} \approx 15.031$	15	15
Libra	879	$\frac{879}{740} \approx 1.188$	1	1
Draco	3518	$\frac{3518}{740} \approx 4.754$	4	4
Cephus	1563	$\frac{1563}{740} \approx 2.112$	2	2
Orion	2917	$\frac{2917}{740} \approx 3.942$	3	3
		Total	25	25

The table below shows how the results of the Hamilton and Jefferson apportionment methods differ. Note that each method assigns a different number of representatives to certain states.

State	Population	Hamilton Plan	Jefferson Plan
Apus	11,123	14	15
Libra	879	1	1
Draco	3518	4	4
Cephus	1563	2	2
Orion	2917	4	3
	Total	25	25

Although we have applied apportionment to allocating representatives to a congress, there are other applications of apportionment. For instance, nurses can be assigned to hospitals according to the number of patients requiring care; police officers can be assigned to precincts on the basis of the number of reported crimes; and math classes can be scheduled on the basis of student demand for those classes. These are only some of the ways apportionment can be used.

Ruben County

City	Population
Cardiff	7020
Solana	2430
Vista	1540
Pauma	3720
Pacific	5290

EXAMPLE 1 ■ Apportioning Board Members Using the Hamilton and Jefferson Methods

Suppose the 18 members on the board of the Ruben County environmental agency are selected according to the populations of the five cities in the county, as shown in the table at the left.

a. Use the Hamilton method to determine the number of board members each city should have.

b. Use the Jefferson method to determine the number of board members each city should have.

Solution

a. First find the total population of the five cities.

$$7020 + 2430 + 1540 + 3720 + 5290 = 20,000$$

Now calculate the standard divisor.

$$\text{Standard divisor} = \frac{\text{population of county}}{\text{number of board members}} = \frac{20,000}{18} \approx 1111.11$$

Use the standard divisor to find the standard quota for each city.

City	Population	Quotient	Standard Quota	Number of Board Members
Cardiff	7020	$\frac{7020}{1111.11} \approx 6.318$	6	6
Solana	2430	$\frac{2430}{1111.11} \approx 2.187$	2	2
Vista	1540	$\frac{1540}{1111.11} \approx 1.386$	1	2
Pauma	3720	$\frac{3720}{1111.11} \approx 3.348$	3	3
Pacific	5290	$\frac{5290}{1111.11} \approx 4.761$	4	5
		Total	16	18

The sum of the standard quotas is 16, so we must add two more members. The two cities with the largest decimal remainders are Pacific and Vista. Each of these two cities gets one additional board member. Thus the composition of the environmental board using the Hamilton method is Cardiff: 6, Solana: 2, Vista: 2, Pauma: 3, Pacific: 5.

CALCULATOR NOTE

Using lists and the IPart function (which returns only the whole number part of a number) of a TI-83/84 Plus calculator can be helpful when trying to find a modified standard divisor. Press STAT ENTER to display the list editor. If a list is present in L1, use the up arrow key to highlight L1, then press CLEAR ENTER. Enter the populations of each city in L1.

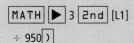

Press 2nd [QUIT]. To divide each number in L1 by a modified divisor (we are using 950 for this example), enter the following.

MATH ▶ 3 2nd [L1]

÷ 950)

STO 2nd [L2] ENTER

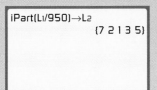

The standard quotas are shown on the screen. The sum of these numbers is 18, the desired number of representatives.

b. To use the Jefferson method, we must find a *modified* standard divisor that is less than the standard divisor we calculated in part a. We must do this by trial and error. For instance, if we choose 925 as the modified standard divisor, we have the following result.

City	Population	Quotient	Standard Quota	Number of Board Members
Cardiff	7020	$\frac{7020}{925} \approx 7.589$	7	7
Solana	2430	$\frac{2430}{925} \approx 2.627$	2	2
Vista	1540	$\frac{1540}{925} \approx 1.665$	1	1
Pauma	3720	$\frac{3720}{925} \approx 4.022$	4	4
Pacific	5290	$\frac{5290}{925} \approx 5.719$	5	5
		Total	19	19

This result yields too many board members. Thus we must increase the modified standard divisor. By experimenting with different divisors, we find that 950 is a possible modified standard divisor. Using 950 as the standard divisor gives the results shown in the table below.

City	Population	Quotient	Standard Quota	Number of Board Members
Cardiff	7020	$\frac{7020}{950} \approx 7.389$	7	7
Solana	2430	$\frac{2430}{950} \approx 2.558$	2	2
Vista	1540	$\frac{1540}{950} \approx 1.621$	1	1
Pauma	3720	$\frac{3720}{950} \approx 3.916$	3	3
Pacific	5290	$\frac{5290}{950} \approx 5.568$	5	5
		Total	18	18

Thus the composition of the environmental board using the Jefferson method is Cardiff: 7, Solana: 2, Vista: 1, Pauma: 3, Pacific: 5.

European Countries'
Populations, 2004

Country	Population
France	60,400,000
Germany	82,400,000
Italy	58,000,000
Spain	40,000,000
Belgium	10,300,000

Source: The World Almanac and Book of Facts, 2005, pages 848–849

CHECK YOUR PROGRESS 1   Suppose the 20 members of a committee from five European countries are selected according to the populations of the five countries, as shown in the table at the left.

a. Use the Hamilton method to determine the number of representatives each country should have.

b. Use the Jefferson method to determine the number of representatives each country should have.

Solution   See page S37.

Suppose that the environmental agency in Example 1 decides to add one more member to the board even though the population of each city remains the same. The total number of members is now 19 and we must determine how the members of the board will be apportioned.

The standard divisor is now $\frac{20,000}{19} \approx 1052.63$. Using Hamilton's method, the calculations necessary to apportion the board members are shown below.

City	Population	Quotient	Standard Quota	Number of Board Members
Cardiff	7020	$\frac{7020}{1052.63} \approx 6.669$	6	7
Solana	2430	$\frac{2430}{1052.63} \approx 2.309$	2	2
Vista	1540	$\frac{1540}{1052.63} \approx 1.463$	1	1
Pauma	3720	$\frac{3720}{1052.63} \approx 3.534$	3	4
Pacific	5290	$\frac{5290}{1052.63} \approx 5.026$	5	5
Total			17	19

The table below summarizes the number of board members each city would have if the board consisted of 18 members (Example 1) or 19 members.

City	Hamilton Apportionment with 18 Board Members	Hamilton Apportionment with 19 Board Members
Cardiff	6	7
Solana	2	2
Vista	2	1
Pauma	3	4
Pacific	5	5
Total	18	19

Notice that although one more board member was added, Vista lost a board member, even though the populations of the cities did not change. This is called the **Alabama paradox** and has a negative effect on fairness. In the interest of fairness, an apportionment method should not exhibit the Alabama paradox. (See the Math Matters below for other paradoxes.)

Math Matters   Apportionment Paradoxes

The Alabama paradox, although it was not given that name until later, was first noticed after the 1870 census. At the time, the House of Representatives had 270 seats. However, when the number of representatives in the House was increased to 280 seats, Rhode Island lost a representative.

After the 1880 census, C. W. Seaton, the chief clerk of the U.S. Census Office, calculated the number of representatives each state would have if the number were set at some number between 275 and 300. He noticed that when the number of representatives was increased from 299 to 300, Alabama lost a representative.

There are other paradoxes that involve apportionment methods. Two of them are the *population paradox* and the *new states paradox*. It is possible for the population of one state to be increasing faster than that of another state and for the state still to lose a representative. This is an example of the **population paradox.**

In 1907, when Oklahoma was added to the Union, the size of the House was increased by five representatives to account for Oklahoma's population. However, when the complete apportionment of the Congress was recalculated, New York lost a seat and Maine gained a seat. This is an example of the **new states paradox.**

Fairness in Apportionment

To decide which plan—the Hamilton or the Jefferson—is better, we might try to determine which plan is fairer. Of course, *fair* can be quite a subjective term, so we will try to state conditions by which an apportionment plan is judged fair. One criterion of fairness for an apportionment plan is that it should satisfy the *quota rule*.

> **Quota Rule**
>
> The number of representatives apportioned to a state is the standard quota or one more than the standard quota.

We can show that the Jefferson plan does not satisfy the quota rule by calculating the standard quota of Apus (see page 604).

$$\text{Standard quota} = \frac{\text{population of Apus}}{\text{standard divisor}} = \frac{11{,}123}{800} \approx 13$$

The standard quota of Apus is 13. However, the Jefferson plan assigns 15 representatives to that state (see page 605), two more than its standard quota. Therefore, the Jefferson method violates the quota rule.

As we have seen, the choice of apportionment method affects the number of representatives a state will have. Given that fact, mathematicians and others have tried to work out an apportionment method that is fair. The difficulty lies in trying to define what is fair.

Another measure of fairness is *average constituency*. This is the population of a state divided by the number of representatives from the state and then rounded to the nearest whole number.

Average Constituency

$$\text{Average constituency} = \frac{\text{population of a state}}{\text{number of representatives from the state}}$$

Consider the two states Hampton and Shasta in the table below.

State	Population	Representatives	Average Constituency
Hampton	16,000	10	$\frac{16{,}000}{10} = 1600$
Shasta	8340	5	$\frac{8340}{5} = 1668$

Because the average constituencies are approximately equal, it seems natural to say that both states are equally represented. See the Take Note at the left.

QUESTION *Although the average constituencies of Hampton and Shasta are approximately equal, which state has the more favorable representation?*

Now suppose that one representative will be added to one of the states. Which state is more deserving of the new representative? In other words, to be fair, which state should receive the new representative?

The changes in the average constituencies are shown below.

State	Average Constituency (old)	Average Constituency (new)
Hampton	$\frac{16{,}000}{10} = 1600$	$\frac{16{,}000}{11} \approx 1455$
Shasta	$\frac{8340}{5} = 1668$	$\frac{8340}{6} = 1390$

From the table, there are two possibilities for adding one representative. If Hampton receives the representative, its average constituency will be 1455 and Shasta's will remain at 1668. The difference in the average constituencies is $1668 - 1455 = 213$. This difference is called the *absolute unfairness of the apportionment.*

ANSWER *Because Hampton's average constituency is smaller than Shasta's, Hampton has the more favorable representation.*

Absolute Unfairness of an Apportionment

The **absolute unfairness of an apportionment** is the absolute value of the difference between the average constituency of state A and the average constituency of state B.

|Average constituency of A − average constituency of B|

If Shasta receives the representative, its average constituency will be 1390 and Hampton's will remain at 1600. The absolute unfairness of apportionment is 1600 − 1390 = 210. This is summarized below.

	Hampton's Average Constituency	Shasta's Average Constituency	Absolute Unfairness of Apportionment
Hampton receives the new representative	1455	1668	213
Shasta receives the new representative	1600	1390	210

Because the smaller absolute unfairness of apportionment occurs if Shasta receives the new representative, it might seem that Shasta should receive the representative. However, this is not necessarily true.

To understand this concept, let's consider a somewhat different situation. Suppose an investor makes two investments, one of $10,000 and another of $20,000. One year later, the first investment is worth $11,000 and the second investment is worth $21,500. This is shown in the table below.

	Original Investment	One Year Later	Increase
Investment A	$10,000	$11,000	$1000
Investment B	$20,000	$21,500	$1500

Although there is a larger increase in investment B, the increase per dollar of the original investment is $\frac{1500}{20,000} = 0.075$. On the other hand, the increase per dollar of investment A is $\frac{1000}{10,000} = 0.10$. Another way of saying this is that each $1 of investment A produced a return of 10 cents (0.10), whereas each $1 of investment B produced a return of 7.5 cents (0.075). Therefore, even though the increase in investment A was less than the increase in investment B, investment A was more productive.

A similar process is used when deciding which state should receive another representative. Rather than look at the difference in the absolute unfairness of apportionment, we determine the *relative unfairness* of adding the representative.

Relative Unfairness of an Apportionment

The **relative unfairness of an apportionment** is the quotient of the absolute unfairness of the apportionment and the average constituency of the state receiving the new representative.

$$\frac{\text{absolute unfairness of the apportionment}}{\text{average constituency of the state receiving the new representative}}$$

EXAMPLE 2 ■ **Determine the Relative Unfairness of an Apportionment**

Determine the relative unfairness of an apportionment that gives a new representative to Hampton rather than Shasta.

Solution

Using the table for Hampton and Shasta shown on the preceding page, we have

Relative unfairness of the apportionment

$$= \frac{\text{absolute unfairness of the apportionment}}{\text{average constituency when Hampton receives a new representative}}$$

$$= \frac{213}{1455} \approx 0.146$$

The relative unfairness of the apportionment is approximately 0.146.

CHECK YOUR PROGRESS 2   Determine the relative unfairness of an apportionment that gives a new representative to Shasta rather than Hampton.

Solution   *See page S38.*

The relative unfairness of an apportionment is used in the following way.

> **Apportionment Principle**
>
> When adding a new representative to a state, the representative is assigned to the state in such a way as to give the smallest relative unfairness of apportionment.

From Example 2, the relative unfairness of adding a representative to Hampton is approximately 0.146. From Check Your Progress 2, the relative unfairness of adding a representative to Shasta is approximately 0.151. Because the smaller relative unfairness results from adding the representative to Hampton, that state should receive the new representative.

Although we have focused on assigning representatives to states, the apportionment principle can be used in many other situations.

EXAMPLE 3 ■ **Use the Apportionment Principle**

The table below shows the numbers of paramedics and the annual numbers of paramedic calls for two cities. If a new paramedic is hired, use the apportionment principle to determine to which city the paramedic should be assigned.

	Paramedics	**Annual Paramedic Calls**
Tahoe	125	17,526
Erie	143	22,461

Solution

Calculate the relative unfairness of the apportionment that assigns the paramedic to Tahoe and the relative unfairness of the apportionment that assigns the paramedic to Erie. In this case, average constituency is the number of annual paramedic calls divided by the number of paramedics.

	Tahoe's Annual Paramedic Calls per Paramedic	Erie's Annual Paramedic Calls per Paramedic	Absolute Unfairness of Apportionment
Tahoe receives a new paramedic	$\dfrac{17,526}{125+1} \approx 139$	$\dfrac{22,461}{143} \approx 157$	$157 - 139 = 18$
Erie receives a new paramedic	$\dfrac{17,526}{125} \approx 140$	$\dfrac{22,461}{143+1} \approx 156$	$156 - 140 = 16$

If Tahoe receives the new paramedic, the relative unfairness of the apportionment is

Relative unfairness of the apportionment

$$= \frac{\text{absolute unfairness of the apportionment}}{\text{Tahoe's average constituency with a new paramedic}}$$

$$= \frac{18}{139} \approx 0.129$$

If Erie receives the new paramedic, the relative unfairness of the apportionment is

Relative unfairness of the apportionment

$$= \frac{\text{absolute unfairness of the apportionment}}{\text{Erie's average constituency with a new paramedic}}$$

$$= \frac{16}{156} \approx 0.103$$

Because the smaller relative unfairness results from adding the paramedic to Erie, that city should receive the paramedic.

CHECK YOUR PROGRESS 3   The table below shows the numbers of first and second grade teachers in a school district and the numbers of students in each of those grades. If a new teacher is hired, use the apportionment principle to determine to which grade the teacher should be assigned.

	Number of Teachers	Number of Students
First grade	512	12,317
Second grade	551	15,439

Solution   *See page S38.*

historical note

According to the U.S. Bureau of the Census, methods of apportioning the House of Representatives have changed over time.

1790–1830: Jefferson method

1840: Webster method (See the paragraph following Exercise 28, page 621.)

1850–1900: Hamilton method

1910, 1930: Webster method

Note that 1920 is missing. In direct violation of the U.S. Constitution, the House of Representatives failed to reapportion the House in 1920.

1940–2000: Method of equal proportions or the Huntington-Hill method

All apportionment plans enacted by the House have created some controversy. For instance, the constitutionality of the Huntington-Hill method was challenged by Montana in 1992 because it lost a seat to Washington after the 1990 census. For more information on this subject, see http://www.census.gov/population/www/censusdata/apportionment/history.html. ∎

Huntington-Hill Apportionment Method

As we mentioned earlier, the members of the House of Representatives are apportioned among the states every 10 years. The present method used by the House is based on the apportionment principle and is called the *method of equal proportions* or the *Huntington-Hill method*. This method has been used since 1940.

The Huntington-Hill method is implemented by calculating what is called a *Huntington-Hill number*. This number is derived from the apportionment principle. See Exercise 39, page 623.

Huntington-Hill Number

The value of $\dfrac{(P_A)^2}{a(a+1)}$, where P_A is the population of state A and a is the current number of representatives from state A, is called the **Huntington-Hill number** for state A.

When the Huntington-Hill method is used to apportion representatives between two states, the state with the greater Huntington-Hill number receives the next representative. This method can be extended to more than two states.

Huntington-Hill Apportionment Principle

When there is a choice of adding one representative to one of several states, the representative should be added to the state with the greatest Huntington-Hill number.

EXAMPLE 4 ■ **Use the Huntington-Hill Apportionment Principle**

The table below shows the numbers of lifeguards that are assigned to three different beaches and the numbers of rescues made by lifeguards at those beaches. Use the Huntington-Hill apportionment principle to determine to which beach a new lifeguard should be assigned.

Beach	Number of Lifeguards	Number of Rescues
Mellon	37	1227
Donovan	51	1473
Ferris	24	889

Solution

Calculate the Huntington-Hill number for each of the beaches. In this case, the population is the number of rescues and the number of representatives is the number of lifeguards.

Mellon: Donovan: Ferris:

$$\frac{1227^2}{37(37+1)} \approx 1071 \qquad \frac{1473^2}{51(51+1)} \approx 818 \qquad \frac{889^2}{24(24+1)} \approx 1317$$

Ferris has the greatest Huntington-Hill number. Thus, according to the Huntington-Hill apportionment principle, the new lifeguard should be assigned to Ferris.

The advantage of using the Huntington-Hill apportionment principle, rather than calculating relative unfairness, occurs when there are many states that could receive the next representative. For instance, if we were to use relative unfairness to determine which of four states should receive a new representative, it would be necessary to compute the relative unfairness for every possible pairing of the states—a total of 24 computations. However, using the Huntington-Hill method, we need only calculate the Huntington-Hill number for each state—a total of four calculations. In a sense, the Huntington-Hill number provides a short cut for applying the relative unfairness method.

CHECK YOUR PROGRESS 4 A university has a president's council that is composed of students from each of the undergraduate classes. If a new student representative is added to the council, use the Huntington-Hill apportionment principle to determine which class the new student council member should represent.

Class	Number of Representatives	Number of Students
First year	12	2015
Second year	10	1755
Third year	9	1430
Fourth year	8	1309

Solution   *See page S39.*

Now that we have looked at various apportionment methods, it seems reasonable to ask which is the best method. Unfortunately, all apportionment methods have some flaws. This was proved by Michael Balinski and H. Peyton Young.

> **Balinski-Young Impossibility Theorem**
>
> Any apportionment method either will violate the quota rule or will produce paradoxes such as the Alabama paradox.

Although there is no perfect apportionment method, Balinski and Young went on to present a strong case that the Webster method (following Exercise 28 on page 621) is the system that most closely satisfies the goal of one person, one vote. However, political expediency sometimes overrules mathematical proof. Some historians have suggested that although the Huntington-Hill apportionment method was better than some of the previous methods, President Franklin Roosevelt chose this method in 1941 because it alloted one more seat to Arkansas and one less to Michigan. This essentially meant that the House of Representatives would have one more seat for the Democrats, Roosevelt's party.

Investigation

Apportioning the 1790 House of Representatives

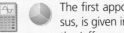

The first apportionment of the House of Representatives, using the 1790 census, is given in the following table. This apportionment was calculated by using the Jefferson method. (See our Online Study Center at **college.hmco.com/pic/aufmannMTQR** for an Excel spreadsheet that will help with the computations.)

(continued)

historical note

According to the U.S. Constitution, each state must have at least one representative to the House of Representatives. The remaining representatives are then assigned to the states using, since 1941, the Huntington-Hill apportionment principle. ■

▼ point of interest

For most states, the populations shown in the table are not the actual populations of the states because each slave was counted as only three-fifths of a person. (See the Historical Note on page 603.) For instance, the actual population of Connecticut was 237,946, of which 2764 were slaves. For apportionment purposes, the number of slaves was subtracted, and then $\frac{3}{5}$ of that population was added back.

$$237,946 - 2764 + \frac{3}{5}(2764) \approx 236,841$$

Apportionment Using the Jefferson Method, 1790

State	Population	Number of Representatives
Connecticut	236,841	7
Delaware	55,540	1
Georgia	70,835	2
Maryland	278,514	8
Massachusetts	475,327	14
Kentucky	68,705	2
New Hampshire	141,822	4
Vermont	85,533	2
New York	331,591	10
New Jersey	179,570	5
Pennsylvania	432,879	13
North Carolina	353,523	10
South Carolina	206,236	6
Virginia	630,560	19
Rhode Island	68,446	2

Source: U.S. Census Bureau

Investigation Exercises

1. Verify this apportionment using the Jefferson method. You will have to experiment with various modified standard divisors until you reach the given representation. See the Calculator Note on page 607.

2. Find the apportionment that would have resulted if the Hamilton method had been used.

3. Give each state one representative. Use the Huntington-Hill method with $a = 1$ to determine the state that receives the next representative. The Calculator Note in this section will help. With the populations stored in L1, enter `2nd` L1 `x²` $\div$ 2 `STO` `2nd` L2 `ENTER`. Now scroll through L2 to find the largest number.

4. Find the apportionment that would have resulted if the Huntington-Hill method (the one used for the 2000 census) had been used in 1790. See our Online Study Center at **college.hmco.com/pic/aufmannMTQR** for a spreadsheet that will help with the calculations.

Exercise Set 9.1

1. Explain how to calculate the standard divisor of an apportionment for a total population p with n items to apportion.

2. **Teacher Aides** A total of 25 teacher aides are to be apportioned among seven classes at a new elementary school. The enrollments in the seven classes are shown in the following table.

Class	Number of Students
Kindergarten	38
First grade	39
Second grade	35
Third grade	27
Fourth grade	21
Fifth grade	31
Sixth grade	33
Total	224

a. Determine the standard divisor. What is the meaning of the standard divisor in the context of this exercise?

b. Use the Hamilton method to determine the number of teacher aides to be apportioned to each class.

c. Use the Jefferson method to determine the number of teacher aides to be apportioned to each class. Is this apportionment in violation of the quota rule?

d. How do the apportionment results produced using the Jefferson method compare with the results produced using the Hamilton method?

3. In the Hamilton apportionment method, explain how to calculate the standard quota for a particular state (group).

4. What is the quota rule?

5. If the average constituency of state A is less than the average constituency of state B, which state has the more favorable representation?

6. If a new representative is added to state A, the relative unfairness of the apportionment is 0.35. If a new representative is added to state B, the relative unfairness of the apportionment is 0.56. According to the apportionment principle, which state should receive the new representative?

7. The following table shows how the average constituency changes for two regional governing boards, Joshua and Salinas, when a new representative is added to each board.

	Joshua's Average Constituency	Salinas's Average Constituency
Joshua receives new board member	1215	1547
Salinas receives new board member	1498	1195

a. Determine the relative unfairness of an apportionment that gives a new board member to Joshua rather than to Salinas. Round to the nearest thousandth.

b. Determine the relative unfairness of an apportionment that gives a new board member to Salinas rather than to Joshua. Round to the nearest thousandth.

c. Using the apportionment principle, determine which regional governing board should receive the new board member.

8. The table below shows how the average constituency changes when two different national parks, Evergreen State Park and Rust Canyon Preserve, add a new forest ranger.

	Evergreen State Park's Average Constituency	Rust Canyon Preserve's Average Constituency
Evergreen receives new forest ranger	466	638
Rust Canyon receives new forest ranger	650	489

a. Determine the relative unfairness of an apportionment that gives a new forest ranger to Evergreen rather than to Rust Canyon. Round to the nearest thousandth.

b. Determine the relative unfairness of an apportionment that gives a new forest ranger to Rust Canyon rather than to Evergreen. Round to the nearest thousandth.

c. Using the apportionment principle, determine which national park should receive the new forest ranger.

9. The table below shows the numbers of sales associates and the average numbers of customers per day at a company's two department stores. The company is planning to add a new sales associate to one of the stores. Use the apportionment principle to determine which store should receive the new employee.

Shopping mall location	Number of Sales Associates	Average Number of Customers per Day
Summer Hill Galleria	587	5289
Seaside Mall Galleria	614	6215

10. The table below shows the numbers of interns and the average numbers of patients admitted each day at two different hospitals. The hospital administrator is planning to add a new intern to one of the hospitals. Use the apportionment principle to determine which hospital should receive the new intern.

Hospital location	Number of Interns	Average Number of Patients Admitted per Day
South Coast Hospital	128	518
Rainer Hospital	145	860

11. **House of Representatives** The U.S. House of Representatives currently has 435 members to represent the 281,424,177 citizens of the United States as determined by the 2000 census.

a. Calculate the standard divisor for the apportionment of these representatives and explain the meaning of this standard divisor in the context of this exercise.

b. According to the 2000 census, the population of Delaware was 785,068. Delaware currently has only one representative in the House of Representatives. Is Delaware currently overrepresented or underrepresented in the House of Representatives? Explain.

c. According to the 2000 census, the population of Vermont was 609,890. Vermont currently has only one representative in the House of Representatives. Is Vermont currently overrepresented or underrepresented in the House of Representatives? Explain.

12. **College Enrollment** The following table shows the enrollment for each of the four divisions of a college. The four divisions are liberal arts, business, humanities, and science. There are 180 new computers that are to be apportioned among the divisions based on the enrollments.

Division	Enrollment
Liberal arts	3455
Business	5780
Humanities	1896
Science	4678
Total	15,809

a. What is the standard divisor for an apportionment of the computers? What is the meaning of the standard divisor in the context of this exercise?

b. Use the Hamilton method to determine the number of computers to be apportioned to each division.

c. If the computers are to be apportioned using the Jefferson method, explain why neither 86 nor 87 can be used as a modified standard divisor. Explain why 86.5 can be used as a modified standard divisor.

d. Explain why the modified standard divisor used in the Jefferson method cannot be larger than the standard divisor.

e. Use the Jefferson method to determine the number of computers to be apportioned to each division. Is this apportionment in violation of the quota rule?

f. How do the apportionment results produced using the Jefferson method compare with the results produced using the Hamilton method?

13. Medical Care A hospital district consists of six hospitals. The district administrators have decided that 48 new nurses should be apportioned based on the number of beds in each of the hospitals. The following table shows the number of beds in each hospital.

Hospital	Number of Beds
Sharp	242
Palomar	356
Tri-City	308
Del Raye	190
Rancho Verde	275
Bel Aire	410
Total	1781

a. Determine the standard divisor. What is the meaning of the standard divisor in the context of this exercise?

b. Use the Hamilton method to determine the number of nurses to be apportioned to each hospital.

c. Use the Jefferson method to determine the number of nurses to be apportioned to each hospital.

d. How do the apportionment results produced using the Jefferson method compare with the results produced using the Hamilton method?

14. What is the Alabama paradox?

15. What is the population paradox?

16. What is the new states paradox?

17. What is the Balinski-Young Impossibility Theorem?

18. Apportionment of Projectors Consider the apportionment of 27 projectors for a school district with four campus locations labeled A, B, C, and D. The following table shows the apportionment of the projectors using the Hamilton method.

Campus	A	B	C	D
Enrollment	840	1936	310	2744
Apportionment of 27 projectors	4	9	1	13

a. If the number of projectors to be apportioned increases from 27 to 28, what will be the apportionment if the Hamilton method is used? Will the Alabama paradox occur? Explain.

b. If the number of projectors to be apportioned using the Hamilton method increases from 28 to 29, will the Alabama paradox occur? Explain.

19. Hotel Management A company operates four resorts. The CEO of the company decides to use the Hamilton method to apportion 115 new flat-screen digital television sets to the resorts based on the number of guest rooms at each resort.

Resort	A	B	C	D
Number of guest rooms	23	256	182	301
Apportionment of 115 televisions	4	39	27	45

a. If the number of television sets to be apportioned by the Hamilton method increases from 115 to 116, will the Alabama paradox occur?

b. If the number of television sets to be apportioned by the Hamilton method increases from 116 to 117, will the Alabama paradox occur?

c. If the number of television sets to be apportioned by the Hamilton method increases from 117 to 118, will the Alabama paradox occur?

20. College Security A college apportions 40 security personnel among three education centers according to their enrollments. The following table shows the present enrollments at the three centers.

Center	A	B	C
Enrollment	356	1054	2590

a. Use the Hamilton method to apportion the security personnel.

b. ✎ After one semester, the centers have the following enrollments.

Center	A	B	C
Enrollment	370	1079	2600

Center A has an increased enrollment of 14 students, which is an increase of $\frac{14}{356} \approx 0.039 = 3.9\%$. Center B

has an increased enrollment of 25 students, which is an increase of $\frac{25}{1054} \approx 0.024 = 2.4\%$.

If the security personnel are reapportioned using the Hamilton method, will the population paradox occur? Explain.

21. Management Scientific Research Corporation has offices in Boston and Chicago. The number of employees at each office is shown in the following table. There are 22 vice presidents to be apportioned between the offices.

Office	Boston	Chicago
Employees	151	1210

a. Use the Hamilton method to find each office's apportionment of vice presidents.

b. ✎ The corporation opens an additional office in San Francisco with 135 employees and decides to have a total of 24 vice presidents. If the vice presidents are reapportioned using the Hamilton method, will the new states paradox occur? Explain.

22. Education The science division of a college consists of three departments: mathematics, physics, and chemistry. The number of students enrolled in each department is shown in the following table. There are 19 clerical assistants to be apportioned among the departments.

Department	Math	Physics	Chemistry
Student enrollment	4325	520	1165

a. Use the Hamilton method to find each department's apportionment of clerical assistants.

b. ✎ The division opens a new computer science department with a student enrollment of 495. The division decides to have a total of 20 clerical assistants. If the clerical assistants are reapportioned using the Hamilton method, will the new states paradox occur? Explain.

23. If two states have the same population but state A has 15 representatives and state B has 16 representatives, which state has the larger Huntington-Hill number?

24. 🗿 If the Huntington-Hill number of state A is less than the Huntington-Hill number of state B, which state should receive the next representative?

25. Education The following table shows the numbers of fifth and sixth grade teachers in a school district and the numbers of students in the two grades. The number of teachers for each of the grade levels was determined by using the Huntington-Hill apportionment method.

	Number of Teachers	Number of Students
Fifth grade	19	604
Sixth grade	21	698

The district has decided to hire a new teacher for either the fifth or the sixth grade.

a. Use the apportionment principle to determine to which grade the new teacher should be assigned.

b. Use the Huntington-Hill apportionment principle to determine to which grade the new teacher should be assigned. How does this result compare with the result in part a?

26. Social Workers The following table shows the numbers of social workers and the numbers of cases (the case load) handled by the social workers for two offices. The number of social workers for each office was determined by using the Huntington-Hill apportionment method.

	Number of Social Workers	Case Load
Hill Street office	20	584
Valley office	24	712

A new social worker is to be hired for one of the offices.

a. Use the apportionment principle to determine to which office the social worker should be assigned.

b. Use the Huntington-Hill apportionment principle to determine to which office the new social worker should be assigned. How does this result compare with the result in part a?

c. The results of part b indicate that the new social worker should be assigned to the Valley office. At this moment the Hill Street office has 20 social workers and the Valley office has 25 social workers. Use the Huntington-Hill apportionment principle to determine to which office the *next* new social worker should be assigned. Assume the case loads remain the same.

27. Computer Usage The table below shows the numbers of computers that are assigned to four different schools and the numbers of students in those schools. Use the Huntington-Hill apportionment principle to determine to which school a new computer should be assigned.

School	Number of Computers	Number of Students
Rose	26	625
Lincoln	22	532
Midway	26	620
Valley	31	754

28. 🖊 **House of Representatives** Currently, the U.S. House of Representatives has 435 members, who have been apportioned by the Huntington-Hill apportionment method. If the number of representatives were to be increased to 436, then, according to the 2000 census figures, Utah would be given the new representative. How must Utah's 2000 census Huntington-Hill number compare with the 2000 census Huntington-Hill numbers for the other 49 states? Explain.

The *Webster method of apportionment* is similar to the Jefferson method except that quotas are rounded up when the decimal remainder is 0.5 or greater and are rounded down when the decimal remainder is less than 0.5. This method of rounding is referred to as **rounding to the nearest integer**. For instance, using the Jefferson method, a quotient of 15.91 would be rounded to 15; using the Webster method, it would be rounded to 16. A quotient of 15.49 would be rounded to 15 in both methods.

To use the Webster method you must still experiment to find a *modified standard divisor* for which the sum of the quotas rounded to the nearest integer equals the number of items to be apportioned. The Webster method is similar to the Jefferson method; however, the Webster method is generally more difficult to apply because the modified divisor may be less than, equal to, or more than the standard divisor.

A calculator can be very helpful in testing a possible modified standard divisor md when applying the Webster method of apportionment. For instance, on a TI-83/84 Plus calculator, first store the populations in L1. Then enter

$$\texttt{iPart(L1/}md\texttt{+.5)}\rightarrow\texttt{L2}$$

where md is the modified divisor. Then press the ENTER key. Now scroll through L2 to view the quotas rounded to the *nearest integer*. If the sum of these quotas equals the total number of items to be apportioned, then you have the Webster apportionment. If the sum of the quotas in L2 is less than the total number of items to be apportioned, try a smaller modified standard divisor. If the sum of the quotas in L2 is greater than the total number of items to be apportioned, try a larger modified standard divisor.

29. **Computer Usage** Use the Webster method to apportion the computers in Exercise 12, page 618. How do the apportionment results produced using the Webster method compare with the results produced using the

   a. Hamilton method?

   b. Jefferson method?

30. **Demographics** The table below shows the populations of five European countries. A committee of 20 people from these countries is to be formed using the Webster method of apportionment.

Country	Population
France	60,400,000
Germany	82,400,000
Italy	58,000,000
Spain	40,000,000
Belgium	10,300,000
Total	251,100,000

Source: World Almanac and Book of Facts, 2005, pages 845–849

   a. Explain why 12,600,000 *cannot* be used as a modified standard divisor.

   b. Explain why 12,700,000 *can* be used as a modified standard divisor.

   c. Use the Webster apportionment method to determine the apportionment of the 20 committee members.

31. Which of the following apportionment methods can violate the quota rule?

   ■ Hamilton method

   ■ Jefferson method

   ■ Webster method

   ■ Huntington-Hill method

32. According to Michael Balinski and H. Peyton Young, which of the apportionment methods most closely satisfies the goal of one person, one vote?

33. What method is presently used to apportion the members of the U.S. House of Representatives?

Extensions

CRITICAL THINKING

34. According to the 2000 census, what is the population of your state? How many representatives does your state have in the U.S. House of Representatives? Is your state underrepresented or overrepresented in the House of Representatives? Explain. $\left(\textit{Hint:}\text{ The result of Exercise 11a on page 618 shows that the ratio of representatives to citizens is about }\frac{1}{646,952}.\right)$ How does your state's current number of representatives compare with the number of representatives it had after the 1990 census?

35. Create an apportionment problem in which the Hamilton, Jefferson, and Webster methods produce the same apportionment.

EXPLORATIONS

36. John Quincy Adams, the sixth president of the United States, proposed an apportionment method. Research this method, which is known as the Adams method of apportionment. Describe how this method works. Also indicate whether it satisfies the quota rule and whether it is susceptible to any paradoxes.

37. In the Huntington-Hill method of apportionment, each state is first given one representative and then additional representatives are assigned, one at a time, to the state currently having the highest Huntington-Hill number. This way of implementing the Huntington-Hill apportionment method is time consuming. Another process for implementing the Huntington-Hill apportionment method consists of using a modified divisor and a special rounding procedure that involves the *geometric mean* of two consecutive integers. Research this process for implementing the Huntington-Hill apportionment method. Apply this process to apportion 22 new security vehicles to each of the following schools, based on their student populations.

School	Number of Students	Number of Security Vehicles
Del Mar	5230	?
Wheatly	12,375	?
West	8568	?
Mountain View	14,245	?

38. It can be shown that the results of the rounding procedure used in the Huntington-Hill method described in Exercise 37 differ only slightly from the results of the rounding procedure used in the Webster apportionment method. Thus both methods often produce the same apportionment. Verify that, for Exercise 37, the Webster method produces the same apportionment as the Huntington-Hill method.

39. The Huntington-Hill number is derived by using the apportionment principle. Let P_A = population of state A, a = number of representatives from state A, P_B = population of state B, and b = number of representatives from state B. Complete the following to derive the Huntington-Hill number.

a. Write the fraction that gives the average constituency of state A when it receives a new representative.

b. Write the fraction that gives the average constituency of state B without a new representative.

c. Express the relative unfairness of apportionment of giving state A the new representative in terms of the fractions from parts a and b.

d. Express the relative unfairness of apportionment of giving state B the new representative.

e. According to the apportionment principle, state A should receive the next representative instead of state B if the relative unfairness of giving the new representative to state A is *less than* the relative unfairness of giving the new representative to state B. Express this inequality in terms of the expressions in parts c and d.

f. Simplify the inequality and you will have the Huntington-Hill number.

SECTION 9.2 # Introduction to Voting

 TAKE NOTE

When an issue requires a **majority vote**, it means that more than 50% of the people voting must vote for the issue. This is not the same as a **plurality**, in which the person or issue with the most votes wins.

Plurality Method of Voting

One of the most revered privileges of those of us who live in a democracy is the right to vote for our representatives. Sometimes, however, we are puzzled by the fact that the best candidate did not get elected. Unfortunately, because of the way our *plurality* voting system works, it is possible to elect someone or pass a proposition that has less than *majority support*. As we proceed through this section, we will look at the problems with plurality voting and alternatives to this system. We start with a definition.

> **The Plurality Method of Voting**
>
> Each voter votes for one candidate, and the candidate with the most votes wins. The winning candidate does not have to have a majority of the votes.

EXAMPLE 1 ■ Determine the Winner Using Plurality Voting

Fifty people were asked to rank their preferences of five varieties of chocolate candy, using 1 for their favorite and 5 for their least favorite. This type of ranking of choices is called a **preference schedule**. The results are shown in the table below.

	Rankings					
Caramel center	5	4	4	4	2	4
Vanilla center	1	5	5	5	5	5
Almond center	2	3	2	1	3	3
Toffee center	4	1	1	3	4	2
Solid chocolate	3	2	3	2	1	1
Number of voters:	17	11	9	8	3	2

According to the table (see the column in blue), three voters ranked solid chocolate first, caramel centers second, almond centers third, toffee centers fourth, and vanilla centers fifth. According to this table, which variety of candy would win the taste test using the plurality voting system?

Solution

To answer the question, we will make a table showing the number of first-place votes for each candy.

	First-Place Votes
Caramel center	0
Vanilla center	17
Almond center	8
Toffee center	11 + 9 = 20
Solid chocolate	3 + 2 = 5

Because toffee centers received 20 first-place votes, this type of candy would win the plurality taste test.

CHECK YOUR PROGRESS 1   According to the table in Example 1, which variety of candy would win second place using the plurality voting system?

Solution   *See page S39.*

Example 1 can be used to show the difference between plurality and majority. There were 20 first-place votes for toffee-centered chocolate, so it wins the taste test. However, toffee-centered chocolate was the first choice of only 40% $\left(\frac{20}{50} = 40\%\right)$ of the people voting. Thus less than half of the people voted for toffee-centered chocolate as number one, so it did not receive a majority vote.

Math Matters   **Gubernatorial and Presidential Elections**

In 1998, in a three-party race, plurality voting resulted in the election of former wrestler Jesse Ventura as governor of Minnesota, despite the fact that more than 60% of the state's voters did not vote for him. In fact, he won the governor's race with only 37% of the voters choosing him. Ventura won not because the majority of voters chose him, but because of the plurality voting method.

There are many situations that can be cited to show that plurality voting can lead to unusual results. When plurality voting is mixed with other voting methods, even the plurality winner may not win. One way this can happen is in U.S. presidential elections. The president of the United States is elected not directly by popular vote but by the Electoral College.

In the 1824 presidential election, the approximate percent of the *popular* vote received by each candidate was: John Quincy Adams, 31%; Andrew Jackson, 43%; William Crawford, 13%; and Henry Clay, 13%. The vote in the Electoral College was John Quincy Adams, 84; Andrew Jackson, 99; William Crawford, 41; and Henry Clay, 37. Because none of the candidates had received 121 electoral votes (the number needed to win in 1824), by the Twelfth Amendment to the U.S. Constitution, the House of Representatives decided the election. The House elected John Quincy Adams, thereby electing a president who had less than one-third of the popular vote.

There have been three other instances in which the candidate with the most popular vote in a presidential election was not elected president: 1876, Hayes versus Tilden; 1888, Harrison versus Cleveland; and 2000, Bush versus Gore.

▼ **point of interest**

Another anomaly in the presidential election of 1824 was that William Crawford actually had fewer popular votes than Henry Clay but more electoral votes. By the Twelfth Amendment, the candidates with the top three *electoral* votes get forwarded to the House of Representatives for consideration. Thus Adams, Jackson, and Crawford were the candidates forwarded to the House for consideration.

historical note

The issue of whether plurality voting methods are fair has been debated for more than 200 years. Jean C. Borda (1733–1799) was a member of the French Academy of Sciences when he first started thinking about the way in which people were elected to the Academy. He was concerned that the plurality method of voting might not result in the best candidate being elected. The Borda Count method was born out of these concerns. It was the first attempt at mathematical quantification of voting systems. ■

Borda Count Method of Voting

The problem with plurality voting is that alternative choices are not considered. For instance, the result of the Minnesota governor's contest might have been quite different if voters had been asked, "Choose the candidate you prefer, but if that candidate does not receive a *majority* of the votes, which candidate would be your second choice?"

To see why this might be a reasonable alternative to plurality voting, consider the following situation. Thirty-six senators are considering an educational funding measure. Because the senate leadership wants an educational funding measure to pass, the leadership first determined that the senators preferred measure A for $50 million over measure B for $30 million. However, because of an unexpected dip in state revenues, measure A was removed from consideration and a new measure C, for $15 million, was proposed. The senate leadership determined that senators favored measure B over measure C. In summary, we have

A majority of senators favor measure A over measure B.

A majority of senators favor measure B over measure C.

From these results, it seems reasonable to think that a majority of senators would prefer measure A over measure C. However, when the senators were asked about their preferences between the two measures, measure C was preferred over measure A. To understand how this could happen, consider the preference schedule for the senators shown in the following table.

	Rankings		
Measure A: $50 million	1	3	3
Measure B: $30 million	3	1	2
Measure C: $15 million	2	2	1
Number of senators:	15	12	9

Notice that 15 senators prefer measure A over measure C, but $12 + 9 = 21$ senators, a majority of the 36 senators, prefer measure C over measure A. This means that if all three measures were on the ballot, A would come in first, B would come in second, and C would come in third. However, if just A and C were on the ballot, C would win over A. This paradoxical result was first discussed by Jean C. Borda in 1770.

In an attempt to remove such paradoxical results from voting, Borda proposed that voters rank their choices by giving each choice a certain number of points.

TAKE NOTE

Paradoxes occur in voting only when there are three or more candidates or issues on a ballot. If there are only two candidates in a race, then the candidate receiving the majority of the votes cast is the winner. In a two-candidate race, the majority and the plurality are the same.

The Borda Count Method of Voting

If there are n candidates or issues in an election, each voter ranks the candidates or issues by giving n points to the voter's first choice, $n - 1$ points to the voter's second choice, and so on, with the voter's least favorite choice receiving 1 point. The candidate or issue that receives the most total points is the winner.

Applying the Borda Count method to the education measures, a measure receiving a first-place vote receives 3 points. (There are three different measures.) Each measure receiving a second-place vote receives 2 points, and each measure receiving a third-place vote receives 1 point. This is summarized in the table below.

point of interest

One way to see the difference between a plurality voting system (sometimes called a "winner-take-all" method) and the Borda Count method is to consider grades earned in school. Suppose a student is going to be selected for a scholarship based on grades. If one student has 10 As and 20 Fs and another student has five As and 25 Bs, it would seem that the second student should receive the scholarship. However, if the scholarship is awarded by the plurality of As, the first student will get the scholarship. The Borda Count method is closely related to the method used to calculate grade-point average (GPA).

Points per vote

Measure A: 15 first-place votes:	$15 \cdot 3 = 45$
0 second-place votes:	$0 \cdot 2 = 0$
21 third-place votes:	$21 \cdot 1 = 21$
Total:	66

Measure B: 12 first-place votes:	$12 \cdot 3 = 36$
9 second-place votes:	$9 \cdot 2 = 18$
15 third-place votes:	$15 \cdot 1 = 15$
Total:	69

Measure C: 9 first-place votes:	$9 \cdot 3 = 27$
27 second-place votes:	$27 \cdot 2 = 54$
0 third-place votes:	$0 \cdot 1 = 0$
Total:	81

Using the Borda Count method, measure C is the clear winner.

EXAMPLE 2 ■ Use the Borda Count Method

The members of a club are going to elect a president from four nominees using the Borda Count method. If the 100 members of the club mark their ballots as shown in the table below, who will be elected president?

	Rankings					
Avalon	2	2	2	2	3	2
Branson	1	4	4	4	2	1
Columbus	3	3	1	3	1	3
Dunkirk	4	1	3	1	4	4
Number of voters:	30	24	18	12	10	6

Solution

Using the Borda Count method, each first-place vote receives 4 points, each second-place vote receives 3 points, each third-place vote receives 2 points, and each last-place vote receives 1 point. The summary for each candidate is shown below.

Avalon: 0 first-place votes $0 \cdot 4 = 0$
          90 second-place votes $90 \cdot 3 = 270$
          10 third-place votes $10 \cdot 2 = 20$
          0 fourth-place votes $\underline{0 \cdot 1 = 0}$
                                  Total   290

Branson: 36 first-place votes $36 \cdot 4 = 144$
          10 second-place votes $10 \cdot 3 = 30$
          0 third-place votes $0 \cdot 2 = 0$
          54 fourth-place votes $\underline{54 \cdot 1 = 54}$
                                  Total   228

Columbus: 28 first-place votes $28 \cdot 4 = 112$
          0 second-place votes $0 \cdot 3 = 0$
          72 third-place votes $72 \cdot 2 = 144$
          0 fourth-place votes $\underline{0 \cdot 1 = 0}$
                                  Total   256

Dunkirk: 36 first-place votes $36 \cdot 4 = 144$
          0 second-place votes $0 \cdot 3 = 0$
          18 third-place votes $18 \cdot 2 = 36$
          46 fourth-place votes $\underline{46 \cdot 1 = 46}$
                                  Total   226

Avalon has the largest total score. By the Borda Count method, Avalon is elected president.

✔ **TAKE NOTE**

Notice in Example 2 that Avalon was the winner even though that candidate did not receive any first-place votes. The Borda Count method was devised to allow voters to say, "If my first choice does not win, then consider my second choice."

CHECK YOUR PROGRESS 2 The preference schedule given in Example 1 for the 50 people who were asked to rank their preferences of five varieties of chocolate candy is shown again below.

	Rankings					
Caramel center	5	4	4	4	2	4
Vanilla center	1	5	5	5	5	5
Almond center	2	3	2	1	3	3
Toffee center	4	1	1	3	4	2
Solid chocolate	3	2	3	2	1	1
Number of voters:	17	11	9	8	3	2

Determine the taste test favorite using the Borda Count method.

Solution   *See page S39.*

Plurality with Elimination

A variation of the plurality method of voting is called *plurality with elimination.* Like the Borda Count method, the method of plurality with elimination considers a voter's alternative choices.

Suppose that 30 members of a regional planning board must decide where to build a new airport. The airport consultants to the regional board have recommended four different sites. The preference schedule for the board members is shown in the table below.

	Rankings			
Apple Valley	3	1	2	3
Bremerton	2	3	3	1
Cochella	1	2	4	2
Del Mar	4	4	1	4
Number of ballots:	12	11	5	2

Using the plurality with elimination method, the board members first eliminate the site with the lowest number of first-place votes. If two or more of these alternatives have the same number of first-place votes, all are eliminated unless that would eliminate all alternatives. In that case, a different method of voting must be used. From the table above, Bremerton is eliminated because it received only two first-place votes. Now a vote is retaken using the following important assumption: *Voters do not change their preferences from round to round.* This means that after Bremerton is deleted, the twelve people in the first column would adjust their preferences so that Apple Valley becomes their second choice, Cochella remains their first

TAKE NOTE

When the second round of voting occurs, the two ballots that listed Bremerton as the first choice must be adjusted. The second choice on those ballots becomes the first, the third choice becomes the second, and the fourth choice becomes the third. The order of preference does not change. Similar adjustments must be made to the 12 ballots that listed Bremerton as the second choice. Because that choice is no longer available, Apple Valley becomes the second choice and Del Mar becomes the third choice. Adjustments must be made to the 11 ballots that listed Bremerton as the third choice. The fourth choice of those ballots, Del Mar, becomes the third choice. A similar adjustment must be made for the five ballots that listed Bremerton as the third choice.

choice, and Del Mar becomes their third choice. For the 11 voters in the second column, Apple Valley remains their first choice, Cochella remains their second choice, and Del Mar becomes their third choice. Similar adjustments are made by the remaining voters. The new preference schedule is

	Rankings			
Apple Valley	2	1	2	2
Cochella	1	2	3	1
Del Mar	3	3	1	3
Number of ballots:	12	11	5	2

The board members now repeat the process and eliminate the site with the fewest first-place votes. In this case it is Del Mar. The new adjusted preference schedule is

	Rankings			
Apple Valley	2	1	1	2
Cochella	1	2	2	1
Number of ballots:	12	11	5	2

From this table, Apple Valley has 16 first-place votes and Cochella has 14 first-place votes. Therefore, Apple Valley is the selected site for the new airport.

EXAMPLE 3 ■ Use the Plurality with Elimination Voting Method

A university wants to add a new sport to its existing program. To help ensure that the new sport will have student support, the students of the university are asked to rank the four sports under consideration. The results are shown in the table below.

	Rankings					
Lacrosse	3	2	3	1	1	2
Squash	2	1	4	2	3	1
Rowing	4	3	2	4	4	4
Golf	1	4	1	3	2	3
Number of ballots:	326	297	287	250	214	197

Use the plurality with elimination method to determine which of these sports should be added to the university's program.

TAKE NOTE

Remember to shift the preferences at each stage of the elimination method. The 297 students who chose rowing as their third choice and golf as their fourth choice now have golf as their third choice. Check each preference schedule and update it as necessary.

Solution

Because rowing received no first-place votes, it is eliminated from consideration. The new preference schedule is shown below.

	Rankings					
Lacrosse	3	2	2	1	1	2
Squash	2	1	3	2	3	1
Golf	1	3	1	3	2	3
Number of ballots:	326	297	287	250	214	197

From this table, lacrosse has 464 first-place votes, squash has 494 first-place votes, and golf has 613 first-place votes. Because lacrosse has the fewest first-place votes, it is eliminated. The new preference schedule is shown below.

	Rankings					
Squash	2	1	2	1	2	1
Golf	1	2	1	2	1	2
Number of ballots:	326	297	287	250	214	197

From this table, squash received 744 first-place votes and golf received 827 first-place votes. Therefore, golf is added to the sports program.

CHECK YOUR PROGRESS 3 A service club is going to sponsor a dinner to raise money for a charity. The club has decided to serve Italian, Mexican, Thai, Chinese, or Indian food. The members of the club were surveyed to determine their preferences. The results are shown in the table below.

	Rankings				
Italian	2	5	1	4	3
Mexican	1	4	5	2	1
Thai	3	1	4	5	2
Chinese	4	2	3	1	4
Indian	5	3	2	3	5
Number of ballots:	33	30	25	20	18

Use the plurality with elimination method to determine the food preference of the club members.

Solution *See page S40.*

Pairwise Comparison Voting Method

The *pairwise comparison* method of voting is sometimes referred to as the "head-to-head" method. In this method, each candidate is compared one-on-one with each of the other candidates. A candidate receives 1 point for a win, 0.5 point for a tie, and 0 points for a loss. The candidate with the greatest number of points wins the election.

A voting method that elects the candidate who wins all head-to-head matchups is said to satisfy the Condorcet criterion.

Condorcet Criterion

A candidate who wins all possible head-to-head matchups should win an election when all candidates appear on the ballot.

This is one of the *fairness criteria* that a voting method should exhibit. We will discuss other fairness criteria later in this section.

EXAMPLE 4 ■ Use the Pairwise Comparison Voting Method

There are four proposals for the name of a new football stadium at a college: Panther Stadium, after the team mascot; Sanchez Stadium, after a large university contributor; Mosher Stadium, after a famous alumnus known for humanitarian work; and Fritz Stadium, after the college's most winning football coach. The preference schedule cast by alumni and students is shown below.

	Rankings				
Panther Stadium	2	3	1	2	4
Sanchez Stadium	1	4	2	4	3
Mosher Stadium	3	1	4	3	2
Fritz Stadium	4	2	3	1	1
Number of ballots:	752	678	599	512	487

Use the pairwise comparison voting method to determine the name of the stadium.

Solution

We will create a table to keep track of each of the head-to-head comparisons. Before we begin, note that a matchup between, say, Panther and Sanchez is the same as the matchup between Sanchez and Panther. Therefore, we will shade the duplicate cells and the cells between the same candidates. This is shown below.

versus	Panther	Sanchez	Mosher	Fritz
Panther				
Sanchez				
Mosher				
Fritz				

To complete the table, we will place the name of the winner in the cell of each head-to-head match. For instance, for the Panther–Sanchez matchup,

    ✓ Panther was favored over Sanchez on 678 + 599 + 512 = 1789 ballots.
        Sanchez was favored over Panther on 752 + 487 = 1239 ballots.

The winner of this matchup is Panther, so that name is placed in the Panther versus Sanchez cell. Do this for each of the matchups.

    ✓ Panther was favored over Mosher on 752 + 599 + 512 = 1863 ballots.
        Mosher was favored over Panther on 678 + 487 = 1165 ballots.

        Panther was favored over Fritz on 752 + 599 = 1351 ballots.
    ✓ Fritz was favored over Panther on 678 + 512 + 487 = 1677 ballots.

        Sanchez was favored over Mosher on 752 + 599 = 1351 ballots.
    ✓ Mosher was favored over Sanchez on 678 + 512 + 487 = 1677 ballots.

        Sanchez was favored over Fritz on 752 + 599 = 1351 ballots.
    ✓ Fritz was favored over Sanchez on 678 + 512 + 487 = 1677 ballots.

        Mosher was favored over Fritz on 752 + 678 = 1430 ballots.
    ✓ Fritz was favored over Mosher on 599 + 512 + 487 = 1598 ballots.

versus	Panther	Sanchez	Mosher	Fritz
Panther		Panther	Panther	Fritz
Sanchez			Mosher	Fritz
Mosher				Fritz
Fritz				

From the above table, Fritz has three wins, Panther has two wins, and Mosher has one win. Using pairwise comparison, Fritz Stadium is the winning name.

CHECK YOUR PROGRESS 4 One hundred restaurant critics were asked to rank their favorite restaurants from a list of four. The preference schedule for the critics is shown in the table below.

	Rankings				
Sanborn's Fine Dining	3	1	4	3	1
The Apple Inn	4	3	3	2	4
May's Steak House	2	2	1	1	3
Tory's Seafood	1	4	2	4	2
Number of ballots:	31	25	18	15	11

Use the pairwise voting method to determine the critics' favorite restaurant.

Solution *See page S40.*

Fairness of Voting Methods and Arrow's Theorem

Kenneth J. Arrow

Now that we have examined various voting options, we will stop to ask which of these options is the *fairest*. To answer that question, we must first determine what we mean by *fair*.

In 1948, Kenneth J. Arrow was trying to develop material for his Ph.D. dissertation. As he studied, it occurred to him that he might be able to apply the principles of order relations to problems in social choice or voting. (An example of an order relation for real numbers is "less than.") His investigation led him to outline various criteria for a fair voting system. A paraphrasing of four fairness criteria is given below.

Fairness Criteria

1. *Majority criterion:* The candidate who receives a majority of the first-place votes is the winner.

2. *Monotonicity criterion:* If candidate A wins an election, then candidate A will also win the election if the only change in the voters' preferences is that supporters of a different candidate change their votes to support candidate A.

3. *Condorcet criterion:* A candidate who wins all possible head-to-head matchups should win an election when all candidates appear on the ballot.

4. *Independence of irrelevant alternatives criterion:* If a candidate wins an election, the winner should remain the winner in any recount in which losing candidates withdraw from the race.

There are other criteria, such as the *dictator criterion*, which we will discuss in the next section. However, what Kenneth Arrow was able to prove is that no matter what kind of voting system we devise, it is impossible for it to satisfy the fairness criteria.

Arrow's Impossibility Theorem

There is no voting method involving three or more choices that satisfies all four fairness criteria.

By Arrow's Impossibility Theorem, none of the voting methods we have discussed is fair. Not only that, but we *cannot* construct a fair voting system for three or more candidates. We will now give some examples of each of the methods we have discussed and show which of the fairness criteria are not satisfied.

EXAMPLE 5 ■ Show That the Borda Count Method Violates the Majority Criterion

Suppose the preference schedule for three candidates, Alpha, Beta, and Gamma, is given by the table below.

	Rankings		
Alpha	1	3	3
Beta	2	1	2
Gamma	3	2	1
Number of ballots:	55	50	3

Show that using the Borda Count method violates the majority criterion.

Solution

The calculations for Borda's method are shown below.

Alpha

55 first-place votes	$55 \cdot 3 = 165$
0 second-place votes	$0 \cdot 2 = 0$
53 third-place votes	$53 \cdot 1 = 53$
	Total $\quad$ 218

Beta

50 first-place votes	$50 \cdot 3 = 150$
58 second-place votes	$58 \cdot 2 = 116$
0 third-place votes	$0 \cdot 1 = 0$
	Total $\quad$ 266

Gamma

3 first-place votes	$3 \cdot 3 = 9$
50 second-place votes	$50 \cdot 2 = 100$
55 third-place votes	$55 \cdot 1 = 55$
	Total $\quad$ 164

From these calculations, Beta should win the election. However, Alpha has the majority (more than 50%) of the first-place votes. This result violates the majority criterion.

CHECK YOUR PROGRESS 5 Using the table in Example 5, show that the Borda Count method violates the Condorcet criterion.

Solution *See page S40.*

QUESTION $\quad$ *Does the pairwise comparison voting method satisfy the Condorcet criterion?*

EXAMPLE 6 ■ Show That Plurality with Elimination Violates the Monotonicity Criterion

Suppose the preference schedule for three candidates, Alpha, Beta, and Gamma, is given by the table below.

	Rankings			
Alpha	2	3	1	1
Beta	3	1	2	3
Gamma	1	2	3	2
Number of ballots:	25	20	16	10

ANSWER $\quad$ *Yes. The pairwise comparison voting method elects the person who wins all head-to-head matchups.*

a. Show that, using plurality with elimination voting, Gamma wins the election.

b. Suppose that the 10 people who voted for Alpha first and Gamma second changed their votes such that they all voted for Alpha second and Gamma first. Show that, using plurality with elimination voting, Beta will now be elected.

c. Explain why this result violates the montonicity criterion.

Solution

a. Beta received the fewest first-place votes, so Beta is eliminated. The new preference schedule is

	Rankings			
Alpha	2	2	1	1
Gamma	1	1	2	2
Number of ballots:	25	20	16	10

From this schedule, Gamma has 45 first-place votes and Alpha has 26 first-place votes, so Gamma is the winner.

b. If the 10 people who voted for Alpha first and Gamma second changed their votes such that they all voted for Alpha second and Gamma first, the preference schedule would be

	Rankings			
Alpha	2	3	1	2
Beta	3	1	2	3
Gamma	1	2	3	1
Number of ballots:	25	20	16	10

From this schedule, Alpha has the fewest first-place votes and is eliminated. The new preference schedule is

	Rankings			
Beta	2	1	1	2
Gamma	1	2	2	1
Number of ballots:	25	20	16	10

From this schedule, Gamma has 35 first-place votes and Beta has 36 first-place votes, so Beta is the winner.

c. This result violates the monotonicity criterion because Gamma, who won the first election, loses the second election even though Gamma received a larger number of first-place votes in the first election.

CHECK YOUR PROGRESS 6   The table below shows the preferences for three new car colors.

	Rankings		
Radiant silver	1	3	3
Electric red	2	2	1
Lightning blue	3	1	2
Number of votes:	30	27	2

Show that the Borda Count method violates the independence of irrelevant alternatives criterion.

Solution   *See page S40.*

Investigation

Variations of the Borda Count Method

Sixty people were asked to select their preferences among plain iced tea, lemon-flavored iced tea, and raspberry-flavored iced tea. The preference schedule is shown in the table below.

	Rankings		
Plain iced tea	1	3	3
Lemon iced tea	2	2	1
Raspberry iced tea	3	1	2
Number of ballots:	25	20	15

Investigation Exercises

1. Using the Borda method of voting, which flavor of iced tea is preferred by this group? Which is second? Which is third?

2. Instead of using the normal Borda method, suppose the Borda method used in Exercise 1 of this Investigation assigned 1 point for first, 0 points for second, and −1 point for third place. Does this alter the preferences you found in Exercise 1?

3. Suppose the Borda method used in Exercise 1 of this Investigation assigned 10 points for first, 5 points for second, and 0 points for third place. Does this alter the preferences you found in Exercise 1?

4. Suppose the Borda method used in Exercise 1 of this Investigation assigned 20 points for first, 5 points for second, and 0 points for third place. Does this alter the preferences you found in Exercise 1?

5. Suppose the Borda method used in Exercise 1 of this Investigation assigned 25 points for first, 5 points for second, and 0 points for third place. Does this alter the preferences you found in Exercise 1?

(continued)

6. Can the assignment of points for first, second, and third place change the preference order when the Borda method of voting is used?

7. Suppose the assignment of points for first, second, and third place for the Borda method of voting consists of consecutive integers. Can the value of the starting integer change the outcome of the preferences?

Exercise Set 9.2

1. What is the difference between a majority and a plurality? Is it possible to have one without the other?

2. Explain why the plurality voting system may not be the best system to use in some situations.

3. Explain how the Borda Count method of voting works.

4. Explain how the plurality with elimination voting method works.

5. Explain how the pairwise comparison voting method works.

6. What does the Condorcet criterion say?

7. Is there a "best" voting method? Is one method more fair than the others?

8. Explain why, if only two candidates are running, the plurality and Borda Count methods will determine the same winner.

9. If a candidate in an election has a majority of the votes, does the candidate have a plurality of the votes? If a candidate has a plurality of the votes, does the candidate have a majority of the votes?

10. Is it possible in an election that uses the Borda Count method that the winner could receive no first-place votes?

11. Presidential Election The table below shows the popular vote and the Electoral College vote for the major candidates in the 2000 presidential election.

Candidate	Popular Vote	Electoral College Vote
George W. Bush	50,456,002	271
Al Gore	50,999,897	266
Ralph Nader	2,882,955	0

Source: *Encyclopaedia Britannica* Online

a. Which candidate received the plurality of the popular vote?

b. Did any candidate receive a majority of the popular vote?

c. Who won the election?

12. Breakfast Cereal Sixteen people were asked to rank three breakfast cereals in order of preference. Their responses are given below.

Corn Flakes	3	1	1	2	3	3	2	2	1	3	1	3	1	2	1	2
Raisin Bran	1	3	2	3	1	2	1	1	2	1	2	2	3	1	3	3
Mini Wheats	2	2	3	1	2	1	3	3	3	2	3	1	2	3	2	1

If the plurality method of voting is used, which cereal is the group's first preference?

13. **Cartoon Characters** A kindergarten class was surveyed to determine the childrens' favorite cartoon characters among Mickey Mouse, Bugs Bunny, and Scooby Doo. The students ranked the characters in order of preference; the results are shown in the preference schedule below.

	Rankings					
Mickey Mouse	1	1	2	2	3	3
Bugs Bunny	2	3	1	3	1	2
Scooby Doo	3	2	3	1	2	1
Number of students:	6	4	6	5	6	8

a. How many students are in the class?

b. How many votes are required for a majority?

c. Using plurality voting, which character is the childrens' favorite?

14. **Catering** A 15-person committee is having lunch catered for a meeting. Three caterers, each specializing in a different cuisine, are available. In order to choose a caterer for the group, each member is asked to rank the cuisine options in order of preference. The results are given in the preference schedule below.

	Rankings				
Italian	1	1	2	3	3
Mexican	2	3	1	1	2
Japanese	3	2	3	2	1
Number of votes:	2	4	1	5	3

Using plurality voting, which caterer should be chosen?

15. **Movies** Fifty consumers were surveyed about their movie watching habits. They were asked to rank the likelihood that they would participate in each listed activity. The results are summarized in the table below.

	Rankings				
Go to a theater	2	3	1	2	1
Rent a video or DVD	3	1	3	1	2
Watch pay-per-view	1	2	2	3	3
Number of votes:	8	13	15	7	7

Using the Borda Count method of voting, which activity is the most popular choice among this group of consumers?

16. **Breakfast Cereal** Use the Borda Count method of voting to determine the preferred breakfast cereal in Exercise 12.

17. **Cartoons** Use the Borda Count method of voting to determine the childrens' favorite cartoon character in Exercise 13.

18. **Catering** Use the Borda Count method of voting to determine which caterer the committee should hire in Exercise 14.

19. **Class Election** A senior high school class held an election for class president. Instead of just voting for one candidate, the students were asked to rank all four candidates in order of preference. The results are shown below.

	Rankings					
Raymond Lee	2	3	1	3	4	2
Suzanne Brewer	4	1	3	4	1	3
Elaine Garcia	1	2	2	2	3	4
Michael Turley	3	4	4	1	2	1
Number of votes:	36	53	41	27	31	45

Using the Borda Count method, which student should be class president?

20. **Cell Phone Usage** A journalist reviewing various cellular phone services surveyed 200 customers and asked each one to rank four service providers in order of preference. The group's results are shown below.

	Rankings				
Verizon	3	4	2	3	4
Sprint PCS	1	1	4	4	3
Cingular	2	2	1	2	1
Nextel	4	3	3	1	2
Number of votes:	18	38	42	63	39

Using the Borda Count method, which provider is the favorite of these customers?

21. **Baseball** A Little League baseball team must choose the colors for its uniforms. The coach offered four different choices, and the players ranked them in order of preference, as shown in the table below.

	Rankings			
Red and white	2	3	3	2
Green and yellow	4	1	4	1
Red and blue	3	4	2	4
Blue and white	1	2	1	3
Number of votes:	4	2	5	4

Using the plurality with elimination method, what colors should be used for the uniforms?

22. **Radio Station** A number of college students were asked to rank four radio stations in order of preference. The responses are given in the table below.

	Rankings				
WNNX	3	1	1	2	4
WKLS	1	3	4	1	2
WWVV	4	2	2	3	1
WSTR	2	4	3	4	3
Number of votes:	57	72	38	61	15

Use plurality with elimination to determine the students' favorite radio station among the four.

23. **Class Election** Use plurality with elimination to choose the class president in Exercise 19.

24. **Cell Phone Usage** Use plurality with elimination to determine the preferred cellular phone service provider in Exercise 20.

25. **Campus Club** A campus club has money left over in its budget and must spend it before the school year ends. The members arrived at five different possibilities, and each member ranked them in order of preference. The results are shown in the table below.

	Rankings				
Establish a scholarship	1	2	3	3	4
Pay for several members to travel to a convention	2	1	2	1	5
Buy new computers for the club	3	3	1	4	1
Throw an end-of-year party	4	5	5	2	2
Donate to charity	5	4	4	5	3
Number of votes:	8	5	12	9	7

   a. Using the plurality voting system, how should the club spend the money?

   b. Use the plurality with elimination method to determine how the money should be spent.

   c. Using the Borda Count method of voting, how should the money be spent?

   d. In your opinion, which of the previous three methods seems most appropriate in this situation? Why?

26. **Recreation** A company is planning its annual summer retreat and has asked its employees to rank five different choices of recreation in order of preference. The results are given in the table below.

	Rankings				
Picnic in a park	1	2	1	3	4
Water skiing at a lake	3	1	2	4	3
Amusement park	2	5	5	1	2
Riding horses at a ranch	5	4	3	5	1
Dinner cruise	4	3	4	2	5
Number of votes:	10	18	6	28	16

a. Using the plurality voting system, what activity should be planned for the retreat?

b. Use the plurality with elimination method to determine which activity should be chosen.

c. Using the Borda Count method of voting, which activity should be planned?

27. **Star Wars Movies** Fans of the *Star Wars* movies have been debating, on a website, regarding which of the films is the best. To see what the overall opinion is, visitors to the website can rank the four films in order of preference. The results are shown in the preference schedule below.

	Rankings			
Star Wars	1	2	1	3
The Empire Strikes Back	4	4	2	1
Return of the Jedi	2	1	3	2
The Phantom Menace	3	3	4	4
Number of votes:	429	1137	384	582

Using pairwise comparison, which film is the favorite of the visitors to the website who voted?

28. **Family Reunion** The Nelson family is trying to decide where to hold a family reunion. They have asked all their family members to rank four choices in order of preference. The results are shown in the preference schedule below.

	Rankings				
Grand Canyon	3	1	2	3	1
Yosemite	1	2	3	4	4
Bryce Canyon	4	4	1	2	2
Yellowstone	2	3	4	1	3
Number of votes:	7	3	12	8	13

Use the pairwise comparison method to determine the best choice of location for the reunion.

29. **School Mascot** A new college needs to pick a mascot for its football team. The students were asked to rank four choices in order of preference; the results are tallied below.

	Rankings				
Bulldog	3	4	4	1	4
Panther	2	1	2	4	3
Hornet	4	2	1	2	2
Bobcat	1	3	3	3	1
Number of votes:	638	924	525	390	673

Using the pairwise comparison method of voting, which mascot should be chosen?

30. **Election** Five candidates are running for president of a charity organization. Interested persons were asked to rank the candidates in order of preference. The results are given below.

	Rankings				
P. Gibson	5	1	2	1	2
E. Yung	2	4	5	5	3
R. Allenbaugh	3	2	1	3	5
T. Meckley	4	3	4	4	1
G. DeWitte	1	5	3	2	4
Number of votes:	16	9	14	9	4

Use the pairwise comparison method to determine the president of the organization.

31. Use the pairwise comparison method to choose the colors for the Little League uniforms in Exercise 21.

32. Use the pairwise comparison method to determine the favorite radio station in Exercise 22.

33. When consumers were given choices between two of three products, product A was preferred to product B and product A was preferred to product C. However, when consumers were given a choice of all three products, product B was preferred. What fairness criterion does this violate?

34. The board of directors of a company decides to use the Borda Count method to elect a new president. In balloting for a new president of the company, 52% of the directors voted for candidate A. However, candidate B won the election. What fairness criterion does this violate?

35. Does the winner in Exercise 13c satisfy the Condorcet criterion?

36. Does the winner in Exercise 14 satisfy the Condorcet criterion?

37. Does the winner in Exercise 25c satisfy the Condorcet criterion?

38. Does the winner in Exercise 26c satisfy the Condorcet criterion?

39. Does the winner in Exercise 19 satisfy the majority criterion?

40. Does the winner in Exercise 22 satisfy the majority criterion?

41. Election Three candidates are running for mayor. A vote was taken in which the candidates were ranked in order of preference. The results are shown in the preference schedule below.

	Rankings		
John Lorenz	1	3	3
Marcia Beasley	3	1	2
Stephen Hyde	2	2	1
Number of votes:	2691	2416	237

a. Use the Borda Count method to determine the winner of the election.

b. Verify that the majority criterion has been violated.

c. Identify a candidate who wins all head-to-head comparisons.

d. ✎ Explain why the Condorcet criterion has been violated.

e. If Marcia Beasley drops out of the race for mayor (and voter preferences remain the same), determine the winner of the election again, using the Borda Count method.

f. ✎ Explain why the independence of irrelevant alternatives criterion has been violated.

42. Film Competition Three films have been selected as finalists in a national student film competition. Seventeen judges have viewed each of the films and have ranked them in order of preference. The results are given in the preference schedule below.

	Rankings			
Film A	1	3	2	1
Film B	2	1	3	3
Film C	3	2	1	2
Number of votes:	4	6	5	2

a. Using the plurality with elimination method, which film should win the competition?

b. Suppose the first vote is declared invalid and a revote is taken. All of the judges' preferences remain the same except for the votes represented by the last column of the table. The judges who cast these votes both decide to switch their first place vote to film C, so their preference now is film C first, then film A, and then film B. Which film now wins using the plurality with elimination method?

c. Has the monotonicity criterion been violated?

Extensions

CRITICAL THINKING

43. **Election** A campus club needs to elect four officers: a president, a vice president, a secretary, and a treasurer. The club has five volunteers. Rather than vote individually for each position, the club members will rank the candidates in order of preference. The votes will then be tallied using the Borda Count method. The candidate receiving the highest number of points will be president, the candidate receiving the next highest number of points will be vice president, the candidate receiving the next highest number of points will be secretary, and the candidate receiving the next highest number of points will be treasurer. For the preference schedule shown below, determine who wins each position in the club.

	Rankings				
Cynthia	4	2	5	2	3
Andrew	2	3	1	4	5
Jen	5	1	2	3	2
Hector	1	5	4	1	4
Medin	3	4	3	5	1
Number of votes:	22	10	16	6	27

44. **Election** Use the plurality with elimination method of voting to determine the four officers in Exercise 43. Is your result the same as the result you arrived at using the Borda Count method?

45. **Scholarship Awards** The members of a scholarship committee have ranked four finalists competing for a scholarship in order of preference. The results are shown in the preference schedule below.

	Rankings			
Francis Chandler	3	4	4	1
Michael Huck	1	2	3	4
David Chang	2	3	1	2
Stephanie Owen	4	1	2	3
Number of votes:	9	5	7	4

If you are one of the voting members and you want David Chang to win the scholarship, which voting method would you suggest that the committee use?

COOPERATIVE LEARNING

46. **Restaurants** Suppose you and three friends, David, Sara, and Cliff, are trying to decide on a pizza restaurant. You like Pizza Hut best, Round Table pizza is acceptable to you, and you definitely do not want to get pizza from Domino's. Domino's is David's favorite, and he also likes Round Table, but he won't eat pizza from Pizza Hut. Sara says she will only eat Round Table pizza. Cliff prefers Domino's, but he will also eat Round Table pizza. He doesn't like Pizza Hut.

   a. Given the preferences of the four friends, which pizza restaurant would be the best choice?

   b. If you use the plurality system of voting to determine which pizza restaurant to go to, which restaurant wins? Does this seem to be the best choice for the group?

   c. If you use the pairwise comparison method of voting, which pizza restaurant wins?

   d. Does one of the four voting methods discussed in this section give the same winner as your choice in part a? Does the method match your reasoning in making your choice?

EXPLORATIONS

47. Another method of voting is to assign a "weight," or score, to each candidate rather than ranking the candidates in order. All candidates must receive a score, and two or more candidates can receive the same score from a voter. A score of 5 represents the strongest endorsement of a candidate. The scores range down to 1, which corresponds to complete disapproval of a candidate. A score of 3 represents indifference. The candidate with the most total points wins the election. The results of a sample election are given in the table.

	Rankings							
Candidate A	2	1	2	5	4	2	4	5
Candidate B	5	3	5	3	2	5	3	2
Candidate C	4	5	4	1	4	3	2	1
Candidate D	1	3	2	4	5	1	3	2
Number of votes:	26	42	19	33	24	8	24	33

   a. Find the winner of the election.

   b. If plurality were used (assuming that a person's vote would go to the candidate that he or she gave the highest score to), verify that a different winner would result.

48. *Approval voting* is a system in which voters may vote for more than one candidate. Each vote counts equally, and the candidate with the most total votes wins the election. Many feel that this is a better system for large elections than simple plurality because it considers a voter's second choices and is a stronger measure of overall voter support for each candidate. Some organizations use

approval voting to elect their officers. The United Nations uses this method to elect the secretary-general.

a. Suppose a math class is going to show a film involving mathematics or mathematicians on the last day of class. The options are *Stand and Deliver, Good Will Hunting, A Beautiful Mind, Pi,* and *Contact.* The students vote using approval voting. The results are as follows.

8 students vote for all five films.

8 students vote for *Good Will Hunting, A Beautiful Mind,* and *Contact.*

8 students vote for *Stand and Deliver, Good Will Hunting,* and *Contact.*

8 students vote for *A Beautiful Mind* and *Pi.*

8 students vote for *Stand and Deliver* and *Pi.*

8 students vote for *Good Will Hunting* and *Contact.*

1 student votes for *Pi.*

Which film will be chosen?

b. Use approval voting in Exercise 46 to determine the pizza restaurant the group of friends should choose. Does your result agree with your answer to part a of Exercise 46?

SECTION 9.3 | **Weighted Voting Systems**

Biased Voting Systems

A **weighted voting system** is one in which some voters have more weight on the outcome of an election. Examples of weighted voting systems are fairly common. A few examples are the stockholders of a company, the Electoral College, the United Nations Security Council, and the European Union.

Math Matters   **The Electoral College**

As mentioned in the Historical Note on the following page, the Electoral College elects the president of the United States. The number of electors representing each state is equal to the sum of the number of senators (2) and the number of members in the House of Representatives for that state. The original intent of the framers of the Constitution was to protect the smaller states. We can verify this by computing the number of people represented by each elector. In the 2004 election, each Vermont elector represented about 203,000 people; each California elector represented about

(continued)

historical note

———

The U.S. Constitution, Article 2, Section 1 states that the members of the Electoral College elect the president of the United States. The original article directed members of the College to vote for two people. However, it did not stipulate that one name was for president and the other name was for vice president. The article goes on to state that the person with the greatest number of votes becomes president and the one with the next highest number of votes becomes vice president. In 1800, Thomas Jefferson and Aaron Burr received exactly the same number of votes even though they were running on a Jefferson for president, Burr for vice president ticket. Thus the House of Representatives was asked to select the president. It took 36 different votes by the House before Jefferson was elected president. In 1804, the Twelfth Amendment to the Constitution was ratified to prevent a recurrence of the 1800 election problems. ∎

616,000 people. To see how this gives a state with a smaller population more *power* (a word we will discuss in more detail later in this section), note that three electoral votes from Vermont represent approximately the same size population as does one electoral vote from California. Not every vote represents the same number of people.

Another peculiarity related to the Electoral College system is that it is very sensitive to small vote swings. For instance, in the 2000 election, if an additional 0.01% of the voters in Florida had cast their votes for Al Gore instead of George Bush, Gore would have won the presidential election.

Consider a small company with a total of 100 shares of stock and three stockholders, A, B, and C. Suppose that A owns 45 shares of the stock (which means A has 45 votes), B owns 45 shares, and C owns 10 shares. If a vote of 51 or greater is required to approve any measure before the owners, then a measure cannot be passed without two of the three owners voting for the measure. Even though C has only 10 shares, C has the same voting power as A and B.

Now suppose that a new stockholder is brought into the company and the shares of the company are redistributed so that A has 27 shares, B has 26 shares, C has 25 shares, and D has 22 shares. Note, in this case, that any two of A, B, or C can pass a measure, but D paired with any of the other shareholders cannot pass a measure. D has virtually no power even though D has only three shares less than C.

The number of votes that are required to pass a measure is called a **quota**. For the two stockholder examples above, the quota was 51. The **weight of a voter** is the number of votes controlled by the voter. In the case of the company whose stock was split A−27 shares, B−26 shares, C−25 shares, and D−22 shares, the weight of A is 27, the weight of B is 26, the weight of C is 25, and the weight of D is 22. Rather than write out in sentence form the quota and weight of each voter, we use the notation

Quota ———┐   Weights

$\{51: 27, 26, 25, 22\}$

• The four numbers after the colon indicate that there are a total of four voters in this system.

This notation is very convenient. We state its more general form in the definition below.

Weighted Voting System

A weighted voting system of n voters is written $\{q: w_1, w_2, \ldots, w_n\}$, where q is the quota and w_1 through w_n represent the weights of each of the n voters.

Using this notation, we can describe various voting systems.

- **One-person, one-vote system:** For instance, $\{5: 1, 1, 1, 1, 1, 1, 1, 1, 1\}$. In this system, each person has one vote and five votes, a majority, are required to pass a measure.

- **Dictatorship:** For instance, $\{20: 21, 6, 5, 4, 3\}$. In this system, the person with 21 votes can pass any measure. Even if the remaining four people get together, their votes do not total the quota of 20.

- **Null system:** For instance, {28: 6, 3, 5, 2}. If all the members of this system vote for a measure, the total number of votes is 16, which is less than the quota. Therefore, no measure can be passed.

- **Veto power system:** For instance, {21: 6, 5, 4, 3, 2, 1}. In this case, the sum of all the votes is 21, the quota. Therefore, if any one voter does not vote for the measure, it will fail. Each voter is said to have **veto power.** In this case, this means that even the voter with one vote can veto a measure (cause the measure not to pass). If at least one voter in a voting system has veto power, the system is a veto power system.

Math Matters   UN Security Council: An Application of Inequalities

UN Security Council Chamber

The United Nations Security Council consists of five permanent members (United States, China, France, Great Britain, and Russia) and 10 members that are elected by the General Assembly for a two-year term. In 2005, the 10 nonpermanent members were Algeria, Argentina, Belize, Brazil, Denmark, Germany, Japan, Philippines, Romania, and United Republic of Tanzania.

For a resolution to pass the Security Council,

1. Nine countries must vote for the resolution; and

2. If one of the five permanent members votes against the resolution, it fails.

This situation can be described using inequalities. Let x be the weight of the vote of one permanent member of the Council and let q be the quota for a vote to pass. Then, by condition 1, $q \leq 5x + 4$. (The weights of the five votes of the permanent members plus the single votes of four nonpermanent members must be greater than or equal to the quota.)

By condition 2, we have $4x + 10 < q$. (If one of the permanent members opposes the resolution, it fails even if all of the nonpermanent members vote for it.)

Combining the inequalities from condition 1 and condition 2, we have

$$4x + 10 < 5x + 4$$
$$10 < x + 4 \qquad \text{• Subtract } 4x \text{ from each side.}$$
$$6 < x \qquad \text{• Subtract 4 from each side.}$$

The smallest whole number greater than 6 is 7. Therefore, the weight of each permanent member is 7. Substituting 7 into $q \leq 5x + 4$ and $4x + 10 < q$, we find that $q = 39$. Thus the weighted voting system of the Security Council is given by {39: 7, 7, 7, 7, 7, 1, 1, 1, 1, 1, 1, 1, 1, 1, 1}.

QUESTION   Is the UN Security Council voting system a veto power system?

In a weighted voting system, a **coalition** is a set of voters each of whom votes the same way, either for or against a resolution. A **winning coalition** is a set of voters the sum of whose votes is greater than or equal to the quota. A **losing**

ANSWER   *Yes. If any of the permanent members votes against a resolution, the resolution cannot pass.*

coalition is a set of voters the sum of whose votes is less than the quota. A voter who leaves a winning coalition and thereby turns it into a losing coalition is called a **critical voter.**

As shown in the next theorem, for large numbers of voters, there are many possible coalitions.

> **Number of Possible Coalitions of *n* Voters**
>
> The number of possible coalitions of n voters is $2^n - 1$.

✔ **TAKE NOTE**

The number of coalitions of n voters is the number of subsets that can be formed from n voters. From Chapter 1, this is 2^n. Because a coalition must contain at least one voter, the empty set is not a possible coalition. Therefore, the number of coalitions is $2^n - 1$.

As an example, if all electors of each state to the Electoral College cast their ballots for one candidate, then there are $2^{51} - 1 \approx 2.25 \times 10^{15}$ possible coalitions (the District of Columbia is included). The number of *winning* coalitions is far less. For instance, any coalition of 10 or fewer states cannot be a winning coalition because the largest 10 states do not have enough electoral votes to elect the president. As we proceed through this section, we will not attempt to list all the coalitions, only the winning coalitions.

EXAMPLE 1 ■ **Determine Winning Coalitions in a Weighted Voting System**

Suppose that the four owners of a company, Ang, Bonhomme, Carmel, and Diaz, own, respectively, 500 shares, 375 shares, 225 shares, and 400 shares. The weighted voting system for this company is {751: 500, 375, 225, 400}.

a. Determine the winning coalitions.

b. For each winning coalition, determine the critical voters.

Solution

a. A winning coalition must represent at least 751 votes. We will list these coalitions in the table below, in which we use A for Ang, B for Bonhomme, C for Carmel, and D for Diaz.

✔ **TAKE NOTE**

The coalition {A, C} is not a winning coalition because the total number of votes for that coalition is 725, which is less than 751.

Winning Coalition	Number of Votes
{A, B}	875
{A, D}	900
{B, D}	775
{A, B, C}	1100
{A, B, D}	1275
{A, C, D}	1125
{B, C, D}	1000
{A, B, C, D}	1500

b. A voter who leaves a winning coalition and thereby creates a losing coalition is a critical voter. For instance, for the winning coalition {A, B, C}, if A leaves, the number of remaining votes is 600, which is not enough to pass a resolution. If B leaves, the number of remaining votes is 725—again, not enough to pass a resolution. Therefore, A and B are critical voters for the coalition {A, B, C} and C is not a critical voter. The table below shows the critical voters for each winning coalition.

Winning Coalition	Number of Votes	Critical Voters
{A, B}	875	A, B
{A, D}	900	A, D
{B, D}	775	B, D
{A, B, C}	1100	A, B
{A, B, D}	1275	None
{A, C, D}	1125	A, D
{B, C, D}	1000	B, D
{A, B, C, D}	1500	None

CHECK YOUR PROGRESS 1   Many countries must govern by forming coalitions from among many political parties. Suppose a country has five political parties named A, B, C, D, and E. The numbers of votes, respectively, for the five parties are 22, 18, 17, 10, and 5.

a. Determine the winning coalitions if 37 votes are required to pass a resolution.

b. For each winning coalition, determine the critical voters.

Solution   *See page S41.*

QUESTION   *Is the voting system in Example 1 a dictatorship? What is the total number of possible coalitions in Example 1?*

Banzhaf Power Index

There are a number of measures of the *power* of a voter. For instance, as we saw from the Electoral College example, some electors represent fewer people and therefore their votes may have more power. As an extreme case, suppose that two electors, A and B, each represent 10 people and that a third elector, C, represents 1000 people. If a measure passes when two of the three electors vote for the measure, then A and B voting together could pass a resolution even though they represent only 20 people.

ANSWER   *No. There is no one shareholder who has 751 or more shares of stock. The number of possible coalitions is $2^4 - 1 = 15$.*

Another measure of power, called the *Banzhaf power index,* was derived by John F. Banzhaf III in 1965. The purpose of this index is to determine the power of a voter in a weighted voting system.

Banzhaf Power Index

The **Banzhaf power index** of a voter, v, symbolized by $BPI(v)$, is given by

$$BPI(v) = \frac{\text{number of times voter } v \text{ is a critical voter}}{\text{number of times any voter is a critical voter}}$$

Consider four people, A, B, C, and D, and the one-person, one-vote system given by $\{3: 1, 1, 1, 1\}$.

Winning Coalition	Number of Votes	Critical Voters
{A, B, C}	3	A, B, C
{A, B, D}	3	A, B, D
{A, C, D}	3	A, C, D
{B, C, D}	3	B, C, D
{A, B, C, D}	4	None

To find $BPI(A)$, we look under the critical voters column and find that A is a critical voter three times. The number of times any voter is a critical voter, the denominator of the Banzhaf power index, is 12. (A is a critical voter three times, B is a critical voter three times, C is a critical voter three times, and D is a critical voter three times. The sum is $3 + 3 + 3 + 3 = 12$.) Thus

$$BPI(A) = \frac{3}{12} = 0.25$$

Similarly, we can calculate the Banzhaf power index for each of the other voters.

$$BPI(B) = \frac{3}{12} = 0.25 \qquad BPI(C) = \frac{3}{12} = 0.25 \qquad BPI(D) = \frac{3}{12} = 0.25$$

In this case, each voter has the same power. This is expected in a voting system in which each voter has one vote.

Now suppose that three people, A, B, and C, belong to a dictatorship given by $\{3: 3, 1, 1\}$.

Winning Coalition	Number of Votes	Critical Voters
{A}	3	A
{A, B}	4	A
{A, C}	4	A
{A, B, C}	5	A

The sum of the numbers of critical voters in all winning coalitions is 4. To find $BPI(A)$, we look under the critical voters column and find that A is a critical voter four times. Thus

$$BPI(A) = \frac{4}{4} = 1 \qquad BPI(B) = \frac{0}{4} = 0 \qquad BPI(C) = \frac{0}{4} = 0$$

Thus A has all the power. This is expected in a dictatorship.

EXAMPLE 2 ■ Compute the BPI for a Weighted Voting System

Suppose the stock in a company is held by five people, A, B, C, D, and E. The voting system for this company is {626: 350, 300, 250, 200, 150}. Determine the Banzhaf power index for A and E.

Solution
Determine all of the winning coalitions and the critical voters in each coalition.

Winning Coalition	Critical Voters	Winning Coalition	Critical Voters
{A, B}	A, B	{B, C, E}	B, C, E
{A, B, C}	A, B	{B, D, E}	B, D, E
{A, B, D}	A, B	{A, B, C, D}	None
{A, B, E}	A, B	{A, B, C, E}	None
{A, C, D}	A, C, D	{A, B, D, E}	None
{A, C, E}	A, C, E	{A, C, D, E}	A
{A, D, E}	A, D, E	{B, C, D, E}	B
{B, C, D}	B, C, D	{A, B, C, D, E}	None

The number of times any voter is critical is 28. To find $BPI(A)$, we look under the critical voters columns and find that A is a critical voter eight times. Thus

$$BPI(A) = \frac{8}{28} \approx 0.29$$

To find $BPI(E)$, we look under the critical voters columns and find that E is a critical voter four times. Thus

$$BPI(E) = \frac{4}{28} \approx 0.14$$

CHECK YOUR PROGRESS 2 Suppose that a government is composed of four political parties, A, B, C, and D. The voting system for this government is {26: 18, 16, 10, 6}. Determine the Banzhaf power index for A and D.

Solution See page S41.

In many cities, the only time the mayor votes on a resolution is when there is a tie vote by the members of the city council. This is also true of the United States Senate. The vice president only votes when there is a tie vote by the senators.

In Example 3, we will calculate the Banzhaf power index for a voting system in which one voter votes only to break a tie.

EXAMPLE 3 ■ Use the BPI to Determine a Voter's Power

Suppose a city council consists of four members, A, B, C, and D, and a mayor, M. The mayor votes only when there is a tie vote among the members of the council. In all cases, a resolution receiving three or more votes passes. Show that the Banzhaf power index for the mayor is the same as the Banzhaf power index for each city council member.

Solution

We first list all of the winning coalitions that do not include the mayor. To this list, we add the winning coalitions in which the mayor votes to break a tie.

Winning Coalition (Without Mayor)	Critical Voters	Winning Coalition (Mayor Voting)	Critical Voters
{A, B, C}	A, B, C	{A, B, M}	A, B, M
{A, B, D}	A, B, D	{A, C, M}	A, C, M
{A, C, D}	A, C, D	{A, D, M}	A, D, M
{B, C, D}	B, C, D	{B, C, M}	B, C, M
{A, B, C, D}	None	{B, D, M}	B, D, M
		{C, D, M}	C, D, M

By examining the table, we see that A, B, C, D, and M each occur in exactly six winning coalitions. The total number of critical voters in all winning coalitions is 30. Therefore, each member of the council and the mayor have the same Banzhaf power index, which is $\frac{6}{30} = 0.2$.

CHECK YOUR PROGRESS 3 The European Economic Community (EEC) was founded in 1958 and originally consisted of Belgium, France, Germany, Italy, Luxembourg, and the Netherlands. The weighted voting system was {12: 2, 4, 4, 4, 1, 2}. Find the Banzhaf power index for each country.

Solution See page S42.

▼ **point of interest**

In 2005, 25 countries were full members of the organization known as the European Union (EU), which at one time was referred to as the Common Market or the European Economic Community (EEC). (See Check Your Progress 3.)

By working Check Your Progress 3, you will find that the Banzhaf power index for the original EEC gave little power to Belgium and the Netherlands and no power to Luxembourg. However, there was an implicit understanding among the countries that a resolution would not pass unless all countries voted for it, thereby effectively giving each country veto power.

Blocking Coalitions and the Banzhaf Power Index

The four members, A, B, C, and D, of an organization adopted the weighted voting system $\{6: 4, 3, 2, 1\}$. The table below shows the winning coalitions.

Winning Coalition	Number of Votes	Critical Voters
{A, B}	7	A, B
{A, C}	6	A, C
{A, B, C}	9	A
{A, B, D}	8	A, B
{A, C, D}	7	A, C
{B, C, D}	6	B, C, D
{A, B, C, D}	10	None

Using the Banzhaf power index, we have $BPI(\text{A}) = \frac{5}{12}$.

A **blocking coalition** is a group of voters who can prevent passage of a resolution. In this case, a critical voter is one who leaves a blocking coalition, thereby producing a coalition that is no longer capable of preventing the passage of a resolution. For the voting system above, we have

Blocking Coalition	Number of Votes	Number of Remaining Votes	Critical Voters
{A, B}	7	3	A, B
{A, C}	6	4	A, C
{A, D}	5	5	A, D
{B, C}	5	5	B, C
{A, B, C}	9	1	None
{A, B, D}	8	2	A
{A, C, D}	7	3	A
{B, C, D}	6	4	B, C

If we count the number of times A is a critical voter in a winning or blocking coalition, we find what is called the *Banzhaf index*. In this case, the Banzhaf index is 10 and we write $BI(\text{A}) = 10$. Using both the winning coalition and the blocking coalition tables, we find that

(continued)

✔ **TAKE NOTE**

The Banzhaf index is always a *whole number*, whereas the Banzhaf power index is often a fraction between zero and one.

$BI(B) = 6$, $BI(C) = 6$, and $BI(D) = 2$. This information can be used to create an alternative definition of the Banzhaf power index.

Banzhaf Power Index—Alternative Definition

$$BPI(A) = \frac{BI(A)}{\text{sum of all Banzhaf indices for the voting system}}$$

Applying this definition to the voting system given above, we have

$$BPI(A) = \frac{BI(A)}{BI(A) + BI(B) + BI(C) + BI(D)} = \frac{10}{10 + 6 + 6 + 2} = \frac{10}{24} = \frac{5}{12}$$

Investigation Exercises

1. Using the data in Example 1 on page 649, list all blocking coalitions.

2. For the data in Example 1 on page 649, calculate the Banzhaf power indices for A, B, C, and D using the alternative definition.

3. Using the data in Check Your Progress 3 on page 653, list all blocking coalitions.

4. For the data in Check Your Progress 3 on page 653, calculate the Banzhaf power indices for Belgium and Luxembourg using the alternative definition.

5. Create a voting system with three members that is a dictatorship. Calculate the Banzhaf power index for each voter in this system using the alternative definition.

6. Create a voting system with four members in which one member has veto power. Calculate the Banzhaf power index for this system using the alternative definition.

7. Create a voting system with five members that satisfies the one-person, one-vote rule. Calculate the Banzhaf power index for this system using the alternative definition.

Exercise Set 9.3

In the following exercises that involve weighted voting systems for voters A, B, C, …, the systems are given in the form $\{q: w_1, w_2, w_3, w_4, …, w_n\}$. The weight of voter A is w_1, the weight of voter B is w_2, the weight of voter C is w_3, and so on.

1. A weighted voting system is given by $\{6: 4, 3, 2, 1\}$.
   a. What is the quota?
   b. How many voters are in this system?
   c. What is the weight of voter B?
   d. What is the weight of the coalition $\{A, C\}$?

   e. Is $\{A, D\}$ a winning coalition?
   f. Which voters are critical voters in the coalition $\{A, C, D\}$?
   g. How many coalitions can be formed?
   h. How many coalitions consist of exactly two voters?

2. A weighted voting system is given by $\{16: 8, 7, 4, 2, 1\}$.
   a. What is the quota?
   b. How many voters are in this system?
   c. What is the weight of voter C?
   d. What is the weight of the coalition $\{B, C\}$?

e. Is {B, C, D, E} a winning coalition?

f. Which voters are critical voters in the coalition {A, B, D}?

g. How many coalitions can be formed?

h. How many coalitions consist of exactly three voters?

In Exercises 3–12, calculate, if possible, the Banzhaf power index for each voter. Round to the nearest hundredth.

3. {6: 4, 3, 2}

4. {10: 7, 6, 4}

5. {10: 7, 3, 2, 1}

6. {14: 7, 5, 1, 1}

7. {19: 14, 12, 4, 3, 1}

8. {3: 1, 1, 1, 1}

9. {18: 18, 7, 3, 3, 1, 1}

10. {14: 6, 6, 4, 3, 1}

11. {80: 50, 40, 30, 25, 5}

12. {85: 55, 40, 25, 5}

13. Which, if any, of the voting systems in Exercises 3 to 12 is

a. a dictatorship?

b. a veto power system? *Note:* A voting system is a veto power system if any of the voters has veto power.

c. a null system?

d. a one-person, one-vote system?

14. Explain why it is impossible to calculate the Banzhaf power index for any voter in the null system {8: 3, 2, 1, 1}.

15. In a one-person, one-vote system, how many voters are necessary to form a winning coalition?

16. What are the losing coalitions in a dictatorship?

17. Music Education A music department consists of a band director and a music teacher. Decisions on motions are made by voting. If both members vote in favor of a motion, it passes. If both members vote against a motion, it fails. In the event of a tie vote, the principal of the school votes to break the tie. For this voting scheme, determine the Banzhaf power index for each department member and for the principal. *Hint:* See Example 3, page 653.

18. Four voters, A, B, C, and D, make decisions by using the voting scheme {4: 3, 1, 1, 1}, except when there is a tie. In the event of a tie, a fifth voter, E, casts a vote to break the tie. For this voting scheme, determine the

Banzhaf power index for each voter, including voter E. *Hint:* See Example 3, page 653.

19. Criminal Justice In a criminal trial, each of the 12 jurors has one vote and all of the jurors must agree to reach a verdict. Otherwise the judge will declare a mistrial.

a. Write the weighted voting system, in the form $\{q: w_1, w_2, w_3, w_4, \ldots, w_{12}\}$, used by these jurors.

b. Is this weighted voting system a one-person, one-vote system?

c. Is this weighted voting system a veto power system?

d. Explain an easy way to determine the Banzhaf power index for each voter.

20. Criminal Justice In California civil court cases, each of the 12 jurors has one vote and at least nine of the jury members must agree to reach a verdict.

a. Write the weighted voting system, in the form $\{q: w_1, w_2, w_3, w_4, \ldots, w_{12}\}$, used by these jurors.

b. Is this weighted voting system a one-person, one-vote system?

c. Is this weighted voting system a veto power system?

d. Explain an easy way to determine the Banzhaf power index for each voter.

A voter who has a weight that is greater than or equal to the quota is called a **dictator.** In a weighted voting system, the dictator has all the power. A voter who is never a critical voter has no power and is referred to as a **dummy.** This term is not meant to be a comment on the voter's intellectual powers. It just indicates that the voter has no ability to influence an election.

In Exercises 21–24, identify any dictator and all dummies for each weighted voting system.

21. {16: 16, 5, 4, 2, 1}

22. {15: 7, 5, 3, 2}

23. {19: 12, 6, 3, 1}

24. {45: 40, 6, 2, 1}

25. **Football** At the beginning of each football season, the coaching staff at Vista High School must vote to decide which players to select for the team. They use the weighted voting system {4: 3, 2, 1}. In this voting system, the head coach A has a weight of 3, the assistant coach B has a weight of 2, and the junior varsity coach C has a weight of 1.

   a. Compute the Banzhaf power index for each of the coaches.

   b. Explain why it seems reasonable that the assistant coach and the junior varsity coach have the same Banzhaf power index in this voting system.

26. **Football** The head coach in Exercise 25 has decided that next year the coaching staff should use the weighted voting system {5: 4, 3, 1}. The head coach is still voter A, the assistant coach is still voter B, and the junior varsity coach is still voter C.

   a. Compute the Banzhaf power index for each coach under this new system.

   b. How do the Banzhaf power indices for this new voting system compare with the Banzhaf power indices in Exercise 25? Did the head coach gain any power, according to the Banzhaf power indices, with this new voting system?

27. Consider the weighted voting system {60: 4, 56, 58}.

   a. Compute the Banzhaf power index for each voter in this system.

   b. Voter B has a weight of 56 compared with only 4 for voter A, yet the results of part a show that voter A and voter B both have the same Banzhaf power index. Explain why it seems reasonable, in this voting system, to assign voters A and B the same Banzhaf power index.

Extensions

CRITICAL THINKING

28. It can be proved that for any natural number constant c, the weighted voting systems $\{q: w_1, w_2, w_3, w_4, \ldots, w_n\}$ and $\{cq: cw_1, cw_2, cw_3, cw_4, \ldots, cw_n\}$ both have the same Banzhaf power index distribution. Verify that this theorem is true for the weighted voting system {14: 8, 7, 6, 3} and the constant $c = 2$.

29. Consider the weighted voting system $\{q: 8, 3, 3, 2\}$, with q an integer and $9 \leq q \leq 16$.

   a. For what values of q is there a dummy?

   b. For what values of q do all voters have the same power?

   c. If a voter is a dummy for a given quota, must the voter be a dummy for all larger quotas?

30. Consider the weighted voting system {17: 7, 7, 7, 2}.

   a. Explain why voter D is a dummy in this system.

   b. Explain an *easy* way to compute the Banzhaf power indices for this system.

31. In a weighted voting system, two voters have the same weight. Must they also have the same Banzhaf power index?

32. In a weighted voting system, voter A has a larger weight than voter B. Must the Banzhaf power index for voter A be larger than the Banzhaf power index for voter B?

33. Consider the one-person, one-vote strict majority system {2: 1, 1, 1} and the weighted voting system {5: 3, 3, 3}. Explain why the systems are essentially equivalent.

34. Consider the voting system {7: 3, 2, 2, 2, 2}. A student has determined that the Banzhaf power index for voter A is $\frac{5}{13}$.

   a. Explain an *easy* way to calculate the Banzhaf power index for each of the other voters.

   b. If the given voting system were changed to {8: 3, 2, 2, 2, 2}, which voters, if any, would lose power according to the Banzhaf power index?

EXPLORATIONS

35. **Shapley-Shubik Power Index** Another index that is used to measure the power of voters in a weighted voting system is the *Shapley-Shubik power*

index. Research, and write a short report on, the Shapley-Shubik power index. Explain how to calculate the Shapley-Shubik power index for each voter in the weighted voting system {6: 4, 3, 2}. How do these Shapley-Shubik power indices compare with the Banzhaf power indices for this voting system? (See your results from Exercise 3.) Explain under what circumstances you would choose to use the Shapley-Shubik power index over the Banzhaf power index.

36.  **UN Security Council** The United Nations Security Council consists of five permanent members and 10 members that are elected for two-year terms. (See the Math Matters on page 648.) The weighted voting system of the Security Council is given by

$$\{39: 7, 7, 7, 7, 7, 1, 1, 1, 1, 1, 1, 1, 1, 1, 1\}$$

In this system, each permanent member has a voting weight of 7 and each elected member has a voting weight of 1.

a. Use the program "BPI" at the website www.math.temple.edu/~cow/bpi.html to compute the Banzhaf power index distribution for the members of the Security Council.

b. According to the Banzhaf power index distribution from part a, each permanent member has about how many times more power than an elected member?

c. Some people think that the voting system used by the United Nations Security Council gives too much power to the permanent members. A weighted voting system that would reduce the power of the permanent members is given by

$$\{30: 5, 5, 5, 5, 5, 1, 1, 1, 1, 1, 1, 1, 1, 1, 1\}$$

In this new system, each permanent member has a voting weight of 5 and each elected member has a voting weight of 1. Use the program "BPI" at www.math.temple.edu/~cow/bpi.html to compute the Banzhaf power indices for the members of the Security Council under this voting system.

d. According to the Banzhaf power indices from part c, each permanent member has about how many times more power than an elected member under this new voting system?

37.  **European Community** The European Community 2005 consists of 25 countries. Issues are decided by the weighted voting system

$$\{232: 29, 29, 29, 29, 27, 27, 13, 12, 12, 12, 12, 12, 10, 10, 7, 7, 7, 7, 7, 7, 4, 4, 4, 4, 4, 4\}$$

The following table shows the weight for each country. (*Source:* http://news.bbc.co.uk)

Country	Weight
Germany, United Kingdom, France, Italy	29
Spain, Poland	27
Netherlands	13
Greece, Portugal, Belgium, Czech Republic, Hungary	12
Sweden, Austria	10
Slovakia, Denmark, Finland, Ireland, Lithuania	7
Latvia, Slovenia, Estonia, Cyprus, Luxembourg, Malta	4

a. Use the program "BPI" on the website at www.math.temple.edu/~cow/bpi.html to compute the Banzhaf power indices for the countries in the European Community 2005.

b. According to the Banzhaf power indices from part a, how many times more voting power does Italy have than Luxembourg? Round to the nearest tenth.

CHAPTER 9   # Summary

Key Terms

absolute unfairness of an apportionment [p. 610]
Alabama paradox [p. 609]
apportionment [p. 603]
apportionment principle [p. 612]
Arrow's Impossibility Theorem [p. 633]
average constituency [p. 610]
Balinski-Young Impossibility Theorem [p. 615]
Banzhaf power index [p. 651]
Borda Count method of voting [p. 626]
coalition [p. 648]
Condorcet criterion [p. 631]
critical voter [p. 649]
dictatorship [p. 647]
fairness criteria [p. 633]
Hamilton plan/method [p. 603]
Huntington-Hill apportionment method [p. 614]
Huntington-Hill apportionment principle [p. 614]
Huntington-Hill number [p. 614]
independence of irrelevant alternatives criterion [p. 633]
Jefferson plan/method [p. 605]
losing coalition [p. 648]
majority [p. 623]
majority criterion [p. 633]
modified standard divisor [p. 605]
monotonicity criterion [p. 633]
new states paradox [p. 609]
null system [p. 648]
one-person, one-vote system [p. 647]
pairwise comparison method of voting [p. 631]
plurality method of voting [p. 623]
plurality with elimination method [p. 628]
population paradox [p. 609]
preference schedule [p. 624]
quota [p. 647]
quota rule [p. 609]
relative unfairness of an apportionment [p. 611]
standard divisor [p. 603]
standard quota [p. 604]
veto power system [p. 648]
Webster apportionment method [p. 621]
weight of a voter [p. 647]
weighted voting system [p. 646]
winning coalition [p. 648]

Essential Concepts

- **Standard Divisor**

$$\text{Standard divisor} = \frac{\text{total population}}{\text{number of items to apportion}}$$

- **Standard Quota**
The *standard quota* is the whole number part of the quotient of a population and the standard divisor.

- **Quota Rule**
The number of representatives apportioned to a state is the standard quota or one more than the standard quota.

- **Huntington-Hill Number**
The value of $\dfrac{(P_A)^2}{a(a+1)}$, where P_A is the population of state A and a is the current number of representatives from state A, is called the Huntington-Hill number for state A.

- **Balinski-Young Impossibility Theorem**
Any apportionment method will either violate the quota rule or will produce paradoxes such as the Alabama paradox.

- **Plurality Method of Voting**
Each voter votes for one candidate, and the candidate with the most votes wins.

- **Borda Count Method of Voting**
With n candidates in an election, each voter ranks the candidates by giving n points to the voter's first choice, $n - 1$ points to the voter's second choice, and so on. The candidate who receives the most total points is the winner.

- **Plurality with Elimination Method of Voting**
Eliminate the candidate with the lowest number of first-place votes. Retake a vote, keeping the same ranking preferences, and eliminate the candidate with the fewest first-place votes. Continue until only one candidate remains.

- **Pairwise Comparison Method of Voting**
Compare each candidate head-to-head with each other candidate. Award 1 point for a win, 0.5 point for a tie, and 0 points for a loss. The candidate with the greatest number of points wins the election.

- **Majority Criterion**
The candidate who receives a majority of the first-place votes is the winner.

- **Monotonicity Criterion**
  If candidate A wins an election, then candidate A will also win the election if the only change in the voters' preferences is that supporters of a different candidate change their votes to support candidate A.

- **Condorcet Criterion**
  A candidate who wins all possible head-to-head matchups should win an election when all candidates appear on the ballot.

- **Independence of Irrelevant Alternatives Criterion**
  If a candidate wins an election, the winner should remain the winner in any recount in which losing candidates withdraw from the race.

- **Arrow's Impossibility Theorem**
  There is no voting method for an election with three or more candidates that satisfies all four fairness criteria.

- **Banzhaf Power Index**
  The Banzhaf power index of a voter, v, symbolized by $BPI(v)$, is given by

$$BPI(v) = \frac{\text{number of times voter } v \text{ is a critical voter}}{\text{number of times any voter is a critical voter}}$$

- **Number of Possible Coalitions of n Voters**
  The number of possible coalitions of n voters is $2^n - 1$.

CHAPTER 9 **Review Exercises**

---

1. **Airline Industry** The table below shows how the average constituencies change when two different airports, High Desert Airport and Eastlake Airport, add a new air traffic controller.

	High Desert Airport Average Constituency	Eastlake Airport Average Constituency
High Desert Airport receives new controller	297	326
Eastlake Airport receives new controller	302	253

   a. Determine the relative unfairness of an apportionment that gives a new air traffic controller to High Desert Airport rather than to Eastlake Airport. Round to the nearest thousandth.

   b. Determine the relative unfairness of an apportionment that gives a new air traffic controller to Eastlake Airport rather than to High Desert Airport. Round to the nearest thousandth.

   c. Using the apportionment principle, determine which airport should receive the new air traffic controller.

2. **Education** The following table shows the numbers of English professors and the numbers of students taking English at two campuses of a state university. The university is planning to add a new English professor to one of the campuses. Use the apportionment principle to determine which campus should receive the new professor.

University Campus	Number of English Professors	Number of Students Taking English
Morena Valley	38	1437
West Keyes	46	1504

3. **Education** The following table shows the enrollments for the four divisions of a college. There are 50 new overhead projectors that are to be apportioned among the divisions based on the enrollments.

Division	Enrollment
Health	1280
Business	3425
Engineering	1968
Science	2936
Total	9609

   a. Use the Hamilton method to determine the number of projectors to be apportioned to each division.

   b. Use the Jefferson method to determine the number of projectors to be apportioned to each division.

   c. Use the Webster method to determine the number of projectors to be apportioned to each division.

4. **Airline Industry** The following table shows the number of ticket agents at five airports for a small airline company. The company has hired 35 new security employees who are to be apportioned among the airports based on the number of ticket agents at each airport.

Airport	Number of Ticket Agents
Newark	28
Cleveland	19
Chicago	34
Philadelphia	13
Detroit	16
Total	110

a. Use the Hamilton method to apportion the new security employees among the airports.

b. Use the Jefferson method to apportion the new security employees among the airports.

c. Use the Webster method to apportion the new security employees among the airports.

5. **Technology** A company has four offices. The president of the company uses the Hamilton method to apportion 66 new computer printers among the offices based on the number of employees at each office.

Office	A	B	C	D
Number of employees	19	195	308	402
Apportionment of 66 printers	1	14	22	29

a. If the number of printers to be apportioned by the Hamilton method increases from 66 to 67, will the Alabama paradox occur? Explain.

b. If the number of printers to be apportioned by the Hamilton method increases from 67 to 68, will the Alabama paradox occur? Explain.

6. **Automobile Sales** Consider the apportionment of 27 automobiles to the sales departments of a business with five regional centers labeled A, B, C, D, and E. The following table shows the Hamilton apportionment of the automobiles based on the number of sales personnel at each center.

Center	A	B	C	D	E
Number of sales personnel	31	108	70	329	49
Apportionment of 27 automobiles	2	5	3	15	2

a. If the number of automobiles to be apportioned increases from 27 to 28, what will be the apportionment if the Hamilton method is used? Will the Alabama paradox occur? Explain.

b. If the number of automobiles to be apportioned using the Hamilton method increases from 28 to 29, will the Alabama paradox occur? Explain.

7. **Music Company** MusicGalore.net has offices in Los Angeles and Newark. The number of employees at each office is shown in the following table. There are 11 new computer file servers to be apportioned between the offices according to their numbers of employees.

Office	Los Angeles	Newark
Employees	1430	235

a. Use the Hamilton method to find each office's apportionment of file servers.

b. The corporation opens an additional office in Kansas City with 111 employees and decides to have a total of 12 file servers. If the file servers are reapportioned using the Hamilton method, will the new states paradox occur? Explain.

8. **Building Inspectors** A city apportions 34 building inspectors among three regions according to their populations. The following table shows the present population of each region.

Region	A	B	C
Population	14,566	3321	29,988

a. Use the Hamilton method to apportion the inspectors.

b. After a year the regions have the following populations.

Region	A	B	C
Population	15,008	3424	30,109

Region A has an increase in population of 442, which is an increase of 3.03%. Region B has an increase in population of 103, which is an increase of 3.10%. Region C has an increase in population of 121, which is an increase of 0.40%. If the inspectors are reapportioned using the Hamilton method, will the population paradox occur? Explain.

9. Is the Hamilton apportionment method susceptible to the population paradox?

10. Is the Jefferson apportionment method susceptible to the new states paradox?

11. **Corporate Security** The Huntington-Hill apportionment method has been used to apportion 86 security guards among three corporate office buildings according to the number of employees at each building. See the following table.

Building	Number of Security Guards	Number of Employees
A	25	414
B	43	705
C	18	293

The corporation has decided to hire a new security guard.

a. Use the Huntington-Hill apportionment principle to determine to which building the new security guard should be assigned.

b. If another security guard is hired, bringing the total number of guards to 88, to which building should this guard be assigned?

12. **Homecoming Queen** Three high school students are running for Homecoming Queen. Students at the school were allowed to rank the candidates in order of preference. The results are shown in the preference schedule below.

	Rankings		
Cynthia L.	3	2	1
Hannah A.	1	3	3
Shannon M.	2	1	2
Number of votes:	112	97	11

a. Use the Borda Count method to find the winner.

b. Find a candidate who wins all head-to-head comparisons.

c. Explain why the Condorcet criterion has been violated.

d. Who wins using the plurality voting system?

e. Explain why the majority criterion has been violated.

13. **Homecoming Queen** In Exercise 12, suppose Cynthia L. withdraws from the Homecoming Queen election.

a. Assuming voter preferences between the remaining two candidates remain the same, who will be crowned Homecoming Queen using the Borda Count method?

b. Explain why the independence of irrelevant alternatives criterion has been violated.

14. **Scholarship Awards** A scholarship committee must choose a winner from three finalists, Jean, Margaret, and Terry. Each member of the committee ranked the three finalists, and Margaret was selected using the plurality with elimination method. This vote was later declared invalid and a new vote was taken. All members voted using the same rankings except one, who changed her first choice from Terry to Margaret. This time Jean won the scholarship using the same voting method. Which fairness criterion was violated, and why?

15. A weighted voting system for voters A, B, C, and D is given by {18: 12, 7, 6, 1}. The weight of voter A is 12, the weight of voter B is 7, the weight of voter C is 6, and the weight of voter D is 1.

a. What is the quota?

b. What is the weight of the coalition {A, C}?

c. Is {A, C} a winning coalition?

d. Which voters are critical voters in the coalition {A, C, D}?

e. How many coalitions can be formed?

f. How many coalitions consist of exactly two voters?

16. A weighted voting system for voters A, B, C, D, and E is given by {35: 29, 11, 8, 4, 2}. The weight of voter A is 29, the weight of voter B is 11, the weight of voter C is 8, the weight of voter D is 4, and the weight of voter E is 2.

a. What is the quota?

b. What is the weight of the coalition {A, D, E}?

c. Is {A, D, E} a winning coalition?

d. Which voters are critical voters in the coalition {A, C, D, E}?

e. How many coalitions can be formed?

f. How many coalitions consist of exactly two voters?

17. Calculate the Banzhaf power indices for voters A, B, and C in the weighted voting system {9: 6, 5, 3}.

18. Calculate the Banzhaf power indices for voters A, B, C, D, and E in the one-person, one-vote system {3: 1, 1, 1, 1, 1}.

19. Calculate the Banzhaf power indices for voters A, B, C, and D in the weighted voting system {31: 19, 15, 12, 10}. Round to the nearest hundredth.

20. Calculate the Banzhaf power indices for voters A, B, C, D, and E in the weighted voting system {35: 29, 11, 8, 4, 2}. Round to the nearest hundredth.

In Exercises 21 and 22, identify any dictator and all dummies for each weighted voting system.

21. Voters A, B, C, D, and E: {15: 15, 10, 2, 1, 1}

22. Voters A, B, C, and D: {28: 19, 6, 4, 2}

23. Four voters, A, B, C, and D, make decisions by using the weighted voting system {5: 4, 2, 1, 1}. In the event of a tie, a fifth voter, E, casts a vote to break the tie. For this voting scheme, determine the Banzhaf power index for each voter, including voter E.

24. Four finalists are competing in an essay contest. Judges have read and ranked each essay in order of preference. The results are shown in the preference schedule below.

	Rankings			
Crystal Kelley	3	2	2	1
Manuel Ortega	1	3	4	3
Peter Nisbet	2	4	1	2
Sue Toyama	4	1	3	4
Number of votes:	8	5	4	6

a. Using the plurality voting system, who is the winner of the essay contest?

b. Does this winner have a majority?

c. Use the Borda Count method of voting to determine the winner of the essay contest.

25. **Ski Club** A campus ski club is trying to decide where to hold its winter break ski trip. The members of the club were surveyed and asked to rank five choices in order of preference. Their responses are tallied in the following table.

	Rankings				
Aspen	1	1	3	2	3
Copper Mountain	5	4	2	4	4
Powderhorn	3	2	5	1	5
Telluride	4	5	4	5	2
Vail	2	3	1	3	1
Number of votes:	14	8	11	18	12

a. Use the plurality method of voting to determine which resort the club should choose.

b. Use the Borda Count method to choose the ski resort the club should visit.

26. **Campus Election** Four students are running for the activities director position on campus. Students were asked to rank the four candidates in order of preference. The results are shown in the table below.

	Rankings			
G. Reynolds	2	3	1	3
L. Hernandez	1	4	4	2
A. Kim	3	1	2	1
J. Schneider	4	2	3	4
Number of votes:	132	214	93	119

Use the plurality with elimination method to determine the winner of the election.

27. **Consumer Preferences** A group of consumers were surveyed about their favorite candy bars. Each participant was asked to rank four candy bars in order of preference. The results are given in the table below.

	Rankings				
Nestle Crunch	1	4	4	2	3
Snickers	2	1	2	4	1
Milky Way	3	2	1	1	4
Twix	4	3	3	3	2
Number of votes:	15	38	27	16	22

Use the plurality with elimination method to determine the group's favorite candy bar.

28. Use the pairwise comparison method of voting to choose the winner of the election in Exercise 26.

29. Use the pairwise comparison method of voting to choose the group's favorite candy bar in Exercise 27.

CHAPTER 9 **Test**

---

1. Postal Service The table below shows the numbers of mail carriers and the populations for two cities, Spring Valley and Summerville. The postal service is planning to add a new mail carrier to one of the cities. Use the apportionment principle to determine which city should receive the new mail carrier.

City	Number of Mail Carriers	Population
Spring Valley	158	67,530
Summerville	129	53,950

2. Computer Allocation The following table shows the numbers of employees in the four divisions of a corporation. There are 85 new computers that are to be apportioned among the divisions based on the number of employees in each division.

Division	Number of Employees
Sales	1008
Advertising	234
Service	625
Manufacturing	3114
Total	4981

a. Use the Hamilton method to determine the number of computers to be apportioned to each division.

b. Use the Jefferson method to determine the number of computers to be apportioned to each division. Does this particular apportionment violate the quota rule?

3. High School Counselors The following table shows the numbers of counselors and the numbers of students at two high schools. The current number of counselors for each school was determined using the Huntington-Hill apportionment method.

School	Number of Counselors	Number of Students
Cedar Falls	9	2646
Lake View	7	1984

A new counselor is to be hired for one of the schools.

a. Calculate the Huntington-Hill number for each of the schools. Round to the nearest whole number.

b. Use the Huntington-Hill apportionment principle to determine to which school the new counselor should be assigned.

4. A weighted voting system for voters A, B, C, D, and E is given by {33: 21, 14, 12, 7, 6}.

a. What is the quota?

b. What is the weight of the coalition {B, C}?

c. Is {B, C} a winning coalition?

d. Which voters are critical voters in the coalition {A, C, D}?

e. How many coalitions can be formed?

f. How many coalitions consist of exactly two voters?

5. Consumer Preference One hundred consumers ranked three brands of bottled water in order of preference. The results are shown in the preference schedule below.

	Rankings				
Arrowhead	2	3	1	3	2
Evian	3	2	2	1	1
Aquafina	1	1	3	2	3
Number of votes:	22	17	31	11	19

a. Using the plurality system of voting, which brand of bottled water is the preferred brand?

b. Does the winner have a majority?

c. Use the Borda Count method of voting to determine the preferred brand of water.

6. Executives' Preferences A company with offices across the country will hold its annual executive meeting in one of four locations. All of the executives were asked to rank the locations in order of preference. The results are shown in the table below.

	Rankings			
New York	3	1	2	2
Dallas	2	3	1	4
Los Angeles	4	2	4	1
Atlanta	1	4	3	3
Number of votes:	19	24	7	35

Use the pairwise comparison method to determine which location should be chosen.

7. Exam Review A professor is preparing an extra review session the Monday before final exams and she wants as many students as possible to be able to attend. She asks all of the students to rank different times of day in order of preference. The results are given below.

	Rankings				
Morning	4	1	4	2	3
Noon	3	2	2	3	1
Afternoon	1	3	1	4	2
Evening	2	4	3	1	4
Number of votes:	12	16	9	5	13

a. Using the plurality with elimination method, for what time of day should the professor schedule the review session?

b. Use the Borda Count method to determine the best time of day for the review.

8. Budget Proposal A committee must vote on which of three budget proposals to adopt. Each member of the committee has ranked the proposals in order of preference. The results are shown below.

	Rankings		
Proposal A	1	3	3
Proposal B	2	2	1
Proposal C	3	1	2
Number of votes:	40	9	39

a. Using the plurality system of voting, which proposal should be adopted?

b. If proposal C is found to be invalid and is eliminated, which proposal wins using the plurality system?

c. Explain why the independence of irrelevant alternatives criterion has been violated.

d. Verify that proposal B wins all head-to-head comparisons.

e. Explain why the Condorcet criterion has been violated.

9. Drama Department The four staff members, A, B, C, and D, of a college drama department use the weighted voting system $\{8: 5, 4, 3, 2\}$ to make casting decisions for a play they are producing. In this voting system, voter A has a weight of 5, voter B has a weight of 4, voter C has a weight of 3, and voter D has a weight of 2. Compute the Banzhaf power index for each member of the department. Round each result to the nearest hundredth.

10. Three voters, A, B, and C, make decisions by using the weighted voting system $\{6: 5, 4, 1\}$. In the case of a tie, a fourth voter, D, casts a single vote to break the tie. For this voting scheme, determine the Banzhaf power index for each voter, including voter D.

CHAPTER 10

The Mathematics of Graphs

10.1	Traveling Roads and Visiting Cities	10.3	Map Coloring and Graphs
10.2	Efficient Routes		

In this chapter, you will learn how to analyze and solve a variety of problems, such as how to find the least expensive route of travel on a vacation, how to determine the most efficient order in which to run errands, and how to schedule meetings at a conference so that no one has two required meetings at the same time.

The methods we will use to study these problems can be traced back to an old recreational puzzle. In the early eighteenth century, the Pregel River in a city called Königsberg (located in modern-day Russia and now called Kaliningrad) surrounded an island before splitting in two. Seven bridges crossed the river and connected four different land areas, similar to the map drawn below.

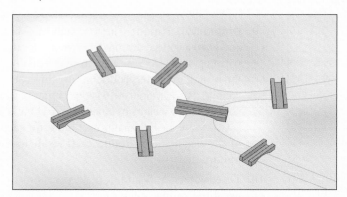

Many citizens of the time attempted to take a stroll that would lead them across each bridge and return them to the starting point without traversing the same bridge twice. None of them could do it, no matter where they chose to start. Try it for yourself with pencil and paper. You will see that it is not easy!

So is it impossible to take such a stroll? Or just difficult? In 1736 the Swiss mathematician Leonhard Euler (1707–1783) proved that it is, in fact, impossible to walk such a path. His analysis of the challenge laid the groundwork for a branch of mathematics known as *graph theory*. We will investigate how Euler approached the problem of the seven bridges of Königsberg in Section 10.1.

Online Study Center

For online student resources, visit this textbook's Online Study Center at **college.hmco.com/pic/aufmannMTQR.**

| # Traveling Roads and Visiting Cities

Introduction to Graphs

The acts of traveling the streets of a city, routing data through nodes on the Internet, and flying between cities have a common link. The goal of each of these tasks is to go from one place to another along a specified path. Efficient strategies for accomplishing these goals can be studied using a branch of mathematics called *graph theory*.

For example, the diagram in Figure 10.1 could represent the flights available on a particular airline among a selection of cities. Each dot represents a city, and a line segment connecting two dots means there is a flight between the two cities. The line segments do not represent the actual paths the planes fly; they simply indicate a relationship between two cities (in this case, that a flight exists). If we wish to travel from San Francisco to New York, the diagram can help us examine the various possible routes. This type of diagram is called a *graph*. Note that this is a very different kind of graph from the graph of a function that we discussed in Chapter 5.

Figure 10.1

> **Graph**
>
> A **graph** is a set of points called **vertices** and line segments or curves called **edges** that connect vertices.

✔️ **TAKE NOTE**

Vertices are always clearly indicated with a "dot." Edges that intersect with no marked vertex are considered to cross over each other without touching.

Graphs can be used to represent many different scenarios. For instance, the three graphs in Figure 10.2 are the exact same graph as in Figure 10.1, but used in different contexts. In Figure 10.2a, each vertex represents a baseball team, and an edge connecting two vertices might mean that the two teams played each other during the current season. Note that the placement of the vertices has nothing to do with geographical location; in fact, the vertices can be shown in any arrangement we choose. The important information is which vertices are connected by edges.

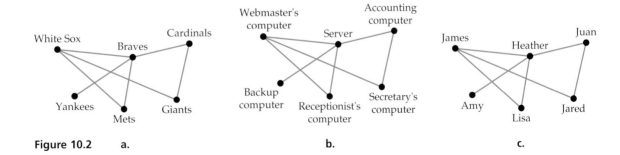

Figure 10.2   **a.**   **b.**   **c.**

Figure 10.2b shows the computer network of a small business. Each vertex represents a computer, and the edges indicate which machines are directly connected to each other. The graph in Figure 10.2c could be used to indicate which students share a class together; each vertex represents a student, and an edge connecting two vertices means that those students share at least one class.

In general, graphs can contain vertices that are not connected to any edges, two or more edges that connect the same vertices (called **multiple edges**), or edges that loop back to the same vertex. We will usually deal with **connected graphs,** graphs in which any vertex can be reached from any other vertex by tracing along edges. (Essentially, the graph consists of only one "piece.") Several examples of graphs are shown below.

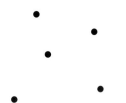

This graph has five vertices but no edges, and is referred to as a **null graph.** It is also an example of a *disconnected graph*.

This is a connected graph that has a pair of *multiple edges*. Notice that two edges cross in the center, but there is no vertex there. Unless a dot is drawn, the edges are considered to pass over each other without touching.

This graph is not connected; it consists of two different sections. It also contains a loop.

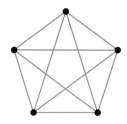

This is a connected graph in which every possible edge is drawn between vertices (without any multiple edges). Such a graph is called a **complete graph.**

Notice that it does not matter whether the edges are drawn straight or curved, and their lengths and positions are not important. Nor is the exact placement of the vertices important. The graph simply illustrates connections between vertices.

Consequently, the three graphs shown below are considered **equivalent graphs** because the edges form the same connections of vertices in each graph.

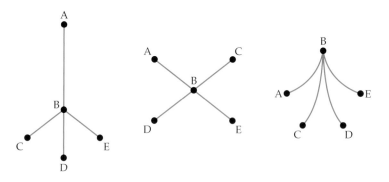

If you have difficulty seeing that these graphs are equivalent, use the labeled vertices to compare each graph. Notice that in each case, vertex B has an edge connecting it to each of the other four vertices, and no other edges exist.

EXAMPLE 1 ▪ Equivalent Graphs

Determine whether the following two graphs are equivalent.

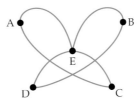
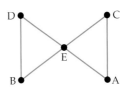

Solution

Despite the fact that the two graphs have different arrangements of vertices and edges, they are equivalent. To illustrate, we examine the edges of each graph. The first graph contains six edges; we can list them by indicating which two vertices they connect. The edges are AC, AE, BD, BE, CE, and DE. If we do the same for the second graph, we get the exact same six edges. Because the two graphs represent the same connections among the vertices, they are equivalent.

CHECK YOUR PROGRESS 1   Determine whether the following two graphs are equivalent.

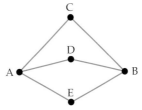
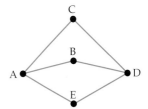

Solution   *See page S42.*

Math Matters   Picture the World Wide Web as a Graph

Graph theory is playing an ever-increasing role in the study of large and complex systems such as the World Wide Web. Internet Cartographer, available at http://www.inventix.com, is a software application that creates graphs from networks of websites. Each vertex is a website, and two vertices are joined by an edge if there is a link from one website to the other. The software was used to generate the image on the following page. You can imagine how large the graph would be if all the billions of existing web pages were included. The Internet search engine Google uses these relationships to rank its search results. Websites are ranked in

(continued)

part according to how many other sites link to them. In effect, the websites represented by the vertices of the graph with the most connected edges would appear first in Google's search results.

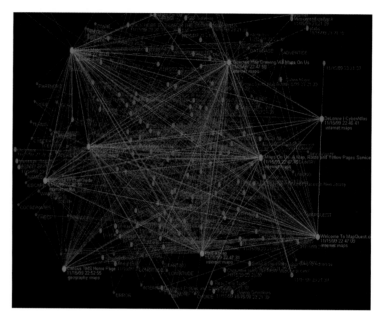

A graph of a network of websites. Each vertex represents a website, and an edge between vertices indicates that one site includes a link to the other.

Another example of an extremely large graph is the graph that uses vertices to represent telephone numbers. An edge connects two vertices if one number called the other. James Abello of AT&T Shannon Laboratories analyzed a graph formed from one day's worth of calls. The graph had more than 53 million vertices and 170 million edges! Interestingly, although the graph was not a connected graph (it contained 3.7 million separate components), more than 80% of the vertices were part of one large connected component. Within this component, any telephone could be linked to any other through a chain of 20 or fewer calls.

Euler Circuits

To solve the Königsberg bridges problem presented on page 666, we can represent the arrangement of land areas and bridges with a graph. Let each land area be represented by a vertex, and then connect two vertices if there is a bridge spanning the corresponding land areas. Then the geographical situation shown in Figure 10.3 on the following page becomes the graph shown in Figure 10.4.

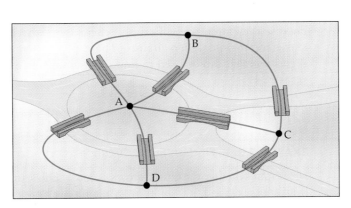

Figure 10.3

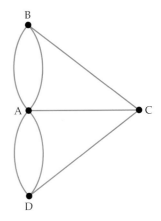

Figure 10.4

In terms of a graph, the original problem can be stated as follows: Can we start at any vertex, move through each edge once (but not more than once), and return to the starting vertex? Again, try it with pencil and paper. Every attempt seems to end in failure.

Before we can examine how Euler proved this task impossible, we need to establish some terminology. A **walk** in a graph can be thought of as a movement from one vertex to another by traversing edges. We can refer to our movement by vertex letters. For example, in the graph in Figure 10.4, one walk would be A–B–A–C. If a walk ends at the same vertex it started at, it is considered a **closed walk,** or **circuit.** A circuit that uses every edge, but never uses the same edge twice, is called an **Euler circuit.** So an Euler circuit is a walk that starts and ends at the same vertex and uses every edge of the graph exactly once. (The walk may cross through vertices more than once.) If we could find an Euler circuit in the graph in Figure 10.4, we would have a solution to the Königsberg bridges problem: a path that crosses each bridge exactly once and returns to the starting point.

Euler essentially proved that the graph in Figure 10.4 cannot have an Euler circuit. He accomplished this by examining the number of edges that met at each vertex. This is called the **degree** of a vertex. He made the observation that in order to complete the desired walk, every time you approached a vertex you would then need to leave that vertex. If you traveled through that vertex again, you would again need an approaching edge and a departing edge. Thus, for an Euler circuit to exist, the degree of every vertex would have to be an even number. Furthermore, he was able to show that any graph that has even degree at every vertex must have an Euler circuit. Consequently, such graphs are called **Eulerian.** If we now look at the graph in Figure 10.4, we can see that it is not Eulerian. No Euler circuit exists because not every vertex is of even degree.

Eulerian Graph Theorem

A connected graph is Eulerian if and only if every vertex of the graph is of even degree.

✔ TAKE NOTE

 For information on and examples of finding Eulerian circuits, search for "Fleury's Algorithm" on the World Wide Web.

The Eulerian Graph Theorem guarantees that when all vertices of a graph are of even degree, an Euler circuit exists, but it does not tell us how to find one. Because the graphs we will examine here are relatively small, we will rely on trial and error to find Euler circuits. There is a systematic method, called *Fleury's Algorithm*, that can be used to find Euler circuits in graphs with large numbers of vertices.

EXAMPLE 2 ▪ Find an Euler Circuit

Determine whether the graph shown below is Eulerian. If it is, find an Euler circuit. If it is not, explain how you know. The number beside each vertex indicates the degree of the vertex.

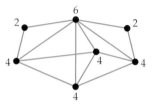

✔ TAKE NOTE

You should verify that the given walk is an Euler circuit. Using your pencil, start at vertex B and trace along edges of the graph, following the vertices in order. Make sure you trace over each edge once (but not twice).

Solution

Each vertex has a degree of 2, 4, or 6, so by the Eulerian Graph Theorem, the graph is Eulerian. There are many possible Euler circuits in this graph. We do not have a formal method of locating one, so we just use trial and error. If we label the vertices as shown below, one Euler circuit is B–A–F–B–E–F–G–E–D–G–B–D–C–B.

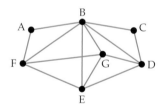

CHECK YOUR PROGRESS 2   Does the following graph have an Euler circuit? If so, find one. If not, explain why.

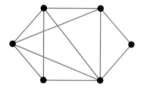

Solution   *See page S42.*

EXAMPLE 3 ■ An Application of Euler Circuits

The subway map below shows the tracks that subway trains traverse as well as the junctions where one can switch trains. Suppose an inspector needs to travel the full length of each track. Is it possible to plan a journey that traverses the tracks and returns to the starting point without traveling through any portion of a track more than once?

▼ point of interest

There are a number of instances in which finding an Euler circuit has practical implications. For example, in cities where it snows in the winter, the highway department must provide snow removal for the streets. The most efficient route is an Euler circuit. In this case, the snow plow leaves the maintenance garage, travels down each street only once, and returns to the garage. The situation becomes more complicated if one-way streets are involved; nonetheless, graph theory techniques can still help find the most efficient route to follow.

Solution

We can consider the subway map a graph, with a vertex at each junction. An edge represents a track that runs between two junctions. In order to find a travel route that does not traverse the same track twice, we need to find an Euler circuit in the graph. Notice, however, that the vertex representing the Civic Center junction is of degree 3. Because a vertex is of odd degree, the graph cannot be Eulerian, and it is impossible for the inspector not to travel at least one track twice.

CHECK YOUR PROGRESS 3 Suppose the city of Königsberg had the arrangement of islands and bridges pictured below instead of the arrangement we introduced previously. Would the citizens be able to complete a stroll across each bridge and return to their starting points without crossing the same bridge twice?

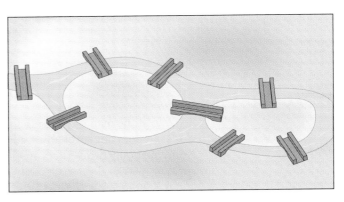

Solution   *See page S42.*

Euler Walks

Perhaps the Königsberg bridge problem would have a solution if we did not need to return to the starting point. Give it a try; you will still find it difficult to find a solution.

If we do not need to return to our starting point, then what we are looking for in Figure 10.4 on page 671 is a walk (not necessarily a circuit) that uses every edge once and only once. We call such a walk an **Euler walk.** Euler showed that even with this relaxed condition, the bridge problem still was not solvable. The general result of his argument is given in the following theorem.

✔ **TAKE NOTE**

Note that an Euler *walk* does not require that we start and stop at the same vertex, whereas an Euler *circuit* does.

Euler Walk Theorem

A connected graph contains an Euler walk if and only if the graph has two vertices of odd degree with all other vertices of even degree. Furthermore, every Euler walk must start at one of the vertices of odd degree and end at the other.

To see why this theorem is true, notice that the only places at which an Euler walk differs from an Euler circuit are the starting and ending vertices. If we never return to the starting vertex, only one edge meets there and the degree of the vertex is 1. If we do return, we cannot stop there. So we depart again, giving the vertex a degree of 3. Similarly, any return trip means that an additional two edges meet at the vertex. Thus the degree of the starting vertex must be odd. By similar reasoning, the ending vertex must also be of odd degree. All other vertices, just as in the case of an Euler circuit, must be of even degree.

EXAMPLE 4 ■ **An Application of Euler Walks**

A photographer would like to travel across all of the roads shown on the map below. The photographer will rent a car that need not be returned to the same city, so the trip can begin in any city. Is it possible for the photographer to design a trip that traverses all of the roads exactly once?

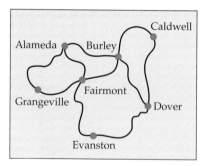

Solution

Looking at the map of roads as a graph, we see that a route that includes all of the roads but does not cover any road twice corresponds to an Euler walk of the graph. Notice that only two vertices are of odd degree, the cities Alameda and Dover. Thus we know that an Euler walk exists, and so it is possible for the photographer to plan a route that travels each road once. Because (abbreviating the cities) A and D are vertices of odd degree, the photographer must start at one of these cities. With a little experimentation, we find that one Euler walk is A−B−C−D−B−F−A−G−F−E−D.

CHECK YOUR PROGRESS 4   A bicyclist wants to mountain bike through all the trails of a national park. A map of the park is shown below. Because the bicyclist will be dropped off in the morning by friends and picked up in the evening, she does not have a preference of where she begins and ends her ride. Is it possible for the cyclist to traverse all of the trails without repeating any portions of her trip?

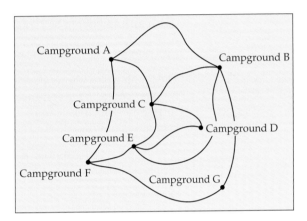

Solution   *See page S42.*

Hamiltonian Circuits

In Example 4, we tried to find a route that traversed every road once without repeating any roads. Suppose our priority is to visit cities rather than travel roads. We do not care whether we use all the roads or not. Is there a route that visits each city once (without repeating any cities)?

You can verify that the route A−B−C−D−E−F−G visits each city in Example 4 just once. After visiting Grangeville, we could return to the starting city to complete the journey. In the language of graph theory, this route corresponds to a walk that uses every vertex of a (connected) graph, does not use any vertex twice, and returns to the starting vertex. Such a walk is called a **Hamiltonian circuit.** (Unlike an Euler circuit, we do not need to use every edge.) If a graph has a Hamiltonian circuit, the graph is called **Hamiltonian.**

QUESTION   *Can a graph be both Eulerian and Hamiltonian?*

Because we found the Hamiltonian circuit A−B−C−D−E−F−G−A in the map of cities in Example 4, we know that the graph is Hamiltonian. Unfortunately, we do not have a straightforward criterion to guarantee that a graph will be Hamiltonian, but we do have the following theorem.

ANSWER   *Yes. For example, the graph shown here has an Euler circuit and a Hamiltonian circuit.*

Dirac's Theorem

Consider a connected graph with at least three vertices and no multiple edges. Let n be the number of vertices in the graph. If every vertex has a degree of at least $n/2$, then the graph must be Hamiltonian.

We must be careful, however; if our graph does not meet the requirements of this theorem, it still might be Hamiltonian. Dirac's Theorem does not help us in this case.

EXAMPLE 5 ■ Apply Dirac's Theorem

The graph below shows the available flights of a popular airline. (An edge between two vertices in the graph means that the airline has direct flights between the two corresponding cities.) Apply Dirac's Theorem to verify that the following graph is Hamiltonian. What does a Hamiltonian circuit represent in terms of flights?

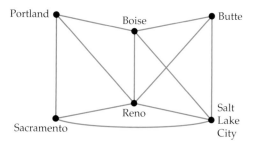

Solution

There are six vertices in the graph, so $n = 6$, and every vertex has a degree of at least $n/2 = 3$. So, by Dirac's Theorem, the graph is Hamiltonian. This means that the graph contains a circuit that visits each vertex once and returns to the starting vertex without visiting any vertex twice. Here, a Hamiltonian circuit represents a sequence of flights that visits each city and returns to the starting city without visiting any city twice. Notice that Dirac's Theorem does not tell us how to find the Hamiltonian circuit; it just guarantees that one exists.

CHECK YOUR PROGRESS 5   A large law firm has offices in seven major cities. The firm has overnight document deliveries scheduled every day between certain offices. In the graph below, an edge between vertices indicates that there is delivery service between the corresponding offices. Use Dirac's Theorem to answer the following question: Using the law firm's existing delivery service, is it possible to route a document to all the offices and return the document to its originating office without sending it through the same office twice?

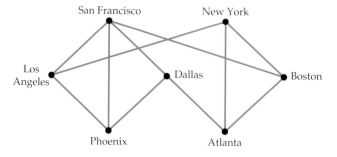

Solution   See page S42.

EXAMPLE 6 ■ An Application of Hamiltonian Circuits

The floor plan of an art gallery is pictured below. Draw a graph that represents the floor plan, where vertices correspond to rooms and edges correspond to doorways. Then use your graph to answer the following questions: Is it possible to take a walking tour through the gallery that visits every room and returns to the starting point without visiting any room twice? Is it possible to take a stroll that passes through every doorway without going through the same doorway twice? If so, does it matter whether we return to the starting point?

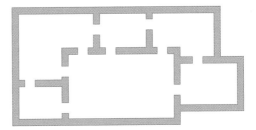

Solution

We can represent the floor plan by a graph if we let a vertex represent each room. Draw an edge between two vertices if there is a doorway between the two rooms, as shown in Figure 10.5.

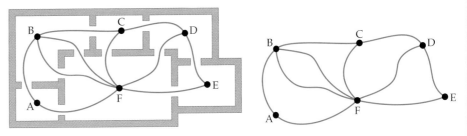

Figure 10.5   **Figure 10.6**

The graph in Figure 10.6 is equivalent to our floor plan. A walk through the gallery that visits each room just once and returns to the starting point corresponds to a visit to each vertex without visiting any vertex twice. In other words, we are looking for a Hamiltonian circuit. There are six vertices in the graph, and not all the vertices have a degree of at least 3, so Dirac's Theorem does not guarantee that the graph is Hamiltonian. However, it is still possible that a Hamiltonian circuit exists. In fact, one is not too hard to find: A–B–C–D–E–F–A.

If we would like to tour the gallery and pass through every doorway once, we must find a walk on our graph that uses every edge once (and no more). Thus we are looking for an Euler walk. In the graph, two vertices are of odd degree and the others are of even degree. So we know that an Euler walk exists, but not an Euler circuit. Therefore, we cannot pass through each doorway once and only once if we want to return to the starting point, but we can do it if we end up somewhere else. Furthermore, we know we must start at a vertex of odd degree—either room C or room D. By trial and error, one such walk is C–B–F–B–A–F–E–D–C–F–D.

✔ **TAKE NOTE**

Recall that a Hamiltonian circuit visits each vertex once (and only once), whereas an Euler circuit uses each edge once (and only once).

CHECK YOUR PROGRESS 6 The floor plan of a warehouse is illustrated below. Use a graph to represent the floor plan and answer the following questions: Is it possible to walk through the warehouse so that you pass through every doorway once but not twice? Does it matter whether you return to the starting point?

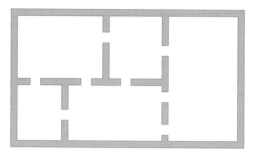

Solution *See page S42.*

You may find it helpful to review the following key definitions from this section before proceeding.

Walk	A path between vertices formed by traversing edges.
Circuit	A walk that begins and ends at the same vertex.
Euler walk	A walk that uses every edge, but does not use any edge more than once.
Euler circuit	A walk that uses every edge, but does not use any edge more than once, and begins and ends at the same vertex.
Eulerian graph	A graph that contains an Euler circuit.
Hamiltonian circuit	A walk that uses every vertex, but does not use any vertex more than once, and begins and ends at the same vertex.
Hamiltonian graph	A graph that contains a Hamiltonian circuit.

Investigation

Pen-Tracing Puzzles

You may have seen puzzles like this one: Can you draw the diagram at the right without lifting your pencil from the paper, and without tracing over the same segment twice?

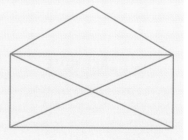

(continued)

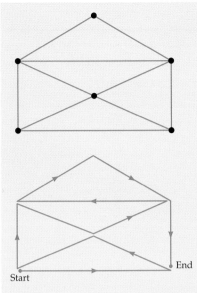

Before reading on, try it for yourself. By trial and error, you may discover a tracing that works. Even though there are several possible tracings, you may notice that only certain starting points seem to allow a complete tracing. How do we know which point to start from? How do we even know that a solution exists?

Puzzles such as this, called "pen-tracing puzzles," are actually problems in graph theory. If we imagine a vertex placed wherever two lines meet or cross over each other, then we have a graph. Our task is to start at a vertex and find a path that traverses every edge of the graph, without repeating any edge. In other words, we need an Euler walk! (An Euler circuit would work as well.)

As we learned in this section, a graph has an Euler walk only if two vertices are of odd degree and the remaining vertices are of even degree. Furthermore, the walk must start at one of the vertices of odd degree and end at the other. In the puzzle in the upper figure at the left, only two vertices are of odd degree—the two bottom corners. So we know that an Euler walk exists, and it must start from one of these two corners. The lower figure at the left shows one possible solution.

Of course, if every vertex is of even degree, we can solve the puzzle by finding an Euler circuit. If more than two vertices are of odd degree, then we know the puzzle cannot be solved.

Investigation Exercises

In Exercises 1–4, a pen-tracing puzzle is given. See if you can find a way to trace the shape without lifting your pen and without tracing over the same segment twice.

1.

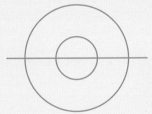

2.

3.

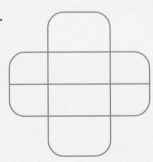

4.

5. Explain why the following pen-tracing puzzle is impossible to solve.

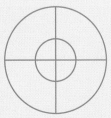

Exercise Set 10.1

In Exercises 1–4, determine (a) the number of edges in the graph, (b) the number of vertices in the graph, (c) the number of vertices that are of odd degree, (d) whether the graph is connected, and (e) whether the graph is a complete graph.

1.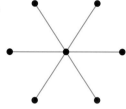

2.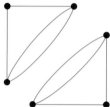

3.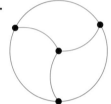

4.

5. **Transportation** An "X" in the table below indicates a direct train route between the corresponding cities. Draw a graph that represents this information, in which each vertex represents a city and an edge connects two vertices if there is a train route between the corresponding cities.

	Springfield	Riverside	Greenfield	Watertown	Midland	Newhope
Springfield	—		X		X	
Riverside		—		X	X	X
Greenfield	X		—	X	X	X
Watertown		X	X	—		
Midland	X	X	X		—	X
Newhope		X	X		X	—

6. **Transportation** The table below shows the nonstop flights offered by a small airline. Draw a graph that represents this information, where each vertex represents a city and an edge connects two vertices if there is a nonstop flight between the corresponding cities.

	Newport	Lancaster	Plymouth	Auburn	Dorset
Newport	—	No	Yes	No	Yes
Lancaster	No	—	Yes	Yes	No
Plymouth	Yes	Yes	—	Yes	Yes
Auburn	No	Yes	Yes	—	Yes
Dorset	Yes	No	Yes	Yes	—

Architecture In Exercises 7 and 8, a floor plan of a museum is shown. Draw a graph that represents the floor plan, where each vertex represents a room and an edge connects two vertices if there is a doorway between the two rooms.

7.

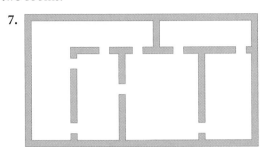

8.

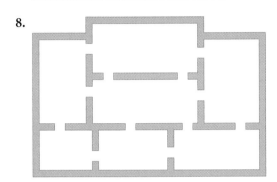

9. **Social Network** A group of friends is represented by the graph below. An edge connecting two names means that the two friends have spoken to each other in the last week.

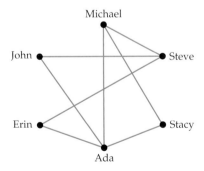

a. Have John and Stacy talked to each other in the last week?

b. How many of the friends in this group has Steve talked to in the last week?

c. Among this group of friends, who has talked to the most people in the last week?

d. Why would it not make sense for this graph to contain a loop?

10. **Baseball** The local Little League baseball teams are represented by the graph below. An edge connecting two teams means that those teams have played a game against each other this season.

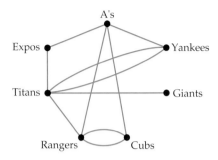

a. Which team has played only one game this season?

b. Which team has played the most games this season?

c. Have any teams played each other twice this season?

11. Are the following two graphs equivalent?

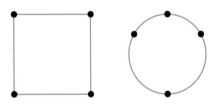

12. Explain why the following two graphs cannot be equivalent.

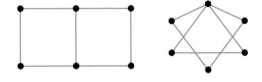

In Exercises 13–16, determine whether the two graphs are equivalent.

13.

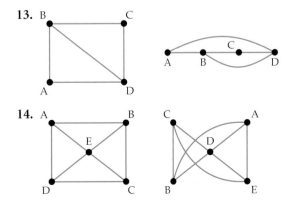

14.

15.

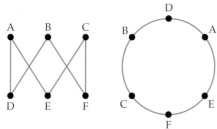

16.

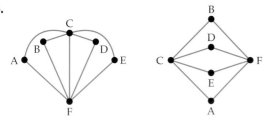

17. Label the vertices of the second graph so that it is equivalent to the first graph.

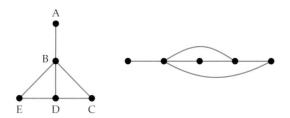

18. Label the vertices of the second graph so that it is equivalent to the first graph.

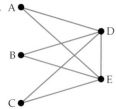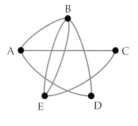

In Exercises 19–26, (a) determine whether the graph is Eulerian. If it is, find an Euler circuit. If it is not, explain why. (b) If the graph does not have an Euler circuit, does it have an Euler walk? If so, find one. If not, explain why.

19.

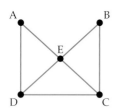

20.

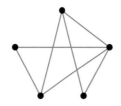

21.

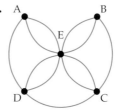

22.

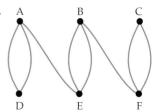

23.

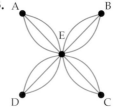

24.

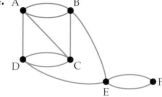

25.

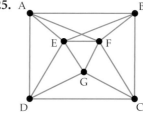

26.

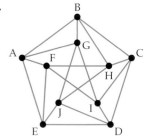

Parks In Exercises 27 and 28, a map of a park is shown with bridges connecting islands in a river to the riverbanks.

a. Represent the map as a graph. See Figures 10.3 and 10.4 on page 671.

b. Is it possible to take a walk that crosses each bridge once and returns to the starting point without crossing any bridge twice? If not, can you do it if you do not end at the starting point? Explain how you know.

27.

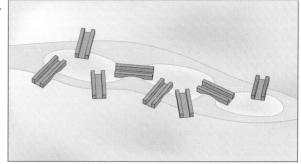

28.

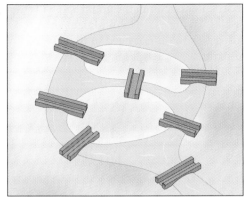

29. ✎ **Transportation** For the train routes given in Exercise 5, is it possible to travel along all of the train routes without traveling along any route twice? Explain how you reached your conclusion.

30. ✎ **Transportation** For the direct air flights given in Exercise 6, is it possible to start at one city and fly every route offered without repeating any flight if you return to the starting city? Explain how you reached your conclusion.

31. Pets The diagram below shows the arrangement of a Habitrail cage for a pet hamster. (Plastic tubes connect different cages.) Is it possible for a hamster to travel through every tube without going through the same tube twice? If so, find a route for the hamster to follow. Can the hamster return to its starting point without repeating any tube passages?

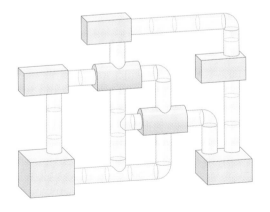

32. Transportation A subway map is shown at the top of the next column. Is it possible for a rider to travel the

length of every subway route without repeating any segments? Justify your conclusion.

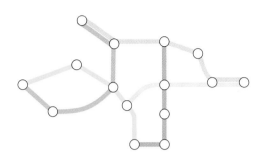

33. ✎ **Architecture** For the floor plan in Exercise 7, is it possible to walk through the museum and pass through each doorway without going through any doorway twice? Does it depend on whether you return to the room in which you started? Justify your conclusion.

34. ✎ **Architecture** For the floor plan in Exercise 8, is it possible to walk through the museum and pass through each doorway without going through any doorway twice? Does it depend on whether you return to your starting point? Justify your conclusion.

35. Explain how we know that the graph shown at the right does not contain an Euler circuit.

36. If a connected graph has six vertices and every vertex is of degree 4, must the graph contain a Hamiltonian circuit?

In Exercises 37–40, use Dirac's Theorem to verify that the graph is Hamiltonian. Then find a Hamiltonian circuit.

37.

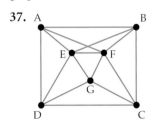

38.

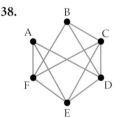

39.

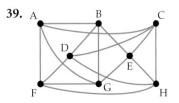

40.

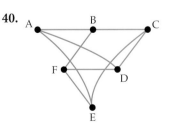

41. **Transportation** For the train routes given in Exercise 5, find a route that visits each city and returns to the starting city without visiting any city twice.

42. **Transportation** For the direct air flights given in Exercise 6, find a route that visits each city and returns to the starting city without visiting any city twice.

43. **Architecture** In Exercise 7, you were asked to draw a graph that represents a museum floor plan. Describe what a Hamiltonian circuit in the graph would correspond to in the museum.

44. **Transportation** Consider a subway map, like the one given in Exercise 32. If we draw a graph in which each vertex represents a train junction, and an edge between vertices means that a train travels between those two junctions, what does a Hamiltonian circuit correspond to in regard to the subway?

Extensions

CRITICAL THINKING

45. ✏️ **Route Planning** A security officer patrolling a city neighborhood needs to drive every street each night. The officer has drawn a graph representing the neighborhood, in which the edges represent the streets and the vertices correspond to street intersections. Would the most efficient way to drive the streets correspond to an Euler circuit, a Hamiltonian circuit, or neither? (The officer must return to the starting location when finished.) Explain your answer.

46. ✏️ **Route Planning** A city engineer needs to inspect the traffic signs at each street intersection of a neighborhood. The engineer has drawn a graph representing the neighborhood, where the edges represent the streets and the vertices correspond to street intersections. Would the most efficient route correspond to an Euler circuit, a Hamiltonian circuit, or neither? (The engineer must return to the starting location when finished.) Explain your answer.

47. Is there an Euler circuit in the graph below? Is there an Euler walk? Is there a Hamiltonian circuit? Justify your answer. (You do not need to find any of the circuits or paths.)

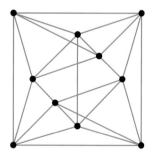

48. Is there an Euler circuit in the graph below? Is there an Euler walk? Is there a Hamiltonian circuit? Justify your answer. (You do not need to find any of the circuits or paths.)

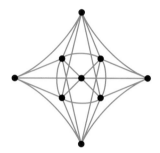

COOPERATIVE LEARNING

49. **a.** Draw a connected graph with six vertices that has no Euler circuit and no Hamiltonian circuit.

   b. Draw a graph with six vertices that has a Hamiltonian circuit but no Euler circuit.

   c. Draw a graph with five vertices that has an Euler circuit but no Hamiltonian circuit.

50. **Travel** A map of South America is shown below.

   a. Draw a graph in which the vertices represent the 13 countries of South America, and two vertices are joined by an edge if the corresponding countries share a common border.

   b. Two friends are planning a driving tour of South America. They would like to drive across every border on the continent. Is it possible to plan such a route that never crosses the same border twice? What would the route correspond to on the graph?

   c. Find a route the friends can follow that will start and end in Venezuela and that crosses every border while recrossing the fewest borders possible. *Hint:* On the graph, add multiple edges corresponding to border crossings that allow an Euler circuit.

| **Efficient Routes**

Weighted Graphs

In Section 10.1, we examined the problem of finding a path that enabled us to visit all of the cities represented on a graph, and we saw that in many cases there were several different paths we could use. What if we are concerned with the distances we must travel between cities? If there is more than one route that brings us through all of the cities, chances are some of the routes will involve a longer total distance than others. If we hope to minimize the number of miles we must travel in order to see the cities, we would be interested in finding the best route among all the possibilities.

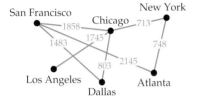

Figure 10.7

We can represent this situation with a *weighted graph*. A **weighted graph** is a graph in which each edge is associated with a value, called a **weight.** The value can represent any quantity we desire. In the case of distances between cities, we can label each edge with the number of miles between the corresponding cities. (Note that the length of an edge does not necessarily correlate to its weight.) For instance, the graph in Figure 10.7 is the graph of the airline flights shown in Figure 10.1, but with the weights of the edges added. We now know at a glance the distances between cities.

We previously looked for routes that visited each city and returned to the starting city—in other words, a Hamiltonian circuit. For each such circuit we find in a weighted graph, we can compute the total weight of the circuit by finding the sum of the weights of the edges we traverse. In this case, the sum gives us the total distance traveled along our route. We can then compare different routes and find the one that requires the least total distance. This is an example of a famous problem called the *traveling salesman problem.*

TAKE NOTE

Remember that a Hamiltonian circuit in a graph is a closed walk that uses each vertex of the graph but does not use any vertex twice.

EXAMPLE 1 ■ **Find Hamiltonian Circuits in a Weighted Graph**

The table below lists the distances in miles between six popular cities to which a particular airline flies. Suppose a traveler would like to start in Chicago, visit the other five cities this airline flies to, and return to Chicago. Find three different routes that the traveler could follow, and find the total distance flown during each route.

	Chicago	New York	Washington, D.C.	Philadelphia	Atlanta	Dallas
Chicago	—	713	597	665	585	803
New York	713	—	No flights	No flights	748	1374
Washington, D.C.	597	No flights	—	No flights	544	1185
Philadelphia	665	No flights	No flights	—	670	1299
Atlanta	585	748	544	670	—	No flights
Dallas	803	1374	1185	1299	No flights	—

Solution

The various options will be simpler to analyze if we first organize the information in a graph. As in Figure 10.7, we can let each city be represented by a vertex. We will draw an edge between two vertices if there is a flight between the corresponding cities, and we will label each edge with a weight that represents the number of miles between the two cities.

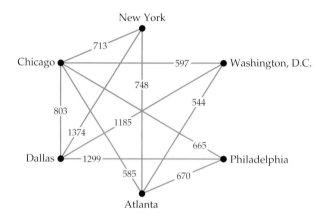

A route that visits each city just once corresponds to a Hamiltonian circuit. Beginning at Chicago, one such circuit is Chicago–New York–Dallas–Philadelphia–Atlanta–Washington, D.C.–Chicago. By adding the weights of each edge in the circuit, we see that the total number of miles traveled is

$$713 + 1374 + 1299 + 670 + 544 + 597 = 5197$$

By trial and error, we can identify two additional routes. One is Chicago–Philadelphia–Dallas–Washington, D.C.–Atlanta–New York–Chicago. The total weight of the circuit is

$$665 + 1299 + 1185 + 544 + 748 + 713 = 5154$$

A third route is Chicago–Washington, D.C.–Dallas–New York–Atlanta–Philadelphia–Chicago. The total mileage is

$$597 + 1185 + 1374 + 748 + 670 + 665 = 5239$$

CHECK YOUR PROGRESS 1   A tourist visiting San Francisco is staying at a hotel near the Moscone Center. The tourist would like to visit five locations by bus tomorrow and then return to the hotel. The numbers of minutes spent traveling by bus between locations are given in the table on the following page. (N/A in the table indicates that no convenient bus route is available.) Find two different routes for the tourist to follow and compare the total travel times.

	Moscone Center	Civic Center	Union Square	Embarcadero Plaza	Fisherman's Wharf	Coit Tower
Moscone Center	—	18	6	22	N/A	N/A
Civic Center	18	—	14	N/A	33	N/A
Union Square	6	14	—	24	28	36
Embarcadero Plaza	22	N/A	24	—	N/A	18
Fisherman's Wharf	N/A	33	28	N/A	—	14
Coit Tower	N/A	N/A	36	18	14	—

Solution   *See page S43.*

Algorithms in Complete Graphs

In Example 1, the second route we found represented the smallest total distance out of the three options. Is there a way we can find the very best route to take? It turns out that this is no easy task. One method is to list every possible Hamiltonian circuit, compute the total weight of each one, and choose the smallest total weight. Unfortunately, the number of different possible circuits can be extremely large. For instance, the graph shown in Figure 10.8, with only six vertices, has 60 unique Hamiltonian circuits. If we have a graph with 12 vertices, and every vertex is connected to every other by an edge, there are almost 20 million different Hamiltonian circuits! Even by using computers, it can take too long to investigate each and every possibility.

Unfortunately, there is no known short cut for finding the best route in a weighted graph. We do, however, have two *algorithms* that we can use to find a pretty good solution to this type of problem. Both of these algorithms apply only to **complete graphs**—graphs in which every possible edge is drawn between vertices (without any multiple edges). For instance, the graph in Figure 10.8 is a complete graph with six vertices. Each algorithm can be applied to find a Hamiltonian circuit. The total weight of this circuit may not be the smallest possible, but it is often smaller than you would find by trial and error.

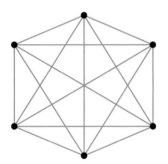

Figure 10.8

✔ **TAKE NOTE**

An **algorithm** is a step-by-step procedure for accomplishing a specific task.

The Greedy Algorithm

1. Choose a vertex to start at, then travel along the connected edge that has the smallest weight. (If two or more edges have the same weight, pick any one.)

2. After arriving at the next vertex, travel along the edge of smallest weight that connects to a vertex not yet visited. Continue this process until you have visited all vertices.

3. Return to the starting vertex.

EXAMPLE 2 ■ The Greedy Algorithm

Use the Greedy Algorithm to find a Hamiltonian circuit in the weighted graph shown in Figure 10.9. Start at vertex A.

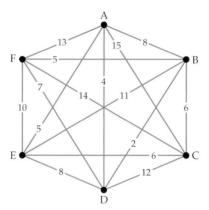

Figure 10.9

Solution

The edge from vertex A with the smallest weight is the edge to D, which has weight 4. From D, the edge with the smallest weight (besides the one we just traveled) has weight 2 and connects to B. The next edge, of weight 5, connects to vertex F. The edge from F of the least weight that has not yet been used is the edge to vertex D. However, we have already visited D, so we choose the edge with the next smallest weight, 10, to vertex E. The only remaining choice is the edge to vertex C, of weight 6, as all other vertices have been visited. Now we are at Step 3 of the algorithm, so we return to starting vertex A by traveling along the edge of weight 15. Thus our Hamiltonian circuit is A−D−B−F−E−C−A, with a total weight of

$$4 + 2 + 5 + 10 + 6 + 15 = 42$$

CHECK YOUR PROGRESS 2 Use the Greedy Algorithm to find a Hamiltonian circuit starting at vertex A in the weighted graph shown below.

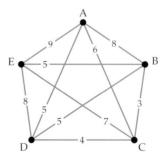

Solution See page S43.

The Greedy Algorithm is so called because it has us choose the "cheapest" option at every chance we get. The next algorithm uses a different strategy to find an efficient route in a traveling salesman problem.

The Edge-Picking Algorithm

1. Mark the edge of smallest weight in the graph. (If two or more edges have the same weight, pick any one.)

2. Mark the edge of next smallest weight in the graph, as long as it does not complete a circuit and does not add a third marked edge to a single vertex.

3. Continue this process until you can no longer mark any edges. Then mark the final edge that completes the Hamiltonian circuit.

In Step 2 of the algorithm, we are instructed not to complete a circuit too early. Also, because a Hamiltonian circuit will always have exactly two edges at each vertex, we are warned not to mark an edge that would allow three edges to meet at one vertex. You can see this algorithm in action in the next example.

EXAMPLE 3 ■ The Edge-Picking Algorithm

Use the Edge-Picking Algorithm to find a Hamiltonian circuit in Figure 10.9.

Solution
The smallest weight appearing in the graph is 2, so we highlight edge B–D of weight 2. The next smallest weight is 4, the weight of edge A–D.

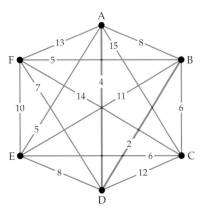

There are two edges of weight 5 (the next smallest weight), A–E and F–B, and we can highlight both of them. Next there are two edges of weight 6. However, we cannot use B–C because it would add a third marked edge to vertex B. We mark E–C.

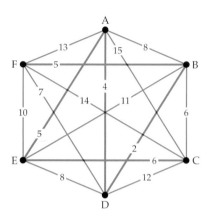

We are now at Step 3 of the algorithm; any edge we mark will either complete a circuit or add a third edge to a vertex. So we mark the final edge to complete the circuit, F–C.

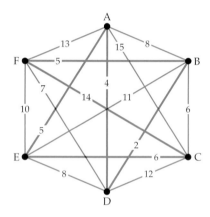

TAKE NOTE

A Hamiltonian circuit forms a complete loop, so we can follow along the circuit starting at any of the vertices. We can also reverse the direction in which we follow the circuit.

If we begin at vertex A, our Hamiltonian circuit is A–D–B–F–C–E–A. (In the reverse direction, an equivalent Hamiltonian circuit is A–E–C–F–B–D–A.) The total weight of the circuit is

$$4 + 2 + 5 + 14 + 6 + 5 = 36$$

CHECK YOUR PROGRESS 3   Use the Edge-Picking Algorithm to find a Hamiltonian circuit in the weighted graph in Check Your Progress 2.

Solution   *See page S43.*

Notice in Examples 2 and 3 that the two algorithms gave different Hamiltonian circuits, and in this case the Edge-Picking Algorithm gave the more efficient route. Is this the best route? We mentioned before that there is no known efficient method for finding the very best circuit. In fact, the Hamiltonian circuit A–D–F–B–C–E–A in Figure 10.9 has a total weight of 33, which is smaller than the weights of both routes given by the algorithms. So when we use the algorithms, we have no guarantee that we are finding the best route, but we can be confident that we are finding a reasonably good solution.

QUESTION *Can the Greedy Algorithm or the Edge-Picking Algorithm be used to identify a route to visit the cities in Example 1?*

Math Matters   Computers and Traveling Salesman Problems

The traveling salesman problem is a long-standing mathematical problem that has been analyzed since the 1920s by mathematicians, statisticians, and, later, computer scientists. The algorithms given in this section can help us find good solutions, but they cannot guarantee that we will find the best solution.

At this point in time, the only way to find the optimal Hamiltonian circuit in a weighted graph is to find each and every possible circuit and compare the total weights of all circuits. Computers are well suited to such a task; however, as the number of vertices increases, the number of possible Hamiltonian circuits increases rapidly, and even computers are not fast enough to handle large graphs. There are so many possible circuits in a graph with a large number of vertices that finding them all could take hundreds or thousands of years, on even the fastest computers.

Computer scientists continue to improve their methods and produce more sophisticated algorithms. In 2004, five researchers (David Applegate, Robert Bixby, Vasek Chvátal, William Cook, and Keld Helsgaun) were able to find the optimal route to visit all 24,978 cities in Sweden. To accomplish this task, they used a network of 96 dual-processor workstation computers. The amount of computing time required was equivalent to a single-processor computer running for 84.8 years. You can see their resulting circuit on the map.

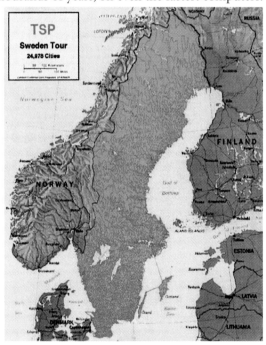

Applications of Weighted Graphs

In Example 1, we examined distances between cities. This is just one example of a weighted graph; the weight of an edge can be used to represent any quantity we like. For example, a traveler might be more interested in the cost of flights than the time

---

ANSWER   *No. Both algorithms apply only to complete graphs.*

or distance between cities. If we labeled each edge of the graph in Example 1 with the cost of traveling between the two cities, the total weight of a Hamiltonian circuit would be the total travel cost of the trip.

EXAMPLE 4 ■ **An Application of the Greedy and Edge-Picking Algorithms**

The costs of flying between various European cities are shown in the following table. Use both the Greedy Algorithm and the Edge-Picking Algorithm to find a low-cost route that visits each city just once and starts and ends in London. Which route is the most economical?

	London, England	Berlin, Germany	Paris, France	Rome, Italy	Madrid, Spain	Vienna, Austria
London, England	—	$325	$160	$280	$250	$425
Berlin, Germany	$325	—	$415	$550	$675	$375
Paris, France	$160	$415	—	$495	$215	$545
Rome, Italy	$280	$550	$495	—	$380	$480
Madrid, Spain	$250	$675	$215	$380	—	$730
Vienna, Austria	$425	$375	$545	$480	$730	—

Solution

First we draw a weighted graph with vertices representing the cities and each edge labeled with the price of the flight between the corresponding cities.

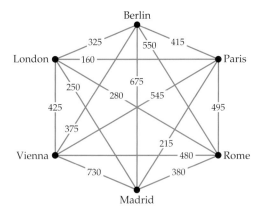

To use the Greedy Algorithm, start at London and travel along the edge with the smallest weight, 160, to Paris. The edge of smallest weight leaving Paris is the edge to Madrid. From Madrid, the edge of smallest weight (that we have not already traversed) is the edge to London, of weight 250. However, we cannot use this edge, because it would bring us to a city we have already seen. We can take the next-smallest-weight edge to Rome. We cannot yet return to London, so the next available edge is to Vienna, then to Berlin, and finally back to London.

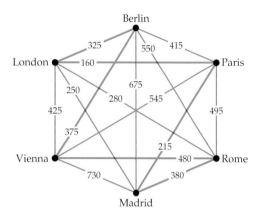

The total weight of the edges, and thus the total airfare for the trip, is

$$160 + 215 + 380 + 480 + 375 + 325 = \$1935$$

If we use the Edge-Picking Algorithm, the edges with the smallest weights that we can highlight are London–Paris and Madrid–Paris. The edge of next smallest weight has a weight of 250, but we cannot use this edge because it would complete a circuit. We can take the edge of next smallest weight, 280, from London to Rome. We cannot take the edge of next smallest weight, 325, because it would add a third edge to the London vertex, but we can take the edge Vienna–Berlin of weight 375. We must skip the edges of weights 380, 415, and 425, but we can take the edge of weight 480, which is the Vienna–Rome edge. There are no more edges we can mark that will meet the requirements of the algorithm, so we mark the last edge to complete the circuit, Berlin–Madrid.

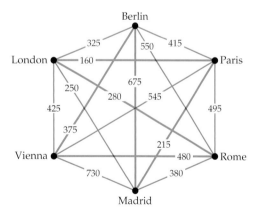

The resulting route is London–Paris–Madrid–Berlin–Vienna–Rome–London, for a total cost of

$$160 + 215 + 675 + 375 + 480 + 280 = \$2185$$

(We could also travel this route in the reverse order.)

CHECK YOUR PROGRESS 4 Susan needs to mail a package at the post office, pick up several items at the grocery store, return a rented video, and make a deposit at her bank. The estimated driving time, in minutes, between each of these locations is given in the table below.

	Home	Post Office	Grocery Store	Video Rental Store	Bank
Home	—	14	12	20	23
Post office	14	—	8	12	21
Grocery store	12	8	—	17	11
Video rental store	20	12	17	—	18
Bank	23	21	11	18	—

Use both of the algorithms from this section to design routes for Susan to follow that will help minimize her total driving time. Assume that she must start from home and return home when her errands are done.

Solution   *See page S43.*

A wide variety of problems are actually traveling salesman problems in disguise, and can be analyzed using the algorithms from this section. An example follows.

EXAMPLE 5 ▪ An Application of the Edge-Picking Algorithm

A toolmaker needs to use one machine to create four different tools. The toolmaker needs to make adjustments to the machine before starting each different tool. However, since the tools have parts in common, the amount of adjustment time required depends on which tool the machine was previously used to create. The table below lists the estimated times (in minutes) required to adjust the machine from one tool to another. The machine is currently configured for Tool A, and should be returned to that state when all the tools are finished.

	Tool A	Tool B	Tool C	Tool D
Tool A	—	25	6	32
Tool B	25	—	18	9
Tool C	6	18	—	15
Tool D	32	9	15	—

Use the Edge-Picking Algorithm to determine a sequence for creating the tools.

Solution

Draw a weighted graph in which each vertex represents a tool configuration of the machine, and the weight of each edge is the number of minutes required to adjust the machine from one tool to another.

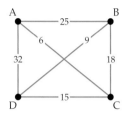

Using the Edge-Picking Algorithm, we first choose edge A–C, of weight 6. The next smallest weight is 9, or edge D–B, followed by edge D–C of weight 15. We end by completing the circuit with edge A–B. The final circuit, of total weight 25 + 9 + 15 + 6 = 55, is A–B–D–C–A. So the machine starts with tool A, is reconfigured for tool B, then for tool D, then for tool C, and finally is returned to the settings used for tool A. (Notice that we can equivalently follow this sequence in the reverse order: from Tool A to C, to D, to B, and back to A.)

CHECK YOUR PROGRESS 5   Businesses often network their various computers. One option is to run cables from a central hub to each computer individually; another is to connect one computer to the next, and that one to the next, and so on until you return to the first computer. Thus the computers are all connected in a large loop. Suppose a company wishes to use the latter method, and the lengths of cable (in feet) required between computers are given in the table below.

	Computer A	Computer B	Computer C	Computer D	Computer E	Computer F	Computer G
Computer A	—	43	25	6	28	30	45
Computer B	43	—	26	40	37	22	25
Computer C	25	26	—	20	52	8	50
Computer D	6	40	20	—	30	24	45
Computer E	28	37	52	30	—	49	20
Computer F	30	22	8	24	49	—	41
Computer G	45	25	50	45	20	41	—

Use the Edge-Picking Algorithm to determine how the computers should be networked if the business wishes to use the least amount of cable possible.

Solution   *See page S43.*

Investigation

Extending the Greedy Algorithm

When we create a Hamiltonian circuit in a graph, it is a closed loop. We can start at any vertex, follow the path, and arrive back at the starting vertex. For instance, if we use the Greedy Algorithm to create a Hamiltonian circuit starting at vertex A, we are actually creating a circuit that could start at any vertex in the circuit.

If we use the Greedy Algorithm and start from different vertices, will we always get the same result? Try it on Figure 10.10 at the left. If we start at vertex A, we get the circuit A–C–E–B–D–A, with a total weight of 26 (see Figure 10.11). However, if we start at vertex B, we get B–E–D–C–A–B, with a total weight of 18 (see Figure 10.12). Even though we found the second circuit by starting at B, we could use the same circuit starting at A, namely A–B–E–D–C–A, or, equivalently, A–C–D–E–B–A. This circuit has a smaller total weight than the one we found by starting at A.

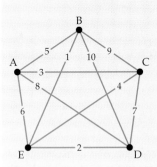

Figure 10.10

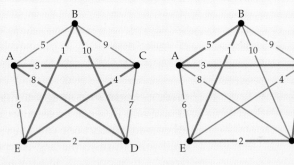

Figure 10.11　　　　　　**Figure 10.12**

In other words, it may pay to try the Greedy Algorithm starting from different vertices, even when we know that we actually want to start from a particular vertex. To be thorough, we can extend the Greedy Algorithm by using it at each and every vertex, generating several different circuits. We then choose from among these the circuit with the smallest total weight, and start at the vertex we want. The following exercises ask you to apply this approach for yourself.

Investigation Exercises

1. Continue investigating Hamiltonian circuits in Figure 10.10 by using the Greedy Algorithm starting at vertices C, D, and E. Then compare the various Hamiltonian circuits to identify the one with the smallest total weight.

2. Use the Greedy Algorithm and the weighted graph at the top of the following page to generate a Hamiltonian circuit starting from each vertex. Then compare the different circuits to find the one of smallest total weight.

(continued)

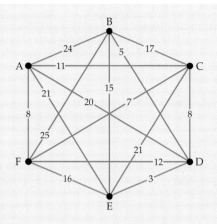

3. Use the Edge-Picking Algorithm to find a Hamiltonian circuit in the graph in Exercise 2. How does the weight of this circuit compare with the weights of the circuits found in Exercise 2?

Exercise Set 10.2

1. The weight of each edge in the graph below represents the cost of traveling by bus between the cities connected by the edge. Which is less expensive, traveling from Billings to Polson directly, or traveling from Billings to Great Falls and then to Polson?

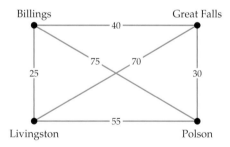

2. From the bus routes represented in the graph in Exercise 1, what would be the cheapest way to travel from Great Falls to Livingston?

In Exercises 3–6, use trial and error to find two Hamiltonian circuits of different total weights, starting at vertex A in the weighted graph. Compute the total weight of each circuit.

3.

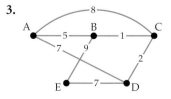

4.

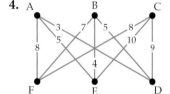

5.

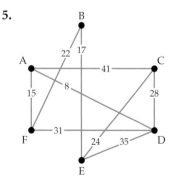

6.

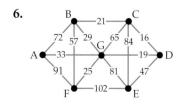

In Exercises 7–10, use the Greedy Algorithm to find a Hamiltonian circuit starting at vertex A in the weighted graph.

7.

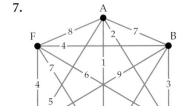

8.

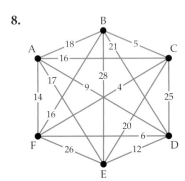

9.

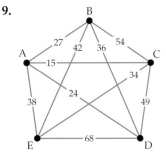

10.

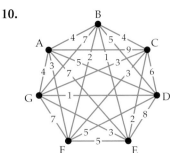

In Exercises 11–14, use the Edge-Picking Algorithm to find a Hamiltonian circuit in the indicated graph.

11. Graph in Exercise 7   **12.** Graph in Exercise 8

13. Graph in Exercise 9   **14.** Graph in Exercise 10

15. Travel A company representative lives in Louisville, Kentucky, and needs to visit offices in five different Indiana cities over the next few days. The representative wants to drive between cities and return to Louisville at the end of the trip. The estimated driving times, in hours, between cities are given in the table below. Represent the driving times by a weighted graph. Use the Greedy Algorithm to design an efficient route for the representative to follow.

	Louisville	Bloomington	Fort Wayne	Indianapolis	Lafayette	Evansville
Louisville	—	3.6	6.4	3.2	4.9	3.1
Bloomington	3.6	—	4.5	1.3	2.4	3.4
Fort Wayne	6.4	4.5	—	3.3	3.0	8.0
Indianapolis	3.2	1.3	3.3	—	1.5	4.6
Lafayette	4.9	2.4	3.0	1.5	—	5.0
Evansville	3.1	3.4	8.0	4.6	5.0	—

16. Travel A tourist is staying in Toronto, Canada, and would like to visit four other Canadian cities by train. The visitor wants to go from one city to the next and return to Toronto while minimizing the total travel distance. The distances between cities, in kilometers, are given in the table on the following page. Represent the distances between the cities using a weighted graph. Use the Greedy Algorithm to plan a route for the tourist.

	Toronto	Kingston	Niagara Falls	Ottawa	Windsor
Toronto	—	259	142	423	381
Kingston	259	—	397	174	623
Niagara Falls	142	397	—	562	402
Ottawa	423	174	562	—	787
Windsor	381	623	402	787	—

17. **Travel** Use the Edge-Picking Algorithm to design a route for the company representative in Exercise 15.

18. **Travel** Use the Edge-Picking Algorithm to design a route for the tourist in Exercise 16.

19. **Travel** Nicole wants to tour Asia. She will start and end her journey in Tokyo and visit Hong Kong, Bangkok, Seoul, and Beijing. The airfares available to her between cities are given in the table. Draw a weighted graph that represents the travel costs between cities and use the Greedy Algorithm to find a low-cost route.

	Tokyo	Hong Kong	Bangkok	Seoul	Beijing
Tokyo	—	$845	$1275	$470	$880
Hong Kong	$845	—	$320	$515	$340
Bangkok	$1275	$320	—	$520	$365
Seoul	$470	$515	$520	—	$225
Beijing	$880	$340	$365	$225	—

20. **Travel** The prices for traveling between five cities in Colorado by bus are given in the table below. Represent the travel costs between cities using a weighted graph. Use the Greedy Algorithm to find a low-cost route that starts and ends in Boulder and visits each city.

	Boulder	Denver	Colorado Springs	Grand Junction	Durango
Boulder	—	$16	$25	$49	$74
Denver	$16	—	$22	$45	$72
Colorado Springs	$25	$22	—	$58	$59
Grand Junction	$49	$45	$58	—	$32
Durango	$74	$72	$59	$32	—

21. **Travel** Use the Edge-Picking Algorithm to find a low-cost route for the traveler in Exercise 19.

22. **Travel** Use the Edge-Picking Algorithm to find a low-cost bus route in Exercise 20.

23. Route Planning Brian needs to visit the pet store, the shopping mall, the local farmers' market, and the pharmacy. His estimated driving times (in minutes) between the locations are given in the table below. Use the Greedy Algorithm and the Edge-Picking Algorithm to find two possible routes, starting and ending at home, that will help Brian minimize his total travel time.

	Home	Pet Store	Shopping Mall	Farmers' Market	Pharmacy
Home	—	18	27	15	8
Pet store	18	—	24	22	10
Shopping mall	27	24	—	20	32
Farmers' market	15	22	20	—	22
Pharmacy	8	10	32	22	—

24. Route Planning A bike messenger needs to deliver packages to five different buildings and return to the courier company. The estimated biking times (in minutes) between the buildings are given in the table below. Use the Greedy Algorithm and the Edge-Picking Algorithm to find two possible routes for the messenger to follow that will help minimize the total travel time.

	Courier Company	Prudential Building	Bank of America Building	Imperial Bank Building	GE Tower	Design Center
Courier company	—	10	8	15	12	17
Prudential building	10	—	10	6	9	8
Bank of America building	8	10	—	7	18	20
Imperial Bank building	15	6	7	—	22	16
GE Tower	12	9	18	22	—	5
Design Center	17	8	20	16	5	—

25. Scheduling A research company has a large supercomputer that is used by different teams for a variety of computational tasks. In between each task, the software must be reconfigured. The time required depends on which tasks follow which, because some settings are shared by different tasks. The times (in minutes) required to reconfigure the machine from one task to another are given in the table on the following page. Use the Greedy Algorithm and the Edge-Picking Algorithm to find time-efficient sequences in which to assign the tasks to the computer. The software configuration must start and end in the home state.

	Home State	Task A	Task B	Task C	Task D
Home state	—	35	15	40	27
Task A	35	—	30	18	25
Task B	15	30	—	35	16
Task C	40	18	35	—	32
Task D	27	25	16	32	—

26. **Computer Networks** A small office wishes to network its six computers in one large loop (see Check Your Progress 5). The lengths of cable, in meters, required between machines are given in the table below. Use the Edge-Picking Algorithm to find an efficient cable configuration in which to network the computers.

	Computer A	Computer B	Computer C	Computer D	Computer E	Computer F
Computer A	—	10	22	9	15	8
Computer B	10	—	12	14	16	5
Computer C	22	12	—	14	9	16
Computer D	9	14	14	—	7	15
Computer E	15	16	9	7	—	13
Computer F	8	5	16	15	13	—

Extensions
CRITICAL THINKING

27. Assign weights to the edges of the complete graph shown below so that the Edge-Picking Algorithm gives a circuit of lower total weight than the circuit given by the Greedy Algorithm. For the Greedy Algorithm, begin at vertex A.

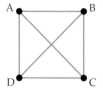

28. Assign weights to the edges of the complete graph in Exercise 27 so that the Greedy Algorithm gives a circuit of lower total weight than the circuit given by the Edge-Picking Algorithm. For the Greedy Algorithm, begin at vertex A.

29. Assign weights to the edges of the complete graph in Exercise 27 so that there is a circuit of lower total weight than the circuits given by the Greedy Algorithm (beginning at vertex A) and the Edge-Picking Algorithm.

30. Route Planning Form a team of five classmates and, using a map, identify a driving route that will visit each home, return to the starting home, and that you believe will result in the least total driving distance. Next determine the driving distances between the homes of the classmates. Represent these distances in a weighted graph, and then use the Edge-Picking Algorithm to determine a driving route that will visit each home and return to the starting home. Does the algorithm find a more efficient route than your first route?

SECTION 10.3 | **Map Coloring and Graphs**

Figure 10.13

Coloring Maps

A map of South America is shown in Figure 10.13. Notice that each country is colored so that no two bordering countries are the same color. This is easy to accomplish with a large number of colors, but what is the minimum number of colors we would need to color the countries in such a way?

It was conjectured that four colors would always be enough, because no one had ever found a map that could not be colored as described with four or fewer colors. Many mathematicians attempted to prove that this would always be the case, but they were unsuccessful for many years. Finally, in 1976, the so-called *Four-Color Theorem* was proved.

EXAMPLE 1 ■ **Coloring a Map**

Color the map of South America in Figure 10.13 using only four colors such that no two neighboring countries are the same color.

Solution
We do not have a systematic way to go about coloring the countries, so we must use trial and error. One possible coloring scheme is shown below.

CHECK YOUR PROGRESS 1   A map of the nine provinces of South Africa is shown below. Color the map using only three colors such that no two neighboring provinces are the same color.

Solution   *See page S44.*

Maps Become Graphs

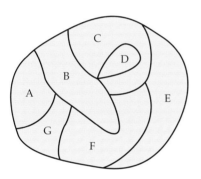

Figure 10.14

What does coloring maps have to do with graph theory? In their attempts to prove the Four-Color Theorem, mathematicians first converted the map-coloring question into a graph theory question. We will use a simple hypothetical map to illustrate.

Suppose the map in Figure 10.14 shows the countries, labeled as letters, of a continent. We will assume that no country is split into more than one piece, and countries that touch just at a corner point will not be considered neighbors. We can represent each country by a vertex, placed anywhere within the boundary of that country. We will then connect two vertices with an edge if the two corresponding countries are neighbors—that is, if they share a common boundary. The result is shown in Figure 10.15.

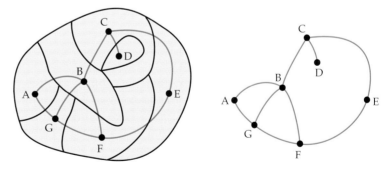

Figure 10.15   **Figure 10.16**

If we erase the boundaries of the countries, we are left with the graph in Figure 10.16. Our map-coloring question then becomes: Can we give each vertex of the graph in Figure 10.16 a color such that no two vertices connected by an edge share the same color? How many different colors will be required? If this can be accomplished using four colors, for instance, we will say that the graph is **4-colorable.** In

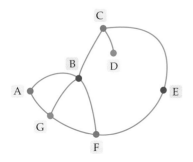

Figure 10.17

fact, the graph in Figure 10.16 is actually *3-colorable*. One possible coloring is given in Figure 10.17.

The graph in Figure 10.16 is an example of a *planar graph*. A **planar graph** is a graph that can be drawn so that no edges intersect each other (except at vertices). The process we just utilized to convert a map to a graph always results in a planar graph. As the following theorem states, four colors will always be enough to color a map in the desired method.

Four-Color Theorem

Every planar graph is 4-colorable.

QUESTION   *The graph shown at the right requires five colors if we wish to color it such that no edge joins two vertices of the same color. Does this contradict the Four-Color Theorem?*

EXAMPLE 2 ■ Represent a Map as a Graph

The map below shows the boundaries of countries on a fictional rectangular continent. Represent the map as a graph, and then find a coloring of the graph that uses the fewest possible colors.

Solution

First draw a vertex in each country and then connect two vertices with an edge if the corresponding countries are neighbors. Now try to color the vertices of the resulting graph so that no edge connects two vertices of the same color. We know we will need at least two colors, so one strategy is simply to pick a starting vertex, give it a color, and then assign colors to the connected vertices one by one. Try to reuse the same colors, and use a new color only when there is no other option. For this graph we will need four colors. (The Four-Color Theorem guarantees that we will not need more than that.) To see why we will need four colors, notice that the one vertex colored green in the graph on the following page connects to a ring of five

---

ANSWER   *No. This graph is not a planar graph, so the Four-Color Theorem does not apply.*

vertices. Three different colors are required to color the five-vertex ring, and the green vertex connects to all these, so it requires a fourth color.

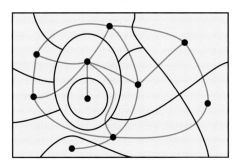

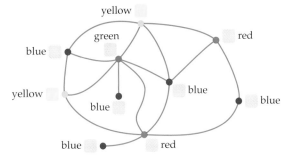

CHECK YOUR PROGRESS 2 Represent the map of fictional countries below as a graph, and then determine whether the graph is 2-colorable, 3-colorable, or 4-colorable by finding a suitable coloring of the graph.

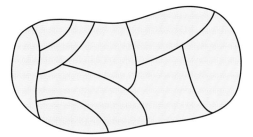

Solution   *See page S44.*

Math Matters   Proving the Four-Color Theorem

The Four-Color Theorem can be stated in a simple, short sentence, but proving it is anything but simple. The theorem was finally proved in 1976 by Wolfgang Haken and Kenneth Appel, two mathematicians at the University of Illinois. Mathematicians had long hunted for a short, elegant proof, but it turned out that the proof had to wait for the advent of computers to help sift through the many possible arrangements that can occur. The final proof used more than 1000 hours of computer time (surely it would be less now) and on paper came to several hundred pages that included some 10,000 diagrams.

The Chromatic Number of a Graph

We mentioned previously that representing a map as a graph always results in a planar graph. The Four-Color Theorem guarantees that we need only four colors to color a *planar* graph; however, if we wish to color a nonplanar graph, we may need quite a few more than four colors. The minimum number of colors needed to color a graph so that no edge connects vertices of the same color is called the **chromatic number** of the graph. In general, there is no efficient method of finding the chromatic number of a graph, but we do have a theorem that can tell us whether a graph is 2-colorable.

2-Colorable Graph Theorem

A graph is 2-colorable if and only if it has no circuits that consist of an odd number of vertices.

EXAMPLE 3 ■ Determine the Chromatic Number of a Graph

Find the chromatic number of the graph below.

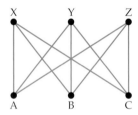

Solution

Notice that the graph contains circuits such as A−Y−C−Z−B−X−A with six vertices and A−Y−B−X−A with four vertices. Any circuit we find, in fact, seems to involve an even number of vertices. It is difficult to determine whether we have looked at all possible circuits, but our observations suggest that the graph may be 2-colorable. A little trial and error confirms this if we simply color vertices A, B, and C one color and the remaining vertices another color. Thus the graph has a chromatic number of 2.

CHECK YOUR PROGRESS 3   Determine whether the following graph is 2-colorable.

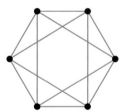

Solution   *See page S44.*

Applications of Graph Coloring

Determining the chromatic number of a graph and finding a corresponding coloring of the graph can solve a wide assortment of practical problems. One common application is in scheduling meetings or events. This is best shown by example.

EXAMPLE 4 ■ A Scheduling Application of Graph Coloring

Eight different school clubs want to schedule meetings on the last day of the semester. Some club members, however, belong to more than one of these clubs, so clubs that share members cannot meet at the same time. How many different time slots are required so that all members can attend all meetings? Clubs that have a member in common are indicated with an "X" in the table below.

	Ski Club	Student Government	Debate Club	Honor Society	Student Newspaper	Community Outreach	Campus Democrats	Campus Republicans
Ski club	—	X		X			X	X
Student government	X	—	X	X	X			
Debate club		X	—	X		X		X
Honor society	X	X	X	—	X	X		
Student newspaper		X		X	—	X	X	
Community outreach			X	X	X	—	X	X
Campus Democrats	X				X	X	—	
Campus Republicans	X		X			X		—

Solution

We can represent the given information by a graph. Each club can be represented by a vertex, and an edge will connect two vertices if the corresponding clubs have at least one common member.

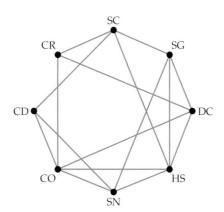

We now must choose time slots for the meetings so that two clubs that share members do not meet at the same time. In the graph, this means that two clubs that are connected by an edge cannot meet simultaneously. If we let a color correspond to a time slot, then we simply need to find a coloring of the graph that uses the fewest possible colors. The graph is not 2-colorable, because we can find circuits of odd length. However, by trial and error, we can find a 3-coloring. One example is shown below. So the chromatic number of the graph is 3, which means we need only three different time slots. Red vertices correspond to the clubs that will meet during the first time slot, green vertices to the clubs that will meet during the second time slot, and blue vertices to the clubs that will meet during the third time slot.

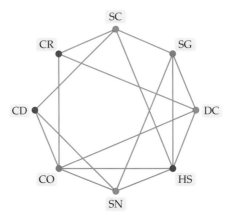

CHECK YOUR PROGRESS 4   Six friends are taking a film history course and, because they have procrastinated, need to view several films the night before the final exam. They have rented a copy of each film on DVD, and they have a total of three DVD players in different dorm rooms. If each film is 2 hours long and they start watching at 8:00 P.M., how soon can they all be finished watching the required films? Create a viewing schedule for the friends.

> Film A needs to be viewed by Brian, Chris, and Damon.
>
> Film B needs to be viewed by Allison and Fernando.
>
> Film C needs to be viewed by Damon, Erin, and Fernando.
>
> Film D needs to be viewed by Brian and Erin.
>
> Film E needs to be viewed by Brian, Chris, and Erin.

Solution   See page S44.

EXAMPLE 5 ■ **A Scheduling Application of Graph Coloring**

Five classes at an elementary school have arranged a tour at a zoo where the students get to feed the animals.

> Class 1 wants to feed the elephants, giraffes, and hippos.
>
> Class 2 wants to feed the monkeys, rhinos, and elephants.
>
> Class 3 wants to feed the monkeys, deer, and sea lions.

Class 4 wants to feed the parrots, giraffes, and polar bears.

Class 5 wants to feed the sea lions, hippos, and polar bears.

If the zoo allows animals to be fed only once a day by one class of students, can the tour be accomplished in two days? (Assume that each class will visit the zoo on only one day.) If not, how many days will be required?

Solution

No animal is listed more than twice in the tour list, so you may be tempted to say that only two days will be required. However, to get a better picture of the problem, we can represent the situation with a graph. Use a vertex to represent each class, and connect two vertices with an edge if the corresponding classes want to feed the same animal. Then we can try to find a 2-coloring of the graph, where a different color represents a different day at the zoo.

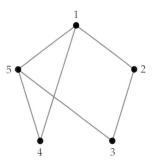

Notice that the graph contains a circuit, $1-4-5-1$, consisting of three vertices. This circuit will require three colors, but the remaining vertices will not need more than that. So the chromatic number of the graph is 3; one possible coloring is given below. Using this coloring, three days are required at the zoo. On the first day classes 2 and 5, represented by the blue vertices, will visit the zoo; on the second day classes 1 and 3, represented by the red vertices, will visit the zoo; and on the third day class 4, represented by the green vertex, will visit the zoo.

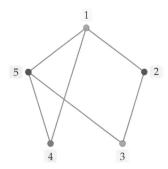

CHECK YOUR PROGRESS 5 Several delis in New York City have arranged deliveries to various buildings at lunchtime. The buildings' managements do not want more than one deli showing up at a particular building in one day, but the delis would like to deliver as often as possible. If they decide to agree on a delivery schedule, how many days will be required before each deli can return to the same building?

> Deli A delivers to the Empire State Building, the Statue of Liberty, and Rockefeller Center.
>
> Deli B delivers to the Chrysler Building, the Empire State Building, and the New York Stock Exchange.
>
> Deli C delivers to the New York Stock Exchange, the American Stock Exchange, and the United Nations Building.
>
> Deli D delivers to New York City Hall, the Chrysler Building, and Rockefeller Center.
>
> Deli E delivers to Rockefeller Center, New York City Hall, and the United Nations Building.

Solution *See page S44.*

Investigation

Modeling Traffic Lights with Graphs

Have you ever watched the cycles that traffic lights go through while you were waiting for a red light to turn green? Some intersections have lights that go through several stages, to allow all the different lanes of traffic to proceed safely.

Ideally, each stage of a traffic light cycle should allow as many lanes of traffic to proceed through the intersection as possible. We can design a traffic light cycle by modeling an intersection with a graph. Figure 10.18 shows a three-way intersection where two two-way roads meet. Each direction of traffic has turn lanes, with left-turn lights. There are six different directions in which vehicles can travel, as indicated in the figure, and we have labeled each possibility with a letter.

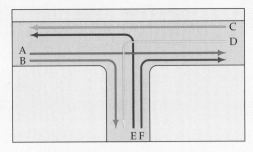

Figure 10.18

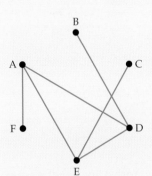

Figure 10.19

We can represent the traffic patterns with a graph; each vertex will represent one of the six possible traffic paths, and we will draw an edge between two vertices if the corresponding paths would allow vehicles to collide. The result is the graph shown in Figure 10.19. Because we do not want to allow vehicles to travel simultaneously along routes on which they could collide, any vertices connected by an edge can allow traffic

(continued)

to move only during different parts of the light cycle. We can represent each portion of the cycle by a color. Our job then is to color the graph using the fewest colors possible.

There is no 2-coloring of the graph because we have a circuit of length 3: A–D–E–A. We can, however, find a 3-coloring. One possibility is given in Figure 10.20.

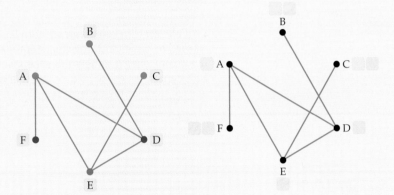

Figure 10.20   **Figure 10.21**

A 3-coloring of the graph means that the traffic lights at the intersection will have to go through a three-stage cycle. One stage will allow the traffic routes corresponding to the red vertices to proceed, the next stage will let the paths corresponding to the blue vertices proceed, and finally, the third stage will let path E, colored green, proceed.

Although safety requires three stages for the lights, we can refine the design to allow more traffic to travel through the intersection. Notice that at the third stage, only one route, path E, is scheduled to be moving. However, there is no harm in allowing path B to move at the same time, since it is a right turn that doesn't conflict with route E. We could also allow path F to proceed at the same time. Adding these additional paths corresponds to adding colors to the graph in Figure 10.20. We do not want to use more than three colors, but we can add a second color to some of the vertices while maintaining the requirement that no edge can connect two vertices of the same color. The result is shown in Figure 10.21. Notice that the vertices in the triangular circuit A–D–E–A can be assigned only a single color, but the remaining vertices can handle two colors.

In summary, our design allows traffic paths A, B, and C to proceed during one stage of the cycle, paths C, D, and F during another, and paths B, E, and F during the third stage.

Investigation Exercises

1. A one-way road ends at a two-way street. The intersection and the different possible traffic routes are shown in the accompanying figure. The one-way road has a left-turn light. Represent the traffic routes with a graph and use graph coloring to determine the minimum number of stages required for a light cycle.

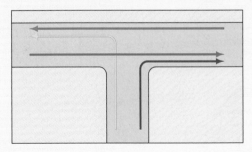

(continued)

2. A one-way road intersects a two-way road in a four-way intersection. Each direction has turn lanes and left-turn lights. Represent the various traffic routes with a graph and use graph coloring to determine the minimum number of stages required for a light cycle. Then refine your design to allow as much traffic as possible to proceed at each stage of the cycle.

3. A two-way road intersects another two-way road in a four-way intersection. One road has left-turn lanes with left-turn lights, but on the other road cars are not allowed to make left turns. Represent the various traffic routes with a graph and use graph coloring to determine the minimum number of stages required for a light cycle. Then refine your design to allow as much traffic as possible to proceed at each stage of the cycle.

Exercise Set 10.3

Map Coloring In Exercises 1–4, a map of the countries of a fictional continent is given. Use four colors to color the graph so that no two bordering countries share the same color.

1.

2.

3.

4.

Map Coloring In Exercises 5–10, represent the map by a graph and then find a coloring of the graph that uses the fewest possible colors.

5. The map in Exercise 1

6. The map in Exercise 2

7. The map in Exercise 3

8. The map in Exercise 4

9. Western portion of the United States

10. Counties of New Hampshire

In Exercises 11–16, show that the graph is 2-colorable by finding a 2-coloring. If the graph is not 2-colorable, explain why.

11.

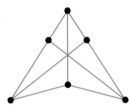

12.

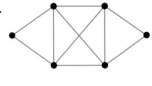

13.

14.

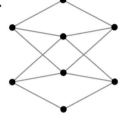

15.

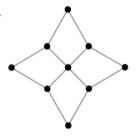

16.

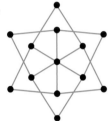

17. Can the vertices of the graph be colored with only two colors so that no edge connects vertices of the same color?

18. Can a complete graph (in which every pair of vertices is connected by an edge) be 2-colorable?

19. What is the chromatic number of the graph?

20. What is the chromatic number of the graph?

In Exercises 21–26, determine (by trial and error) the chromatic number of the graph.

21.

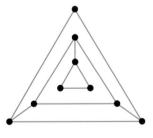

22.

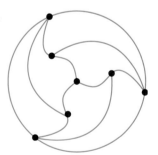

23.

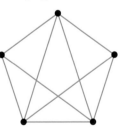

24.

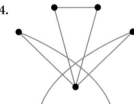

25.

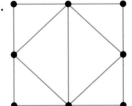

26.

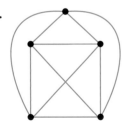

27. **Scheduling** Six student clubs need to hold meetings on the same day, but some students belong to more than one club. In order to avoid members missing meetings, the meetings need to be scheduled during different time slots. An "X" in the table below indicates that the two corresponding clubs share at least one member. Use graph coloring to determine the minimum number of time slots necessary to ensure that all club members can attend all meetings.

	Student Newspaper	Honor Society	Biology Assoc.	Gaming Club	Debate Team	Engineering Club
Student newspaper	—		X		X	
Honor society		—	X		X	X
Biology assoc.	X	X	—	X		
Gaming club			X	—	X	X
Debate team	X	X		X	—	
Engineering club		X		X		—

28. **Scheduling** Eight political committees must meet on the same day, but some members are on more than one committee. Thus any committees that have members in common cannot meet at the same time. An "X" in the table below indicates that the two corresponding committees share a member. Use graph coloring to determine the minimum number of meeting times that will be necessary so that all members can attend the appropriate meetings.

	Appropriations	Budget	Finance	Judiciary	Education	Health	Foreign Affairs	Housing
Appropriations	—		X			X	X	
Budget		—		X		X		
Finance	X		—	X			X	X
Judiciary		X	X	—		X		X
Education					—		X	X
Health	X	X		X		—		
Foreign affairs	X		X		X		—	
Housing			X	X	X			—

29. **Scheduling** Six different groups of children would like to visit the zoo and feed different animals. (Assume each group will visit the zoo on only one day.)

Group 1 would like to feed the bears, dolphins, and gorillas.

Group 2 would like to feed the bears, elephants, and hippos.

Group 3 would like to feed the dolphins and elephants.

Group 4 would like to feed the dolphins, zebras, and hippos.

Group 5 would like to feed the bears and hippos.

Group 6 would like to feed the gorillas, hippos, and zebras.

Use graph coloring to find the minimum number of days that are required so that all groups can feed the animals they would like to feed but no animals will be fed twice on the same day. Design a schedule to accomplish this goal.

30. **Scheduling** Five different charity organizations send trucks on various routes to pick up donations that residents leave on their doorsteps.

> Charity A covers Main St., First Ave., and State St.
>
> Charity B covers First Ave., Second Ave., and Third Ave.
>
> Charity C covers State St., City Dr., and Country Lane.
>
> Charity D covers City Dr., Second Ave., and Main St.
>
> Charity E covers Third Ave., Country Lane, and Fourth Ave.

Each charity has its truck travel down all three streets on its route on the same day, but no two charities wish to visit the same streets on the same day. Use graph coloring to design a schedule for the charities. Arrange their pickup routes so that no street is visited twice on the same day by different charities. The schedule should use the smallest possible number of days.

31. **Scheduling** Students in a film class have volunteered to form groups and create several short films. The class has three digital video cameras that may be checked out for one day only, and it is expected that each group will need the entire day to finish shooting. All members of each group must participate in the film they volunteered for, so a student cannot work on more than one film on any given day.

> Film 1 will be made by Brian, Angela, and Kate.
>
> Film 2 will be made by Jessica, Vince, and Brian.
>
> Film 3 will be made by Corey, Brian, and Vince.
>
> Film 4 will be made by Ricardo, Sarah, and Lupe.
>
> Film 5 will be made by Sarah, Kate, and Jessica.
>
> Film 6 will be made by Angela, Corey, and Lupe.

Use graph coloring to design a schedule for lending the cameras, using the fewest possible days, so that each group can shoot its film and all members can participate.

Extensions

CRITICAL THINKING

Maps In Exercises 32–34, draw a map of a fictional continent with country boundaries corresponding to the given graph.

32.

33.

34.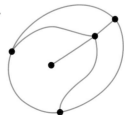

35. Map Coloring Draw a map of a fictional continent consisting of four countries, for which the map cannot be colored with three or fewer colors without adjacent countries sharing a color.

36. If the chromatic number of a graph with five vertices is 1, what must the graph look like?

EXPLORATIONS

37. Edge Coloring In this section, we colored vertices of graphs so that no edge connected two vertices of the same color. We can also consider coloring edges, rather than vertices, so that no vertex connects two or more edges of the same color. In parts a to d, assign each edge in the graph a color so that no vertex connects two or more edges of the same color. Use the fewest colors possible.

a.  b. 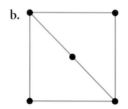 c. 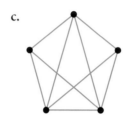 d.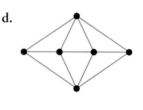

  e. Explain why the number of colors required will always be at least the number of edges that meet at the vertex of highest degree in the graph.

38. Scheduling Edge colorings, as explained in Exercise 37, can be used to solve scheduling problems. For instance, suppose five players are competing in a tennis tournament. Each player needs to play every other player in a match (but not more than once). Each player will participate in no more than one match per day, and two matches can occur at the same time when possible. How many days will be required for the tournament? Represent the tournament as a graph, in which each vertex corresponds to a player and an edge joins two vertices if the corresponding players will compete against each other in a match. Next color the edges, where each different color corresponds to a different day of the tournament. Because one player will not be in more than one match per day, no two edges of the same color can meet at the same vertex. If we can find an edge coloring of the graph that uses the minimum number of colors possible, it will correspond to the minimum number of days required for the tournament. Sketch a graph that represents the tournament, find an edge coloring that uses the fewest colors possible, and use your graph to design a schedule of matches for the tournament that minimizes the number of days required.

CHAPTER 10   # Summary

Key Terms

4-colorable [p. 703]
algorithm [p. 687]
chromatic number [p. 706]
circuit or closed walk [p. 671]
complete graph [p. 668]
connected graph [p. 668]
degree [p. 671]
edge [p. 667]
equivalent graphs [p. 668]
Euler circuit [p. 671]
Euler walk [p. 674]
Eulerian graph [p. 671]
graph [p. 667]
Hamiltonian circuit [p. 675]
Hamiltonian graph [p. 675]
multiple edges [p. 668]
null graph [p. 668]
planar graph [p. 704]
traveling salesman problem [p. 685]
vertex [p. 667]
walk [p. 671]
weight [p. 685]
weighted graph [p. 685]

Essential Concepts

- **Eulerian Graph Theorem**
  A connected graph is *Eulerian* if and only if every vertex of the graph is of even degree.

- **Euler Walk Theorem**
  A connected graph contains an Euler walk if and only if the graph has two vertices of odd degree with all other vertices of even degree. Furthermore, every Euler walk must start at one of the vertices of odd degree and end at the other.

- **Hamiltonian Graphs and Dirac's Theorem**
  Dirac's Theorem states that in a connected graph with at least three vertices and with no multiple edges, if n is the number of vertices in the graph and every vertex has a degree of at least $n/2$, then the graph must be Hamiltonian. If it is not the case that every vertex has a degree of at least $n/2$, then the graph may or may not be Hamiltonian.

- **The Greedy Algorithm**
  One method of finding a Hamiltonian circuit in a complete weighted graph is given by the following procedure.

  1. Choose a vertex to start at, then travel along the connected edge that has the smallest weight. (If two or more edges have the same weight, pick any one.)

  2. After arriving at the next vertex, travel along the edge of smallest weight that connects to a vertex not yet visited. Continue this process until you have visited all vertices.

  3. Return to the starting vertex.

  The Greedy Algorithm attempts to find a circuit of minimal total weight, although it does not always succeed.

- **The Edge-Picking Algorithm**
  A second method of finding a Hamiltonian circuit in a complete weighted graph consists of the following steps.

  1. Mark the edge of smallest weight in the graph. (If two or more edges have the same weight, pick any one.)

  2. Mark the edge of next smallest weight in the graph, as long as it does not complete a circuit and does not add a third marked edge to a single vertex.

  3. Continue this process until you can no longer mark any edges. Then mark the final edge that completes the Hamiltonian circuit.

- **The Four-Color Theorem**
  Every planar graph is 4-colorable. (Note that fewer than four colors may be sufficient; if the graph is not planar, more than four colors may be necessary.)

- **Representing Maps as Graphs**
  Draw a vertex in each region (country, state, etc.) of the map. Connect two vertices if the corresponding regions share a common border.

- **2-Colorable Graph Theorem**
  A graph is 2-colorable if and only if it has no circuits that consist of an odd number of vertices.

| CHAPTER 10 | **Review Exercises** |

In Exercises 1 and 2, (a) determine the number of edges in the graph, (b) find the number of vertices in the graph, (c) list the degree of each vertex, and (d) determine whether the graph is connected.

1.

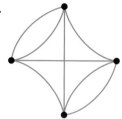

2.

3. **Soccer** In the table below, an "X" indicates teams from a junior soccer league that have played each other in the current season. Draw a graph representing the games by using a vertex to represent each team. Connect two vertices with an edge if the corresponding teams played a game against each other this season.

	Mariners	**Scorpions**	**Pumas**	**Stingrays**	**Vipers**
Mariners	—	X		X	X
Scorpions	X	—	X		
Pumas		X	—	X	X
Stingrays	X		X	—	
Vipers	X		X		—

4. Each vertex in the graph at the right represents a freeway in the Los Angeles area. An edge connects two vertices if the corresponding freeways have interchanges allowing drivers to transfer from one freeway to the other.

   a. Can drivers transfer from the 105 freeway to the 10 freeway?

   b. Among the freeways represented, how many have interchanges with the 5 freeway?

   c. Which freeways have interchanges to all the other freeways in the graph?

   d. Of the freeways represented in the graph, which has the fewest interchanges to the other freeways?

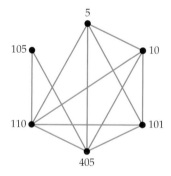

In Exercises 5 and 6, determine whether the two graphs are equivalent.

5.

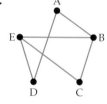

6.

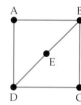

In Exercises 7–10, (a) find an Euler walk if possible, and (b) find an Euler circuit if possible.

7.
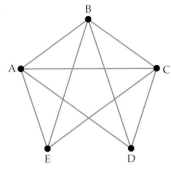

8.
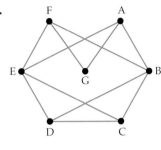

9.
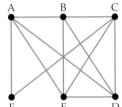

10.
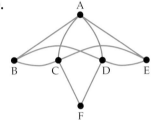

11. Parks The figure shows an arrangement of bridges connecting land areas in a park. Represent the map as a graph and then determine whether it is possible to stroll across each bridge exactly once and return to the starting position.

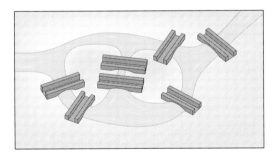

12. Architecture The floor plan of a sculpture gallery is shown. Is it possible to walk through each doorway exactly once? Is it possible to walk through each doorway exactly once and return to the starting point?

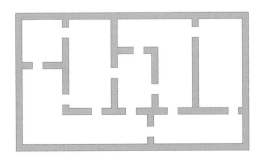

In Exercises 13 and 14, use Dirac's Theorem to verify that the graph is Hamiltonian, and then find a Hamiltonian circuit.

13.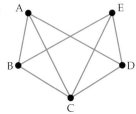

14.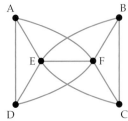

15. **Travel** An "X" in the table indicates a direct flight offered by a small airline. Draw a graph that represents the direct flights, and use your graph to find a route that visits each city exactly once and returns to the starting city.

	Casper	Rapid City	Minneapolis	Des Moines	Topeka	Omaha	Boulder
Casper	—	X					X
Rapid City	X	—	X				
Minneapolis		X	—	X	X		X
Des Moines			X	—	X		
Topeka			X	X	—	X	X
Omaha					X	—	X
Boulder	X		X		X	X	—

16. **Travel** For the direct flights given in Exercise 15, find a route that travels each flight exactly once and returns to the starting city. (You may visit cities more than once.)

In Exercises 17 and 18, use the Greedy Algorithm to find a Hamiltonian circuit starting at vertex A in the weighted graph.

17.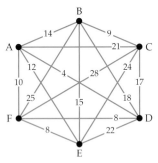

18.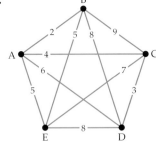

In Exercises 19 and 20, use the Edge-Picking Algorithm to find a Hamiltonian circuit starting at vertex A in the weighted graph.

19. The graph in Exercise 17   **20.** The graph in Exercise 18

21. Efficient Route The distances, in miles, between five different cities are given in the table. Sketch a weighted graph that represents the distances, and then use the Greedy Algorithm to design a route that starts in Memphis, visits each city, and returns to Memphis while attempting to minimize the total distance traveled.

	Memphis	Nashville	Atlanta	Birmingham	Jackson
Memphis	—	210	394	247	213
Nashville	210	—	244	189	418
Atlanta	394	244	—	148	383
Birmingham	247	189	148	—	239
Jackson	213	418	383	239	—

22. Computer Networking A small office needs to network five computers by connecting one computer to another and forming a large loop. The lengths of cable needed (in feet) between pairs of machines are given in the table. Use the Edge-Picking Algorithm to design a method of networking the computers while attempting to use the smallest possible amount of cable.

	Computer A	Computer B	Computer C	Computer D	Computer E
Computer A	—	85	40	55	20
Computer B	85	—	35	40	18
Computer C	40	35	—	60	50
Computer D	55	40	60	—	30
Computer E	20	18	50	30	—

Map Coloring In Exercises 23 and 24, a map is given showing the states of a fictional country. Represent the map by a graph and find a coloring of the graph, using the minimum number of colors possible, such that no two neighboring countries share the same color.

23.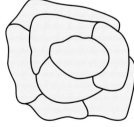

24.

In Exercises 25 and 26, show that the graph is 2-colorable by finding a 2-coloring, or explain why the graph is not 2-colorable.

In Exercises 27 and 28, determine (by trial and error) the chromatic number of the graph.

25. **26.**

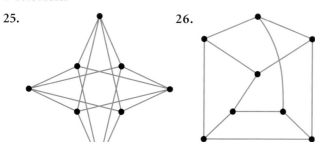

27. **28.**

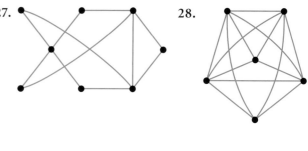

29. Scheduling A company has scheduled a retreat at a hotel resort. It needs to hold meetings the first day there, and several conference rooms are available. Some employees must attend more than one meeting, however, so the meetings cannot all be scheduled at the same time. An "X" in the table below indicates that at least one employee must attend both of the corresponding meetings, and so those two meetings must be held at different times. Draw a graph in which each vertex represents a meeting, and an edge joins two vertices if the corresponding meetings require the attendance of the same employee. Then use graph coloring to design a meeting schedule that uses the minimum number of different time slots possible.

	Budget Meeting	Marketing Meeting	Executive Meeting	Sales Meeting	Research Meeting	Planning Meeting
Budget meeting	—	X	X		X	
Marketing meeting	X	—		X		
Executive meeting	X		—		X	X
Sales meeting		X		—		X
Research meeting	X		X		—	X
Planning meeting			X	X	X	—

CHAPTER 10 **Test**

---

1. Social Network Each vertex in the graph at the right represents a student. An edge connects two vertices if the corresponding students have at least one class in common during the current term.

 a. Is there a class that both Jacob and Cheung are enrolled in?

 b. Who shares the most classes among the members of this group of students?

 c. How many classes does Victor have in common with the other students?

 d. Is this graph connected?

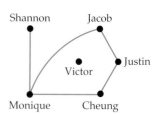

2. Determine whether the following two graphs are equivalent. Explain your reasoning.

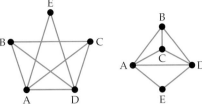

3. Answer the following questions for the graph shown at the right.

   a. Is the graph Eulerian? If so, find an Euler circuit. If not, explain how you know.

   b. Does the graph have an Euler walk? If so, find one. If not, explain how you know.

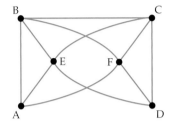

4. **Recreation** The illustration below depicts bridges connecting islands in a river to the riverbanks. Can a person start at one location and take a stroll so that each bridge is crossed once, but no bridge is crossed twice? (The person does not need to return to the starting location.) Draw a graph that represents the land areas and bridges. Answer the question, using the graph to explain your reasoning.

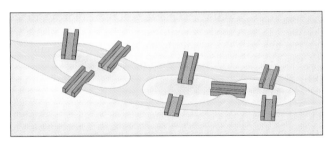

5. a. What does Dirac's Theorem state? Explain how it guarantees that the graph at the right is Hamiltonian.

   b. Find a Hamiltonian circuit in the graph.

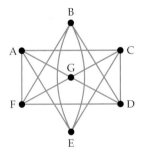

6. **Low-Cost Route** The table below shows the costs of direct train travel between various cities.

   a. Draw a weighted graph that represents the train fares.

   b. Use the Edge-Picking Algorithm to find a route that begins in Angora, visits each city, and returns to Angora. What is the total cost of this route?

   c. Is it possible to travel along each train route and return to the starting city without traveling any route more than once? Explain how you know.

	Angora	Bancroft	Chester	Davenport	Elmwood
Angora	—	$48	$52	$36	$90
Bancroft	$48	—	$32	$42	$98
Chester	$52	$32	—	$76	$84
Davenport	$36	$42	$76	—	$106
Elmwood	$90	$98	$84	$106	—

7. Use the Greedy Algorithm to find a Hamiltonian circuit beginning at vertex A in the weighted graph shown.

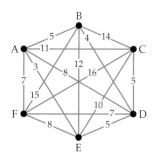

8. Map Coloring A map of the countries of a fictional continent is shown below.

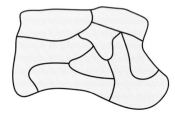

   a. Represent the map by a graph in which the vertices correspond to countries and the edges indicate which countries share a common border.

   b. What is the chromatic number of your graph?

   c. Use your work in part b to color the countries of the map with the fewest colors possible so that no two neighboring countries share the same color.

9. For the graph shown below, find a 2-coloring of the vertices or explain why a 2-coloring is impossible.

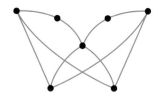

10. A group of eight friends is planning a vacation in Las Vegas, where they will split into different groups each evening to see various shows.

   Karen, Ryan, and Ruby want to see Cirque du Soleil together.

   Ruby, Anthony, and Jay want to see a magic show together.

   Anthony, Ricardo, and Heather want to see a comedy show together.

   Jenna, Ryan, and Ricardo want to see a tribute band play in concert together.

   Karen, Jay, and Jenna want to see a musical together.

   Ricardo, Jay, and Heather want to see a play together.

Draw a graph in which each vertex represents one of the shows, and connect vertices with an edge if at least one person wants to see both corresponding shows. Then use graph coloring to determine the fewest evenings needed so that all the friends can see the shows they would like to see, and design a schedule for the group.

CHAPTER 1

SECTION 1.1

CHECK YOUR PROGRESS 1, *page 2*

a. Each successive number is 5 larger than the preceding number. Thus we predict that the next number in the list is 5 larger than 25, which is 30.

b. The first two numbers differ by 3. The second and third numbers differ by 5. It appears that the difference between any two numbers is always 2 more than the preceding difference. Since 17 and 26 differ by 9, we predict that the next number will be 11 more than 26, which is 37.

CHECK YOUR PROGRESS 2, *page 3*

If the original number is 2, then $\dfrac{2 \times 9 + 15}{3} - 5 = 6$, which is three times the original number.

If the original number is 7, then $\dfrac{7 \times 9 + 15}{3} - 5 = 21$, which is three times the original number.

If the original number is -12, then $\dfrac{-12 \times 9 + 15}{3} - 5 = -36$, which is three times the original number.

It appears, by inductive reasoning, that the procedure produces a number that is three times the original number.

CHECK YOUR PROGRESS 3, *page 4*

a. It appears that when the velocity of a tsunami is doubled, its height is quadrupled.

b. A tsunami with a velocity of 30 feet per second will have a height that is four times that of a tsunami with a speed of 15 feet per second. Thus, we predict a height of $4 \times 25 = 100$ feet for a tsunami with a velocity of 30 feet per second.

CHECK YOUR PROGRESS 4, *page 5*

a. Let $x = 0$. Then $\dfrac{x}{x} \neq 1$, because division by 0 is undefined.

b. Let $x = 1$. Then $\dfrac{x + 3}{3} = \dfrac{1 + 3}{3} = \dfrac{4}{3}$, whereas $x + 1 = 1 + 1 = 2$.

c. Let $x = 3$. Then $\sqrt{x^2 + 16} = \sqrt{3^2 + 16} = \sqrt{25} = 5$, whereas $x + 4 = 3 + 4 = 7$.

CHECK YOUR PROGRESS 5, *page 6*

Let n represent the original number.
Multiply the number by 6: $6n$
Add 10 to the product: $6n + 10$

Divide the sum by 2: $\dfrac{6n + 10}{2} = 3n + 5$

Subtract 5: $3n + 5 - 5 = 3n$

The procedure always produces a number that is three times the original number.

CHECK YOUR PROGRESS 6, *page 7*

a. The conclusion is a specific case of a general assumption, so the argument is an example of deductive reasoning.

b. The argument reaches a conclusion based on specific examples, so the argument is an example of inductive reasoning.

CHECK YOUR PROGRESS 7, *page 9*

From clue 1, we know that Ashley is not the president or the treasurer. In the following chart, write X1 (which stands for "ruled out by clue 1") in the President and Treasurer columns of Ashley's row.

	Pres.	V. P.	Sec.	Treas.
Brianna				
Ryan				
Tyler				
Ashley	X1			X1

From clue 2, Brianna is not the secretary. We know from clue 1 that the president is not the youngest, and we know from clue 2 that Brianna and the secretary are the youngest members of the group. Thus Brianna is not the president. In the chart, write X2 for these two conditions. Also we know from clues 1 and 2 that Ashley is not the secretary, because she is older than the treasurer. Write an X2 in the Secretary column of Ashley's row.

	Pres.	V. P.	Sec.	Treas.
Brianna	X2		X2	
Ryan				
Tyler				
Ashley	X1		X2	X1

At this point we see that Ashley must be the vice president and that none of the other members is the vice president. Thus we can update the chart as shown below.

	Pres.	V. P.	Sec.	Treas.
Brianna	X2	X2	X2	
Ryan		X2		
Tyler		X2		
Ashley	X1	√	X2	X1

Now we can see that Brianna must be the treasurer and that neither Ryan nor Tyler is the treasurer. Update the chart as shown below.

	Pres.	V. P.	Sec.	Treas.
Brianna	X2	X2	X2	√
Ryan		X2		X2
Tyler		X2		X2
Ashley	X1	√	X2	X1

From clue 3, we know that Tyler is not the secretary. Thus we can conclude that Tyler is the president and Ryan must be the secretary. See the chart below.

	Pres.	V. P.	Sec.	Treas.
Brianna	X2	X2	X2	√
Ryan	X3	X2	√	X2
Tyler	√	X2	X3	X2
Ashley	X1	√	X2	X1

Tyler is the president, Ashley is the vice president, Ryan is the secretary, and Brianna is the treasurer.

SECTION 1.2

CHECK YOUR PROGRESS 1, *page 18*

Understand the Problem   In order to go past Starbucks, Allison must walk along Third Avenue from Board Walk to Park Avenue.

Devise a Plan   Label each intersection that Allison can pass through with the number of routes to that intersection. If she can reach an intersection from two different routes, then the number of routes to that intersection is the sum of the numbers of routes to the two adjacent intersections.

Carry Out the Plan   The following figure shows the number of routes to each of the intersections that Allison could pass through. Thus there are a total of nine routes that Allison can take if she wishes to walk directly from point A to point B and pass by Starbucks.

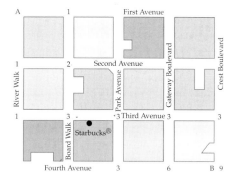

Review the Solution   The total of nine routes seems reasonable. We know from Example 1 that if Allison can take any route, the total number of routes is 35. Requiring Allison to go past Starbucks eliminates several routes.

CHECK YOUR PROGRESS 2, *page 19*

Understand the Problem   There are several ways to answer the questions so that two answers are "false" and three answers are "true." One way is TTTFF and another is FFTTT.

Devise a Plan   Make an organized list. Try the strategy of listing a T unless doing so will produce too many Ts or a duplicate of one of the previous orders in your list.

Carry Out the Plan   (Start with three Ts in a row.)

TTTFF	(1)
TTFTF	(2)
TTFFT	(3)
TFTTF	(4)
TFTFT	(5)
TFFTT	(6)
FTTTF	(7)
FTTFT	(8)
FTFTT	(9)
FFTTT	(10)

Review the Solution   Each entry in the list has two F's and three T's. Since the list is complete and has no duplications, we know that there are 10 ways for a student to mark two questions with "false" and the other three with "true."

CHECK YOUR PROGRESS 3, *page 20*

Understand the Problem   There are six people, and each person shakes hands with each of the other people.

Devise a Plan   Each person will shake hands with five other people (a person won't shake his or her own hand; that would be silly). Since there are six people, we could multiply 6 times 5 to get the total number of handshakes. However, this procedure would count each handshake exactly twice, so we must divide this product by 2 for the actual answer.

Carry Out the Plan   6 times 5 is 30. 30 divided by 2 is 15.

Review the Solution   Denote the people by the letters A, B, C, D, E, and F. Make an organized list. Remember that AB and BA represent the same people shaking hands, so do not list both AB and BA.

$$AB \quad AC \quad AD \quad AE \quad AF$$
$$BC \quad BD \quad BE \quad BF$$
$$CD \quad CE \quad CF$$
$$DE \quad DF$$
$$EF$$

The method of making an organized list verifies that if six people shake hands with each other there will be a total of 15 handshakes.

CHECK YOUR PROGRESS 4, *page 21*

Understand the Problem   We need to find the ones digit of 4^{200}.

Devise a Plan   Compute a few powers of 4 to see if there are any patterns. $4^1 = 4$, $4^2 = 16$, $4^3 = 64$, and $4^4 = 256$. It appears that the last digit (ones digit) of 4^{200} must be either a 4 or a 6.

Carry Out the Plan   If the exponent n is an even number, then 4^n has a ones digit of 6. If the exponent n is an odd number, then 4^n has a ones digit of 4. Because 200 is an even number, we conjecture that 4^{200} has a ones digit of 6.

Review the Solution   You could try to check the answer by using a calculator, but you would find that 4^{200} is too large to be displayed. Thus we need to rely on the patterns we have observed to conclude that 6 is indeed the ones digit of 4^{200}.

CHECK YOUR PROGRESS 5, *page 22*

Understand the Problem   We are asked to find the possible numbers that Melody could have started with.

Devise a Plan   Work backward from 18 and do the inverse of each operation that Melody performed.

Carry Out the Plan   To get 18, Melody subtracted 30 from a number, so that number was $18 + 30 = 48$. To get 48, she divided a number by 3, so that number was $48 \times 3 = 144$. To get 144, she squared a number. She could have squared either 12 or -12 to produce 144. If the number she squared was 12, then she must have doubled 6 to get 12. If the number she squared was -12, then the number she doubled was -6.

Review the Solution   We can check by starting with 6 or -6. If we do exactly as Melody did, we end up with 18. The operation that prevents us from knowing with 100% certainty which number she started with is the squaring operation. We have no way of knowing whether the number she squared was a positive number or a negative number.

CHECK YOUR PROGRESS 6, *page 23*

Understand the Problem   We need to find Diophantus's age when he died.

Devise a Plan   Read the hint and then look for clues that will help you make an educated guess. You know from the given information that Diophantus's age must be divisible by 6, 12, 7, and 2. Find a number divisible by all of these numbers and check to see if it is a possible solution to the problem.

Carry Out the Plan   All multiples of 12 are divisible by 6 and 2, but the smallest multiple of 12 that is divisible by 7 is $12 \times 7 = 84$. Thus we conjecture that Diophantus's age when he died was $x = 84$ years. If $x = 84$, then $\frac{1}{6}x = 14$, $\frac{1}{12}x = 7$, $\frac{1}{7}x = 12$, and $\frac{1}{2}x = 42$. Then $\frac{1}{6}x + \frac{1}{12}x + \frac{1}{7}x + 5 + \frac{1}{2}x + 4 = 14 + 7 + 12 + 5 + 42 + 4 = 84$. It seems that 84 years is a correct solution to the problem.

Review the Solution   After 84, the next multiple of 12 that is divisible by 7 is 168. The number 168 also satisfies all the conditions of the problem, but it is unlikely that Diophantus died at the age of 168 years or at any age older than 168 years. Hence the only reasonable solution is 84 years.

CHECK YOUR PROGRESS 7, *page 25*

Understand the Problem   We need to determine two U.S. coins that have a total value of 35¢, given that one of the coins is not a quarter.

Devise a Plan   Experiment with different coins to try to produce 35¢. After a few attempts, you should conclude that one of the coins must be a quarter. Consider that the problem may be a *deceptive problem.*

Carry Out the Plan   A total of 35¢ can be produced by using a dime and a quarter. One of the coins is a quarter, but it is also true that *one of the coins, the dime, is not a quarter.*

Review the Solution   A dime and a quarter satisfy all the conditions of the problem. No other combination of coins satisfies the conditions of the problem. Thus the only solution is a dime and a quarter.

CHECK YOUR PROGRESS 8, *page 27*

a. The maximum of the average yearly ticket prices is displayed by the tallest vertical bar in Figure 1.2. Thus the maximum of the average yearly U.S. movie theatre ticket prices for the years from 1996 to 2004 was $6.21, in the year 2004.

b. Figure 1.3 indicates that in 2005, 9% of the automobile accidents in Twin Falls were accidents involving lane changes. Thus $0.09 \cdot 4300 = 387$ of the accidents were accidents involving lane changes.

c. To estimate the average age at which women married for the first time in the year 1975, locate 1975 on the horizontal axis of Figure 1.4 and then move directly upward to the point on the green broken-line graph. The height of this point represents the average age at first marriage for women in the year 1975, and it can be estimated by moving horizontally to the vertical axis on the left. Thus the average age at first marriage for women in the year 1975 was 21 years, rounded to the nearest quarter of a year. This same procedure shows that in the year 1975, the average age at which men first married was 23.5 years, rounded to the nearest quarter of a year.

SECTION 1.3

CHECK YOUR PROGRESS 1, *page 34*

a. {0, 1, 2, 3}   **b.** {12, 13, 14, 15, 16, 17, 18, 19}

c. {−4, −3, −2, −1}

CHECK YOUR PROGRESS 2, *page 35*

a. $n(C) = 5$ **b.** $n(D) = 1$

CHECK YOUR PROGRESS 3, *page 36*

a. Set S' contains all the elements of
$U = $ {sea bass, halibut, salmon, tuna, cod, rockfish}
that are not elements of $S = $ {sea bass, salmon}. Thus,
$S' = $ {halibut, tuna, cod, rockfish}.

b. Set M' contains all the elements of
$U = $ {sea bass, halibut, salmon, tuna, cod, rockfish}
that are not elements of $M = $ {halibut, salmon, cod, rockfish}.
Thus, $M' = $ {sea bass, tuna}.

CHECK YOUR PROGRESS 4, *page 37*

a. False. The number 3 is an element of the first set but not an element of the second set. Therefore, the first set is not a subset of the second set.

b. True. The set of counting numbers is the same set as the set of natural numbers, and every set is a subset of itself.

c. True. The empty set is a subset of every set.

d. True. Each element of the first set is an integer.

CHECK YOUR PROGRESS 5, *page 39*

Subsets with zero elements: { }

Subsets with one element: {a}, {b}, {c}, {d}, {e}

Subsets with two elements: {a, b}, {a, c}, {a, d}, {a, e}, {b, c}, {b, d}, {b, e}, {c, d}, {c, e}, {d, e}

Subsets with three elements: {a, b, c}, {a, b, d}, {a, b, e}, {a, c, d}, {a, c, e}, {a, d, e}, {b, c, d}, {b, c, e}, {b, d, e}, {c, d, e}

Subsets with four elements: {a, b, c, d}, {a, b, c, e}, {a, b, d, e}, {a, c, d, e}, {b, c, d, e}

Subsets with five elements: {a, b, c, d, e}

CHECK YOUR PROGRESS 6, *page 40*

a. $D \cap E = \{0, 3, 8, 9\} \cap \{3, 4, 8, 9, 11\}$
$= \{3, 8, 9\}$

b. $D \cap F = \{0, 3, 8, 9\} \cap \{0, 2, 6, 8\}$
$= \{0, 8\}$

CHECK YOUR PROGRESS 7, *page 40*

a. $D \cup E = \{0, 4, 8, 9\} \cup \{1, 4, 5, 7\}$
$= \{0, 1, 4, 5, 7, 8, 9\}$

b. $D \cup F = \{0, 4, 8, 9\} \cup \{2, 6, 8\}$
$= \{0, 2, 4, 6, 8, 9\}$

CHECK YOUR PROGRESS 8, *page 41*

The following Venn diagrams show that $(A \cap B)'$ is equal to $A' \cup B'$.

The white region represents $A \cap B$.
The shaded region represents $(A \cap B)'$.

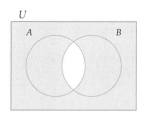

The grey shaded region below represents A'.
The diagonal patterned region below represents B'.
$A' \cup B'$ is the union of the grey shaded region and the diagonal patterned region.

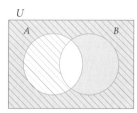

CHECK YOUR PROGRESS 9, *page 42*
The following Venn diagrams show that
$A \cup (B \cap C) = (A \cup B) \cap (A \cup C)$.

The grey shaded region represents $A \cup (B \cap C)$.

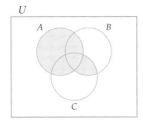

The grey shaded region below represents $A \cup B$.
The diagonal patterned region below represents $A \cup C$.
The intersection of the grey shaded region and the diagonal patterned region represents $(A \cup B) \cap (A \cup C)$.

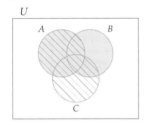

CHECK YOUR PROGRESS 10, *page 45*

The intersection of the two sets includes the 85 students who like both volleyball and basketball.

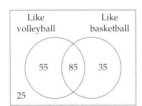

a. Because 140 students like volleyball and 85 like both sports, there must be $140 - 85 = 55$ students who like only volleyball.

b. Because 120 students like basketball and 85 like both sports, there must be $120 - 85 = 35$ students who like only basketball.

c. The Venn diagram shows that the number of students who like only volleyball plus the number who like only basketball plus the number who like both sports is $55 + 35 + 85 = 175$. Thus of the 200 students surveyed, only $200 - 175 = 25$ do not like either of the sports.

CHECK YOUR PROGRESS 11, *page 46*

The intersection of the three sets includes the 15 people who like all three activities.

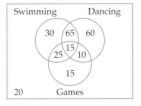

a. There are 25 people who like dancing and games. This includes the 15 people who like all three activities. Thus there must be another $25 - 15 = 10$ people who like only dancing and games. There are 40 people who like swimming and games. Thus there must be another $40 - 15 = 25$ people who like only swimming and games. There are 80 people who like swimming and dancing. Thus there must be another $80 - 15 = 65$ people who like only swimming and dancing. Hence $10 + 25 + 65 = 100$ people who like exactly two of the three activities.

b. There are 135 people who like swimming. We have determined that 15 people like all three activities, 25 like only swimming and games, and 65 like only swimming and dancing. This means that $135 - (15 + 25 + 65) = 30$ people like only swimming.

c. There are a total of 240 passengers surveyed. The Venn diagram shows that $15 + 25 + 10 + 15 + 30 + 65 + 60 = 220$ passengers like at least one of the activities. Thus $240 - 220 = 20$ passengers like none of the activities.

CHAPTER 2

SECTION 2.1

CHECK YOUR PROGRESS 1, *page 63*

a. The sentence "Open the door" is a command. It is not a statement.

b. The word *large* is not a precise term. It is not possible to determine whether the sentence "7055 is a large number" is true or false and thus the sentence is not a statement.

c. The sentence $4 + 5 = 8$ is a false statement.

d. At this time we do not know whether the given sentence is true or false, but we know that the sentence is either true or false and that it is not both true and false. Thus the sentence is a statement.

e. The sentence $x > 3$ is a statement because for any given value of x, the inequality $x > 3$ is true or false, but not both.

CHECK YOUR PROGRESS 2, *page 65*

a. 1001 is not divisible by 7. **b.** 5 is not an even number.

c. The fire engine is red.

CHECK YOUR PROGRESS 3, *page 66*

a. $\sim p \wedge r$ **b.** $\sim s \wedge \sim r$ **c.** $r \leftrightarrow q$ **d.** $p \rightarrow \sim r$

CHECK YOUR PROGRESS 4, *page 66*

$e \wedge \sim t$: All men are created equal and I am not trading places.

$a \vee \sim t$: I get Abe's place or I am not trading places.

$e \rightarrow t$: If all men are created equal, then I am trading places.

$t \leftrightarrow g$: I am trading places if and only if I get George's place.

CHECK YOUR PROGRESS 5, *page 67*

a. True. A conjunction is true provided both components are true.

b. True. A disjunction is true provided at least one component is true.

c. False. If both components of a disjunction are false, then the disjunction is false.

CHECK YOUR PROGRESS 6, *page 68*

a. Some bears are not brown.

b. Some math classes are fun.

c. All vegetables are green.

SECTION 2.2

CHECK YOUR PROGRESS 1, *page 75*

a.

p	q	$\sim p$	$\sim q$	$p \wedge \sim q$	$\sim p \vee q$	$(p \wedge \sim q) \vee (\sim p \vee q)$	
T	T	F	F	F	T	T	Row 1
T	F	F	T	T	F	T	Row 2
F	T	T	F	F	T	T	Row 3
F	F	T	T	F	T	T	Row 4
		1	2	3	4	5	

b. p is true and q is false in row 2 of the above truth table. The truth value of $(p \wedge \sim q) \vee (\sim p \vee q)$ in row 2 is T (true).

CHECK YOUR PROGRESS 2, *page 75*

a.

p	q	r	$\sim p$	$\sim r$	$\sim p \wedge r$	$q \wedge \sim r$	$(\sim p \wedge r) \vee (q \wedge \sim r)$	
T	T	T	F	F	F	F	F	Row 1
T	T	F	F	T	F	T	T	Row 2
T	F	T	F	F	F	F	F	Row 3
T	F	F	F	T	F	F	F	Row 4
F	T	T	T	F	T	F	T	Row 5
F	T	F	T	T	F	T	T	Row 6
F	F	T	T	F	T	F	T	Row 7
F	F	F	T	T	F	F	F	Row 8

b. p is false, q is true, and r is false in row 6 of the above truth table. The truth value of $(\sim p \wedge r) \vee (q \wedge \sim r)$ in row 6 is T (true).

CHECK YOUR PROGRESS 3, *page 77*

The given statement has two simple statements. Thus you should use a standard form that has $2^2 = 4$ rows.

Step 1 Enter the truth values for each simple statement and their negations. See columns 1, 2, and 3 in the table following step 3.

Step 2 Use the truth values in columns 2 and 3 to determine the truth values to enter under the "and" connective. See column 4 in the table following step 3.

Step 3 Use the truth values in columns 1 and 4 to determine the truth values to enter under the "or" connective. See column 5 in the table below.

p	q	~p	∨	(p	∧	q)
T	T	F	T	T	T	T
T	F	F	F	T	F	F
F	T	T	T	F	F	T
F	F	T	T	F	F	F

| | 1 | 5 | 2 | 4 | 3 |

The truth table for $\sim p \vee (p \wedge q)$ is displayed in column 5.

CHECK YOUR PROGRESS 4, *page 78*

p	q	p	∨	(p	∧	~q)
T	T	T	T	T	F	F
T	F	T	T	T	T	T
F	T	F	F	F	F	F
F	F	F	F	F	F	T

| 1 | 5 | 2 | 4 | 3 |

The above truth table shows that $p \equiv p \vee (p \wedge \sim q)$.

CHECK YOUR PROGRESS 5, *page 79*

Let d represent "I am going to the dance." Let g represent "I am going to the game." The original sentence in symbolic form is $\sim(d \wedge g)$. Applying one of De Morgan's laws, we find that $\sim(d \wedge g) \equiv \sim d \vee \sim g$. Thus an equivalent form of "It is not true that I am going to the dance and I am going to the game" is "I am not going to the dance or I am not going to the game."

CHECK YOUR PROGRESS 6, *page 79*

The following truth table shows that $p \wedge (\sim p \wedge q)$ is always false. Thus $p \wedge (\sim p \wedge q)$ is a self-contradiction.

p	q	p	∧	(~p	∧	q)
T	T	T	F	F	F	T
T	F	T	F	F	F	F
F	T	F	F	T	T	T
F	F	F	F	T	F	F

| 1 | 5 | 2 | 4 | 3 |

SECTION 2.3

CHECK YOUR PROGRESS 1, *page 86*

a. Because the antecedent is true and the consequent is false, the statement is a false statement.

b. Because the antecedent is false, the statement is a true statement.

c. Because the consequent is true, the statement is a true statement.

CHECK YOUR PROGRESS 2, *page 87*

p	q	[p	∧	(p	→	q)]	→	q
T	T	T	T	T	T	T	T	T
T	F	T	F	T	F	F	T	F
F	T	F	F	F	T	T	T	T
F	F	F	F	F	T	F	T	F

| 1 | 6 | 2 | 5 | 3 | 7 | 4 |

CHECK YOUR PROGRESS 3, *page 88*

a. If it is a square, then it is a rectangle.

b. If I am older than 30, then I am at least 21.

CHECK YOUR PROGRESS 4, *page 89*

Converse: If we are not going to have a quiz tomorrow, then we will have a quiz today.

Inverse: If we don't have a quiz today, then we will have a quiz tomorrow.

Contrapositive: If we have a quiz tomorrow, then we will not have a quiz today.

CHECK YOUR PROGRESS 5, *page 90*

a. The second statement is the inverse of the first statement. Thus the statements are not equivalent. This can also be demonstrated by the fact that the first statement is true for $c = 0$ and the second statement is false for $c = 0$.

b. The second statement is the contrapositive of the first statement. Thus the statements are equivalent.

CHECK YOUR PROGRESS 6, *page 91*

a. *Contrapositive:* If x is an odd integer, then $3 + x$ is an even integer. The contrapositive is true and so the original statement is also true.

b. *Contrapositive:* If two triangles are congruent triangles, then the two triangles are similar triangles. The contrapositive is true and so the original statement is also true.

c. *Contrapositive:* If tomorrow is Thursday, then today is Wednesday. The contrapositive is true and so the original statement is also true.

CHECK YOUR PROGRESS 7, *page 91*

a. Let $x = 6.5$. Then the first component of the biconditional is false and the second component of the biconditional is true. Thus the given biconditional statement is false.

b. Both components of the biconditional are true for $x > 2$, and both components are false for $x \leq 2$. Because both components have the same truth value for any real number x, the given biconditional is true.

SECTION 2.4

CHECK YOUR PROGRESS 1, *page 98*

Let p represent the statement "She got on the plane." Let r represent the statement "She will regret it." Then the symbolic form of the argument is

$$\sim p \rightarrow r$$
$$\underline{\sim r}$$
$$\therefore p$$

CHECK YOUR PROGRESS 2, *page 100*

Let r represent the statement "The stock market rises." Let f represent the statement "The bond market will fall." Then the symbolic form of the argument is

$$r \rightarrow f$$
$$\underline{\sim f}$$
$$\therefore \sim r$$

The truth table for this argument is as follows:

		First Premise	Second Premise	Conclusion	
r	f	$r \rightarrow f$	$\sim f$	$\sim r$	
T	T	T	F	F	Row 1
T	F	F	T	F	Row 2
F	T	T	F	T	Row 3
F	F	T	T	T	Row 4

Row 4 is the only row in which all the premises are true, so it is the only row that we examine. Because the conclusion is true in row 4, the argument is valid.

CHECK YOUR PROGRESS 3, *page 101*

Let a represent the statement "I arrive before 8 A.M." Let f represent the statement "I will make the flight." Let p represent the statement "I will give the presentation." Then the symbolic form of the argument is

$$a \rightarrow f$$
$$\underline{f \rightarrow p}$$
$$\therefore a \rightarrow p$$

The truth table for this argument is as follows:

a	f	p	First Premise $a \rightarrow f$	Second Premise $f \rightarrow p$	Conclusion $a \rightarrow p$	
T	T	T	T	T	T	Row 1
T	T	F	T	F	F	Row 2
T	F	T	F	T	T	Row 3
T	F	F	F	T	F	Row 4
F	T	T	T	T	T	Row 5
F	T	F	T	F	T	Row 6
F	F	T	T	T	T	Row 7
F	F	F	T	T	T	Row 8

The only rows in which all the premises are true are rows 1, 5, 7, and 8. In each of these rows the conclusion is also true. Thus the argument is a valid argument.

CHECK YOUR PROGRESS 4, *page 102*
Let *f* represent "I go to Florida for spring break." Let ~*s* represent "I will not study." Then the symbolic form of the argument is

$$f \rightarrow \sim s$$
$$\underline{\sim f}$$
$$\therefore s$$

This argument has the form of the fallacy of the inverse. Thus the argument is invalid.

CHECK YOUR PROGRESS 5, *page 104*
Let *r* represent "I read a math book." Let *f* represent "I start to fall asleep." Let *d* represent "I drink a soda." Let *e* represent "I eat a candy bar." Then the symbolic form of the argument is

$$r \rightarrow f$$
$$f \rightarrow d$$
$$\underline{d \rightarrow e}$$
$$\therefore r \rightarrow e$$

The argument has the form of the extended law of syllogism. Thus the argument is valid.

SECTION 2.5

CHECK YOUR PROGRESS 1, *page 109*
The following Euler diagram shows that the argument is valid.

CHECK YOUR PROGRESS 2, *page 110*
From the given premises we can conclude that 7 may or may not be a prime number. Thus the argument is invalid.

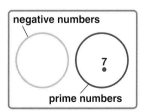
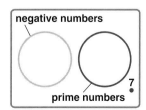

CHECK YOUR PROGRESS 3, *page 111*

From the given premises we can construct two possible Euler diagrams.

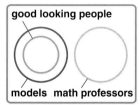 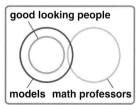

From the rightmost Euler diagram we can determine that the argument is invalid.

CHECK YOUR PROGRESS 4, *page 112*

The following Euler diagram illustrates that all squares are quadrilaterals, so the argument is a valid argument.

CHECK YOUR PROGRESS 5, *page 113*

The following Euler diagrams illustrate two possible cases. In both cases we see that all white rabbits like tomatoes.

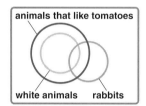 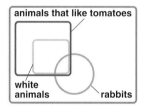

CHAPTER 3

SECTION 3.1

CHECK YOUR PROGRESS 1, *page 125*

a.
$$c - 6 = -13$$
$$c - 6 + 6 = -13 + 6$$
$$c = -7$$

The solution is -7.

b.
$$4 = -8z$$
$$\frac{4}{-8} = \frac{-8z}{-8}$$
$$-\frac{1}{2} = z$$

The solution is $-\frac{1}{2}$.

c.
$$22 + m = -9$$
$$22 - 22 + m = -9 - 22$$
$$m = -31$$

The solution is -31.

d. $5x = 0$
$$\frac{5x}{5} = \frac{0}{5}$$
$$x = 0$$

The solution is 0.

CHECK YOUR PROGRESS 2, *page 127*

a.
$$4x + 3 = 7x + 9$$
$$4x - 7x + 3 = 7x - 7x + 9$$
$$-3x + 3 = 9$$
$$-3x + 3 - 3 = 9 - 3$$
$$-3x = 6$$
$$\frac{-3x}{-3} = \frac{6}{-3}$$
$$x = -2$$

The solution is -2.

b.
$$7 - (5x - 8) = 4x + 3$$
$$7 - 5x + 8 = 4x + 3$$
$$15 - 5x = 4x + 3$$
$$15 - 5x - 4x = 4x - 4x + 3$$
$$15 - 9x = 3$$
$$15 - 15 - 9x = 3 - 15$$
$$-9x = -12$$
$$\frac{-9x}{-9} = \frac{-12}{-9}$$
$$x = \frac{4}{3}$$

The solution is $\frac{4}{3}$.

c.
$$\frac{3x - 1}{4} + \frac{1}{3} = \frac{7}{3}$$
$$12\left(\frac{3x - 1}{4} + \frac{1}{3}\right) = 12\left(\frac{7}{3}\right)$$
$$12 \cdot \frac{3x - 1}{4} + 12 \cdot \frac{1}{3} = 12 \cdot \frac{7}{3}$$
$$9x - 3 + 4 = 28$$
$$9x + 1 = 28$$
$$9x + 1 - 1 = 28 - 1$$
$$9x = 27$$
$$\frac{9x}{9} = \frac{27}{9}$$
$$x = 3$$

The solution is 3.

CHECK YOUR PROGRESS 3, *page 128*

a. $P = 0.05Y - 95$

$P = 0.05(1990) - 95$

$P = 99.5 - 95$

$P = 4.5$

The amount of garbage was about 4.5 pounds per day.

b. $\qquad P = 0.05Y - 95$

$\qquad 5.6 = 0.05Y - 95$

$5.6 + 95 = 0.05Y - 95 + 95$

$\qquad 100.6 = 0.05Y$

$$\frac{100.6}{0.05} = \frac{0.05Y}{0.05}$$

$\qquad 2012 = Y$

The year will be 2012.

CHECK YOUR PROGRESS 4, *page 129*

| $17.50 for the first three lines + $2.50 for each additional line | = | $30 |

Let L = the number of lines in the ad.

$17.50 + 2.50(L - 3) = 30$

$17.50 + 2.50L - 7.50 = 30$

$\qquad 10.00 + 2.50L = 30$

$10.00 - 10.00 + 2.50L = 30 - 10.00$

$\qquad 2.50L = 20$

$$\frac{2.50L}{2.50} = \frac{20}{2.50}$$

$\qquad L = 8$

You can place an eight-line ad.

CHECK YOUR PROGRESS 5, *page 130*

Let n = the number of years.

| The 1990 population of Vermont plus an annual increase times n | = | The 1990 population of North Dakota minus an annual decrease times n |

$562{,}576 + 5116n = 638{,}800 - 1370n$

$562{,}576 + 5116n + 1370n = 638{,}800 - 1370n + 1370n$

$562{,}576 + 6486n = 638{,}800$

$562{,}576 - 562{,}576 + 6486n = 638{,}800 - 562{,}576$

$6486n = 76{,}224$

$$\frac{6486n}{6486} = \frac{76{,}224}{6486}$$

$n \approx 12$

$1990 + 12 = 2002$

The populations would have been the same in 2002.

CHECK YOUR PROGRESS 6, *page 132*

a. $\qquad s = \dfrac{A + L}{2}$

$\qquad 2 \cdot s = 2 \cdot \dfrac{A + L}{2}$

$\qquad 2s = A + L$

$\qquad 2s - A = A - A + L$

$\qquad 2s - A = L$

b. $\qquad L = a(1 + ct)$

$\qquad \dfrac{L}{a} = \dfrac{a(1 + ct)}{a}$

$\qquad \dfrac{L}{a} = 1 + ct$

$\qquad \dfrac{L}{a} - 1 = 1 - 1 + ct$

$\qquad \dfrac{L}{a} - 1 = ct$

$\qquad \dfrac{\dfrac{L}{a} - 1}{t} = \dfrac{ct}{t}$

$\qquad \dfrac{\dfrac{L}{a} - 1}{t} = c$

$\qquad \left(\dfrac{L}{a} - 1\right)\left(\dfrac{1}{t}\right) = c$

$\qquad \dfrac{L}{at} - \dfrac{1}{t} = c$

SECTION 3.2

CHECK YOUR PROGRESS 1, *page 140*

$4.92 \div 1.5 = 3.28$

$$\frac{\$4.92}{1.5 \text{ pounds}} = \frac{\$3.28}{1 \text{ pound}} = \$3.28/\text{pound}$$

The hamburger costs $3.28 per pound.

CHECK YOUR PROGRESS 2, *page 141*

Find the difference in the hourly wage.

$\$6.75 - \$5.15 = \$1.60$

Multiply the difference in the hourly wage by 35.

$\$1.60(35) = \$56$

An employee's pay for working 35 hours and earning the California minimum wage is $56 greater.

CHECK YOUR PROGRESS 3, *page 141*

$$\frac{\$2.99}{32 \text{ ounces}} \approx \frac{\$.093}{1 \text{ ounce}} \qquad \frac{\$3.99}{48 \text{ ounces}} \approx \frac{\$.083}{1 \text{ ounce}}$$

$$\$.093 > \$.083$$

The more economical purchase is 48 ounces of detergent for \$3.99.

CHECK YOUR PROGRESS 4, *page 143*

a. $20,000(1.1616) = 23,232$
23,232 Canadian dollars would be needed to pay for an order costing \$20,000.

b. $25,000(0.8540) = 21,350$
21,350 euros would be exchanged for \$25,000.

CHECK YOUR PROGRESS 5, *page 144*

a. $\dfrac{24 \text{ hours}}{1 \text{ day}} \cdot 7 \text{ days} = (24 \text{ hours})(7) = 168 \text{ hours}$

$$\frac{120 \text{ hours}}{1 \text{ week}} = \frac{120 \text{ hours}}{168 \text{ hours}} = \frac{120}{168} = \frac{5}{7}$$

The ratio is $\frac{5}{7}$.

b. $\dfrac{60 \text{ hours}}{(168 - 60) \text{ hours}} = \dfrac{60 \text{ hours}}{108 \text{ hours}} = \dfrac{60}{108} = \dfrac{5}{9}$

The ratio is 5 to 9.

CHECK YOUR PROGRESS 6, *page 145*

$$6742 + 7710 = 14,452$$

$$\frac{14,452}{798} \approx \frac{18.11}{1} \approx \frac{18}{1}$$

The ratio is 18 to 1.

CHECK YOUR PROGRESS 7, *page 147*

$$\frac{42}{x} = \frac{5}{8}$$

$$42 \cdot 8 = x \cdot 5$$

$$336 = 5x$$

$$\frac{336}{5} = \frac{5x}{5}$$

$$67.2 = x$$

The solution is 67.2.

CHECK YOUR PROGRESS 8, *page 148*

$$\frac{15 \text{ kilometers}}{2 \text{ centimeters}} = \frac{x \text{ kilometers}}{7 \text{ centimeters}}$$

$$\frac{15}{2} = \frac{x}{7}$$

$$15 \cdot 7 = 2 \cdot x$$

$$105 = 2x$$

$$\frac{105}{2} = \frac{2x}{2}$$

$$52.5 = x$$

The distance between the two cities is 52.5 kilometers.

CHECK YOUR PROGRESS 9, *page 149*

$$\frac{7}{5} = \frac{\$28,000}{x \text{ dollars}}$$

$$\frac{7}{5} = \frac{28,000}{x}$$

$$7 \cdot x = 5 \cdot 28,000$$

$$7x = 140,000$$

$$\frac{7x}{7} = \frac{140,000}{7}$$

$$x = 20,000$$

The other partner receives \$20,000.

CHECK YOUR PROGRESS 10, *page 150*

$$\frac{10.1 \text{ deaths}}{1,000,000 \text{ people}} = \frac{d \text{ deaths}}{4,000,000 \text{ people}}$$

$$10.1(4,000,000) = 1,000,000 \cdot d$$

$$40,400,000 = 1,000,000d$$

$$\frac{40,400,000}{1,000,000} = \frac{1,000,000d}{1,000,000}$$

$$40.4 = d$$

Approximately 40 people aged 5 to 34 die from asthma each year in New York City.

SECTION 3.3

CHECK YOUR PROGRESS 1, *page 160*

a. $74\% = 0.74$

b. $152\% = 1.52$

c. $8.3\% = 0.083$

d. $0.6\% = 0.006$

CHECK YOUR PROGRESS 2, *page 160*

a. $0.3 = 30\%$

b. $1.65 = 165\%$

c. $0.072 = 7.2\%$

d. $0.004 = 0.4\%$

CHECK YOUR PROGRESS 3, *page 161*

a. $8\% = 8\left(\dfrac{1}{100}\right) = \dfrac{8}{100} = \dfrac{2}{25}$

b. $180\% = 180\left(\dfrac{1}{100}\right) = \dfrac{180}{100} = 1\dfrac{80}{100} = 1\dfrac{4}{5}$

c. $2.5\% = 2.5\left(\dfrac{1}{100}\right) = \dfrac{2.5}{100} = \dfrac{25}{1000} = \dfrac{1}{40}$

d. $66\dfrac{2}{3}\% = \dfrac{200}{3}\% = \dfrac{200}{3}\left(\dfrac{1}{100}\right) = \dfrac{2}{3}$

CHECK YOUR PROGRESS 4, *page 162*

a. $\dfrac{1}{4} = 0.25 = 25\%$

b. $\dfrac{3}{8} = 0.375 = 37.5\%$

c. $\dfrac{5}{6} = 0.83\overline{3} = 83.\overline{3}\%$

d. $1\dfrac{2}{3} = 1.66\overline{6} = 166.\overline{6}\%$

CHECK YOUR PROGRESS 5, *page 163*

$\dfrac{\text{Percent}}{100} = \dfrac{\text{amount}}{\text{base}}$

$\dfrac{70}{100} = \dfrac{22{,}400}{B}$

$70 \cdot B = 100(22{,}400)$

$70B = 2{,}240{,}000$

$\dfrac{70B}{70} = \dfrac{2{,}240{,}000}{70}$

$B = 32{,}000$

The Blazer cost $32,000 when it was new.

CHECK YOUR PROGRESS 6, *page 164*

$\dfrac{\text{Percent}}{100} = \dfrac{\text{amount}}{\text{base}}$

$\dfrac{p}{100} = \dfrac{416{,}000}{1{,}300{,}000}$

$p \cdot 1{,}300{,}000 = 100(416{,}000)$

$1{,}300{,}000p = 41{,}600{,}000$

$\dfrac{1{,}300{,}000p}{1{,}300{,}000} = \dfrac{41{,}600{,}000}{1{,}300{,}000}$

$p = 32$

32% of the enlisted people are over the age of 30.

CHECK YOUR PROGRESS 7, *page 165*

$\dfrac{\text{Percent}}{100} = \dfrac{\text{amount}}{\text{base}}$

$\dfrac{3.5}{100} = \dfrac{A}{32{,}500}$

$3.5(32{,}500) = 100(A)$

$113{,}750 = 100A$

$\dfrac{113{,}750}{100} = \dfrac{100A}{100}$

$1137.5 = A$

The customer would receive a rebate of $1137.50.

CHECK YOUR PROGRESS 8, *page 165*

$PB = A$

$0.05(32{,}685) = A$

$1634.25 = A$

The teacher contributes $1634.25.

CHECK YOUR PROGRESS 9, *page 166*

$PB = A$

$0.03B = 14{,}370$

$\dfrac{0.03B}{0.03} = \dfrac{14{,}370}{0.03}$

$B = 479{,}000$

The selling price of the home was $479,000.

CHECK YOUR PROGRESS 10, *page 166*

$PB = A$

$P \cdot 90 = 63$

$\dfrac{P \cdot 90}{90} = \dfrac{63}{90}$

$P = 0.7$

$P = 70\%$

You answered 70% of the questions correctly.

CHECK YOUR PROGRESS 11, *page 167*

$PB = A$

$0.90(21{,}262) = A$

$19{,}135.80 = A$

$21{,}262 - 19{,}135.80 = 2126.20$

The difference between the cost of the remodeling and the increase in value of your home is $2126.20.

CHECK YOUR PROGRESS 12, *page 169*

$5.67 - 1.82 = 3.85$

$PB = A$

$P \cdot 1.82 = 3.85$

$\dfrac{P \cdot 1.82}{1.82} = \dfrac{3.85}{1.82}$

$P \approx 2.115$

The percent increase in the federal debt from 1985 to 2000 was 211.5%.

CHECK YOUR PROGRESS 13, *page 171*

$\dfrac{\text{Percent}}{100} = \dfrac{\text{amount}}{\text{base}}$

$\dfrac{3.81}{100} = \dfrac{A}{20{,}416}$

$3.81(20{,}416) = 100(A)$

$77{,}784.96 = 100A$

$\dfrac{77{,}784.96}{100} = \dfrac{100A}{100}$

$778 \approx A$

$20{,}416 - 778 = 19{,}638$

There were 19,638 passenger car fatalities in the United States in 2003.

SECTION 3.4

CHECK YOUR PROGRESS 1, *page 183*

$y = kx$	• Write the general form of a direct variation equation.
$120 = k \cdot 8$	• Substitute the given values for y and x.
$15 = k$	• Solve for k, the constant of variation.
$y = 15x$	• Write the specific direct variation equation that relates y and x by substituting 15 for k.

The direct variation equation is $y = 15x$.

CHECK YOUR PROGRESS 2, *page 184*

$T = kx$	• Write the general form of a direct variation equation.
$8 = k \cdot 2$	• Substitute 8 for T and 2 for x.
$4 = k$	• Solve for k, the constant of variation.
$T = 4x$	• Write the specific direct variation equation by substituting 4 for k.
$T = 4 \cdot 4$	• Find T when $x = 4$.
$T = 16$	

The tension is 16 pounds when the distance stretched is 4 inches.

CHECK YOUR PROGRESS 3, *page 185*

$d = kv^2$	• Write the general form of a direct variation equation.
$130 = k \cdot 40^2$	• Substitute 130 for d and 40 for v.
$130 = k \cdot 1600$	
$0.08125 = k$	• Solve for k, the constant of variation.
$d = 0.08125v^2$	• Write the specific direct variation equation by substituting 0.08125 for k.
$d = 0.08125 \cdot 60^2$	• Find d when $v = 60$.
$d = 0.08125 \cdot 3600$	
$d = 292.5$	

The stopping distance is 292.5 feet.

CHECK YOUR PROGRESS 4, *page 186*

$P = \dfrac{k}{R}$	• Write the general form of an inverse variation equation.
$20 = \dfrac{k}{5}$	• Substitute 20 for P and 5 for R.
$100 = k$	• Solve for k by multiplying each side of the equation by 5.

The constant of variation is 100.

$P = \dfrac{100}{R}$	• Write the specific inverse variation equation that relates P and R by substituting 100 for k.

The inverse variation equation is $P = \dfrac{100}{R}$.

CHECK YOUR PROGRESS 5, *page 186*

$h = \dfrac{k}{m}$	• Write the general form of an inverse variation equation.
$9 = \dfrac{k}{5}$	• Substitute 9 for h and 5 for m.
$45 = k$	• Solve for k, the constant of variation.
$h = \dfrac{45}{m}$	• Write the specific inverse variation equation by substituting 45 for k.
$h = \dfrac{45}{4}$	• Find h when $m = 4$.
$h = 11.25$	

It takes 11.25 hours for four assembly machines to complete the daily quota of plastic molds.

CHECK YOUR PROGRESS 6, *page 188*

$$I = \frac{k}{d^2}$$ • Write the general form of an inverse variation equation.

$$20 = \frac{k}{8^2}$$ • Substitute 20 for *I* and 8 for *d*.

$$20 = \frac{k}{64}$$

$$1280 = k$$ • Solve for , the constant of variation.

$$I = \frac{1280}{d^2}$$ • Write the specific inverse variation equation by substituting 1280 for *k*.

$$I = \frac{1280}{5^2}$$ • Find *I* when *d* = 5.

$$I = \frac{1280}{25}$$

$$I = 51.2$$

The intensity is 51.2 foot-candles at a distance of 5 feet.

CHAPTER 4

SECTION 4.1

CHECK YOUR PROGRESS 1, *page 205*

a. 1 295 m = 1.295 km

b. 7 543 g = 7.543 kg

c. 6.3 L = 6 300 ml

d. 2 kl = 2 000 L

CHECK YOUR PROGRESS 2, *page 207*

12(274 mg) = 3 288 mg

3 288 mg = 3.288 g

There are 3.288 g of cholesterol in one dozen eggs.

SECTION 4.2

CHECK YOUR PROGRESS 1, *page 214*

a. $40 \text{ in.} = 40 \text{ in.} \cdot 1 = \frac{40 \text{ in.}}{1} \cdot \frac{1 \text{ ft}}{12 \text{ in.}} = \frac{40 \text{ ft}}{12} = 3\frac{1}{3} \text{ ft}$

b. $1 \text{ mi} = 1 \text{ mi} \cdot 1 = \frac{1 \text{ mi}}{1} \cdot \frac{5280 \text{ ft}}{1 \text{ mi}} \cdot \frac{1 \text{ yd}}{3 \text{ ft}}$

$= \frac{5280 \text{ yd}}{3} = 1760 \text{ yd}$

CHECK YOUR PROGRESS 2, *page 214*

$2880 \text{ min} = 2880 \text{ min} \cdot 1 \cdot 1 = \frac{2880 \text{ min}}{1} \cdot \frac{1 \text{ h}}{60 \text{ min}} \cdot \frac{1 \text{ day}}{24 \text{ h}}$

$= \frac{2880 \text{ days}}{1440} = 2 \text{ days}$

CHECK YOUR PROGRESS 3, *page 216*

Find the volume of water in the fish tank.

$V = LWH$

$V = (36)(23)(16)$

$V = 13{,}248$

The volume of water is 13,248 in³.

Multiply the volume of water by the conversion rate $\frac{1 \text{ gal}}{231 \text{ in}^3}$ to convert the volume to gallons.

$13{,}248 \text{ in}^3 = \frac{13{,}248 \text{ in}^3}{1} \cdot \frac{1 \text{ gal}}{231 \text{ in}^3} \approx 57.4 \text{ gal}$

There are 57.4 gal of water in the tank.

CHECK YOUR PROGRESS 4, *page 217*

a. $45 \text{ cm} = \frac{45 \text{ cm}}{1} \cdot \frac{1 \text{ in.}}{2.54 \text{ cm}}$

$= \frac{45 \text{ in.}}{2.54} \approx 17.72 \text{ in.}$

b. $\frac{75 \text{ km}}{h} \approx \frac{75 \text{ km}}{h} \cdot \frac{1 \text{ mi}}{1.61 \text{ km}}$

$= \frac{75 \text{ mi}}{1.61 \text{ h}} \approx 46.58 \text{ mph}$

CHECK YOUR PROGRESS 5, *page 217*

$\$2.39 \text{ per gallon} = \frac{\$2.39}{\text{gal}} \approx \frac{\$2.39}{\text{gal}} \cdot \frac{1 \text{ gal}}{3.79 \text{ L}}$

$= \frac{\$2.39}{3.79 \text{ L}} \approx \frac{\$.63}{\text{L}}$

The price is approximately $.63 per liter.

CHECK YOUR PROGRESS 6, *page 218*

$\$1.75 \text{ per liter} = \frac{\$1.75}{\text{L}} \approx \frac{\$1.75}{\text{L}} \cdot \frac{3.79 \text{ L}}{1 \text{ gal}}$

$= \frac{\$6.6325}{1 \text{ gal}} \approx \$6.63/\text{gal}$

The price is approximately $6.63 per gallon.

SECTION 4.3

CHECK YOUR PROGRESS 1, *page 223*

$QR + RS + ST = QT$

$28 + 16 + 10 = QT$

$54 = QT$

$QT = 54$ cm

CHECK YOUR PROGRESS 2, *page 223*

$AB + BC = AC$

$\frac{1}{4}(BC) + BC = AC$

$\frac{1}{4}(16) + 16 = AC$

$4 + 16 = AC$

$20 = AC$

$AC = 20$ ft

CHECK YOUR PROGRESS 3, *page 226*

$m\angle G + m\angle H = 127° + 53° = 180°$

The sum of the measures of $\angle G$ and $\angle H$ is $180°$. Angles G and H are supplementary angles.

CHECK YOUR PROGRESS 4, *page 227*

Supplementary angles are two angles the sum of whose measures is $180°$. To find the supplement, let x represent the supplement of a $129°$ angle.

$x + 129 = 180$

$x = 51$

The supplement of a $129°$ angle is a $51°$ angle.

CHECK YOUR PROGRESS 5, *page 228*

$m\angle a + 68° = 118°$

$m\angle a = 50°$

The measure of $\angle a$ is $50°$.

CHECK YOUR PROGRESS 6, *page 229*

$m\angle b + m\angle a = 180°$

$m\angle b + 35° = 180°$

$m\angle b = 145°$

$m\angle c = m\angle a = 35°$

$m\angle d = m\angle b = 145°$

$m\angle b = 145°$, $m\angle c = 35°$, and $m\angle d = 145°$.

CHECK YOUR PROGRESS 7, *page 231*

$m\angle b = m\angle g = 124°$

$m\angle d = m\angle g = 124°$

$m\angle c + m\angle b = 180°$

$m\angle c + 124° = 180°$

$m\angle c = 56°$

$m\angle b = 124°$, $m\angle c = 56°$, and $m\angle d = 124°$.

CHECK YOUR PROGRESS 8, *page 233*

$m\angle b + m\angle d = 180°$

$m\angle b + 105° = 180°$

$m\angle b = 75°$

$m\angle a + m\angle b + m\angle c = 180°$

$m\angle a + 75° + 35° = 180°$

$m\angle a + 110° = 180°$

$m\angle a = 70°$

$m\angle e = m\angle a = 70°$

CHECK YOUR PROGRESS 9, *page 233*

Let x represent the measure of the third angle.

$x + 90° + 27° = 180°$

$x + 117° = 180°$

$x = 63°$

The measure of the third angle is $63°$.

SECTION 4.4

CHECK YOUR PROGRESS 1, *page 244*

$P = a + b + c$

$P = 4\frac{3}{10} + 2\frac{1}{10} + 6\frac{1}{2}$

$P = 4\frac{3}{10} + 2\frac{1}{10} + 6\frac{5}{10}$

$P = 12\frac{9}{10}$

The total length of the bike trail is $12\frac{9}{10}$ mi.

CHECK YOUR PROGRESS 2, *page 245*

$P = 2L + 2W$

$P = 2(12) + 2(8)$

$P = 24 + 16$

$P = 40$

You will need 40 ft of molding to edge the tops of the walls.

CHECK YOUR PROGRESS 3, *page 246*

$P = 4s$

$P = 4(24)$

$P = 96$

The homeowner should purchase 96 ft of fencing.

CHECK YOUR PROGRESS 4, *page 246*

$P = 2b + 2s$

$P = 2(5) + 2(7)$

$P = 10 + 14$

$P = 24$

24 m of plank is needed to surround the garden.

CHECK YOUR PROGRESS 5, *page 248*

$C = \pi d$

$C = 9\pi$

The circumference of the circle is 9π km.

CHECK YOUR PROGRESS 6, *page 248*

12 in. = 1 ft

$C = \pi d$

$C = \pi(1)$

$C = \pi$

$12C = 12\pi \approx 37.70$

The tricycle travels approximately 37.70 ft when the wheel makes 12 revolutions.

CHECK YOUR PROGRESS 7, *page 250*

$A = LW$

$A = 308(192)$

$A = 59{,}136$

59,136 cm^2 of fabric is needed.

CHECK YOUR PROGRESS 8, *page 251*

$A = s^2$

$A = 24^2$

$A = 576$

The area of the floor is 576 ft^2.

CHECK YOUR PROGRESS 9, *page 252*

$A = bh$

$A = 14(8)$

$A = 112$

The area of the patio is 112 m^2.

CHECK YOUR PROGRESS 10, *page 253*

$A = \dfrac{1}{2}bh$

$A = \dfrac{1}{2}(18)(9)$

$A = 9(9)$

$A = 81$

81 in^2 of felt is needed.

CHECK YOUR PROGRESS 11, *page 254*

$A = \dfrac{1}{2}h(b_1 + b_2)$

$A = \dfrac{1}{2} \cdot 9(12 + 20)$

$A = \dfrac{1}{2} \cdot 9(32)$

$A = \dfrac{9}{2} \cdot (32)$

$A = 144$

The area of the patio is 144 ft^2.

CHECK YOUR PROGRESS 12, *page 255*

$r = \dfrac{1}{2}d = \dfrac{1}{2}(12) = 6$

$A = \pi r^2$

$A = \pi(6)^2$

$A = 36\pi$

The area of the circle is 36π km^2.

CHECK YOUR PROGRESS 13, *page 256*

$r = \dfrac{1}{2}d = \dfrac{1}{2}(4) = 2$

$A = \pi r^2$

$A = \pi(2)^2$

$A = \pi(4)$

$A \approx 12.57$

Approximately 12.57 ft^2 of material is needed.

SECTION 4.5

CHECK YOUR PROGRESS 1, *page 266*

$\dfrac{AC}{DF} = \dfrac{CH}{FG}$

$\dfrac{10}{15} = \dfrac{7}{FG}$

$10(FG) = (15)7$

$10(FG) = 105$

$FG = 10.5$

The height *FG* of triangle *DEF* is 10.5 m.

CHECK YOUR PROGRESS 2, *page 269*

$$\frac{AO}{DO} = \frac{AB}{DC}$$

$$\frac{AO}{3} = \frac{10}{4}$$

$$4(AO) = 3(10)$$

$$4(AO) = 30$$

$$AO = 7.5$$

$$A = \frac{1}{2}bh$$

$$A = \frac{1}{2}(10)(7.5)$$

$$A = 5(7.5)$$

$$A = 37.5$$

The area of triangle *AOB* is 37.5 cm^2.

CHECK YOUR PROGRESS 3, *page 271*

$a^2 + b^2 = c^2$ • Use the Pythagorean Theorem.

$2^2 + b^2 = 6^2$ • $a = 2$, $c = 6$

$4 + b^2 = 36$

$b^2 = 32$ • Solve for b^2. Subtract 4 from each side.

$\sqrt{b^2} = \sqrt{32}$ • Take the square root of each side of the equation.

$b \approx 5.66$ • Use a calculator to approximate $\sqrt{32}$.

The length of the other leg is approximately 5.66 m.

SECTION 4.6

CHECK YOUR PROGRESS 1, *page 281*

$V = LWH$

$V = 5(3.2)(4)$

$V = 64$

The volume of the solid is 64 m^3.

CHECK YOUR PROGRESS 2, *page 281*

$$V = \frac{1}{3}s^2h$$

$$V = \frac{1}{3}(15)^2(25)$$

$$V = \frac{1}{3}(225)(25)$$

$$V = 1875$$

The volume of the pyramid is 1875 m^3.

CHECK YOUR PROGRESS 3, *page 282*

$$r = \frac{1}{2}d = \frac{1}{2}(16) = 8$$

$V = \pi r^2 h$

$V = \pi(8)^2(30)$

$V = \pi(64)(30)$

$V = 1920\pi$

$$\frac{1}{4}(1920\pi) = 480\pi$$

$$\approx 1507.96$$

Approximately 1507.96 ft^3 is not being used for storage.

CHECK YOUR PROGRESS 4, *page 285*

$$r = \frac{1}{2}d = \frac{1}{2}(6) = 3$$

$S = 2\pi r^2 + 2\pi rh$

$S = 2\pi(3)^2 + 2\pi(3)(8)$

$S = 2\pi(9) + 2\pi(3)(8)$

$S = 18\pi + 48\pi$

$S = 66\pi$

$S \approx 207.35$

The surface area of the cylinder is approximately 207.35 ft^2.

CHECK YOUR PROGRESS 5, *page 285*

Surface area of the cube $= 6s^2$

$= 6(8)^2$

$= 6(64)$

$= 384$ cm^2

Surface area of the sphere $= 4\pi r^2$

$= 4\pi(5)^2$

$= 4\pi(25)$

≈ 314.16 cm^2

The cube has a larger surface area.

SECTION 4.7

CHECK YOUR PROGRESS 1, *page 294*

Use the Pythagorean Theorem to find the length of the hypotenuse.

$a^2 + b^2 = c^2$

$3^2 + 4^2 = c^2$

$9 + 16 = c^2$

$25 = c^2$

$\sqrt{25} = \sqrt{c^2}$

$5 = c$

$\sin \theta = \dfrac{\text{opp}}{\text{hyp}} = \dfrac{3}{5}$,

$\cos \theta = \dfrac{\text{adj}}{\text{hyp}} = \dfrac{4}{5}$,

$\tan \theta = \dfrac{\text{opp}}{\text{adj}} = \dfrac{3}{4}$

CHECK YOUR PROGRESS 2, *page 295*

$\tan 37.1° = 0.7563$

CHECK YOUR PROGRESS 3, *page 296*

We are given the measure of $\angle B$ and the hypotenuse. We want to find the length of side a. The cosine function involves the side adjacent and the hypotenuse.

$$\cos B = \frac{\text{adj}}{\text{hyp}}$$

$$\cos 48° = \frac{a}{12}$$

$$12(\cos 48°) = a$$

$$8.03 \approx a$$

The length of side a is approximately 8.03 ft.

CHECK YOUR PROGRESS 4, *page 297*

$\tan^{-1}(0.3165) \approx 17.6°$

CHECK YOUR PROGRESS 5, *page 297*

$\theta \approx \tan^{-1}(0.5681)$

$\theta \approx 29.6°$

CHECK YOUR PROGRESS 6, *page 298*

We want to find the measure of $\angle A$, and we are given the length of the side opposite $\angle A$ and the hypotenuse. The sine function involves the side opposite an angle and the hypotenuse.

$$\sin A = \frac{\text{opp}}{\text{hyp}}$$

$$\sin A = \frac{7}{11}$$

$$A = \sin^{-1}\frac{7}{11}$$

$$A \approx 39.5°$$

The measure of $\angle A$ is approximately 39.5°.

CHECK YOUR PROGRESS 7, *page 299*

Let d be the distance from the center of the base of the lighthouse to the boat.

$$\tan 25° = \frac{20}{d}$$

$$d(\tan 25°) = 20$$

$$d = \frac{20}{\tan 25°}$$

$$d \approx 42.9$$

The boat is approximately 42.9 m from the base of the lighthouse.

CHAPTER 5

SECTION 5.1

CHECK YOUR PROGRESS 1, *page 314*

x	$-2x + 3 = y$	(x, y)
-2	$-2(-2) + 3 = 7$	$(-2, 7)$
-1	$-2(-1) + 3 = 5$	$(-1, 5)$
0	$-2(0) + 3 = 3$	$(0, 3)$
1	$-2(1) + 3 = 1$	$(1, 1)$
2	$-2(2) + 3 = -1$	$(2, -1)$
3	$-2(3) + 3 = -3$	$(3, -3)$

CHECK YOUR PROGRESS 2, *page 315*

x	$-x^2 + 1 = y$	(x, y)
-3	$-(-3)^2 + 1 = -8$	$(-3, -8)$
-2	$-(-2)^2 + 1 = -3$	$(-2, -3)$
-1	$-(-1)^2 + 1 = 0$	$(-1, 0)$
0	$-(0)^2 + 1 = 1$	$(0, 1)$
1	$-(1)^2 + 1 = 0$	$(1, 0)$
2	$-(2)^2 + 1 = -3$	$(2, -3)$
3	$-(3)^2 + 1 = -8$	$(3, -8)$

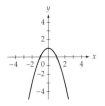

CHECK YOUR PROGRESS 3, *page 318*

$$f(z) = z^2 - z$$
$$f(-3) = (-3)^2 - (-3)$$
$$= 12$$

The value of the function is 12 when $z = -3$.

CHECK YOUR PROGRESS 4, *page 318*

$$N(s) = \frac{s^2 - 3s}{2}$$

$$N(12) = \frac{(12)^2 - 3(12)}{2}$$

$$= \frac{144 - 36}{2}$$

$$= 54$$

A polygon with 12 sides has 54 diagonals.

CHECK YOUR PROGRESS 5, *page 320*

x	$f(x) = 2 - \dfrac{3}{4}x$	(x, y)
-3	$f(-3) = 2 - \dfrac{3}{4}(-3) = 4\dfrac{1}{4}$	$\left(-3, 4\dfrac{1}{4}\right)$
-2	$f(-2) = 2 - \dfrac{3}{4}(-2) = 3\dfrac{1}{2}$	$\left(-2, 3\dfrac{1}{2}\right)$
-1	$f(-1) = 2 - \dfrac{3}{4}(-1) = 2\dfrac{3}{4}$	$\left(-1, 2\dfrac{3}{4}\right)$
0	$f(0) = 2 - \dfrac{3}{4}(0) = 2$	$(0, 2)$
1	$f(1) = 2 - \dfrac{3}{4}(1) = 1\dfrac{1}{4}$	$\left(1, 1\dfrac{1}{4}\right)$
2	$f(2) = 2 - \dfrac{3}{4}(2) = \dfrac{1}{2}$	$\left(2, \dfrac{1}{2}\right)$
3	$f(3) = 2 - \dfrac{3}{4}(3) = -\dfrac{1}{4}$	$\left(3, -\dfrac{1}{4}\right)$

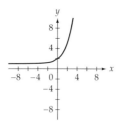

SECTION 5.2

CHECK YOUR PROGRESS 1, *page 326*

$$f(x) = \frac{1}{2}x + 3 \qquad\qquad f(x) = \frac{1}{2}x + 3$$

$$0 = \frac{1}{2}x + 3 \qquad\qquad f(0) = \frac{1}{2}(0) + 3$$

$$-3 = \frac{1}{2}x \qquad\qquad\qquad = 3$$

$$-6 = x$$

The x-intercept is $(-6, 0)$.　　The y-intercept is $(0, 3)$.

CHECK YOUR PROGRESS 2, *page 327*

$$g(t) = -20t + 8000$$
$$g(0) = -20(0) + 8000 = 8000$$

The intercept on the vertical axis is $(0, 8000)$. This means that the plane is at an altitude of 8000 feet when it begins its descent.

$$g(t) = -20t + 8000$$
$$0 = -20t + 8000$$
$$-8000 = -20t$$
$$400 = t$$

The intercept on the horizontal axis is $(400, 0)$. This means that the plane reaches the ground 400 seconds after beginning its descent.

CHECK YOUR PROGRESS 3, *page 329*

a. $(x_1, y_1) = (-6, 5), (x_2, y_2) = (4, -5)$

$$m = \frac{y_2 - y_1}{x_2 - x_1} = \frac{-5 - 5}{4 - (-6)} = \frac{-10}{10} = -1$$

The slope is -1.

b. $(x_1, y_1) = (-5, 0), (x_2, y_2) = (-5, 7)$

$$m = \frac{y_2 - y_1}{x_2 - x_1} = \frac{7 - 0}{-5 - (-5)} = \frac{7}{0}$$

The slope is undefined.

c. $(x_1, y_1) = (-7, -2), (x_2, y_2) = (8, 8)$

$$m = \frac{y_2 - y_1}{x_2 - x_1} = \frac{8 - (-2)}{8 - (-7)} = \frac{10}{15} = \frac{2}{3}$$

The slope is $\frac{2}{3}$.

d. $(x_1, y_1) = (-6, 7), (x_2, y_2) = (1, 7)$

$$m = \frac{y_2 - y_1}{x_2 - x_1} = \frac{7 - 7}{1 - (-6)} = \frac{0}{7} = 0$$

The slope is 0.

CHECK YOUR PROGRESS 4, *page 330*

For the linear function $d(t) = 50t$, the slope is the coefficient of t. Therefore, the slope is 50. This means that a homing pigeon can fly 50 miles for each 1 hour of flight time.

CHECK YOUR PROGRESS 5, *page 331*

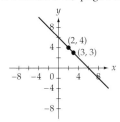

CHECK YOUR PROGRESS 6, *page 332*

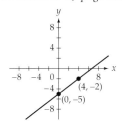

SECTION 5.3

CHECK YOUR PROGRESS 1, *page 339*

$f(a) = ma + b$

$f(a) = -3.5a + 100$

The linear function is $f(a) = -3.5a + 100$, where $f(a)$ is the boiling point of water in degrees Celcius at an altitude of a kilometers above sea level.

CHECK YOUR PROGRESS 2, *page 339*

$y - y_1 = m(x - x_1)$

$y - 2 = -\dfrac{1}{2}[x - (-2)]$

$y - 2 = -\dfrac{1}{2}x - 1$

$y = -\dfrac{1}{2}x + 1$

CHECK YOUR PROGRESS 3, *page 340*

$C - C_1 = m(t - t_1)$

$C - 191 = 3.8(t - 50)$

$C - 191 = 3.8t - 190$

$C = 3.8t + 1$

A linear function that models the number of calories burned after t minutes is $C(t) = 3.8t + 1$.

CHECK YOUR PROGRESS 4, *page 340*

$m = \dfrac{y_2 - y_1}{x_2 - x_1} = \dfrac{1 - 3}{4 - (-2)} = \dfrac{-2}{6} = -\dfrac{1}{3}$

$y - y_1 = m(x - x_1)$

$y - 3 = -\dfrac{1}{3}[x - (-2)]$

$y - 3 = -\dfrac{1}{3}x - \dfrac{2}{3}$

$y = -\dfrac{1}{3}x + \dfrac{7}{3}$

CHECK YOUR PROGRESS 5, *page 344*

a. We can simplify the data by letting $x = 0$ correspond to 1990. Then $x = 2$ represents 1992, $x = 4$ represents 1994, and so on.

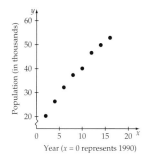

b. Looking at the scatter diagram, it appears that a line through the points $(4, 26.31)$ and $(14, 49.87)$ will fit the data points well.

The slope of the line through these points is

$m = \dfrac{y_2 - y_1}{x_2 - x_1} = \dfrac{49.87 - 26.31}{14 - 4} = \dfrac{23.56}{10} = 2.356$

and the equation of the line is

$y - y_1 = m(x - x_1)$

$y - 26.31 = 2.356(x - 4)$

$y - 26.31 = 2.356x - 9.424$

$y = 2.356x + 16.886$

The city's population (in thousands) is approximately $f(x) = 2.356x + 16.886$, where $x = 0$ corresponds to 1990.

c. Evaluate the function when $x = 35$:

$f(35) = 2.356(35) + 16.886 = 99.346$. The population in 2025 is predicted to be approximately 99,346.

SECTION 5.4

CHECK YOUR PROGRESS 1, *page 351*

a. The regression equation is approximately $y = 5.6333x - 252.8667$.

b. When $x = 63$, we have $y = 5.6333(63) - 252.8667 \approx 102$. The estimated weight of a woman swimmer who is 63 inches tall is approximately 102 pounds.

CHECK YOUR PROGRESS 2, *page 352*

a. The equation of the regression line is approximately
$y = 3.130x - 5.55$.

b. When $x = 2.7$, we have $y = 3.130(2.7) - 5.55 \approx 2.9$, and
when $x = 4.5$, we have $y = 3.130(4.5) - 5.55 \approx 8.5$. The
predicted average speed of a camel with a stride length of
2.7 meters is about 2.9 meters per second, and the predicted
average speed for a camel with a stride length of 4.5 meters is
approximately 8.5 meters per second.

CHECK YOUR PROGRESS 3, *page 354*

A graphing calculator gives $r = 0.998497842$, so the linear
correlation coefficient is approximately 1.00.

CHAPTER 6

SECTION 6.1

CHECK YOUR PROGRESS 1, *page 370*

a. $g(t) = 3t^2 - 4t + 5$

b. The degree is 2, the largest exponent on the variable.

c. $g(t) = 3t^2 - 4t + 5$
$g(3) = 3(3)^2 - 4(3) + 5$
$\quad = 27 - 12 + 5$
$\quad = 20$

CHECK YOUR PROGRESS 2, *page 372*

$a = 1, b = 0; x = -\dfrac{b}{2a} = -\dfrac{0}{2(1)} = 0$

$y = x^2 - 2$

$y = (0)^2 - 2$

$y = -2$

The vertex is $(0, -2)$.

CHECK YOUR PROGRESS 3, *page 373*

$a = 2, b = -3; x = -\dfrac{b}{2a} = -\dfrac{-3}{2(2)} = \dfrac{3}{4}$

$f(x) = 2x^2 - 3x + 1$

$f\left(\dfrac{3}{4}\right) = 2\left(\dfrac{3}{4}\right)^2 - 3\left(\dfrac{3}{4}\right) + 1$

$f\left(\dfrac{3}{4}\right) = -\dfrac{1}{8}$

The vertex is $\left(\dfrac{3}{4}, -\dfrac{1}{8}\right)$. The minimum value of the function is $-\dfrac{1}{8}$,
the y-coordinate of the vertex.

CHECK YOUR PROGRESS 4, *page 374*

$a = -16, b = 64; x = -\dfrac{b}{2a} = -\dfrac{64}{2(-16)} = 2$

The ball reaches its maximum height in 2 seconds.

$s(t) = -16t^2 + 64t + 4$

$s(2) = -16(2)^2 + 64(2) + 4$

$s(2) = 68$

The maximum height of the ball is 68 feet.

CHECK YOUR PROGRESS 5, *page 375*

Perimeter: $w + l + w + l = 44$
$\qquad\qquad\qquad 2w + 2l = 44$
$\qquad\qquad\qquad\quad w + l = 22$
$\qquad\qquad\qquad\qquad\quad l = -w + 22$

Area: $A = lw$
$\qquad\quad A = (-w + 22)w$
$\qquad\quad A = -w^2 + 22w$

$w = -\dfrac{b}{2a} = -\dfrac{22}{2(-1)} = 11$

The width is 11 feet.

$l = -w + 22$

$l = -(11) + 22 = 11$

The length is 11 feet.

The dimensions of the rectangle with maximum area are 11 feet by
11 feet.

CHECK YOUR PROGRESS 6, *page 376*

The surface area is the sum of the areas of the four sides of the box
and the area of its bottom.

4 sides   Area of each side   Area of the base

$S(x) = 4(50 - 2x)x + (50 - 2x)(50 - 2x)$

$\qquad = 200x - 8x^2 + 2500 - 200x + 4x^2$

$\qquad = -4x^2 + 2500$

CHECK YOUR PROGRESS 7, *page 377*

a. $[d(t)]^2 = 2^2 + (400t)^2$
$\quad [d(t)]^2 = 4 + 160{,}000t^2$
$\qquad d(t) = \sqrt{4 + 160{,}000t^2}$

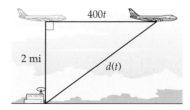

b. $d(t) = \sqrt{4 + 160{,}000t^2}$
$\quad d(3) = \sqrt{4 + 160{,}000(3)^2}$
$\quad d(3) \approx 1200$

CHECK YOUR PROGRESS 8, *page 378*

a. Let $A(x)$ be the average cost per chair.

$$A(x) = \frac{\text{total cost}}{x} = \frac{C(x)}{x}$$

$$= \frac{35x + 500}{x}$$

b. $A(x) = \frac{35x + 500}{x}$

$$A(40) = \frac{35(40) + 500}{40} = 47.50$$

The average cost is $47.50 per chair.

SECTION 6.2

CHECK YOUR PROGRESS 1, *page 387*

$$g(x) = \left(\frac{1}{2}\right)^x$$

$$g(3) = \left(\frac{1}{2}\right)^3 = \frac{1}{8}$$

$$g(-1) = \left(\frac{1}{2}\right)^{-1} = \frac{1}{\frac{1}{2}} = 2$$

$$g(\sqrt{3}) = \left(\frac{1}{2}\right)^{\sqrt{3}} \approx \left(\frac{1}{2}\right)^{1.732} \approx 0.301$$

CHECK YOUR PROGRESS 2, *page 389*

Because the base $\frac{3}{2}$ is greater than 1, f is an exponential growth function.

x	$f(x) = \left(\frac{3}{2}\right)^x$	(x, y)
−3	$f(-3) = \left(\frac{3}{2}\right)^{-3} = \frac{8}{27}$	$\left(-3, \frac{8}{27}\right)$
−2	$f(-2) = \left(\frac{3}{2}\right)^{-2} = \frac{4}{9}$	$\left(-2, \frac{4}{9}\right)$
−1	$f(-1) = \left(\frac{3}{2}\right)^{-1} = \frac{2}{3}$	$\left(-1, \frac{2}{3}\right)$
0	$f(0) = \left(\frac{3}{2}\right)^0 = 1$	$(0, 1)$
1	$f(1) = \left(\frac{3}{2}\right)^1 = \frac{3}{2}$	$\left(1, \frac{3}{2}\right)$
2	$f(2) = \left(\frac{3}{2}\right)^2 = \frac{9}{4}$	$\left(2, \frac{9}{4}\right)$
3	$f(3) = \left(\frac{3}{2}\right)^3 = \frac{27}{8}$	$\left(3, \frac{27}{8}\right)$

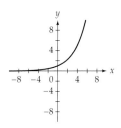

CHECK YOUR PROGRESS 3, *page 390*

x	−2	−1	0	1	2
$f(x) = e^{-x} + 2$	9.4	4.7	3	2.4	2.1

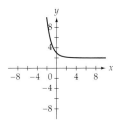

CHECK YOUR PROGRESS 4, *page 391*

$$N(t) = 1.5\left(\frac{1}{2}\right)^{t/193.7}$$

$$N(24) = 1.5\left(\frac{1}{2}\right)^{24/193.7}$$

$$\approx 1.5(0.9177) \approx 1.3766$$

After 24 hours, there is approximately 1.3766 grams of the isotope in the body.

CHECK YOUR PROGRESS 5, *page 392*

$$A(t) = 200e^{-0.014t}$$

$$A(45) = 200e^{-0.014(45)}$$

$$\approx 107$$

After 45 minutes, there is approximately 107 milligrams of aspirin in the patient's bloodstream.

CHECK YOUR PROGRESS 6, *pages 393–394*

The regression equation is $P(a) \approx 10.1468(0.8910)^a$.

The atmospheric pressure at an altitude of 24 kilometers is approximately 0.6 newton per square centimeter.

SECTION 6.3

CHECK YOUR PROGRESS 1, *page 400*

a. $2^{10} = 4x$

b. $\log_{10} 2x = 3$

CHECK YOUR PROGRESS 2, *page 401*

a. $\log_{10} 0.001 = x$
$\qquad 10^x = 0.001$
$\qquad 10^x = 10^{-3}$
$\qquad\quad x = -3$
$\quad \log_{10} 0.001 = -3$

b. $\log_5 125 = x$
$\qquad 5^x = 125$
$\qquad 5^x = 5^3$
$\qquad\ x = 3$
$\log_5 125 = 3$

CHECK YOUR PROGRESS 3, *page 401*

$\log_2 x = 6$
$\quad 2^6 = x$
$\quad 64 = x$

CHECK YOUR PROGRESS 4, *page 402*

a. $\log x = -2.1$
$\quad 10^{-2.1} = x$
$\quad 0.008 \approx x$

b. $\ln x = 2$
$\quad e^2 = x$
$\quad 7.389 \approx x$

CHECK YOUR PROGRESS 5, *page 403*

$y = \log_5 x$
$5^y = x$

$x = 5^y$	$\dfrac{1}{25}$	$\dfrac{1}{5}$	1	5	25
y	-2	-1	0	1	2

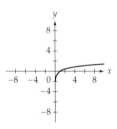

CHECK YOUR PROGRESS 6, *page 404*

a. $S(0) = 5 + 29 \ln(0 + 1) = 5$
The average typing speed when the student first started to type was 5 words per minute.
$S(3) = 5 + 29 \ln(3 + 1) \approx 45$
The average typing speed after 3 months was about 45 words per minute.

b. $S(3) - S(0) = 45 - 5 = 40$
The typing speed increased by 40 words per minute during the 3 months.

CHECK YOUR PROGRESS 7, *page 405*

$I = 2 \cdot (12,589,254 I_0) = 25,178,508 I_0$

$M = \log\left(\dfrac{I}{I_0}\right) = \log\left(\dfrac{25,178,508 I_0}{I_0}\right) = \log(25,178,508) \approx 7.4$

The Richter scale magnitude of an earthquake whose intensity is twice that of the Amazonas, Brazil earthquake is 7.4.

CHECK YOUR PROGRESS 8, *page 405*

$\log\left(\dfrac{I}{I_0}\right) = 4.6$

$\qquad \dfrac{I}{I_0} = 10^{4.6}$

$\qquad\ I = 10^{4.6} I_0$

$\qquad\ I \approx 39,811 I_0$

The April 29, 2003 earthquake had an intensity that was approximately 40,000 times the intensity of a zero-level earthquake.

CHECK YOUR PROGRESS 9, *page 406*

a. $pH = -\log[H^+] = -\log(2.41 \times 10^{-13}) \approx 12.6$
The cleaning solution has a pH of 12.6.

b. $pH = -\log[H^+] = -\log(5.07 \times 10^{-4}) \approx 3.3$
The cola soft drink has a pH of 3.3.

c. $pH = -\log[H^+] = -\log(6.31 \times 10^{-5}) \approx 4.2$
The rainwater has a pH of 4.2.

CHECK YOUR PROGRESS 10, *page 407*

$\qquad\quad pH = -\log[H^+]$
$\qquad\ 10.0 = -\log[H^+]$
$\qquad -10.0 = \log[H^+]$
$\qquad 10^{-10.0} = H^+$
$1.0 \times 10^{-10} = H^+$

The hydronium-ion concentration of the water in the Great Salt Lake in Utah is 1.0×10^{-10} mole per liter.

CHAPTER 7

SECTION 7.1

CHECK YOUR PROGRESS 1, *page 416*
$P = 500$, $r = 4\% = 0.04$, $t = 1$

$I = Prt$
$I = 500(0.04)(1)$
$I = 20$

The simple interest earned is $20.

CHECK YOUR PROGRESS 2, *page 417*
$P = 1500$, $r = 5.25\% = 0.0525$

$t = \dfrac{4 \text{ months}}{1 \text{ year}} = \dfrac{4 \text{ months}}{12 \text{ months}} = \dfrac{4}{12}$

$I = Prt$

$I = 1500(0.0525)\left(\dfrac{4}{12}\right)$

$I = 26.25$

The simple interest due is $26.25.

CHECK YOUR PROGRESS 3, *page 417*
$P = 700, r = 1.25\% = 0.0125, t = 5$

$I = Prt$
$I = 700(0.0125)(5)$
$I = 43.75$

The simple interest due is $43.75.

CHECK YOUR PROGRESS 4, *page 418*
$P = 7000, r = 5.25\% = 0.0525$

$t = \dfrac{\text{number of days}}{360} = \dfrac{120}{360}$

$I = Prt$

$I = 7000(0.0525)\left(\dfrac{120}{360}\right)$

$I = 122.5$

The simple interest due is $122.50.

CHECK YOUR PROGRESS 5, *page 419*

$I = Prt$

$462 = 12{,}000(r)\left(\dfrac{6}{12}\right)$

$462 = 6000r$
$0.077 = r$
$r = 7.7\%$

The simple interest rate on the loan is 7.7%.

CHECK YOUR PROGRESS 6, *page 420*
Find the interest.

$P = 4000, r = 8.75\% = 0.0875, t = \dfrac{9}{12}$

$I = Prt$

$I = 4000(0.0875)\left(\dfrac{9}{12}\right)$

$I = 262.50$

Find the maturity value.
$A = P + I$
$A = 4000 + 262.50$
$A = 4262.50$

The maturity value of the loan is $4262.50.

CHECK YOUR PROGRESS 7, *page 421*
$P = 6700, r = 8.9\% = 0.089, t = 1$

$A = P(1 + rt)$
$A = 6700[1 + 0.089(1)]$
$A = 6700(1 + 0.089)$
$A = 6700(1.089)$
$A = 7296.30$

The maturity value of the loan is $7296.30.

CHECK YOUR PROGRESS 8, *page 421*
$P = 680, r = 6.4\% = 0.064, t = 1$

$A = P(1 + rt)$
$A = 680[1 + 0.064(1)]$
$A = 680(1 + 0.064)$
$A = 680(1.064)$
$A = 723.52$

After 1 year, $723.52 is in the account.

CHECK YOUR PROGRESS 9, *page 422*
$I = A - P$
$I = 9240 - 9000$
$I = 240$

$I = Prt$

$240 = 9000(r)\left(\dfrac{4}{12}\right)$

$240 = 3000r$
$0.08 = r$
$r = 8.0\%$

The simple interest rate on the loan is 8.0%.

SECTION 7.2

CHECK YOUR PROGRESS 1, *page 431*
$A = P(1 + rt)$

$A = 2000\left[1 + 0.04\left(\dfrac{1}{12}\right)\right]$

$A \approx 2006.67$

$A = P(1 + rt)$

$A \approx 2006.67\left[1 + 0.04\left(\dfrac{1}{12}\right)\right]$

$A \approx 2013.36$

$A = P(1 + rt)$

$A \approx 2013.36\left[1 + 0.04\left(\dfrac{1}{12}\right)\right]$

$A \approx 2020.07$

$A = P(1 + rt)$

$A = 2020.07\left[1 + 0.04\left(\dfrac{1}{12}\right)\right]$

$A \approx 2026.80$

$A = P(1 + rt)$

$A = 2026.80\left[1 + 0.04\left(\dfrac{1}{12}\right)\right]$

$A \approx 2033.56$

$A = P(1 + rt)$

$A = 2033.56\left[1 + 0.04\left(\dfrac{1}{12}\right)\right]$

$A \approx 2040.34$

The total amount in the account at the end of 6 months is $2040.34.

CHECK YOUR PROGRESS 2, *page 434*

$r = 6\% = 0.06$, $n = 12$, $t = 2$, $i = \dfrac{r}{n} = \dfrac{0.06}{12} = 0.005$,

$N = nt = 12(2) = 24$

$A = P(1 + i)^N$

$A = 4000(1 + 0.005)^{24}$

$A = 4000(1.005)^{24}$

$A \approx 4000(1.127160)$

$A \approx 4508.64$

The compound amount after 2 years is approximately $4508.64.

CHECK YOUR PROGRESS 3, *page 434*

$r = 9\% = 0.09$, $n = 360$, $t = 4$, $i = \dfrac{r}{n} = \dfrac{0.09}{360} = 0.00025$,

$N = nt = 360(4) = 1440$

$A = P(1 + i)^N$

$A = 2500(1 + 0.00025)^{1440}$

$A = 2500(1.00025)^{1440}$

$A \approx 2500(1.4332649)$

$A \approx 3583.16$

The future amount after 4 years is approximately $3583.16.

CHECK YOUR PROGRESS 4, *page 435*

$r = 9\% = 0.09$, $n = 12$, $t = 6$, $i = \dfrac{r}{n} = \dfrac{0.09}{12} = 0.0075$,

$N = nt = 12(6) = 72$

$A = P(1 + i)^N$

$A = 8000(1 + 0.0075)^{72}$

$A = 8000(1.0075)^{72}$

$A \approx 8000(1.7125527)$

$A \approx 13,700.42$

$I = A - P$

$I = 13,700.42 - 8000$

$I = 5700.42$

The amount of interest earned is approximately $5700.42.

CHECK YOUR PROGRESS 5, *page 435*

The following solution utilizes the finance feature of a TI-83/84 Plus calculator.

Press $\boxed{\text{2nd}}$ [Finance] to display the FINANCE CALC menu or press $\boxed{\text{APPS}}$ $\boxed{\text{ENTER}}$.

Press $\boxed{\text{ENTER}}$ to select 1: TVM Solver.

After N =, enter 10.

After I% =, enter 6.

After PV =, enter −3500.

After PMT =, enter 0.

After P/Y =, enter 2.

After C/Y =, enter 2.

Use the up arrow key to place the cursor at FV =.

Press $\boxed{\text{ALPHA}}$ [Solve].

The solution is displayed to the right of FV =.

The compound amount is $4703.71.

CHECK YOUR PROGRESS 6, *page 437*

$r = 9\%$, $n = 2$, $t = 5$, $i = \dfrac{r}{n} = \dfrac{9\%}{2} = 4.5\% = 0.045$,

$N = nt = 2(5) = 10$

$P = \dfrac{A}{(1 + i)^N}$

$P = \dfrac{20{,}000}{(1 + 0.045)^{10}}$

$P \approx \dfrac{20{,}000}{1.5529694}$

$P \approx 12{,}878.55$

$12,878.55 should be invested in the account.

CHECK YOUR PROGRESS 7, *page 438*

The following solution utilizes the finance feature of a TI-83/84 Plus calculator.

Press $\boxed{\text{2nd}}$ [Finance] to display the FINANCE CALC menu or press $\boxed{\text{APPS}}$ $\boxed{\text{ENTER}}$.

Press $\boxed{\text{ENTER}}$ to select 1: TVM Solver.

After N =, enter 5400 (15 × 360).

After I% =, enter 6.

After PMT =, enter 0.

After FV =, enter 25000.

After P/Y =, enter 360.
After C/Y =, enter 360.
Use the up arrow key to place the cursor at PV =.
Press ALPHA [Solve].
The solution is displayed to the right of PV =.

$10,165.00 should be invested in the account.

CHECK YOUR PROGRESS 8, *page 439*

$r = 5\% = 0.05, n = 1, t = 17, i = \dfrac{r}{n} = \dfrac{0.05}{1} = 0.05,$

$N = nt = 1(17) = 17$

$A = P(1 + i)^N$
$A = 28{,}000(1 + 0.05)^{17}$
$A = 28{,}000(1.05)^{17}$
$A \approx 28{,}000(2.2920183)$
$A \approx 64{,}176.51$

The average new car sticker price in 2025 will be approximately
$64,176.51.

CHECK YOUR PROGRESS 9, *page 439*

$r = 7\%, n = 1, t = 40, i = \dfrac{r}{n} = \dfrac{7\%}{1} = 7\% = 0.07,$

$N = nt = 1(40) = 40$

$P = \dfrac{A}{(1 + i)^N}$

$P = \dfrac{500{,}000}{(1 + 0.07)^{40}}$

$P \approx \dfrac{500{,}000}{14.9744578}$

$P \approx 33{,}390.19$

In 2050, the purchasing power of $500,000 will be approximately
$33,390.19.

CHECK YOUR PROGRESS 10, *page 441*

$r = 4\% = 0.04, n = 4, t = 1, i = \dfrac{r}{n} = \dfrac{0.04}{4} = 0.01,$

$N = nt = 4(1) = 4$

$A = P(1 + i)^N$
$A = 100(1 + 0.01)^4$
$A = 100(1.01)^4$
$A \approx 100(1.040604)$
$A \approx 104.06$

$I = A - P$
$I = 104.06 - 100$
$I = 4.06$

The effective interest rate is 4.06%.

CHECK YOUR PROGRESS 11, *page 442*

$i = \dfrac{r}{n} = \dfrac{0.05}{4}$

$N = nt = 4(1) = 4$

$(1 + i)^N = \left(1 + \dfrac{0.05}{4}\right)^4$

≈ 1.050945

$i = \dfrac{r}{n} = \dfrac{0.0525}{2}$

$N = nt = 2(1) = 2$

$(1 + i)^N = \left(1 + \dfrac{0.0525}{2}\right)^2$

≈ 1.053189

An investment that earns 5.25% compounded semiannually has a
higher annual yield than an investment that earns 5%
compounded quarterly.

SECTION 7.3

CHECK YOUR PROGRESS 1, *page 451*

Date	Payments or Purchases	Balance Each Day	Number of Days Until Balance Changes	Unpaid Balance Times Number of Days
July 1–6		$1024	6	$6144
July 7–14	$315	$1339	8	$10,712
July 15–21	−$400	$939	7	$6573
July 22–31	$410	$1349	10	$13,490
Total				$36,919

Average daily balance $= \dfrac{\text{sum of the total amounts owed each day of the month}}{\text{number of days in the billing period}}$

$= \dfrac{36{,}919}{31} \approx \$1190.94$

$I = Prt$

$I = 1190.94(0.012)(1)$

$I \approx 14.29$

The finance charge on the August 1 bill is \$14.29.

CHECK YOUR PROGRESS 2, *page 454*

a. Down payment = Percent down × purchase price

$= 0.20 \times 750 = 150$

Amount financed = purchase price − down payment

$= 750 - 150 = 600$

Interest owed = finance rate × amount financed

$= 0.08 \times 600 = 48$

The finance charge is \$48.

b. $APR \approx \dfrac{2Nr}{N+1}$

$\approx \dfrac{2(12)(0.08)}{12+1} \approx \dfrac{1.92}{13} \approx 0.148$

The annual percentage rate is approximately 14.8%.

CHECK YOUR PROGRESS 3, *page 456*

Sales tax amount = sales tax rate × purchase price

$= 0.0425 \times 1499 \approx 63.71$

Amount financed = purchase price + sales tax amount

$= 1499 + 63.71 = 1562.71$

$i = \dfrac{\text{annual interest rate}}{\text{number of payments per year}} = \dfrac{0.084}{12} = 0.007$

$n = 3(12) = 36$

$PMT = A\left(\dfrac{i}{1-(1+i)^{-n}}\right)$

$PMT = 1562.71\left(\dfrac{0.007}{1-(1+0.007)^{-36}}\right)$

$PMT \approx 49.26$

The monthly payment is \$49.26.

CHECK YOUR PROGRESS 4, *page 457*

a. Sales tax $= 0.0525(26{,}788) = 1406.37$

b. Loan amount

$=$ purchase price + sales tax + license fee − down payment

$= 26{,}788 + 1406.37 + 145 - 2500$

$= 25{,}839.37$

c. $i = \dfrac{APR}{12} = \dfrac{0.081}{12} = 0.00675;\ n = 12 \times 5 = 60$

$PMT = A\left(\dfrac{i}{1-(1+i)^{-n}}\right)$

$PMT = 25{,}839.37\left(\dfrac{0.00675}{1-(1+0.00675)^{-60}}\right)$

$PMT \approx 525.17$

The monthly payment is \$525.17.

CHECK YOUR PROGRESS 5, *page 458*

$i = \dfrac{APR}{12} = \dfrac{0.084}{12} = 0.007$

$A = PMT\left(\dfrac{1-(1+i)^{-n}}{i}\right)$

$A = 592.57\left(\dfrac{1-(1+0.007)^{-24}}{0.007}\right)$

$A \approx 13{,}049.34$

The loan payoff is \$13,049.34.

CHECK YOUR PROGRESS 6, *page 460*

Residual value $= 0.40(33{,}395) = 13{,}358$

Money factor $= \dfrac{\text{annual interest rate as a percent}}{2400} = \dfrac{8}{2400}$

≈ 0.00333333

Average monthly finance charge

$=$ (net capitalized cost + residual value) × money factor

$= (31{,}900 + 13{,}358) \times 0.00333333$

≈ 150.86

Average monthly depreciation

$= \dfrac{\text{net capitalized cost} - \text{residual value}}{\text{term of the lease in months}}$

$= \dfrac{31{,}900 - 13{,}358}{60}$

≈ 309.03

Monthly lease payment
= average monthly finance charge + average monthly depreciation
= 150.86 + 309.03
= 459.89

The monthly lease payment is $459.89.

SECTION 7.4

CHECK YOUR PROGRESS 1, *page 467*

($.72 per share) × (550 shares) = $396

The shareholder receives $396 in dividends.

CHECK YOUR PROGRESS 2, *page 468*

$I = Prt$

$0.82 = 51.25r(1)$ • Let I = the annual dividend and P = the stock price. The time is 1 year.

$0.82 = 51.25r$

$0.016 = r$ • Divide each side of the equation by 51.25.

The dividend yield is 1.6%.

CHECK YOUR PROGRESS 3, *page 470*

a. From Table 7.2, the 52-week low is $29.43, and the 52-week high is $38.89.

Profit = selling price − purchase price

= 300($38.89) − 300($29.43)

= $11,667 − $8829 = $2838

The profit on the sale of the stock was $2838.

b. Commission = 2.1%(selling price)

= 0.021($11,667) ≈ $245.01

The broker's commission was $245.01.

CHECK YOUR PROGRESS 4, *page 471*

Use the simple interest formula to find the annual interest payment. Substitute the following values into the formula: $P = 15,000$, $r = 3.5\% = 0.035$, and $t = 1$.

$I = Prt$

$I = 15,000(0.035)(1)$

$I = 525$

Multiply the annual interest payment by the term of the bond.

$525(4) = 2100$

The total of the interest payments paid to the bondholder is $2100.

CHECK YOUR PROGRESS 5, *page 472*

a. $A - L = (750 \text{ million} + 0.75 \text{ million}$
$+ 1.5 \text{ million}) - 1.5 \text{ million}$
$= 750.75 \text{ million}$

$N = 20 \text{ million}$

$NAV = \dfrac{A - L}{N} = \dfrac{750.75 \text{ million}}{20 \text{ million}} = 37.5375$

The NAV of the fund is $37.5375.

b. $\dfrac{10,000}{37.5375} \approx 266$ • Divide the amount invested by the *NAV* of the fund. Round down to the nearest whole number.

You will purchase 266 shares of the mutual fund.

SECTION 7.5

CHECK YOUR PROGRESS 1, *page 478*

Down payment = 25% of 410,000 = 0.25(410,000)
= 102,500

Mortgage = selling price − down payment
= 410,000 − 102,500
= 307,500

Points = 1.75% of 307,500 = 0.0175(307,500)
= 5381.25

Total = 102,500 + 375 + 5381.25 = 108,256.25

The total of the down payment and the closing costs is $108,256.25.

CHECK YOUR PROGRESS 2, *page 479*

a. $i = \dfrac{0.07}{12} \approx 0.00583333$

$n = 25(12) = 300$

$PMT = A\left(\dfrac{i}{1 - (1 + i)^{-n}}\right)$

$PMT \approx 223,000\left(\dfrac{0.00583333}{1 - (1 + 0.00583333)^{-300}}\right)$

$PMT \approx 1576.12$

The monthly payment is $1576.12.

b. Total = 1576.12(300) = 472,836

The total of the payments over the life of the loan is $472,836.

c. Interest = 472,836 − 223,000 = 249,836

The amount of interest paid over the life of the loan is $249,836.

CHECK YOUR PROGRESS 3, *page 482*

Down payment $= 0.25(295{,}000) = 73{,}750$

Mortgage $= 295{,}000 - 73{,}750 = 221{,}250$

$i = \dfrac{0.0675}{12} = 0.005625$

$n = 30(12) = 360$

$PMT = A\left(\dfrac{i}{1 - (1 + i)^{-n}}\right)$

$PMT = 221{,}250\left(\dfrac{0.005625}{1 - (1 + 0.005625)^{-360}}\right)$

$PMT \approx 1435.02$

The monthly payment is \$1435.02.

$I = Prt$

$ = 221{,}250(0.0675)\left(\dfrac{1}{12}\right)$

$ \approx 1244.53$

The interest paid on the first payment is \$1244.53.

Principal $= 1435.02 - 1244.53 = 190.49$

The principal paid on the first payment is \$190.49.

CHECK YOUR PROGRESS 4, *page 483*

$i = \dfrac{0.069}{12} = 0.00575$

$n = 25(12) - 4(12) = 300 - 48 = 252$

$A = PMT\left(\dfrac{1 - (1 + i)^{-n}}{i}\right)$

$A = 846.82\left(\dfrac{1 - (1 + 0.00575)^{-252}}{0.00575}\right)$

$A \approx 112{,}548.79$

The mortgage payoff is \$112,548.79.

CHECK YOUR PROGRESS 5, *page 485*

Monthly property tax $= 2332.80 \div 12 = 194.40$

Monthly fire insurance $= 450 \div 12 = 37.50$

Total monthly payment $= 1492.89 + 194.40 + 37.50 = 1724.79$

The total monthly payment for mortgage, property tax, and fire insurance is \$1724.79.

CHAPTER 8

SECTION 8.1

CHECK YOUR PROGRESS 1, *page 495*

Q: quarter; D: dime; N: nickel; P: penny; H: head; T: tail

{(QH, DH, NH, PH), (QH, DH, NH, PT), (QH, DH, NT, PH), (QH, DT, NH, PH),
(QT, DH, NH, PH), (QH, DH, NT, PT), (QH, DT, NT, PH), (QT, DT, NH, PH), (QH, DT, NH, PT),
(QT, DH, NT, PH), (QT, DT, NH, PH), (QT, DT, NH, PT), (QT, DT, NT, PH), (QT, DH, NT, PT),
(QT, DT, NT, PH), (QT, DT, NT, PT)}

CHECK YOUR PROGRESS 2, *page 496*

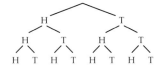

There are eight possible outcomes.

CHECK YOUR PROGRESS 3, *page 497*

$9 \cdot 8 \cdot 7 = 504$

CHECK YOUR PROGRESS 4, *page 498*

a. $7! + 4! = 7 \cdot 6 \cdot 5 \cdot 4 \cdot 3 \cdot 2 \cdot 1 + 4 \cdot 3 \cdot 2 \cdot 1 = 5064$

b. $\dfrac{8!}{4!} = \dfrac{8 \cdot 7 \cdot 6 \cdot 5 \cdot 4!}{4!} = 1680$

CHECK YOUR PROGRESS 5, *page 499*

$$P(8, 5) = \frac{8!}{(8-5)!} = \frac{8!}{3!} = \frac{8 \cdot 7 \cdot 6 \cdot 5 \cdot 4 \cdot 3!}{3!} = 8 \cdot 7 \cdot 6 \cdot 5 \cdot 4 = 6720$$

CHECK YOUR PROGRESS 6, *page 501*

$$C(16, 10) = \frac{16!}{10!6!} = \frac{16 \cdot 15 \cdot 14 \cdot 13 \cdot 12 \cdot 11 \cdot 10!}{10!6!} = \frac{16 \cdot 15 \cdot 14 \cdot 13 \cdot 12 \cdot 11}{6 \cdot 5 \cdot 4 \cdot 3 \cdot 2 \cdot 1} = 8008$$

CHECK YOUR PROGRESS 7, *page 501*

$$C(4, 3) \cdot C(6, 2) = 4 \cdot 15 = 60$$

CHECK YOUR PROGRESS 8, *page 502*

Begin by considering just the spade suit. There are $C(13, 4)$ ways to choose four spades from 13 and $C(48, 1)$ ways to choose the remaining card. Therefore, there are $C(13, 4) \cdot C(48, 1)$ ways to choose four spades and one nonspade card. This analysis can be applied to the other suits. Because there are four suits, the total number of ways of choosing four cards of the same suit is $4 \cdot C(13, 4) \cdot C(48, 1) = 137,280$

SECTION 8.2

CHECK YOUR PROGRESS 1, *page 507*

{HH, HT, TH, TT}

CHECK YOUR PROGRESS 2, *page 509*

The sample space is $S = \{1, 2, 3, 4, 5, 6\}$. The event is $E = \{1, 3, 5\}$. Therefore,

$$P(E) = \frac{n(E)}{n(S)} = \frac{3}{6} = \frac{1}{2}$$

CHECK YOUR PROGRESS 3, *page 510*

There are 36 outcomes from the roll of two dice. The event E, a sum of 7 is rolled, can occur in six ways. Therefore,

$$P(E) = \frac{6}{36} = \frac{1}{6}$$

CHECK YOUR PROGRESS 4, *page 510*

The sample space is the number of ways of choosing five professors from the 11 professors without regard to the disciplines of the professors and is $C(11, 5)$. The number of elements in the event E that the committee contain two mathematicians and three economists is $C(5, 2) \cdot C(6, 3)$. Therefore,

$$P(E) = \frac{C(5, 2) \cdot C(6, 3)}{C(11, 5)} = \frac{10 \cdot 20}{462} = \frac{200}{462} \approx 0.433$$

CHECK YOUR PROGRESS 5, *page 511*

The sample space is the total number of people surveyed. Thus $n(S) = 3228$.

The event E is that the person selected is between 39 and 49. Thus $n(E) = 773$. Therefore,

$$P(E) = \frac{n(E)}{n(S)} = \frac{773}{3228} \approx 0.24$$

CHECK YOUR PROGRESS 6, *page 512*

There are 30 favorable outcomes (not rolling a sum of 7) and six unfavorable outcomes (rolling a sum of 7). The odds against rolling a sum of 7 are $\frac{30}{6} = \frac{5}{1}$. The odds against rolling a sum of 7 are 5 to 1.

CHECK YOUR PROGRESS 7, *page 513*

$A = \{(1, 6), (2, 5), (3, 4), (4, 3), (5, 2), (6, 1)\}$
$B = \{(5, 6), (6, 5)\}$

The events are mutually exclusive. Therefore

$$P(A \text{ or } B) = P(A) + P(B) \qquad \text{• Addition formula for the probability of mutually exclusive events}$$

$$= \frac{6}{36} + \frac{2}{36} \qquad \text{• } P(A) = \frac{6}{36}, P(B) = \frac{2}{36}$$

$$= \frac{8}{36} = \frac{2}{9}$$

CHECK YOUR PROGRESS 8, *page 516*

$B = \text{sum of } 6 = \{(1, 5), (2, 4), (3, 3), (4, 2), (5, 1)\}$
$A = \text{sum is not } 7 = \{(1, 1), (1, 2), (1, 3), (1, 4), (1, 5), (2, 1), (2, 2), (2, 3),$
$(2, 4), (2, 6), (3, 1), (3, 2), (3, 3), (3, 5), (3, 6), (4, 1), (4, 2), (4, 4), (4, 5),$
$(4, 6), (5, 1), (5, 3), (5, 4), (5, 5), (5, 6), (6, 2), (6, 3), (6, 4), (6, 5), (6, 6)\}$
$A \cap B = \{(1, 5), (2, 4), (3, 3), (4, 2), (5, 1)\}$

$$P(B \mid A) = \frac{P(A \cap B)}{P(A)} = \frac{\dfrac{6}{36}}{\dfrac{30}{36}} \qquad \begin{array}{l} \text{• } P(A \cap B) = \dfrac{n(A \cap B)}{n(S)} = \dfrac{6}{36} \\[2mm] \text{• } P(A) = \dfrac{n(A)}{n(S)} = \dfrac{30}{36} \end{array}$$

$$= \frac{6}{30} = \frac{1}{5}$$

CHECK YOUR PROGRESS 9, *page 517*

Let $A = \{\text{first card is a spade}\}$, $B = \{\text{second card is a heart}\}$, and $C = \{\text{third card is a spade}\}$. Then
$P(A \cap B \cap C) = P(A) \cdot P(B \mid A) \cdot P(C \mid A \cap B)$

$$= \frac{13}{52} \cdot \frac{13}{51} \cdot \frac{12}{50} = \frac{13}{850}$$

CHECK YOUR PROGRESS 10, *page 519*

The tosses of coins are independent; the probability of a head on a toss does not depend on the previous toss. Let $A = \{\text{first toss is a head}\}$, $B = \{\text{second toss is a head}\}$, and $C = \{\text{third toss is a head}\}$. Then

$$P(A \cap B \cap C) = P(A) \cdot P(B) \cdot P(C) = \frac{1}{2} \cdot \frac{1}{2} \cdot \frac{1}{2} = \frac{1}{8}$$

CHECK YOUR PROGRESS 11, *page 519*

Let S be the event that a person has the genetic defect and let T be the event that the test for the defect is positive. Then the probability we wish to determine is $P(S \mid T)$. Using the conditional probability formula, we have

$$P(S \mid T) = \frac{P(S \cap T)}{P(T)}$$

A tree diagram, shown at the right, can be used to calculate this probability. A positive test result can occur in two ways: either the person has the genetic defect and correctly tests positive, or the person does not have the defect and incorrectly tests positive. The probability of a positive test result, $P(T)$, corresponds to path ST or path $S'T$ in the tree diagram. (S' symbolizes no genetic defect.) $P(S \cap T)$, the probability of the genetic defect and a positive test, is path ST. Thus,

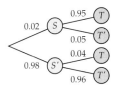

$$P(S|T) = \frac{P(S \cap T)}{P(T)}$$

$$= \frac{0.02 \cdot 0.95}{0.02 \cdot 0.95 + 0.98 \cdot 0.04} \approx 0.326$$

CHECK YOUR PROGRESS 12, *page 521*

Make a Punnett square.

Parents	C	c
C	CC	Cc
c	Cc	cc

For an offspring to be white, both alleles must be c. There are four possibilities, of which one is cc. Therefore, the probability of a white hamster is $\frac{1}{4}$.

SECTION 8.3

CHECK YOUR PROGRESS 1, *page 528*

$$\bar{x} = \frac{\Sigma x}{n} = \frac{245 + 235 + 220 + 210}{4} = \frac{910}{4} = 227.5$$

The mean of the patient's blood cholesterol levels is 227.5.

CHECK YOUR PROGRESS 2, *page 529*

a. The list 14, 27, 3, 82, 64, 34, 8, 51 contains eight numbers. The median of a list of data with an even number of numbers is found by ranking the numbers and computing the mean of the two middle numbers. Ranking the numbers from smallest to largest gives 3, 8, 14, 27, 34, 51, 64, 82. The two middle numbers are 27 and 34. The mean of 27 and 34 is 30.5. Thus 30.5 is the median of the data.

b. The list 21.3, 37.4, 11.6, 82.5, 17.2 contains five numbers. The median of a list of data with an odd number of numbers is found by ranking the numbers and finding the middle number. Ranking the numbers from smallest to largest gives 11.6, 17.2, 21.3, 37.4, 82.5. The middle number is 21.3. Thus 21.3 is the median.

CHECK YOUR PROGRESS 3, *page 530*

a. In the list 3, 3, 3, 3, 3, 4, 4, 5, 5, 5, 8, the number 3 occurs more often than the other numbers. Thus 3 is the mode.

b. In the list 12, 34, 12, 71, 48, 93, 71, the numbers 12 and 71 both occur twice and the other numbers occur only once. Thus 12 and 71 are both modes of the data.

CHECK YOUR PROGRESS 4, *page 532*

Weighted mean
$$= \frac{(70 \times 1) + (55 \times 1) + (90 \times 2) + (72 \times 2) + (68 \times 2) + (85 \times 6)}{14}$$

$$= \frac{1095}{14} \approx 78.2$$

Larry's weighted mean is approximately 78.2.

CHECK YOUR PROGRESS 5, *page 534*

$$\text{Mean} = \frac{\Sigma(x \cdot f)}{\Sigma f}$$

$$= \frac{(2 \cdot 5) + (3 \cdot 25) + (4 \cdot 10) + (5 \cdot 5)}{45}$$

$$= \frac{150}{45}$$

$$= 3\frac{1}{3}$$

The mean number of bedrooms per household for the homes in the subdivision is $3\frac{1}{3}$.

SECTION 8.4

CHECK YOUR PROGRESS 1, *page 541*

Tara's highest test score is 84 and her lowest test score is 76. The range of Tara's test scores is $84 - 76 = 8$.

CHECK YOUR PROGRESS 2, *page 543*

$$\mu = \frac{5 + 8 + 16 + 17 + 18 + 20}{6} = \frac{84}{6} = 14$$

x	$x - \mu$	$(x - \mu)^2$
5	$5 - 14 = -9$	$(-9)^2 = 81$
8	$8 - 14 = -6$	$(-6)^2 = 36$
16	$16 - 14 = 2$	$2^2 = 4$
17	$17 - 14 = 3$	$3^2 = 9$
18	$18 - 14 = 4$	$4^2 = 16$
20	$20 - 14 = 6$	$6^2 = 36$
		Sum: 182

$$\sigma = \sqrt{\frac{\Sigma(x - \mu)^2}{n}} = \sqrt{\frac{182}{6}} \approx \sqrt{30.33} \approx 5.51$$

The standard deviation for this population is approximately 5.51.

CHECK YOUR PROGRESS 3, *page 544*

The rope from Trustworthy has a breaking point standard deviation of

$$s_1 = \sqrt{\frac{(122 - 130)^2 + (141 - 130)^2 + \cdots + (125 - 130)^2}{6}}$$

$$= \sqrt{\frac{1752}{6}} \approx 17.1 \text{ pounds}$$

The rope from Brand X has a breaking point standard deviation of

$$s_2 = \sqrt{\frac{(128 - 130)^2 + (127 - 130)^2 + \cdots + (137 - 130)^2}{6}}$$

$$= \sqrt{\frac{3072}{6}} \approx 22.6 \text{ pounds}$$

The rope from NeverSnap has a breaking point standard deviation of

$$s_3 = \sqrt{\frac{(112 - 130)^2 + (121 - 130)^2 + \cdots + (135 - 130)^2}{6}}$$

$$= \sqrt{\frac{592}{6}} \approx 9.9 \text{ pounds}$$

The rope from NeverSnap has the smallest breaking point standard deviation.

CHECK YOUR PROGRESS 4, *page 545*

The mean is approximately 46.577.

The population standard deviation is approximately 2.876.

L1	L2	L3	1
54.2	------	------	
49.4			
49.2			
53.2			
50.0			
48.2			
49.6			
L1(1) = 54.2			

```
1-Var Stats
 x̄=46.57653846          ← Mean
 Σx=1210.99
 Σx²=56618.7787
 Sx=2.932960201        ← Sample standard deviation
 σx=2.876004095        ← Population standard deviation
 ↓n=26
```

CHECK YOUR PROGRESS 5, *page 546*

In Check Your Progress 2 we found $\sigma \approx \sqrt{30.33}$. Variance is the square of the standard deviation. Thus the variance is $\sigma^2 \approx \left(\sqrt{30.33}\right)^2 = 30.33$.

SECTION 8.5

CHECK YOUR PROGRESS 1, *page 553*

$$z_{15} = \frac{15 - 12}{2.4} = 1.25 \qquad z_{14} = \frac{14 - 11}{2.0} = 1.5$$

These z-scores indicate that in comparison with her classmates, Cheryl did better on the second quiz than she did on the first quiz.

CHECK YOUR PROGRESS 2, *page 553*

$$z_x = \frac{x - \mu}{\sigma}$$

$$0.6 = \frac{70 - 65.5}{\sigma}$$

$$\sigma = \frac{4.5}{0.6} = 7.5$$

The standard deviation for this set of test scores is 7.5.

CHECK YOUR PROGRESS 3, *page 554*

a. By definition, the median is the 50th percentile. Therefore, 50% of the police dispatchers earned less than $28,288 per year.

b. Because $25,640 is the 30th percentile, 100% − 30% = 70% of all police dispatchers made more than $25,640.

c. From parts a and b, 50% − 30% = 20% of the police dispatchers earned between $25,640 and $28,288.

CHECK YOUR PROGRESS 4, *page 555*

$$\text{Percentile} = \frac{\text{number of data values less than 405}}{\text{total number of data values}} \cdot 100$$

$$= \frac{3952}{8600} \cdot 100$$

$$\approx 46$$

Hal's score of 405 places him at the 46th percentile.

CHECK YOUR PROGRESS 5, *page 556*

Rank the data.

7.5 9.8 10.2 10.8 11.4 11.4 12.2 12.4 12.6 12.8 13.1 14.2 14.5 15.6 16.4

The median of these 15 data values has a rank of 8. Thus the median is 12.4. The second quartile, Q_2, is the median of the data, so $Q_2 = 12.4$.

The first quartile is the median of the seven values less than Q_2. Thus Q_1 has a rank of 4, so $Q_1 = 10.8$.

The third quartile is the median of the values greater than Q_2. Thus Q_3 has a rank of 12, so $Q_3 = 14.2$.

CHECK YOUR PROGRESS 6, *page 557*

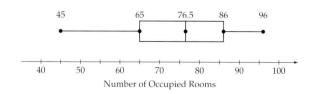

SECTION 8.6

CHECK YOUR PROGRESS 1, *page 566*

a. The percent of data in all classes with an upper bound of 25 seconds or less is the sum of the percents for the first five classes in Table 8.5. Thus the percent of subscribers who required less than 25 seconds to download the file is 30.9%.

b. The percent of data in all the classes with a lower bound of at least 10 seconds and an upper bound of 30 seconds or less is the sum of the percents in the third through sixth classes in Table 8.5. Thus the percent of subscribers who required from 10 to 30 seconds to download the file is 47.8%. The probability that a subscriber chosen at random will require from 10 to 30 seconds to download the file is 0.478.

CHECK YOUR PROGRESS 2, *page 570*

a. 0.76 pound is one standard deviation above the mean of 0.61 pound. In a normal distribution, 34.1% of all data lie between the mean and one standard deviation above the mean, and 50% of all data lie below the mean. Thus 34.1% + 50% = 84.1% of the tomatoes weigh less than 0.76 pound.

b. 0.31 pound is two standard deviations below the mean of 0.61 pound. In a normal distribution, 47.7% of all data lie between the mean and two standard deviations below the mean, and 50% of all data lie above the mean. This gives a total of 47.7% + 50% = 97.7% of the tomatoes that weigh more than 0.31 pound. Therefore

$$(97.7\%)(6000) = (0.977)(6000) = 5862$$

of the tomatoes can be expected to weigh more than 0.31 pound.

c. 0.31 pound is two standard deviations below the mean of 0.61 pound and 0.91 pound is two standard deviations above the mean of 0.61 pound. In a normal distribution, 95.4% of all data lie within two standard deviations of the mean. Therefore

$$(95.4\%)(4500) = (0.954)(4500) = 4293$$

of the tomatoes can be expected to weigh from 0.31 pound to 0.91 pound.

CHECK YOUR PROGRESS 3, *page 573*

The area of the standard normal distribution between $z = -0.67$ and $z = 0$ is equal to the area between $z = 0$ and $z = 0.67$. The entry in Table 8.7 associated with $z = 0.67$ is 0.249. Thus the area of the standard normal distribution between $z = -0.67$ and $z = 0$ is 0.249 square unit.

CHECK YOUR PROGRESS 4, *page 573*

Table 8.7 indicates that the area from $z = 0$ to $z = -1.47$ is 0.429 square unit. The area to the left of $z = 0$ is 0.500 square unit. Thus the area to the left of $z = -1.47$ is $0.500 - 0.429 = 0.071$ square unit.

CHECK YOUR PROGRESS 5, *page 575*

Round z-scores to the nearest hundredth so you can use Table 8.7.

a. $z_9 = \dfrac{9 - 6.1}{1.8} \approx 1.61$

Table 8.7 indicates that 0.446 (44.6%) of the data in the standard normal distribution are between $z = 0$ and $z = 1.61$. The percent of the data to the right of $z = 1.61$ is $50\% - 44.6\% = 5.4\%$.

Approximately 5.4% of professional football players have careers of more than 9 years.

b. $z_3 = \dfrac{3 - 6.1}{1.8} \approx -1.72$ $z_4 = \dfrac{4 - 6.1}{1.8} \approx -1.17$

From Table 8.7:
$A_{1.72} = 0.457$ $A_{1.17} = 0.379$

$$0.457 - 0.379 = 0.078$$

The probability that a professional football player chosen at random will have a career of between 3 and 4 years is about 0.078.

SECTION 8.7

CHECK YOUR PROGRESS 1, *page 582*

The null hypothesis is the accepted standard, which in this case is that the dice are fair. The alternative hypothesis is that the dice are loaded, producing more 7s than expected.

Thus, H_0: $p = \dfrac{1}{6}$ and H_1: $p > \dfrac{1}{6}$.

CHECK YOUR PROGRESS 2, *page 584*

1. The null hypothesis is that the proportion of people with this type of cancer is the same as the national proportion: H_0: $\hat{p} = 0.015$. The alternative hypothesis is that the proportion of people with this cancer in this region is greater than the national proportion: H_1: $(\hat{p} > 0.015)$.

2. This is a proportion hypothesis test. $p = \dfrac{4}{200} = 0.02$, $\hat{p} = 0.015$, and $n = 200$.

$$z = \dfrac{p - \hat{p}}{\sqrt{\dfrac{\hat{p}(1 - \hat{p})}{n}}}$$

$$z = \dfrac{0.02 - 0.015}{\sqrt{\dfrac{0.015(1 - 0.015)}{200}}}$$

$$z = \dfrac{0.005}{\sqrt{\dfrac{0.015(0.985)}{200}}} \approx 0.58$$

3. The significance level is 5% and the alternative hypothesis uses a greater than symbol. The critical z-value is $z_{0.05} = 1.65$. Compare the test statistic value with the critical z-value: $z \approx 0.582 < 1.65$.

4. Because the value of the test statistic is less than the critical z-value, we cannot reject the null hypothesis. This means that on the basis of the sample, researchers cannot conclude that the region is a cancer cluster.

CHECK YOUR PROGRESS 3, *page 585*

1. The null hypothesis is that the hybrid car's fuel efficiency is 42 mpg: H_0: $\mu = 42$. The alternative hypothesis is that the fuel efficiency is less than 42 mpg: H_1: $\mu < 42$.

2. This is a hypothesis about a mean. $\bar{x} = 40$, $\mu = 42$, $\sigma = 2.5$, and $n = 30$.

$$z = \frac{\bar{x} - \mu}{\dfrac{\sigma}{\sqrt{n}}}$$

$$z = \frac{40 - 42}{\dfrac{2.5}{\sqrt{30}}}$$

$$z \approx -4.38$$

3. The significance level is 1% and the alternative hypothesis uses a less than symbol. The critical z-value is $z_{0.01} = -2.33$. Compare the test statistic value with the critical z-value: $z \approx -4.38 < -2.33$.

4. Because the value of the test statistic is less than the critical z-value, we reject the null hypothesis. This means that, on the basis of the sample, the fuel efficiency of the hybrid car is not 42 mpg.

CHECK YOUR PROGRESS 4, *page 587*

Use the formula for the 99% confidence interval for a proportion. From the survey, $n = 200$ and $p = \dfrac{15}{200} = 0.075$.

$$p - 2.58\sqrt{\frac{p(1 - p)}{n}} < \hat{p} < p + 2.58\sqrt{\frac{p(1 - p)}{n}}$$

$$0.075 - 2.58\sqrt{\frac{0.075(1 - 0.075)}{200}} < \hat{p} < 0.075 + 2.58\sqrt{\frac{0.075(1 - 0.075)}{200}}$$

$$0.075 - 0.048 < \hat{p} < 0.075 + 0.048$$

$$0.027 < \hat{p} < 0.123$$

The 99% confidence interval for the true proportion of zinc in the reservoir is $0.027 < \hat{p} < 0.123$.

CHECK YOUR PROGRESS 5, *page 587*

The sample mean is $\bar{x} = 0.81$ COR, and the standard deviation is $\sigma = 0.04$ COR. The sample size is $n = 60$. Use the formula for the 95% confidence interval for the mean.

$$\bar{x} - 1.96\frac{\sigma}{\sqrt{n}} < \mu < \bar{x} + 1.96\frac{\sigma}{\sqrt{n}}$$

$$0.81 - 1.96\frac{0.04}{\sqrt{60}} < \mu < 0.81 + 1.96\frac{0.04}{\sqrt{60}}$$

$$0.81 - 0.01 < \mu < 0.81 + 0.01$$

$$0.80 < \mu < 0.82$$

The 95% confidence interval for the population mean is $0.80 < \mu < 0.82$.

CHECK YOUR PROGRESS 6, *page 588*

Use the formula for the margin of error at the 99% confidence level with $p = \dfrac{350}{500} = 0.70$ and $n = 500$.

$$\text{Margin of error} = 2.58 \sqrt{\frac{p(1-p)}{n}}$$

$$= 2.58 \sqrt{\frac{0.7(0.3)}{500}} \approx 0.053$$

The margin of error, to the nearest tenth of a percent, is 5.3%.

CHECK YOUR PROGRESS 7, *page 589*

Use the formula for the greatest margin of error at the 99% confidence level.

$$\text{Greatest margin of error} = \frac{1.29}{\sqrt{n}}$$

$$= \frac{1.29}{\sqrt{500}} \approx 0.058$$

The greatest margin of error at the 99% confidence level is 5.8%.

CHAPTER 9

SECTION 9.1

CHECK YOUR PROGRESS 1, *page 608*

a. The standard divisor is the sum of all the populations (251,100,000) divided by the number of representatives (20).

$$\text{Standard divisor} = \frac{251,100,000}{20} = 12,555,000$$

Country	Population	Quotient	Standard Quota	Number of Representatives
France	60,400,000	$\dfrac{60,400,000}{12,555,000} \approx 4.811$	4	5
Germany	82,400,000	$\dfrac{82,400,000}{12,555,000} \approx 6.563$	6	6
Italy	58,000,000	$\dfrac{58,000,000}{12,555,000} \approx 4.620$	4	5
Spain	40,000,000	$\dfrac{40,000,000}{12,555,000} \approx 3.186$	3	3
Belgium	10,300,000	$\dfrac{10,300,000}{12,555,000} \approx 0.820$	0	1
		Total	17	20

Because the sum of the standard quotas is 17 and not 20, we add one representative to each of the three countries with the largest decimal remainders. These are Belgium, France, and Italy. Thus the composition of the committee is France: 5, Germany: 6, Italy: 5, Spain: 3, Belgium: 1.

b. To use the Jefferson method, we must find a modified divisor such that the sum of the standard quotas is 20. This modified divisor is found by trial and error, but is always less than or equal to the standard divisor. We are using 11,000,000 for the modified standard divisor.

Country	Population	Quotient	Standard Quota	Number of Representatives
France	60,400,000	$\dfrac{60,400,000}{11,000,000} \approx 5.491$	5	5
Germany	82,400,000	$\dfrac{82,400,000}{11,000,000} \approx 7.491$	7	7
Italy	58,000,000	$\dfrac{58,000,000}{11,000,000} \approx 5.273$	5	5
Spain	40,000,000	$\dfrac{40,000,000}{11,000,000} \approx 3.636$	3	3
Belgium	10,300,000	$\dfrac{10,300,000}{11,000,000} \approx 0.936$	0	0
		Total	20	20

Thus the composition of the committee is France: 5, Germany: 7, Italy: 5, Spain: 3, Belgium: 0.

CHECK YOUR PROGRESS 2, *page 612*

$$\text{Relative unfairness of the apportionment} = \frac{\text{absolute unfairness of the apportionment}}{\text{average constituency of Shasta with a new representative}}$$

$$= \frac{210}{1390} \approx 0.151$$

The relative unfairness of the apportionment is approximately 0.151.

CHECK YOUR PROGRESS 3, *page 613*

Calculate the relative unfairness of the apportionment that assigns the teacher to the first grade and the relative unfairness of the apportionment that assigns the teacher to the second grade. In this case, the average constituency is the number of students divided by the number of teachers.

	First Grade Number of Students per Teacher	Second Grade Number of Students per Teacher	Absolute Unfairness of Apportionment
First grade receives teacher	$\dfrac{12,317}{512 + 1} \approx 24$	$\dfrac{15,439}{551} \approx 28$	$28 - 24 = 4$
Second grade receives teacher	$\dfrac{12,317}{512} \approx 24$	$\dfrac{15,439}{551 + 1} \approx 28$	$28 - 24 = 4$

If the first grade receives the new teacher, then the relative unfairness of the apportionment is

$$\text{Relative unfairness of the apportionment} = \frac{\text{absolute unfairness of the apportionment}}{\text{first grade's average constituency with a new teacher}}$$

$$= \frac{4}{24} \approx 0.167$$

If the second grade receives the new teacher, then the relative unfairness of the apportionment is

$$\text{Relative unfairness of the apportionment} = \frac{\text{Absolute unfairness of the apportionment}}{\text{Second grade's average constituency with a new teacher}}$$

$$= \frac{4}{28} \approx 0.143$$

Because the smaller relative unfairness results from adding the teacher to the second grade, that class should receive the new teacher.

CHECK YOUR PROGRESS 4, *page 615*

Calculate the Huntington-Hill number for each of the classes. In this case, the population is the number of students.

First year:

$$\frac{2015^2}{12(12 + 1)} \approx 26{,}027$$

Second year:

$$\frac{1755^2}{10(10 + 1)} \approx 28{,}000$$

Third year:

$$\frac{1430^2}{9(9 + 1)} \approx 22{,}721$$

Fourth year:

$$\frac{1309^2}{8(8 + 1)} \approx 23{,}798$$

Because the second-year class has the greatest Huntington-Hill number, the new representative should represent the second-year class.

SECTION 9.2

CHECK YOUR PROGRESS 1, *page 624*

To answer the question, we will make a table showing the number of second-place votes for each candy.

	Second-Place Votes
Caramel center	3
Vanilla center	0
Almond center	17 + 9 = 26
Toffee center	2
Solid chocolate	11 + 8 = 19

The largest number of second-place votes (26) were for almond centers. Almond centers would win second place using the plurality voting system.

CHECK YOUR PROGRESS 2, *page 628*

Using the Borda Count method, each first-place vote receives 5 points, each second-place vote receives 4 points, each third-place vote receives 3 points, each fourth-place vote receives 2 points, and each last-place vote receives 1 point. The summaries for the five varieties are shown in the next column.

Caramel:

0 first-place votes	$0 \cdot 5 =$ 0
3 second-place votes	$3 \cdot 4 = 12$
0 third-place votes	$0 \cdot 3 =$ 0
30 fourth-place votes	$30 \cdot 2 = 60$
17 fifth-place votes	$17 \cdot 1 = 17$
	Total 89

Vanilla:

17 first-place votes	$17 \cdot 5 = 85$
0 second-place votes	$0 \cdot 4 =$ 0
0 third-place votes	$0 \cdot 3 =$ 0
0 fourth-place votes	$0 \cdot 2 =$ 0
33 fifth-place votes	$33 \cdot 1 = 33$
	Total 118

Almond:

8 first-place votes	$8 \cdot 5 = 40$
26 second-place votes	$26 \cdot 4 = 104$
16 third-place votes	$16 \cdot 3 = 48$
0 fourth-place votes	$0 \cdot 2 =$ 0
0 fifth-place votes	$0 \cdot 1 =$ 0
	Total 192

Toffee:

20 first-place votes	$20 \cdot 5 = 100$
2 second-place votes	$2 \cdot 4 =$ 8
8 third-place votes	$8 \cdot 3 = 24$
20 fourth-place votes	$20 \cdot 2 = 40$
0 fifth-place votes	$0 \cdot 1 =$ 0
	Total 172

Chocolate:

5 first-place votes	$5 \cdot 5 = 25$
19 second-place votes	$19 \cdot 4 = 76$
26 third-place votes	$26 \cdot 3 = 78$
0 fourth-place votes	$0 \cdot 2 =$ 0
0 fifth-place votes	$0 \cdot 1 =$ 0
	Total 179

Using the Borda Count method, almond centers is the first choice.

CHECK YOUR PROGRESS 3, *page 630*

	Rankings				
Italian	2	5	1	4	3
Mexican	1	4	5	2	1
Thai	3	1	4	5	2
Chinese	4	2	3	1	4
Indian	5	3	2	3	5
Number of ballots:	33	30	25	20	18

Indian food received no first place votes, so it is eliminated.

	Rankings				
Italian	2	4	1	3	3
Mexican	1	3	4	2	1
Thai	3	1	3	4	2
Chinese	4	2	2	1	4
Number of ballots:	33	30	25	20	18

In this ranking, Chinese food received the fewest first-place votes, so it is eliminated.

	Rankings				
Italian	2	3	1	2	3
Mexican	1	2	3	1	1
Thai	3	1	2	3	2
Number of ballots:	33	30	25	20	18

In this ranking, Italian food received the fewest first-place votes, so it is eliminated.

	Rankings				
Mexican	1	2	2	1	1
Thai	2	1	1	2	2
Number of ballots:	33	30	25	20	18

In this ranking, Thai food received the fewest first-place votes, so it is eliminated. The preference for the banquet food is Mexican.

CHECK YOUR PROGRESS 4, *page 632*

Do a head-to-head comparison for each of the restaurants and enter each winner in the table below. For instance, in the Sanborn's versus Apple Inn comparison, Sanborn's was favored by $31 + 25 + 11 = 67$ critics. In the Apple Inn versus Sanborn's comparison, Apple Inn was favored by $18 + 15 = 33$ critics. Therefore, Sanborn's wins this head-to-head match. The completed table is shown below.

versus	Sanborn's	Apple Inn	May's	Tory's
Sanborn's		Sanborn's	May's	Sanborn's
Apple Inn			May's	Tory's
May's				May's
Tory's				

From the table, May's has the most points, so it is the critics' choice.

CHECK YOUR PROGRESS 5, *page 634*

Do a head-to-head comparison for each of the candidates and enter each winner in the table below.

versus	Alpha	Beta	Gamma
Alpha		Alpha	Alpha
Beta			Beta
Gamma			

From this table, Alpha is the winner. However, using the Borda Count method (See Example 5), Beta is the winner. Thus the Borda Count method violates the Condorcet criterion.

CHECK YOUR PROGRESS 6, *page 636*

	Rankings		
Radiant silver	1	3	3
Electric red	2	2	1
Lightning blue	3	1	2
Number of votes:	30	27	2

Using the Borda Count method, we have

Silver:

30 first-place votes	$30 \cdot 3 =$	90
0 second-place votes	$0 \cdot 2 =$	0
29 third-place votes	$29 \cdot 1 =$	29
	Total	119

Red:

2 first-place votes	$2 \cdot 3 =$	6
57 second-place votes	$57 \cdot 2 =$	114
0 third-place votes	$0 \cdot 1 =$	0
	Total	120

Blue:

27 first-place votes	$27 \cdot 3 =$	81
2 second-place votes	$2 \cdot 2 =$	4
30 third-place votes	$30 \cdot 1 =$	30
	Total	115

Using this method, electric red is the preferred color. Now suppose we eliminate the third-place choice (lightning blue). This gives the following table.

	Rankings		
Radiant silver	1	2	2
Electric red	2	1	1
Number of votes:	30	27	2

Recalculating the results, we have

Silver:

30 first-place votes	$30 \cdot 2 = 60$
29 second-place votes	$29 \cdot 1 = 29$
	Total 89

Red:

29 first-place votes	$29 \cdot 2 = 58$
30 second-place votes	$30 \cdot 1 = 30$
	Total 88

Now radiant silver is the preferred color. By deleting an alternative, the result of the voting changed. This violates the irrelevant alternatives criterion.

SECTION 9.3

CHECK YOUR PROGRESS 1, *page 650*

a. and **b.**

Winning Coalition	Number of Votes	Critical Voters
{A, B}	40	A, B
{A, C}	39	A, C
{A, B, C}	57	A
{A, B, D}	50	A, B
{A, B, E}	45	A, B
{A, C, D}	49	A, C
{A, C, E}	44	A, C
{A, D, E}	37	A, D, E
{B, C, D}	45	B, C, D
{B, C, E}	40	B, C, E
{A, B, C, D}	67	None
{A, B, C, E}	62	None
{A, B, D, E}	55	A
{A, C, D, E}	54	A
{B, C, D, E}	50	B, C
{A, B, C, D, E}	72	None

CHECK YOUR PROGRESS 2, *page 652*

Winning Coalition	Number of Votes	Critical Voters
{A, B}	34	A, B
{A, C}	28	A, C
{B, C}	26	B, C
{A, B, C}	44	None
{A, B, D}	40	A, B
{A, C, D}	34	A, C
{B, C, D}	32	B, C
{A, B, C, D}	50	None

The number of times any voter is critical is 12.

$$BPI(A) = \frac{4}{12} = \frac{1}{3}$$

$$BPI(D) = \frac{0}{12} = 0$$

CHECK YOUR PROGRESS 3, *page 653*

The countries are represented as follows: B, Belgium; F, France; G, Germany; I, Italy; L, Luxembourg; N, Netherlands.

Winning Coalition	Number of Votes	Critical Voters
{F, G, I}	12	F, G, I
{B, F, G, I}	14	F, G, I
{B, F, G, I, L}	15	F, G, I
{B, F, G, I, N}	16	None
{B, F, G, I, L, N}	17	None
{B, F, G, N}	12	B, F, G, N
{B, F, I, N}	12	B, F, I, N
{B, G, I, N}	12	B, G, I, N
{B, G, I, N, L}	13	B, G, I, N
{B, F, I, N, L}	13	B, F, I, N
{B, F, G, N, L}	13	B, F, G, N
{F, G, I, L}	13	F, G, I
{F, G, I, L, N}	15	F, G, I
{F, G, I, N}	14	F, G, I

The number of times any vote is critical is 42. The BPIs of the nations are:

$$BPI(B) = \frac{6}{42} = \frac{1}{7}$$

$$BPI(F) = \frac{10}{42} = \frac{5}{21}$$

$$BPI(G) = \frac{10}{42} = \frac{5}{21}$$

$$BPI(I) = \frac{10}{42} = \frac{5}{21}$$

$$BPI(L) = \frac{0}{42} = 0$$

$$BPI(N) = \frac{6}{42} = \frac{1}{7}$$

CHAPTER 10

SECTION 10.1

CHECK YOUR PROGRESS 1, *page 669*

Because the second graph has edge AB and the first graph does not, the two graphs are not equivalent.

CHECK YOUR PROGRESS 2, *page 672*

One vertex in the graph is of degree 3, and another is of degree 5. Because not all vertices are of even degree, the graph is not Eulerian.

CHECK YOUR PROGRESS 3, *page 673*

Represent the land areas and bridges with a graph, as we did for the Königsberg bridges earlier in the section. The vertices of the resulting graph, shown at the right below, all are of even degree. Thus we know that the graph has an Euler circuit. An Euler circuit corresponds to a stroll that crosses each bridge and returns to the starting point without crossing any bridge twice.

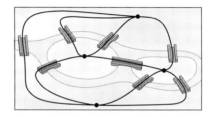 

CHECK YOUR PROGRESS 4, *page 675*

Consider the campground map as a graph. A route through all the trails that does not repeat any trail corresponds to an Euler walk. Because only two vertices (A and F) are of odd degree, we know that an Euler walk exists. Furthermore, the walk must begin at A and end at F or begin at F and end at A. By trial and error, one Euler walk is A–B–C–D–E–B–G–F–E–C–A–F.

CHECK YOUR PROGRESS 5, *page 676*

The graph has seven vertices, so $n = 7$ and $n/2 = 3.5$. Several vertices are of degree less than $n/2$, so Dirac's Theorem does not apply. Still, it may be possible to route the document as specified. By trial and error, one such route is Los Angeles–New York–Boston–Atlanta–Dallas–Phoenix–San Francisco–Los Angeles.

CHECK YOUR PROGRESS 6, *page 678*

Represent the floor plan with a graph, as in Example 6.

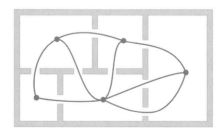

A stroll passing through each doorway just once corresponds to an Euler circuit or walk. Because four vertices are of odd degree, no Euler circuit or walk exists, so it is not possible to take such a stroll.

SECTION 10.2

CHECK YOUR PROGRESS 1, *page 686*
Draw a graph in which the vertices represent locations and the edges indicate available bus routes between locations. Each edge should be given a weight corresponding to the number of minutes for the bus ride.

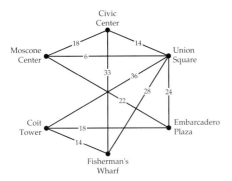

A route that visits each location and returns to the Moscone Center corresponds to a Hamiltonian circuit. Using the graph, we find that one such route is Moscone Center–Civic Center–Union Square–Fisherman's Wharf–Coit Tower–Embarcadero Plaza–Moscone Center, with a total weight of $18 + 14 + 28 + 14 + 18 + 22 = 114$. Another route is Moscone Center–Union Square–Embarcadero Plaza–Coit Tower–Fisherman's Wharf–Civic Center–Moscone Center, with a total weight of $6 + 24 + 18 + 14 + 33 + 18 = 113$. The travel time is one minute less for the second route.

CHECK YOUR PROGRESS 2, *page 688*
Starting at vertex A, the edge of smallest weight is the edge to D, with weight 5. From D, take the edge of weight 4 to C, then the edge of weight 3 to B. From B, the edge of least weight to a vertex not yet visited is the edge to vertex E (with weight 5). This is the last vertex, so we return to A along the edge of weight 9. Thus the Hamiltonian circuit is A–D–C–B–E–A, with a total weight of 26.

CHECK YOUR PROGRESS 3, *page 690*
The smallest weight appearing in the graph is 3, so we mark edge BC. The next smallest weight is 4, on edge CD. Three edges have weight 5, but we cannot mark edge BD, because it would complete a circuit. We can, however, mark edge AD. The next valid edge of smallest weight is BE, also of weight 5. No more edges can be marked without completing a circuit or adding a third edge to a vertex, so we mark the final edge, AE, to complete the Hamiltonian circuit. In this case, the Edge-Picking Algorithm generated the same circuit as the Greedy Algorithm did in Check Your Progress 2.

CHECK YOUR PROGRESS 4, *page 694*
Represent the times between locations with a weighted graph.

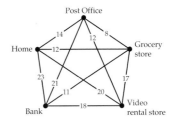

Starting at the home vertex and using the Greedy Algorithm, we first use the edge to the grocery store (of weight 12) followed by the edge of weight 8 to the post office and then the edge of weight 12 to the video rental store. The edge of next smallest weight is to the grocery store, but that vertex has already been visited, so we take the edge to the bank, with weight 18. All vertices have now been visited, so we select the last edge, of weight 23, to return home. The total weight is 73, corresponding to a total driving time of 73 minutes.

For the Edge-Picking Algorithm, we first select the edge of weight 8, followed by the edge of weight 11. Two edges have weight 12, but one adds a third edge to the grocery store vertex, so we must choose the edge from the post office to the video rental store. The next smallest weight is 14, but that edge would add a third edge to a vertex, as would the edge of weight 17. The edge of weight 18 would complete a circuit too early, so the next edge we can select is that of weight 20, the edge from home to the video rental store. The final step is to select the edge from home to the bank to complete the circuit. The resulting route is home–video rental store–post office–grocery store–bank–home (we could travel the same route in the reverse order) with a total travel time of 74 minutes.

CHECK YOUR PROGRESS 5, *page 695*
Represent the computer network by a graph in which the weights of the edges indicate the distances between computers.

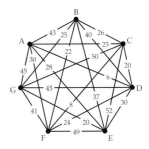

The edges with the smallest weights, which can all be chosen, are those of weights 6, 8, 20, 20, and 22. The edge of next smallest weight, 24, cannot be selected. There are two edges of weight 25; edge AC would add a third edge to vertex C, but edge BG can be chosen. All that remains is to complete the circuit with edge AE. The computers should be networked in this order: A, D, C, F, B, G, E, and back to A.

SECTION 10.3

CHECK YOUR PROGRESS 1, *page 703*
One possible coloring is 1-blue, 2-green, 3-red, 4-green, 5-blue, 6-red, 7-green, 8-red, 9-blue.

CHECK YOUR PROGRESS 2, *page 705*
Draw a graph on the map as in Example 2. More than two colors are required to color the resulting graph, but, by experimenting, it can be determined that the graph can be colored with three colors. Thus the graph is 3-colorable.

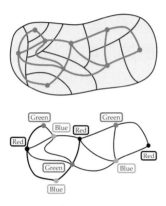

CHECK YOUR PROGRESS 3, *page 706*
There are several locations in the graph at which three edges form a triangle. Because a triangle is a circuit with an odd number of vertices, the graph is not 2-colorable.

CHECK YOUR PROGRESS 4, *page 708*
Draw a graph in which each vertex corresponds to a film and an edge joins two vertices if one person needs to view both of the corresponding films. We can use colors to represent the different times at which the films can be viewed. No two vertices connected by an edge can share the same color, because that would mean one person would have to watch two films at the same time.

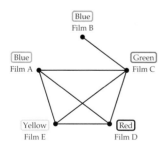

It is not possible to color the vertices with only three colors; one possible 4-coloring is shown. This means that four different time slots will be required to view the films, and the soonest all the friends can finish watching them is 4:00 A.M. A schedule can be set

using the coloring in the graph. From 8 to 10, the films labeled blue, film A and film B, can be shown in two different rooms. The remaining films are represented by unique colors and thus will require their own viewing times. Film C can be shown from 10 to 12, film D from 12 to 2, and film E from 2 to 4.

CHECK YOUR PROGRESS 5, *page 709*
Draw a graph in which each vertex represents a deli, and an edge connects two vertices if the corresponding delis deliver to a common building. Try to color the vertices using the fewest colors possible; each color can correspond to a day of the week that the delis can deliver.

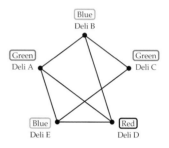

As shown, a 3-coloring is possible (but a 2-coloring is not). Therefore three different delivery days will be necessary—delis A and C deliver on one day, delis B and E on another day, and deli D on a third day.

CHAPTER 1

EXERCISE SET 1.1   *page 12*

1. 28   **3.** 45   **5.** 64   **7.** $\dfrac{15}{17}$ **9.** -13 **11.** Correct   **13.** Correct   **15.** Incorrect
17. No effect   **19.** 150 inches   **21.** The distance is quadrupled.   **23.** 0.5 second   **25.** Inductive
27. Deductive   **29.** Deductive   **31.** Inductive
For Exercises 33–43, only one possible answer is given. Your answers may vary from the given answers.

33. $x = \dfrac{1}{2}$ **35.** $x = \dfrac{1}{2}$ **37.** $x = -3$ **39.** Consider 1 and 3. $1 + 3$ is even, but $1 \cdot 3$ is odd.

41. It does not work for 121.   **43.** n
$$6n + 8$$
$$\frac{6n + 8}{2} = 3n + 4$$
$$3n + 4 - 2n = n + 4$$
$$n + 4 - 4 = n$$

45. Maria: the utility stock; Jose: the automotive stock; Anita: the technology stock; Tony: the oil stock   **47.** Atlanta: stamps;
Chicago: baseball cards; Philadelphia: coins; Seattle: comic books   **49.** Home, bookstore, supermarket, credit union, home; or
home, credit union, supermarket, bookstore, home   **51.** N, because the first letter of Nine is N.   **53.** d
55. Answers will vary.

EXERCISE SET 1.2   *page 28*

1. 195   **3.** 91   **5.** $40   **7.** 18   **9.** $2^{12} = 4096$ **11.** 28   **13.** 21 ducks, 14 pigs   **15.** 12
17. 6   **19.** 7   **21. a.** 80,200   **b.** 151,525   **c.** 1892   **23.** Yes   **25.** Four times as large

27. a. 121, 484, and 676   **b.** 1331   **29.** $1\dfrac{1}{2}$ inches   **31. a.** 1.3 billion; 1.5 billion; 1.6 billion   **b.** 1995

c. 2002   **d.** 2001 to 2002   **33. a.** 1994   **b.** 2004   **c.** The number of theatre admissions in 2003 was less than
the number of admissions in 2002.   **35.** 2601 tiles   **37.** Four more sisters than brothers   **39.** The 11th day
41. 91   **43. a.** Place four coins on the left balance pan and the other four on the right balance pan. The pan that is higher
contains the fake coin. Take the four coins from the higher pan and use the balance scale to compare the weight of two of these coins with
the weight of the other two. The pan that is higher contains the fake coin. Take the two coins from the higher pan and use the balance scale
to compare the weight of one of these coins with the weight of the other. The pan that is higher contains the fake coin. This procedure
enables you to determine the fake coin in three weighings.   **b.** Place three of the coins on one of the balance pans and another three
coins on the other. If the pans balance, then the fake coin is one of the two remaining coins. You can use the balance scale to determine
which of the remaining coins is the fake coin because it will be lighter than the other coin. If the three coins on the left pan do not balance
with the three coins on the right pan, then the fake coin must be one of the three coins on the higher pan. Pick any two coins from these
three and place one on each balance pan. If these two coins do not balance, then the one that is higher is the fake. If the coins balance, then
the third coin (the one that you did not place on the balance pan) is the fake. In any case, this procedure enables you to determine the fake
coin in two weighings.   **45. a.** 1600. Sally likes perfect squares.   **47.** d.   64. Each number is the cube of a term in the
sequence 1, 2, 3, 4, 5, 6.   **49.** Adding 83 is the same as adding 100 and subtracting 17. Thus, after you add 83, you will have a
number that has 1 as the hundreds digit. The number formed by the tens digit and the units digit will be 17 less than your original
number. After you add the hundreds digit, 1, to the other two digits of this new number, you will have a number that is 16 less than

your original number. If you subtract this number from your original number, you must get 16.   **51.** Answers will vary.

53. a. The Collatz problem (the $3n + 1$ problem): Start with any counting number $n > 1$. Now generate a sequence of numbers using the following rules.

■ If n is even, divide n by 2.

■ If n is odd, multiply n by 3 and add 1.

Repeat the above procedure on the new number you have just generated. Keep applying the same procedure until you obtain the number 1. Collatz conjectured that the procedure would always generate a sequence of numbers that would eventually reach the number 1, regardless of the starting number n. Thus far no one has been able to prove that Collatz's conjecture is true or to show that it is false. The sequences generated are sometimes called "hailstone" sequences because the numbers in the sequences tend to bounce up and down, much like a hailstone in a storm.

b. Here are the Collatz sequences for the counting numbers 2 through 10.

   2: 2, 1
   3: 3, 10, 5, 16, 8, 4, 2, 1
   4: 4, 2, 1
   5: 5, 16, 8, 4, 2, 1
   6: 6, 3, 10, 5, 16, 8, 4, 2, 1
   7: 7, 22, 11, 34, 17, 52, 26, 13, 40, 20, 10, 5, 16, 8, 4, 2, 1
   8: 8, 4, 2, 1
   9: 9, 28, 14, 7, 22, 11, 34, 17, 52, 26, 13, 40, 20, 10, 5, 16, 8, 4, 2, 1
  10: 10, 5, 16, 8, 4, 2, 1

55. a.

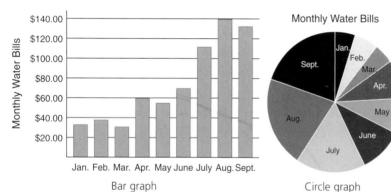

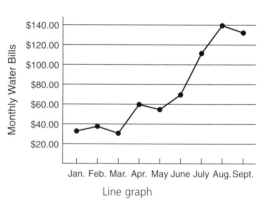

Bar graph   Circle graph   Line graph

b. Answers will vary.

EXERCISE SET 1.3   *page 47*

1. {penny, nickel, dime, quarter}   **3.** {Mercury, Mars}   **5.** $\{-5, -4, -3, -2, -1\}$ **7.** $\{-5\}$ **9.** The set of days of the week that begin with the letter T.   **11.** The set consisting of the two planets in our solar system that are closest to the sun.   **13.** The set of single-digit natural numbers.   **15.** {California, Arizona}   **17.** {2000, 2002, 2004}
19. {1985, 1986, 1987, 1989}   **21.** 11   **23.** 0   **25.** 121   **27.** 45   **29.** 28   **31.** {0, 1, 3, 5, 8}
33. {0, 2, 4, 6, 8}   **35.** Subset   **37.** Subset   **39.** Not a subset   **41.** 18 hours   **43. a.** 256   **b.** 11
45. a. 4096   **b.** 14   **47.** {1, 2, 4, 5, 6, 8}   **49.** {4, 6}   **51.** {3, 7}   **53.** $U = \{1, 2, 3, 4, 5, 6, 7, 8\}$
55. ∅   **57.** {2, 5, 8}   **59.** {2, 5, 8}
61.   *U*   **63.**   *U*   **65.**   *U*

67. Not equal   **69.** Equal   **71.** Not equal   **73.** Not equal   **75.** Yellow   **77.** Cyan   **79.** Red

81.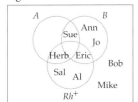

83. 7 **85.** 8 **87.** 8 **89.** 12 **91. a.** 180 **b.** 200

93. a. 109 **b.** 328 **c.** 104 **95. a.** 450 **b.** 140 **c.** 130 **97. a.** 6% **b.** 38%
c. 7% **99. a.** 200 **b.** 271 **c.** 16

CHAPTER 1 REVIEW EXERCISES *page 54*

1. Deductive [Sec 1.1] **2.** Inductive [Sec. 1.1] **3.** Inductive [Sec. 1.1] **4.** Deductive [Sec. 1.1]
5. $x = 0$ provides a counterexample because $0^4 = 0$ and 0 is not greater than 0. [Sec. 1.1]

6. $x = 4$ provides a counterexample because $\dfrac{(4)^3 + 5(4) + 6}{6} = 15$, which is not an even number. [Sec. 1.1]

7. $x = 1$ provides a counterexample because $[(1) + 4]^2 = 25$, but $(1)^2 + 16 = 17$. [Sec. 1.1] **8.** Let $a = 1$ and $b = 1$. Then
$(a + b)^3 = (1 + 1)^3 = 2^3 = 8$. However, $a^3 + b^3 = 1^3 + 1^3 = 2$. [Sec. 1.1] **9.** 320 feet by 1600 feet [Sec. 1.2]
10. $3^{15} = 14{,}348{,}907$ [Sec. 1.2] **11.** 48 skyboxes [Sec. 1.2] **12.** On the first trip the rancher takes the rabbit across the
river. The rancher returns alone. The rancher takes the dog across the river and returns with the rabbit. The rancher next takes the carrots
across the river and returns alone. On the final trip the rancher takes the rabbit across the river. [Sec. 1.2] **13.** $300 [Sec. 1.2]
14. 105 [Sec. 1.2] **15.** Answers will vary. [Sec. 1.2] **16.** Answers will vary. [Sec. 1.2] **17.** Michael: biology
major; Clarissa: business major; Reggie: computer science major; Ellen: chemistry major [Sec. 1.1]
18. Dodgers: drug store; Pirates: supermarket; Tigers: bank; Giants: service station [Sec. 1.1]
19. a. Yes. Answers will vary. **b.** No. The countries of India, Bangladesh, and Myanmar all share borders with each other. Thus at
least three colors are needed to color the map. [Sec. 1.1]
20. a. The following figure shows a route that starts from North Bay **b.** No. [Sec. 1.1]
and passes over each bridge once and only once.

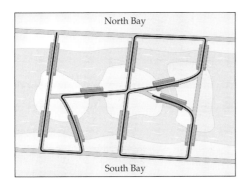

21. 1 square inch; 4 square inches; 25 square inches [Sec. 1.2] **22.** A represents 1, B represents 9, and D represents 0. [Sec. 1.2]
23. 5 [Sec. 1.2] **24.** 10 [Sec. 1.2] **25.** 1 [Sec. 1.2] **26.** 3 [Sec. 1.2]
27. n
$\quad$ $4n$
$\quad$ $4n + 12$
$\quad$ $\dfrac{4n + 12}{2} = 2n + 6$
$\quad$ $2n + 6 - 6 = 2n$ [Sec. 1.1]
28. Each nickel is worth 5 cents. Thus 2004 nickels are
worth $2004 \times 5 = 10{,}020$ cents, or $100.20. [Sec. 1.2]

29. a. 1970 to 1980 **b.** 5.5% [Sec. 1.2] **30. a.** 16 times as many **b.** 61 times as many [Sec. 1.2]
31. a. 2002 **b.** 2004 [Sec. 1.2] **32.** 5005 [Sec. 1.2] **33.** There are no narcissistic numbers. [Sec. 1.2]
34. a. 10 **b.** Yes [Sec. 1.1] **35.** $\{0, 1, 2, 3, 4, 5, 6, 7\}$ [Sec. 1.3] **36.** $\{-8, 8\}$ [Sec. 1.3]
37. $\{1, 2, 3, 4\}$ [Sec. 1.3] **38.** $\{1, 2, 3, 4, 5, 6\}$ [Sec. 1.3] **39.** $\{6, 10\}$ [Sec. 1.3] **40.** $\{2, 6, 10, 16, 18\}$ [Sec. 1.3]

41. $C = \{14, 16\}$ [Sec. 1.3]   **42.** $\{2, 6, 8, 10, 12, 16, 18\}$ [Sec. 1.3]   **43.** $\{2, 6, 10, 16\}$ [Sec. 1.3]
44. $\{8, 12\}$ [Sec. 1.3]   **45.** $\{6, 8, 10, 12, 14, 16, 18\}$ [Sec. 1.3]   **46.** $\{8, 12\}$ [Sec. 1.3]   **47.** Proper subset   [Sec. 1.3]
48. Proper subset   [Sec. 1.3]   **49.** Not a proper subset   [Sec. 1.3]   **50.** Not a proper subset   [Sec. 1.3]
51. $\varnothing, \{I\}, \{II\}, \{I, II\}$ [Sec. 1.3]   **52.** $\varnothing, \{s\}, \{u\}, \{n\}, \{s, u\}, \{s, n\}, \{u, n\}, \{s, u, n\}$ [Sec. 1.3]
53. $\varnothing$, {penny}, {nickel}, {dime}, {quarter}, {penny, nickel}, {penny, dime}, {penny, quarter}, {nickel, dime}, {nickel, quarter},
{dime, quarter}, {penny, nickel, dime}, {penny, nickel, quarter}, {penny, dime, quarter}, {nickel, dime, quarter},
{penny, nickel, dime, quarter}   [Sec. 1.3]   **54.** $\varnothing$, {A}, {B}, {C}, {D}, {E}, {A, B}, {A, C}, {A, D}, {A, E}, {B, C}, {B, D}, {B, E},
{C, D}, {C, E}, {D, E}, {A, B, C}, {A, B, D}, {A, B, E}, {A, C, D}, {A, C, E}, {A, D, E}, {B, C, D}, {B, C, E}, {B, D, E}, {C, D, E}, {A, B, C, D},
{A, B, C, E}, {A, B, D, E}, {A, C, D, E}, {B, C, D, E}, {A, B, C, D, E}   [Sec. 1.3]   **55.** $2^4 = 16$ [Sec. 1.3]
56. $2^{26} = 67,108,864$ [Sec. 1.3]   **57.** $2^{15} = 32,768$ [Sec. 1.3]   **58.** $2^7 = 128$ [Sec. 1.3]

59.
[Sec. 1.3]

60.
[Sec. 1.3]

61.
[Sec. 1.3]

62.
[Sec. 1.3]

63. Equal   [Sec. 1.3]

64. Not equal   [Sec. 1.3]   **65.** Not equal   [Sec. 1.3]   **66.** Not equal   [Sec. 1.3]

67.
[Sec. 1.3]

68.
[Sec. 1.3]

69. 5   [Sec. 1.3]   **70.** 10   [Sec. 1.3]   **71. a.** 135   **b.** 59   **c.** 71   [Sec. 1.3]   **72.** 391   [Sec. 1.3]
73. a. 42   **b.** 31   **c.** 20   **d.** 142   [Sec. 1.3]   **74. a.** 79   **b.** 37   **c.** 4   [Sec. 1.3]

CHAPTER 1 TEST   *page 58*

1. a. Deductive   **b.** Inductive   **c.** Inductive   **d.** Deductive   [Sec. 1.1]   **2.** 15   [Sec. 1.2]
3. 3   [Sec. 1.2]   **4.** $672   [Sec. 1.2]   **5.** 126   [Sec. 1.2]   **6.** 36   [Sec. 1.2]   **7.** Reynaldo is 13, Ramiro is 5,
Shakira is 15, and Sasha is 7.   [Sec. 1.1]   **8.** 606   [Sec. 1.2]   **9.** $x = 4$ provides a counterexample because division by zero
is undefined.   [Sec. 1.1]   **10. a.** 2002 to 2003   **b.** 250,000   [Sec. 1.2]   **11.** $\{0, 1, 2, 3, 4, 5\}$ [Sec. 1.3]
12. a. $\{2, 3, 5, 7, 8, 9, 10\}$ **b.** $\{2, 9, 10\}$ **c.** $\{5, 7, 8\}$ [Sec. 1.3]   **13. a.** 7   **b.** 1   [Sec. 1.3]
14. $\varnothing, \{a\}, \{d\}, \{p\}, \{a, d\}, \{a, p\}, \{d, p\}, \{a, d, p\}$ [Sec. 1.3]   **15.** $2^{21} = 2,097,152$ [Sec. 1.3]
16.
[Sec. 1.3]

17. $A' \cap B'$ [Sec. 1.3]   **18. a.** $2^5 = 32$ **b.** 8   [Sec. 1.3]

19. 541   [Sec. 1.3]   **20. a.** 232   **b.** 102   **c.** 857   **d.** 79   [Sec. 1.3]

CHAPTER 2

EXERCISE SET 2.1   *page 71*

1. Statement   **3.** Statement   **5.** Not a statement   **7.** Statement   **9.** Not a statement   **11.** One component is "The principal will attend the class on Tuesday." The other component is "The principal will attend the class on Wednesday."   **13.** One component is "A triangle is an acute triangle." The other component is "It has three acute angles."   **15.** One component is "I ordered a salad." The other component is "I ordered a cola."   **17.** One component is $5 + 2 > 6$. The other component is $5 + 2 = 6$.   **19.** The Giants did not lose the game.   **21.** The game went into overtime.   **23.** $w \rightarrow t$; conditional   **25.** $s \rightarrow r$; conditional   **27.** $l \leftrightarrow a$; biconditional   **29.** $d \rightarrow f$; conditional   **31.** $m \vee c$; disjunction   **33.** The tour goes to Italy and the tour does not go to Spain.   **35.** If we go to Venice, then we will not go to Florence.   **37.** We will go to Florence if and only if we do not go to Venice.   **39.** q is a true statement.   **41.** $p \vee q$ is a true statement.   **43.** All cats have claws.   **45.** Some classic movies were not first produced in black and white.   **47.** Some of the numbers were even numbers.   **49.** Some irrational numbers can be written as terminating decimals.   **51.** Some cars do not run on gasoline.   **53.** Some items are not on sale.   **55.** True   **57.** True   **59.** True   **61.** True   **63.** True   **65.** True   **67.** True   **69.** $p \rightarrow q$, where p represents "you can count your money" and q represents "you don't have a billion dollars."   **71.** $p \rightarrow q$, where p represents "you do not learn from history" and q represents "you are condemned to repeat it."   **73.** $p \rightarrow q$, where p represents "people concentrated on the really important things in life" and q represents "there'd be a shortage of fishing poles."   **75.** $p \leftrightarrow q$, where p represents "an angle is a right angle" and q represents "its measure is 90°."   **77.** $p \rightarrow q$, where p represents "two sides of a triangle are equal in length" and q represents "the angles opposite those sides are congruent."   **79.** $p \rightarrow q$, where p represents "it is a square" and q represents "it is a rectangle."   **81.** Raymond Smullyan was born in 1919 in Far Rockaway, New York. Smullyan dropped out of high school because he found it boring and wanted to pursue his own mathematical interests. He attended several universities and eventually received a degree in mathematics from the University of Chicago. After earning a Ph.D. at Princeton in 1959, he taught mathematics and logic at Dartmouth, Princeton, and the City University of New York. In 1981 he became a member of the faculty at Indiana University. Smullyan has written several popular puzzle books and many books on mathematical logic.

EXERCISE SET 2.2   *page 83*

1. It is a true statement.   **3.** Self-contradiction   **5.** True   **7.** False   **9.** False   **11.** False   **13.** False   **15. a.** If p is false, then $p \wedge (q \vee r)$ must be a false statement.   **b.** For a conjunctive statement to be true, it is necessary that all components of the statement be true. Because it is given that one of the components (p) is false, $p \wedge (q \vee r)$ must be a false statement.

p	q	17.	19.	21.
T	T	T	F	F
T	F	F	T	T
F	T	T	F	T
F	F	T	F	T

p	q	r	23.	25.	27.	29.	31.
T	T	T	F	T	T	T	F
T	T	F	F	T	F	T	T
T	F	T	F	T	F	T	T
T	F	F	F	T	F	T	F
F	T	T	F	F	T	T	F
F	T	F	F	F	F	T	F
F	F	T	F	T	T	F	T
F	F	F	T	T	F	T	F

See the *Online Student's Solutions Manual* for the solutions to Exercises 33–39.   **41.** It did not rain and it did not snow.
43. She did not visit either France or Italy.   **45.** She did not get a promotion and she did not receive a raise.   **47.** Tautology
49. Tautology   **51.** Tautology   **53.** Self-contradiction   **55.** Self-contradiction   **57.** Not a self-contradiction
59. The symbol $\leq$ means "less than or equal to."   **61.** $2^5 = 32$ **63.** F F F T T T F T F F F T F F F
65. a. $(p \wedge q \wedge r) \vee (p \wedge \sim q \wedge r) \vee (\sim p \wedge \sim q \wedge r)$ **b.** The disjunctive normal form is a valuable concept because it
provides an easy mechanical method for naming any proposition defined by a truth table.

EXERCISE SET 2.3   *page 95*

1. *Antecedent:* I had the money
 Consequent: I would buy the painting

3. *Antecedent:* they had a guard dog
 Consequent: no one would trespass on their property

5. True   **7.** True   **9.** False

p	q	11.	13.
T	T	T	T
T	F	T	T
F	T	T	T
F	F	T	T

p	q	r	15.	17.	19.
T	T	T	T	T	T
T	T	F	T	F	T
T	F	T	T	T	T
T	F	F	T	T	T
F	T	T	T	T	T
F	T	F	T	F	T
F	F	T	T	T	T
F	F	F	T	T	T

21. $p \rightarrow v$ **23.** $t \rightarrow \sim v$ **25.** Not equivalent   **27.** Not equivalent   **29.** Equivalent   **31.** If we take the
aerobics class, then we will be in good shape for the ski trip.   **33.** If he has the talent to play a keyboard, then he can join the band.
35. If Education 147 is offered in the spring semester, then I will be able to receive my credential.   **37.** $s \rightarrow r$ **39.** Yes
41. a. If I quit this job, then I am rich.   **b.** If I were not rich, then I would not quit this job.   **c.** If I would not quit this job,
then I would not be rich.   **43. a.** If we are not able to attend the party, then she did not return soon.   **b.** If she returns soon,
then we will be able to attend the party.   **c.** If we are able to attend the party, then she returned soon.   **45. a.** If a figure is a
quadrilateral, then it is a parallelogram.   **b.** If a figure is not a parallelogram, then it is not a quadrilateral.   **c.** If a figure is
not a quadrilateral, then it is not a parallelogram.   **47. a.** If I am able to get current information about astronomy, then I have
access to the Internet.   **b.** If I do not have access to the Internet, then I will not be able to get current information about astronomy.
c. If I am not able to get current information about astronomy, then I don't have access to the Internet.   **49. a.** If we don't have
enough money for dinner, then we took a taxi.   **b.** If we did not take a taxi, then we will have enough money for dinner.
c. If we have enough money for dinner, then we did not take a taxi.   **51. a.** If she can extend her vacation for at least two days, then
she will visit Kauai.   **b.** If she does not visit Kauai, then she could not extend her vacation for at least two days.   **c.** If she
cannot extend her vacation for at least two days, then she will not visit Kauai.   **53. a.** If two lines are parallel, then the two lines
are perpendicular to a given line.   **b.** If two lines are not perpendicular to a given line, then the two lines are not parallel.
c. If two lines are not parallel, then the two lines are not both perpendicular to a given line.   **55.** Not equivalent
57. Equivalent   **59.** If $x = 7$, then $3x - 7 \neq 11$. The original statement is true.   **61.** If $|a| = 3$, then $a = 3$. The original
statement is false.   **63.** If $a + b = 25$, then $\sqrt{a + b} = 5$. The original statement is true.   **65.** False   **67.** True
69. True   **71. a. and b.** Answers will vary.   **73.** If you can dream it, then you can do it.   **75.** If I were a dancer, then
I would not be a singer.   **77.** A conditional statement and its contrapositive are equivalent. They always have the same truth values.
79. The Hatter is telling the truth.   **81.** Student demonstration

---

EXERCISE SET 2.4   *page 106*

1. $r \rightarrow c$
  r
  $\therefore c$

3. $g \rightarrow s$
  $\sim g$
  $\therefore \sim s$

5. $s \rightarrow i$
  s
  $\therefore i$

7. $\sim p \rightarrow \sim a$
  a
  $\therefore p$

9. Invalid   **11.** Invalid   **13.** Valid

15. Invalid   **17.** Invalid   **19.** Invalid   **21.** Valid   **23.** Valid   **25.**
$$h \to r$$
$$\underline{\sim h}$$
$$\therefore \sim r$$
Invalid

27.
$$\sim b \to d$$
$$\underline{b \lor d}$$
$$\therefore b$$
Invalid

29.
$$c \to t$$
$$\underline{t}$$
$$\therefore c$$
Invalid

31. Valid argument; modus ponens   **33.** Invalid argument; fallacy of the converse

35. Valid argument; modus tollens   **37.** Invalid argument; fallacy of the inverse   **39.** Valid argument; law of syllogism
41. Valid argument; modus ponens   **43.** Valid argument; modus tollens   See the *Online Instructor's Solutions Manual* for the solutions to Exercises 45–50.   **51.** Valid

EXERCISE SET 2.5   *page 115*

1. All cats (*C*) are nimble (*N*).

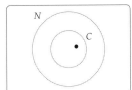

3. Some actors (*A*) are not famous (*F*).

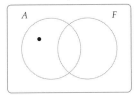

5. Valid   **7.** Valid   **9.** Valid   **11.** Valid   **13.** Invalid   **15.** Invalid   **17.** Invalid
19. Valid   **21.** Valid   **23.** Invalid   **25.** All Reuben sandwiches need mustard.   **27.** 1001 ends with a 5.
29. Some horses are grey.   **31. a.** Invalid   **b.** Invalid   **c.** Invalid   **d.** Invalid   **e.** Valid   **f.** Valid
33.

1	2
2	1
3	4
8	4

CHAPTER 2 REVIEW EXERCISES   *page 118*

1. Not a statement [Sec. 2.1]   **2.** Statement [Sec. 2.1]   **3.** Statement [Sec. 2.1]   **4.** Statement [Sec. 2.1]
5. Not a statement [Sec. 2.1]   **6.** Statement [Sec. 2.1]   **7.** $m \land b$; conjunction [Sec. 2.1]
8. $d \to e$; conditional [Sec. 2.1]   **9.** $g \leftrightarrow d$; biconditional [Sec. 2.1]   **10.** $t \to s$; conditional [Sec. 2.1]
11. No dogs bite. [Sec. 2.1]   **12.** Some desserts at the Cove restaurant are not good. [Sec. 2.1]   **13.** Some winners do
not receive prizes. [Sec. 2.1]   **14.** All cameras use film. [Sec. 2.1]   **15.** Some of the students received As. [Sec. 2.1]
16. Nobody enjoyed the story. [Sec. 2.1]   **17.** True [Sec. 2.1]   **18.** True [Sec. 2.1]   **19.** True [Sec. 2.1]
20. True [Sec. 2.1]   **21.** True [Sec. 2.1]   **22.** True [Sec. 2.1]   **23.** False [Sec. 2.2]
24. False [Sec. 2.2/2.3]   **25.** True [Sec. 2.2]   **26.** True [Sec. 2.2/2.3]   **27.** True [Sec. 2.2/2.3]
28. False [Sec. 2.2/2.3]

p	*q*	**29.** [Sec. 2.2/2.3]	**30.** [Sec. 2.2/2.3]	**31.** [Sec. 2.2/2.3]	**32.** [Sec. 2.2/2.3]
T	T	T	F	F	T
T	F	T	F	F	T
F	T	T	T	F	F
F	F	F	F	F	T

p	q	r	33. [Sec. 2.2/2.3]	34. [Sec. 2.2/2.3]	35. [Sec. 2.2/2.3]	36. [Sec. 2.2/2.3]
T	T	T	T	F	F	T
T	T	F	T	T	F	T
T	F	T	T	T	T	T
T	F	F	F	T	T	T
F	T	T	T	F	F	F
F	T	F	T	F	F	T
F	F	T	T	T	F	T
F	F	F	T	T	F	T

37. Bob passed the English proficiency test or he did not register for a speech course. [Sec. 2.2]   **38.** It is not true that Ellen went to work this morning or she took her medication. [Sec. 2.2]   **39.** It is not the case that Wendy will not go to the store this afternoon and she will be able to prepare her fettuccine al pesto recipe. [Sec. 2.2]   **40.** It is not the case that Gina did not enjoy the movie or she enjoyed the party. [Sec. 2.2/2.3]   See the *Online Instructor's Solutions Manual* for solutions to Exercises 41–44.   **45.** Self-contradiction [Sec. 2.2]   **46.** Tautology [Sec. 2.2/2.3]   **47.** Tautology [Sec. 2.2/2.3]   **48.** Tautology [Sec. 2.2/2.3]   **49.** *Antecedent:* he has talent   **50.** *Antecedent:* I had a teaching credential   *Consequent:* he will succeed [Sec. 2.3]   *Consequent:* I could get the job [Sec. 2.3]   **51.** *Antecedent:* I join the fitness club   **52.** *Antecedent:* I will attend   *Consequent:* I will follow the exercise program [Sec. 2.3]   *Consequent:* it is free [Sec. 2.3]   **53.** False [Sec. 2.3]   **54.** True [Sec. 2.3]   **55.** True [Sec. 2.3]   **56.** False [Sec. 2.3]   **57.** False [Sec. 2.3]   **58.** False [Sec. 2.3]   **59.** If a real number has a nonrepeating, nonterminating decimal form, then the real number is irrational. [Sec. 2.3]   **60.** If you are a politician, then you are well known. [Sec. 2.3]   **61.** If I could sell my condominium, then I could buy the house. [Sec. 2.3]   **62.** If a number is divisible by 9, then the number is divisible by 3. [Sec. 2.3]   **63. a.** *Converse:* If $x > 3$, then $x + 4 > 7$.   **b.** *Inverse:* If $x + 4 \le 7$, then $x \le 3$.   **c.** *Contrapositive:* If $x \le 3$, then $x + 4 \le 7$. [Sec. 2.3]   **64. a.** *Converse:* If the recipe can be prepared in less than 20 minutes, then the recipe is in this book.   **b.** *Inverse:* If the recipe is not in this book, then the recipe cannot be prepared in less than 20 minutes. **c.** *Contrapositive:* If the recipe cannot be prepared in less than 20 minutes, then the recipe is not in this book.   [Sec. 2.3]   **65. a.** *Converse:* If $(a + b)$ is divisible by 3, then a and b are both divisible by 3.   **b.** *Inverse:* If a and b are not both divisible by 3, then $(a + b)$ is not divisible by 3.   **c.** *Contrapositive:* If $(a + b)$ is not divisible by 3, then a and b are not both divisible by 3. [Sec. 2.3]   **66. a.** *Converse:* If they come, then you built it.   **b.** *Inverse:* If you do not build it, then they will not come. **c.** *Contrapositive:* If they do not come, then you did not build it. [Sec. 2.3]   **67. a.** *Converse:* If it has exactly two parallel sides, then it is a trapezoid.   **b.** *Inverse:* If it is not a trapezoid, then it does not have exactly two parallel sides.   **c.** *Contrapositive:* If it does not have exactly two parallel sides, then it is not a trapezoid. [Sec. 2.3]   **68. a.** *Converse:* If they returned, then they liked it. **b.** *Inverse:* If they do not like it, then they will not return.   **c.** *Contrapositive:* If they do not return, then they did not like it. [Sec. 2.3]   **69.** $q \to p$, the converse of the original statement [Sec. 2.3]   **70.** $p \to q$, the original statement [Sec. 2.3]   **71.** If x is an odd prime number, then $x > 2$. [Sec. 2.3]   **72.** If the senator attends the meeting, then she will vote on the motion. [Sec. 2.3]   **73.** If their manager contacts me, then I will purchase some of their products. [Sec. 2.3]   **74.** If I can rollerblade, then Ginny can rollerblade. [Sec. 2.3]   **75.** Valid [Sec. 2.4]   **76.** Valid [Sec. 2.4]   **77.** Invalid [Sec. 2.4]   **78.** Valid [Sec. 2.4]   **79.** Valid argument; disjunctive syllogism [Sec. 2.4]   **80.** Valid argument; law of syllogism [Sec. 2.4]   **81.** Invalid argument; fallacy of the inverse [Sec. 2.4]   **82.** Valid argument; disjunctive syllogism [Sec. 2.4]   **83.** Valid argument; modus tollens [Sec. 2.4]   **84.** Invalid argument; fallacy of the inverse [Sec. 2.4]   **85.** Valid [Sec. 2.5]   **86.** Invalid [Sec. 2.5]   **87.** Invalid [Sec. 2.5]   **88.** Valid [Sec. 2.5]

CHAPTER 2 TEST   *page 120*

1. a. Not a statement   **b.** Statement [Sec. 2.1]   **2. a.** All trees are green.   **b.** Some of the kids had seen the movie. [Sec. 2.1]   **3. a.** False   **b.** True [Sec. 2.1]   **4. a.** False   **b.** True [Sec. 2.2/2.3]

p	q	5. [Sec. 2.2/2.3]
T	T	T
T	F	T
F	T	T
F	F	T

p	q	r	6. [Sec. 2.2/2.3]
T	T	T	F
T	T	F	T
T	F	T	F
T	F	F	F
F	T	T	F
F	T	F	T
F	F	T	T
F	F	F	F

7. It is not the case that Elle ate breakfast or took a lunch break. [Sec. 2.2]   **8.** A tautology is a statement that is always true. [Sec. 2.2]   **9.** The statements are equivalent. [Sec. 2.3]   **10. a.** False   **b.** False [Sec. 2.2/2.3]
11. a. *Converse:* If $x > 4$, then $x + 7 > 11$.   **b.** *Inverse:* If $x + 7 \le 11$, then $x \le 4$.   **c.** *Contrapositive:* If $x \le 4$, then $x + 7 \le 11$.   [Sec. 2.3]

12.
$$\begin{array}{l} p \to q \\ \underline{ p } \\ \therefore q \end{array}$$ [Sec. 2.4]

13.
$$\begin{array}{l} p \to q \\ \underline{ q \to r } \\ \therefore p \to r \end{array}$$ [Sec. 2.4]

14. Valid [Sec. 2.4]

15. Invalid [Sec. 2.4]   **16.** Invalid argument; the argument is a fallacy of the inverse. [Sec. 2.4]   **17.** Valid argument; the argument is a disjunctive syllogism. [Sec. 2.4]   **18.** Invalid argument, as shown by an Euler diagram. [Sec. 2.5]
19. Invalid argument, as shown by an Euler diagram. [Sec. 2.5]   **20.** Invalid argument; the argument is a fallacy of the converse. [Sec. 2.4]

CHAPTER 3

EXERCISE SET 3.1   *page 134*

1. An equation expresses the equality of two mathematical expressions. An equation contains an equals sign; an expression does not.
3. Substitute the solution back into the original equation and confirm the equality.   **5.** 12   **7.** 22   **9.** -8

11. -20 **13.** 8   **15.** Positive   **17.** $y > 30$ **19.** -1 **21.** 1   **23.** $-\dfrac{1}{3}$ **25.** $\dfrac{2}{3}$

27. -2 **29.** 4   **31.** 2   **33.** 2   **35.** $\dfrac{1}{4}$ **37.** -2 **39.** 4   **41.** 8   **43.** 1

45. -32 **47.** $16,859.34   **49.** 1350 inches   **51.** 60 feet   **53.** 1952   **55.** 168 feet
57. 18.6 degrees Celsius   **59.** 175   **61.** $22,000   **63. a.** $2395   **b.** $4694   **65.** $117.75
67. a. 163,000 kilograms   **b.** 1930 kilograms   **69.** $12.50   **71.** 1280 horizontal pixels   **73.** More than

16 minutes but not over 17 minutes   **75.** $b = P - a - c$ **77.** $R = \dfrac{E}{I}$ **79.** $r = \dfrac{I}{Pt}$ **81.** $C = \dfrac{5}{9}(F - 32)$

83. $t = \dfrac{A - P}{Pr}$ **85.** $f = \dfrac{T + gm}{m}$ **87.** $S = C - Rt$ **89.** $b_2 = \dfrac{2A}{h} - b_1$ **91.** $h = \dfrac{S}{2\pi r} - r$

93. $y = 2 - \dfrac{4}{3}x$ **95.** $x = \dfrac{y - y_1}{m} + x_1$ **97.** 0   **99.** Every real number is a solution.   **101.** $x = \dfrac{-b}{a - c}$. $a \ne c$

or the denominator equals zero and the expression is undefined.   **103. a.** $4125   **b.** 4325 passenger cars
c. $T = 325m + 0.04x$, where T is the total cost, m is the number of months, and x is the number of copies   **d.** $T = 750m + 30,000$, where T is the total number of miles driven and m is the number of months you have driven the car   **e.** $C = 2.50 + 1.75(h - 1)$, where C is the parking charge and h is the number of hours parked in the garage   **f.** $C = 19.95d + 0.25(m - 100)$, where C is the total cost, d is the number of days, and m is the number of miles driven   **g.** Answers will vary.

EXERCISE SET 3.2   *page 152*

1. Examples will vary.   **3.** The purpose is to allow currency from one country to be converted into the currency of another country.   **5.** Explanations will vary.   **7.** The cross-products method is a shortcut for multiplying each side of the proportion by the least common multiple of the denominators.   **9.** 23 miles per gallon   **11.** 12.5 meters per second   **13.** 400 square feet per gallon   **15.** c   **17.** 272,160 kilograms   **19.** A 24-ounce jar of mayonnaise for $2.09   **21.** $16.50 per hour   **23. a.** Australia   **b.** 859 more people per square mile   **25.** 63,652 krona   **27.** 397,936 pesos   **29.** For each state, the ratio is 3.125 to 1.   **31.** $13:1$ or 13 to 1. There is one faculty member for every 13 students at Syracuse University.   **33.** University of Connecticut   **35.** No   **37.** 17.14   **39.** 25.6   **41.** 20.83   **43.** 2.22   **45.** 13.71   **47.** 39.6   **49.** 0.52   **51.** 6.74   **53.** $t > 3$ **55.** $45,000   **57.** 5.5 milligrams   **59.** 24 feet; 15 feet   **61.** 160,000 people   **63.** 63,000 miles   **65.** 11.25 grams. Explanations will vary.   **67. a.** True   **b.** True   **c.** True   **d.** False   **69. a.–c.** Explanations will vary.   **71.** Answers will vary.

EXERCISE SET 3.3   *page 173*

1. Answers will vary.   **3.** 3   **5.** Employee B   **7.** 0.5; 50%   **9.** $\frac{2}{5}$; 0.4   **11.** $\frac{7}{10}$; 70%   **13.** 0.55; 55%   **15.** $\frac{5}{32}$; 0.15625   **17.** $x > y$ **19. a.** 73 fans   **b.** More fans approved.   **c.** 7%; $100\% - 73\% - 20\% = 7\%$ **21.** $26.6 billion   **23.** 23.7%   **25.** $1846.64   **27. a.** 48.2 hours   **b.** 33.6 hours   **c.** 3.4 hours   **29. a.** 1998 to 2000   **b.** 2004 to 2006   **c.** From 1998 to 2002   **31. a.** Arizona: 118,153; California: 625,041; Colorado: 165,038; Maryland: 142,718; Massachusetts: 246,833; Minnesota: 229,543; New Hampshire: 51,714; New Jersey: 229,543; Vermont: 17,274; Virginia: 185,900   **b.** New Hampshire; California   **c.** More than half   **d.** Massachusetts: 7.5%; New Jersey: 5.5%; Virginia: 5.2%   **e.** 5.3%   **f.** The rate is 10 times the percent.   **33. a.** 900%   **b.** 60%   **c.** 1500%   **d.** Explanations will vary.   **35. a.** Home health aides   **b.** Software engineers (applications)   **c.** 63,200 people   **d.** 320,580 people   **e.** Computer support specialists   **f.** The percent increases are based on different original employment figures.   **g.** Answers will vary.   **37.** Less than   **39.** 26 months   **41. a.** 10,700,000 TV households; 62,700,000 TV households   **b.** 5,900,000 TV households; 53,700,000 TV households   **c.** 2.5 people   **d.** Answers will vary.

EXERCISE SET 3.4   *page 190*

1. a. When an increase in one quantity leads to a proportional increase in the other quantity   **b.** When an increase in one quantity leads to a proportional decrease in the other quantity   **3.** 7.5   **5.** 1.6   **7.** 6   **9.** 24   **11.** 288   **13.** 200   **15.** Larger   **17.** a   **19.** $307.50   **21.** 5.4 pounds per square inch   **23.** 3 amperes   **25.** 287.3 feet   **27.** 80 feet   **29.** 10 feet   **31.** $1.\overline{6}$ ohms   **33.** 45 cubic feet   **35.** 12.8 foot-candles   **37.** $55.\overline{5}$ decibels   **39.** 2.25 seconds   **41. a.** 8   **b.** $\frac{1}{8}$ **43. a.** $f = kaw^2$ **b.** It doubles the force of the wind.   **c.** It quadruples the force of the wind.   **d.** 180 pounds   **e.** 200 watts   **f.** 22.5 pounds per square inch

CHAPTER 3 REVIEW EXERCISES   *page 195*

1. 4   [Sec. 3.1]   **2.** $\frac{1}{8}$ [Sec. 3.1]   **3.** -2 [Sec. 3.1]   **4.** $\frac{10}{3}$ [Sec. 3.2]   **5.** $-\frac{9}{4}$ [Sec. 3.1]   **6.** $y = -\frac{4}{3}x + 4$ [Sec. 3.1]   **7.** $t = \frac{f - v}{a}$ [Sec. 3.1]   **8.** 3   [Sec. 3.4]   **9.** 100   [Sec. 3.4]   **10.** 2450 feet   [Sec. 3.1]   **11.** 3 seconds   [Sec. 3.1]   **12.** 60°C   [Sec. 3.1]   **13.** 39 minutes   [Sec. 3.1]   **14.** 28.4 miles per gallon   [Sec. 3.2]   **15.** $\frac{1}{4}$ [Sec. 3.2]   **16.** $1260   [Sec. 3.1]   **17. a.** New York, Chicago, Philadelphia, Los Angeles, Houston   **b.** 21,596 more people per square mile   [Sec. 3.2]   **18. a.** $15:1$, 15 to 1. There are 15 students for every one faculty member at the university.   **b.** Grand Canyon University, Arizona State University   **c.** Embry-Riddle Aeronautical University, Northern Arizona University, and University of Arizona   [Sec. 3.2]   **19.** Department A: $105,000;

Department B: $245,000   [Sec. 3.2]   **20.** 7.5 tablespoons   [Sec. 3.2]   **21. a.** No   **b.** 41:9   **c.** $1,828.6 billion
d. $490.6 billion   [Sec. 3.2]   **22. a.** 51.0%   **b.** More than   [Sec. 3.3]   **23.** 735,000 people   [Sec. 3.3]
24. 69.0%   [Sec. 3.3]   **25. a.** 83.$\overline{3}$%   **b.** 100%   **c.** 50%   [Sec. 3.3]   **26.** 16.4%   [Sec. 3.3]
27. a. Ages 9–10   **b.** Ages 15–16   **c.** 46.5%; less than   **d.** 7230 boys   **e.** 549,400 young people   [Sec. 3.3]
28. $80,000 and $60,000   [Sec. 3.2]   **29.** $410   [Sec. 3.4]   **30.** 44 miles per hour   [Sec. 3.4]

CHAPTER 3 TEST   *page 199*

1. 14   [Sec. 3.1]   **2.** 10   [Sec. 3.1]   **3.** $\dfrac{21}{4}$ [Sec. 3.2]   **4.** 2   [Sec. 3.1]   **5.** $y = \dfrac{1}{2}x - \dfrac{15}{2}$ [Sec. 3.1]

6. $F = \dfrac{9}{5}C + 32$ [Sec. 3.1]   **7.** 2.5 minutes   [Sec. 3.1]   **8.** 10 days   [Sec. 3.1]   **9.** 54.8 miles per hour   [Sec. 3.2]

10. 843 acres   [Sec. 3.1]   **11. a.** 2.727, 2.905, 2.777, 2.808, 2.901, 2.904   **b.** Ty Cobb, Rogers Hornsby, Joe Jackson,

Tris Speaker, Ted Williams, Billy Hamilton   [Sec. 3.2]   **12.** $\dfrac{4}{7}$ [Sec. 3.2]   **13.** $112,500 and $67,500   [Sec. 3.2]

14. 2.75 pounds   [Sec. 3.2]   **15. a.** Miami-Dade   **b.** 9420 violent crimes   [Sec. 3.2]   **16.** 14.4%   [Sec. 3.3]
17. a. 20.6%   **b.** Between 2004 and 2005   **c.** 39.1%   [Sec. 3.3]   **18. a.** 20%   **b.** 1.6 million working farms
c. Answers will vary.   [Sec. 3.3]   **19.** 54 feet   [Sec. 3.4]   **20.** 167 decibels   [Sec. 3.4]

CHAPTER 4

EXERCISE SET 4.1   *page 208*

1. Meter, liter, gram   **3.** Kilometer   **5.** Centimeter   **7.** Gram   **9.** Meter   **11.** Gram
13. Milliliter   **15.** Milligram   **17.** Millimeter   **19.** Milligram   **21.** Gram   **23.** Kiloliter

25. Milliliter   **27. a.** Column 2: k, c, m; column 3: 10^9, 10^3, 10^2, $\dfrac{1}{10^3}$, $\dfrac{1}{10^{12}}$; column 4: 1 000 000, 10, 0.1, 0.01, 0.000 001,

0.000 000 001   **b.** Answers will vary.   **29.** 910   **31.** 1.856   **33.** 7.285   **35.** 8 000   **37.** 0.034
39. 29.7   **41.** 7.530   **43.** 9 200   **45.** 36   **47.** 2 350   **49.** 83   **51.** 0.716   **53.** 6.302
55. 458   **57.** 9.2   **59.** 2 g   **61.** 4.4 km   **63.** 24 L   **65.** $20.77   **67.** 16 servings
69. 4 tablets   **71.** The case of 12 one-liter bottles   **73.** 500 s   **75.** $17,430   **77.** $66.50

EXERCISE SET 4.2   *page 219*

1. Answers will vary.   **3.** $5\dfrac{1}{3}$ ft   **5.** $2\dfrac{5}{8}$ lb   **7.** $1\dfrac{1}{2}$ mi   **9.** $\dfrac{1}{4}$ ton   **11.** $2\dfrac{1}{2}$ gal   **13.** 20 fl oz

15. 11,880 ft   **17.** $\dfrac{5}{8}$ ft   **19.** $7\dfrac{1}{2}$ c   **21.** $3\dfrac{3}{4}$ qt   **23.** 2640 yd   **25.** 1,104,537,600 s   **27.** $12\dfrac{1}{2}$ gal

29. $7\dfrac{1}{2}$ gal   **31.** 25 qt   **33.** $800   **35.** $65,340   **37.** 65.91 kg   **39.** 8.48 c   **41.** 54.20 L
43. 189.2 lb   **45.** 1.38 in.   **47.** 6.33 gal   **49.** 18.29 m/s   **51.** $1.30   **53.** 49.69 mph   **55.** $3.16
57. 40 068.07 km   **59. a.** False   **b.** True   **c.** False   **d.** True   **e.** True   **f.** True
61. Answers will vary.   **63.** Answers will vary.

EXERCISE SET 4.3   *page 236*

1. $\angle O$, $\angle AOB$, and $\angle BOA$ **3.** 40°, acute   **5.** 30°, acute   **7.** 120°, obtuse   **9.** Yes   **11.** No
13. A 28° angle   **15.** An 18° angle   **17.** 14 cm   **19.** 28 ft   **21.** 30 m   **23.** 86°   **25.** 71°
27. 30°   **29.** 36°   **31.** 127°   **33.** 116°   **35.** 20°   **37.** 20°   **39.** 20°   **41.** 141°
43. 106°   **45.** 11°   **47.** $m\angle a = 38°$, $m\angle b = 142°$ **49.** $m\angle a = 47°$, $m\angle b = 133°$ **51.** 20°   **53.** 47°

55. $m\angle x = 155°$, $m\angle y = 70°$ **57.** $m\angle a = 45°$, $m\angle b = 135°$ **59.** $90° - x$ **61.** $60°$ **63.** $35°$
65. $102°$ **67.** The three angles form a straight angle. The sum of the measures of the angles of a triangle is 180°.
69. Adjacent angles   **71.** The distances differ because you can travel in a straight line in the air but you cannot travel in a straight line on the roads. The shortest distance between two points is a straight line.   **73.** The line segments are the same length.
75. The sum of the measures of the interior angles of a triangle is 180°; therefore, $m\angle a + m\angle b + m\angle c = 180°$. The sum of the measures of an interior and an exterior angle is 180°; therefore, $m\angle c + m\angle x = 180°$. Solving this equation for $m\angle c$, $m\angle c = 180° - m\angle x$. Substitute $180° - m\angle x$ for $m\angle c$ in the equation $m\angle a + m\angle b + m\angle c = 180°$: $m\angle a + m\angle b + 180° - m\angle x = 180°$. Add $m\angle x$ to each side of the equation, and subtract 180° from each side of the equation: $m\angle a + m\angle b = m\angle x$. The measure of an exterior angle of a triangle is equal to the sum of the measures of the two opposite interior angles: $m\angle a + m\angle c = m\angle z$.

EXERCISE SET 4.4   *page 258*

1.    **3. a.** Perimeter is not measured in square units.   **b.** Area is measured in square units.

5. Heptagon   **7.** Quadrilateral   **9.** Isosceles   **11.** Scalene   **13.** Right   **15.** Obtuse
17. a. 30 m   **b.** 50 m^2 **19. a.** 16 cm   **b.** 16 cm^2 **21. a.** 40 km   **b.** 100 km^2 **23. a.** 40 ft
b. 72 ft^2 **25. a.** 8π cm; 25.13 cm   **b.** 16π cm^2; 50.27 cm^2 **27. a.** 11π mi; 34.56 mi   **b.** 30.25π mi^2; 95.03 mi^2
29. a. 17π ft; 53.41 ft   **b.** 72.25π ft^2; 226.98 ft^2 **31.** $29\frac{1}{2}$ ft   **33.** $10\frac{1}{2}$ mi   **35.** 68 ft   **37.** 20 in.
39. 214 yd   **41.** 10 mi   **43.** 2 packages   **45.** 144 m^2 **47.** 9 in.   **49.** 10 in.   **51.** 8 m
53. 96 m^2 **55.** 607.5 m^2 **57.** 2 bags   **59.** 2 qt   **61.** 20 tiles   **63.** $40   **65.** $34
67. 120 ft^2 **69.** 13.19 ft   **71.** 1256.6 ft^2 **73.** 94.25 ft   **75.** 144π in^2 **77.** 113.10 in^2
79. 266,281 km   **81.** $8r^2 - 2\pi r^2$ **83.** Rectangle   **85.** A square whose side is 1 ft   **87. a.** Sometimes true
b. Sometimes true   **c.** Always true   **d.** Always true   **e.** Always true   **f.** Always true
89. $(2x + 3)$ by $(2x - 3)$ **91. a.** 12 units   **b.** 10 units   **c.** 14 units   **d.** 10 units

---

EXERCISE SET 4.5   *page 273*

1. $\frac{1}{2}$ **3.** $\frac{3}{4}$ **5.** 7.2 cm   **7.** 3.3 m   **9.** 12 m   **11.** 12 in.   **13.** 56.3 cm^2 **15.** 18 ft
17. 16 m   **19.** $14\frac{3}{8}$ ft   **21.** 15 m   **23.** 8 ft   **25.** 13 cm   **27.** 35 m   **29. a.** Always true
b. Sometimes true   **c.** Always true   **d.** Always true   **31.** 13 in.   **33.** 11.4 cm   **35.** 8.7 ft   **37.** 7.9 m
39. 7.4 m   **41.** 21.6 mi   **43.** 24 in.   **45.** Yes. Explanations will vary.

EXERCISE SET 4.6   *page 288*

1. 840 in^3 **3.** 15 ft^3 **5.** 4.5π cm^3; 14.14 cm^3 **7.** 94 m^2 **9.** 56 m^2 **11.** 96π in^2; 301.59 in^2
13. yd^2, mi   **15. a.** Feet   **b.** Cubic feet   **c.** Square feet   **17.** 20.25 ft^3 **19.** 343 cm^3
21. 7.24 m^3 **23.** 115.2π m^3 **25.** 392.70 cm^3 **27.** 216 m^3 **29.** 184 ft^2 **31.** 69.36 m^2
33. 225π cm^2 **35.** 402.12 in^2 **37.** 6π ft^2 **39.** 297 in^2 **41.** 2.5 ft   **43.** 3 cm   **45.** 11 cans
47. 456 in^2 **49.** 22.53 cm^2 **51.** 5 m^3 **53.** 69.12 in^3 **55.** 192 in^3 **57.** 208 in^2 **59.** 204.57 cm^3
61. 165.99 cm^2 **63.** $3515.00   **65.** 158 cans   **67.** $498.75   **69.** 256,000 gal   **71.** $V = \frac{2}{3}\pi r^3$;
$SA = 3\pi r^2$ **73.** Surface area of the sphere $= 4\pi r^2$. Surface area of the side of the cylinder $= 2\pi rh = 2\pi r(2r) = 4\pi r^2$.
75. a. For example, make a cut perpendicular to the top and bottom faces and parallel to two of the sides.   **b.** For example,

beginning at an edge that is perpendicular to the bottom face, cut at an angle through to the bottom face.   **c.** For example, beginning at the top face at a distance d from the vertex, cut at an angle to the bottom face, ending at a distance greater than d from the opposite vertex.   **d.** For example, beginning on the top face at a distance d from a vertex, cut across the cube to a point just above the opposite vertex.

EXERCISE SET 4.7   *page 300*

1. a. $\dfrac{a}{c}$ **b.** $\dfrac{b}{c}$ **c.** $\dfrac{b}{c}$ **d.** $\dfrac{a}{c}$ **e.** $\dfrac{a}{b}$ **f.** $\dfrac{b}{a}$ **3.** $\sin\theta = \dfrac{5}{13}$, $\cos\theta = \dfrac{12}{13}$, $\tan\theta = \dfrac{5}{12}$

5. $\sin\theta = \dfrac{24}{25}$, $\cos\theta = \dfrac{7}{25}$, $\tan\theta = \dfrac{24}{7}$ **7.** $\sin\theta = \dfrac{8}{\sqrt{113}}$, $\cos\theta = \dfrac{7}{\sqrt{113}}$, $\tan\theta = \dfrac{8}{7}$ **9.** $\sin\theta = \dfrac{1}{2}$, $\cos\theta = \dfrac{\sqrt{3}}{2}$,

$\tan\theta = \dfrac{1}{\sqrt{3}}$ **11.** 0.6820   **13.** 1.4281   **15.** 0.9971   **17.** 1.9970   **19.** 0.8878   **21.** 0.8453

23. 0.8508   **25.** 0.6833   **27.** 38.6°   **29.** 41.1°   **31.** 21.3°   **33.** 38.0°   **35.** 72.5°   **37.** 0.6°
39. 66.1°   **41.** 29.5°   **43.** Sine   **45.** Tangent   **47.** Larger; smaller   **49.** 841.79 ft   **51.** 13.6°
53. 29.14 ft   **55.** 52.92 ft   **57.** 13.59 ft   **59.** 1056.63 ft   **61.** 29.58 yd   **63.** No. Explanations will vary.

65. $\dfrac{\sqrt{5}}{3}$ **67.** $\dfrac{\sqrt{7}}{3}$ **69.** $\sqrt{1-a^2}$ **71.** 4 radians   **73.** $\dfrac{2}{3}$ radian   **75.** $\left(\dfrac{180}{\pi}\right)^{\circ}$ **77.** $\dfrac{\pi}{4}$ radian;

0.7854 radian   **79.** $\dfrac{7\pi}{4}$ radians; 5.4978 radians   **81.** $\dfrac{7\pi}{6}$ radians; 3.6652 radians   **83.** 60°   **85.** 240°

87. $\left(\dfrac{540}{\pi}\right)^{\circ}$; 171.8873°

CHAPTER 4 REVIEW EXERCISES   *page 307*

1. 1 240 m   [Sec. 4.1]   **2.** 450 mg   [Sec. 4.1]   **3.** $4\dfrac{1}{2}$ lb   [Sec. 4.2]   **4.** 6600 ft   [Sec. 4.2]

5. 46.58 mph   [Sec. 4.2]   **6.** $m\angle x = 22°$, $m\angle y = 158°$ [Sec. 4.3]   **7.** 24 in.   [Sec. 4.5]   **8.** 240 in³   [Sec. 4.6]
9. 68°   [Sec. 4.3]   **10.** 220 ft²   [Sec. 4.6]   **11.** 40π m²   [Sec. 4.6]   **12.** 44 cm   [Sec. 4.3]   **13.** 19°   [Sec. 4.3]
14. 27 in²   [Sec. 4.4]   **15.** 96 cm³   [Sec. 4.6]   **16.** 14.14 m   [Sec. 4.4]   **17.** $m\angle a = 138°$, $m\angle b = 42°$ [Sec. 4.3]
18. A 148° angle   [Sec. 4.3]   **19.** 39 ft³   [Sec. 4.6]   **20.** 95°   [Sec. 4.3]   **21.** 8 cm   [Sec. 4.4]   **22.** 288π mm³
[Sec. 4.6]   **23.** 21.5 cm   [Sec. 4.4]   **24.** 4 cans   [Sec. 4.6]   **25.** 208 yd   [Sec. 4.4]   **26.** 90.25 m²   [Sec. 4.4]
27. 12 oz   [Sec. 4.2]   **28.** 127.6 ft/s   [Sec. 4.2]   **29.** 50 L   [Sec. 4.1]   **30.** 276 m²   [Sec. 4.4]

31. 9.75 ft   [Sec. 4.5]   **32.** $\sin\theta = \dfrac{5}{\sqrt{89}}$, $\cos\theta = \dfrac{8}{\sqrt{89}}$, $\tan\theta = \dfrac{5}{8}$ [Sec. 4.7]

33. $\sin\theta = \dfrac{\sqrt{3}}{2}$, $\cos\theta = \dfrac{1}{2}$, $\tan\theta = \sqrt{3}$ [Sec. 4.7]   **34.** 25.7°   [Sec. 4.7]   **35.** 29.2°   [Sec. 4.7]   **36.** 53.8°

[Sec. 4.7]   **37.** 1.9°   [Sec. 4.7]   **38.** 100.1 ft   [Sec. 4.7]   **39.** 153.2 mi   [Sec. 4.7]   **40.** 56.0 ft   [Sec. 4.7]

CHAPTER 4 TEST   *page 309*

1. 46.5 m   [Sec. 4.1]   **2.** 4 100 ml   [Sec. 4.1]   **3.** 126 ft   [Sec. 4.2]   **4.** 20 fl oz   [Sec. 4.2]   **5.** 340.2 g
[Sec. 4.2]   **6.** 169.65 m³   [Sec. 4.6]   **7.** 6.8 m   [Sec. 4.4]   **8.** A 58° angle   [Sec. 4.3]   **9.** 3.14 m²   [Sec. 4.4]
10. 150°   [Sec. 4.3]   **11.** $m\angle a = 45°$; $m\angle b = 135°$ [Sec. 4.3]   **12.** 5.0625 ft²   [Sec. 4.4]   **13.** 448π cm³   [Sec. 4.6]

14. $1\dfrac{1}{5}$ ft   [Sec. 4.5]   **15.** 90° and 50°   [Sec. 4.3]   **16.** 125°   [Sec. 4.3]   **17.** 32 m²   [Sec. 4.4]

18. 25 ft   [Sec. 4.5]   **19.** 113.10 in²   [Sec. 4.4]   **20.** 7.55 cm   [Sec. 4.5]   **21.** $\sin\theta = \dfrac{4}{5}$, $\cos\theta = \dfrac{3}{5}$, $\tan\theta = \dfrac{4}{3}$

[Sec. 4.7]   **22.** 127 ft   [Sec. 4.7]   **23.** 103.87 ft²   [Sec. 4.4]   **24.** 889.3125 in³   [Sec. 4.6]   **25.** $114   [Sec. 4.2]

CHAPTER 5

EXERCISE SET 5.1 *page 321*

1.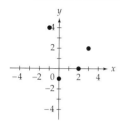

3.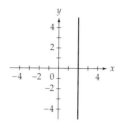

5.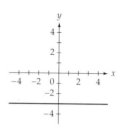

7.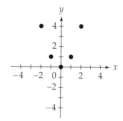

9.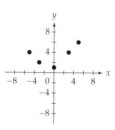

11.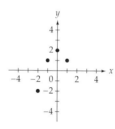

13.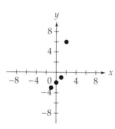

15. a. Quadrant IV
 b. Quadrant II

17.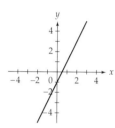

19.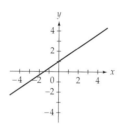

21.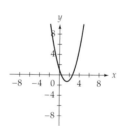

23.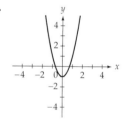

25.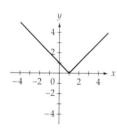

27. No **29.** 3 **31.** 3 **33.** 10 **35.** 0 **37.** $P(2000)$

39. a. 16 meters **b.** 20 feet **41. a.** 100 feet **b.** 68 feet **43. a.** 1087 feet per second **b.** 1136 feet per second **45. a.** 10% **b.** 40% **47.** 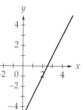 **49.**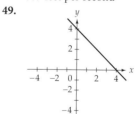

51. 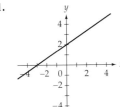   **53.** 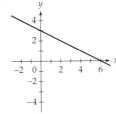   **55.** 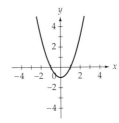   **57.**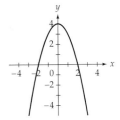

59. 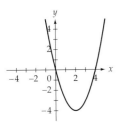   **61.** 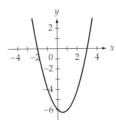   **63.** 48 square units   **65.** No. A function cannot have different elements in the range corresponding to one element in the domain.

67. 2   **69.** 17   **71.** $x = 0$ **73.** **a.** $M(10, 8) = 10; M(-3, -1) = -1; M(12, -13) = 12; M(-11, 15) = 15$
b. Answers will vary.   **c.** The value of the function is equal to the larger of the two input values.   **d.** It is a good name for the function because the value of the function is equal to the maximum of x and y.   **e.** Answers will vary. For example,

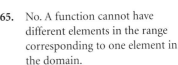

EXERCISE SET 5.2   *page 333*

1. $(2, 0), (0, -6)$ **3.** $(6, 0), (0, -4)$ **5.** $(-4, 0), (0, -4)$ **7.** $(4, 0), (0, 3)$ **9.** $\left(\dfrac{9}{2}, 0\right), (0, -3)$

11. $(2, 0), (0, 3)$ **13.** $(1, 0), (0, -2)$ **15.** $\left(\dfrac{30}{7}, 0\right)$; At $\dfrac{30}{7}$ °C the cricket stops chirping.   **17.** The intercept on the vertical axis is $(0, -15)$. This means that the temperature of the object is -15°F before it is removed from the freezer. The intercept on the horizontal axis is $(5, 0)$. This means that it takes 5 minutes for the temperature of the object to reach 0°F.   **19.** $f(x)$ **21.** -1

23. $\dfrac{1}{3}$ **25.** $-\dfrac{2}{3}$ **27.** $-\dfrac{3}{4}$ **29.** Undefined   **31.** $\dfrac{7}{5}$ **33.** 0   **35.** $-\dfrac{1}{2}$ **37.** Undefined

39. Positive   **41.** The slope is 40, which means the motorist was traveling at 40 miles per hour.   **43.** The slope is 0.25, which means the tax rate for an income range of \$29,050 to \$70,350 is 25%.   **45.** The slope is approximately 343.9, which means the runner traveled at a rate of 343.9 meters per minute.

47. 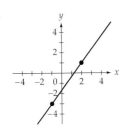   **49.** 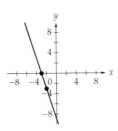   **51.** 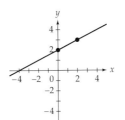   **53.**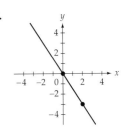

55. 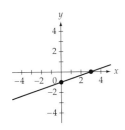   **57.** $y = \dfrac{2}{3}x - 216$ **59.** 2.5   **61.** Line *A* represents the distance traveled by Lois in x hours, line *B* represents the distance traveled by Tanya in x hours, and line *C* represents the distance between them after x hours.

63. No **65.** 7 **67.** 2 **69.** It rotates the line counterclockwise. **71.** It raises the line on the rectangular coordinate system. **73.** No. A vertical line through $(4, 0)$ does not have a y-intercept. **75. a.** $0.91\overline{6}$ **b.** $0.\overline{63}$
c. New design. $0.\overline{6}$ is closer to $0.\overline{63}$ than to $0.91\overline{6}$. **d.** Designs will vary. **e.** Answers will vary.

EXERCISE SET 5.3 *page 345*

1. $y = 2x + 5$ **3.** $y = -3x + 4$ **5.** $y = -\dfrac{2}{3}x + 7$ **7.** $y = -3$ **9.** $y = x + 2$ **11.** $y = -\dfrac{3}{2}x + 3$

13. $y = \dfrac{1}{2}x - 1$ **15.** $y = -\dfrac{5}{2}x$ **17.** Positive **19.** $R(x) = -\dfrac{3}{5}x + 545$; 485 rooms **21.** $D(t) = 415t$;
1867.5 miles **23.** $N(x) = -20x + 230{,}000$; 60,000 cars **25.** Green **27. a.**

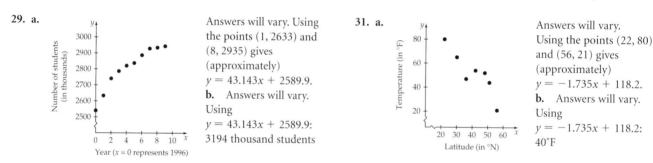

$y = 0.65x + 34.3$
b. 89.6 millimeters of mercury

29. a.

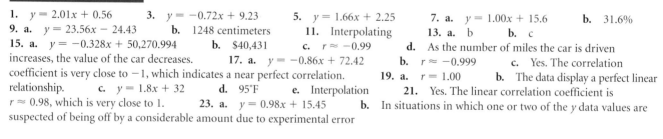

Answers will vary. Using the points $(1, 2633)$ and $(8, 2935)$ gives (approximately) $y = 43.143x + 2589.9$. **b.** Answers will vary. Using $y = 43.143x + 2589.9$: 3194 thousand students

31. a.

Answers will vary. Using the points $(22, 80)$ and $(56, 21)$ gives (approximately) $y = -1.735x + 118.2$. **b.** Answers will vary. Using $y = -1.735x + 118.2$: 40°F

33. Answers will vary. For example, $(0, 3)$, $(1, 2)$, and $(3, 0)$. **35.** 7 **37.** No. The three points do not lie on a straight line.
39. 3°. The car is climbing. **41. a.** The parametric equations giving the plane's position after t minutes are $x = 9000t$ and
$y = 100t + 5000$. **b.** 5500 feet **c.** Approximately 5133 feet

---

EXERCISE SET 5.4 *page 358*

1. $y = 2.01x + 0.56$ **3.** $y = -0.72x + 9.23$ **5.** $y = 1.66x + 2.25$ **7. a.** $y = 1.00x + 15.6$ **b.** 31.6%
9. a. $y = 23.56x - 24.43$ **b.** 1248 centimeters **11.** Interpolating **13. a.** b **b.** c
15. a. $y = -0.328x + 50{,}270.994$ **b.** \$40,431 **c.** $r \approx -0.99$ **d.** As the number of miles the car is driven increases, the value of the car decreases. **17. a.** $y = -0.86x + 72.42$ **b.** $r \approx -0.999$ **c.** Yes. The correlation coefficient is very close to -1, which indicates a near perfect correlation. **19. a.** $r = 1.00$ **b.** The data display a perfect linear relationship. **c.** $y = 1.8x + 32$ **d.** 95°F **e.** Interpolation **21.** Yes. The linear correlation coefficient is $r \approx 0.98$, which is very close to 1. **23. a.** $y = 0.98x + 15.45$ **b.** In situations in which one or two of the y data values are suspected of being off by a considerable amount due to experimental error

CHAPTER 5 REVIEW EXERCISES *page 364*

1.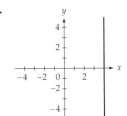
[Sec. 5.1]

2.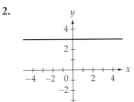
[Sec. 5.1]

3.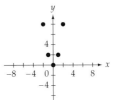
[Sec. 5.1]

4.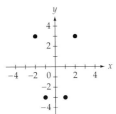
[Sec. 5.1]

5.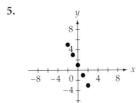
[Sec. 5.1]

6.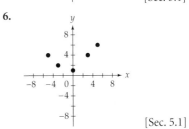
[Sec. 5.1]

7.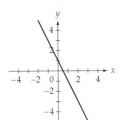
[Sec. 5.1]

8.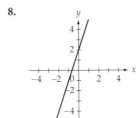
[Sec. 5.1]

9.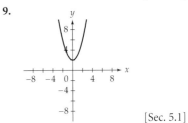
[Sec. 5.1]

10.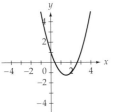
[Sec. 5.1]

11.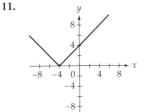
[Sec. 5.1]

12.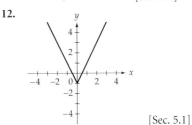
[Sec. 5.1]

13. -14 [Sec. 5.1] **14.** -20 [Sec. 5.1] **15.** 1 [Sec. 5.1] **16.** 0 [Sec. 5.1] **17. a.** 78.5 square feet **b.** 804.2 square centimeters [Sec. 5.1] **18.** 144 feet [Sec. 5.1] **19.** 1.76 seconds [Sec. 5.1] **20.** $(-5, 0), (0, 10)$ [Sec. 5.2] **21.** $(12, 0), (0, -9)$ [Sec. 5.2] **22.** $(5, 0), (0, -3)$ [Sec. 5.2] **23.** $(6, 0), (0, 8)$ [Sec. 5.2] **24.** The intercept on the vertical axis is $(0, 25{,}000)$. This means that the value of the truck was \$25,000 when it was new. The intercept on the horizontal axis is $(5, 0)$. This means that after 5 years the truck will be worth \$0. [Sec. 5.2] **25.** 5 [Sec. 5.2] **26.** $\dfrac{5}{2}$ [Sec. 5.2] **27.** 0 [Sec. 5.2] **28.** Undefined [Sec. 5.2] **29.** The slope is -0.6, which means that the revenue from home video rentals is decreasing \$600,000,000 annually. [Sec. 5.2]

30.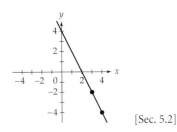
[Sec. 5.2]

31.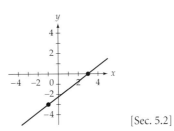
[Sec. 5.2]

32. $y = 2x + 7$ [Sec. 5.3]

33. $y = x - 5$ [Sec. 5.3]   **34.** $y = \frac{2}{3}x + 3$ [Sec. 5.3]   **35.** $y = \frac{1}{4}x$ [Sec. 5.3]

36.

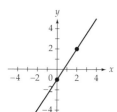

37. a. $f(x) = 25x + 100$ **b.** $900   [Sec. 5.3]
38. a. $A(p) = -25,000p + 10,000$ **b.** 10,750 gallons   [Sec. 5.3]

[Sec. 5.2]

39. a.

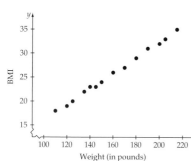

Weight (in pounds)

Answers will vary. Using the points (120, 19) and (205, 33) gives approximately $y = 0.165x - 0.765$.
b. Answers will vary. Using $y = 0.165x - 0.765$: 25 [Sec. 5.3]

40. a.

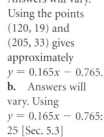

Year (x = 0 represents 1990)

Answers will vary. Using the points (0, 5) and (13, 70) gives $y = 5x + 5$.
b. Answers will vary. Using $y = 5x + 5$: 45% [Sec. 5.3]

41. a. $y = -1.08x + 19.44$ **b.** 10.8   **c.** $r \approx -0.948$ [Sec. 5.4]   **42. a.** $r \approx 0.999$ **b.** $y = 0.073x + 0.288$
c. 14.5 inches   [Sec. 5.4]   **43. a.** $y = 0.018x + 0.001$ **b.** Yes. $r \approx 0.999$, which is very close to 1.   **c.** 1.80 seconds
[Sec. 5.4]   **44. a.** $y = 12.9x + 8.15$ **b.** $y = -4.5x + 71$ **c.** Greater than [Sec. 5.4]
45. a. $y = 13.67x + 108.24, r \approx 0.998$ **b.** 668.7   [Sec. 5.4]

CHAPTER 5 TEST   *page 367*

1. -21 [Sec. 5.1]   **2.** 5   [Sec. 5.1]   **3.**

[Sec. 5.1]

4.

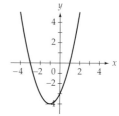

[Sec. 5.1]

5. $\left(\frac{8}{5}, 0\right), \left(0, -\frac{8}{3}\right)$ [Sec. 5.2]   **6.** $\frac{3}{5}$ [Sec. 5.2]   **7.** $y = \frac{2}{3}x + 3$ [Sec. 5.3]   **8.** The vertical intercept is (0, 250).
This means that the plane starts 250 miles from its destination. The horizontal intercept is (2.5, 0). This means that it takes the plane
2.5 hours to reach its destination.   [Sec. 5.2]   **9.** $y = -1.5x + 740$ [Sec. 5.3]

10. a.

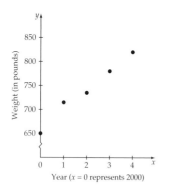

Year (x = 0 represents 2000)

Answers will vary. Using the points (0, 650) and (4, 820) gives $y = 42.5x + 650$.
b. Answers will vary. Using $y = 42.5x + 650$: 990 pounds   [Sec. 5.3]

11. $y = 238.4x - 427,372; r \approx 0.998$. The regression line fits the data well because the linear correlation coefficient is close to 1.   [Sec. 5.4]
12. a. $y = -7.98x + 767.12$ **b.** 57 calories   [Sec. 5.4]

CHAPTER 6

1. a. Standard form   **b.** 2   **c.** 6   **3. a.** $g(x) = 2x^3 + x^2 - 3x - 1$ **b.** 3   **c.** 1
5. a. $y(z) = -z^5 + 2z^3 - 3z^2 + 4z + 6$ **b.** 5   **c.** 2   **7.** Positive   **9.** $(0, -2)$ **11.** $(0, -1)$

13. $(0, 2)$ **15.** $(0, -1)$ **17.** $\left(\dfrac{1}{2}, -\dfrac{9}{4}\right)$ **19.** $\left(\dfrac{1}{4}, -\dfrac{41}{8}\right)$ **21.** Minimum: 2   **23.** Maximum: -3

25. Minimum: $-\dfrac{13}{4}$ **27.** Maximum: $\dfrac{9}{4}$ **29.** Not necessarily. The coordinates of the vertex for $f(x) = x^2$ and $g(x) = -x^2$
are $(0, 0)$, but the graph of f opens up while the graph of g opens down.   **31.** 150 feet   **33.** 100 lenses   **35.** 24.36 feet
37. Yes   **39.** 141.6 feet   **41.** 41 miles per hour, 33.658 miles per gallon   **43. a.** $V(x) = 4x^3 - 240x^2 + 3600x$
b. 14,812 cubic inches   **45.** $S(x) = -4x^2 + 3600$ square inches   **47. a.** $V(x) = 4x^3 - 150x^2 + 1350x$
b. 3564 cubic inches   **49.** $S(x) = -4x^2 + 1350$ square inches   **51.** $V(x) = -16x^2 + 24x$ cubic feet

53. a. $d(t) = \sqrt{10,000 + 625t^2}$ **b.** 388.10 feet   **55. a.** $C(x) = 1000x + \dfrac{15,000x + 20,000}{x}$ **b.** $40,800

57. a. $T(s) = \dfrac{\sqrt{s}}{4} + \dfrac{s}{1130}$ **b.** 2.6 seconds   **59.** 7   **61.** Yes   **63.** $F(2) = 2^2 + 2 - 6 = 4 + 2 - 6 = 0$
65. $-1, 5$ **67.** $-0.32, 0.59$ **69.** $-1.77, 1.06$ **71.** $-4, -1, 3$

1. a. 9   **b.** 1   **c.** $\dfrac{1}{9}$ **3. a.** 16   **b.** 4   **c.** $\dfrac{1}{4}$ **5. a.** 1   **b.** $\dfrac{1}{8}$ **c.** 16   **7.** No.

$2^{-x} = \dfrac{1}{2^x}$ and $2^x > 0$ for all values of x.   **9.** Exponential growth. Because $e > 1$, $f(x) = e^x$ represents an exponential growth function.

11. a. 7.3891   **b.** 0.3679   **c.** 1.2840   **13. a.** 54.5982   **b.** 1   **c.** 0.1353   **15. a.** 16   **b.** 16
c. 1.4768   **17. a.** 0.1353   **b.** 0.1353   **c.** 0.0111

19. 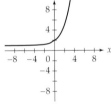   **21.** 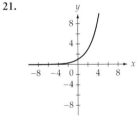   **23.** 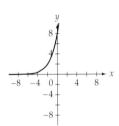   **25.**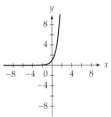

27. 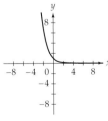   **29.** $11,202.50   **31.** $3210.06   **33.** 5.2 micrograms   **35.** $F(n) = 440(2^{n/12})$

37. $y = 8000(1.1066^t)$; 552,200,000 automobiles   **39.** $y = 100(0.99^t)$ **41. a.** $y = 4.959(1.063^x)$ **b.** 12.4 milliliters
43. a. 5.9 billion people   **b.** 7.2 billion people   **c.** The maximum population that Earth can support is 70.0 billion people.

45. a. 0.0075   **b.** $P = \dfrac{Ar(1 + r)^n}{(1 + r)^n - 1}$ **c.** $107.33   **d.** $193.33   **e.** There is no residual value on the car when
it is purchased.   **f.** $3999.92; $6000.08   **g.** $10,000.14; approximately $0   **h.** The $6000 remaining on the leased car is
its residual value.

EXERCISE SET 6.3   *page 408*

1. $\log_7 49 = 2$ **3.** $\log_5 625 = 4$ **5.** $\log 0.0001 = -4$ **7.** $\log x = y$ **9.** $3^4 = 81$ **11.** $5^3 = 125$
13. $4^{-2} = \dfrac{1}{16}$ **15.** $e^y = x$ **17.** 4   **19.** 2   **21.** -2 **23.** 6   **25.** 9   **27.** $\dfrac{1}{7}$ **29.** $\dfrac{1}{9}$
31. 1   **33.** 316.23   **35.** 7.39   **37.** 2.24   **39.** 14.39   **41.** Always true   **43.** $a > b$

45. 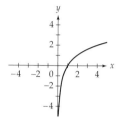   **47.** 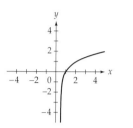   **49.** 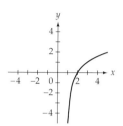   **51.** 79%

53. 65 decibels   **55.** 1.35   **57.** 6.8   **59.** 794,328,235I_0 **61.** 100 times as strong   **63.** 6.0 parsecs
65. 7805.5 billion barrels   **67.** $x = \dfrac{10^{0.47712}}{10^{0.30103}} = 10^{0.17609} \approx 1.5; \log x = 0.17609$ **69.** Answers will vary.   **71.** 4.9

CHAPTER 6 REVIEW EXERCISES   *page 412*

1. -13 [Sec. 6.1]   **2.** 13 [Sec. 6.1]   **3.** $-\dfrac{1}{2}$ [Sec. 6.1]   **4.** -21 [Sec. 6.1]   **5.** 4 [Sec. 6.2]
6. $\dfrac{4}{9}$ [Sec. 6.2]   **7.** 15.78 [Sec. 6.2]   **8.** -2.26 [Sec. 6.2]

9. 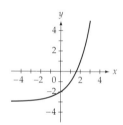 [Sec. 6.2]   **10.** 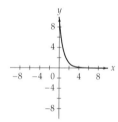 [Sec. 6.2]   **11.** 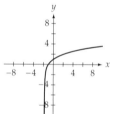 [Sec. 6.3]

12. 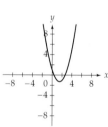 [Sec. 6.1]   **13. a.** 113.1 cubic inches   **b.** 7238.2 cubic centimeters [Sec. 6.1]

14. a. 133 feet   **b.** 133 feet [Sec. 6.1]   **15. a.** 5%   **b.** 36.7%   **c.** nonlinear [Sec. 6.1]
16. a. $V(x) = 4x^3 - 170x^2 + 1750x$ **b.** 5168 cubic centimeters [Sec. 6.1]   **17.** $(-4x^2 + 1750)$ square centimeters [Sec. 6.1]
18. a. $d(t) = \sqrt{100 + 1600t^2}$ **b.** 2400 feet [Sec. 6.1]   **19. a.** $C(x) = 55x + \dfrac{550x + 750}{x}$ **b.** \$3315 [Sec. 6.1]
20. $(-1, 3)$ [Sec. 6.1]   **21.** $\left(-\dfrac{3}{2}, \dfrac{11}{2}\right)$ [Sec. 6.1]   **22.** $(1, 2)$ [Sec. 6.1]   **23.** $\left(-\dfrac{5}{2}, -\dfrac{29}{4}\right)$ [Sec. 6.1]
24. 5, maximum [Sec. 6.1]   **25.** $-\dfrac{15}{2}$, minimum [Sec. 6.1]   **26.** -5, minimum [Sec. 6.1]   **27.** $\dfrac{1}{8}$, maximum
[Sec. 6.1]   **28.** 125 feet [Sec. 6.1]   **29.** 2000 CD-RWs [Sec. 6.1]   **30.** \$12,297.11 [Sec. 6.2]

31. 7.94 milligrams [Sec. 6.2]   **32.** 3.36 micrograms [Sec. 6.2]   **33. a.** $H(n) = 6\left(\dfrac{2}{3}\right)^n$; 0.79 foot [Sec. 6.2]

34. a. $N = 250.2056(0.6506)^t$ **b.** 19 thousand people [Sec. 6.2]   **35.** 5 [Sec. 6.3]   **36.** -4 [Sec. 6.3]

37. -1 [Sec. 6.3]   **38.** 6 [Sec. 6.3]   **39.** 64 [Sec. 6.3]   **40.** 32 [Sec. 6.3]   **41.** $\dfrac{1}{3}$ [Sec. 6.3]

42. $\dfrac{1}{16}$ [Sec. 6.3]   **43.** 43.7 parsecs [Sec. 6.3]   **44.** 140 decibels [Sec. 6.3]

CHAPTER 6 TEST   *page 414*

1. a. -21 [Sec. 6.1]   **b.** $\dfrac{1}{9}$ [Sec. 6.2]   **2.** $\log_2 b = a$ [Sec. 6.3]   **3.** 3 [Sec. 6.3]   **4.** 36 [Sec. 6.3]

5.

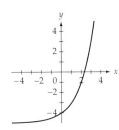

[Sec. 6.2]

6.

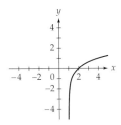

[Sec. 6.3]

7. $(-3, -10)$ [Sec. 6.1]

8. a. 10%   **b.** 70% [Sec. 6.1]   **9. a.** $V(x) = 4x^3 - 132x^2 + 1064x$ **b.** 2352 cubic inches [Sec. 6.1]

10. 148 feet [Sec. 6.1]   **11.** 404°F [Sec. 6.2]   **12.** 32 times greater [Sec. 6.3]

CHAPTER 7

EXERCISE SET 7.1   *page 425*

1. Divide the number of months by 12.   **3.** I is the interest, P is the principal, r is the interest rate, and t is the time period.
5. $560   **7.** $227.50   **9.** $202.50   **11.** $16.80   **13.** $159.60   **15.** $125   **17.** $168
19. $15,667.50   **21.** $7390.80   **23.** $2864.40   **25.** 7.5%   **27.** 10.5%   **29.** 9.3%   **31.** A period
of 8 months   **33.** Equal to   **35.** Equal to   **37.** $161   **39.** $78   **41.** $18   **43.** $7406
45. $5730.40   **47.** $804.75   **49.** 10.2%   **51.** 10%   **53.** $133.32   **55.** Exact method: $306.85. Ordinary
method: $311.11. The ordinary method yields the greater maturity value. The lender benefits.   **57.** There are fewer days in
September than there are in August.   **59. a.** $5, $10, $15, $20, $25   **b.** $30   **c.** $35   **d.** $40   **e.** $45
f. Multiply the interest due by 8.   **g.** Twice as large   **h.** Three times as large

EXERCISE SET 7.2   *page 444*

1. $2739.99   **3.** $852.88   **5.** $20,836.54   **7.** $12,575.23   **9.** $3532.86   **11.** $5450.09
13. $2213.84   **15.** $13,625.23   **17.** $14,835.46   **19.** $41,210.44   **21.** $7641.78   **23.** $3182.47
25. $10,094.57   **27.** The account paying 7.25% compounded quarterly   **29.** The loan at 7.5% compounded semiannually
31. Less than   **33.** $11,887.58   **35.** $3583.16   **37.** $8188.40   **39.** $391.24   **41.** $18,056.35
43. $11,120.58   **45. a.** $450   **b.** $568.22   **c.** $118.22   **47. a.** $20,528.54   **b.** $20,591.79
c. $63.25   **49. a.** $1698.59   **b.** $1716.59   **c.** $18.00   **51. a.** $10,401.63   **b.** $3401.63
53. $9612.75   **55. a.** $67,228.19   **b.** $94,161.98   **57.** $6789.69   **59.** $4651.73   **61.** $56,102.07
63. $12,152.77   **65.** $2495.45   **67.** 7.40%   **69.** 7.76%   **71.** 8.44%   **73.** 6.10%
75. $2.68, $3.58, $6.41   **77.** $3.60, $4.82, $8.63   **79.** $12.04, $16.12, $28.86   **81.** $368,012.03, $492,483.12,
$881,962.25   **83.** $25,417.46   **85.** $25,841.90   **87.** $53,473.96   **89. a.** 4.0%, 4.04%, 4.06%, 4.07%, 4.08%
b. Increase   **91.** 3.04%   **93.** 6.25% compounded semiannually   **95.** 5.8% compounded quarterly   **97.** 10%
99. $561.39

EXERCISE SET 7.3   *page 461*

1. $1.48   **3.** $152.32   **5.** $335.87   **7.** $15.34   **9.** $5.00   **11.** $0   **13.** $20.37   **15.** 13.7%
17. 12.9%   **19.** $56.03   **21. a.** $696.05   **b.** $174.01   **c.** $88.46   **23. a.** $68,569.73
b. $13,713.95   **c.** $641.17   **25.** $874.88   **27.** $571.31   **29. a.** $621.19   **b.** $4372.12
31. $13,575.25   **33.** $3472.57   **35. a.** $22,740   **b.** $101.90   **c.** $161.25   **d.** $263.15
37. a. $21,100   **b.** 0.003375   **c.** $121.84   **d.** $169.44   **e.** $291.28   **39.** The monthly payment for the
loan is *PMT*. The interest rate per period *i* is the annual interest rate divided by 12. The term of the loan *n* is the number of years of the
loan times 12. Substitute these values into the payment formula for an APR loan and solve for *A*, the selling price of the car. The selling
price of the car is $9775.72.   **41. a.** $168.48   **b.** $2669   **c.** $11,391.04   **43. a.** $336.04   **b.** $339.29
c. $1.40   **d.** $22.63

EXERCISE SET 7.4   *page 474*

1. $382.50   **3.** $535.50   **5.** 2.51%   **7.** 1.83%   **9. a.** $8.73   **b.** $67.50   **c.** 3,750,600 shares
d. Decrease   **e.** $22.59   **11.** 50 shares   **13. a.** Profit of $290   **b.** $78.66   **15. a.** Profit of $9096
b. $472.33   **17.** $40   **19.** Double the PE ratio of last year   **21.** No   **23.** $562.50   **25.** $2832
27. $16.50   **29.** 250 shares   **31.** 500 shares   **33.** Answers will vary.

EXERCISE SET 7.5   *page 486*

1. $64,500; $193,500   **3.** $5625   **5.** $99,539   **7.** $34,289.38   **9.** $974.37   **11.** $2155.28
13. A 25% down payment   **15.** Paying a 20% down payment and a fee of 2 points   **17.** Less than   **19. a.** $1569.02
b. $470,706   **c.** $271,706   **21.** $664,141.60   **23.** Interest: $1297.13; principal: $37.49   **25.** Interest: $986.59;
principal: $110.44   **27.** $126,874.00   **29.** $96,924.63   **31.** $1903.71   **33.** $1827.28   **35. a.** $804.08
b. $325,058.40   **37. a.** $343.07   **b.** $188,254.80   **39.** $120,000   **41.** 260th payment   **43.** Yes. If the
interest rate is lower, it will take fewer months.   **45. a.** $65,641.88   **b.** $138,596.60   **c.** $28,881.52   **d.** 44%

CHAPTER 7 REVIEW EXERCISES   *page 490*

1. $61.88   [Sec. 7.1]   **2.** $782   [Sec. 7.1]   **3.** $90   [Sec. 7.1]   **4.** $7218.40   [Sec. 7.1]
5. 7.5%   [Sec. 7.1]   **6.** $3654.90   [Sec. 7.2]   **7.** $11,609.72   [Sec. 7.2]   **8.** $7859.52   [Sec. 7.2]
9. $200.23   [Sec. 7.2]   **10.** $10,683.29   [Sec. 7.2]   **11. a.** $11,318.23   **b.** $3318.23   [Sec. 7.2]
12. $19,225.50   [Sec. 7.2]   **13.** 1.1%   [Sec. 7.4]   **14.** $9000   [Sec. 7.4]   **15.** $2.31   [Sec. 7.2]
16. $43,650.68   [Sec. 7.2]   **17.** 6.06%   [Sec. 7.2]   **18.** 5.4% compounded semiannually   [Sec. 7.2]
19. $431.16   [Sec. 7.3]   **20.** $6.12   [Sec. 7.3]   **21. a.** $259.38   **b.** 12.9%   [Sec. 7.3]
22. a. $36.03   **b.** 12.9%   [Sec. 7.3]   **23.** $45.41   [Sec. 7.3]   **24. a.** $10,092.69   **b.** $2018.54
c. $253.01   [Sec. 7.3]   **25.** $704.85   [Sec. 7.3]   **26. a.** $540.02   **b.** $12,196.80   [Sec. 7.3]
27. a. $29,450   **b.** $181.80   **c.** $224.17   **d.** $436.42   [Sec. 7.3]   **28. a.** Profit of $5325
b. $256.10   [Sec. 7.4]   **29.** 200 shares   [Sec. 7.4]   **30.** $99,041   [Sec. 7.5]
31. a. $1659.11   **b.** $597,279.60   **c.** $341,479.60   [Sec. 7.5]   **32. a.** $1396.69
b. $150,665.74   [Sec. 7.5]   **33.** $2658.53   [Sec. 7.5]

CHAPTER 7 TEST   *page 492*

1. $108.28   [Sec. 7.1]   **2.** $202.50   [Sec. 7.1]   **3.** $8408.89   [Sec. 7.1]   **4.** 9%   [Sec. 7.1]
5. $7340.87   [Sec. 7.2]   **6.** $312.03   [Sec. 7.2]   **7. a.** $15,331.03   **b.** $4831.03   [Sec. 7.2]
8. $21,949.06   [Sec. 7.2]   **9.** 1.2%   [Sec. 7.4]   **10.** $1900   [Sec. 7.4]   **11.** $612,184.08   [Sec. 7.2]
12. 6.40%   [Sec. 7.2]   **13.** 4.6% compounded semiannually   [Sec. 7.2]   **14.** $7.79   [Sec. 7.3]   **15. a.** $48.56
b. 16.6%   [Sec. 7.3]   **16.** $60.61   [Sec. 7.3]   **17. a.** Loss of $4896   **b.** $226.16   [Sec. 7.4]
18. 208 shares   [Sec. 7.4]   **19. a.** $6985.94   **b.** $1397.19   **c.** $174.62   [Sec. 7.3]
20. $60,083.50   [Sec. 7.5]   **21. a.** $1530.69   **b.** $221,546.46   [Sec. 7.5]   **22.** $2595.97   [Sec. 7.5]

CHAPTER 8

EXERCISE SET 8.1   *page 503*

1. {0, 2, 4, 6, 8}   **3.** {Monday, Tuesday, Wednesday, Thursday, Friday, Saturday, Sunday}   **5.** {HH, HT, TH, TT}
7. {1H, 1T, 2H, 2T, 3H, 3T, 4H, 4T, 5H, 5T, 6H, 6T}   **9.** {Andy Bob Cassidy, Andy Cassidy Bob, Bob Andy Cassidy, Bob Cassidy Andy, Cassidy Bob Andy, Cassidy Andy Bob}   **11.** Salads: S_1, S_2; Entrees: E_1, E_2, E_3; Desserts: D_1, D_2

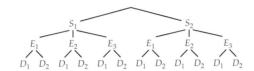

13.

15.

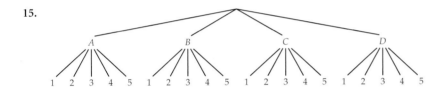

17. 25   **19.** 4^{15} **21.** 6000   **23.** 24   **25.** 18   **27.** 40,320   **29.** 362,760   **31.** 60
33. 6720   **35.** 210   **37.** 60,480   **39.** 1   **41.** 36   **43.** 1   **45.** 525   **47.** 336
49. 40,320   **51.** No   **53.** 21   **55.** 6840   **57.** 240   **59.** $9 \cdot 26^3 \cdot 10^3$ **61.** 5040
63. a. 9!   **b.** 8!   **65.** 210   **67. a.** 518,400   **b.** 720   **69.** 3,628,800   **71.** 35   **73.** 455
75. a. 495   **b.** 11,880   **77.** 756,756   **79. a.** 36^8 **b.** 36^9 **c.** 89,457 years   **d.** 3,220,445 years
81. a. 1,048,576   **b.** 12,600   **83.** Answers will vary.   **85.** 7.9×10^{28}

EXERCISE SET 8.2   *page 522*

1. {HHH, HHT, HTH, HTT, THH, THT, TTH, TTT}   **3.** {♠A, ♥A, ♦A, ♣A}   **5.** $\dfrac{3}{8}$ **7.** $\dfrac{1}{16}$ **9.** $\dfrac{1}{18}$

11. $\dfrac{5}{18}$ **13.** $\dfrac{19}{36}$ **15.** $\dfrac{4}{13}$ **17.** $\dfrac{3}{13}$ **19.** $\dfrac{58}{293}$ **21.** $\dfrac{36}{293}$ **23.** $\dfrac{2492}{5111}$ **25.** 1 to 3

27. 1 to 101   **29.** $\dfrac{137}{425}$ **31.** $\dfrac{103}{850}$ **33.** No   **35.** $\dfrac{2}{9}$ **37.** 0.000092   **39.** 0.0000015   **41.** $\dfrac{9}{19}$

43. $\dfrac{1}{38}$ **45.** $\dfrac{3}{19}$ **47.** $\dfrac{173}{778}$ **49.** $\dfrac{341}{1642}$ **51.** $\dfrac{1}{6}$ **53.** $\dfrac{3}{51}$ **55.** $\dfrac{6}{1045}$ **57.** $\dfrac{3}{1045}$

59. $\dfrac{13}{102}$ **61.** $\dfrac{8}{5525}$ **63.** 0.00048   **65.** 0.0014   **67.** Independent   **69.** Dependent   **71.** $\dfrac{1}{4}$

73. 1   **75.** 0.46   **77.** 0.11   **79. a.** $\dfrac{1}{8}$ **b.** $\dfrac{3}{8}$ **c.** $\dfrac{3}{4}$ **81.** 0.0000000068   **83.** 0.000000024

85. 0.000208   **87. a.** {(1, 1), (1, 2), (1, 2), (1, 3), (1, 3), (1, 4), (3, 1), (3, 2), (3, 2), (3, 3), (3, 3), (3, 4), (4, 1), (4, 2), (4, 3), (4, 4), (5, 1), (5, 2), (5, 2), (5, 3), (5, 3), (5, 4), (6, 1), (6, 2), (6, 2), (6, 3), (6, 3), (6, 4), (8, 1), (8, 2), (8, 2), (8, 3), (8, 3), (8, 4)}
b. $P(2) = \dfrac{1}{36}$, $P(3) = \dfrac{1}{18}$, $P(4) = \dfrac{1}{12}$, $P(5) = \dfrac{1}{9}$, $P(6) = \dfrac{5}{36}$, $P(7) = \dfrac{1}{6}$, $P(8) = \dfrac{5}{36}$, $P(9) = \dfrac{1}{9}$, $P(10) = \dfrac{1}{12}$, $P(11) = \dfrac{1}{18}$, $P(12) = \dfrac{1}{36}$
c. Answers will vary.

EXERCISE SET 8.3 *page 536*

1. 7; 7; 7 **3.** 22; 14; no mode **5.** 18.8; 8.1; no mode **7.** 192.4; 190; 178 **9.** 0.1; −3; −5
11. a. Yes. The mean is computed by using the sum of all the data. **b.** No. The median is not affected unless the middle value, or one of the two middle values, in a data set is changed. **13.** Greater than M; equal to N **15.** 38; 35; 33; 35
17. a. Answers will vary. **b.** Answers will vary. **19.** ≈82.9 **21.** 82 **23.** 2.5 **25.** ≈6.1 points;
5 points; 2 points and 5 points **27.** ≈7.2; 7; 7 **29.** 64° **31.** −6°F **33.** 92 **35. a.** ≈0.275
b. ≈0.273 **c.** No **37.** 81
39. $d_1 = d_2$

$$r_1 = \frac{d_1}{t_1} \qquad r_2 = \frac{d_2}{t_2} = \frac{d_1}{t_2}$$

$$t_1 = \frac{d_1}{r_1} \qquad t_2 = \frac{d_1}{r_2}$$

$$r = \frac{d_1 + d_2}{t_1 + t_2} = \frac{d_1 + d_1}{t_1 + t_2} = \frac{2d_1}{t_1 + t_2}$$

$$= \frac{2d_1}{\dfrac{d_1}{r_1} + \dfrac{d_1}{r_2}} = \frac{2d_1}{d_1\left(\dfrac{1}{r_1} + \dfrac{1}{r_2}\right)}$$

$$= \frac{2}{\dfrac{1}{r_1} + \dfrac{1}{r_2}} \cdot \left(\frac{r_1 r_2}{r_1 r_2}\right) = \frac{2r_1 r_2}{r_1 + r_2}$$

41. Yes. Joanne has lower averages for the first month and the second month, but she has a higher average for both months combined.
43. Answers will vary.

EXERCISE SET 8.4 *page 548*

1. 84°F **3.** 21; 8.2; 67.1 **5.** 3.3; 1.3; 1.7 **7.** 52; 17.7; 311.6 **9.** 0; 0; 0 **11.** 23; 8.3; 69.6
13. No **15.** Opinions will vary. However, many climbers would consider rope B to be safer because of its small standard
deviation. **17.** The students in the college statistics course because the range of weights would probably be greater.
19. a. 30.1; 10.1 **b.** 15.6; 6.5 **c.** Winning scores; winning scores **21. a.** 44.9; 9.3 **b.** 42.8; 7.7
c. National League; National League **23.** 54.8 years; 6.2 years **25. a.** Answers will vary. **b.** The population
standard deviation remains the same. **27. a.** 0 **b.** Yes **c.** No **29. a.** Identical **b.** Identical
c. Identical **d.** They will be identical.

EXERCISE SET 8.5 *page 561*

1. a. ≈0.87 **b.** ≈1.74 **c.** ≈−2.17 **d.** 0.0 **3. a.** ≈−0.32 **b.** ≈0.21 **c.** ≈1.16
d. ≈−0.95 **5.** There is insufficient information to answer the question. **7. a.** ≈−0.67 **b.** 147.78 mm Hg
9. a. ≈0.72 **b.** ≈112.16 mg/dl **11.** The score in part a. **13.** ≈59th percentile **15.** 6396 students
17. a. 50% **b.** 12% **c.** 38% **19.** Median, range **21.** $Q_1 = 5, Q_2 = 10, Q_3 = 26$
23.

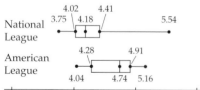

25.

789.5 971.5
638 891   1280
           Women
Men
849   1095 / 1305.5   1736
           1219.5

600 800 1000 1200 1400 1600 1800

The ERAs in the National League tend to be lower than the ERAs in the American League. Also, the range of the ERAs is larger for the National League.

The median salary for men was approximately as high as the highest salary for women. The lowest salary for men was approximately the median salary for women.

27. a. $\mu = 0, \sigma = 1$ **b.** $\mu = 0, \sigma = 1$ **c.** $\mu = 0, \sigma = 1$ **29.** A value is an outlier for a set of data provided the
number is less than $Q_1 - 1.5(Q_3 - Q_1)$ or greater than $Q_3 + 1.5(Q_3 - Q_1)$. For the given data, 85 is the only outlier.

EXERCISE SET 8.6 *page 577*

1. Greater   **3.** a   **5.** Answers will vary.   **7. a.** 95.4%   **b.** 15.9%   **c.** 81.8%   **9. a.** 81.8%
b. 0.15%   **11. a.** 1272 vehicles   **b.** 12 vehicles   **13.** 0.433 square unit   **15.** 0.468 square unit
17. 0.130 square unit   **19.** 0.878 square unit   **21.** 0.097 square unit   **23.** 0.013 square unit
25. 0.926 square unit   **27.** 0.997 square unit   **29.** $z = 0.84$ **31.** $z = -0.90$ **33.** $z = 0.35$
35. a. 30.9%   **b.** 33.4%   **37. a.** 13.6%   **b.** 35.2%   **39. a.** 10.6%   **b.** 98.8%   **41. a.** 0.106
b. ≈ 0.460 **43. a.** ≈ 0.749 **b.** 0.023   **45.** Answers will vary.   **47.** True   **49.** False   **51.** True
53. False   **55.** False   **57.** -0.84 and 0.84 **59.** For any positive constant k, the probability that a random variable will

take on a value within k standard deviations of the mean is at least $1 - \dfrac{1}{k^2}$. According to Chebyshev's Theorem, a minimum of 75% of the

data in any data set must lie within two standard deviations of the mean.

EXERCISE SET 8.7 *page 590*

1. No   **3.** Yes, this event is highly unlikely compared with what is expected by chance.   **5.** $H_0: \mu = 48$ mpg, $H_1: \mu < 48$ mpg
7. a. $\dfrac{1}{38}$ **b.** $H_0: \hat{p} = \dfrac{1}{38}, H_1: \hat{p} > \dfrac{1}{38}$ **c.** $z \approx 2.02$ **d.** Yes   **e.** No   **9.** Yes
11. a. $H_0: \mu = \$6.91, H_1: \mu > \$6.91$ **b.** $z \approx 2.89$ **c.** Yes   **13.** No   **15.** $0.49 < \hat{p} < 0.63$
17. $0.40 < \hat{p} < 0.47$ **19.** 15.25 ounces $< \mu <$ 15.55 ounces   **21.** $\$747 < \mu < \$783$ **23.** 3.9%
25. 4.1%   **27.** It is reduced to half of its previous value.   **29.** The 95% confidence interval   **31.** Answers will vary.

CHAPTER 8 REVIEW EXERCISES *page 595*

1. $\{(1, 1), (1, 2), (1, 3), (2, 1), (2, 2), (2, 3), (3, 1), (3, 2), (3, 3)\}$ [Sec. 8.1]   **2.** $\{(2, 6), (2, 8), (6, 2), (6, 8), (8, 2), (8, 6)\}$ [Sec. 8.1]
3.   **4.** 64   [Sec. 8.1]   **5.** 5040   [Sec. 8.1]

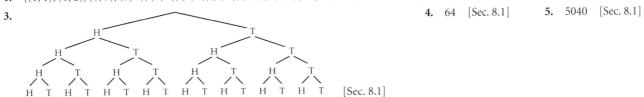

[Sec. 8.1]

6. 40,296   [Sec. 8.1]   **7.** 1260   [Sec. 8.1]   **8.** 151,200   [Sec. 8.1]   **9.** 336   [Sec. 8.1]   **10.** 0.42   [Sec. 8.1]
11. 5040   [Sec. 8.1]   **12.** 2520   [Sec. 8.1]   **13.** 180   [Sec. 8.1]   **14.** 495   [Sec. 8.1]

15. 3,268,760   [Sec. 8.1]   **16.** 165   [Sec. 8.1]   **17.** $\dfrac{3}{8}$ [Sec. 8.2]   **18.** $\dfrac{1}{8}$ [Sec. 8.2]   **19.** 0.56   [Sec. 8.2]

20. 0.37   [Sec. 8.2]   **21.** 0.85   [Sec. 8.2]   **22.** $\dfrac{1}{9}$ [Sec. 8.2]   **23.** $\dfrac{5}{9}$ [Sec. 8.2]   **24.** $\dfrac{2}{9}$ [Sec. 8.2]

25. $\dfrac{1}{6}$ [Sec. 8.2]   **26.** $\dfrac{3}{4}$ [Sec. 8.2]   **27.** $\dfrac{4}{13}$ [Sec. 8.2]   **28.** $\dfrac{12}{13}$ [Sec. 8.2]   **29.** $\dfrac{2}{3}$ [Sec. 8.2]

30. $\dfrac{1}{2}$ [Sec. 8.2]   **31.** 0.036   [Sec. 8.2]   **32.** $\dfrac{173}{1000}$ [Sec. 8.2]   **33.** $\dfrac{5}{8}$ [Sec. 8.2]   **34.** 1 to 5   [Sec. 8.2]

35. $\dfrac{7}{16}$ [Sec. 8.2]   **36.** $\dfrac{646}{1771}$ [Sec. 8.2]   **37.** $\dfrac{7}{253}$ [Sec. 8.2]   **38.** 0.37   [Sec. 8.2]   **39.** 0.62   [Sec. 8.2]

40. 0.07   [Sec. 8.2]   **41.** 0.47   [Sec. 8.2]   **42.** 0.29   [Sec. 8.2]   **43.** 0.648   [Sec. 8.2]   **44.** 0.0525   [Sec. 8.2]
45. Mean $= 5.58$, median $= 5$, mode $= 5$ [Sec. 8.3]   **46.** 14   [Sec. 8.3]   **47.** Answers will vary.   [Sec. 8.3]   **48.**
a. Median   **b.** Mode   **c.** Mean   [Sec. 8.3]   **49.** Mean $= 331.125$, median $= 223.5$, mode $= 200$ [Sec. 8.3]
50. 3.10   [Sec. 8.4]   **51. a.** 1.25   **b.** 89%   [Sec. 8.4]   **52.** 3.16, 9.99   [Sec. 8.4]   **53.** 5.219, 5.225, 0.69
[Sec. 8.4]   **54. a.** The mean of the second student is 5 points higher than the mean of the first student.   **b.** The standard
deviations are the same.   [Sec. 8.3/8.4]   **55. a.** -1.73 **b.** 0.58   [Sec. 8.5]

56.   **57. a.** 135   **b.** 356   **c.** 9

[Sec. 8.5]

d. 43.0% [Sec. 8.6] **58. a.** 58.9% **b.** 0.236 [Sec. 8.6] **59.** No. It is not known whether the distribution of scores was a normal distribution. [Sec. 8.6] **60. a.** 0.067 **b.** 0.309 [Sec. 8.6] **61. a.** 6.7% **b.** 84% [Sec. 8.6]

62. a. 6.7% **b.** 64.5% **c.** 78.8% [Sec. 8.6] **63. a.** $H_0: \hat{p} = \dfrac{32}{663}, H_1: \hat{p} < \dfrac{32}{663}$ **b.** -1.535

c. No [Sec. 8.7] **64. a.** $H_0: \mu = 40{,}000, H_1: \mu < 40{,}000$ **b.** -6.44 **c.** Yes [Sec. 8.7] **65.** Yes, $z = -4.47 < 2.58$ [Sec. 8.7] **66.** No, $z = -1.18 > -1.96$ [Sec. 8.7] **67.** Between 3.4 and 4.0 minutes [Sec. 8.7]
68. Between 27.7 mpg and 28.3 mpg [Sec. 8.7] **69.** 4.6% [Sec. 8.7]

CHAPTER 8 TEST   *page 600*

1. 72 [Sec. 8.1] **2.** 125,970 [Sec. 8.1] **3.** $\dfrac{7}{12}$ [Sec. 8.2] **4.** $\dfrac{1}{17}$ [Sec. 8.2] **5.** 1 to 7 [Sec. 8.2]

6. 0.63 [Sec. 8.2] **7.** 0.47 [Sec. 8.2] **8.** $\dfrac{1}{2}$ [Sec. 8.2] **9.** Mean = 45.4, median = 46, mode = 51 [Sec. 8.3]

10. 88.2 [Sec. 8.3] **11. a.** 1.25 **b.** -2.08 [Sec. 8.5] **12. a.** 0.345 **b.** 0.682 **c.** 0.767 [Sec. 8.5]
13.

[Sec. 8.6]

14. a. 100 **b.** 60 **c.** 19% [Sec. 8.6]

15. a. $H_0: \mu = 15, H_1: \mu > 15$ **b.** 0.44 **c.** No [Sec. 8.6] **16. a.** $0.25 < \hat{p} < 0.38$ **b.** 9.8% [Sec. 8.7]

CHAPTER 9

EXERCISE SET 9.1   *page 617*

1. To calculate the standard divisor, divide the total population p by the number of items to apportion n. **3.** The standard quota for a state is the whole number part of the quotient of the state's population divided by the standard divisor. **5.** State A
7. a. 0.273 **b.** 0.254 **c.** Salinas **9.** Seaside Mall **11. a.** 646,952; There is one representative for every 646,952 citizens in the U.S. **b.** Underrepresented; The average constituency is greater than the standard divisor.
c. Overrepresented; The average constituency is less than the standard divisor. **13. a.** 37.10; There is one new nurse for every 37.10 beds. **b.** Sharp: 7; Palomar: 10; Tri-City: 8; Del Raye: 5; Rancho Verde: 7; Bel Aire: 11 **c.** Sharp: 7; Palomar: 10; Tri-City: 8; Del Raye: 5; Rancho Verde: 7; Bel Aire: 11 **d.** They are identical. **15.** The population paradox occurs when the population of one state is increasing faster than that of another state, yet the first state still loses a representative. **17.** The Balinski-Young Impossibility Theorem states that any apportionment method will either violate the quota rule or will produce paradoxes such as the Alabama paradox. **19. a.** Yes **b.** No **c.** Yes **21. a.** Boston: 2; Chicago: 20 **b.** Yes. Chicago will lose a vice president while Boston will gain one. **23.** State A **25. a.** Sixth grade **b.** Sixth grade; same
27. Valley **29. a.** They are the same. **b.** Using the Jefferson method, the humanities division gets one less computer and the sciences division gets one more computer compared with using the Webster method. **31.** The Jefferson and Webster methods
33. The Huntington-Hill method **35.** Answers will vary. **37.** Del Mar: 3; Wheatly: 7; West: 5; Mountain View: 7

39. a. $\dfrac{P_A}{a+1}$ **b.** $\dfrac{P_B}{b}$ **c.** $\dfrac{\dfrac{P_B}{b} - \dfrac{P_A}{a+1}}{\dfrac{P_A}{a+1}}$ **d.** $\dfrac{\dfrac{P_A}{a} - \dfrac{P_B}{b+1}}{\dfrac{P_B}{b+1}}$ **e.** $\dfrac{\dfrac{P_A}{a} - \dfrac{P_B}{b+1}}{\dfrac{P_B}{b+1}} < \dfrac{\dfrac{P_B}{b} - \dfrac{P_A}{a+1}}{\dfrac{P_A}{a+1}}$

f. $\dfrac{(P_B)^2}{b(b+1)} < \dfrac{(P_A)^2}{a(a+1)}$

EXERCISE SET 9.2   *page 637*

1. A majority means that a choice receives more than 50% of the votes. A plurality means that the choice with the most votes wins. It is possible to have a plurality without a majority when there are more than two choices. **3.** If there are n choices in an election, each voter ranks the choices by giving n points to the voter's first choice, $n - 1$ points to the voter's second choice, and so on, with the voter's least favorite choice receiving 1 point. The choice with the most points is the winner.   **5.** In the pairwise comparison voting method, each choice is compared one-on-one with each of the other choices. A choice receives 1 point for a win, 0.5 point for a tie, and 0 points for a loss. The choice with the greatest number of points is the winner.   **7.** No; no   **9.** Yes; not necessarily **11. a.** Al Gore   **b.** No   **c.** George W. Bush   **13. a.** 35   **b.** 18   **c.** Scooby Doo   **15.** Go to a theater   **17.** Bugs Bunny   **19.** Elaine Garcia   **21.** Blue and white   **23.** Raymond Lee   **25. a.** Buy new computers for the club.   **b.** Pay for several members to travel to a convention.   **c.** Pay for several members to travel to a convention.   **d.** Answers will vary.   **27.** *Return of the Jedi*   **29.** Hornet   **31.** Blue and white **33.** Condorcet criterion   **35.** No   **37.** Yes   **39.** No   **41. a.** Stephen Hyde   **b.** Stephen Hyde received the fewest number of first-place votes.   **c.** John Lorenz   **d.** The candidate that wins all head-to-head matches does not win the election.   **e.** John Lorenz   **f.** The candidate winning the original election (Stephen Hyde) did not remain the winner in a recount after a losing candidate withdrew from the race.   **43.** Medin is president, Jen is vice president, Andrew is secretary, and Hector is treasurer.   **45.** The Borda Count method   **47. a.** Candidate B   **b.** Candidate A

EXERCISE SET 9.3   *page 655*

1. a. 6   **b.** 4   **c.** 3   **d.** 6   **e.** No   **f.** A and C   **g.** 15   **h.** 6   **3.** 0.60, 0.20, 0.20 **5.** 0.50, 0.30, 0.10, 0.10   **7.** 0.36, 0.28, 0.20, 0.12, 0.04   **9.** 1.00, 0.00, 0.00, 0.00, 0.00, 0.00   **11.** 0.44, 0.20, 0.20, 0.12, 0.04   **13. a.** Exercise 9   **b.** Exercises 3, 5, 6, 9, and 12   **c.** None   **d.** Exercise 8   **15.** One more than 50% of the number of voters   **17.** 0.33, 0.33, 0.33   **19. a.** {12: 1, 1, 1, 1, 1, 1, 1, 1, 1, 1, 1, 1}   **b.** Yes **c.** Yes   **d.** Divide the voting power, 1, by the quota, 12.   **21.** Dictator: A; dummies: B, C, D, E   **23.** None **25. a.** 0.60, 0.20, 0.20   **b.** Answers will vary.   **27. a.** 0.33, 0.33, 0.33   **b.** This system has the same effect as a one-person, one-vote system.   **29. a.** 11 and 14   **b.** 15 and 16   **c.** No   **31.** Yes   **33.** In the weighted system, all voters have equal weight and the quota is equal to 50% of the sum of the weights rounded up to the nearest whole number, or a strict majority.   **35.** Answers will vary.   **37. a.** $BPI(29) \approx 0.0855$, $BPI(27) \approx 0.081$, $BPI(13) \approx 0.0422$, $BPI(12) \approx 0.039$, $BPI(10) \approx 0.0326$, $BPI(7) \approx 0.0229$, $BPI(4) \approx 0.0131$ **b.** 6.5 times

CHAPTER 9 REVIEW EXERCISES   *page 660*

1. a. 0.098   **b.** 0.194   **c.** High Desert   [Sec. 9.1]   **2.** Morena Valley   [Sec. 9.1]   **3. a.** Health: 7; business: 18; engineering: 10; science: 15   **b.** Health: 6; business: 18; engineering: 10; science: 16   **c.** Health: 7; business: 18; engineering: 10; science: 15   [Sec. 9.1]   **4. a.** Newark: 9; Cleveland: 6; Chicago: 11; Philadelphia: 4; Detroit: 5 **b.** Newark: 9; Cleveland: 6; Chicago: 11; Philadelphia: 4; Detroit: 5   **c.** Newark: 9; Cleveland: 6; Chicago: 11; Philadelphia: 4; Detroit: 5   [Sec. 9.1]   **5. a.** No. None of the offices will lose a new printer.   **b.** Yes. Office A will drop from two new printers to only one new printer.   [Sec. 9.1]   **6. a.** A: 2; B: 5; C: 3; D: 16; E: 2. No. None of the centers will lose an automobile. **b.** No. None of the centers will lose an automobile.   [Sec. 9.1]   **7. a.** Los Angeles: 9; Newark: 2   **b.** Yes. Newark will lose a computer file server.   [Sec. 9.1]   **8. a.** A: 10; B: 3; C: 21   **b.** Yes. The population of region B grew at a higher rate than the population of region A, yet region B will lose an inspector to region A.   [Sec. 9.1]   **9.** Yes   [Sec. 9.1]   **10.** Yes   [Sec. 9.1] **11. a.** Building A   **b.** Building B   [Sec. 9.1]   **12. a.** Shannon M.   **b.** Hannah A.   **c.** Hannah A. won all head-to-head comparisons, but lost the overall election.   **d.** Hannah A.   **e.** Hannah A. received a majority of the first-place votes, but lost the overall election.   [Sec. 9.2]   **13. a.** Hannah A.   **b.** Cynthia L., a losing candidate, withdrew from the race and caused a change in the overall winner of the election.   [Sec. 9.2]   **14.** The monotonicity criterion was violated because the only change was that the supporter of a losing candidate changed his or her vote to support the original winner, but the original winner did not win the second vote.   [Sec. 9.2]   **15. a.** 18   **b.** 18   **c.** Yes   **d.** A and C   **e.** 15 **f.** 6   [Sec. 9.3]   **16. a.** 35   **b.** 35   **c.** Yes   **d.** A   **e.** 31   **f.** 10   [Sec. 9.3]   **17.** 0.60, 0.20, 0.20   [Sec. 9.3]   **18.** 0.20, 0.20, 0.20, 0.20, 0.20   [Sec. 9.3]   **19.** 0.42, 0.25, 0.25, 0.08   [Sec. 9.3]   **20.** 0.62, 0.14, 0.14, 0.05, 0.05   [Sec. 9.3]   **21.** Dictator: A; dummies: B, C, D, E   [Sec. 9.3]   **22.** Dummy: D   [Sec. 9.3] **23.** 0.50, 0.125, 0.125, 0.125, 0.125   [Sec. 9.3]   **24. a.** Manuel Ortega   **b.** No   **c.** Crystal Kelley   [Sec. 9.2] **25. a.** Vail   **b.** Aspen   [Sec. 9.2]   **26.** A. Kim   [Sec. 9.2]   **27.** Snickers   [Sec. 9.2]   **28.** A. Kim   [Sec. 9.2] **29.** Snickers   [Sec. 9.2]

CHAPTER 9 TEST   *page 664*

1. Spring Valley   [Sec. 9.1]   **2. a.** Sales: 17; advertising: 4; service: 11; manufacturing: 53   **b.** Sales: 17; advertising: 4; service: 10; manufacturing: 54; no   [Sec. 9.1]   **3. a.** Cedar Falls $\approx$ 77,792; Lake View $\approx$ 70,290   **b.** Cedar Falls   [Sec. 9.1] **4. a.** 33   **b.** 26   **c.** No   **d.** A and C   **e.** 31   **f.** 10   [Sec. 9.3]   **5. a.** Aquafina

b. No   **c.** Evian [Sec. 9.2]   **6.** New York [Sec. 9.2]   **7. a.** Afternoon   **b.** Noon [Sec. 9.2]
8. a. Proposal A   **b.** Proposal B   **c.** Eliminating a losing choice changed the outcome of the vote.   **d.** Answers will
vary.   **e.** Proposal B won all head-to-head comparisons but lost the vote when all the choices were on the ballot. [Sec. 9.2]
9. 0.42, 0.25, 0.25, 0.08 [Sec. 9.3]   **10.** 0.40, 0.20, 0.20, 0.20 [Sec. 9.3]

CHAPTER 10

EXERCISE SET 10.1   *page 680*

1. a. 6   **b.** 7   **c.** 6   **d.** Yes   **e.** No   **3. a.** 6   **b.** 4   **c.** 4   **d.** Yes   **e.** Yes
5.   **7.**   **9. a.** No   **b.** Three   **c.** Ada

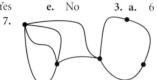

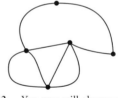

d. A loop would correspond to a friend speaking to himself or herself.   **11.** Yes   **13.** Equivalent   **15.** Not equivalent
17.   **19. a.** D–A–E–B–D–C–E–D

21. a. Not Eulerian   **b.** A–E–A–D–E–D–C–E–C–B–E–B   **23. a.** Not Eulerian   **b.** No   **25. a.** Not
Eulerian   **b.** E–A–D–E–G–D–C–G–F–C–B–F–A–B–E–F   **27. a.**   **b.** Yes

29. Yes   **31.** Yes, but the hamster cannot return to its starting point.   **33.** Yes; you will always return to the starting room.
35. Not every vertex is of even degree.   **37.** A–B–C–D–E–G–F–A   **39.** A–B–E–C–H–D–F–G–A
41. Springfield–Greenfield– Watertown–Riverside–Newhope–Midland–Springfield   **43.** A route through the museum that visits
each room once and returns to the starting room without visiting any room twice   **45.** Euler circuit   **47.** There is not an
Euler circuit or an Euler walk; there is a Hamiltonian circuit.   **49. a.**

There are
many possible
answers.

b.   **c.**

There are
many possible
answers.

There are
many possible
answers.

EXERCISE SET 10.2   *page 697*

1. From Billings to Great Falls, then to Polson   **3.** A–B–E–D–C–A, total weight 31; A–D–E–B–C–A, total weight 32
5. A–D–C–E–B–F–A, total weight 114; A–C–D–E–B–F–A, total weight 158   **7.** A–D–B–C–F–E–A

9. A–C–E–B–D–A   **11.** A–D–B–F–E–C–A   **13.** A–C–E–B–D–A   **15.** Louisville–Evansville–
Bloomington–Indianapolis–Lafayette–Fort Wayne–Louisville   **17.** Louisville–Evansville–Bloomington–Indianapolis–
Lafayette–Fort Wayne–Louisville   **19.** Tokyo–Seoul–Beijing–Hong Kong–Bangkok–Tokyo   **21.** Tokyo–Seoul–Beijing–
Hong Kong–Bangkok–Tokyo   **23.** Home–pharmacy–pet store–farmers' market–shopping mall–home; home–pharmacy–
pet store–shopping mall–farmers' market–home   **25.** Home state–task B–task D–task A–task C–home state; Edge-Picking
Algorithm gives the same sequence.   **27.**    **29.**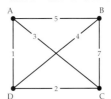

EXERCISE SET 10.3 *page 712*

1.    **3.** 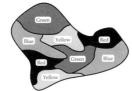   **5.**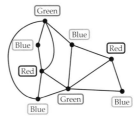

7. 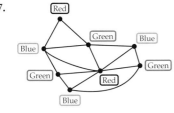   **9.** 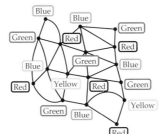   **11.**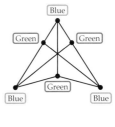

13. Not 2-colorable   **15.** 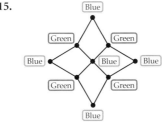   **17.** Yes   **19.** 3   **21.** 3   **23.** 4

25. 3   **27.** Two time slots   **29.** Five days; one possible schedule: group 1, group 2, groups 3 and 5, group 4, group 6
31. Three days: films 1 and 4, films 2 and 6, films 3 and 5   **33.**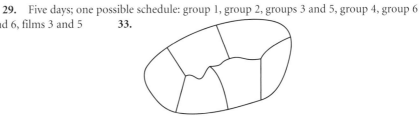

35. 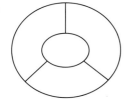   **37. a.** 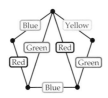   **b.** 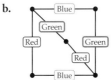   **c.**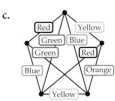

d.

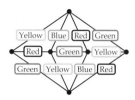

e.   Answers will vary.

CHAPTER 10 REVIEW EXERCISES   *page 718*

1. a.   8   **b.**   4   **c.**   All vertices are of degree 4.   **d.**   Yes   [Sec. 10.1]   **2. a.**   6   **b.**   7
c.   1, 1, 2, 2, 2, 2, 2   **d.**   No   [Sec. 10.1]   **3.**   **4. a.**   No   **b.**   4

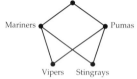

[Sec. 10.1]

c.   110, 405   **d.**   105   [Sec. 10.1]   **5.**   Equivalent   [Sec. 10.1]   **6.**   Equivalent   [Sec. 10.1]
7. a.   E–A–B–C–D–B–E–C–A–D   **b.**   Not possible   [Sec. 10.1]   **8. a.**   Not possible   **b.**   Not possible   [Sec. 10.1]
9. a. and b.   F–A–E–C–B–A–D–B–E–D–C–F   [Sec. 10.1]   **10. a.**   B–A–E–C–A–D–F–C–B–D–E
b.   Not possible   [Sec. 10.1]   **11.**   Yes   **12.**   Yes; no   [Sec. 10.1]

[Sec. 10.1]

13.   A–B–C–E–D–A   [Sec. 10.1]   **14.**   A–D–F–B–C–E–A   [Sec. 10.1]
15.

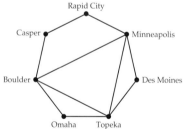

Casper–Rapid City–Minneapolis–Des Moines–Topeka–Omaha–Boulder–Casper
[Sec. 10.1]

16.   Casper–Boulder–Topeka–Minneapolis–Boulder–Omaha–Topeka–Des Moines–Minneapolis–Rapid City–Casper   [Sec. 10.1]
17.   A–D–F–E–B–C–A   [Sec. 10.2]   **18.**   A–B–E–C–D–A   [Sec. 10.2]   **19.**   A–D–F–E–C–B–A   [Sec. 10.2]
20.   A–B–E–D–C–A   [Sec. 10.2]   **21.**

Memphis–Nashville–Birmingham–Atlanta–
Jackson–Memphis   [Sec. 10.2]

22. A–E–B–C–D–A [Sec. 10.2] **23.** Requires four colors

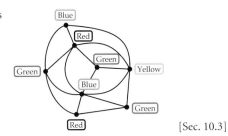

[Sec. 10.3]

24. Requires four colors

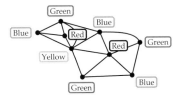

[Sec. 10.3]

25. 2-colorable

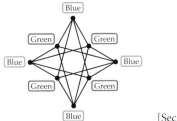

[Sec. 10.3]

26. Not 2-colorable [Sec. 10.3] **27.** 3 [Sec. 10.3]

28. 5 [Sec. 10.3] **29.** Three time slots: Budget and Planning, Marketing and Executive, Sales and Research

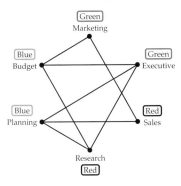

[Sec. 10.3]

CHAPTER 10 TEST *page 722*

1. a. No **b.** Monique **c.** 0 **d.** No [Sec. 10.1] **2.** Equivalent [Sec. 10.1]
3. a. No **b.** A–B–E–A–F–D–C–F–B–C–E–D [Sec. 10.1] **4.** No

[Sec. 10.1]

5. a. See page 676. **b.** A–G–C–D–F–B–E–A [Sec. 10.1]

6. a.

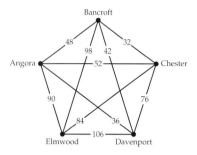

b.   Angora–Elmwood–Chester–Bancroft–Davenport–Angora; $284
c.   Yes   [Sec. 10.2]

7.   A–E–D–B–C–F–A   [Sec. 10.2]

8. a.

b.   3

c.

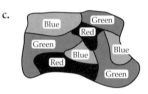

[Sec. 10.3]

9.

[Sec. 10.3]

10.   Three evenings: Cirque du Soleil and a play on one evening; a magic show and a tribute band concert on another evening; and a comedy show and a musical on a third evening   [Sec. 10.3]

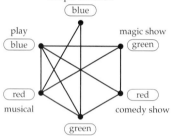

PHOTO CREDITS

INDEX